T0134572

Lecture Notes in Computer Science 13741

The series Lecture Notes in Computer Science (LNCS), including its subseries Lecture Notes in Artificial Intelligence (LNAI) and Lecture Notes in Bioinformatics (LNBI), has established itself as a medium for the publication of new developments in computer science and information technology research, teaching, and education.

LNCS enjoys close cooperation with the computer science R & D community, the series counts many renowned academics among its volume editors and paper authors, and collaborates with prestigious societies. Its mission is to serve this international community by providing an invaluable service, mainly focused on the publication of conference and workshop proceedings and postproceedings. LNCS commenced publication in 1973.

Hakimjon Zaynidinov · Madhusudan Singh ·
Uma Shanker Tiwary · Dhananjay Singh
Editors

Intelligent Human Computer Interaction

14th International Conference, IHCI 2022
Tashkent, Uzbekistan, October 20–22, 2022
Revised Selected Papers

Editors
Hakimjon Zaynidinov
Tashkent University Information Technologies
Tashkent, Uzbekistan

Madhusudan Singh
Oregon Institute of Technology
Klamath Falls, USA

Uma Shanker Tiwary
Indian Institute of Information Technology
Allahabad, India

Dhananjay Singh
Hankuk University of Foreign Studies
Yongin, Korea (Republic of)

ISSN 0302-9743 ISSN 1611-3349 (electronic)
Lecture Notes in Computer Science
ISBN 978-3-031-27198-4 ISBN 978-3-031-27199-1 (eBook)
https://doi.org/10.1007/978-3-031-27199-1

This Springer imprint is published by the registered company Springer Nature Switzerland AG
The registered company address is: Gewerbestrasse 11, 6330 Cham, Switzerland

Preface

The science and technology of Human Computer Interaction (HCI) has taken a giant leap forward in the last few years. This has given impetus to two opposing trends. One divergent trend is to organize separate conferences on focused topics such as 'Interaction Design and User-Centered Design', etc., which earlier would have been covered under HCI. The other convergent trend is to assimilate new areas in HCI conferences, such as 'Computing with Words', 'Prosocial Agents Development', 'Attention-based Applications', and 'Signals and Systems for HCI' etc. IHCI-2022 is one of the rare conferences focusing on those issues of 'Intelligence' and 'Human Computer Interaction' which exist at the crossroads of the above-mentioned trends. It is a privilege to present the proceedings of the 14th International Conference on Intelligent Human Computer Interaction (IHCI-2022) organized by the Tashkent University of Information Technologies during October 20–22, 2022 at Tashkent, Uzbekistan. IHCI is an annual international conference in the Human Computer Interaction field, where we explore research challenges emerging in the complex interaction between machine intelligence and human intelligence. This is the fourteenth event in the series and had the theme of "Interactive Technologies for the post-Covid Era", having eleven special tracks related to the main theme of the conference as well as general topics in the HCI field.

Out of 148 submitted papers, 60 papers were accepted for oral presentation and publication by the program committee based on the recommendations of at least 2 expert reviewers in a double-blind review process. The proceedings are organized in five sections corresponding to the five tracks of the conference. The 14th IHCI conference included five keynote speakers and ten invited talks, five workshops, four tutorial lectures, and twenty-one powerful expert session chairs who have worked in both industry and academia, to attract more than 200 participants and has emerged as the foremost worldwide (more than 20 countries) gathering of academic researchers, graduate students, top research think tanks and industry technology developers. Therefore, we do believe that the biggest benefit to the participant is the actualization of their goals in the field of HCI. That will ultimately lead to greater success in business, which is ultimately beneficial to society. Moreover, our warm gratitude should be given to all the authors who submitted their work to IHCI-2022. During the submission, review, and editing stages, the Easychair conference system proved very helpful. We are grateful to the technical program committee (TPC) and local organizing committee for their immeasurable efforts to ensure the success of this conference. Finally we would like to thank our speakers, authors, and participants for their contribution to making IHCI-2022 a stimulating and productive conference. This IHCI conference series cannot achieve yearly milestones without their continued support in future.

November 2022

Hakimjon Zaynidinov
Madhusudan Singh
Uma Shanker Tiwary
Dhananjay Singh

Organization

General Chairs

Hakimjon Zaynidinov	Tashkent University of Information Technologies, Uzbekistan
Dae-Ki Kang	Dongseo University, South Korea
Ajay Gupta	Massachusetts Institute of Technology, USA

Technical Program Chairs

Hakimjon Zaynidinov	Tashkent University of Information Technologies, Uzbekistan
Madhusudan Singh	Oregon Institute of Technology (OregonTech.), USA
Uma Shanker Tiwary	Indian Institute of Information Technology Allahabad, India
Dhananjay Singh	Hankuk University of Foreign Studies, South Korea

Steering Committee

Uma Shanker Tiwary	IIIT Allahabad, India
Santanu Chaudhury	IIT Jodhpur, India
Tom D. Gedeon	Australian National University, Australia
Debasis Samanta	IIT Kharagpur, India
Atanendu Sekhar Mandal	CSIR-CEERI, India
Tanveer Siddiqui	University of Allahabad, India
Jaroslav Pokorny	Charles University, Czech Republic
Sukhendu Das	IIT Madras, India
Samit Bhattacharya	IIT Guwahati, India
Wan-Young Chung	Pukyong National University, South Korea
Javed Khan	Kent State University, USA
Mriganka Sur	Massachusetts Institute of Technology, USA
Dhananjay Singh	Hankuk University of Foreign Studies, South Korea

Session Chairs

Uma Shanker Tiwary	IIIT-Allahabad, India
Madhusudan Singh	Oregon Institute of Technology, USA
Mark D. Whitaker	State University of New York, South Korea
Mohd Helmy Abd Wahab	Universiti Tun Hussein Onn, Malaysia
Masoud Mohammadian	University of Canberra, Australia
Eui-Chul Lee	Sangmyung University, South Korea
Jee Hang Lee	Sangmyung University, South Korea
Hakimjon Zaynidinov	Tashkent University of IT, Uzbekistan
Elmira Nazirova	Tashkent University of IT, Uzbekistan
Ibrohimbek Yusupov	Tashkent University of IT, Uzbekistan
Sarvarbek Makhmudjanov	Tashkent University of IT, Uzbekistan
Irish Singh	Ajou University, Korea
Preeti Parwekar	SRM-IST, Ghaziabad, India
Nagamani Molakatala	University of Hyderabad, India
David (Bong Jun) Choi	Soongsil University, South Korea
Hanumant Singh Shekhawat	IIT Guwahati, India
Dhananjay Singh	Hankuk University of Foreign Studies, South Korea
Nagarajan Prabakar	Florida International University, USA
Elena Novak	Kent State University, USA
Ikechi Ukaegbu	Nazarbayev University, Kazakhstan
Jong-Hoon Kim	Kent State University, USA
Ajit Kumar	Soongsil University, South Korea
Mukesh Saini	IIT Ropar, India
Abhishek Shrivastava	IIT Guwahati, India

Local Organizing Committee

Jasur Rizayev	Samarkand State Medical University, Uzbekistan
Yaxshibayev Doniyor Sultonbayevich	Tashkent University of IT, Uzbekistan
Anvar Kabulov	National University of Uzbeistan, Uzbekistan
Halimjon Khujamatov	Tashkent University of IT, Uzbekistan
Fakhriddin Nuraliev	Tashkent University of IT, Uzbekistan
Zafar Juraev	Andijan Machine-Building Institute, Uzbekistan
Elmira Nazirova	Tashkent University of IT, Uzbekistan
Akbarali Rasulov	Tashkent University of IT, Uzbekistan
Jonibek Usmonov	Tashkent University of IT, Uzbekistan

Kholmatvay Makhkambaevich Shadimetov	Tashkent University of IT, Uzbekistan
Farkhod Abduganievich Nuraliyev	Tashkent University of IT, Uzbekistan
Abdullo Rakhmonovich Hayotov	V.I. Romanovsky Institute of Mathematics, Uzbekistan
Ortik Baxtiyorovich Ruzibaev	Tashkent University of IT, Uzbekistan
Ulugbek Abdumurodovich Berdanov	Tashkent University of IT, Uzbekistan
Oybek Mallayev	Tashkent University of IT, Uzbekistan
Bunyod Azimov	Tashkent University of IT, Uzbekistan
Abdulaziz Gaybulayev	Tashkent University of IT, Uzbekistan
Temurbek Kuchkorov	Tashkent University of IT, Uzbekistan
Fayzullajon Botirov	Tashkent University of IT, Uzbekistan
Saydiaxad Pasriyev	Tashkent University of IT, Uzbekistan
Utkir Khamdamov	Tashkent University of IT, Uzbekistan
Saloxiddin Toshmatov	Tashkent University of IT, Uzbekistan
Shahzoda Anarova	Tashkent University of IT, Uzbekistan
Marat Rakhmatullaev	Tashkent University of IT, Uzbekistan
Mirsaid Aripov	National University of Uzbekistan, Uzbekistan
Normohammad Ravshanov	Digital Technologies and Artificial Intelligence Research Institute, Uzbekistan
Jonibek Jurayev	Samarkand State University, Uzbekistan
Shokir Urakov	Tashkent University of IT, Uzbekistan
Abdulhaq Kholikov	Tashkent State Transport University, Uzbekistan
Gayrat Jurayev	National University of Uzbekistan, Uzbekistan
Botir Usmonov	Tashkent Chemical-Technological Institute, Uzbekistan
Orif Zaripov	Tashkent State Technical University, Uzbekistan
Narzulla Mamatov	Tashkent institute of Irrigation and Agricultural Mechanization Engineers, Uzbekistan
Furqatbek Nurjanov	Tashkent State Agrarian University, Uzbekistan
Nietbay Utevliev	Tashkent University of IT, Uzbekistan

Publicity Chairs

Jamshid Sultanov	Tashkent University of IT, Uzbekistan
Koumudi Patil	IIT-Kanpur, India

Program Chairs

Ibrohimbek Yusupov	Tashkent University of IT, Uzbekistan
Sarvarbek Makhmudjanov	Tashkent University of IT, Uzbekistan

Operation Chairs

Jong-Hoon Kim	Kent State University, USA
Dhananjay Singh	Hankuk University of Foreign Studies, South Korea

Industry Forum Chairs

Garima Bajpai	DevOps Community of Practice, Canada
Temurbek Kuchkorov	TUIT, Uzbekistan
Mario Jose Divan	Intel Corporation, USA

Graduate Student Forum Chairs

Kamoliddin Shukurov	Tashkent University of IT, Uzbekistan
Ajit Kumar	Soongsil University, South Korea
Naagmani Molakatala	University of Hyderabad, India

Technical Committee Members

Tao Shen	Kent State University, USA
Hakimjon Zaynidinov	Tashkent University of Information Technologies, Uzbekistan
Elmira Nazirova	Tashkent University of Information Technologies, Uzbekistan
Ibrohimbek Yusupov	Tashkent University of Information Technologies, Uzbekistan
Sarvarbek Makhmudjanov	Tashkent University of Information Technologies, Uzbekistan
Mina Choi	Kent State University, USA
Lauren Copeland	Kent State University, USA
Jaehyun Park	Incheon National University, South Korea
Hyun K. Kim	Kwangwoon University, South Korea

Prima Sanjaya	University of Helsinki, Finland
Thierry Oscar Edoh	Bonn University, Germany
Garima Agrawal	Vulcan-ai, Singapore
David (Bong Jun) Choi	Soongsil University, South Korea
Jia Uddin	Woosong University, Korea
Alex Wong Ming Hui	Osaka University, Japan
Bharat Rawal	Gannon University, USA
Wesley De Neve	Ghent University Global Campus, South Korea
Satish Kumar L. Varma	Pillai College of Engineering, India
Alex Kuhn	State University of New York, South Korea
Mark Whitaker	State University of New York, South Korea
Satish Srirama	University of Hyderabad, India
Nagamani Molakatala	University of Hyderabad, India
Shyam Perugu	National Institute of Technology, Warangal, India
Neeraj Parolia	Towson University, USA
Stella Tomasi	Towson University, USA
Marcelo Marciszack	National University of Technology, Argentina
Andres Navarro- Newball	Pontificia Universidad Javeriana – Cali, Colombia
Indranath Chatterjee	Tongmyong University, South Korea
Gaurav Tripathi	BEL, India
Bernardo Nugroho Yahya	HUFS, South Korea
Carlene Campbell	University of Wales Trinity Saint David, UK
Himanshu Chaudhary	University of Gothenburg, Sweden
Rajiv Mishra	iOligos Technologies, India
Rakesh Pandey	Banaras Hindu University, India
Anwar Alruwaili	Stevens Institute of Technology, USA
Sayed Chhattan Shah	Hankuk University of Foreign Studies (HUFS), South Korea
Rajesh Mishra	Gautam Buddha University, India
Akshay Bhardwaj	IIT Guwahati, India
Angela Guercio	Kent State University, USA
Yujin Fu	Alabama A&M University, USA
Frank Zhu	University of Alabama at Huntsville, USA
Rajesh Sankaran	Argonne National Laboratory, USA
Ricardo Calix	Purdue University Northwest, USA
Hyun K. Kim	Kwangwoon University, South Korea
Jaehyun Park	Incheon National University, South Korea
Ismael Ali	University of Zakho, Iraq
Amal Babour	King Abdul Aziz University, Saudi Arabia
Durdona Irgasheva	Tashkent University of Information Technologies, Uzbekistan

Nitesh Funde	Visvesvaraya National Institute of Technology, Nagpur, India
Utkir Khamdamov	Tashkent University of Information Technologies, Uzbekistan
Mekhriddin Rakhimov	Tashkent University of Information Technologies, Uzbekistan
Smriti Agarwal	Motilal Nehru National Institute of Technology Allahabad, India
Mukhriddin Mukhiddinov	Tashkent University of Information Technologies, Uzbekistan
Irish Singh	Ajou University, South Korea
Longhui Zou	Kent State University, USA
Pamul Yadav	Yonsei University, South Korea
Bong Jun Choi	Soongsil University, South Korea
Kamoliddin Shukurov	Tashkent University of Information Technologies, Uzbekistan
Seungjik Lee	Korea Institute of Energy Technology (KENTECH), South Korea
Rustam Sadikov	Tashkent University of Information Technologies, Uzbekistan
Alok Chauhan	VIT, India
Satishkumar Varma	Pillai College of Engineering, India
Jan-Willem van't Klooster	University of Twente, The Netherlands
Md. Milon Islam	University of Waterloo, Canada
Nureize Arbaiy	University Tun Hussein Onn, Malaysia
Hafiza Abas	UTM Kuala Lumpur, Malaysia
Benazarova Saida	Tashkent University of Information Technologies, Uzbekistan
Indranath Chatterjee	Tongmyong University, South Korea
Rushali Deshmukh	Jayawant Shikshan Prasarak Mandal's Rajarshi Shahu College of Engineering, India
Temurbek Kuchkorov	Tashkent University of Information Technologies, Uzbekistan
Mario José Diván	National University of La Pampa, Argentina
Mukesh Saini	IIT Ropar, India
Sreeja S. R.	IIIT Sri City, India
Gyanendra Verma	NIT Raipur, India

Keynote Speakers

Mukhamadjan Makhmudovich Musaev	TUIT, Uzbekistan
Venkatasubramanian Ganesan	NIMHANS, India
Ajay Gupta	Western Michigan University, USA
Jan Treur	Vrije Universiteit Amsterdam, The Netherlands
Shiho Kim	Yonsei University, South Korea
Ingmar Weber	Qatar Computing Research Institute, Qatar

Invited Speakers

Mukesh Saini	IIT Ropar, India
Abhishek Shrivastava	IIT Guwahati, India
Nagarajan Prabakar	Florida International University, USA
Jan-Willem van't Klooster	University of Twente, The Netherlands
Irish Singh	Ajou University, South Korea
Antonio J. Jara	Libelium, Spain
Koumudi Patil	IIT Kanpur, India
Rodrigo da Rosa Righi	UNISINOS, Brazil
Muhammad Taqi Raza	University of Arizona, USA
Chintan Amrit	University of Amsterdam, The Netherlands
Jitendra P. Khatait	IIT Delhi, India
Sophie C. F. Hendrikse	Vrije Universiteit Amsterdam, The Netherlands

Tutorial Speakers

Hanumant Singh Shekhawat	Indian Institute of Technology, Guwahati, India
Shodhan Rao	Ghent University Global Campus, South Korea
Sandeep Kumar Pandey	Samsung Research and Development Institute, Bangalore, India
Gaurav Tripathi	Bharat Electronics Limited, India
Naagmani Molakatala	University of Hyderabad, India
Shankhanil Ghosh	University of Hyderabad, India
Bong Jun Choi	Soongsil University, South Korea
Ajit Kumar	Soongsil University, South Korea
Ankit Kumar Singh	Soongsil University, South Korea

Workshop Speakers

Shiho Kim	Yonsei University, South Korea
Pamul Yadav	Yonsei University, South Korea
Garima Bajpai	Canada DevOps Community of Practice, Canada
Eui-Chul Jung	Seoul National University, South Korea
Hyewon Kim	Seoul National University, South Korea
Younhee Cho	Seoul National University, South Korea

Contents

Deepfake Video Detection Using the Frequency Characteristic of Remote Photoplethysmography 1
Su Min Jeon, Hyeon Ah Seong, and Eui Chul Lee

A Multi-layered Deep Learning Approach for Human Stress Detection 7
Jayesh Soni, Nagarajan Prabakar, and Himanshu Upadhyay

Digital Processing Algorithms of Biomedical Signals Using Cubic Base Splines 18
Mukhriddin Abduganiev, Rakhimjon Azimov, and Lazizbek Muydinov

Methods for Creating a Morphological Analyzer 27
Elov Botir Boltayevich, Hamroyeva Shahlo Mirdjonovna, and Axmedova Xolisxon Ilxomovna

Uzbek Speech Synthesis Using Deep Learning Algorithms 39
M. I. Abdullaeva, D. B. Juraev, M. M. Ochilov, and M. F. Rakhimov

Speech Recognition Technologies Based on Artificial Intelligence Algorithms 51
Muhammadjon Musaev, Ilyos Khujayarov, and Mannon Ochilov

Multimodal Human Computer Interaction Using Hand Gestures and Speech ... 63
Mohammed Ridhun, Rayan Smith Lewis, Shane Christopher Misquith, Sushanth Poojary, and Kavitha Mahesh Karimbi

Emotion Recognition in VAD Space During Emotional Events Using CNN-GRU Hybrid Model on EEG Signals 75
Mohammad Asif, Majithia Tejas Vinodbhai, Sudhakar Mishra, Aditya Gupta, and Uma Shanker Tiwary

Multiclass Classification of Online Reviews Using NLP & Machine Learning for Non-english Language 85
Priyanka Sharma and Pritee Parwekar

A Higher Performing DARTS Model for CIFAR-10 95
Jie Yong Shin and Dae-Ki Kang

Automatic Speech Recognition on the Neutral Network Based on Attention Mechanism 100
N. S. Mamatov, N. A. Niyozmatova, Yu. Sh. Yuldoshev, Sh. Sh. Abdullaev, and A. N. Samijonov

Equal Temperament and Just Intonation Feature Based Analysis of Indian Music 109
D. V. K. Vasudevan, Nagamani Molakatala, Nikil Priyatham, Ravikant Gautam, and M. Rajender

Emotion Classification Through Facial Expressions Using SVM and Convolutional Neural Classifier 121
Varsha Singh, Ravi Kumar Singh, and Uma Shanker Tiwary

On the Evaluation of Generated Stylised Lyrics Using Deep Generative Models: A Preliminary Study 132
Hye-Jin Hong, So-Hyeon Kim, and Jee-Hang Lee

GWD: Graded Word Drop Model for When Type Questions for Hindi QA 140
Vani, Sumit Singh, Puja Burman, Anmol Jain, and Uma Shanker Tiwary

Masked Face Recognition Model with Explainable AI 154
Hyeon Ah Sung, Seunghyun Kim, and Eui Chul Lee

UX Design Workshop for Building Relationships Between Humans and Intelligent Objects Using 'T + e = B' Toolkit 160
Eui-chul Jung, Younhee Cho, and Hyewon Kim

A Longitudinal Study of the Emotional Content in Indian Political Speeches ... 166
Sandeep Kumar Pandey, Mohit Manohar Nirgulkar, and Hanumant Singh Shekhawat

Building a Local Classifier for Component-Based Face Recognition 177
Shavkat Kh. Fazilov, Olimjon N. Mirzaev, and Shukrullo S. Kakharov

Dominance Submissiveness Predisposition Scale (DSPS): Development and Validation 188
Ankita Shah and Uma Shanker Tiwary

Design of a Mixed Reality-Based Immersive Virtual Environment System for Social Interaction and Behavioral Studies 201
Sophia Matar, Alfred Shaker, Saifuddin Mahmud, Jong-Hoon Kim, and Jan-Willem van't Klooster

Privacy-Preserving Digital Intervention for Mental Health Using Federated Learning 213
Ankit Kumar Singh, Ajit Kumar, and Bong Jun Choi

How Can Humans and Robots Live Together?: The 5 Types of Human-Robot Relationship 225
Karam Park and Eui-Chul Jung

Effectiveness of Deep Learning Based Filtering Algorithm in Separation of Human Objects from Images 230
S. P. Khalilov, I. Yusupov, M. G. Mannapova, N. B. Nasrullayev, and F. Botirov

Building the Groundwork for a Natural Search, to Make Accurate and Trustworthy Filtered Searches: The Case of a New Educational Platform with a Global Heat Map to Geolocate Innovations in Renewable Energy 239
Seongyun Ku, Sunghwan Kim, Minji You, and Mark D. Whitaker

Implementation of Virtual Sea Environment with 3D Whale Animation 251
Uipil Chong and Shokhzod Alimardanov

A Constant-Factor Approximation Algorithm for Online Coverage Path Planning with Energy Constraint 257
Ayan Dutta and Gokarna Sharma

Real-Time Image Based Plant Phenotyping Using Tiny-YOLOv4 271
Sonal Jain, Dwarikanath Mahapatra, and Mukesh Saini

Influence of Packet Switching and Routing Methods on the Reliability of the Data Transmission Network and the Application of Artificial Neural Networks 284
D. Davronbekov, J. Aripov, Sh. Jabbarov, R. Djuraev, and D. Matkurbonov

Do Users' Values Influence Trust in Automation? 297
Liang Tang, Priscilla Ferronato, and Masooda Bashir

Automation of Calibration Procedure for Milk Non Automatic Weighing Instrument (NAWI) Process Using AI Methods 312
Nagamani Molakatala, Vimal Babu Undru, Shalem Raju Tambala, M. Tejaswini, M. Teja Kiran, M. Tejo Seshadri, and Venkateswara Sagar Juturi

Low-Cost Entry-Level Educational Drone with Associated K-12 Education Strategy 323
Bailey Wimer, Justin Dannemiller, Saifuddin Mahmud, and Jong-Hoon Kim

Creating a Modular and Decentralized Smart Mailbox System Using LoRaWan Networks 336
Prakash Shekhar, Abdolla Hegazy, Ajay Gupta, and Ammar Kamel

A Converting Model 3D Gaze Direction to 2D Gaze Position 348
Chaewon Lee, Seunghyun Kim, and Eui Chul Lee

Application of Fiber Optic Sensors in Aircraft Fuel Management System 354
Azizbek Umarov, Oripjon Zaripov, and Ruslan Zakirov

Intelligent Multi-tariff Payment Collection System for Inter-Municipal Buses in the Department of Atlántico – Colombia 361
Paola-Patricia Ariza-Colpas, Guillermo Hernandez-Sánchez, Guillermo Serrano-Torné, Marlon Alberto Piñeres-Melo, Shariq Butt-Aziz, and Roberto-Cesar Morales-Ortega

Agent-Based Modelling and Simulation of Public Transport to Identify Effects of Network Changes on Passenger Flows 373
Sophie Ensing and Chintan Amrit

Exploiting Security and Privacy Issues in Human-IoT Interaction Through the Virtual Assistant Technology in Amazon Alexa 386
Amrth Ashok Shenava, Saifuddin Mahmud, Jong-Hoon Kim, and Gokarna Sharma

User Experience in Virtual Reality Using Threshold Space in Between Different Physical Laws 396
Lori Minyoung Kim, Jung-Ryun Kwon, and Eui-Chul Jung

Monitoring Pollination by Honeybee Using Computer Vision 406
Vinit Kujur, Anterpreet Kaur Bedi, and Mukesh Saini

Modeling the Problem of Integral Geometry on the Family of Broken Lines Based on Tikhonov Regularization 417
N. U. Uteuliev, G. M. Djaykov, and A. O. Pirimbetov

A Framework for Privacy-Preserved Collaborative Learning in Smart Factory Environment 428
Ericka Pamela Bermudez Pillado, Tori Bukit, Sean Yonathan Tanjung, Hyun-Woo Lim, Ignatius Iwan, Bernardo Nugroho Yahya, and Seok-Lyong Lee

Co-creating Computer Supported Collective Intelligence in Citizen Science Hubs 435
Aelita Skarzauskiene and Monika Mačiulienė

Gaze Detection Using Encoded Retinomorphic Events 442
Abeer Banerjee, Shyam Sunder Prasad, Naval Kishore Mehta, Himanshu Kumar, Sumeet Saurav, and Sanjay Singh

Extremely Lightweight Skin Segmentation Networks to Improve Remote Photoplethysmography Measurement 454
Kunyoung Lee, Hojoon You, Jaemu Oh, and Eui Chul Lee

Privacy-Friendly Phishing Attack Detection Using Personalized Federated Learning 460
Jun Yong Yoon and Bong Jun Choi

CovidMis20: COVID-19 Misinformation Detection System on Twitter Tweets Using Deep Learning Models 466
Aos Mulahuwaish, Manish Osti, Kevin Gyorick, Majdi Maabreh, Ajay Gupta, and Basheer Qolomany

Development of School Library Network Based on Cloud Technologies in Uzbekistan 480
Marat Rakhmatullaev and Sherbek Normatov

Parallel Resource Defined Fitness Sharing 493
Blayne Rogers, Ajay Gupta, and Pranjal Minocha

EEG-Based Key Generation Cryptosystem for Strengthening Security of Blockchain Transactions 504
Ngoc-Dau Mai, Ha-Trung Nguyen, and Wan-Young Chung

Improving Gaze Estimation Performance Using Ensemble Loss Function 510
Seung Hyun Kim, Seung Gun Lee, Jee Hang Lee, and Eui Chul Lee

Non-overlayed Guidance in Augmented Reality: User Study in Radio-Pharmacy 516
Yves Simmen, Tabea Eggler, Alexander Legath, Doris Agotai, and Hilko Cords

Development of a Novel Method for Image Resizing Using Artificial Neural Network 527
Mukhriddin Arabboev, Shohruh Begmatov, Khabibullo Nosirov, Shakhzod Tashmetov, Saydiakhrol Saydiakbarov, Jean Chamberlain Chedjou, and Kyandoghere Kyamakya

Algorithms for Selecting and Comparing Features of Digital Image Vectors Based on the Analysis of Local Extrema 540
Mumtozali Tuktasinov

Calculation of Spectral Coefficients of Signals on the Basis of Haar by the Method of Machine Learning 547
Yusupov Ibrohimbek, Nurmurodov Javohir, Ibragimov Sanjarbek, Gofurjonov Muhammadali, and Qobilov Sirojiddin

Comparison for Polyp Segmentation Models: Focusing on Inference Speed 559
Seung Gun Lee, Seung Min Jeong, Chae Lin Seok, Jin Man Kim, and Eui Chul Lee

Custom Object Segmentation by Training R-CNN 565
Javlon Tursunov, Aziza Narimonova, Hamroev Bekzod, Djabbarov Bakhtiyor, Rustam Rakhmonov, and Shakhzod Bobokulov

Optimization of Fractal Structure Pattern Colors in Carpet Design Using Genetic Algorithm 575
Fakhriddin Nuraliev, Oybek Narzulloyev, and Saida Tastanova

Algorithm for Digital Processing of Seismic Signals in Distributed Systems 586
Oybek Mallaev, Bunyodbek Azimov, Kuchkarov Muslimjon, and Ahmadova Kamola

Author Index 595

Deepfake Video Detection Using the Frequency Characteristic of Remote Photoplethysmography

Su Min Jeon[1], Hyeon Ah Seong[1], and Eui Chul Lee[2](✉)

[1] Department of AI & Informatics, Graduate School, Sangmyung University, Seoul, Republic of Korea
{202132044,202132040}@sangmyung.kr

[2] Department of Human-Centered Artificial Intelligence, Sangmyung University, Seoul, Republic of Korea
eclee@smu.ac.kr

Abstract. Photoplethysmography is a technique for measuring the blood flow per unit time of an artery. Remote photoplethysmography is a method for obtaining photoplethysmography signals in a non-contact manner through a sensor such as a camera and has been recently applied to various fields. In this study, we propose a method for detecting Deepfake modulated color video based on remote photoplethysmography concept. As experimental data, 50 real videos and their 50 Deepfake videos using Face Swapping Generative Adversarial Networks were used. The photoplethysmography signals of face and neck regions were extracted, respectively, and the signals were preprocessed by detrending and performing Butterworth bandpass filtering. The 80 power values in the frequency domain were defined as feature vectors. As a result of analyzing the L2 Norm between the two vectors extracted from the face region and the neck region, the L2 Norms of the real video and the fake video were 0.0000307 and 0.0001332, respectively, confirming that the distributions were clearly separated. It was confirmed that there is a significant difference between the real and the fake videos. Also, as a result of calculating the degree of separation of distributions with d-prime, 2.32 was derived.

Keywords: Deepfake detection · Remote photoplethysmography · Face substitution · Bio-signals · Face recognition

1 Introduction

Photoplethysmography (PPG) is a simple optical technique to detect changes in blood volume in the microvascular layer of tissues [1]. Recently, a remote PPG (rPPG) method that senses a PPG signal using a camera without a sensor has been developed [2]. Since this method has a great advantage of being a non-contact method, research on estimating bio-signals such as heart rate and oxygen saturation is in progress and is being applied in various fields [3, 4]. In this study, we also propose a method for detecting Deepfake modulated videos.

H. Zaynidinov et al. (Eds.): IHCI 2022, LNCS 13741, pp. 1–6, 2023.
https://doi.org/10.1007/978-3-031-27199-1_1

Deepfake is a compound word of deep learning and fake that refers to image synthesis technology based on artificial intelligence. Recently, concerns are growing as it is used in digital crimes, and in particular, it is frequently used in fake news and illegal video production crimes that synthesize celebrities' faces [5]. To prevent the damage of these crimes, research has been conducted to detect tampered videos of Deepfakes. Among previous studies, Güera *et al.* proposed a method for automatically detecting Deepfakes. They used a Convolutional Neural Network (CNN) to extract features and train a Recurrent Neural Network (RNN) to classify Deepfakes. As a result, it showed an accuracy of 97% [6]. Hernandez *et al.* proposed Deepfakeson-Phys to detect Deepfakes using rPPG. Both Celeb-DF and DFDC databases showed more than Area Under the Curve (AUC) of 98% [7].

However, techniques using these deep learning models cannot explain why this video is a modulated one. In contrast, the method proposed in this paper can explain the difference between real video and deepfake modulated video by using the frequency characteristics of the rPPG signal. As a result of deriving the d-prime value to express the degree of quantitative separation of L2 Norms between frequency characteristics, the reliability is increased by showing that there is a clear difference between the two.

2 Materials and Method

Figure 1 shows the flow of the Deepfake modulated video detection method proposed in this paper. The input data is a 20-s video and uses the face and neck regions. Then, the color information extracted from the corresponding region is converted into a YCbCr color model which separates brightness values and color information. Then, the algorithm for extracting the rPPG signal is applied, and a detailed description is given in Sect. 2.2. Then, the extracted rPPG signal is frequency analyzed and feature vectors are extracted. After that, the L2 Norm between 80-dimensional features is calculated to determine the real video and the Deepfake modulated video. The details arc included in Sect. 2.3.

2.1 Dataset

This study was conducted using the Deepfake modulated video constructed with the support of the Korea Intelligent Information Society Promotion Agency with the funds of the Ministry of Science and ICT. The utilized data can be downloaded from AI HUB [8]. The data used in the experiment were 50 original FHD (1920 × 1080) videos of 90 s or longer (30 frames per second) obtained from the subject, 50 FHD-modulated videos of 15 s or longer generated by learning these original videos with Face Swapping Generative Adversarial Networks (FSGAN).

2.2 Remote Photoplethysmography Signal Extraction

The method of extracting rPPG uses an algorithm made in our laboratory [2]. Figure 2 shows the rPPG signal extraction process, and a more detailed description of the algorithm is as follows.

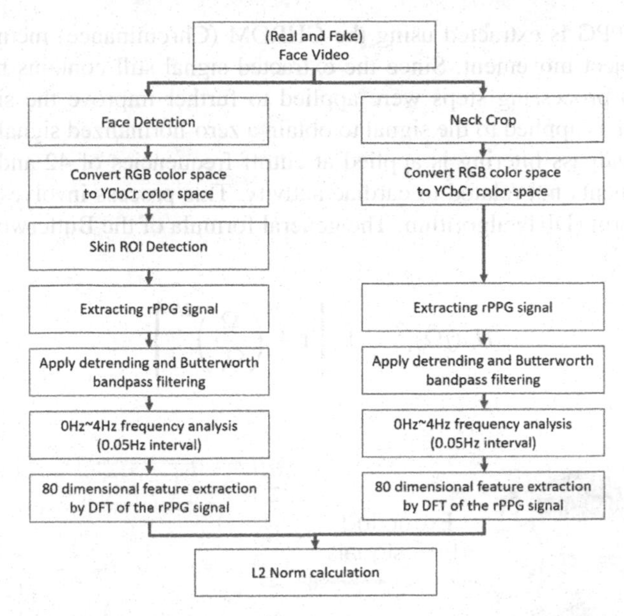

Fig. 1. The flow chart of the proposed Deepfake modulated video detection method

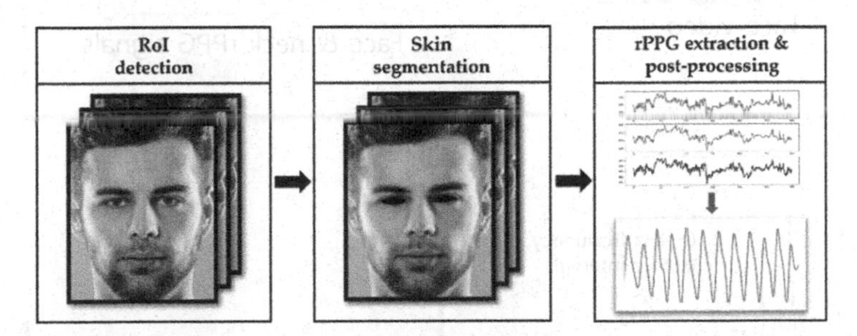

Fig. 2. The process of extracting the rPPG signal from the skin region

Experimental data includes clothes, background in addition to the face. In this study, rPPG signals in the face and neck regions were compared. The neck region was directly detected by the mouse click event, and the face region was detected using the face detector of the OpenCV DNN module [9]. To select only the skin region from the detected region, the RGB color space was converted into the YCbCr color space, and the ranges of $0 \leq Y' \leq 235$, $77 \leq C\_B \leq 127$, $133 \leq C\_R \leq 173$ were used [10]. The transformation formula is the same as Eq. (1).

$$\begin{aligned} Y' &= 16 + \left(65.481 \cdot R' + 128.553 \cdot G' + 24.966 \cdot B'\right) \\ C_B &= 128 + \left(-37.797 \cdot R' + 74.203 \cdot G' + 112.0 \cdot B'\right) \\ C_R &= 128 + (112.0 \cdot R' + 93.786 \cdot G' + 18.214 \cdot B') \end{aligned} \tag{1}$$

Then, the rPPG is extracted using the CHROM (Chrominance) method, which is resistant to subject movement. Since the extracted signal still contains noise components, two post-processing steps were applied to further improve the signal quality. First, detrending is applied to the signal to obtain a zero-normalized signal. In addition, Butterworth bandpass filtering is applied at cutoff frequencies of 42 and 240 bpm to remove components not related to cardiac activity. This process involves the Discrete Fourier Transform (DFT) algorithm. The general formula of the Butterworth bandpass filter is as Eq. (2).

$$|H_a(j\Omega)|^2 = 1/\left\{1 + \left(\frac{\Omega}{\Omega_c}\right)^{2N}\right\} \tag{2}$$

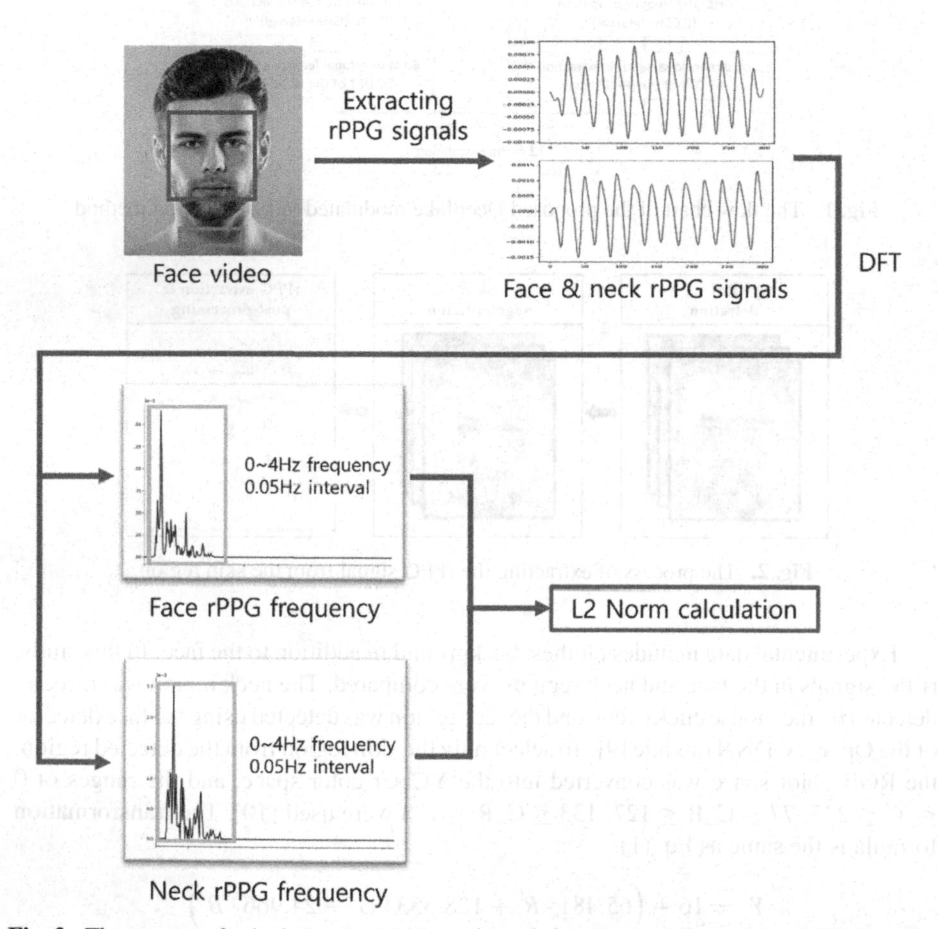

Fig. 3. The process of calculating the L2 Norm through frequency analysis of the rPPG signal in the face and neck

2.3 Distance Calculation

The signal obtained in Sect. 2.2 was subjected to frequency analysis because it was difficult to see a visual difference between the real video and the fake video. The frequency of 0–4 Hz was analyzed at 0.05 Hz intervals, and the power spectrum of each frequency domain was extracted as a feature vector. For each data, an 80-dimensional feature vector for each face and neck region was extracted, and the L2 Norm between features was calculated using this vector. Figure 3 schematically shows all the above processes.

3 Results

In this experiment, a total of 100 videos such as 50 real videos and 50 fake videos, were used to discriminate the real face video and the Deepfake modulated video. Figure 4 shows two L2 Norm distributions of real and fake videos by applying our proposed method. In Fig. 4, the blue-colored distribution is generated by the 50 L2 Norms between two feature vectors extracted from face and neck regions of the real videos. In contrast, the orange-colored distribution is generated by the 50 L2 Norms between two feature vectors extracted from face and neck regions of the fake videos. Consequently, we qualitatively confirmed that the two distributions are clearly separated. To express quantitatively the degree of separation, we calculated the d-prime value. The distribution means of the real video and the fake video were 0.0000307 and 0.0001332, respectively, and the standard deviation was 0.0000193 and 0.0000957, respectively. Therefore, as a result of calculating the d-prime, 2.32 was derived. The general formula of d-prime is as Eq. (3).

$$d' = \frac{|\mu_1 + \mu_2|}{\sqrt{(\sigma_1^2 + \sigma_2^2)/2}} \tag{3}$$

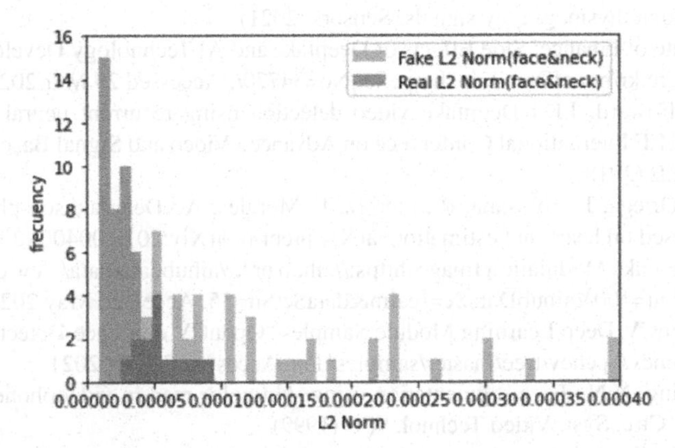

Fig. 4. L2 Norm of face and neck feature vectors in real video and fake video

4 Conclusion

We aimed to differentiate between real face video and Deepfake modulated video using rPPG signals. While AI technology-based contents such as face recognition and Deepfakes are spreading recently, cases of digital crimes such as defamation and financial fraud using 'Deepfake' technology are increasing significantly. This study was conducted to propose a method to prevent this. Through the experiment, we confirmed that the L2 Norm of the power spectrum of the face and neck region was different between the real video and the fake video. Therefore, 2.32 was derived as a result of calculating d-prime to see a quantitative numerical value. This supports our assertion that there is a significant difference between real face video and deepfake video. It seems that our proposed method can reduce the risk of exploitation of Deepfakes. In future research, we plan to increase the amount of data to increase the reliability of the research results, and we plan to conduct research with datasets built using multiple models.

Acknowledgement. This paper was supported by Field-oriented Technology Development Project for Customs Administration through National Research Foundation of Korea (NRF) funded by the Ministry of Science & ICT and Korea Customs Service (2022M3I1A1095155).

References

1. Allen, J.: "Photoplethysmography and its application in clinical physiological measurement. Physiol. Meas. **28**(3) (2007)
2. Suh, K.H., Eui, C.L.: Contactless physiological signals extraction based on skin color magnification. J. Electron. Imaging **26**(6) (2017)
3. Kim, N.H., Yu, S.G., Kim, S.E., Lee, E.C.: Non-contact oxygen saturation measurement using YCgCr color space with an RGB camera. Sensors (2021)
4. Yu, S.G., Kim, S.E., Kim, N.H., Suh, K.H., Lee, E.C.: Pulse rate variability analysis using remote photoplethysmography signals. Sensors (2021)
5. Hana Institute of Finance: Side Effects of Deepfake and AI Technology Development. http://www.hanaif.re.kr/boardDetail.do?hmpeSeqNo=34770/. Accessed 29 Mar 2021
6. Güera, D., Edward, J.D.: Deepfake video detection using recurrent neural networks. In: 2018 15th IEEE International Conference on Advanced Video and Signal Based Surveillance (AVSS). IEEE (2018)
7. Hernandez-Ortega, J., Tolosana, R., Fierrez, J., Morales, A.: Deepfakeson-phys: deepfakes detection based on heart rate estimation. arXiv preprint arXiv:2010.00400 (2020)
8. Dataset: Deepfake Modulation Image. https://aihub.or.kr/aihubdata/data/view.do?currMenu=115&topMenu=100&aihubDataSe=realm&dataSetSn=55. Accessed May 2022
9. Github: OpenCV Deep Learning Module Samples "OpenCV dnn Face Detector". https://github.com/opencv/opencv/tree/master/samples/dnn. Accessed 01 Oct 2021
10. Chai, D., King, N.N.: Face segmentation using skin-color map in videophone applications. IEEE Trans. Circ. Syst. Video Technol. **9**(4) (1999)

A Multi-layered Deep Learning Approach for Human Stress Detection

Jayesh Soni[1](✉), Nagarajan Prabakar[1], and Himanshu Upadhyay[2]

[1] Knight Foundation School of Computing and Information Sciences, Florida International University, Miami, FL, USA
jsoni@fiu.edu, prabakar@cis.fiu.edu
[2] Applied Research Center, Florida International University, Miami, FL, USA
upadhyay@fiu.edu

Abstract. In today's fast-paced world, stress is common on various occasions in everyday life. However, long-term stress hinders normal lives. Detection of such mental stress at an earlier stage can prevent many associated health problems. There are significant changes in the multiple bio-signals, such as electrical, thermal, optical, etc., when an individual is under stress. Such bio-signals can be utilized to identify stress. In this paper, we propose a multi-layered deep learning-based approach for detecting human stress using the multimodal dataset. We use an open-source dataset, namely Wearable Stress and Affect Detection (WESAD), which contains data from wearable physiological and motion sensors. The modalities of these sensors include axis acceleration, body temperature, electrocardiogram, and electrodermal activity with three conditions: baseline, amusement, and stress. In the first layer of our multi-layered approach, we train and compare AutoEncoder and Variational AutoEncoder to learn the normal emotional state of the subject. In the second layer, we train and compare LSTM and Transformer models to classify the subjects as either in an amused or stressed state. This multi-layered approach helps to achieve a higher stress detection rate of 98%.

Keywords: Stress Detection · WESAD · Long Short Term Memory · AutoEncoder

1 Introduction

With the ongoing economic recession, pandemics, and natural calamities, the mental health of people is affected negatively. This leads to stress in one's daily life that directly causes psychological illnesses such as depression, anxiety, disorders in the regular functions of the organs, etc. Sometimes the existing symptoms get even worse, affecting physical health. Stress is a reaction to complex situations pertaining to a heavy workload or something that demands in-depth thinking psychologically. In the current fast pace world, every other person is seen to be under stress. The health might deteriorate if it is not resolved at an earlier stage. Even more, in the worst-case scenarios, stress may lead to death. A recent study has shown a high correlation between the psychological and

H. Zaynidinov et al. (Eds.): IHCI 2022, LNCS 13741, pp. 7–17, 2023.
https://doi.org/10.1007/978-3-031-27199-1_2

physical health of an individual [1]. Their observations indicate that the measurement of the physiological signals can detect the emotional state of an individual. Thus, such signals can be used for periodical observation of our mental health, including depression, stress, and anxiety. Stress detection systems have applications in a wide variety of domains that, includes students, employees, job applicants, etc.

The methods proposed by researchers in the past include the self-reporting questionnaire that needs to be filled by the individual to detect the stress. The limitation of such an approach is that people under stress lack interest in filling up such questionnaires. Thus, to improve health, an accurate system is required that can detect stress efficiently. Automation in detecting stress has been proposed by some researchers [2–4]. Even with the increase in the research for detecting stress automatically, it is still not enough to provide a considerable amount of accuracy in the case of real scenarios. In this paper, a physiological dataset, namely WESAD, is used to develop an efficient stress detection model. To increase the detection rate, we propose a multi-level deep learning-based approach.

The rest of the chapter is as follows. We provide a literature review in the current problem domain in Sect. 2. The dataset is explained in Sect. 3. Next, Sect. 4 explains the proposed multi-level stress detection model in depth. Analysis of the experimental results is performed in Sect. 5. Finally, we conclude in Sect. 6 with future work discussion.

2 Related Work

A considerable amount of research has been done in detecting stress in people. Traditionally, smart devices are being used to detect stress during the daytime and nighttime. Individual data is being recorded using smartphone devices. Different variations in self's perception are recorded, and subsequently, the subject was asked to fill the questionnaires. Many times, researchers analyze the voice recordings to measure the stress. Readings from the biological sensors have a major impact on the detection of stress. There is numerous research done on the automatic detection of stress. Lin et al. [4] use a neural network to detect stress. Siirtola et al.[5] uses the commercially available sensors in smartphones for the same purpose. They provide a comparison between the sensors in multiple operating systems. Furthermore, they used a support vector machine and achieved an accuracy of 84%. Queyam et al. [6] developed a method for stress detection in automobile drivers. They collected the ECG and EMG of ten subjects and classified them into three class classifications, namely high, moderate and low stress. An Artificial Neural Network was used to develop the model, achieving an 88% accuracy rate. Padmaja et al. [7] used FITBIT activity trackers. They analyzed the BMI, sleeping patterns, and physical activities of the subjects. Giannakakis et al. [8] used the recording of the facial signals to evaluate the emotional states. SVM, Adaboost, and K-nearest neighbors were trained to develop the system. They achieved an accuracy rate of 91.68% with the AdaBoost classifier. The author in [9] used wearable sensors to collect the signals. They recorded a total of 19 physiological signals. The feature set was condensed to 7 vectors by using principal component analysis. They achieved an accuracy of 80% with the K-nearest classifier. In [10], various signal processing methods were applied to extract the features from the physiological signals. Next, they train decision trees and SVM to distinguish the relaxed state subject from the stressed state subject. A KNN-based classifier is used for stress prediction from the signals collected from the subject trying to solve the famous Tower of Hanoi problem [11]. Werner et al. [12] extracted the gradient-based and the facial-based features from the recorded video frames. They combined these features with the statistical features generated from the biological signals. The final hybrid feature vectors were used to train a random forest model for stress detection. However, the author considered features highly correlated only within the individual modalities. Rostgoo et al. [13] used Long Short Term Memory for detecting the stress of drivers and achieved 92.8% accuracy. Umematsu et al. [14] predicted the stress level on the next day based on the previous one-week data of the students. Though considerable research has been conducted in automatic stress detection, the current research is far from achieving the optimal model. Thus, in this paper, we propose a multi-level deep learning-based approach. Our research indicates that a sequential deep learning-based algorithm can significantly improve the stress detection rate more than conventional machine learning-based algorithms.

3 Stress Detection Dataset

We use the Wearable Stress and Affect dataset, an open-source benchmark dataset from the UCI machine learning repository. It is made available publicly by Schmidt et al. [15]. It contains a multimodal dataset of subjects who experiences both normal and emotional stress. This dataset records data from the chest and wrist-worn devices of a total of 15 subjects. There are numerous modalities extracted from each subject. There are a total of three states in the dataset, namely, stress, baseline, and amusement. During the baseline condition, the subjects sat or stood at the table while they read neutral magazines. It aims at extracting the subject's affective state in neutral conditions. During the amusement condition, the subjects were set to watch a total of seven video clips. Each video clip was followed by five seconds of short neutral video sequences. The total length of the video in the amusement condition is 392 s. During the stress condition, the subject was set toward the well-researched Social Stress Test. It consists of solving the arithmetic problem and public speaking. It generates high mental stress on the individual subject. In the WESAD dataset collection, the author asked each subject to provide a five-minute speech regarding their personal behavior in front of a panel consisting of three people. Here, each subject was asked to focus on their individual strength and weaknesses. Subjects were informed that the members in the panel are specialized in human resources from their facility. Subjects were also told to present as efficiently as possible since that can impact their career options. The participants were given only three minutes to prepare the five-minute speech; furthermore, they were not allowed to have any notes while presenting the speech. The panel members asked each subject to count inversely from 2023 to zero with the step of 17 after the speech delivery. The subject has to start from the beginning if they make any mistakes. Next, the same panel member gave a total of five minutes for performing both tasks; hence, the total duration of the social stress test was ten minutes. After the test, each participant was given a rest period of five minutes. Table 1 shows the dataset descriptors for WESAD. Accelerometer (ACC), Electrocardiogram (ECG), Electrodermal Activity (EDA), and Temperature (TEMP) are a few of the physiological modalities being recorded.

Table 1. WESAD Dataset Descriptors

Total participants	Modalities	States	Methods	Devices
15	ECG EDA TEMP ACC	Baseline, Amusement, Stress	Social Stress Test, Video Clips	Empatica E4, RespiBAN

4 Proposed Framework

The proposed multi-layered deep learning-based framework is depicted in Fig. 1. It consists of the following four phases: Dataset Collection, Data Pre-Processing, Training of the algorithm, and finally, the testing with deployment.

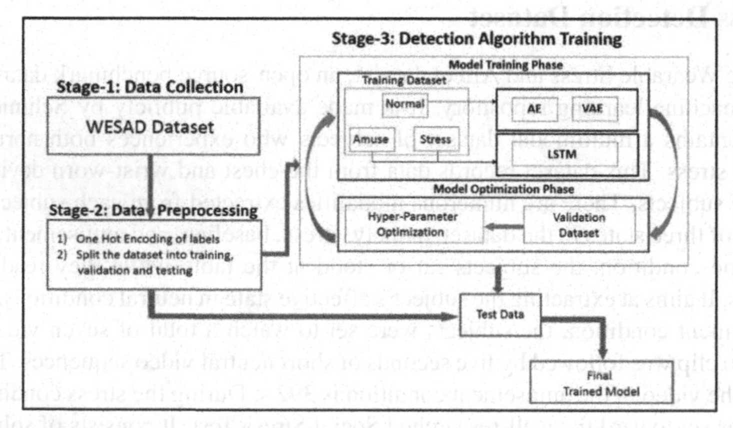

Fig. 1. Proposed Multi-Layered Deep Learning-based Framework

We provide an in-depth explanation of the four stages of the proposed framework below.

Stage-1: Data Collection
We utilized the WESAD dataset described in brief in the previous section.

Stage-2: Data Pre-Processing
The following transformation is applied during the pre-processing step:

1) Encoding the target labels as One-Hot
2) Divide the whole dataset into training for model building, validation for hyper-parameter optimization, and testing for final deployment

Stage-3: Detection Algorithm Training
This is the most crucial stage of the whole proposed multi-layered deep learning-based approach for stress detection. The two layered are trained in the following ways:
3.1) Model Training Phase: Layer 1
The primary purpose of the first layer is to train a model that learns the patterns of the subjects in their normal emotional state. We employed AutoEncoder (AE) and Variational AutoEncoder (VAE) deep learning-based algorithms to achieve this. Since these algorithms adds non-linearity, it makes them learn complex representations in lower dimensions with less loss of information. This algorithm takes the normal patterns as input and encodes them into lower dimensional space by performing the feature representation embedding. Next, it tries to reconstruct the pattern from such embedding. The difference between the original and generated input is called the reconstruction error. The significant difference between AE and VAE is that the latter uses the mean and standard deviation while encoding the input feature space into latent space. Figure 2 shows the VAE model.

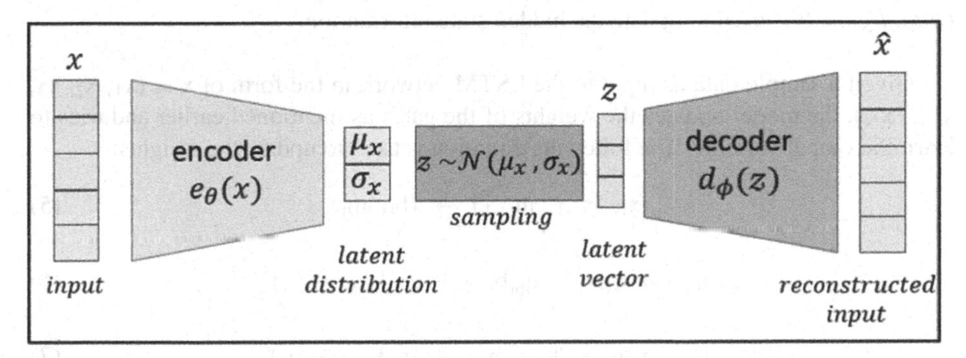

Fig. 2. VAE Model

We use the KL-Divergence Loss for updating the weights through backpropagation using the following equations.

$$\textit{reconstruction loss} = \|x - \hat{x}\|_2 = \|x - d_\phi(\ddagger)\|_2 = \|x - d_\phi(\mu_x + \sigma_x \epsilon)\|_2 \tag{1}$$

$$\mu_x, \sigma_x = e_\theta(x), \epsilon \sim \mathcal{N}(0, I) \tag{2}$$

$$\textit{similarity loss} = \textit{K L Divergence} = D_{KL}(\mathcal{N}(\mu_x, \sigma_x) \| \mathcal{N}(0, I)) \tag{3}$$

$$\textit{loss} = \textit{reconstruction loss} + \textit{similarity loss} \tag{4}$$

The dataset is divided into training and testing as 80% and 20%, respectively. Furthermore, we employed K-Fold Cross Validation to obtain the optimal hyper-parameter values. Here, we applied the value of K as 10. The proposed VAE model has three layers in the encoder and decoder part, each with 20, 10, and 5 neurons, respectively. Next, to avoid overfitting, we use the dropout layer with a dropout probability of 0.1.

3.2) Model Training Phase: Layer 2

In the second layer of the proposed multi-layered approach, we train the Long Short Term Memory (LSTM) and Transformer model to distinguish the subject's emotional state as being stressed or amused. LSTM is one of the variants of the Recurrent Neural Network [16]. The significant advantage of LSTM over RNN is that the former has memory cells [17]. With the reading of the input by the LSTM cell, the memory is updated. LSTM has the following gates to retain the memory.

Forget gate: It regulates the information that needs to be discarded.
Input gate: It controls the amount of new info that requires an updation.
Memory gate: It creates a new memory based on the input gate information.
Output gate: It is used to update the hidden state information.

Given a sample data as input to the LSTM network in the form of $x = (x_1, x_2, x_3, x_{N-1}, x_N)$, the model updates the weights of the gates as mentioned earlier and tries to learn the output variable. The following equation is used to update the weights:

$$(x_t,\ b_{t-1},\ m_{t-1}) \rightarrow (b_t,\ m_t) \tag{5}$$

$$i_t = \sigma(q_{xi}x_t + q_{bi}b_{t-1} + q_{mi}k_{t-1} + z_i) \tag{6}$$

$$f_t = \sigma\ (q_{xf}x_t + q_{bf}b_{t-1} + q_{kf}k_{t-1} + z_f) \tag{7}$$

$$k_t = f_t * k_{t-1} + i_t * \tanh\big(q_{xk}x_t + q_{bk}b_{t-1} + z_k\big) \tag{8}$$

$$g_t = \sigma\ (q_{xg}x_t + q_{bg}b_{t-1} + q_{kg}k_t + z_g) \tag{9}$$

$$b_t = g_t * \tanh(k_t) \tag{10}$$

where z_i, z_f,z_g, and z_k represent the bias units for the input, forget, output, and memory gate, respectively. Next, we use *b* to represent the output from the hidden layer, *k* as the current state of the memory, and *q* as the weight matrix developed using the sigmoid and tanh activation functions. These gates mitigate the inaccuracies that occur during the training of RNN by acting as a memory to retain long-term information. The major difference between LSTM [18, 19] and Transformer is that the latter has the attention mechanism.

The proposed architecture has 3 LSTM layers with 50 neurons each. Softmax activation is used at the dense layer to classify the subject as either in a stressed or amused state.

The following evaluation metrics are used for determining the efficiency of the model.

$$Recall = \frac{TP}{TP + FN} \tag{11}$$

$$Precision = \frac{TP}{TP + FP} \tag{12}$$

$$Specificity = \frac{TN}{TN + FP} \tag{13}$$

$$False\ Negative\ Rate\ (FNR) = \frac{FN}{TP + FN} \tag{14}$$

$$False\ Positive\ Rate\ (FPR) = \frac{FP}{FP + TN} \tag{15}$$

$$Accuracy = \frac{TP + TN}{TP + FP + FN + TN} \tag{16}$$

where TP: True Positive, FN: False Negative, TN: True Negative, and FP: False Positive.

3.3) Model Optimization

For the optimization of the hyper-parameters, we use the validation dataset. The following parameters are tuned:

BatchSize: Total number of sample data points used to update the model's weights through backpropagation.

Epochs: Total number of times the model is trained on the entire training dataset.

Upon successfully optimizing the parameters, we use the test data for the final evaluation of the model. This trained model can be used for the precise detection of the emotional state of the subjects.

5 Experimental Results

We trained the models in both the layers with Rectifier Linear Unit (ReLu) activation function in the hidden layers. Next, we employed the Adam optimizer for the gradient updates to obtain the optimal weights. We obtained the optimal accuracy rate for LSTM and Transformer models at a batch size of 128.

We trained the model with the batch size range of {16, 32, 64, 128, 256, and 512}. From the experimental results, we found out that the highest accuracy of 0.98 was achieved with a batch size of 128.

Figure 3 depicts the transformer model's accuracy rate and loss value at each epoch. We achieved an accuracy of 98% with a loss of 0.07% with the transformer model and an accuracy of 88% with a loss of 0.25% with the LSTM Model. Figure 4 depicts the loss value at each epoch for the VAE model. Based on literature review, Liapis et al. [20] achieved an accuracy of 91.1%. We achieved a loss of 0.015% with VAE and 0.12% with the AE model. Figure 5 shows the accuracy rate at each batch size.

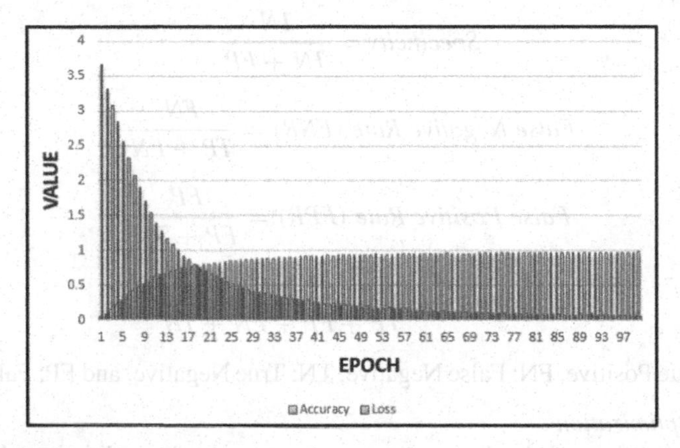

Fig. 3. Accuracy and Loss for Transformer Model w.r.t Epoch

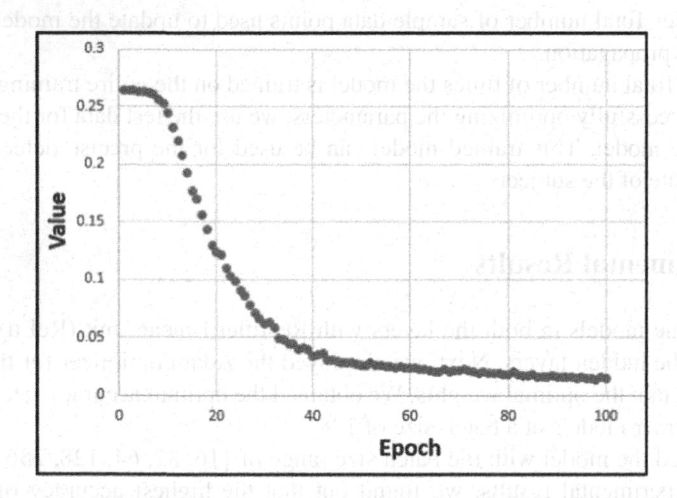

Fig. 4. Loss for VAE Model w.r.t Epoch

Table 2 shows the metrics achieved while evaluating the model on the test dataset for LSTM and transformer models. We notice that the trained Transformer model could distiniguish the subject as being in the amused or stress state with a 98% accuracy rate.

All our experiments are performed on Google Colaboratory. Here, we use Keras as the wrapper on top of the Tensorflow for model training purposes. Furthermore, we use seaborn and matplotlib for visualization purposes.

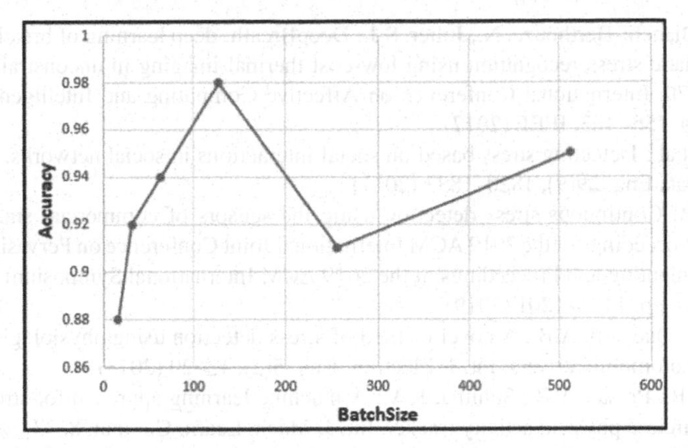

Fig. 5. Accuracy w.r.t BatchSize

Table 2. Evaluated Metrics on LSTM and Transformer Models

Model	Recall	Specificity	Precision	FPR	FNR	Accuracy
Transformer	0.882	0.978	0.889	0.005	0.132	0.981
LSTM	0.863	0.857	0.824	0.105	0.357	0.884

6 Conclusion

This paper employs a multi-level deep learning-based approach for stress detection using physiological parameters. The model uses the UCI-based open source WESAD dataset. In the first level of the proposed framework, we train AE and VAE to learn the pattern of the subject while they are in a normal, stress-free emotional state. In the second level, we train LSTM and Transformer model that can distinguish the subject as being in an amused or stressful emotional state. From the experimental results, we found out that the Transformer achieves an accuracy rate of 98% in distinguishing the emotional state of the subject. Furthermore, the VAE model achieved the least reconstruction loss of 0.015% in learning the pattern of normal behavior.

This work can be extended by applying an ensemble approach to improve the accuracy rate. Furthermore, we can also employ the Convolution Neural Network based approach first to reduce the dimensionality of the dataset, which further reduces the training time of the algorithm.

References

1. Liapis, A., Katsanos, C., Karousos, N., Xenos, M., Orphanoudakis, T.: User experience evaluation: a validation study of a tool-based approach for automatic stress detection using physiological signals. Int. J. Human-Comput. Interact. 1–14 (2020)
2. Gjoreski, M., et al.: Datasets for cognitive load inference using wearable sensors and psychological traits. Appl. Sci. **10**, 3843 (2020)

3. Cho, Y., Bianchi-Berthouze, N., Julier, S.J.: DeepBreath: deep learning of breathing patterns for automatic stress recognition using low-cost thermal imaging in unconstrained settings. In: 2017 7th International Conference on Affective Computing and Intelligent Interaction (ACII), pp. 456- 463. IEEE (2017)
4. Lin, H., et al.: Detecting stress based on social interactions in social networks. IEEE Trans. Knowl. Data Eng. **29**(9), 1820–1833 (2017)
5. Siirtola, P.: Continuous stress detection using the sensors of commercial smartwatch. In: Adjunct Proceedings of the 2019 ACM International Joint Conference on Pervasive and Ubiquitous Computing and Proceedings of the 2019 ACM International Symposium on Wearable Computers, pp. 1198–1201 (2019)
6. Singh, M., Queyam, A.B.: A novel method of stress detection using physiological measurements of automobile drivers. Int. J. Electron. Eng. **5**(2), 13–20 (2013)
7. Padmaja, B., Prasad, V.R., Sunitha, K.V.: A machine learning approach for stress detection using a wireless physical activity tracker. Int. J. Mach. Learn. Comput. **8**, 33–38 (2018)
8. Giannakakisa, G., Pediaditisa, M., Manousos, D.: Stress and anxiety detection using facial cues from videos. Elsevier (2016)
9. Wijsman, J., Grundlehner, B., Liu, H.: Towards mental stress detection using wearable physiological sensors. IEEE (2011)
10. Barreto, A., Zhai, J., Adjouadi, M.: Non-intrusive physiological monitoring for automated stress detection in human-computer interaction. In: Lew, M., Sebe, N., Huang, T.S., Bakker, E.M. (eds.) HCI 2007. LNCS, vol. 4796, pp. 29–38. Springer, Heidelberg (2007). https://doi.org/10.1007/978-3-540-75773-3_4
11. Ciabattoni, L., Ferracuti, F., Longhi, S., Pepa, L., Romeo, L., Verdini, F.: Real-time mental stress detection based on smartwatch. In: 2017 IEEE International Conference on Consumer Electronics (ICCE), pp. 110–111. IEEE (2017)
12. Werner, P., Al-Hamadi, A., Niese, R., Walter, S., Gruss, S., Traue, H.C.: Automatic pain recognition from video and biomedical signals. In: 2014 22nd International Conference on Pattern Recognition, pp. 4582–4587 (2014)
13. Rastgoo, M.N., Nakisa, B., Maire, F., Rakotonirainy, A., Chandran, V.: Automatic driver stress level classification using multimodal deep learning. Expert Syst. Appl. **138**, 112793 (2019)
14. Umematsu, T., Sano, A., Taylor, S., Picard, R.W.: Improving students' daily life stress forecasting using LSTM neural networks. In: 2019 IEEE EMBS IC on Biomedical & Health Informatics (BHI), pp. 1–4. IEEE (2019)
15. Schmidt, P., Reiss, A., Duerichen, R., et al.: Introducing wesad, a multimodal dataset for wearable stress and affect detection. In: Proceedings of the 20th ACM International Conference on Multimodal Interaction, 400–408 (2018)
16. Zaremba, W., Sutskever, I., Vinyals, O.: Recurrent neural network regularization. arXiv preprint arXiv:1409.2329. 8 Sep 2014
17. Greff, K., Srivastava, R.K., Koutník, J., Steunebrink, B.R., Schmidhuber, J.: LSTM: a search space odyssey. IEEE Trans. Neural Netw. Learn. Syst. **28**(10), 2222–2232 (2016)
18. Soni, J., Prabakar, N., Upadhyay, H.: Behavioral analysis of system call sequences using lstm seq-seq, cosine similarity and jaccard similarity for real-time anomaly detection In: International Conference on Computational Science and Computational Intelligence (CSCI), pp. 214–219 (2019).https://doi.org/10.1109/CSCI49370.2019.00043

19. Soni, J., Prabakar, N., Upadhyay, H.: Visualizing high-dimensional data using t-distributed stochastic neighbor embedding algorithm. In: Arabnia, H.R., Daimi, K., Stahlbock, R., Soviany, C., Heilig, L., Brüssau, K. (eds.) Principles of Data Science. TCSCI, pp. 189–206. Springer, Cham (2020). https://doi.org/10.1007/978-3-030-43981-1_9
20. Liapis, A., Faliagka, E., Katsanos, C., Antonopoulos, C., Voros, N.: Detection of subtle stress episodes during ux evaluation: assessing the performance of the wesad bio-signals dataset. In: Ardito, C., et al. (eds.) INTERACT 2021. LNCS, vol. 12934, pp. 238–247. Springer, Cham (2021). https://doi.org/10.1007/978-3-030-85613-7_17

Digital Processing Algorithms of Biomedical Signals Using Cubic Base Splines

Mukhriddin Abduganiev[1(✉)], Rakhimjon Azimov[1], and Lazizbek Muydinov[2]

[1] Department of Information Technologies, Andijan State University named after Zahiriddin Muhammad Babur, Andijan, Uzbekistan
mr.muhriddin.20@gmail.com

[2] Department of Methodology of Primary Education, Andijan State University named after Zahiriddin Muhammad Babur, Andijan, Uzbekistan

Abstract. In this article, the issues of digital processing and restoration of electroencephalogram (EEG) signals from biomedical signals are considered, the location of 21 sensors in the EEG apparatus along the brain, the naming of the sensors, their connection types, the use of bipolar coupling in the detection of disease symptoms, interpolation of received signals, disease symptoms. The processes of separating parts into scales have been studied. During the work, the B-spline function was selected as the most convenient mathematical model for digital processing of EEG signals, and the construction of the B-spline function was presented. Based on the constructed mathematical model, an algorithm for restoring the electroencephalogram signals by dividing the problematic parts into scales was developed, and the absolute error in the restoration of EEG signals was estimated.

Keywords: EEG signal · EEG apparatus · Bipolar coupling · Monopolar coupling · Spline functions · Cubic B-spline · Piece-polynomial methods

1 Introduction

Today, extensive scientific and research work is being carried out on digital processing of biomedical signals. We know that all the organs of our body are controlled by the brain. That is, hands and feet determine the principles of working of all our internal organs. Electrochemical impulses detected from brain tissue take the form of different types of waves. Sensors that measure electrochemical impulses placed around the brain of patients who are expected to be analyzed have different names depending on the area where they are located. Their location differs depending on specific areas. Currently, many devices are equipped with sensors as shown in the image below. The EEG device can have 4, 8, 10, 16, 19 and 21 sensors [1–3]. The more sensors there are, the higher the result and the easier it is to identify the symptoms of the disease. Sensors are named according to the area of the brain where they are located. For example, Fp1, Fp2 and FpZ in the forehead area, F3, F7, T3, T5, T3, C3, T5 and etc. in the left ear area. The image below shows how the sensors of the 21-channel EEG machine are placed across the brain [3] (Fig. 1).

H. Zaynidinov et al. (Eds.): IHCI 2022, LNCS 13741, pp. 18–26, 2023.
https://doi.org/10.1007/978-3-031-27199-1_3

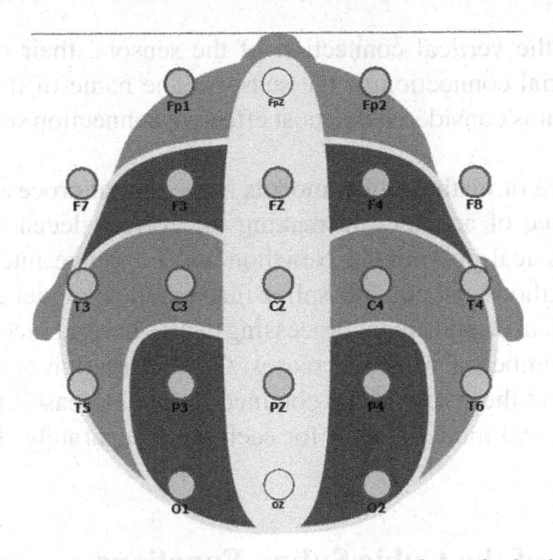

Fig. 1. The arrangement of the sensors along the brain.

From the picture above, we can see that the sensors detect changes in certain parts of the brain tissue. In the EEG device, the data from all 21 sensors is displayed on the screen at the same time.

2 Methods of Connecting the EEG Devices

In the EEG device's sensors have bipolar and monopolar connections. In many cases, monopolar connection hides the initial symptoms of the disease, that is, low-frequency signals. Taking this situation into account, we chose bipolar connections in our work. In this method, the sensors are connected in series (Fig. 2).

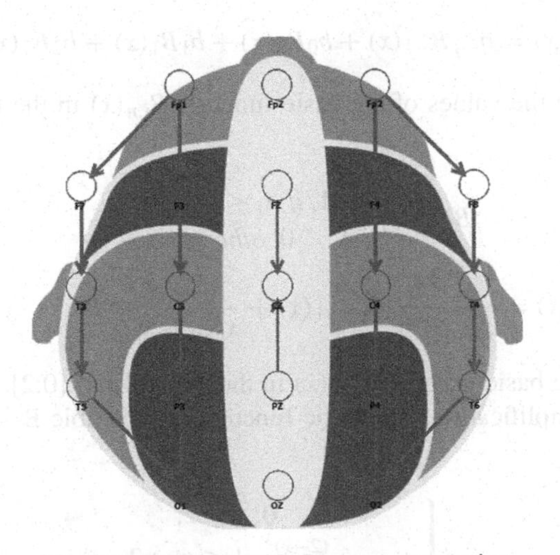

Fig. 2. Vertical (Double Banana) connection.

Figure 2 shows the vertical connection of the sensors, their difference from the monopolar is the serial connection of the sensors. The name of this method is called Double Banana, and it is considered the most effective connection scheme in diagnosing patients.

The correct choice of mathematical models in the digital processing of such signals leads to a high degree of accuracy in making the correct decision of the experts in the field. Today, classical polynomial Newthon and Lagrange interpolation methods, piece-polynomial methods and cubic B-spline mathematical model are used. The use of classical polynomials in digital signal processing causes several inconveniences. One of them is that as the number of values increases, the construction of the model becomes more complicated, and the errors of the obtained results increase. In piece-polynomial methods, a mathematical model is built for each piece separately. Therefore, the level of accuracy is high.

3 Construction of the Cubic Spline-Functions

We chose B-spline for digital processing of signals obtained by bipolar coupling. Below we will look at constructing a cubic B-spline function. If the considered function is not smooth enough, then it is appropriate to smooth this function using spline functions. Approximation of the spline function using third degree B-spline gives good results [4, 5]. Below is the formula of the B-spline in the interval [a:b], in which the interpolated function $f(x)$ of arbitrary $S_n(x)$ of n - degree has the following form (1):

$$f(x) \cong S_n(x) = \sum_{i=-1}^{n-1} b_i B_{i,n}(x), a \leq x \leq b, \tag{1}$$

Here n is the level of the spline, b_i are the coefficients, $B_{i,n}(x)$ represents the basic function. Based on the above formula, we will construct the cubic B-spline function in the case of n = 3 (2).

$$S_3(x) = b_{-1}B_{-1}(x) + b_0B_0(x) + b_1B_1(x) + b_2B_2(x), \tag{2}$$

(2) we calculate the values of the basic function $B_{i,n}(x)$ in the formula as follows (3), (4):

$$B_{i,0}(x) = \begin{cases} 1, \text{ } if \text{ } x_i \leq x < x_{i+1}, \\ 0, \text{ } otherwise. \end{cases} \tag{3}$$

$$B_{i,n}(x) = \frac{x - x_i}{x_{i+n} - x_i} B_{i,n-1}(x) + \frac{x_{i+n} - x}{x_{i+n} - x_{i+1}} B_{i+1,n-1}(x), \tag{4}$$

We calculate the basic function $B_{i,3}(x)$ in the interval $x \in [0:2]$. After a number of calculations and simplifications, the basic function of the cubic B-spline will have the following form (5).

$$B_3(x) = \begin{cases} 0, x \geq 2, \\ \frac{(2-x)^3}{6}, 1 \leq x < 2, \\ \frac{1+3(1-x)+3(1-x)^2-3x^3}{6}, 0 \leq x < 1, \\ B(-x), 0 < x \end{cases} \tag{5}$$

The graph of the basic function of the cubic B-spline in the interval [−2:2] is as follows in the Matlab working environment (Fig. 3).

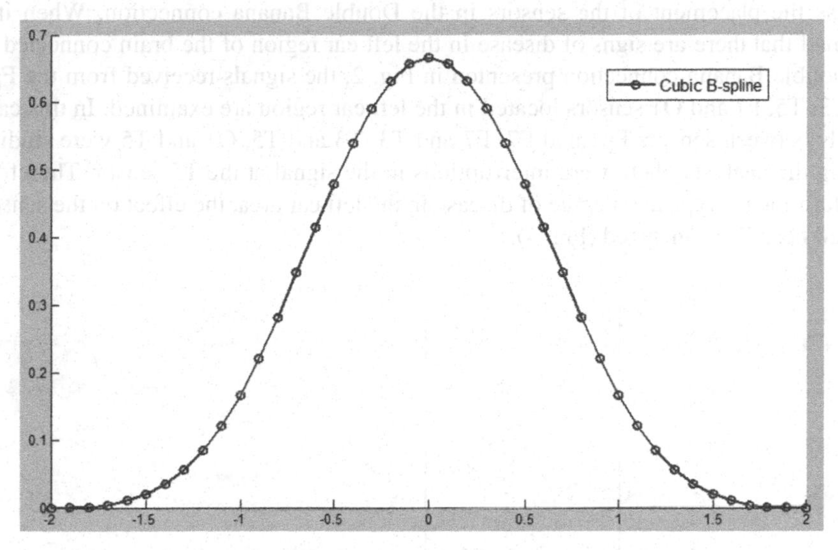

Fig. 3. Cubic B-spline plot.

During the calculation of $S_3(x)$, signal recovery is achieved using all values of the B-spline in the range [0:1].

b_i in the formula (1) is calculated differently depending on how many of the given values are taken in the process of calculating the coefficients [4–8]. We determine the coefficients b_i using the given values using the three-, five-, seven-point formula (6), (7), (8).

3-point formula for calculating b_i coefficients:

$$b_i = \left(\frac{1}{6}\right)(-f_{i-1} + 8f_i - f_{i+1}), \tag{6}$$

5-point formula for calculating b_i coefficients:

$$b_i = \left(\frac{1}{36}\right)(f_{i-2} - 10f_{i-1} + 54f_i - 10f_{i+1} + f_{i+2}), \tag{7}$$

7-point formula for calculating b_i coefficients:

$$b_i = \left(\frac{1}{216}\right)\left(-f_{i-3} + 12f_{i-2} - 75f_{i-1} + 344f_i - 75f_{i+1} + 12f_{i+2} - f_{i+3}\right), \tag{8}$$

where, is the value of the initial given points. Where f_i is the value of the initial given points.

4 Identify Symptoms of the Disease

Before the digital processing of the EEG signal based on the B-spline model, we briefly discuss the placement of the sensors in the Double Banana connection. When it is assumed that there are signs of disease in the left ear region of the brain connected by the Double Banana connection presented in Fig. 2, the signals received from the Fp1, C3, T3, T5, F7 and O1 sensors located in the left ear region are examined. In this case, signals between sensors Fp1 and F7, F7 and T3, T3 and T5, O1 and T5 were studied. During the analysis, there were interruptions in the signal at the T5 sensor. Therefore, to determine the type and degree of disease in the left ear area, the effect on the sensors located near T5 is analyzed (Fig. 4).

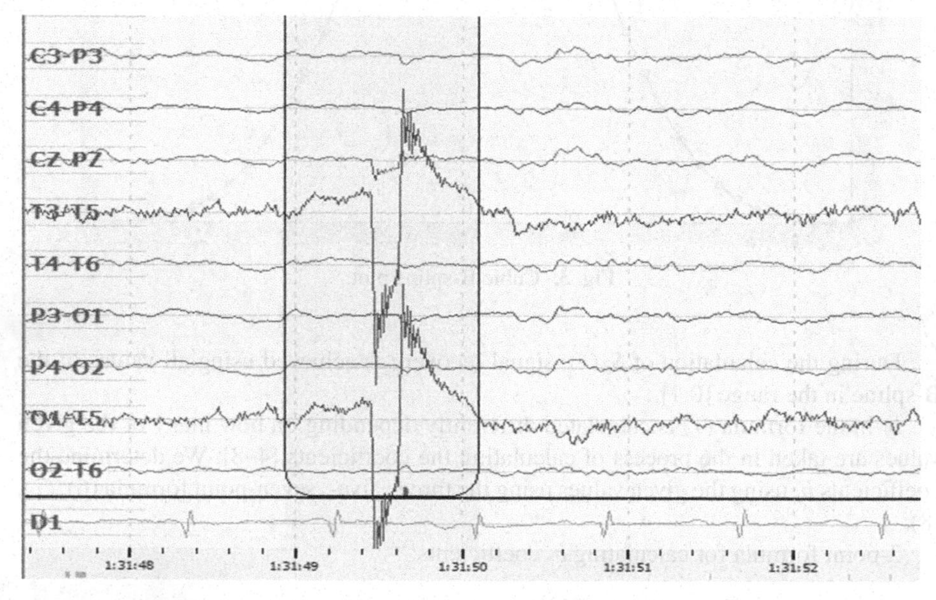

Fig. 4. In a bipolar connection scheme. A sign of artifactual disease.

The figure above shows the graphical representation of the signal frequencies detected by the sensors T3-T5 and O1-T5. Such images make it easier to identify diseases in the brain tissue. But it takes a lot of time to isolate the intervals with disease symptoms as shown in Fig. 4.

5 Way to Create Algorithm for Separation and Restoration of the Values

We know that it is important to diagnose and treat patients in a short period of time. Algorithms for EEG signal segmentation and reconstruction using mathematical models are good to solve this problem. Below is the algorithm for identifying disease symptoms using B-spline [9–13] (Fig. 5).

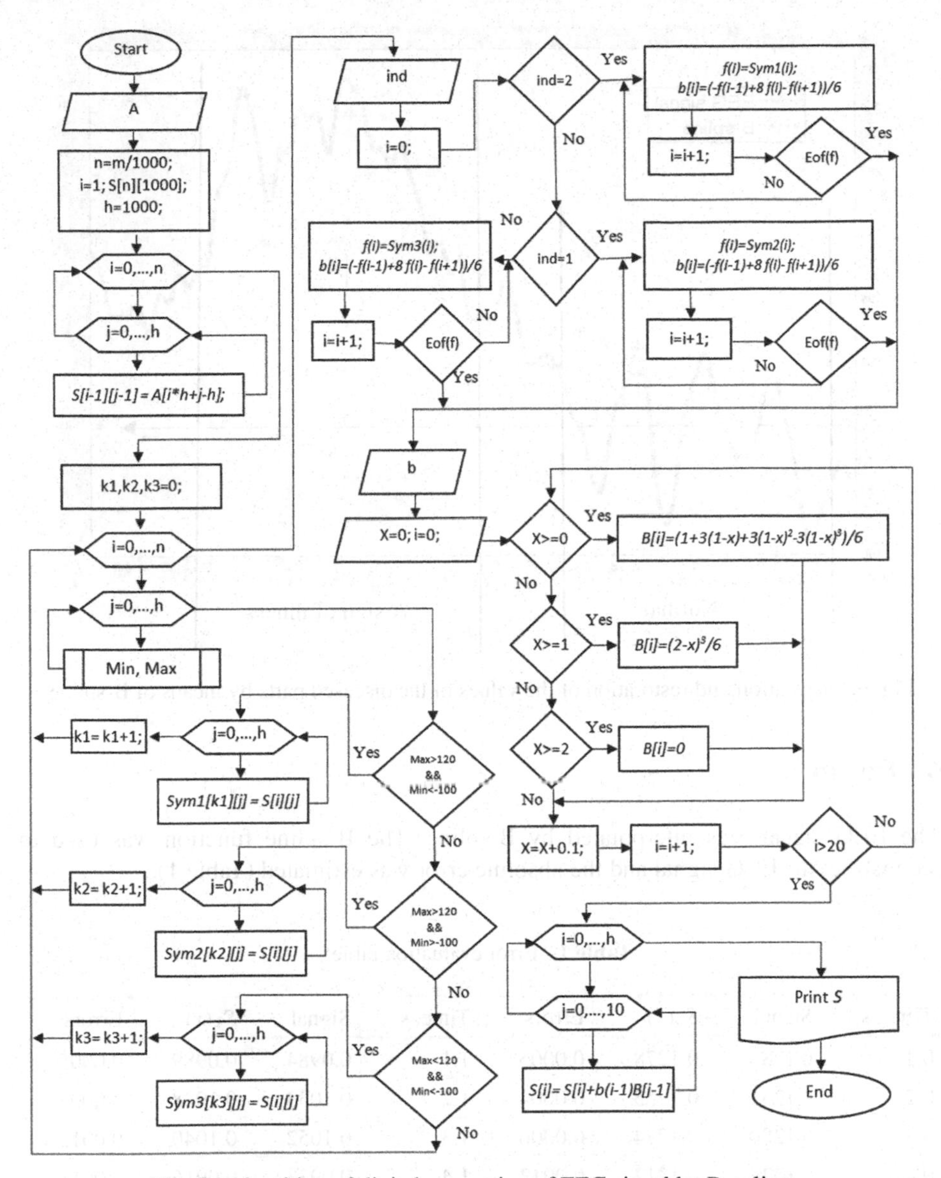

Fig. 5. Algorithm of digital processing of EEG signal by B-spline.

The digital processing of the EEG signal is described in the algorithm, in which the parts with an amplitude higher than 120 are recorded separately in the arrays (Sym1, Sym2, Sym3) and their values are restored using the selected B-spline. This allows the doctor to identify the primary symptoms of the disease in a short time and make an accurate diagnosis. On the basis of the above-mentioned algorithm, the parts with disease symptoms were separated and their values were restored [4, 8] (Fig. 6).

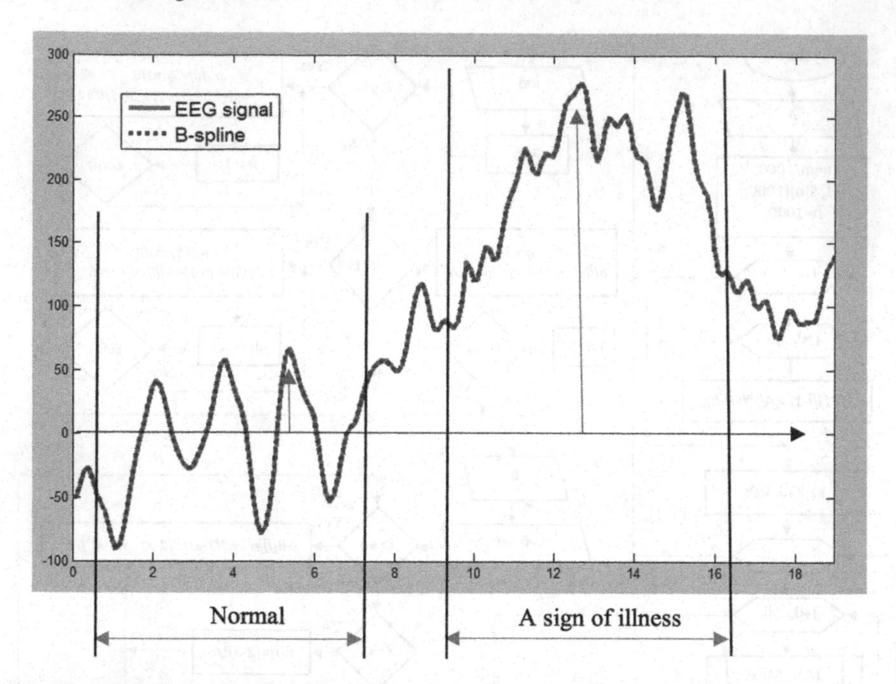

Fig. 6. Separation and restoration of the values of the diseased parts by means of B-spline.

6 Result

The EEG signal was interpolated by B-spline. The B-spline function was used to reconstruct the EEG signal and the absolute error was estimated (Table 1).

Table 1. Error evaluation table.

Time, s	Signal	$S_3(x)$	Errors	Time, s	Signal	$S_3(x)$	Errors
0.1	0.1283	0.1278	0.0005	1.1	0.0984	0.0989	0.0005
0.2	0.1251	0.1255	0.0004	1.2	0.1033	0.1038	0.0005
0.3	0.1280	0.1274	0.0006	1.3	0.1052	0.1040	0.0012
0.4	0.1224	0.1212	0.0012	1.4	0.0933	0.0916	0.0017
0.5	0.1123	0.1121	0.0002	1.5	0.0780	0.0774	0.0006
0.6	0.1112	0.1120	0.0008	1.6	0.0748	0.0757	0.0009
0.7	0.1183	0.1188	0.0005	1.7	0.0850	0.0863	0.0012
0.8	0.1211	0.1205	0.0006	1.8	0.0962	0.0965	0.0003
0.9	0.1136	0.1124	0.0012	1.9	0.0976	0.0972	0.0004
1	0.1025	0.1019	0.0006	2	0.0921	0.0915	0.0006
An absolute error							0.0017

7 Conclusion

In conclusion, it can be said that the issues of digital processing and restoration from biomedical signals to electroencephalogram (EEG) signals have been considered. During the work, the placement of 21 sensors in the EEG apparatus along the brain, their connection types, interpolation of the received signals, and separation into parts indicating disease symptoms were carried out. The B-spline function was selected as a mathematical model convenient for segmentation of EEG signals, and the construction of the B-spline function was presented. The absolute error in EEG signal recovery was found to be 0.0017. Based on the constructed mathematical model, an algorithm for restoring electroencephalogram signals by dividing them into problematic parts has been developed. Based on the algorithm, it is concluded that if the frequency of the signal is in the range of 5–100 Hz, it is normal, and if the frequency is in the range of 120–300 Hz, there are signs of illness. This will definitely help doctors to diagnose patients correctly.

References

1. Zhang, H., Zhao, M., Wei, C., Mantini, D., Li, Z., Liu, Q.: EEGdenoiseNet: a benchmark dataset for deep learning solutions of EEG denoising. J. Neural Eng. **18**(5) (2021). https://doi.org/10.1088/1741-2552/ac2bf8
2. Dadebayev, D., Goh, W.W., Tan, E.X.: EEG-based emotion recognition: review of commercial EEG devices and machine learning techniques. J. King Saud Univ. – Comput. Inf. Sci. **34**(7), 4385–4401 (2022). https://doi.org/10.1016/j.jksuci.2021.03.009
3. Soufineyestani, Malisa, Dowling, Dale, Khan, Arshia: Electroencephalography (EEG) technology applications and available devices. Appl. Sci. **10**(21), 7453 (2020). https://doi.org/10.3390/app10217453
4. Singh, D., Singh, M., Hakimjon, Z.: B-Spline approximation for polynomial splines. In: Signal Processing Applications Using Multidimensional Polynomial Splines. SAST, pp. 13–19. Springer, Singapore (2019). https://doi.org/10.1007/978-981-13-2239-6_2
5. Zaynidinov, H.N.: Cubic basic splines and parallel algorithms. Int. J. Adv. Trends Comput. Sci. Eng. **9**(3), 3957–3960 (2020). https://doi.org/10.30534/ijatcse/2020/219932020
6. Zaynidinov, H., Mallayev, O., Nurmurodov, J.: Parallel algorithm for constructing a cubic spline on multi-core processors in a cluster (2020). https://doi.org/10.1109/AICT50176.2020.9368680
7. Hidayov, O., Ukaegbu, I.A., Zaynidinov, H., Lee, S.G.: Comparative analysis of piecewise-polynomial of local bases. In: International Conference on Advanced Communication Technology, ICACT, vol. 2 (2010)
8. Xakimjon, Z., Bunyod, A.: Biomedical signals interpolation spline models (2019). https://doi.org/10.1109/ICISCT47635.2019.9011926
9. Zaynidinov, H., Bakhromov, S., Azimov, B., Makhmudjanov, S.: Comparative analysis spline methods in digital processing of signals. Adv. Sci. Technol. Eng. Syst. J. **5**(6), 1499–1510 (2020). https://doi.org/10.25046/aj0506180
10. Hakimjon, Z., Artikova, M.: Analysis of the use of wavelets for processing signals of a seismological nature (2021). https://doi.org/10.1109/ICISCT52966.2021.9670254
11. Zaynidinov, X.N., Turakulov, A.A., Mullajonova, F.T.: Using the wi-fi technology and devices to transmit results of human body biosignals processing. IARJSET **8**(10),(2021). https://doi.org/10.17148/IARJSET.2021.81001

12. Zaynidinov, H., Singh, D., Makhmudjanov, S., Yusupov, I.: Methods for determining the optimal sampling step of signals in the process of device and computer integration. In: Kim, J.-H., Singh, M., Khan, J., Tiwary, U.S., Sur, M., Singh, D. (eds.) IHCI 2021. LNCS, vol. 13184, pp. 471–482. Springer, Cham (2022). https://doi.org/10.1007/978-3-030-98404-5_44
13. Zaynidinov, H., Bahromov, S., Azimov, B., Kuchkarov, M.: Lacol interpolation bicubic spline method in digital processing of geophysical signals. Adv. Sci. Technol. Eng. Syst. J. **6**(1), 487–492 (2021). https://doi.org/10.25046/aj060153
14. Singh, M., Zaynidinov, H., Zaynutdinova, M., Singh, D.: Bi-cubic spline based temperature measurement in the thermal field for navigation and time system. J. Appl. Sci. Eng. **22**(3) (2019). https://doi.org/10.6180/jase.201909_22(3).0019
15. Zaynidinov, Khakimjon Nasridinovich, Juraev, Jonibek Uktamovich, Boytemirov, Asror Mahmadostovich: Digital processing of biomedical signalsin haar's part-wavelet models. Asian J. Multidimens. Res. **10**(9), 130–139 (2021). https://doi.org/10.5958/2278-4853.2021.00656.X
16. Zaynidinov, K.N., Anarova, S.A., Jabbarov, J.S.: Determination of dimensions of complex geometric objects with fractal structure. In: Kim, J.-H., Singh, M., Khan, J., Tiwary, U.S., Sur, M., Singh, D. (eds.) IHCI 2021. LNCS, vol. 13184, pp. 437–448. Springer, Cham (2022). https://doi.org/10.1007/978-3-030-98404-5_41
17. Zaynidinov, H.N., Singh, D., Yusupov, I., Makhmudjanov, S.U.: Algorithms and service for digital processing of two-dimensional geophysical fields using octave method. In: Kim, J.-H., Singh, M., Khan, J., Tiwary, U.S., Sur, M., Singh, D. (eds.) IHCI 2021. LNCS, vol. 13184, pp. 460–470. Springer, Cham (2022). https://doi.org/10.1007/978-3-030-98404-5_43
18. Jumaniyozov, D., Omirov, B., Redjepov, S., Zaynidinov, K.: Irreversibility of 2D linear ca on pentagonal lattice over periodic boundary condition and garden of eden. SSRN Electron. J. (2022). https://doi.org/10.2139/ssrn.4047903
19. Zaynidinov, H., Sayfiddin, B., Bunyod, A., Umidjon, J.: Parallel processing of signals in local spline methods (2021). https://doi.org/10.1109/ICISCT52966.2021.9670409
20. Zaynidinov, H., Mallayev, O., Kuchkarov, M.: Parallel algorithm for modeling temperature fields using the splines method (2021). https://doi.org/10.1109/IEMTRONICS52119.2021.9422645
21. Zaynidinov, H., Ibragimov, S., Tojiboyev, G.: Comparative analysis of the architecture of dual-core blackfin digital signal processors (2021). https://doi.org/10.1109/ICISCT52966.2021.9670135
22. Zaynidinov, H., Ibragimov, S., Tojiboyev, G., Nurmurodov, J.: Efficiency of parallelization of haar fast transform algorithm in dual-core digital signal processors (2021). https://doi.org/10.1109/ICCCE50029.2021.9467190

Methods for Creating a Morphological Analyzer

Elov Botir Boltayevich(✉), Hamroyeva Shahlo Mirdjonovna, and Axmedova Xolisxon Ilxomovna

Tashkent State University named after Alisher Navoi University of Uzbek Language and Literature, Tashkent O'qituvchi Street 103, Tashkent, Uzbekistan

{elov,a.xolisa}@navoiy-uni.uz

Abstract. The morphological analysis process is an important component of natural language processing systems such as spelling correction tools, parsers, machine translation systems, and electronic dictionaries. This article describes the stages of a text analyzer, methods for creating a morphological analyzer and a morphological generator. Ways to use the NLTK package tools in Python when creating a morphological analyzer, examples of software codes are given. Also, morphological analyzer structure and architecture are presented on the basis of the morphological analysis process (flexion, derivative, affixpids detection, compound forms).

Keywords: Natural language processing · NLP · Python · Morphological analyzer · Token · Lemmatization · Stemming · Morphological generator · Search engine · Stemming algorithm · PorterStemmer

1 Introduction

Morphology is a section that studies the grammatic meanings of words through morpheme. Morpheme is the smallest unit of language that can not be divided into meaningful parts. This article analyzes the issue of creating morphological analyzer and morphological generator for languages other than English using stemming and lemmatization, stemmer and lemmatizer, machine learning tools, search engines. Today, a number of scientists around the world are conducting scientific research on the creation of a morphological analyzer. In particular, Adnan Öztürel, Tolga Kayadelen and ISIN Demirsahin presented a broad coverage model of Turkish language morphology and an open source morphological analyzer that implements it [1]. This model covers the subtle aspects of Turkish morphology, syntax, from which it can be used as a guide in the development of a language model. The Model performs Turkish morphotactic using OpenFst as finite state transducers and Morphophonemic processes Thrax grammatics. Arabic linguists Y. Jeefer and K. Bouzoubaa Arabic Morphological Analyzers presented the methods of use in syntactic analysis programs, search engines and machine translation systems.

2 Materials and Methods

Natural Language understanding the first version of the Arabic language corps dedicated to the evaluation of Natural Language was created [2]. The four most common and

H. Zaynidinov et al. (Eds.): IHCI 2022, LNCS 13741, pp. 27–38, 2023.
https://doi.org/10.1007/978-3-031-27199-1_4

most advanced morphological analyzers (Hunmorf-Okamorf, Hunmorf-Foma, Humor and Hunspell) for the Hungarian language were analyzed, compared by G. Szabó, L. Kovács. They compared the token systems of annotation instead of the lemmatization properties of analyzers [3].

1. NLPni ikkita asosiy komponentga ajratish Understanding, NLU).
2. Tabiiy tini generatsiya qilish (Natural Language Generation, NLG).

The steps required in the performance of NLP tasks will be considered. The source of the natural language can be speech (sound) or text. Stages of text processing are presented in Fig. 1.

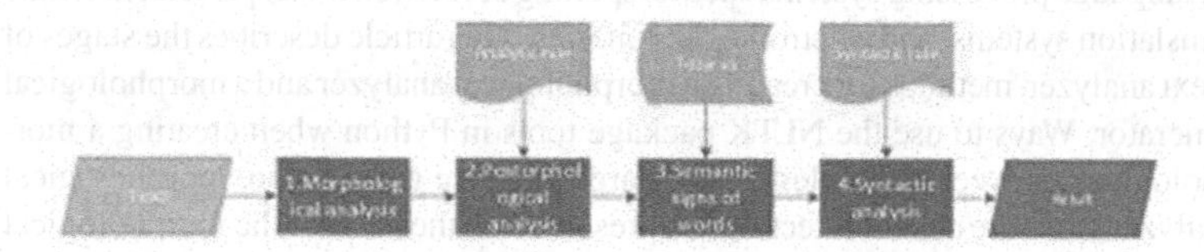

Fig. 1. Stages of text analyzer

In the field of Uzbek computer linguistics, morphological analyzer and morphological generator has not yet been created. In this article, we will consider ways to solve this issue based on world experience.

When creating a morphological Analyzer, It is worthwhile to study the following concepts:

1. morphological units;
2. stemmer;
3. lemmatization;
4. development of Stemmer in Uzbek language;
5. Morphological Analyzer;
6. Morphological Generator;
7. search engine development.

3 Morphological Units

Morpheme is the smallest meaningful part of the language. Morpheme is divided into two types: the limbs are also referred to as free morphemes, since they can also be used without adding affix (suffix) in sentences. (Additional) are not used separately in the language. Because they can not have a lexical meaning in an independent forms and always exists together with independent morphemes. Let's look at the word unbelievable in English. Here the believe is a predicate or a free morpheme. It means quot; independently; flour and able morphemes are affiks (addition) or paired morphemes. These morphemes can not exist in an independent form, they do not mean, they are used together with the core [22, 23].

Natural languages according to the formation of the grammatic form are divided into the following groups:

1. (isolating languages) – Languages of the peoples of South East Asian countries [24];
2. (agglutinative languages) – Turkic, bantu, Mongolian, Finnish-Ugric languages [22, 25];
3. (inflecting languages) – Indo-European and families of som languages [26, 27].

1. The dictionary composition of these languages consists mainly of singlestringed limbs, which do not have features of punctuation, punctuation. Therefore, in these languages word-building exercises, there are only loads that perform the function of auxiliary words. In amorphous languages, words consist only of independent morphemes; they do not include the data of the time (past, present and future) and the number (Unit or plural).
2. In agglyutinative languages, words will consist of a suffix and a suffix attached to it; the morphological composition of the word (suffix) is clearly distinguished. Bunda represents each additional separate meaning, task. For example, in Turkic languages, including Uzbek, the forms of words and words are formed by the addition of suffixes to the basis with a certain consistency.
3. Flektiv languages – it is characterized by the fact that the appendages merge with the core and are absorbed into it. In such languages, thematikmatic meanings are expressed by inflection, book in Arabic (unit) (plural). In Russian, the drug (unit) is a druzya (plural). Flektiv languages include IndoEuropean and som language famllles. Flektlv languages are dıvıded ınto synthetıc and analytical languages. In synthetic languages, the grammatic meanings (interaction of words in the sentence) are expressed by means of form-forming suffixes (eg., Russian, German). In analytical languages, the grammatic meanings are expressed not by means of word forms (form- forming suffixes), but by means of auxiliary words, word order, tone(English, French, Spanish). In Flektiv languages, words are divided into simple units, all these simple units show different meanings.
4. Polisynthetic languages– the main unit of speech is the word-sentence. There can not be a strict limit between the classified languages, since some language phenomena that occur in one language can also occur in others. For example, Oceanic languages can be described as both amorphous and agglyutinative languages [28].

Morphological (typological) classification of languages it is important to group the languages of the world according to certain morphological features, to create a general drawing of them [29, 30, 31].

Morphological processes are divided into the following others [32]:

1. (inflection);
2. (derivation);
3. (semiaffixes);;
4. (combining forms);
5. (cliticization).

On the basis of this stage, the structure of the work of the morphological analyzer of the Uzbek language is shown in Fig. 2 below:

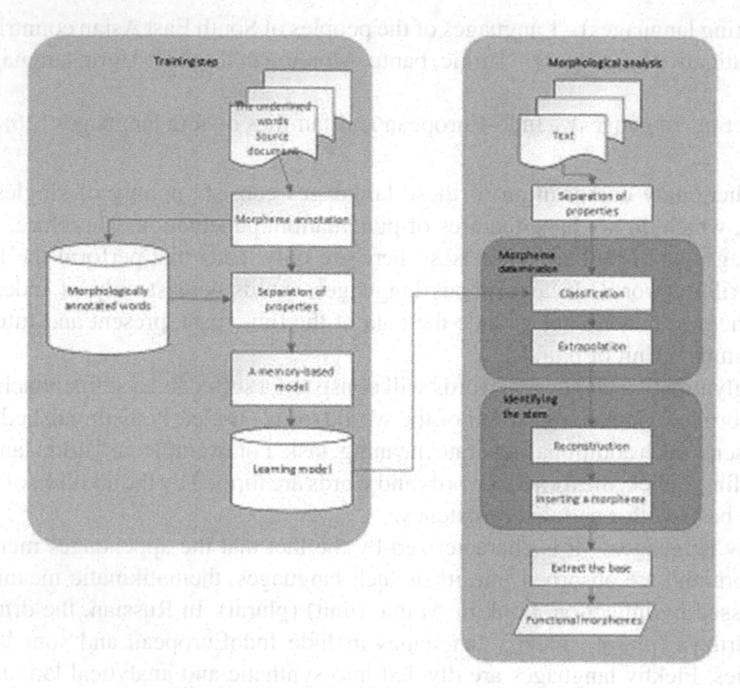

Fig. 2. Morphological analyzer structure

At the stage of infection, the word is transformed into a form that expresses the person, Number, time, gender and other grammatic categories. At this stage, the syntactic category of the lexeme remains unchanged.

At the stage of derivation, the syntactic category of the word changes. At the stage of determining semifixes, morphemes are formed in the form of compound words, abbreviations. For example: noteworthy, antisocial, anticlockwise, etc.

Stemmer: In the process of stemmatization, the determination of the word core is carried out by removing suffixes from the word [33–35]. For example, the question from nature is analyzed as follows: from [nature] (conjugate suffixes Yas forming a syntactic form)

Search engines (Google, Yahoo, Yandex) use stemming mainly in identifying cores and storing them as indexed words in order to increase the accuracy of the search for information. Search engines call words that have the same meaning as synonyms, and this query can be some kind of more accurate implementation of the results.

Currently, many moroanalysts are using the stemming algorithm, developed by Martin Porter. This algorithm is designed to replace and analyze suffixes that are present mainly in English words. To implement stemming in the NLTK package, we can create a copy of the PorterStemmer class, and then execute stemming by calling the stem() method

(Fig. 3). In NLTK, we will consider the stemming process using the PorterStemmer class [36, 37]:

```
    import nltk
from nltk.stem
import PorterStemmer
stemmerporter = PorterStemmer()
print(stemmerporter.stem('removing'))
print(stemmerporter.stem('happies'))
result
    remov
    happi
```

PorterStemmer the class has a lot of vocabulary and word forms in English. The process of determining the cores consists of several stages: the word shorter becomes a word or a form with a similar meaning. The Stemmer I interface defines the stem () method and all Stemmers are sampled from the Stemmer I interface [36].

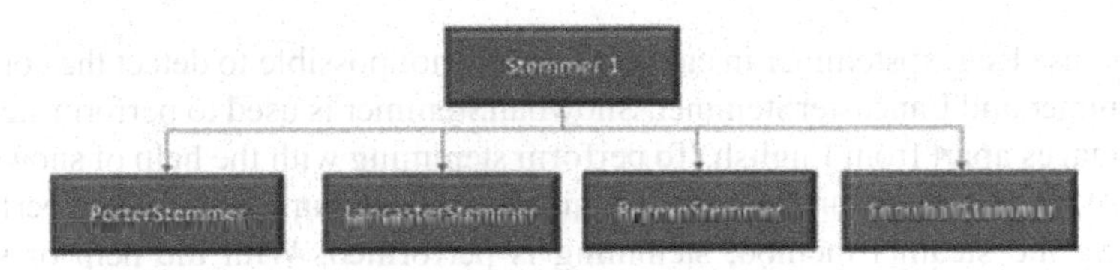

Fig. 3. Stemming methods in the NLTK package

Another stemming algorithm, known as the Lancaster stemming algorithm, was developed at Lancaster University. Similar to the PorterStemmer class, the Lancaster Stemmer class is also used to implement Lancaster stemming in NLTK. But the main difference between the two algorithms is that Lancaster stemming involves the use of more words of different characteristics in comparison with Porter Stemming.

```
from nltk.stem import LancasterStemmer
stemmerlan=LancasterStemmer()
print(stemmerlan.stem('remov' ))
print(stemmerlan.stem( 'happies'))
```

---

remov
happy

We can also build our own Stemmer in NLTK using Regexpstemmer. It works by taking a line and removing the prefix or suffix of the word when compatibility is found. Let's look at an example of stemming using Regexpstemmer in NLTK:

```
from nltk.stem import RegexpStemmer
stemmerregexp=RegexpStemmer('ing')
print(stemmerregexp.stem('removing' ))
print(stemmerregexp.stem( 'happiness'))
print(stemmerregexp.stem( 'pairing'))
```

---

remov
happiness
pair

We can use Regexpstemmer in cases where it is not possible to detect the core using PorterStemmer and LancasterStemmer. SnowballStemmer is used to perform stemming in 13 languages apart from English. To perform stemming with the help of snowststemmer, first of all, it is necessary to specify tilni where the stemming should be performed. Then, using the steam() method, stemming is performed. With the help of snowststemmer, the process of stemming in NLTK in Spanish and French is carried out as follows:

```
import nltk
from nltk.stem import SnowballStemmer
print(" ".join(SnowballStemmer.languages)) //see which languages are supported
spanishstemmer=SnowballStemmer('spanish') // language selection
print(spanishstemmer.stem( 'comiendo'))
frenchstemmer=SnowballStemmer( 'french') //language selection
print(frenchstemmer.stem('manger')) // word steam
```

Let's look at the following code available, which allows us to implement stemming:

```
Class StemmerI(object):
"""
"""
def stem(self, token):
"""
"""
raise NotImplementedError()
```

With the help of several Stemmers, we will consider the code that will be used to perform stemming:

```
import nltk
from nltk.stem.porter import PorterStemmer
from nltk.stem.lancaster import LancasterStemmer
from nltk.stem import SnowballStemmer
def obtain-tokens():
text-file = open("D:/Examples/Examples.txt", "r")
stem = text-file.read()
text-file.close()
tok = nltk.word-tokenize(stem)
return tok
def stemming(filtered):
stem=[]
for x in filtered:
stem.append(PorterStemmer().stem(x))
return stem
tok=obtain-tokens()
print("tokens is %s") stem-tokens= stemming(tok)
print(stem-tokens)
print("After stemming is %s"')
res=dict(zip(tok,stem-tokens))
print(res)
```

3.1 Lemmatization

Lemmatization is the process of determining the shape of the head (lexical appearance) of the word. The word formed after the lemmatization plays an important role. Through the morphy () method, the lemmatization process is performed in Wordnetlemmatizer. If the entered word is not found in the WordNet, it remains unchanged. Pos is formed part of the word category of the word entered in the argument. Let's look at the process of lemmatization in nltk [36, 38–40]: in the argument. Let's look at the process of lemmatization in nltk [36, 38–40]:

```
import nltk
from nltk.stem import WordNetLemmatizer
lemmatizer-output=WordNetLemmatizer()
print(lemmatizer-output.lemmatize('working'))
print(lemmatizer-output.lemmatize('working',pos='v'))
print(lemmatizer-output.lemmatize('works'))
```

WordNetLemmatizer the library generates Lemma through the Murphy () method using an information system (ontology) called WordNet Corps. If the lemma is not formed, the word is returned in the initial (original) form. For example, the returned lemma unit forms for works: work. In the Uzbek language, the word lemmasi stemi of books will be a book. The following program code shows the difference between the stemming and lemmatization processes:

```
import nltk
from nltk.stem import PorterStemmer
stemmer-output=PorterStemmer()
print(stemmer-output.stem(happiness)) from nltk.stem import WordNetLem-
matizer
lemmatizer-output=WordNetLemmatizer()
print(lemmatizer-output.lemmatize('happiness'))
```

```
happi
happiness
```

In the previous code, happiness became happi as a result of the stemming process. Lemmatization can not find the core of the word happiness. Therefore, he returns the word happiness. The process of stemmatization in the Uzbek language is relatively easy, since in the Uzbek language lemmings often correspond to stem. But in the Uzbek language, too, cases of Flexion are flying. For example: in words such as: achievement, interrogative, the predicate is not an achievement, but an achievement, in the form of an interrogative. In such cases, the method of finding stem is used as above.

Polyglot – the software used to provide models called morphessor models, which are used to identify morphemes from tokens [41, 42]. The purpose of the Morpho project is to generate unmanaged processes for processing information. On the basis of the Morpho project, morphemes with the smallest unit of the syntax are created. Natural language morphemes play an important role in processing. Morphemes are used in automatic recognition and creation. With the help of polyglot's dictionaries, morphessor models were developed in 50000 tokens of different languages.

3.2 Morphological Analyzer

NIn the process of morphological analysis, the acquisition ofmatikmatic information is carried out taking into account the meanings of tokens based on attachments. Morphological analysis can be carried out in three ways:

1. morphology based on morpheme;
2. morphology based on lexeme;
3. ordinal-based morphology.

Morphological Analyzer is interpreted as a program that is responsible for analyzing the morphological composition of a particular token. Morphological analyzer analyzes the given token and form data such as Category, variety of meanings. To carry out morphological analysis on a token without a given space, a pyEnchant dictionary is used. To determine the type of word, a set of rules is required. We can determine the word category by the following rules:

1. Morphological rules. Information about suffixes will help to determine the word category. For example, in English, the suffixes-ness and-ment are combined with nouns. So in English it is possible to determine its category, depending on the suffixes that make up the word chord. But in Uzbek it is not possible to determine the category of the word by this method. Therefore, the Uzbek language morpholexicon should be developed.
2. Syntactic rules. Contextual information helps to determine the word category. For example, if we find a word belonging to the category of nouns, then the syntactic rule is useful to determine whether the adjective comes before or after the noun category.
3. Semantic rules. Semantic rules are significant in determining the word category. For example, if it is determined that the word represents the place name, then it can be concluded that the noun belongs to the category of speech.
4. Open class (group). Every day a new word is added to the list of words in this group, their number increases. Words in the open class are usually words belonging to the category of nouns. Prepositions, in principle, belong to the closed class. For example, in the list there can be an infinite number of words.
5. Definition of word categories (POS, Part of Speech): a set of tags of word categories contains information that helps to determine the morphological feature. For example, oynadi Sor comes with a word belonging to the noun category in the third person unit.
6. Omorf package is licensed by GNU GPL version 3, it is used for many functions such as modeling, morphological analysis, rules-based machine translation, and data search, statistical machine translation, morphological segmentation, ontologies, spell checking and Correction.

3.3 Morphological Generator

Morphological generator is a program that performs the function of morphological generalization; it can be considered a dependent function of morphological analysis. Here the original word is formed if the description of the word according to the number, Category, core and other information is given. For example, **ozak=bormoq, gap bolagi = kesim, zamon=hozirgi and when a third person comes along with the subject of the unit, morphologically generator forms its becoming form.** There are many Python-based software applications that perform morphological analysis and generation:

1. **ParaMorfo:** It is a noun in Spanish and guarani, used for morphological formation and analysis of adjectives and verbs.
2. **HornMorpho:** It is used for morphological formation and analysis of nouns and verbs in the Oromo and Amharic languages, as well as Tigrinya verbs.
3. **AntiMorfo:** It is used for morphological creation and analysis of adjectives, verbs and nouns in the night language, as well as Spanish verbs.
4. **MorfoMelayu:** It is used for morphological analysis of words in the Malay language.
5. **Morph** morphological generator and analyzer for English.

6. **Morphy** morphological generator for German language, analyzer and POS tagger.
7. **Morphisto** it is a morphological generator and analyzer for the German language.
8. **Morfette** performs controlled learning (flexion morphology) of Spanish and French.

From the above feedback, it is possible to formulate the architecture of the text analyzer software (Fig. 4) as follows.

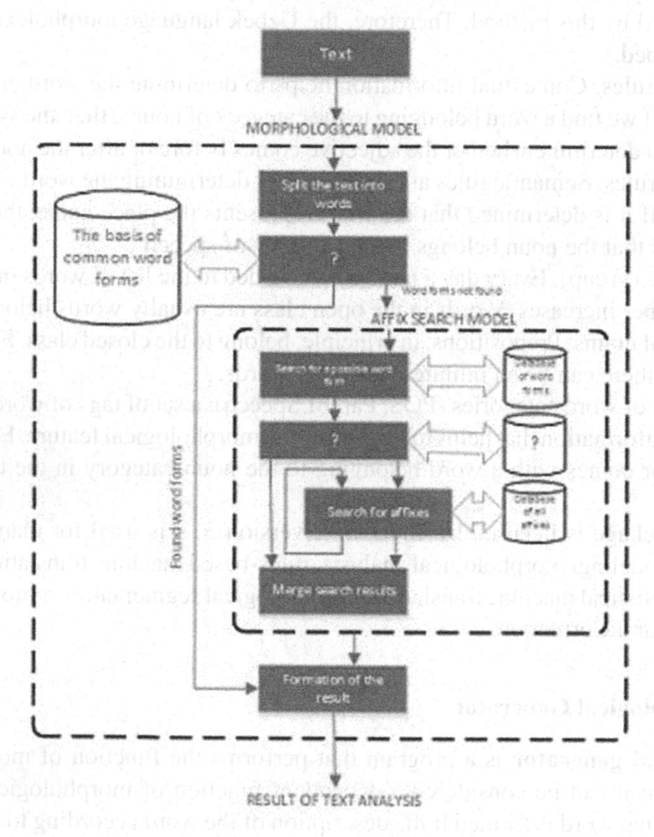

Fig. 4. Text Analyzer Software Architecture

4 Conclusion

There are many networks of computer linguistics. For processing (analysis) or generalization of the text, it is necessary to perform a number of linguistic actions on the given text. This article considered the implementation of stemming, lemmatization and morphological analysis and generalization activities with the help of NLTK package tools. Also, search engines and methods of their implementation were discussed.

References

1. Öztürel, A., Kayadelen, T., Demir¸sahin, I.: A syntactically expressive morphological analyzer for Turkish. In: FSMNLP 2019 14th International Conference on Finite-State Methods and Natural Language Processing, Proceedings (2019). https://doi.org/10.18653/v1/w19-3110
2. Jabbar, A., Iqbal, S., Akhunzada, A., Abbas, Q.: An improved Urdu stemming algorithm for text mining based on multi-step hybrid approach. J. Exp. Theor. Artif. Intell. **30**(5), 1–21 (2018). https://doi.org/10.1080/0952813X.2018.1467495
3. Szabó, G., Kovács, L.: Benchmarking morphological analyzers for the Hungarian language. Annales Mathematicae et Informaticae **49** (2018). https://doi.org/10.33039/ami.2018.05.001
4. Jurafsky, D.S., Martin, J.H., Kehler, A., Linden, K.V., Ward, N.: Speech and Langauge Processing, P. 950. Prentice Hall, Englewood Cliffs (2000)
5. Mohri, M.A.: Finite-state transducers in language and speech processing. Comput. Linguist. **23**, 269–311 (1997)
6. Alfred, V.A., Monika, S.L., Ravi, S., Djeffri, D.U.: Kompilyatory: prinsipy, texnologii i instrumentariy. OOO "I.D. Vilyamc", Per. sangl
7. Toldova, S.Yu.: Bonch-Osmolovskaya A.A. Avtomaticheskiy morfologicheskiy analiz. Fond znaniy "Lomonosov" (2011). www.lomonosov-fund.ru/enc/ru/encyclopedia:0127430
8. Sadykov, T., Kochkonbaeva, B.: Ob optimizatsii algoritma morfologicheskogo analiza. In: Shestaya Mejdunarodnaya konferensiya po kompyuternoy obrabotke tyurkskix yazykov. Turklang 2018. Trudy konferensii, Tashkent (2018)
9. Rodolfe, D.: Computational Linguistic Text Processing: Lexicon, Grammar, Parsing and Anaphora Resolution, p. 4–5. Nova Science Publishers, Inc., NewYork (2008)
10. Yermakov, A.: Morfologicheskiy analizator - osnova poiskovyx sistem. https://www.kv.by/archive/index2004154301.htm
11. Nojov, I.M.: Morfologicheskaya i sintaksicheskaya obrabotka teksta (modeli i programmy): dissertatsiya kand, p. 190. Moskva, nauk (2003)
12. Dybo, A.V., Sheymovich, A.V.: Avtomaticheskiy morfologicheskiy analiz dlya korpusov xakasskogo i drevnetyurkskogo yazykov. In: Nauchnoe obozrenie sayano-altaya retsenziruemyy nauchnyy jurnal Nomer **2**(08), 9–31 (2014)
13. Suleymanov, D.S., Gatiatullin, A.R.: Model tatarskoy affiksalnoy morfemy i yee realizatsiya, pp. 113–127. Intellekt. Yazyk. Kompyuter. – Vyp.4, Kazan, Seriya (1996)
14. Suleymanov, D.Sh., Gilmullin, R.A., Gataullin, R.R.: Morfologicheskiy analizator tatarskogo yazyka na osnove dvuxurovnevoy modeli morfologii. In: Pyataya Mejdunarodnaya konferensiya po kompyuternoy obrabotke tyurkskix yazykov TurkLang 2017, p. 327. Trudy konferensii. V 2-x tomax. T 2. Izdatelstvo Akademii nauk Respubliki Tatarstan, Kazan (2017)
15. Jeltov, P.V.: Razrabotka morfologicheskogo analizatora chuvashskogo yazyka. In: Pyataya Mejdunarodnaya konferensiya po kompyuternoy obrabotke tyurkskix yazykov TurkLang 2017. Trudy konferensii. V 2-x tomax. T 2, p. 327. Izdatelstvo Akademii nauk Respubliki Tatarstan, Kazan (2017)
16. Israilova, N.A., Bakasova, P.S.: Morfologicheskiy analizator kyrgyzskogo yazyka. In: Pyataya Mejdunarodnaya konferensiya po kompyuternoy obrabotke tyurkskix yazykov "TurkLang 2017". Trudy konferensii. V 2-x tomax. T 2, p. 327. Izdatelstvo Akademii nauk Respubliki Tatarstan, Kazan (2017)
17. Leontev, N.A.: Morfologicheskiy analizator yakutskogo yazyka. In: Shestaya Mejdunarodnaya konferensiya po kompyuternoy obrabotke tyurkskix yazykov "TurkLang-2018", vol. 320, pp. 276–279. (Trudy konferensii), Tashkent (2018)

18. Kukanova, V.V., Kadjiev, A.Y.: Algoritm raboty morfologicheskogo parsera kalmyskogo yazyka. In: V sbornike: Pismenoto nasledstvo i informatsionnite texnologii. El'Manuscript-2014 Materiali ot V mejdunarodna nauchnoy konferensii, pp. 116–119 (2014)
19. Orxun, M.: Computational analysis of uzbek nouns. In: Shestaya Mejdunarodnaya konferensiya po kompyuternoy obrabotke tyurkskix yazykov "TurkLang-2018", p. 320. Trudy konferensii, Tashkent (2018)

Uzbek Speech Synthesis Using Deep Learning Algorithms

M. I. Abdullaeva[1(✉)], D. B. Juraev[2(✉)], M. M. Ochilov[2(✉)], and M. F. Rakhimov[2(✉)]

[1] Department of Computer Systems, Tashkent University of Information Technologies named after Muhammad al-Khwarizmi, Tashkent, Uzbekistan
malika.ilkhamovna@gmail.com

[2] Department of Artificial Intelligence, Tashkent University of Information Technologies named after Muhammad al-Khorazmi, Tashkent, Uzbekistan
dilsamtuit@gmail.com, raximov022@gmail.com

Abstract. This paper presents modern architectures for effective speech synthesis. Since each language has its own subtleties, the task of applying the world methods for the Uzbek language was relevant, due to the lack of research in this direction. The paper presents a method consisting of the acoustic model Tacotron and the neural vocoder parallel waveGAN. The formed speech corpus with the volume of 31 h of Uzbek speech is described. The quality of the synthesized speech was evaluated using the MOS scale, according to that the intelligibility and accuracy of the synthesized speech was 4.36 points out of five.

Keywords: neural vocoder · tts system · Tacotron · parallel waveGAN · speech corpus

1 Introduction

Text-to-speech (TTS) is the computer simulation of human speech from a textual representation using machine learning techniques. The first speech synthesis system, called "têtes parlantes" (talking heads), appeared in the 18th century and was a pioneer, but was an imperfect imitation of the human voice.

There are many speech synthesizers around the world that synthesize electronic texts written in different languages into speech signals, and all the methods and tools used in their synthesis differ in the lexical and phonetic features of the chosen language. Currently, one of the important issues is to conduct research on the organization of speech synthesis and preprocessing of electronic texts, for full and perfect linguistic expression of the existing features of the chosen language, as well as to achieve synthesis of speech signals close to the natural pronunciation.

Automatic speech synthesis technology may be useful in a variety of industries and areas, such as telecommunications, mobile devices, industrial and consumer electronics, automotive industry, educational systems, computerized

H. Zaynidinov et al. (Eds.): IHCI 2022, LNCS 13741, pp. 39–50, 2023.
https://doi.org/10.1007/978-3-031-27199-1_5

systems, Internet services, access restriction systems, aerospace industry, the military-industrial complex. Speech synthesis technology offers great opportunities for people with physical disabilities. Speaking machines have been developed for the blind and visually impaired. For the dumb there are portable speech synthesis devices in which a message is typed on a keyboard, allowing communication with other people.

Nowadays, research on the recognition and processing of Uzbek speech is being carried out, and many scientific papers have been published on the results of this research. However, studies on the synthesis of Uzbek electronic texts into speech signals and the development of computational linguistics for the Uzbek language are insufficient. In particular, there is a lack of research on text analysis and processing, syllabic representation of texts, detection and correction of grammatical errors in the text, and real-time speech synthesis systems.

This research paper describes the general scheme of TTS systems, its constituent stages, their description and sequence. The classification of methods of text-to-speech conversion systems is given. In this paper defined families of acoustic models and given the capabilities and goals of neural vocoders. The paper includes the proposed method as a sequence of application of Tacotron 2 as an acoustic model and Parallel WaveGAN as neural vocoder. The Tacotron 2 and Parallel WaveGAN architectures are also described below. The peculiarity of the proposed method is that this sequence is applicable to the synthesis of Uzbek speech with its peculiarities.

2 Related Works

TTS system overcame great development and the first synthesizers were mechanical, which could generate separate sounds or small fragments of fused human-like speech like musical instruments [2]. Due to the age of the field there is a large number of methods of speech synthesis (Fig. 1). In scientific papers [1–4] are detailed reviews on existing synthesis technologies, where described their advantages and disadvantages, as well as their clear differences from each other.

The most simple and yet effective method is concatenative method. In works [1,5,6] the concatenative method algorithm is described and given its results in solving speech synthesis problems. This method is based on combining segments of recorded speech. The most important disadvantage of the method is the need for a large storage and the inability to apply various changes to the voice.

Studying the works [1,3,5,6,16] it can be concluded that speech synthesis can be achieved on the basis of a small amount of data. On the other hand, speech synthesis based on the selection of units proved that it is possible to reconstruct all the nuances and characteristics of the voice, if there is a large database. Hence, combining the two HMM synthesis and unit selection-based synthesis methods in one hybrid approach is another solution and method for high-fidelity speech synthesis [23,24].

The most advanced methods are those that are based on deep learning [11,15, 17,18]. Such speech synthesizers, are trained on recorded speech data. Common deep learning based synthesizers are WaveNet from Google DeepMind, Tacotron from Google, and DeepVoice from Baidu.

One of the modern and high-quality acoustic models is the Tacotron model from Google. In [15,17] the method based on the acoustic model of the Tacotron2 family is described. The key points of the method, which are worth paying attention to, are given. General architecture of the acoustic model Tacotron2 is described too.

In scientific articles [12,18], there is information about the Parallel WaveGAN neural vocoder, which is the final step in modern TTS systems. This vocoder converts the acoustic features, which is received at the input of the acoustic model, into a speech signal.

3 Description of Modern TTS Systems

Speech synthesis is a long-developing area and in the process of development it has opened many methods that differ from each other as the quality of the synthesized speech, as well as the complexity of the algorithm, the amount of memory occupied. Figure 1 shows and classifies the basic and most effective methods of speech synthesis.

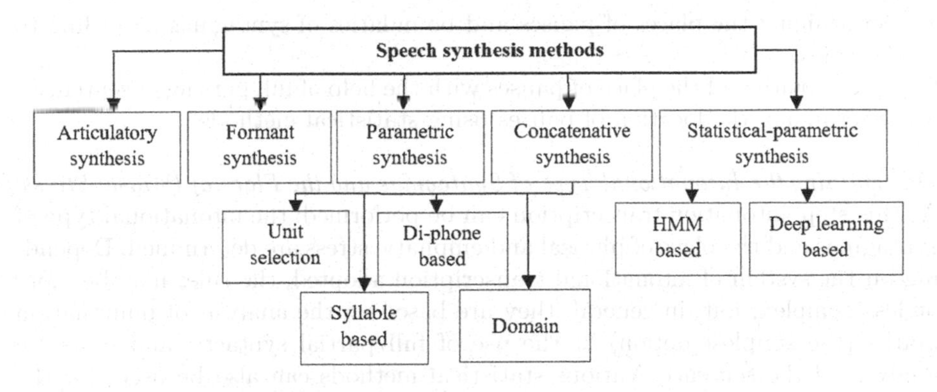

Fig. 1. Classification of text-to-speech synthesis methods

Speech synthesis technology in modern TTS systems consists of stages front-end and back-end, which in turn consist of a number of steps (Fig. 2). Each of these steps are described below [7].

Linguistic Analysis. This stage consists of text preprocessing and the main task is normalization of non-standard words. Normalization is the process of identifying numbers, abbreviations, acronyms, and idioms and converting them to full text, usually based on the context of the sentence.

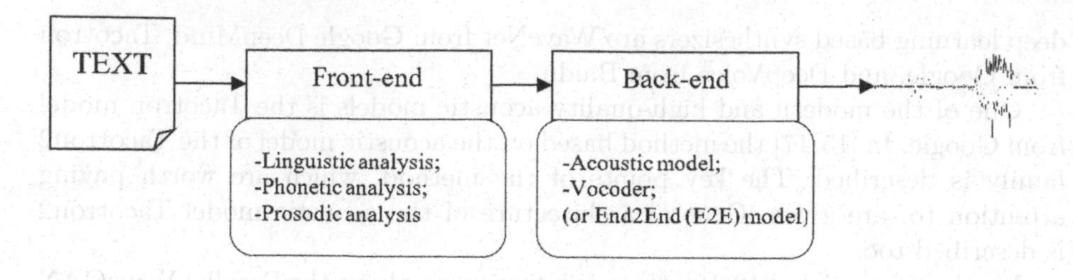

Fig. 2. General scheme of TTS systems

Phonetic Analysis. This step is a conversion of a grapheme into a phoneme. It is known that the grapheme is the minimal unit of writing and the phoneme is the minimal unit of oral speech.

Prosodic Analysis. At this stage, the boundaries of syntagms, localization and duration of pauses are determined, and the intonational type of phrases and the place of phrase emphasis in them are selected.

The existing methods for choosing the place of pauses can be divided into the following groups

1. Determining the places of pauses and boundaries of syntagmas according to the rules.
2. Determination of the place of pauses with the help of full parsing of sentences.
3. Determining the location of pauses using statistical methods.

Determining the Intonational Type of Syntagmas and the Place of Phrasal Stress. At this step, intonation transcription can be performed: the intonational type of syntagmas and the place of phrasal and emphatic stress are determined. Depending on the system of intonational transcription adopted, the rules may be more or less complex, but, in general, they are based on the analysis of punctuation marks (the simplest option) or the use of full/partial syntactic and semantic analysis of the sentence. Various statistical methods can also be used, for the training of which a text base is required, pre-marked with intonational transcription.

In modern TTS systems, the back-end environment is a synthesizer. It generates speech by converting each unit of transcription into sound using a selected method, algorithm or vocoder [15–17]. Thus Back-end consists of an acoustic model and a neural vocoder to approximate the parameters and relations between the input text and the waveform that constitute speech (Fig. 3).

Acoustic Model. The acoustic model algorithms are optimized to convert the pre-processed/normalized text into Mel spectrograms, thus converting the vector of linguistic features into acoustic features [19,20]. It is known that the spectrogram ensures that all significant sound features are taken into account and

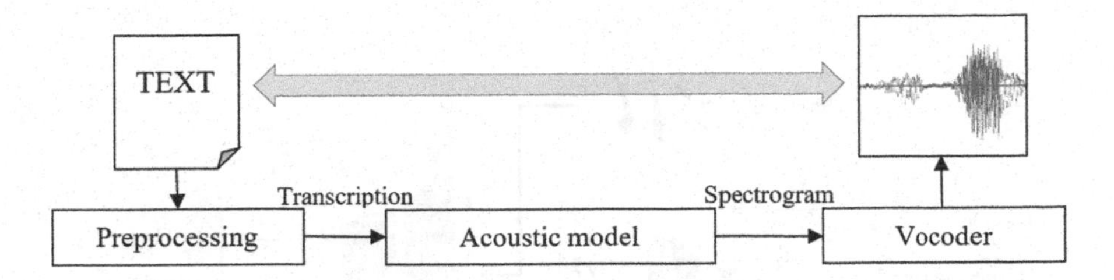

Fig. 3. Block diagram of the two-stage TTS system

carry high-level features. It is on the mel spectrogram that accents, features of interphoneme transitions and speaker's pronunciation are determined. In TTS systems the acoustics, with which the speech case is assembled, plays a very important role. For today there are various speech corpus in open access for the world languages. The most famous and quality ones are listed below:

- LJ Speech - EN, single speaker, 24 h
- Libri-TTS - EN, multi-speaker, 585 h
- RUSLAN - RU, single speaker, 29 h
- NATASHA - RU, single speaker, 13 h
- M-AILABS - multi language, 1000 h, 47 h of Russian

There are two main types of acoustic models - Tacotron family and Fast family.

Neural Vocoder. The input data for the latter stage are Mel spectrograms, which are converted into a waveform using a neural vocoder. Although there are many different types of neural vocoders, among them a special place belongs to vocoders with GAN(Generative Adversarial Networks) basis. For example, Parallel WaveGAN, Multi-band MelGAN, HiFiGAN, Style MelGAN.

Vocoders with GAN basis are based on a generator and a dicriminator, between which there is a constant interaction and struggle. The purpose of the generator is to generate high-quality speech, which will be close to the natural one, and the discriminator is focused on whether the generated speech is natural from the speech corpus or generated by the generator.

4 Method Description

The technology applicable for speech synthesis of the world languages, alas, is not suitable for the Uzbek language. This is due to the peculiarities of the language, such as unique letters, syllables and words. For this reason, the task of speech synthesis for the Uzbek language is relevant and unsolved to this day.

To develop the method of Uzbek speech synthesis, we have proposed using Tacotron2 architecture as an acoustic model for transition from transcription to

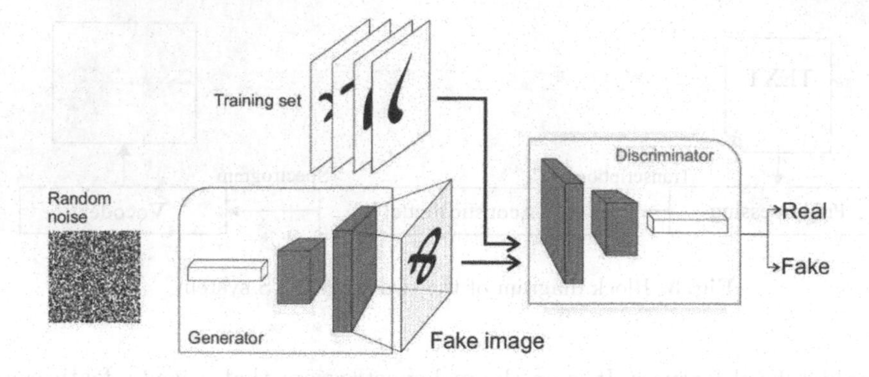

Fig. 4. Neural vocoder architecture based on Generative Adversarial Networks

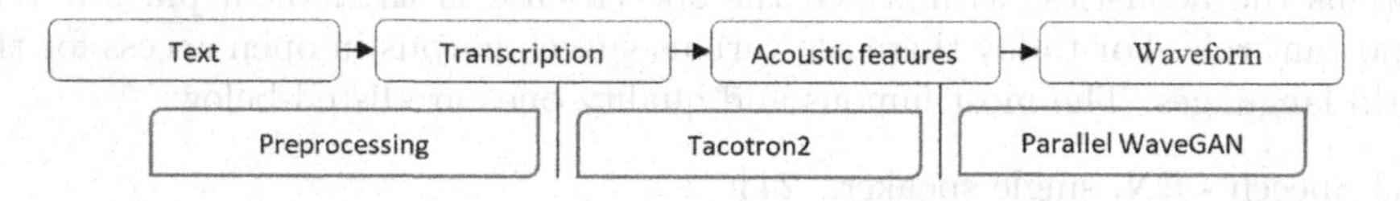

Fig. 5. General scheme of the sequence of operations for the Uzbek speech synthesis system

mel spectrogram, and Parallel WaveGAN architecture as a neural vocoder for mel spectrogram vocalization, as shown in Fig. 5.

Consider the architecture of Tacotron2. In architecture Tacotron2 has encoder, which is designed to work with embedding phonemes, has decoder with two heads - predicts next mel and Stop Token. Stop Token learns a binary classification of whether to stop producing speech. The output of the other linear projection goes to 3 points: in post-Net, residual connection and pre-net, which consists of two layers with dropout.

Parallel WaveGAN has a WaveNet generator and differs from the original WaveNet in that:

- Uses non-causal convolutions instead of causal convolutions;
- Takes random noise as input.
- The model is not autoregressive.

The detailed architecture of a simple and efficient method of generating parallel signals based on the Parallel WaveGAN is shown in Fig. 7.

Assessing the difference between the features of true and generated speech presents a particular difficulty. Without introducing such an estimate into the loss function, the convergence process will be extremely slow and unstable [22]. To solve this problem, Multi-resolution STFT loss functions were first proposed, which for brevity we will call STFT loss.

Let x is audio and $\hat{x}$ the generated one corresponding to x mel-spectrogram.

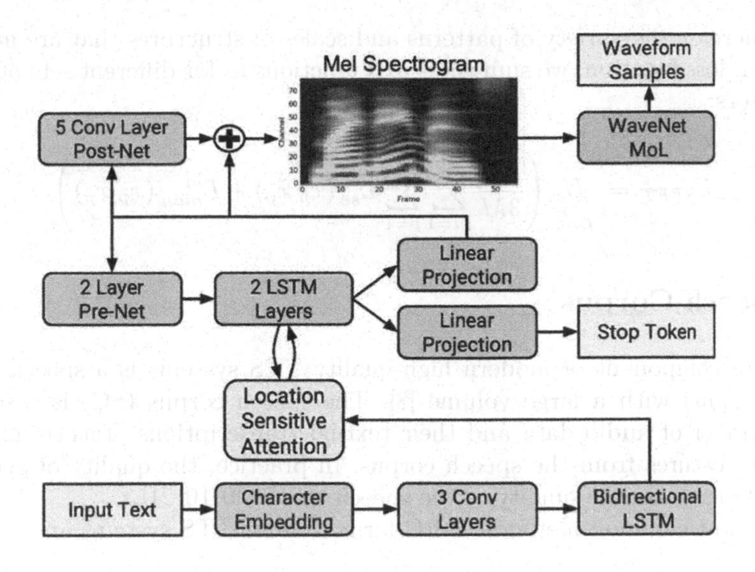

Fig. 6. Tacotron2 architecture

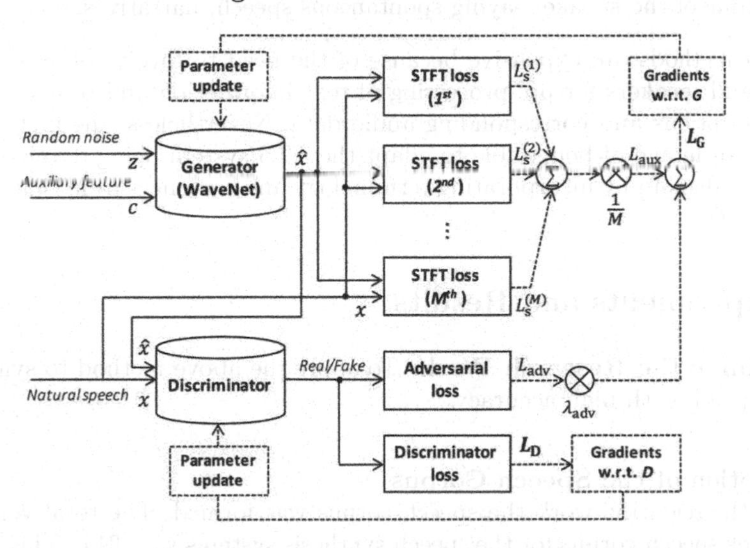

Fig. 7. Parallel WaveGAN architecture with STFT losses

Then for some chosen uniform STFT parameters:

$$L_{sc}(x, \hat{x}) = \frac{\| |STFT(x)| - |STFT(\hat{x})| \|_F}{\| |STFT(\hat{x})| \|_F}$$

$$L_{mag}(x, \hat{x}) = \frac{1}{N} \| \log |STFT(x)| - \log |STFT(\hat{x})| \|_1$$

where $\| \cdot \|_F$- denote the Frobenius and $\| \cdot \|_1 - L_1$ norms;

To increase the variety of patterns and scales of structures that are involved in a given loss function, we sum these loss functions as for different sets of STFT parameters:

$$L_{STFT} = \underset{x,\tilde{x}}{E}\left(\frac{1}{3M}\sum_{m=1}^{M}\sum_{p=1}^{3} L_{sc}^{m}\left(x_p,\tilde{x}_p\right) + L_{mag}^{m}\left(x_p,\tilde{x}_p\right)\right)$$

5 Speech Corpus

The main component of modern high-quality TTS systems is a speech corpus (voice corpus) with a large volume [8]. The speech corpus (SC) is a set of a large number of audio data and their textual transcriptions. Tacotron2 learns language features from the speech corpus. In practice, the quality of synthetic speech depends on the quality of the speech corpus [9,10,21].

The most common methods of RC formation for TTS systems are:

- Recording of the speaker reading a pre-prepared text material;
- Recording of the speaker saying spontaneous speech, narratives, etc.

Both methods are expensive because of the need to involve additional specialists and speakers for pre-processing of text information and post-processing of transcriptions and corresponding audio data. Nevertheless, the first method has the advantage of being able to adapt the TTS system being developed to a particular domain by incorporating terminology and sentences from that domain into the SC.

6 Experiments and Results

The Aim of the Research Work. To apply the above method to synthesize Uzbek speech with high accuracy.

Description of the Speech Corpus

Within the scientific work the speech corpus was formed. The total volume of the Uzbek speech corpus for the speech synthesis systems was 31 h. This volume of the speech corpus was voiced by two speakers separately. The speech signals were recorded in a studio environment in .wav audio format with a sampling rate 22050 Hz, quantization of 16 bits and mono type.

A total of over 14,523 utterances were used in the texts provided for reading. There are 170 thousand words in the sentences, and 25 thousand of them are not repeated words in Uzbek.

Using the above proposed algorithms of linguistic analysis, the experts checked the quality of each statement and compliance of the statements with the audio data. This expert procedure was conducted manually. As a result, only verified transcriptions and their quality audio soundings were stored in the speech corpus.

Statistics of the speech corpus based on the parameter length of utterances in seconds vs the number of utterances with the current length in the speech corpus is shown in Fig. 8. According to Fig. 8, we can argue that the speech corpus mostly consists of utterances of 10 s and similar utterances in the speech corpus 794 audio. And the most rare were expressions with 39 s, which in the speech corpus in total 13.

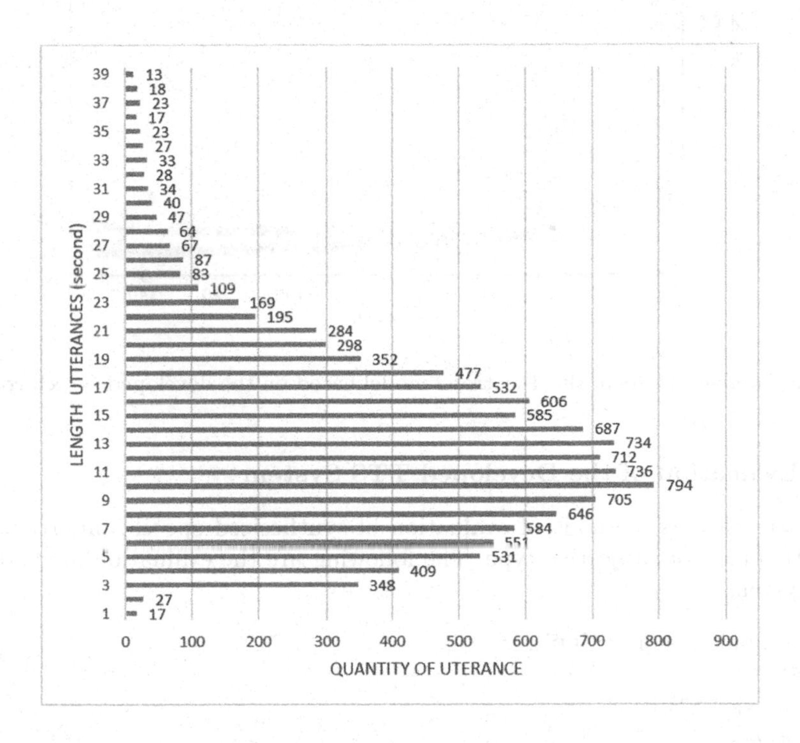

Fig. 8. Statistics of the Uzbek speech corpus

Description and Training of the Acoustic Model Tacotron2

The scientific work uses logarithmic spectrograms with a Hann window, a frame length of 50 ms, a frame shift of 12.5 ms, and a Fourier transform of 2048 points. In the paper, the sampling rate is defined as 22 kHz.

The Tacotron 2 was trained using the word sequence as input and the mel spectrogram extracted from the recorded speech. The model contained 5 encoder layers and 8 decoder layers.

The model was trained on an NVIDIA DGX-2 server with a 32G NVIDIA TESLA V100 GPU. Figure 9 shows the results of training on the Tacotron2 model. The trained data were pairs of audio data and their transcriptions from the speech case. For the speech corpus with a volume of 31 h.

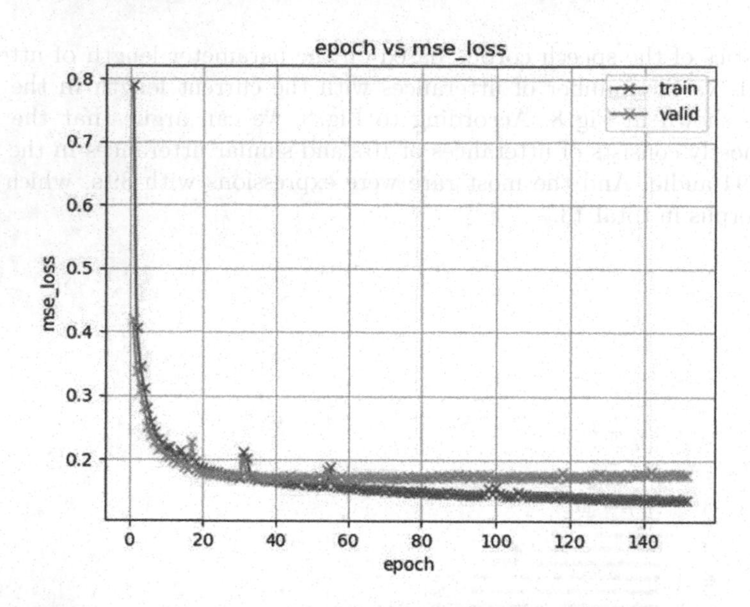

Fig. 9. Training results of the Tacotron2 model based on the developed speech corpus

6.1 Evaluation of the Developed TTS System

There are various methods of evaluation of synthesized speech, unfortunately, all of them are of subjective type. The following are the evaluation methods for TTS systems:

- Mean opinion score (MOS).
- MUSHRA
- Side by side SBS
- Robotness

Among them, MOS is particularly widely used, because of the availability of the evaluation method in different understandings.

To evaluate developed TTS system 12 Uzbek linguists (further experts) from Uzbek language department were involved. Each expert was given synthesized signals by the developed system, as well as a textual representation of the audio data. The synthesized speech was rated from 1 to 5, with 1 being very poor, 2 being unsatisfactory, 3 being satisfactory, 4 being good, and 5 being excellent. Finally, all the expert opinions were compiled into a table and an analysis was performed on them.

Table 1. MOS results for the proposed method

Model	MOS
Tacotron2 + Parallel WaveGAN	4.36 ± 0.09
Recording	4.72 ± 0.01

When testing the applicability and quality of the proposed combination of Tacotron 2 as a model to calculate the acoustic parameters and Parallel WaveGAN for speech synthesis, the results were obtained 4.36 ± 0.09. This figure is high enough, and the proposed method is applicable to the synthesis of Uzbek speech.

7 Conclusion

The results of the synthesized speech showed that the synthesis of Uzbek speech is achieved with a score of 4.36 (excellent) according to the MOS evaluation method. These scores were achieved using the Tacotron2 acoustic model and the Parallel WaveGAN neural vocoder.

While training the acoustic model of the speech corpus, it became known that the quality of the synthesized speech is fully dependent on the quality of the voice corpus. In order to form a high-quality Uzbek speech corpus, a method was developed that includes text preparation, audio recording, text-to-audio synchronization, and a final check to match the audio data to their transcriptions. According to the results of MOS we can state that the chosen method of speech corpus formation is effective and efficient. In addition to assessing the quality of the synthesized speech, the feature and importance of STFT loss and its impact on the quality of the synthesized speech were identified.

References

1. Kireev, N.S., Ilyushin, E.A.: Review of existing algorithms for text-to-speech conversion. Int. J. Open Inf. Technol. **8**(7) (2020). ISSN: 2307-8162
2. Rybin, S.V.: Textbook on the Discipline "Speech Synthesis". SPb: ITMO University, 92 p., (2014)
3. Prahallad, K.: Automatic building of synthetic voices from audio books. School of Computer Science, Carnegie Mellon University, Pittsburgh, PA (2010). CMU-LTI-10-XXX, July 26
4. Allen, J., Hunnicutt, M.S., Klatt, D.H., Armstrong, R.C., Pisoni, D.B.: From Text to Speech: The MITalk System. Cambridge University Press, NY (1987)
5. Hunt, A., Black, A.: Unit selection in a concatenative speech synthesis system using a large speech database. In: Proc. ICASSP, pp. 373–376 (1996)
6. Sawant, R., Virani, H.G., Desai, C.: Database selection for Concatenative speech synthesis With novel endpoint detection algorithm. IJAIEM **2**(5), 173–180 (2013)
7. Psutka, J.: Communication with Computer by Speech (in Czech). Academia, Prague (1995)

8. Radová, V.: UWB S01 corpus - a Czech read-speech corpus. In: Proceedings of ICSLP2000, vol. IV, pp. 732–735. Beijing (2000)
9. Anguera, X., Luque, J., Gracia, C.: Audio-to-text alignment for speech recognition with very limited resources. In: Fifteenth Annual Conference of the International Speech Communication Association (2014)
10. Sproat, R., et al.: Normalization of non-standard words. Comput. Speech Lang. **15**(3), 287–333 (2001)
11. Musaev, M., Khujayorov, I., Ochilov, M.: The use of neural networks to improve the recognition accuracy of explosive and unvoiced phonemes in Uzbek language. Inf. Commun. Technol. Conf. **2020**, 231–234 (2020). https://doi.org/10.1109/ICTC49638.2020.9123309
12. Ping, W., Peng, K., Chen, J.: ClariNet: parallel wave generation in end-to-end text-to-speech. In: Proc. ICLR (2019)
13. Sutskever, I., Vinyals, O., Le, Q.V.: Sequence to sequence learning with neural networks. In: Advances in Neural Information Processing Systems, pp. 3104–3112 (2014)
14. Musaev, M., Khujayorov, I., Ochilov, M.: Automatic recognition of Uzbek speech based on integrated neural networks. In: Aliev, R.A., Yusupbekov, N.R., Kacprzyk, J., Pedrycz, W., Sadikoglu, F.M. (eds.) WCIS 2020. AISC, vol. 1323, pp. 215–223. Springer, Cham (2021). https://doi.org/10.1007/978-3-030-68004-6_28
15. Gonzalvo, X., Tazari, S., Chan, C.A., Becker, M., Gutkin, A., Silen, H.: Recent advances in Google real-time HMM-driven unit selection synthesizer. In: Proc. Interspeech, pp. 2238–2242 (2016)
16. Ze, H., Senior, A., Schuster, M.: Statistical parametric speech synthesis using deep neural networks. In: Proc. ICASSP, pp. 7962–7966 (2013)
17. Wang, Y., et al.: Tacotron: towards end-to-end speech synthesis. In: Proc. Interspeech (2017). arXiv:1703.10135
18. Oord, A.V.D., et al.: WaveNet: a generative model for raw audio (2016). arXiv:1609.03499
19. Musaev, M., Mussakhojayeva, S., Khujayorov, I., Khassanov, Y., Ochilov, M., Atakan Varol, H.: USC: an open-source Uzbek speech corpus and initial speech recognition experiments. In: Karpov, A., Potapova, R. (eds.) SPECOM 2021. LNCS (LNAI), vol. 12997, pp. 437–447. Springer, Cham (2021). https://doi.org/10.1007/978-3-030-87802-3_40
20. Abdullaeva, M., Khujayorov, I., Ochilov, M.: Formant set as a main parameter for recognizing vowels of the Uzbek language. Int. Conf. Inf. Sci. Commun. Technol. **2021**, 1–5 (2021). https://doi.org/10.1109/ICISCT52966.2021.9670268
21. Raximov, R., Primova, H., Ruziyeva, Z.: Methods of recognizing texts in different images. In: International Conference on Information Science and Communications Technologies: Applications, Trends and Opportunities (2021). http://www.icisct2021.org/
22. Fazliddinovich, R.M., Abdumurodovich, B.U.: Parallel processing capabilities in the process of speech recognition. Int. Conf. Inf. Sci. Commun. Technol. **2017**, 1–3 (2017). https://doi.org/10.1109/ICISCT.2017.8188585
23. Musaev, M., Rakhimov, M.: A method of mapping a block of main memory to cache in parallel processing of the speech signal. Int. Conf. Inf. Sci. Commun. Technol. **2019**, 1–4 (2019). https://doi.org/10.1109/ICISCT47635.2019.9011946
24. Rakhimov, M., Ochilov, M.: Distribution of operations in heterogeneous computing systems for processing speech signals. In: 2021 IEEE 15th International Conference on Application of Information and Communication Technologies (AICT), pp. 1–4 (2021). https://doi.org/10.1109/AICT52784.2021.9620451

Speech Recognition Technologies Based on Artificial Intelligence Algorithms

Muhammadjon Musaev[1], Ilyos Khujayarov[2], and Mannon Ochilov[1](✉)

[1] Artificial Intelligence, Tashkent University of Information Technology named after Muhammad Al-Khwarizmi, Tashkent, Uzbekistan
mm.musaev@rambler.ru, ochilov.mannon@mail.ru

[2] Information Technologies, Samarkand branch of Tashkent University of Information Technology named after Muhammad Al-Khwarizmi, Tashkent, Uzbekistan

Abstract. In this article, research was conducted on the development of automatic Uzbek speech recognition technology based on integral models. Methods of continuous speech recognition technology in Uzbek were studied at all stages and suitable ones were selected. A 200-hour speech corpus was trained on the DNN-CTC architecture for acoustic modeling. The accuracy of the developed speech recognition system achieved WER = 17.3%, CER = 7.5% on the test data set.

Keywords: AI · Uzbek speech recognition · Speech-text · E2E · Speech corpus · MFCC · Perplexity · CNN · RNN · LSTM · CTC

1 Introduction

Today artificial intelligence (AI) is the most strategically important technology for the emergence of a new digital economy. Computer tools are a new inexpensive and powerful resource for a new digital technological breakthrough. Unlike other resources, it is developing very rapidly, and today it is the main potential source of economic growth. Three main aspects of the growth of modern information technologies are known: global networks and mobile communications (Net), promising human-machine interfaces (HMI), and intelligent technologies (IT). As is known, the systemic impact is observed in multidisciplinary and interdisciplinary scientific research. This means that areas of research and development that cross these areas are of great interest: promising interfaces in global mobile networks, intelligent interfaces in IT data processing technologies, and the use of artificial intelligence technologies on the Internet. Artificial intelligence in combination with speech technologies provides direct interaction between computer resources and the user through programs for processing text and speech requests. Examples of the most successful application of speech recognition technology are software products used in call centers, in minutes of meetings, journalism, and medical diagnostics [18,23]. Speech recognition technology is widely used in education, and in various software products aimed at learning foreign languages.

H. Zaynidinov et al. (Eds.): IHCI 2022, LNCS 13741, pp. 51–62, 2023.
https://doi.org/10.1007/978-3-031-27199-1_6

These technologies can be of great help to people with hearing and speech disorders, disorders of the speech apparatus as a result of injuries or since birth, in the rehabilitation of patients who have lost all or part of spoken language, children who have been deaf since childhood with hearing implants. However, if the interaction with the computer is conducted using a procedural programming language, this does not satisfy the majority of end users, who usually do not know how to program and do not want to program. The development of artificial intelligence systems involves the use of natural language for user interaction with a computer. The creation of speech recognition technologies with high accuracy at the pace of speech pronunciation is an urgent task and requires the widespread use of artificial intelligence algorithms [1]. In this article, we are discussing Uzbek speech recognition algorithms. The history of the development of speech recognition technologies in Uzbekistan is 5–6 years old, however, the experience in researching and creating artificial intelligence algorithms in other areas made it possible to quickly apply these algorithms in combination with neural networks to the tasks of analyzing and recognizing Uzbek speech [2].

2 Related Works

One of the most natural ways to ensure interaction in human and machine life is speech. Automatic speech recognition is an important component of the human machine interface, the developing of such systems is one of the pressing issues. As we know, the quality and volume of the speech corpus is the main factor of the effectiveness of the speech recognition system. Unfortunately, most of the datasets are developed for popular languages such as English, Russian, Spanish, and Chinese, while less common languages such as agglutinative languages (Uzbek, Kazakh, Turkish, etc.) are not given much attention. But at the same time, speech corpus is being developed for these languages. For example, in Uzbekistan, in order to develop speech processing research, researchers have developed an open-source Uzbek language corpus for speech recognition and speech synthesis [17] applications. Also for the Kazakh language [19], the scientific research paper presents many recommendations for the development of discourse corpus and overcoming the encountered difficulties. This may benefit other researchers who are planning to build a speech corpus for a low-resource language. A corpus of 332 h of speech was developed in the work, and many studies on automatic speech recognition were conducted based on this corpus. In the last 5 years, many researches have been conducted by scientific researchers on Uzbek language speech recognition. Most of the studies have a limited dataset and are done for specific areas. For example, work [16] developed a speech recognition system using 10 h of transcribed audio and focused on limited vocabulary voice command recognition systems. In [18], methods of rehabilitation of speech-impaired children based on the speech recognition system are presented. In another study [17], research was conducted based on a corpus of 105 h of speech recorded by 958 different speakers. In particular, 18.1 % and 17.4 % errors in words were achieved on the validation and test sets, respectively. Currently, work on increasing the size of the Uzbek speech corpus is being carried out at the "Images and speech signal processing" laboratory at the Tashkent University of Information Technologies named after Muhammad al-Khwarizmi. Based on this speech corpus, we can conduct research on the use, testing and comparison of speech recognition technologies and methods.

3 Stages of Speech Recognition Processing

The main requirements for software tools such as "speech to text" one are that the algorithms that implement the processes of automatic speech recognition provide the required processing speed with the required recognition accuracy, and can work at the real pace of speech pronunciation [3]. Before starting work on creating a set of algorithms and speech recognition programs, the requirements for the recognition system should be formulated, taking into account the features of the construction and sound of a particular type of speech [4]. Table 1 presents and highlights the requirements in accordance with which automatic speech recognition technologies are developed. The work we have done in this article corresponds to the types listed in italics in Table 1.

Table 1. Speech recognition system classification.

Type of analyzed speech	Dependence on speaker	Volume of speech vocabulary	Selected technology of processing
Recognition of individual words	Recognition, dependent on speaker	Limited vocabulary	Based on HMM (Hidden Markov model)
Recognition of Word combinations	With customizing for speaker	*With high-volume vocabulary*	Based on HMM +ANN (Artificial neural networks)
Recognition continuous (confluent) speech	*Recognition, independent of speaker*	Vocabulary without limitations	*Based on End to End - E2E model*

The whole process of developing automatic speech recognition programs [4] is a complete technology for preparing and processing information; it includes the following main steps:

- developing of a speech corpus;
- selection of informative features of the speech signal;
- developing of a language model;
- development of neural network architecture for speech recognition;
- developing of automatic speech recognition modules.

Figure 1 shows he main components of speech recognition technology.

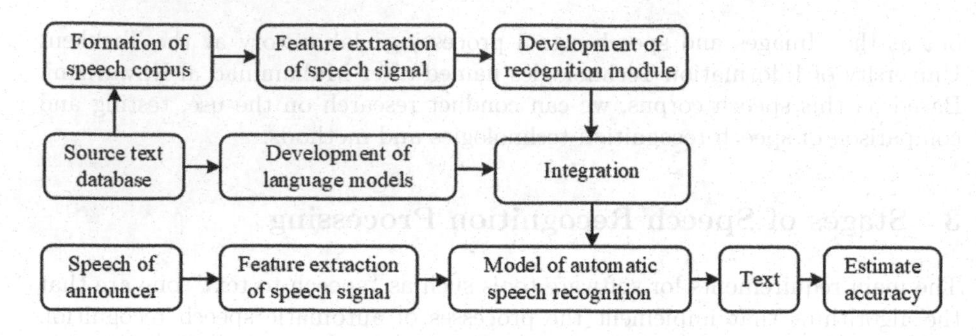

Fig. 1. Stages of speech recognition processing.

4 Development of the Speech Corpus of Uzbek Speech

The developing of a speech corpus is a long process of reading a text by different groups of speakers of different ages, genders, and pronunciations. At that, both office and web technologies for accumulating audio files are widely used to speed up the process. Figure 2 shows a functional diagram of the formation of a speech corpus.

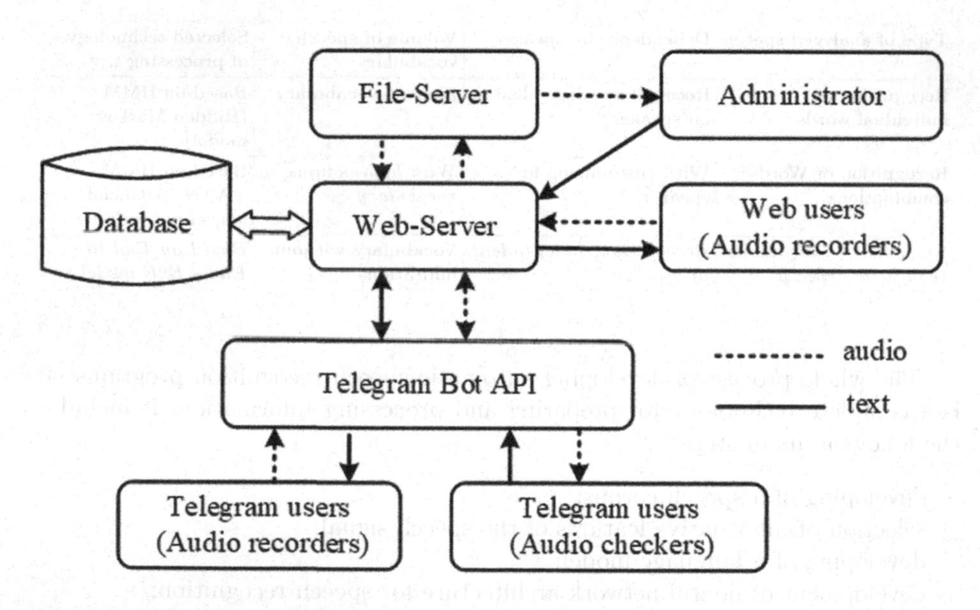

Fig. 2. Speech corpus formation modules.

The speech corpus is formed in advance; it is a laborious stage of data preparation and requires a lot of human resources [5]. Audio data from announcers can be entered both through the Telegram Bot API service and directly through the web server. A special operator checks the correctness of the Telegram audio input and its compliance with the text read. All input audio data is controlled by the administrator, whose task is to remove characters, commas, and pauses and finalize the audio for entering into the database. The database also stores the texts of audio files, data about announcers (gender, full name, age, address) and operators. In the speech corpus, in addition to the audio files, their parameters are stored: the number of the text fragment, the numbers of the announcer and operator, the check time, and duration). Office premises or studios are chosen as the environment for recording audio files. If violations (interference or distortion) are detected in the speech corpus formation technology, the recorded audio file can be deleted or overwritten. The validation algorithm consists in comparing the recorded audio file and its transcript (text analog). In addition, the quality of the audio recording itself is checked. If necessary, the recording is entered into the desired speech range using bandpass filters [6,21]. The volume of the speech corpus depends on the purpose of the recognition system. For speech control systems, the volume of the speech corpus is up to 40 h, and for continuous speech recognition systems, the volume of the corpus is 1000 or more hours. Preliminary preparation of texts for the work of voice-over announcers in the form of audio files includes the following procedures:

- convert into the Latin alphabet;
- inserting spaces between sentences;
- removal of incomprehensible characters in the language;
- converting digital data into text;
- selection of sentences no longer than 2–15 words.

Parameters of the speech corpus used in this version of the system:

- duration - 200 h;
- number of speakers - 1140;
- number of words - 630 thousand.

5 Identification of Informative Features of Speech

The sequence of algorithms for extracting informative features from a speech signal is applied twice: after developing a speech corpus and when recognizing speech from a microphone coming to the input of the system (Fig. 1). Based on the selected informative features (spectrograms), the learning process of the neural network of the E2E model is implemented [7]. Based on the training results, a text transcript of the audio file is created. A feature of the proposed approach to automatic recognition of continuous Uzbek speech is the use of spectrograms of speech fragments as input data.

Algorithms for calculating the neural network parameters of the signal spectrograms [8] based on fast Fourier transform (FFT), Mel-transform (Mel Frequency Cepstral Coefficients - MFCC), and Discrete Cosine Transform (DCT)

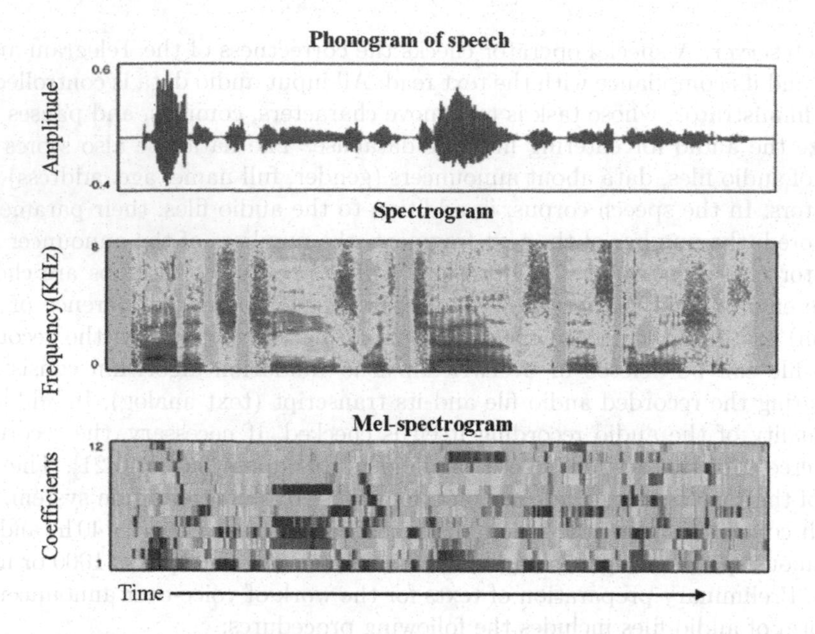

Fig. 3. Start, intermediate and end representation format of a signal

are presented in Fig. 4. Formulas for calculating fast Fourier transform, Discrete Cosine Transform, and Mel-transform are, respectively:

$$X_k = \sum_{n=0}^{N-1} x_n w_n e^{\frac{-2\pi i}{N} kn}, k = 0, \ldots\ldots, N-1. \quad (1)$$

$$c[k] = \sum_{m=0}^{I-1} S[m] \cos\left(\frac{\pi}{I}\left(m + \frac{1}{2}\right)k\right), 0 \le k \le I \quad (2)$$

$$f[k] = \frac{N}{F_s} m^{-1}\left(m\left(f_{\min}\right) + k\frac{m\left(f_{\max} - f_{\min}\right)}{N+1}\right) \quad (3)$$

Here k is the signal frequency, Fs is the sampling frequency, N is the number of variables, m are the transition functions to the Mel scale. The use of Mel spectrograms is due to the fact that, while maintaining the phonetic characteristics of the signal, compression and logarithm are provided to obtain a cepstral signal. In the set of Mel frequency cepstral features, the spectrum is manipulated to imitate the processing features of the auditory system: the spectrum components are collected in accordance with the Mel frequency scale and the energy values in each channel are logarithmized. The Mel scale is a pseudo-logarithmic frequency scale experimentally obtained as a result of psychoacoustic experiments. Its importance lies not only in the fact that it corresponds to our ideas about the operation of the auditory system but also in the fact that, combining the spectral components of high frequencies in ever wider zones, it allows us to substantially reduce the dimension of the feature vector. The logarithm models the amplitude compression characteristic of signal propagation along the nerve canals. Additionally, an inverse cosine transform is performed, passing to the cepstrum, and only the first few components (coefficients) are left [9,20].

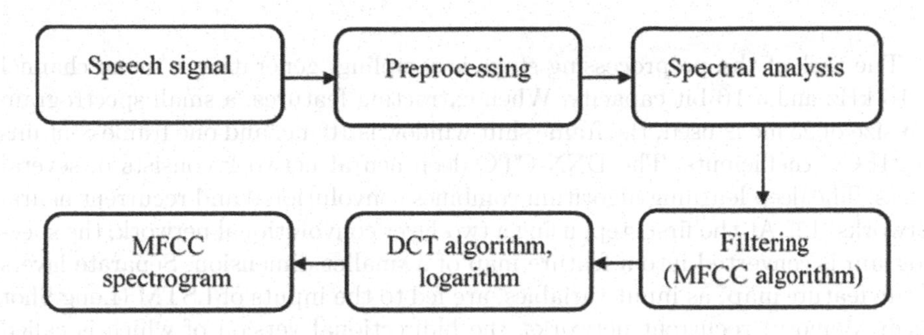

Fig. 4. Algorithms for calculating MFCC spectrograms.

6 Neural Network Architectures Based on the End-to-End (E2E) Model

In the E2E model, due to the absence of the problem of forced segmentation during training, high efficiency in terms of speed and accuracy is ensured. Compared with other standards, the choice of the E2E model and the CTC (Connection temporal classification) method [22] is justified by the fact that it is easier to build a recognition model with them, and the recognition accuracy is related to the volume and quality of the speech corpus intended for training. Language models play an important role in automatic recognition systems of any type of speech. The difficulties in developing a language model of the Uzbek language lie in the fact that there is no single standard for phonemes, diphones, triphones, and other sound units. In addition, for high accuracy recognition, speech corpora of large volume are required. There are several types of models that differ in the length of the context, the complexity of construction, and the accuracy of recognition. According to the results of preliminary assessments, a model based on statistics was selected. In the field of language processing, statistical models are used to predict the sequence of sounds, syllables, words, or sounds. The selected N-gram model calculates the probability of a word position depending on the preceding words. The IRSTLM toolkit [10,11,13] was used as a tool for creating a 3-gram model. In this case, the sequence of input speech recognition steps is as follows (Fig. 5).

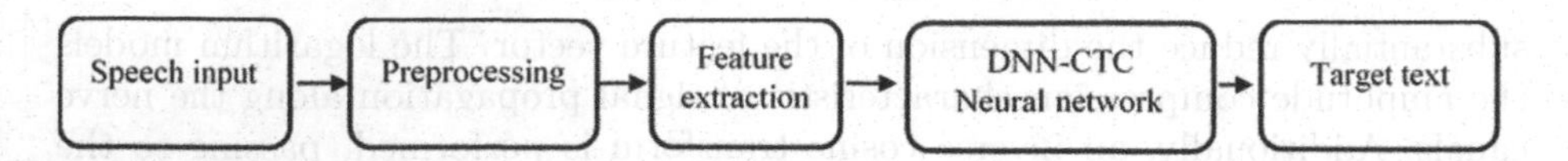

Fig. 5. The architecture of the E2E model

The task of the preprocessing stage is sampling, generating a mono channel at 16 kHz and a 16-bit capacity. When extracting features, a small spectrogram of a size of 25 mc is used, the frame shift window is 10 mc, and one frame contains 40 MFCC coefficients. The DNN-CTC deep neural network consists of several layers. The deep learning algorithm combines convolutional and recurrent neural networks [12]. At the first step, using a two-layer convolutional network, the spectrogram is converted into a feature map of a smaller dimension. Separate layers of the feature map, as input variables, are fed to the inputs of LSTM (Long Shot Term Memory) recurrent networks, the bidirectional version of which is called BLSTM (Bidirectional LSTM based RNN), at the output of the hidden layers of which, in accordance with the previously conducted training, an intermediate series of vectors is formed, denoting probable positions of the current frames. These vectors are fed to the inputs of the fully connected multi-layer perceptron 1-FC (Fully connected). The network transforms incoming vectors according to the number of letters in the alphabet. After the Softmax (0,1) normalization procedure, a matrix of probable feature values for each of the 26 letters and 3 special characters of the Uzbek alphabet is formed. Then frame symbols with the maximum probability value are selected, and the position number of the probability matrix is translated into a sequence of alphabetic symbols. The decoding algorithm CTC-Decoding using the CTC Loss function translates a sequence of characters into the text. The process of combining the speech model and the automatic recognition module is implemented in the decoding module. Thus, the complete architecture and main components of the proposed neural network are shown in Fig. 6.

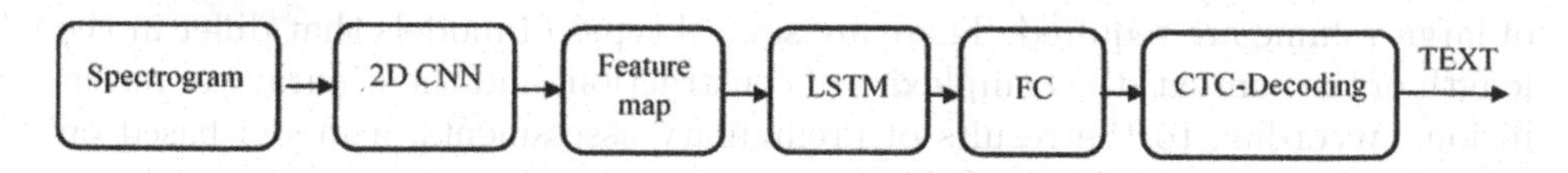

Fig. 6. Main components of the proposed deep neural network

7 Parameters of the Designed Neural Network

The input data are the coefficients of the Mel spectrogram. Two layers of the CNN convolutional network have the same parameters: filter size 11×22, ReLU activation function. Two layers of the recurrent network have parameters: RNN - BLSTM type, layer size - 512, ReLU activation function. Fully connected FC network: first layer size - 600, ReLU activation function, second layer size - 29, Softmax activation function. Transcription layer - CTC Loss - function. Training parameters: Adam optimization algorithm, step length - 0.001, Batch size - 64. Software complex for recognition of Uzbek speech. The developed software package consists of interacting programs that process the data in the following stages.

1. Developing of a speech corpus.
2. Selection of informative features of audio data.
3. Development of a language model.
4. Developing of a deep learning neural network.
5. Programs for testing the recognition system.

Building, training and testing of models is implemented using the capabilities of the python programming language. The network built on the basis of the CTC classification method was created using the tensorflow and keras libraries of the python language. Programming is done in the Linux operating system. During pre-processing of audio signals and developing test modules, special bash scripts were used. The models were trained on the Nvidia Tesla V100 graphics processors (GPU) of the NVIDIA DGX-2 server 32GB. To evaluate the language model, PP coefficients (Perplexity) were used as a measure. They are the inverse probability of the normalized number of words in the tested text.

$$PP(\mathrm{W}) = P(\mathrm{w}_1\mathrm{w}_2 \ldots \mathrm{w}_N)^{-\frac{1}{N}} - \sqrt[N]{\prod_{i=1}^{N} \frac{1}{P(\mathrm{w}_i \mid \mathrm{w}_i \ldots \mathrm{w}_{i-1})}} \tag{4}$$

When testing a statistically based language model using the IRSTLM tool and building a 3-gram model for groups of speech corpus during training and testing, the PP score for *train* and *test* was 6.2 and 7.9, respectively. For a neural network based on the LSTM speech model, these values were 2.56 and 3.12, respectively. Characteristics of evaluations of speech recognition systems are WER (Word Error Rate) - the frequency of erroneous words, CER (Character Error Rate) - the frequency of character errors. The WER and CER parameters are calculated using the following formula

$$WER/CER = \frac{D+U+I}{N} * 100\% \tag{5}$$

Here N is the number of words or characters in the presented sequence, D is the number of words or characters, U is the number of words or characters, I is the number of words or characters. During the experiments, the following methods for expanding the audio data of the speech corpus were used: adding noise (AN), changing the audio reading speed perturbation (SP) and spectrum masking (SpecAugument-SA) along the time and frequency axes. In addition, the effectiveness of using the language model was evaluated during decoding. Gaussian noise of an amplitude of $\sigma = 0.01$ was used as noise. The audio reading speed was changed within 0.9, 1.0 and 1.1. When masking in the spectral domain, the maximum time in width was chosen T = 40 for two masks, and in the frequency domain, the maximum time in frequency was chosen F = 30 for two masks. Good result of the model in terms of WER and CER can be seen. For the test group with 6243 samples, the errors were WER = 17.3% and CER = 7.5%. The results of experiments are given in Table 2.

Table 2. Results of experiments

LM	SP	SA	AN	WER	CER
–	–	–	–	25.3	11.5
+	–	–	–	21.8	9.2
+	+	–	–	19.3	8.3
+	–	+	–	18.1	8.1
+	–	–	+	20.4	8.9
+	+	+	+	**17.3**	**7.5**

The developed speech recognition system works in soft real-time mode. This means that the response time to the request depends on the duration of the input speech fragment. Long sentences require more time, short ones require less time.

8 Conclusion

In conclusion, DNN-CTC neural network based on Ent-to-End model was developed using the existing 200-hour speech corpus for the purpose of recognizing continuous Uzbek speech. To develop a speech recognition model, audio augmentation methods such as adding noise, speed purtubution, specaugment were used. To increase the accuracy of the recognition system, the Uzbek language model was developed. When a speech recognition system was created using augmentation methods and a language model, the error rates of the system were WER = 17.3% and CER = 7.5%, respectively.

References

1. Alhawiti, K.M.: Advances in artificial intelligence using speech recognition. Int. J. Comput. Inf. Eng. **9**, 1432–1435 (2015)
2. Musaev, M., Khujayorov, I., Ochilov, M.: Automatic recognition of Uzbek speech based on integrated neural networks. In: Aliev, R.A., Yusupbekov, N.R., Kacprzyk, J., Pedrycz, W., Sadikoglu, F.M. (eds.) WCIS 2020. AISC, vol. 1323, pp. 215–223. Springer, Cham (2021). https://doi.org/10.1007/978-3-030-68004-6_28
3. Prasad, V.: Voice recognition system: speech-to-text. J. Appl. Fundam. Sci. **1**(2), 191 (2015)

4. Serizel, R., Giuliani, D.: Vocal tract length normalization approaches to DNN-based children's and adults' speech recognition. In: IEEE Workshop on Spoken Language Technology, pp. 135–140 (2014)
5. Kipyatkova, I., Karpov, A.: An analytical survey of large vocabulary Russian speech recognition systems. SPIIRAS Proc. **1**(12), 7–20 (2010). https://doi.org/10.15622/sp.12.1
6. Parada-Cabaleiro, E., Costantini, G., Batliner, A., Schmitt, M., Schuller, B.W.: DEMoS: an Italian emotional speech corpus. Lang. Resour. Eval. **54**(2), 341–383 (2019). https://doi.org/10.1007/s10579-019-09450-y
7. Musaev, M.M., Ochilov, M.M., Khujayarov, I.S.: E2E models of continuous speech recognition with large vocabulary size. TATU Bull. **2**(58), 19–40 (2021)
8. Khujayorov, I., Ochilov, M.: Parallel signal processing based-on graphics processing units. Int. Conf. Inf. Sci. Commun. Technol. **2019**, 1–4 (2019). https://doi.org/10.1109/ICISCT47635.2019.9011976
9. Musaev, M., Khujayorov, I., Ochilov, M.: The use of neural networks to improve the recognition accuracy of explosive and unvoiced phonemes in Uzbek language. Inf. Commun. Technol. Conf. **2020**, 231–234 (2020). https://doi.org/10.1109/ICTC49638.2020.9123309
10. Abdel-Hamid, O., Mohamed, A., Jiang, H., Deng, L., Penn, G., Yu, D.: Convolutional neural networks for speech recognition. IEEE/ACM Trans. Audio Speech Lang. Process. **22**(10), 1533–1545 (2014). https://doi.org/10.1109/TASLP.2014.2339736
11. Zhang, Y., Pezeshki, M., Brakel, P., Zhang, S., Bengio, C.L.Y., Courville, A.: Towards end-to-end speech recognition with deep convolutional neural networks. arXiv preprint arXiv:1701.02720 (2017)
12. Musaev, M., Khujayorov, I., Ochilov, M.: Image approach to speech recognition on CNN. In: Proceedings of the 2019 3rd International Symposium on Computer Science and Intelligent Control (ISCSIC 2019). Association for Computing Machinery, New York, Article 57, pp. 1–6 (2019). https://doi.org/10.1145/3386164.3389100
13. Heafield, K.: KenLM: faster and smaller language model queries. In: Proceedings of the Sixth Workshop on Statistical Machine Translation, pp. 187–197 (2011)
14. Kumar, A., Vembu, S., Menon, A.K., Elkan, C.: Beam search algorithms for multi-label learning. Mach. Learn. **92**(1), 65–89 (2013). https://doi.org/10.1007/s10994-013-5371-6
15. Dong, L., Xu, S., Xu, B.: Speech-transformer: a no-recurrence sequence-to-sequence model for speech recognition. In: 2018 IEEE International Conference on Acoustics, Speech and Signal Processing (ICASSP), pp. 5884–5888. IEEE (2018)
16. Musaev, M., Khujayorov, I., Ochilov, M.: Development of integral model of speech recognition system for Uzbek language. In: 2020 IEEE 14th International Conference on Application of Information and Communication Technologies (AICT), pp. 1–6. IEEE (2020). https://doi.org/10.1109/AICT50176.2020.9368719
17. Musaev, M., Mussakhojayeva, S., Khujayorov, I., Khassanov, Y., Ochilov, M., Atakan Varol, H.: USC: an open-source Uzbek speech corpus and initial speech recognition experiments. In: Karpov, A., Potapova, R. (eds.) SPECOM 2021. LNCS (LNAI), vol. 12997, pp. 437–447. Springer, Cham (2021). https://doi.org/10.1007/978-3-030-87802-3_40
18. Abdullaeva, M., Khujayorov, I., Ochilov, M.: Formant set as a main parameter for recognizing vowels of the Uzbek language. Int. Conf. Inf. Sci. Commun. Technol. **2021**, 1–5 (2021). https://doi.org/10.1109/ICISCT52966.2021.9670268

19. Khassanov, Y., Mussakhojayeva, S., Mirzakhmetov, A., Adiyev, A., Nurpeiissov, M., Varol, H.A.: A crowdsourced open-source Kazakh speech corpus and initial speech recognition baseline. In: Proc. of the Conference of the European Chapter of the Association for Computational Linguistics, pp. 697–706. Association for Computational Linguistics (2021)
20. Rakhimov, M., Ochilov, M.: Distribution of operations in heterogeneous computing systems for processing speech signals. In: 2021 IEEE 15th International Conference on Application of Information and Communication Technologies (AICT), pp. 1–4 (2021). https://doi.org/10.1109/AICT52784.2021.9620451
21. Fazliddinovich, R.M., Abdumurodovich, B.U.: Parallel processing capabilities in the process of speech recognition. Int. Conf. Inf. Sci. Commun. Technol. **2017**, 1–3 (2017). https://doi.org/10.1109/ICISCT.2017.8188585
22. Graves, A., Fernández, S., Gomez, F., Schmidhuber, J.: Connectionist temporal classification: labelling unsegmented sequence data with recurrent neural networks. In: Proceedings of the 23rd International Conference on Machine Learning, pp. 369–376. Association for Computing Machinery, New York (2006)
23. Nasimova, N., Muminov, B., Nasimov, R., Abdurashidova, K., Abdullaev, M.: Comparative analysis of the results of algorithms for dilated cardiomyopathy and hypertrophic cardiomyopathy using deep learning. Int. Conf. Inf. Sci. Commun. Technol. **2021**, 1–5 (2021). https://doi.org/10.1109/ICISCT52966.2021.9670134

Multimodal Human Computer Interaction Using Hand Gestures and Speech

Mohammed Ridhun, Rayan Smith Lewis, Shane Christopher Misquith, Sushanth Poojary, and Kavitha Mahesh Karimbi(✉)

St Joseph Engineering College Mangaluru, Visvesvaraya Technological University, Belagavi 575028, Karnataka, India
kavitham@sjec.ac.in

Abstract. The paper presents multimodal human-computer interaction using speech and gesture recognition to develop a system for mouse movement and operation. The approach allows users to perform mouse navigation and various mouse operations without the need for physical contact with the system. Splitting up the task of mouse navigation and operations with gesture and speech recognition respectively led to a user-friendly and seamless experience for the user. Since no physical contact is required between the user and the system, it could be used by doctors while performing surgery, mechanics while they are handling their instruments from a distance, and casual users if circumstance arise. Unlike a unimodal gesture recognition system the proposed multimodal system allows mouse pointer control using speech and employs gestures to perform mouse operations.

Keywords: Human Computer Interaction · hand gestures · speech recognition

1 Introduction

With the rapid developments in technology (specifically processors), capabilities of computer systems are improving exponentially every year. Inspite of high efficiency and improved performance of modern systems, the way users can interact with their systems remains restricted. Current systems only exploit unimodal human-computer interaction via a mouse, keyboard or touch based systems (which uses the sense of touch), speech recognition system etc. There is a huge potential for multimodal systems which would result in a much more natural interaction with the system.

The developments made with regard to multimodal systems till date have been specifically restricted to highly specialized systems. These systems demand expensive hardware and are often employed in performing highly specialized tasks used by niche fields. Studies analyzing the performance difference between unimodal and multimodal systems for a range of tasks reveal that multimodal

H. Zaynidinov et al. (Eds.): IHCI 2022, LNCS 13741, pp. 63–74, 2023.
https://doi.org/10.1007/978-3-031-27199-1_7

systems are better compared to unimodal systems when the task to be done is complex enough, needing additional parallel control made available by the said multimodal system.

As applications become complex, a single modality limits effective interactions across different tasks and environments. A Multimodal interface typically offers users the freedom to use a combination of modalities or to switch to a better-suited modality, depending on the specifics of the task or environment. Unimodal systems leverage on a single modality which can limit its functionality and usability. Traditional input modalities such as the keyboard are majorly designed for the English-speaking population and are not easily adaptable to other languages. Multimodal systems on the other hand facilitate multiple input modalities to increase its usability and the weaknesses of one modality are offset by the strengths of another. Compared to a unimodal system, a multimodal system can offer a flexible, efficient, and usable environment allowing users to interact through input modalities, such as speech and hand gesture. For instance, patient information in an operating room can be accessed verbally by the members of the surgical team to maintain an antiseptic environment and presented in near real time aurally and visually to maximize comprehension. Multimodal interfaces have the potential to accommodate a broader range of users than traditional graphical user interfaces (GUIs) and unimodal interfaces, including users of different ages, skill levels, native language status and cognitive styles. The growing interest in multimodal interface design is inspired largely by the goal of supporting more transparent, flexible, efficient, and powerfully expressive means of human-computer interaction. Multimodal interfaces also are expected to be easier to learn and use and are preferred by users for many applications. They have the potential to expand computing to more challenging applications, to be used by a broader spectrum of everyday people and to accommodate more adverse usage conditions than in the past.

We present a multimodal system that combines speech and gesture recognition to perform mouse-related tasks. Gesture system is used to control the mouse pointer while speech commands are given to perform various operations. The approach requires a webcam and mic which is available in most modern laptops and hence does not involve additional development cost. The realised system shows great potential but the occasional delay between the issue of a speech command and its execution remains an issue as speech recognition relies on internet connection.

2 Related Work

Table 1. Comparison of Tools, Techniques, Benefits and Limitations in Multimodal Human Computer Interaction

Related Work	Tools	Techniques	Benefits	Limitations
Multimodal Natural Interaction for 3D Images [1]	For both speech and gesture commands, the Microsoft Kinect device was chosen for implementation	Speech commands are used to initiate and close operations of the system, while gestures are used to interact with 3-D Graphics	A much more natural interaction with the computer	Additional hardware and software requirements which drives up cost
Comparing the use of single versus multiple combined abilities in conducting complex computer tasks hands-free [2]	The multimodal tongue drive system (mTDS) provides proportional cursor control via head motion, discrete clicks via tongue movements, and simultaneous typing via speech recognition	Four tasks were used for evaluation which were Center-out Tapping (pointing device evaluation), Maze Navigation (Constrait navigation evaluation), Game(Unconstraint navigation evaluation) and Email(Complex task evaluation)	MTDS results in a noticeable improvement in complex task performance compared to unimodal tongue gestures for people with physical disabilities like tetraplegia	Unimodal tongue gestures were found to be better than mTDS when it came to simple tasks
Voice to Text transcription using CMUSphinx [3]	Toolkit for CMUSphinx	Acoustic model, phonetic dictionary, and language model are all used	Offline usage with reasonable accuracy	Less accuracy compared to online speech recognition libraries like google speech recognition
Multimodal interaction: A review [4]	CUBRICON, The Koons et al	Integrating speech, gesture, and eye gaze	Better flexibility and dependability can provide interaction options that better match the needs of a wide range of users with different usage habits and preferences	Approaches and structures for multimodal integration must consider a broader range of methods and modality combinations
PocketSphinx: A Free, Real-Time Continuous Speech Recognition System for Hand-Held Devices [5]	PocketSphnix, CMU Sphnix -II	Combination of memory optimization, machine optimization and algorithm optimization	Since this was lite system, it could be used onsmall and mobile devices	Running on other system required more powerful hardware

(*continued*)

Table 1. (*continued*)

Related Work	Tools	Techniques	Benefits	Limitations
The OpenInterface Framework: A tool for multimodal interaction [6]	OI Kernal, OpenInterface Interaction Development Environment	Device and Task Components and Transformation components	As a result of the OpenInterface framework, we take a more realistic hybrid approach, allowing the designer/developer to use both generic and customised components	Multimodal system implementation is still a difficult task
Study of Deep learning and CMU sphinx in automatic speech recognition [7]	CMU Sphinx, Deep Learning Algorithm using CNN (Convoluted Neural Networks)	Pattern Recognition And Deep Learning	Helps determine the best speech recognition system for maximum accuracy in word and pattern detection	Deep Learning algorithms require large amounts of data. CMU Sphinx is outperformed by Deep Learning in terms of accuracy and word error rate
Mouse Cursor Control System based on Hand Gestures [8]	External webcam and hand pad	Image Processing	Simple to use, feels natural and is intuitive	Result is dependent on the webcam quality and fps
Benefit, Design and Evaluation of Multimodal Interaction [9]	Speech recognition and tactile feedback	Multimodal Interaction	MMI provides a user-friendly UI model for a user that caters to users need	Difficult to implement
Integration Themes in Multimodal Human-Computer [10]	Speech and typing recognition	Semi-automatic simulation technique	People's utilisation of spoken and written input during multimodal human-computer interactions is specified	Does not applicable for images. Accuracy depends on the presentation format

3 Methodology

3.1 Architecture of a Multimodal User Interface

We follow the Maybury and Wahlster high-level architecture of intelligent user interfaces. As their notion of intelligent user interfaces includes multimodal interaction, the model is suitable for modelling multimodal interfaces. Figure 1 (left) shows an adaptation of their model and presents the different levels of processes that are contained in the proposed architecture [11].

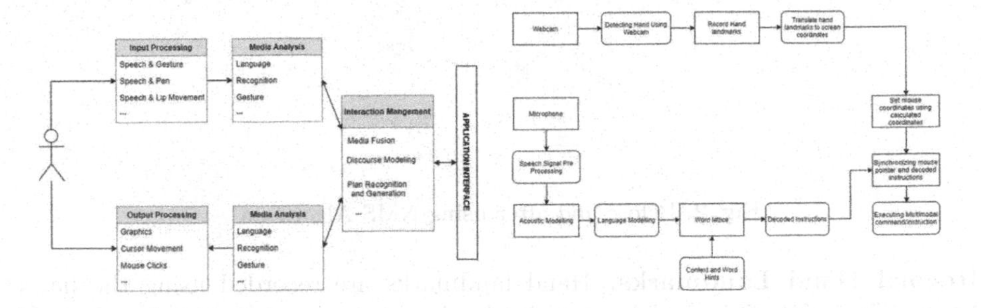

Fig. 1. Architecture of a Multimodal User Interface (left) and UML Design for Multimodal Human Computer Interaction

The Fig. 1 (left) depicts several models for different processes in the system. Each of these models can be refined to fulfill the requirements of a given system. The user provides a pair of inputs which are analyzed and integrated. The actions from these inputs take place on the application interface. These actions are visualized as outputs by the user.

The Fig. 1 (right) shows the approach employed in multimodal interaction involving a combination of gesture and voice control modalities. While these tasks take place simultaneously, the webcam is used to detect the gestures and the microphones pick up the voice command. The gestures are translated to the mouse pointer co-ordinates, and speech signals to instructions. These are synchronized and the multimodal instruction is executed.

3.2 Gesture Recognition System

We used the Gesture recognition system to detect user's hands. Using its position at any given time we determine the mouse pointer location with respect to the screen coordinates. We used the library MediaPipe[1] for detecting hands and getting hand landmarks. AutoPy[2] was used for extracting the screen resolution information and for controlling the mouse. The following steps were employed in the gesture recognition.

Detecting Hand Using Webcam. The palm detection model was used to first detect palm that operates on the full image captured from the webcam and returns an oriented hand bounding box. Non Maximum Suppression (NMS) Algorithm, an unsupervised machine learning algorithm was used for the purpose which whittles down large number of detected rectangles (bounding box) to a few. The hand bounding box (output) was later given to the landmark module. The crops could also be generated based on the hand landmarks identified in the previous frame, and only when the landmark model could no longer identify hand presence was palm detection invoked to re-localize the hand.

[1] https://google.github.io/mediapipe/solutions/hands#python-solution-api/.

[2] Michael Sanders, AutoPy Introduction and Tutorial, Available: https://pypi.org/project/autopy/.

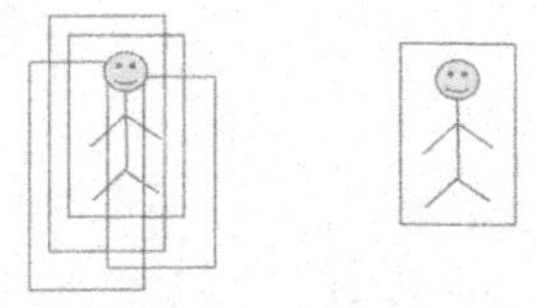

Fig. 2. Before and after using NMS Algorithm

Record Hand Landmarks. Hand landmarks are recorded using the hand landmark model. This model was trained using approximately 30,000 real world images with 21 3D coordinates. It performs precise key point localization of 21 3D hand-knuckle coordinates (landmarks) inside the detected hand regions as seen in Fig. 3 via regression. Providing the accurately cropped hand image to the hand landmark model by the palm detection model drastically reduces the need for data augmentation (Eg. rotation, translation, and scale) and instead allows the network to dedicate most of its capacity towards coordinate prediction accuracy. Landmark 9 is chosen for tracking since it is the centremost landmark.

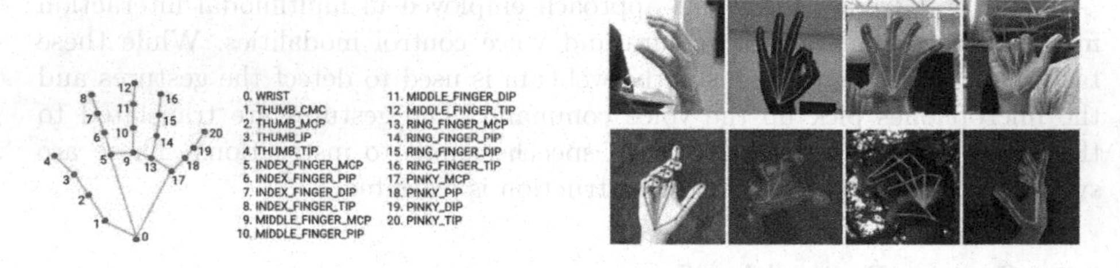

Fig. 3. Hand Landmarks (left) and Crops (right)

Translate Hand Landmarks to Screen Coordinates. Each hand is represented as a list of 21 hand landmarks and each landmark is composed of x, y and z coordinate. Hand landmarks are normalised values in the range [0.0, 1.0] with respect to the image (webcam frame) width and height (x and y coordinate). z coordinate represents the landmark depth with the depth at the wrist being the origin. The smaller the value of the z coordinate, the closer the landmark is to the camera. z coordinate is not used in this project. This normalized value is converted to webcam frame coordinates by multiplying it with frame width and frame height for x and y coordinate, respectively. NumPy[3] function interp (interpolate) is used for the conversion of the frame coordinates to screen coordinates. This function converts the point passed from the range of [0, frame_width] to [0, screen_width] for x coordinate and [0, frame_height] to [0, screen_height] for y coordinate. It uses screen resolution information (provided by AutoPy screen.size() function) and the x and y values of the landmarks to calculate screen coordinates which ensures that the program will work in all devices regardless of their screen resolutions.

[3] NumPy, Available: https://numpy.org/.

Set Mouse Coordinates Using the Calculated Coordinates. The coordinates can that was calculated in the previous step can be used by the AutoPy[4] library to set mouse coordinates. This is done for every frame which results in the mouse pointer moving according how the user moves their hand in real time.

3.3 Speech Recognition System

Speech recognition is the task of detecting spoken words. There are many techniques to do Speech Recognition. To implement speech recognition, we have used Google Cloud speech-to-text API[5] and pyaudio[6] as a device driver for the device microphone. Following are the main steps in the Speech Recognition System:

Speech Signal Pre-processing. The Speech signal is detected using the microphone of the device. Then the waveform of the speech signal is split at utterances by silences and then tries to recognize what is being said in each utterance. Google Cloud speech-to-text API offers Streaming speech recognition which offers real-time speech recognition results as the API processes the audio input streamed from your microphone (inline or through Cloud Storage).

Acoustic Model. An acoustic model is used in automatic speech recognition to represent the relationship between an audio signal and the phonemes or other linguistic units that make up speech. The model is learned from a set of audio recordings and their corresponding transcripts. It is created by taking audio

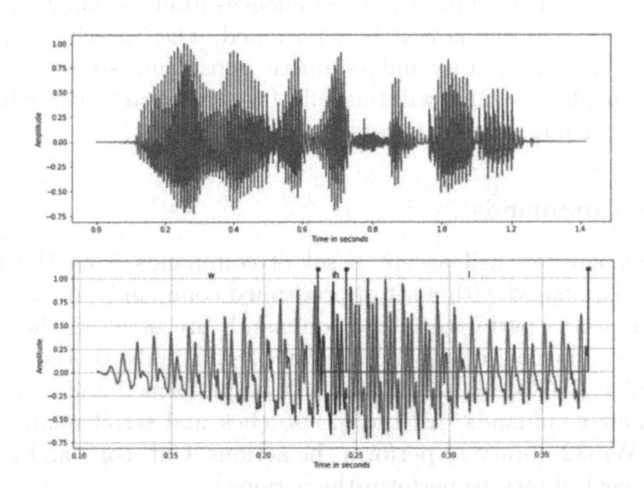

Fig. 4. Speech signal with the phrase "will we ever forget it" (top) and Phoneme level segments for 'will' as 'w', 'ih' and 'l' (bottom).

---

[4] Michael Sanders, AutoPy Introduction and Tutorial, Available: https://pypi.org/project/autopy/.

[5] Google Cloud speech-to-text, Available: https://cloud.google.com/speech-to-text.

[6] PyAudio, Available: https://pypi.org/project/PyAudio/.

recordings of speech, and their text transcriptions, and using software to create statistical representations of the sounds that make up each word. The acoustic model models the relationship between the audio signal and the phonetic units in the language. Figure 4 (top) illustrates a speech signal mapped on a Amplitude vs Time axis representing the phrase "will we ever forget it".

Figure 4 (bottom) shows how the speech signal is broken down. The speech signal is converted into sounds that the model interprets as phoneme level segments. Here the word 'will' in the phrase 'will we ever forget it' is broken down into phoneme level segments as 'w', 'ih' and 'l'.

Language Modelling. A language model is used to restrict word search. It defines which word could follow previously recognized words (matching is a sequential process) and helps to significantly restrict the matching process by stripping words that are not probable. The language model provides context to distinguish between words and phrases that sound similar. For example, in American English, the phrases "recognize speech" and "wreck a nice beach" sound similar but mean different things. In speech recognition, sounds are matched with word sequences. Ambiguities are easier to resolve when evidence from the language model is integrated with a pronunciation model and an acoustic model.

Word Lattice. The language model produces a lattice of word alternatives and the Speech Adaptation API[7] allows to set context and word hints about what the audio might contain. This information is then used to operate on the word lattice, based on these hints and confidences from the language model, the highest confidence hypothesis text is determined. The Speech Adaptation API allows to set a particular phrase and a numeric boost value so that we can specify a weight for that phrase which will take effect on the word lattice which is useful if you have proper nouns or some rare words.

3.4 Speech Commands

The speech recognition will accept a set of commands from the user. These commands are compared with a set of predefined commands, if they match then a respective action is performed. The commands are listed in the table below. If the commands do not match the following commands, or the system is not able to recognize the speech, no actions will be performed and will wait for the user to give new commands. All the mouse click and scroll related commands utilize the PyWin32 library to perform the actions. Cut, copy and paste use the Python Keyboard library to perform the actions.

Click or Left Click. To perform a click or a left click, we use the mouse_event() function from the win32api package. To simulate the mouse clicking down the left button, we pass MOUSEEVENTF_LEFTDOWN as the parameter to the

[7] Cloud speech-to-text Documentation, Available: https://cloud.google.com/speech-totext/docs/basics.

mouse_event() function. Then a time delay for 0.01 s is introduced to ensure the system registers the click. Then to simulate the release of the left button, we pass MOUSEEVENTF_LEFTUP as the parameter to another mouse_event() function[8].

Right Click. To perform a right click, we use the mouse_event() function from the win32api package. To simulate the mouse clicking down the right button, we pass MOUSEEVENTF_RIGHTDOWN as the parameter to the mouse_event() function. Then a time delay for 0.01 s is introduced to ensure the system registers the click. Then to simulate the release of the right button, we pass MOUSEEVENTF_RIGHTUP as the parameter to another mouse_event() function.

Open or Double Click. The mouse double click action is performed, when the speech recognition system recognizes the input as "open" or "double click". Then it calls the mouse_left_click() method twice. This emulates the mouse left click being pressed twice.

Copy. To perform the copy command, we first need the pointer to perform a click action on the object to be copied. The click action ensures the correct object is selected. Next, use the send() function that is a part of the Keyboard library. "ctrl+c", is sent as the parameter for the send() function[9].

Cut or Move. To perform the cut or move command, we first need the pointer to perform a click action on the object to be moved. The click action ensures the correct object is selected. Next, use the send() function that is a part of the Keyboard library. "ctrl+x", is sent as the parameter for the send() function.

Paste. To perform the paste command, we first need to navigate to the destination where the object needs to be placed. Use the send() function that is a part of the Keyboard library. "ctrl+v", is sent as the parameter for the send() function. This will paste the file to the destination.

Scroll or Zoom. To perform scroll up, scroll down, zoom in and zoom out commands, we utilize the MOUSEEVENTF_WHEEL attribute of the win32con function. The scroll() function takes input for the number of scroll clicks in negative or positive value. If the value is positive, then it will scroll up or zoom in the page, and if the value is negative, then it will scroll down or zoom out the page. The scroll clicks are represented as the WHEEL_DELTA attribute.

[8] mouse_event function (winuser.h), Available: https://docs.microsoft.com/en-us/windows/win32/api/winuser/nf-winuser-mouse_event.

[9] Keyboard module in Python, Available: https://www.geeksforgeeks.org/keyboard-module-in-python/.

3.5 Integration

The synchronization of gesture and speech is performed via python multithreading package. Separate threads are made for gesture and, speech recognition system and then they are run parallelly. For inter-thread communication, the concept of global variables was used to indicate the gesture recognition system to end its execution once a speech command to stop the speech recognition system was given. This multimodal system is designed to be run at the system level. Hence, it is not dependent on specific applications. Application developers may choose to add in more features which is specific to their application.

4 Experimental Setup and Evaluation

4.1 Gesture Recognition System

Table 2 highlights the difference in performance between a modality (hand gesture) in a unimodal and a multimodal system. Here we focus on the performance difference between the gesture recognition system in a unimodal and multimodal system (which includes both speech and gesture recognition system). A similar comparison is done for speech recognition system in the following sections. For the performance test, 10 trials were conducted on each system. The metric used for the performance is average framerate (frames per second). After the conduction of the test, it is observed that the performance of the gesture recognition is very similar in both the unimodal and multimodal system. The small difference seen in the performance fall under the margin of error while testing. The similarity is performance is because most of the modern systems has at least a dual core processor and hence the speech and gesture recognition system in the multimodal system could independently run in the two separate cores.

Table 2. Performance comparison of Gesture Recognition System

Trial	**Average Framerate in Unimodal System (Frames/Second)**	**Average Framerate in Multimodal System (Frames/Second)**
1	31.86	29.98
2	26.69	31.17
3	32.61	31.12
4	47.61	32.68
5	32.62	25.76
6	33.13	33.02
7	32.78	32.18
8	33.35	31.97
9	33.47	31.32
10	33.37	32.21
Average	**33.74**	**31.14**

4.2 Speech Recognition System

The graphs in Fig. 5 compare the accuracy between the unimodal (left) and multimodal (right) speech recognition system. The test was performed for 300 s, 37 instructions. Both the systems had an accuracy of 67.57%.

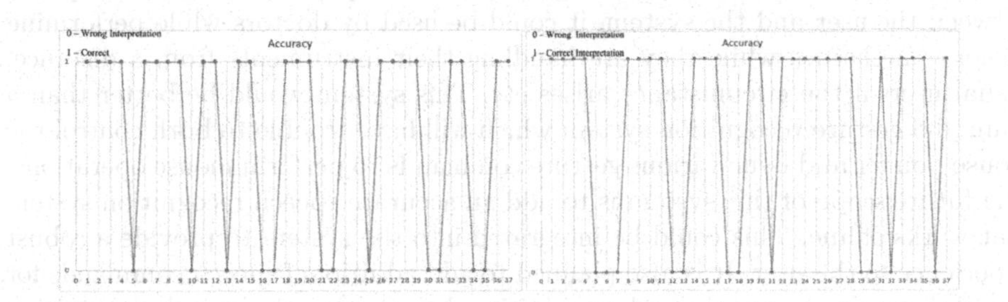

Fig. 5. Accuracy of Unimodal (left) and Multimodal (right) Speech Recognition System

The Fig. 6 represent the average latency in execution of each command for the speech recognition system over a series of 10 commands. The latency observed between each unimodal and multimodal command is negligible. From these observations we can say, the modality of the system does not have an effect on its performance in a significant manner.

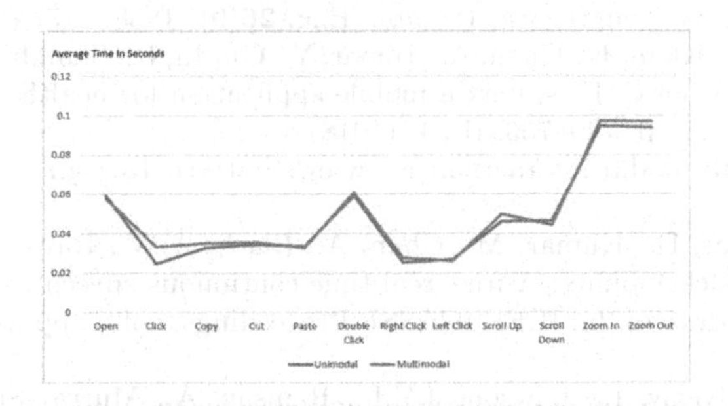

Fig. 6. Latency Graph for each Command in Unimodal and Multimodal System

5 Conclusion and Future Scope

The main objective of the work was to explore the concept of multimodal human-computer interaction and to compare it with the established unimodal human-computer interaction. The realized system combines speech and gesture recognition allowing complete control over mouse movement and operation. The system allows users to perform mouse navigation and various mouse operations without a need for physical contact with the system. Splitting up the task of mouse navigation and operations with Gesture and Speech Recognition respectively led to a user-friendly and seamless experience for the user. Gesture and Speech Recognition system uses two independent modalities (Kinesthetic modality -hand, and Auditory modality - speech). Hence, complexity is reduced as the end users can easily manage the two modalities together. Since no physical contact is required between the user and the system, it could be used by doctors while performing surgery, mechanics while they are handling their instruments from a distance, casual users if the circumstance arises etc. This system would be better than a unimodal gesture recognition system which will have trouble to both control the mouse pointer and issue various gesture commands to perform mouse operations. The future scope of this system is to add an accurate speech recognition system that works offline. This could be integrated into the system to provide a robust experience to the user. It is also planned to add additional speech commands for complex mouse operations and an additional feature to the speech recognition system to accurately recognize lengthy natural sentences that could be used to type instead of using a keyboard.

References

1. Ergüner, F., Durdu, P.O.: Multimodal natural interaction for 3d images. In: 9th International Conference on AICT, pp. 305–309. IEEE (2015)
2. Sahadat, M.N., Alreja, A., Mikail, N., Ghovanloo, M.: Comparing the use of single versus multiple combined abilities in conducting complex computer tasks hands-free. IEEE Trans. Neural Syst. Rehabil. Eng. **26**(9), 1868–1877 (2018)
3. Lakdawala, B., Khan, F., Khan, A., Tomar, Y., Gupta, R., Shaikh, A.: Voice to text transcription using CMU sphinx a mobile application for healthcare organization. In: 2nd ICICCT, pp. 749–753. IEEE (2018)
4. Turk, M.: Multimodal interaction: a review. Pattern Recogn. Lett. **36**, 189–195 (2014)
5. Huggins-Daines, D., Kumar, M., Chan, A., Black, A.W., Ravishankar, M., Rudnicky, A.I.: Pocketsphinx: A free, real-time continuous speech recognition system for hand-held devices. In: IEEE ICASSP Proceedings, vol. 1, pp. I-185-I-188. IEEE (2006)
6. Serrano, M., Nigay, L., Lawson, J.Y.L., Ramsay, A., Murray-Smith, R., Denef, S.: The openinterface framework: a tool for multimodal interaction. In: CHI'08 Extended Abstracts on Human Factors in Computing Systems, pp. 3501–3506 (2008)
7. Dhankar, A.: Study of deep learning and CMU sphinx in automatic speech recognition. In: ICACCI 2017, pp. 2296–2301. IEEE (2017)
8. Grif, H.S., Farcas, C.C.: Mouse cursor control system based on hand gesture. Procedia Technol. **22**, 657–661 (2016)
9. Schaffer, S., Reithinger, N.: Benefit, design and evaluation of multimodal interaction. In: Proceedings of the 2016 DSLI Workshop. ACM CHI (2016)
10. Oviatt, S., Olsen, E.: Integration themes in multimodal human-computer interaction. In: 3rd ICSLP (1994)
11. Raisamo, R.: Multimodal Human-Computer Interaction: a constructive and empirical study. Tampere University Press (1999)

Emotion Recognition in VAD Space During Emotional Events Using CNN-GRU Hybrid Model on EEG Signals

Mohammad Asif(✉), Majithia Tejas Vinodbhai(✉), Sudhakar Mishra(✉), Aditya Gupta, and Uma Shanker Tiwary

Indian Institute of Information Technology, Allahabad, Uttar Pradesh, India
{pse2017001,mit2020058,rs163}@iiita.ac.in

Abstract. Emotion recognition from brain signals is an emerging area of interest in the scientific community. We used EEG signals to classify emotional events on different combinations of valence(V), arousal(A) and dominance(D) dimensions and compared their results. DENS data is used for this purpose which is primarily recorded on the Indian population. STFT is used for feature extraction and used in the classification model consisting of CNN-GRU hybrid layers. Two classification models were evaluated to classify emotional feelings in valence-arousal-dominance space (eight classes) and valence-arousal space (four classes). The results show that VAD space's accuracy is 97.50% and VA space is 96.93%. We conclude that having precise information about emotional feelings improves the classification accuracy in comparison to long-duration EEG signals which might be contaminated by mind-wandering. In addition, our results suggest the importance of considering the dominance dimension during the emotion classification.

Keywords: Affective Computing · CNN · Deep Learning · EEG · Electroencephalography · Emotion Recognition · GRU

1 Introduction

Emotion classification has been a challenging and emerging topic in AI and especially in affective computing. There are several methods for detecting emotions by the intelligent systems, e.g.- detecting emotions from a given image, video or from a text sequences. It is also possible to detect emotions from the brain signals and we can trust the results as these brain signals are infallible, so no manipulation there on this front. But brain signals come with their own challenges which mostly related to noise present in the signal. These signals consist of several artefacts including but not limited to line noise, noise from the muscle activities and eye movements, interference from other cognitive activities etc.

M. Asif, M. T. Vinodbhai, S. Mishra, A. Gupta—These authors contributed equally to this work and author sequence is random.

H. Zaynidinov et al. (Eds.): IHCI 2022, LNCS 13741, pp. 75–84, 2023.
https://doi.org/10.1007/978-3-031-27199-1_8

There are several methods for recording brain signals e.g., functional Magnetic Resonance Imaging(fMRI), Electroencephalography (EEG) and Magnetoencephalography (MEG). For emotion recognition EEG is widely used as it is reliable, relatively less expensive and offers better temporal information. Some famous studies to recognize emotion from EEG data are [1–4]. We have used our own data collected in our lab which follows a modified paradigm of collecting emotion information from the participants which we call 'Emotional Event', more information in Method section.

After selecting which type of data to use, next comes an important decision of feature engineering and deep learning model architecture. As EEG data contains time series data, it is beneficial to convert it into cosine functions via a Fourier Transformation hence we used Short time fourier transform (STFT) for feature extraction. As these signals are converted into 3D spectrograms, Convolution neural networks (CNNs) are best suited to handle this type of data and Recursive neural networks (RNNs) are useful for handling the temporal information of the data. So we built a hybrid architecture for the emotion classification task.

Convolution Neural Networks are embodied with neurons capable of optimizing on their own through learning. CNNs are primarily used in images for pattern recognition, so we can extract key features from images to make the network stronger for accurate results [13]. The architecture of CNN consists of three dimension data as input height, width and depth. CNN can extract features, and CNN kernels are used to determine which part of input data we need to extract features [14]. The major advantage of using CNN is that its layer is not necessarily connected to all the previous neurons. It majorly focuses on the part of data where useful information is recited. And also, in CNN many connections can use similar weights to reduce the complexity of the network.

Gated recurrent unit (GRU) is the latest form of the recurrent neural network, which has the capability to resolve vanishing exploding problems of RNN similar to LSTM but is basically a lighter version of LSTM with similar or more efficiency [15]. The GRU controls the flow of information like the LSTM unit, but without using a memory unit. It just exposes the full hidden content without any control [16]. The major change in GRU is fusing the inner cell and the hidden state into a single one, and this collective data get passed on to the next GRU.

2 Method

2.1 Participants

Forty participants participated in the study, including 3 females (Mean age = 23.3, SD = 1.4). The institutional review board approved the study. The participants are taken from the engineering institution which predominantly has higher ratio of male students. In total, forty three participants have participated in the experiment. Three participants were removed due to excessive movement. Since we had skewed gender-wise sampling, we didn't perform any gender based analysis.

2.2 Dataset on Emotions with Naturalistic Stimuli (DENS)

The EEG data which we have used in this analysis was collected from Indian samples for different emotion categories. EGI 128 dense array system was used for recording the brain activity during emotional activity. The complete experiment paradigm is available elsewhere [10]. Subjects are shown emotional videos selected from an affective film stimuli dataset validated on the Indian population [9]. Subjects were asked to mark the moments of emotional feelings by performing left mouse clicks while looking at the film stimulus. At the end of the stimulus watching, participants categorized their emotional feelings into suitable emotional categories. They were also asked to provide their feedback on valence, arousal, and dominance scales.

2.3 Data Preprocessing

The complete preprocessing details are available elsewhere [10]. The data is filtered using the fifth order Butterworth bandpass filter with 1 Hz 40 Hz. Independent component analysis (ICA) was used to remove artefacts. After the preprocessing, a seven seconds segment of EEG signal around the subjective marking of emotional feeling is extracted. Since no information about time duration of emotional experience is available in the literature, we considered different time duration (say, 10s, 9s, and so on). With seven seconds we observed that only 18 emotional events were overlapping. We removed these 18 emotional events from our analysis. We will refer to this seven seconds segment as the emotional events. Likewise, 465 emotional events from 40 participants were collected.

2.4 Feature Extraction and Input to the Model

The Short-Time Fourier Transform, also called STFT, is a logical continuation of the Fourier transform that uses windows to do segmented analysis to deal with non-stationarity of the signals [11]. The time-localized frequency information that is provided by the STFT is superior to the frequency information that is provided by the standard Fourier transform, which is an average of frequency information over the entirety of the signal's time interval [12]. Within the realms of signal processing and signal synthesis, STFT has been put to work in a wide variety of applications over the years. STFT is more effective on unimodal and univariate signals as it involves very less complexity, moreover the signal artifacts and noise are low. Since the fixed window limits how many non-stationary features can be extracted, STFT features might need a threshold level when they are extracted and fitted. On the other hand, in comparison to the Fourier transform, the short-time Fourier transform (STFT) possesses superior temporal and frequency localization capabilities. The Space-Time Frequency Transform (STFT) divides the space of time and frequency into grids of the same size. Spectrogram is a two-dimensional image with time on the horizontal axis and frequency bins on the vertical axis. Frequency bin count = (framesize/2) + 1, while ((signal size - framesize)/hopsize) + 1 represents the number of time frames. The Gabor transform serves as the mathematical foundation for the spectrogram. Utilizing the Gabor transform, the spectrogram is computed. A special form of the short-time Fourier transform known as the Gabor transform is used to isolate the sinusoidal frequency and phase content of a signal in a certain region. Formally, we can extract the frequency components that make up any signal after applying a Fourier transform to it.

We have a total of 465 mat (MATLAB formatted data) files with emotional events in the DENS dataset. All 465 of the files have been chosen for the experiment. Each mat files have the following format: (128, 1751), where 128 is the total number of channels and 1751 is the EEG data. After that, we changed the data from (465,128,1751) into (59520, 1751) so that we could extract features with a 0.5-second window and a 0.25-second overlap. Following the feature extraction, 59,520 spectrograms each have shape (63,26).

2.5 Architecture of the Model

Our model consists of 32 channels CNN followed by Dropout layer and Max Pooling layer. CNN architecture is a combination of different layers like convolution layer and pooling layer, connected layer and many more, and it works in a feed-forward fashion to execute various layers. The input shape of CNN is in 3 dimensions which compromise of height, width and features, usually 3 to represent R, G and B of the image.

Then we used two layers of GRU having 256 and 128 units, respectively, each followed by a dropout layer. The two important gates of GRU are updated and reset gates. The update gate basically determines how past information needs to be passed on future whereas the reset gate determines how much past information needs to be deleted (Fig. 2). The advantage of using GRU is that it requires less run time compared to LSTM, with quite a similar result.

At last, we used a layer of 64 fully-connected nodes followed by the output layer. The detailed architecture of the model is given in Fig. 1

Formulation of GRU

$$r_t = \sigma(U^r x_t + W^r s_{t-1} + b^r)$$
$$z_t = \sigma(U^z x_t + W^z s_{t-1} + b^z)$$
$$s_t = Z_t o s_{t-1} + (1 - z_t) o (\sim s_t)$$
$$\sim s_t = g(U^s x^t + r_t o W^s s_{t-1}) + b^s$$

Where, t is the time instance; x_t is the input at time t; U^* $\&W^*$ are weight matrices; σ $\&g$ are sigmoid and tangent activation function, p^* $\&b^*$ are peephole connection (i.e., all the gates are having an input along with the cell state) and biases, and o means element wise product.

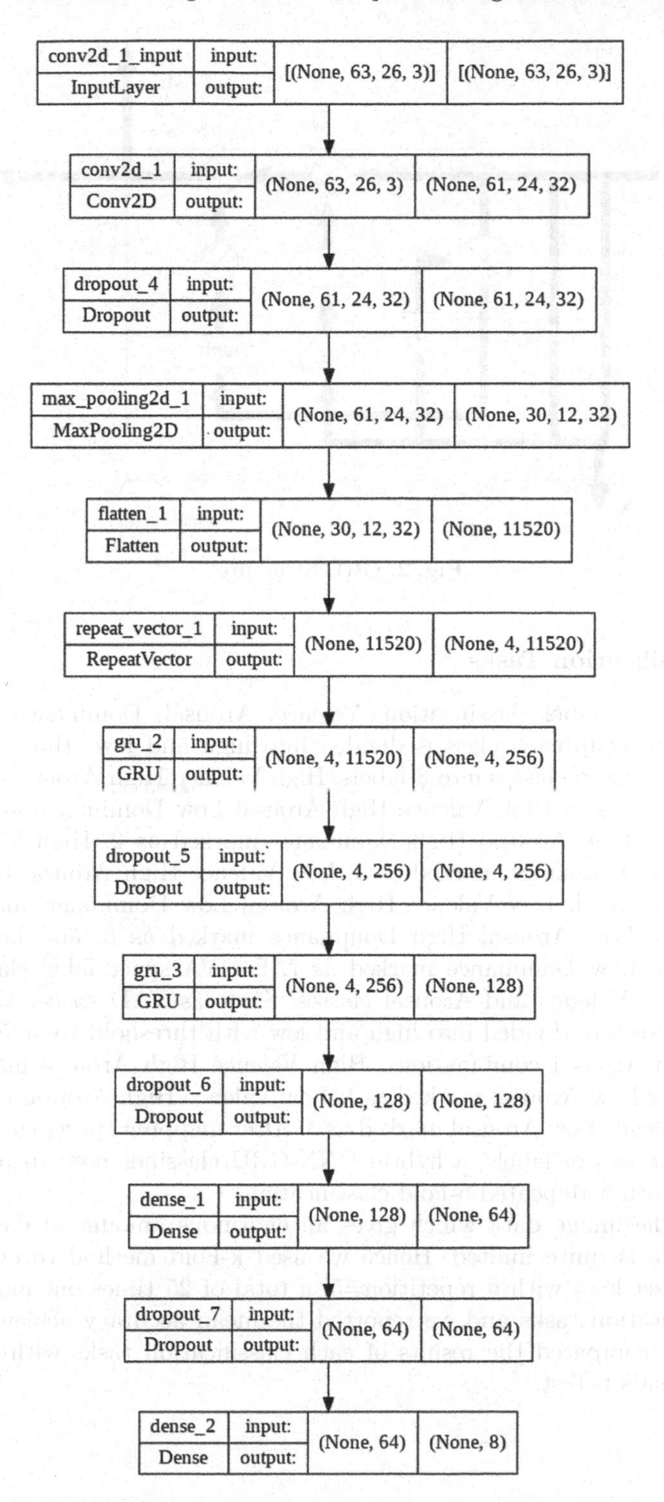

Fig. 1. Proposed Architecture of the Model

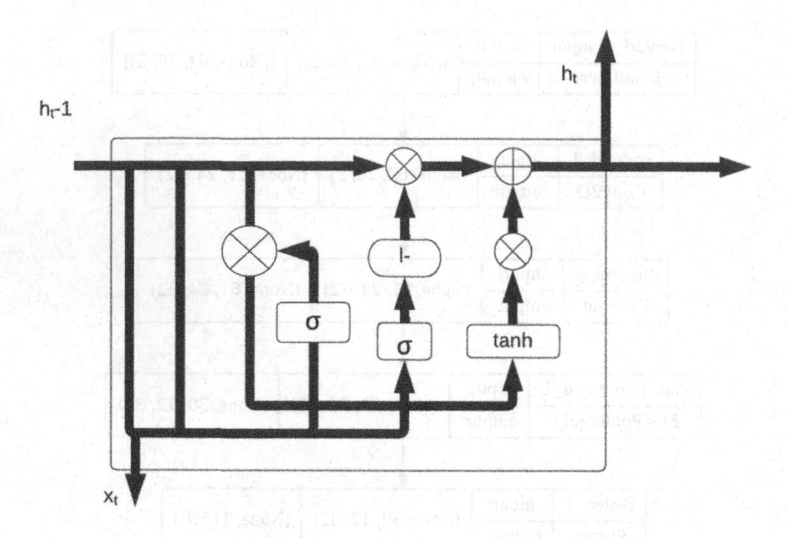

Fig. 2. GRU Structure

2.6 Classification Tasks

For VAD space label classification, Valence, Arousal, Dominance classes are used. In this setup each class is divided into high and low (threshold 5) and by combining these classes into 8 labels: High Valence High Arousal High Dominance marked as 0, High Valence High Arousal Low Dominance marked as 1, High Valence Low Arousal High Dominance marked as 2, High Valence Low Arousal Low Dominance marked as 3, Low Valence High Arousal High Dominance marked as 4, Low Valence High Arousal Low Dominance marked as 5, Low Valence Low Arousal High Dominance marked as 6, and Low Valence Low Arousal Low Dominance marked as 7. For VA space label classification, we have used Valence and Arousal classes. Same as VAD space, Valence and Arousal classes are divided into high and low with threshold value 5 and using Valence and Arousal combinations: High Valence High Arousal marked as 0, High Valence Low Arousal marked as 1, Low Valence High Arousal marked as 2 and Low Valence Low Arousal marked as 3. After mapping spectrogram features to particular sets of labels, a hybrid CNN-GRU classifier used to populate 25 accuracies with 5-Repeated 5-Fold classification.

Unlike the image data which gives an enormous amount of data to train on, our data is quite limited. Hence we used k-Fold method to evaluate our model and set k=5 with a repetition=5, a total of 25 times our model run on each classification tasks and we reported the mean accuracy achieved from it. Further, we compared the results of each classification tasks with each other based student's t-Test.

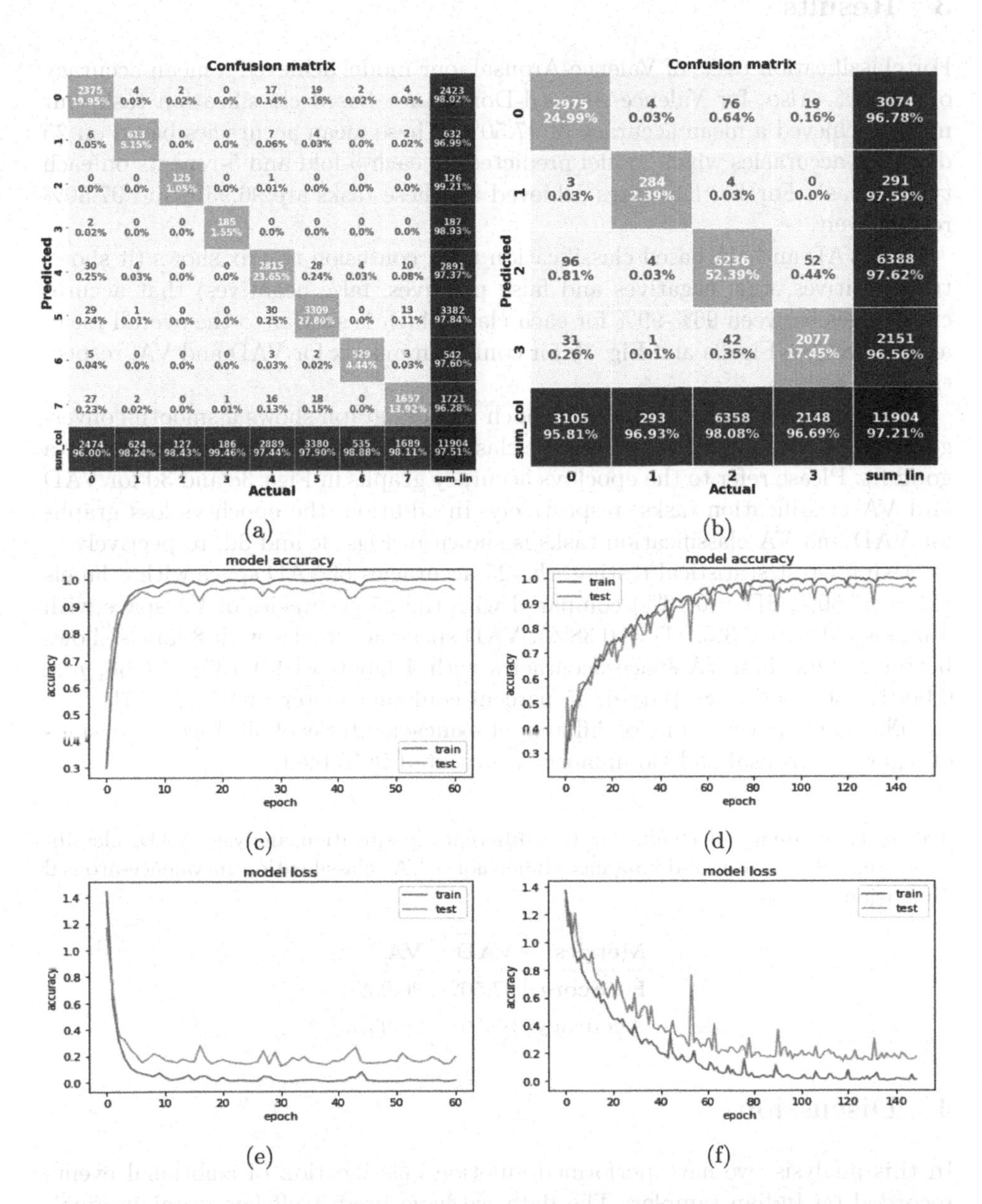

Fig. 3. (a) Confusion Matrix for Valence-Arousal-Dominance Based Classification. (b) Confusion Matrix for Valence-Arousal Based Classification. (c) Accuracy graph for classification in valence-arousal-dominance dimension. (d) Accuracy graph for classification in valence-arousal dimension. (e) Model loss for classification in valence-arousal-dominance dimension. (f) Model loss for classification in valence-arousal dimension.

3 Results

For classification tasks of Valence-Arousal, our model achieved a mean accuracy of 96.93%. Also, for Valence-Arousal-Dominance based classification task, our model achieved a mean accuracy of 97.50%. These mean accuracies based on 25 different accuracies which model predicted on each 5-fold and 5-repeats on each type of task. Further F1-scores achieved for these tasks are 96.93% and 97.50% respectively.

For VAD and VA based classification task, confusion matrix shows (it shows true positives, true negatives and false positives, false negatives) that accuracies ranges between 96%-99% for each class which is similar to the overall mean accuracies (see Fig. 3a and Fig. 3b for confusion matrix for VAD and VA, respectively).

Epoch vs Accuracy graph and Epoch vs Loss graph shows a smooth convergence to an optimal point for both the classification tasks, which is desirable for a good fit. Please refer to the epoch vs accuracy graphs in Figs. 3c and 3d for VAD and VA classification tasks, respectively. In addition, the epoch vs loss graphs for VAD and VA classification tasks is shown in Figs. 3c and 3d, respectively.

Using t-test statistical testing, the 25 accuracies of VAD space with 8 labels (M = 97.50%, SD = 0.16%) compared with the 25 accuracies of VA space with 4 labels (M = 96.93%, SD = 0.38%), VAD space accuracies with 8 labels shows better results than VA space accuracies with 4 labels with t (32) = 6.68, $p <$ 0.0001, cohen's d=1.89 (large), 95 percent confidence interval=[1.20 2.57].

The summary of results of different classification tasks of all three dimensions of Valence, Arousal and Dominance is presented in Table 1.

Table 1. Summary of results for two different classification analysis. VAD: classification in valence-arousal-dominance dimensions, VA: classification in valence-arousal dimension.

Metrics	VAD	VA
F1 Score	97.50%	96.93%
Accuracy	97.50%	96.93%

4 Discussion

In this analysis, we have performed emotion classification of emotional events recorded on Indian samples. The data we have used itself has a unique quality. For the first time, the spontaneous information about emotional feelings was captured while subjects are watching the naturalistic emotional multimedia stimuli (in comparison to previous datasets with multimedia stimuli, which lack this temporal information). This data is recorded in our lab. Hence, one of our aim in this analysis is to show that if the spontaneous emotional feelings, situated in the context, is captured, we don't need complicated models to perform emotion recognition. Because the recorded spontaneous emotional events are not mixed with the mind-wandering activity. We would have recorded the mind-wandering activity mixed with emotional feelings if we had taken the EEG recording for the whole stimulus length (which is a general trend in EEG based emotion experiments [1–4]).

Another unique observation in this study is related to the importance of the dimension of dominance. The literature is overwhelmingly filled with the classification of emotion in the quadrants taken only in valence-arousal space. Our analysis found that including the dimension of dominance provided significantly better accuracy than if only valence and arousal dimensions were considered.

Generally, in literature other than valence and arousal, dominance is proposed as the third dimension [5,6] but it is rarely considered in the classification tasks. The reason is that consideration of dominance as a dimension rarely explains more than 15% of the variance in subjective rating [5] hence it is still debatable to consider dominance as the third dimension. However, it is acknowledged in the circumplex model of affect [7] that emotional episodes cannot be merely defined using core affect dimensions (valence and arousal), and there is a need to include third dimension for the attribution of cause and meta-cognitive judgement. For example, experientially, fear and anger are distinct emotional states. However, the two-dimensional model will project it as a high-arousal and low-valence state, which does not reflect the subjective experience of these emotions. In addition, time is also a contributing variable in the perception of emotion as emotions are dynamic in nature. The temporal dynamics of emotion perception is mainly dominated by subjective meta-cognitive evaluation of the situation [8]. Based on our results, we believe that the essential information related to meta-cognitive evaluation in emotional processing is contributed by the dimension of dominance. We encourage future researchers to consider dominance as an essential dimension when performing emotion classification tasks.

References

1. Koelstra, S., et al.: Deap: a database for emotion analysis; using physiological signals. IEEE Trans. Affect. Comput. **3**, 18–31 (2011)
2. Zheng, W., Lu, B.: Investigating critical frequency bands and channels for EEG-based emotion recognition with deep neural networks. IEEE Trans. Auton. Ment. Dev. **7**, 162–175 (2015)
3. Katsigiannis, S., Ramzan, N.: DREAMER: a database for emotion recognition through EEG and ECG signals from wireless low-cost off-the-shelf devices. IEEE J. Biomed. Health Inform. **22**, 98–107 (2018)
4. Miranda-Correa, J., Abadi, M., Sebe, N., Patras, I.: AMIGOS: a dataset for affect, personality and mood research on individuals and groups. IEEE Trans. Affect. Comput. **12**, 479–493 (2021)
5. Jerram, M., Lee, A., Negreira, A., Gansler, D.: The neural correlates of the dominance dimension of emotion. Psychiatry Res. Neuroimaging **221**, 135–141 (2014)
6. Verma, G., Tiwary, U.: Multimodal fusion framework: a multiresolution approach for emotion classification and recognition from physiological signals. Neuroimage **102**, 162–172 (2014)

7. Barrett, L., Russell, J.: The structure of current affect: controversies and emerging consensus. Curr. Dir. Psychol. Sci. **8**, 10–14 (1999)
8. Blascovich, J., Mendes, W.: Challenge and Threat Appraisals: The Role of Affective Cues. Cambridge University Press, Cambridge (2000)
9. Mishra, S., Srinivasan, N., Tiwary, U.: Affective film dataset from India (AFDI): creation and validation with an Indian sample. (PsyArXiv 2021)
10. Mishra, S., Srinivasan, N., Tiwary, U.: Cardiac–brain dynamics depend on context familiarity and their interaction predicts experience of emotional arousal. Brain Sci. **12** (2022). https://www.mdpi.com/2076-3425/12/6/702
11. Kehtarnavaz, N.: CHAPTER 7 - Frequency Domain Processing. Digital Signal Processing System Design (2nd Edn), pp. 175–196 (2008). https://www.sciencedirect.com/science/article/pii/B9780123744906000076
12. Krishnan, S.: 5 - Advanced analysis of biomedical signals. Biomed. Signal Anal. Connected Healthcare, pp. 157–222 (2021). https://www.sciencedirect.com/science/article/pii/B9780128130865000037
13. Li, Z., Liu, F., Yang, W., Peng, S., Zhou, J.: A survey of convolutional neural networks: analysis, applications, and prospects. IEEE Trans. Neural Netw. Learn. Syst. (2021)
14. O'Shea, K., Nash, R.: An introduction to convolutional neural networks. ArXiv Preprint ArXiv:1511.08458 (2015)
15. Golmohammadi, M., et al.: Gated recurrent networks for seizure detection. 2017 IEEE Signal Processing In Medicine and Biology Symposium (SPMB), pp. 1–5 (2017)
16. Rana, R.: Gated recurrent unit (GRU) for emotion classification from noisy speech. ArXiv Preprint ArXiv:1612.07778 (2016)

Multiclass Classification of Online Reviews Using NLP & Machine Learning for Non-english Language

Priyanka Sharma and Pritee Parwekar(✉)

SRM Institute of Science and Technology, Modinagar, Ghaziabad 201204, UP, India
{ps9627,priteep}@srmist.edu.in

Abstract. The classification of reviews or comments provided by the customers after shopping has a wide scope in terms of the categories it can be classified. Big companies like Walmart, Tesco and Amazon have customers from all over the world with a variety of product range and can have reviews written in any language. Sometimes customers intend to provide reviews not only on the same platform but on various other platforms like Facebook, Twitter. To get an overall picture of the products it's required to check the reviews from all these platforms at a single place. This paper classifies the comments\reviews written in Spanish language and category names are taken in English language for 30 product categories. The purpose is to get the product categorized from comments/reviews on different platforms in non-English language, to gather insights of that product and to reduce the dependency faced during the manual process of classification and barrier to have command on that language. The approach used reduces the chances of manual errors during prediction of new reviews/comments to a particular category. A multiclass Classification model is trained using traditional Machine Learning algorithms & NLP with an accuracy of 90%. It is envisioned that the proposed methodology is scalable for other non-English languages as well.

Keywords: Machine Learning · SVM · Logistic Regression · Collocations · Feature Extraction · NLP

1 Introduction

1.1 Background

In current times, the feedback and reviews gathered from customers after each purchase helps in getting insights of the market trends and to find the pain points to focus on to retain the customers and better understanding of business. From Walmart to Amazon everyone is trying to make their customer's online and offline shopping experience smooth and to retain their customers across the globe. When asked for reviews or feedback the data being received is mostly text. Based on the content of the reviews one can map that comment or feedback to its associated category or product if it is written on different platforms like Twitter or Facebook. It also helps in having insights about that

H. Zaynidinov et al. (Eds.): IHCI 2022, LNCS 13741, pp. 85–94, 2023.
https://doi.org/10.1007/978-3-031-27199-1_9

product and if any problem is faced by customers related to the product can be resolved. The reviews can be written in any language and is currently being categorized manually using human efforts and skills. But doing this process of manual categorization has its own drawbacks. First the process needs command & knowledge of that language and the categories. Second, the manual categorization, depending on the number of categories will be a time taking process. Sometimes product related experience is required to classify the products which create dependency on a person. Lastly chances of human error are there which can lead to misclassifications.

Machine learning and Natural Language processing have provided a strong platform in last few years to solve such type of classification problem. As machine learning problem can be classified into supervised and unsupervised techniques. In Unsupervised Machine Learning techniques, clustering is used which works on un-labeled data. On the other hand supervised machine learning techniques can be divided into Classification and Regression Techniques. In Regression problem the output is generated in form of numerical values. But in Classification Technique, the data is classified into categorical and discrete values like Yes or No. Classification Techniques can further be divided into Binary classification and Multiclass Classification problem. If the prediction categories are just two classes e.g. classifying an email as Spam or not Spam can be considered as Binary classification, but if the number of classes are greater than two then it is known as multi-class classification. For multi-class classification problems one should have enough samples per category to train the model as shown in Fig. 1.

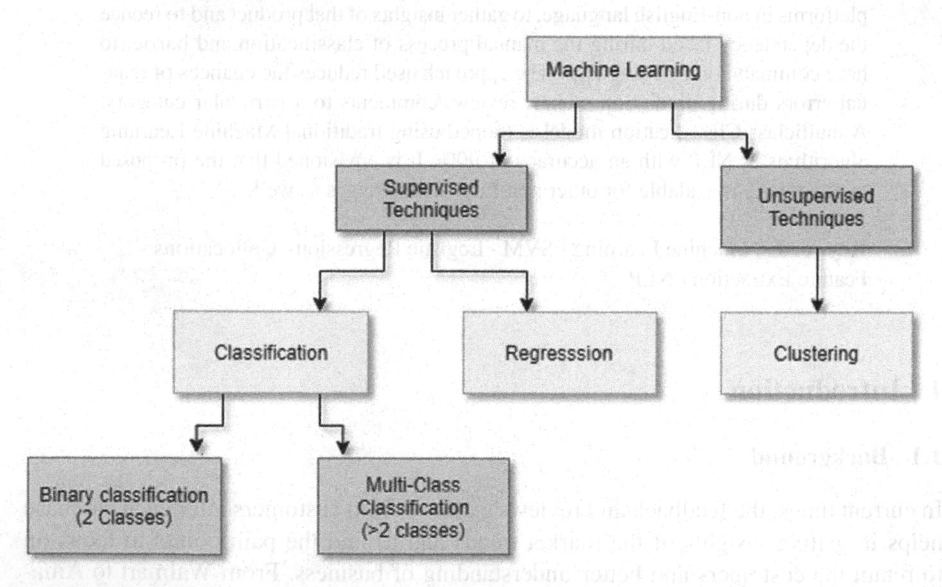

Fig. 1. Machine Learning categorization

Multiclass Classification can provide solution for online reviews/comments for multiple categories, but it becomes difficult when customers write these comments in language which is non-English. Many techniques have been defined in Machine learning to handle text in English languages but if the language is non-English, one need proficiency in that language to understand the language and then classify it based on the features into pre-defined product range. So, a scalable solution is needed to this problem of multiclass classification of non-English language using Machine Learning so that manual efforts in classifying these reviews can be minimized. With the evolution of various cloud platforms, it's easy to store the data in form of reviews/feedback and then process it based on the requirements to get business insights. Classification of the data in form of text needs feature extraction and then prediction can be done on text based on the trained model. Remaining part of the paper is organized as follows. Section 2 conveys a thorough description of the model. Sections 3–5 uses the NLP techniques and Machine learning model to label the data and other prospects for future research. Section 6 concludes.

1.2 Related Work

Several studies have been considered for this, "Multi-Class Classification of Turkish Texts with Machine Learning Algorithms in IEEE in year 2018" [16], by Fatih Gürcan worked on the text in Turkish language and presented the problem of classifying text written in Turkish language into five predefined categories which are "Sport, Economy, Technology, Politics and Health". It also uses Machine learning model.

Next study is on paper named "English languages enrich scientific knowledge: The example of economic costs of biological invasions" by Elena Angulo and at all in Elsevier in June, 2021.

Next [14] by Arun Babhulgaonkar and Shefali, this paper provides a precise summary of necessity & encounters which are involved in the automatic identification of language for machine translation task. It also mentioned that "Language identification" & "machine translation" are quite important for availability of cross lingual information. As Hindi, Marathi and Sanskrit are quite close to each other and the task is to have distinctive features which can classify them is a problematic task. The paper mentioned "Segmentation and translation of individual languages in a multilingual document really improved the quality of machine translation". Hindi and Sanskrit languages were used in the paper giving a room to recognize other languages as well.

Another related research is "Entity Linking: A Problem to Extract Corresponding Entity with Knowledge Base" [15], by Gongqing Wu and at all, Member IEEE. In this work, the researchers familiarize the complications and submissions of the entity linking task & it emphases proceeding the important approaches for addressing this issue. At last, the researchers list the "knowledge bases, datasets, evaluation criterion & certain challenges of entity linking. The current approaches are relevant for linking similar languages.

Most of the time, the labeling and classification of Non-English Language can be erroneous process for humans. Classification of the reviews/feedback or data from customer based on language, features, product range or some pre-defined multiple categories can save hours. Many times the reviews from different platforms are missed during the manual process of classification which if classified can make a difference and provide a big picture.

2 Proposed Methodology

The methodology proposed here is to make the multi-classification of non-English language reviews, written on various platforms by using a single model based on the correlation within the words being used along with the TF-IDF and factorization method. It will be started from collecting the data, wrangling data, applying proposed approach and getting the results which is scalable for other non-English languages as well. The approach used is highly cost effective & time saving and model can be trained in very less time in comparison to other approaches used.

2.1 Dataset Collection

The data was gathered from marketplace in Spain for Spanish language reviews/comments. The purchases were included for around 30 product categories. The product categories selected can have reviews on platform other than amazon. For such reviews proposed model can easily identify the category. In Fig. 2, the sample dataset of amazon reviews is shown for better understanding. Initially in the dataset 21697 samples were collected with their categories.

	product_category	review_body
0	electronics	Nada bueno se me fue ka pantalla en menos de 8...
1	electronics	Horrible, nos tuvimos que comprar otro porque ...
2	drugstore	Te obligan a comprar dos unidades y te llega s...
3	wireless	No entro en descalificar al vendedor, solo pue...
4	shoes	Llega tarde y co la talla equivocada
...	...	...
21692	automotive	El motivo de ponerle una estrella no es porque...
21693	baby_product	Una cuna para usar como cuna solamente o como ...
21694	beauty	No me llegaron las plantillas y algunos frasco...
21695	home	Muy mal acabado te cortas y pinchas con solo t...
21696	personal_care_appliances	Al principio cumplía su función, eso si, creo ...

21697 rows × 2 columns

Fig. 2. Sample Dataset of Amazon Reviews in Spanish Language

2.2 Data Pre-processing

The dataset as mentioned in Fig. 2 is having two columns named "product_category" which is the label. Another column named "review_body" has the Spanish text/reviews by customers on which the data will be trained. On a global platform these reviews can be written in their native languages as well. Here we are taking Spanish language as non-English language.

	Category	Reviews
0	electronics	nada bueno se me fue ka pantalla en menos de 8...
1	electronics	horrible, nos tuvimos que comprar otro porque ...
2	drugstore	te obligan a comprar dos unidades y te llega s...
3	wireless	no entro en descalificar al vendedor, solo pue...
4	shoes	llega tarde y co la talla equivocada
...	...	...
21692	automotive	el motivo de ponerle una estrella no es porque...
21693	baby_product	una cuna para usar como cuna solamente o como ...
21694	beauty	no me llegaron las plantillas y algunos frasco...
21695	home	muy mal acabado te cortas y pinchas con solo t...
21696	personal_care_appliances	al principio cumplía su función, eso si, creo ...

21055 rows × 3 columns

Fig. 3. Dataset after Pre-processing

As seen in Fig. 3, as part of data pre-processing all the null values as well as the duplicate values were dropped from the data. Next all the text is converted into lower case. The same type of comments related to product categories can be found on twitter or Facebook where some pre-processing can be required.

After preprocessing we get 21055 records and then factorization process is applied to provide unique numbers to each category and then process the result dataset for further Natural language processing steps of Tokenization, Lemmatization and collocations. Collocations are implemented here to get the most co-occurring words that exist and it will also make it language dependent. Next "stopwords" for Spanish Language are removed from the data. For the current process of Multi-class classification 'stopwords' removal will help in getting the meaningful features. Next step is to apply Named Entity Relationship (NER) on the data. The Fig. 4 below explains the proposed model of Multiclass classification for Spanish Language of Amazon Product reviews.

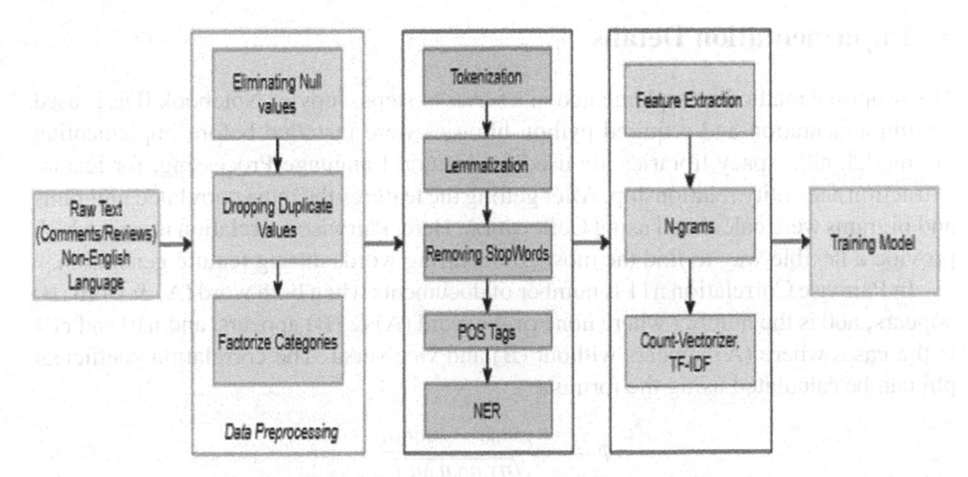

Fig. 4. Proposed model for Multiclass Classification for Spanish Language of Amazon Product reviews.

In Fig. 5 all the 30 categories with their count can be seen; it can be seen that few of the categories have more reviews than others. This condition can relate in future as well so leaving the data intact.

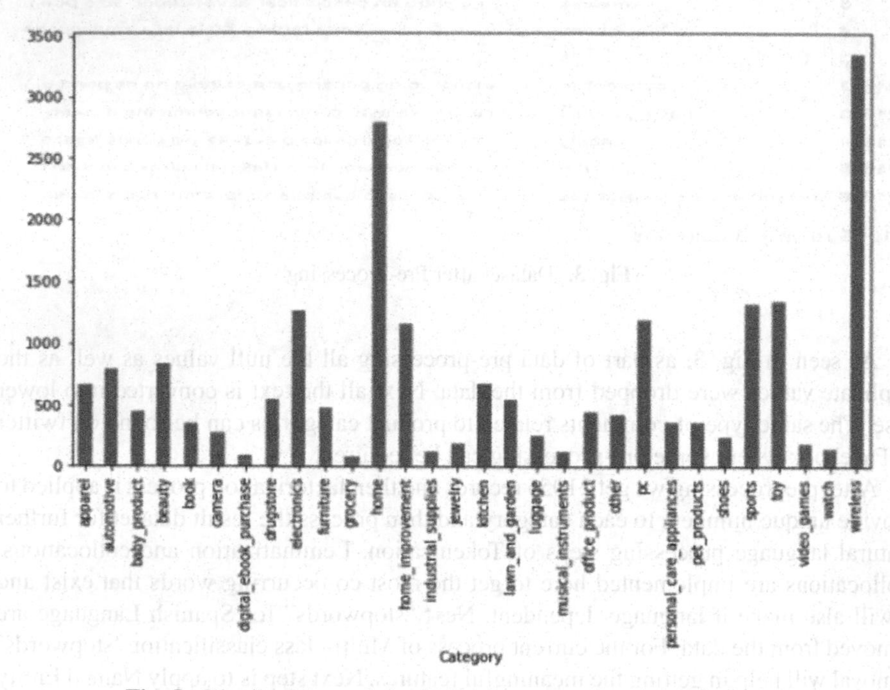

Fig. 5. Distribution graph for 30 Product categories and their count

3 Implementation Details

The proposed method is implemented in a series of steps. Jupyter Notebook IDE is used for implementation and required python libraries were installed before implementing the model. nltk, spacy libraries are used for Natural Language Processing, for feature extraction and entity relationship. After getting the features the most correlated unigrams and bigrams were calculated using Collocation. Here, Pairwise correlation is used which provide a flexible way to find the most co-occurring words during feature generation.

In Pairwise Correlation n11 is number of documents when both word (A) & word (B) appears, n00 is the number where none of the word (A) & (B) appears, and n10 and n01 is the cases where (A) appears without (B) and vice versa. The correlation coefficient phi can be calculated using the formula:

$$\phi = \frac{n_{11}n_{00} - n_{10}n_{01}}{\sqrt{n_{1}.n_{0}.n_{.0}n_{.1}}}$$

Here as a part of proposed model n-grams selected will not be based on the frequency but by finding the correlation between the most co-related words. This is used so that it becomes independent of language and same approach can be used to classify the product categories reviews written in different languages. Here, pairwise correlation method is being used for getting the most correlated unigrams ad bigrams. These correlated unigrams and bigrams generated after applying this approach are then given as input for the next step i.e. TF-IDF to get the required result. This proposed approach helps in getting the better correlated unigrams and bigrams in Spanish Language in comparison to the frequency based unigrams and bigrams. Figure 6 shows a sample of the result of applied process of collocations.

```
# 'apparel':
  . Most correlated unigrams:
. camiseta
. talla
  . Most correlated bigrams:
. tela horrible
. tejido mala
# 'automotive':
  . Most correlated unigrams:
. moto
. coche
  . Most correlated bigrams:
. pila gastada
. aconsejo vendedor
# 'baby_product':
  . Most correlated unigrams:
. cuna
. bebé
```

Fig. 6. Most correlated unigrams and bigrams for each Category.

4 Results

The result of the proposed model is measured using performance matrix with accuracy, Precision, Recall, F1 Score, by calculating confusion matrix. Various Machine Learning Algorithms were used on dataset. The data is divided into two parts. First is training data set and another is testing data set. 80% of data was taken for training and 20% for testing.

As seen in Table 1 the Stochastic Gradient Descent Algorithm with perceptron gave the highest accuracy i.e. 90.10% and model take very less amount of time to train the model and similarly predictions on unseen data can be made in a very less time. Algorithms like Multinomial Naïve Bayes, initially gave 24.5% accuracy on the training data which as seen is very less. Next Logistic Regression was implemented which increased the accuracy to 45.4%. Then SVM was applied with different kernels. Out of all kernels applied Radial Bias Function performed quite well and accuracy was increased to 72%. Then for further increase the accuracy of the model optimizers were applied by measuring the losses. The result seems to be highly efficient, scalable and optimized in terms of time, cost and efforts.

Table 1. Performance table with accuracy

S. No	Algorithm Used	Accuracy
1	Multinomial Naïve Bayes	24.50%
2	Logistic Regression	45.40%
3	Support Vector Machine (Linear Kernel)	54.60%
4	Support Vector Machine (RBF Kernel)	72.00%
5	Stochastic Gradient Descent (Hinge)	83.20%
6	Stochastic Gradient Descent (Modified Huber)	88.70%
7	Stochastic Gradient Descent (Perceptron)	90.10%

The classification report for the Model on Non-English Language is provided in Fig. 7 with Precision, Recall, F1-Score and support for SGD optimizer.

	precision	recall	f1-score	support
apparel	0.78	0.92	0.84	90
automotive	0.93	0.96	0.94	115
baby_product	0.88	0.89	0.88	72
beauty	0.82	0.96	0.88	125
book	0.94	0.98	0.96	65
camera	0.95	0.98	0.96	41
digital_ebook_purchase	1.00	1.00	1.00	11
drugstore	0.95	0.92	0.94	79
electronics	0.87	0.89	0.88	193
furniture	0.79	0.91	0.85	82
grocery	1.00	0.91	0.95	11
home	0.94	0.84	0.88	421
home_improvement	0.89	0.87	0.88	170
industrial_supplies	1.00	0.92	0.96	13
jewelry	0.83	0.92	0.87	26
kitchen	0.94	0.95	0.95	107
lawn_and_garden	0.90	0.91	0.91	91
luggage	0.91	0.91	0.91	35
musical_instruments	1.00	1.00	1.00	19
office_product	0.98	0.93	0.96	61
other	0.96	0.97	0.96	69
pc	0.86	0.91	0.88	159
personal_care_appliances	0.86	0.96	0.91	46
pet_products	0.95	0.93	0.94	58
shoes	0.95	0.97	0.96	39
sports	0.85	0.90	0.88	190
toy	0.94	0.90	0.92	210
video_games	0.94	0.88	0.91	34
watch	0.93	0.93	0.93	15
wireless	0.93	0.89	0.91	512

Fig. 7. Classification Report with Precision, Recall, f1score & support for all 30 categories

Other parameters like accuracy, macro average and weighted average for the optimizer Stochastic Gradient Descent with perceptron can be seen in Fig. 8.

accuracy			0.90	3159
macro avg	0.92	0.93	0.92	3159
weighted avg	0.91	0.90	0.90	3159

Fig. 8. Accuracy, Macro average and weighted average

5 Future Work

In the current implementation just one non-English language i.e. Spanish Language is used. As future work the number of languages can be increased to check the result. Some other correlation technique can also be used to enhance the scalability and usability for multiclass classification in simple manner. Other approaches to increase the performance and accuracy of the models can also be considered as a part of future work.

6 Conclusion

The proposed model of using the collocations increases the accuracy with good rate for non-English language i.e. Spanish Language. The approach is scalable to other languages as well and highly optimized and efficient in terms of saving time, efforts and is cost saving. It can be easily used to get the reviews/comments predicted from various other platforms where the product category is not present.

References

1. Keung, P., Lu, Y., Szarvas, G., Smith, N.A.: The multilingual Amazon reviews corpus. arXiv2010.02573v1 (2020)
2. Amazon Inc. Amazon customer reviews dataset. https://registry.opendata.aws/amazon-reviews/ (2015)
3. Artetxe, M., Schwenk, H.: Massively multilingual sentence embeddings for zeroshot cross-lingual transfer and beyond. Trans. Assoc. Comput. Linguist. **7**, 597–610 (2019)
4. Bel, N., Koster, C.H.A., Villegas, M.: Cross-lingual text categorization. In: Koch, T. (ed.) Research and Advanced Technology for Digital Libraries. Lecture Notes in Computer Science, vol. 2769, pp. 126–139. Springer, Heidelberg (2003). https://doi.org/10.1007/978-3-540-45175-4_13
5. Bojanowski, P., Grave, E., Joulin, A., Mikolov, T.: Enriching word vectors with subword information. Trans. Assoc. Comput. Linguis. **5**, 135–146 (2017)
6. Kingma, D.P., Ba, J.: Adam: a method for stochastic optimization. arXiv preprint arXiv:1412.6980 (2014)
7. Joulin, A., Grave, E., Bojanowski, P., Mikolov, T.: Bag of tricks for efficient text classification. In: Proceedings of the 15th Conference of the European Chapter of the Association for Computational Linguistics, Vol. 2, Short Papers, pp 427–431, Valencia, Spain (2017)
8. Singh, R.P., Haque, R., Hasanuzzaman, M., Way, A.: Identifying complaints from product reviews: a case study on Hindi, CEUR-WS.org, Vol. 2771, Paper 28, Ireland (2020)
9. Bowman, S.R., Angeli, G., Potts, C., Manning, C.D.: A large annotated corpus for learning natural language inference. In: Proceedings of the Conference on Empirical Methods in Natural Language Processing (EMNLP). Association for Computational Linguistics in (2015)
10. Conneau, A., et al.: Unsupervised cross-lingual representation learning at scale. arXiv preprint arXiv:1911.02116 (2019)

11. Conneau, A.: Xnli: evaluating crosslingual sentence representations. In: Proceedings of the 2018 Conference on Empirical Methods in Natural Language Processing. Association for Computational Linguistics (2018)
12. de Melo, G., Siersdorfer, S.: Multilingual text classification using ontologies. In: Amati, G., Carpineto, C., Romano, G. (eds.) Advances in Information Retrieval. Lecture Notes in Computer Science, vol. 4425, pp. 541–548. Springer, Heidelberg (2007). https://doi.org/10.1007/978-3-540-71496-5_49
13. Yu, S., Su, J., Luo, D.: Improving bert-based text classification with auxiliary sentence and domain knowledge. IEEE Access, **7**, 176600176612 (2019)
14. Babhulgaonkar, A., Sonavane, S.: Language identification for multilingual machine translation. In: IEEE International Conference on Communication and Signal Processing, pp. 0401–0405 (2020)
15. Wu, G., He, Y., Hu, X.: Entity linking: a problem to extract corresponding entity with knowledge base IEEE Access, 6220 – 6231 (2016)
16. Gürcan, F.: Multi-class classification of turkish texts with machine learning algorithms In: IEEE (2018)

A Higher Performing DARTS Model for CIFAR-10

Jie Yong Shin and Dae-Ki Kang(✉)

Dongseo University, Busan, KN 47011, Republic of Korea
dkkang@dongseo.ac.kr

Abstract. Machine Learning experts spend much time on fine-tuning. A methodology that automatically searches neural architectures has been to solve this problem. Differentiable Architecture Search (DARTS) is an algorithm that solves a Neural Architecture Search problem using a gradient-based approach. We found an architecture that shows higher test accuracy than the existing DARTS architecture with the DARTS algorithm on the CIFAR-10 dataset. The architecture performed the DARTS algorithm several times and recorded the highest test accuracy of 97.62%. This result exceeds the test accuracy of 97.24 ± 0.09 shown in the existing DARTS paper. These results are expected to raise the baseline for making a practical difference in the study of Neural Architecture Search.

Keywords: Automated Machine Learning · Neural Architecture Search · Differentiable Architecture Search · Hyperparameter Optimization · Continuous Search Space · Fine-tuning

1 Introduction

The development of machine learning has had a positive impact on the development of many fields other than artificial intelligence. Ironically, the more machine learning contributes to society, the more it causes a lack of machine learning experts. One of these reasons will be fine-tuning. Fine-tuning is one of the most time-consuming parts of designing a neural network model. When designing a neural network model, the designer must carefully decide how to design the architecture and what hyperparameter values to put in. This is because even in a single model, the performance of a model can vary greatly depending on how the neural architecture configuration and hyperparameters are configured. On the other hand, such a configuration is sometimes unreasonably resource-intensive. Since the search space can be infinite, sometimes it is impossible to try every configuration. Automated Machine Learning (AutoML) has emerged to solve these problems [1]. AutoML aims to automate the machine learning procedures, including fine-tuning steps, so that machine learning experts do not spend too much resources and time on fine-tuning. For this purpose, AutoML divided the fine-tuning procedure into two major categories. One is Hyperparameter Optimization (HPO), which finds the optimal hyperparameter configuration, and the other is Neural Architecture Search (NAS), which finds the optimal Neural Architecture. This paper deals with the NAS among them.

H. Zaynidinov et al. (Eds.): IHCI 2022, LNCS 13741, pp. 95–99, 2023.
https://doi.org/10.1007/978-3-031-27199-1_10

2 Neural Architecture Search

NAS is a field that studies how to stack the layers constituting the neural network to maximize the model's performance. That is, automatically finding the configuration of Neural Architecture is essential because the configuration of the resulting model varies depending on which algorithm is used to find the Neural Architecture. Basically, the NAS problem is considered a discrete search space because problems such as which operation to put in a specific layer in one model and where to place the skip connection are similar to the problem of choosing among several options rather than finding the optimal value within the real number. NAS algorithms have been developed while leaving the discrete NAS problem itself at first [2–3]. At the beginning of NAS research, some algorithms even demand thousands of GPU days [4].

3 Differentiable Architecture Search

In the case of Differentiable Architecture Search (DARTS), researchers have approached this discrete problem with a gradient-based method by making it differentiable [5]. Compared to the state-of-the-art NAS algorithms at that time, DARTS had a test error of 2.76 ± 0.09 and a search cost of 4 GPU days, widely publicizing the possibility of a gradient-based algorithm [5–7]. This was possible because one-shot Architecture Search was applied to continuous search space [5, 8, 9]. Thanks to the efforts of the NAS researchers, DARTS is more advanced than the original. P-DARTS reduced search costs by reducing repeated exploration [10]. β-DARTS improved the low robustness and generalization performance, which are the problems of DARTS, by regularizing the architectural parameter values passed through the softmax function [11]. We tested how much better models can be found with original DARTS algorithms on the CIFAR-10 dataset and found architectures with better performance than those found.

4 Experiments and Results

Our experiment is the model that achieved the highest test accuracy among the models found by running the DARTS algorithm ten times on the CIFAR-10 dataset. The model was searched through one-step unrolled validation loss (second order derivative) as in DARTS V2, and a cutout was applied in the training process [8, 9, 12]. We've set the hyperparameters such as batch size, learning rate scheduler, inner training epoch, weight decay, cutout length, drop path probability, gradient clipping range, and momentum the same as DARTS. We ran our experiment independently with NVIDIA GeForce RTX 2080 Ti with a GPU memory of 12 Giga bytes. For the CPU, we used Intel Core i9-10900KF.

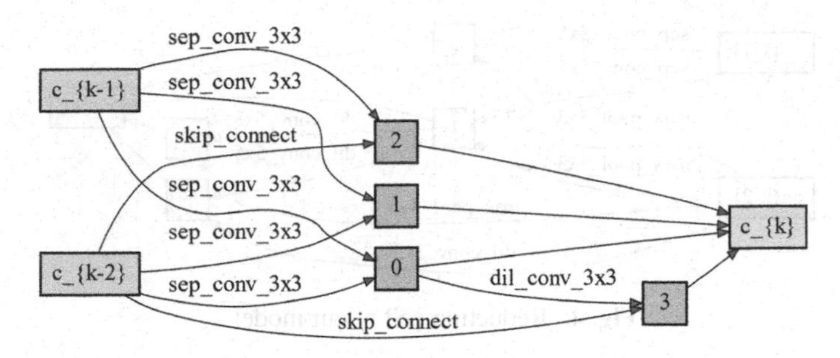

Fig. 1. Normal cell of DARTS V2.

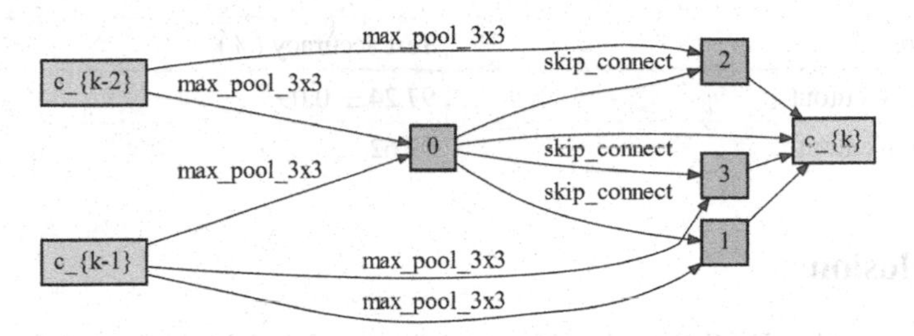

Fig. 2. Reduction cell of DARTS V2.

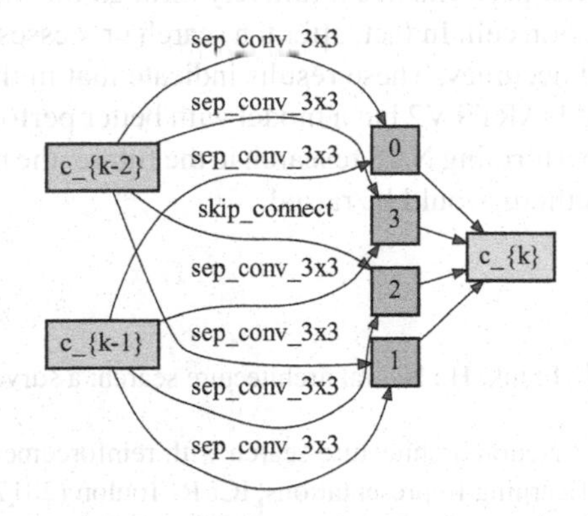

Fig. 3. Normal cell of our model.

Figure 1, 2, 3, and 4, shows that the occurrence of skip connection differs a total of 4 times in DARTS V2 and one time in our model [13]. The frequent skip connection is one of the problems of DARTS, and since we saw it appear three times in the reduction cell, we expected that we could find a model with better performance even with the same algorithm.

As can be seen in Table 1, our test accuracy in CIFAR-10 outperforms DARTS V2 by 0.38% points.

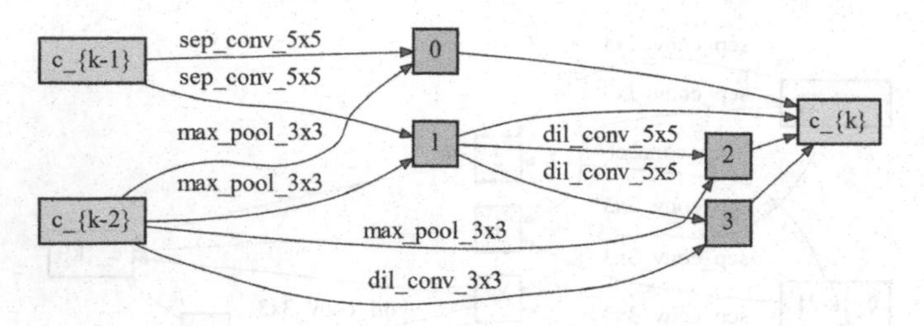

Fig. 4. Reduction cell of our model.

Table 1. Comparison of the results of DARTS V2 and our model on the CIFAR-10 dataset.

Architecture	Test accuracy (%)
DARTS V2 + cutout	97.24 ± 0.09
Our Model + cutout	97.62

5 Conclusion

We have shown that DARTS's algorithm can find models with higher performance on the CIFAR-10 dataset. In the DARTS V2 model, we hypothesized that we would be able to find a model with higher performance intuitively through the high occurrence of skip connections of the reduction cell. In fact, after ten search processes, we were able to find a model with higher test accuracy. These results indicate that in the DARTS algorithm, not only DARTS V1 and DARTS V2 but a model with better performance can be found. This implies that when performing NAS research in the future, the baseline of algorithms using gradient-based methods would be raised.

References

1. Thomas, E., Jan, H.M., Frank, H.: Neural architecture search: a survey. J. Mach. Learn. Res. **20**, 1–21 (2019)
2. Barret, Z., Quoc, V.L.: Neural architecture search with reinforcement learning. In: International Conference on Learning Representations, ICLR, Toulon (2017)
3. Hieu, P., Melody, Y.G., Barret, Z., Quoc, V.L., Jeff, D.: Efficient neural architecture search via parameter sharing. In: International Conference on Machine Learning, ICML, Stockholm (2018)
4. Xin, Y.: Evolving artificial neural networks. IEEE Trans. Neural Netw. **8**, 694–713 (1997)
5. Hanxiao, L., Karen, S., Yiming, Y.: Differentiable architecture search. In: International Conference on Learning Representations, ICLR, New Orleans (2019)
6. Barret, Z., Vijay, V., Jonathon, S., Quoc, V.L.: Learning transferable architectures. In: Computer Vision and Pattern Recognition, CVPR, Salt Lake City (2018)
7. Esteban, R., Alok, A., Yanping, H., Quoc, V.L.: Regularized evolution for image classifier architecture search. In: AAAI Conference on Artificial Intelligence, AAAI, Hawaii (2019)
8. Gabriel, B., Pieter-Jan, K., Barret, Z., Vijay, V., Quoc, L.: Understanding and simplifying one-shot architecture search. In: International Conference on Machine Learning, ICML, Stockholm, pp. 549–558 (2018)

9. Andrew, B., Theodore, L., James, M.R., Nick, W.: Smash: one-shot model architecture search through hypernetworks. In: International Conference on Learning Representations, ICLR, Vancouver (2018)
10. Chen, X., Xie, L., Wu, J., Tian, Q.: Progressive DARTS: bridging the optimization gap for NAS in the wild. Int. J. Comput. Vis. **129**(3), 638–655 (2020). https://doi.org/10.1007/s11263-020-01396-x
11. Peng, Y., Baopu, L., Yikang, L., Tao, C., Jiayuan, F., Wanli, O.: b-DARTS: beta-decay regularization for differentiable architecture search. In: Proceedings of the IEEE/CVF Computer Vision and Pattern Recognition Conference, CVPR, New Orleans, pp. 10874–10883 (2022)
12. Terrance, D., Graham, W.T.: Improved regularization of convolutional neural networks with cutout. arXiv preprint https://arxiv.org/abs/1708.04552 (2017)
13. Kaiming, H., Xiangyu, Z., Shaoqing, R., Jian, S.: Deep residual learning for image recognition. In: Computer Vision and Pattern Recognition, CVPR, Boston (2015)

Automatic Speech Recognition on the Neutral Network Based on Attention Mechanism

N. S. Mamatov[1](✉), N. A. Niyozmatova[1], Yu. Sh. Yuldoshev[1], Sh. Sh. Abdullaev[1], and A. N. Samijonov[2]

[1] Tashkent Institute of Irrigation and Agricultural Mechanization Engineers National Research University, Tashkent, Uzbekistan
m_narzullo@mail.ru

[2] Tashkent University of Information Technology After Named Muhammad Al-Khwarizmi, Tashkent, Uzbekistan

Abstract. This article focuses on the problems that arise in the recognition of speech through machine learning and the methods based on in-depth learning used to overcome them, which outlines approaches to the transition to a coding-decoding architecture system based on the attention mechanism. It also describes the hybrid CTC/Attention architecture, which is now widely used in speech recognition. In recent years, models of neural network architectures and neural network model based on attention mechanism, which are widely used in automatic speech recognition, have been proposed, which are taught on the basis of Uzbek and Russian speech corpuscles and the results obtained are comparatively analyzed.

Keywords: Automatic Speech Recognition · Attention Mechanism · Encoder · Decoder · Loss Function · Neural Network Model

1 Introduction

Machine learning/deep learning problems require a set of predefined and known outputs that match the inputs. It is assumed that there is some function *f* () representing the outputs corresponding to the inputs. In speech recognition, a character vector is generated based on short time spectra of the speech signal, that is, every 20–25 ms or so.

One of the main problems of speech recognition is that the boundary between one sound (phoneme) and another is not known in advance. If the speech signal is shortened to ~ 200–300 ms by some speech processing software, then it cannot be understood by ear. Typically, the training sample is formed in the form of a speech signal and a sequence of words corresponding to it. However, it does not store information about the end and beginning of a word or phoneme.

Some error functions, including cross -entropy, mean quadratic error, require that the input and output parts of the neural network match together. In some cases, however, such compatibility is not required. For example, speech recognition, machine translation, manuscript recognition, and so on. In such cases, it is recommended to implement compatibility on the basis of CTC [5], Attention, seq2seq models.

H. Zaynidinov et al. (Eds.): IHCI 2022, LNCS 13741, pp. 100–108, 2023.
https://doi.org/10.1007/978-3-031-27199-1_11

2 The Main Part

Many systems that are fully based on neural networks will have their own encoding network module. If a two-way LSTM encoder is used in the identification, then the previous hidden state character vectors sequence is taken as the encoder network, and based on a certain encoding function they are converted into a hidden (encoded) sequence of vectors [1].

Each encoded description h_t focuses on the t-entry of the input sequence and stores information about the entire sequence. The hidden state function at time t is defined as:

$$h_t = Encoder(x_t, h_{t-1}) \tag{1}$$

where Encoder () is a specific function that performs encoding when updating a hidden description.

These types of encoders can be multi-layered, meaning that each subsequent layer has a deep BLSTM encoder that changes the output of the previous layers. Since how a word is pronounced depends on the previous and subsequent phonemes, it is advisable to use a two-way network in speech recognition. In consonant, many consonant sounds depend on the presence of a vowel or consonant sound after them, and they are pronounced differently. For example, the pronunciation of the letter "t" in the words "student" and "passport" is different. In such cases, the use of a two-way LSTM encoder allows better modeling of the speech signal [2].

The LSTM encoder, like other deep neural networks, x receives a specific vector at the input and y generates a vector at the output. However, this vector does not depend on the full input value, and it also depends to some extent on the previous (next if two-way) series values.

In Connectionist Temporal Classification, alignment is done by integrating all possible time-character adjustments as follows:

$$L_{ctc}(X, W) = \sum_{C:k(C)=W} p(C|X)$$

$$L_{ctc}(X, W) = \sum_{C:k(C)=W} \prod_{t=1}^{T} p(c_t|X)$$

Each dimension of the LSTM encoding target vector corresponds to the probability that it will appear in the current step. In step 17 of the algorithm, the probability of the formation of the letter "B" can be seen (Fig. 1).

The CTC probability calculation is done by counting each possible character sequence and adding them. If the word "Hello" is taught, the sequence of characters corresponding to it is considered a separate probability and their sum is obtained. This requires maximizing the value of this parameter [3].

To maximize the probability of the correct sequence of characters, the network parameters are updated using the following formula [1].

$$\theta^* = \arg\max_{\theta} \sum_i \log P\left(y^{*(i)}|x^i\right) = \arg\max_{\theta} \sum_i \log \sum_{c:\beta(c)=y^*(i)} P\left(c|x^i\right)$$

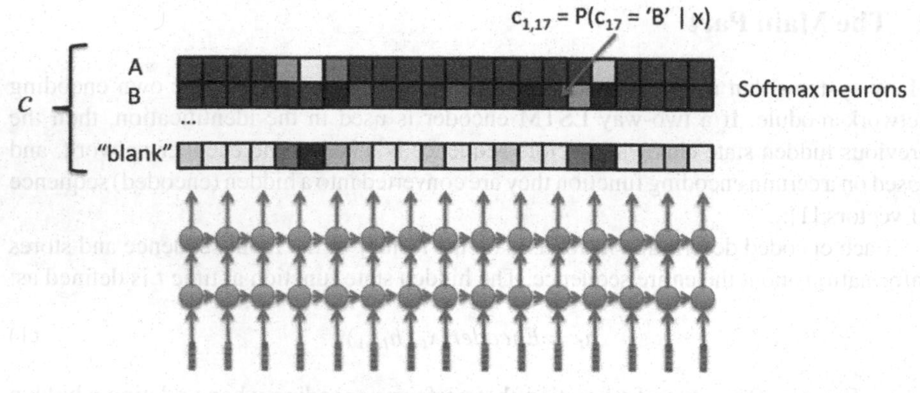

Fig. 1. Encoder output.

In the Encoder-Decoder structure, the Encoder attempts to sum the entire input sequence into a vector h_t of constant size. As a coder, it uses a recurrent neural network (RNN/LSTM/BLSTM/GRU), h_t which itself receives each input x_t character vector and h_t serves to display (collect) the sequence so far within its internal state.

When predicting at each step, h_t can be obtained, but it is desirable to wait until the end of the sequence at time T and obtain h_T descriptions in order to start generating the output sequence. This is due to the fact that the word/letter/phoneme boundaries are fuzzy. The encoder provides a relatively better result if the input sequence is fully generalized to h_T.

At the <sos> input, the starting token of the sequence is passed to the decoder and the generation of output characters is started. The decoder uses a different recurrent neural network (not bidirectional) that changes its internal state each time to predict this outcome [4].

At each step of the time, the output of the previous time step is used to estimate the current output, i.e.:

$$s_i = Decoder(s_{i-1},\ y_{i-1}) \tag{2}$$

(2) The formula $i-$ represents the latent state of the decoder in character prediction. Here is $Decoder()$ a specific function that uses the LSTM decoder to update its hidden internal state.

An example of how an encoder/decoder network works is shown in Fig. 2.

Decoder <eos> - When generating a token at the end of a sequence, the generation of the outgoing character sequence is stopped.

Given $s(i-1)$ the latent state of the decoder (at the previous output time) and $y(i-1)$ the output symbol (previous output symbol), the prediction of the output symbol at the current time stage can be determined as follows:

$$p(y_i|\{y_1,\ y_2,\ \dots,\ y_{i-1}\}) = g(y_{i-1},\ s_i), \tag{3}$$

where $g()$- decoder function.

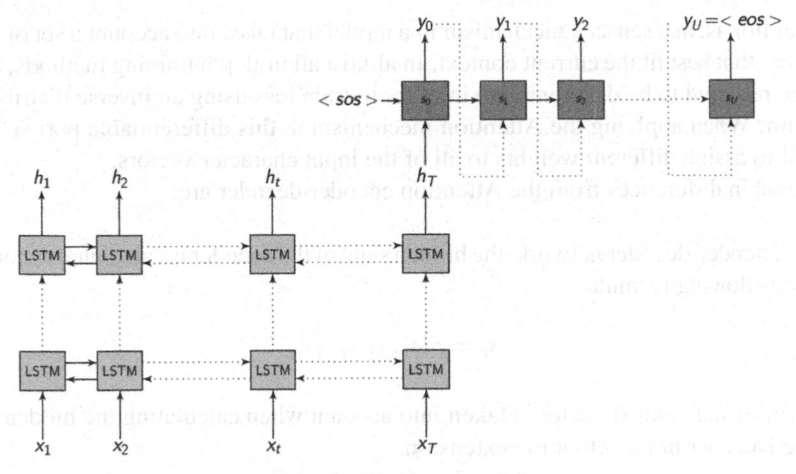

Fig. 2. Encoder-decoder.

complete output sequence y is calculated by the following formula:

$$p(y) = \prod_{i=1}^{U} p(y_i | \{y_1, y_2, \ldots, y_{i-1}\}, s_i) \tag{4}$$

There are the following problems with the encoder-decoder:

- neural network vectors have the ability to compress all important information of the input sequence into a constant vector;
- if the sequence is long, especially if the introductory sequence in the test is longer than the training sequence, then the performance of the encoder/decoder network will be impaired;
- the aggregation of the whole sequence of encoding character vectors into a constant-dimensional vector depends on the size of the vector, i.e., the longer the sentence, the longer the vector is desirable. However, this cannot be done in this network, as the length of the sequence can vary significantly.

The use of the Attention mechanism is important in solving the problems listed above. Attention encoder is an extension of the decoder framework.

When a model needs to generate an output character, it (programmatically) looks for the set of positions in the input sequence of character vectors with the most significant information, and it is important that the set is chosen exactly. The main difference from the encoder-decoder scheme is that there is no need to collect the entire input sequence into a fixed-size vector. In this case, instead of providing a hidden state h_T, the subset h closest to the given context is selected to generate output to the network of decoders. This relevant h_T is linearly transformed to form a vector C_i called the context vector.

$$C_i = q(\{h_1, h_2, \ldots, h_T\}, \alpha_i) \tag{5}$$

Attention is, in a sense, a mechanism in a model that takes into account a set of input characters that best fit the current context. In almost all in-depth training methods, functions are required to be differentiable in order to train loss using an inverse distribution algorithm. When applying the Attention mechanism to this differentiable part set, it is required to assign different weights to all of the input character vectors.

The main differences from the Attention encoder-decoder are:

- In the encoder-decoder network, the hidden state of the decoder is calculated according to the following formula:

$$s_i = f(s_{i-1}, y_{i-1}),$$

- The following context vector is taken into account when calculating the hidden state of the Decoder in the Attention extension:

$$s_i = f(s_{i-1}, y_{i-1}, C_i).$$

The context vector contains only the input character vector with high relevance. To evaluate this relevance, the variable α is introduced. α_i is the hidden state weight, encoded as h_i in the context vector C_i, which serves to predict the output at time i. Given α, the context vector C_i is computed as:

$$C_i = \sum_{j=1}^{T} \alpha_{i,j} \cdot h_j$$

$$\sum_{j=1}^{T} \alpha_{i,j} = 1$$

- To calculate $\alpha_{i,j}$, it is necessary to determine the value $e_{i,j}$, which is the level of significance of the j-comment vector in predicting the i-th character. The weight $\alpha_{i,j}$ of each comment h_j is calculated using the following formula.

$$\alpha_{i,j} = Soft\max\left(e_{i,j}\right) = \frac{e^{e_{i,j}}}{\sum_{k=1}^{T} e^{e_{i,k}}}$$

$$\sum_{j=1}^{T} e_{i,j} \neq 1$$

where $e_{i,j} = AttentionFunction\left(s_{i-1}, h_j\right)$ is the hidden state $s(i-1)$ of the decoder, and the fitness function calculates the importance of each comment h_j.

Prior to model training, i.e. in the 0-epoch, the values of attention weights are obtained randomly and therefore the C_i context vector contains an unnecessary interference in the insignificant input character vectors. This reduces the efficiency of the model. A good attention model provides a good context vector, which leads to an increase in model efficiency.

Abstract this can be expressed as follows. Each sequence of character vectors is mixed with a certain weight and then transmitted to a decoding network for decision making. Characters and the weights attached to them are determined by AttentionFunction (Fig. 3).

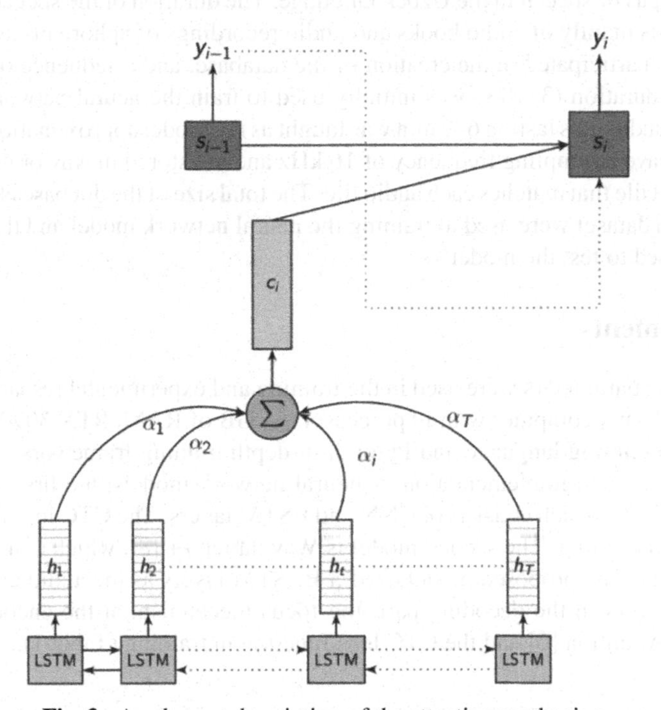

Fig. 3. An abstract description of the attention mechanism.

To gain a deeper understanding of how the model is taught, it is possible to observe how Attention weights change over time (in epochs). Many attention models have been described in detail in various literatures. A set of tools, such as ESPnet, includes more than 10 attention mechanisms designed for easy use in experimental research.

Attention -based speech recognition systems are prone to shutdown and input errors due to their flexible alignment feature. This is due to the consideration of an arbitrary part of the sequence of encoder states to predict the next character. Because the Attention is generated by the decoder network, it is able to predict the end of the sequence earlier, even if it does not pay attention to all the encoder frames. This makes the sentence very short. On the other hand, he can predict the next character with a high probability by focusing on the same parts as the previous ones. In this case, the hypothesis is too long and can repeat the same character sequence. For this and many other reasons, a collaborative hybrid CTC/Attention architecture has been proposed that can be used to train and decode the two architectural advantages mentioned. Multi-criteria training is used during training to increase reliability and achieve rapid convergence. Joint decoding is performed in a one-time beam search algorithm by combining attention-based and CTC-scores to reduce uneven distribution during decoding.

3 Dataset

To train the proposed neural network model in the laboratory "Data Processing Systems" created a corpus of speech in the Uzbek language. The duration of the speeches is 632 h, which consists mainly of audio books and audio recordings of aphorisms and proverbs. 168 speakers participated in the creation of the database, and a sequence of audio files with a short duration (3–10 s) was initially used to train the neural network model. A sequence of audio files lasting 6–9 min was taught as the model approximation increased. Audio files have a sampling frequency of 16 kHz and are stored in wav or mp3 formats. There is a text file that matches each audio file. The total size of the dot base is 46 Gb. 90% of the speech dataset were used to training the neural network model and the remaining 10% were used to test the model.

4 Experiments

The following parameters were used in the training and experimental research of neural network models: a computer with i9 processor, 64 GB of RAM, RTX 3090 video card. Python programming language and Pytorch in-depth training framework were used in the construction and implementation of neural network models, the first model being DeepSpeech2 [7], which consists of CNN and LSTM layers. The CTC loss function was used in model training. The second model is Wav2latter ++ [8], which consists entirely of CNN layers. The proposed model used 3 BLSTM layers as the audio encoding part and LSTM layers in the decoding part, the focus mechanism at the encoder output - Multi-Head Attention [6] and the CTC loss function in training (Table 1).

Table 1. Results of experimental research on speech recognition in Uzbek language.

Neural network architecture	Training time (hours)	Training accuracy (%)	Test accuracy (%)	Test error (WER) (%)
DeepSpeech2	42	85	75	25
Wav2latter ++	41	87	78	22
BLSTM + Attention + LSTM + CTC loss	**45**	**89**	**82**	**18**

In addition, experimental studies on speech recognition in English and Russian were conducted on these models. English speech data - LibriSpeech dataset and Russian speech data - Open STT dataset were used. It can be seen that a lot of time was spent on training because the data on training in Russian was significantly larger, and at the same time relatively high results were obtained (Table 2).

Table 2. Results of application of well-known open-source ASR models and proposed models for Russian speech.

Neural network architecture	Training time (hours)	Training accuracy (%)	Testing accuracy (%)	Testing error (WER) (%)
DeepSpeech2	62	89	81	19
Wave2letter ++	46	88	80	20
Transformers based on model	**51**	**92**	**85**	**15**

5 Conclusion

The problems that arise in the recognition of speech through machine learning and the existing methods based on in-depth training used to solve them have been thoroughly studied. In addition, an overview of the transition to a focus-based encoder-decoder architecture system was provided, some shortcomings of traditional architectures and ways to overcome them using Encoder-Attention-Decoder architecture were suggested. It also provides a brief overview of the new hybrid CTC/Attention architecture, based on the Uzbek and Russian speech corpuses developed in recent years based on neural network architectures and attention-grabbing neural network models used in automatic speech recognition, and a comparative analysis of the results.

References

1. Bahl, L.R., Jelinek, F., Mercer, R.L.: A maximum likelihood approach to continuous speech recognition. IEEE Trans. Pattern Anal. Mach. Intell. **PAMI-5**(2), 179–190 (1983)
2. Noisy channel model (2020). Page was last edited on 12 April 2020, at 09:02 (UTC). https://en.wikipedia.org/wiki/Noisy_channel_model
3. Watanabe, S., et al.: Hybrid CTC/attention architecture for end-to-end. IEEE J. Sel. Top. Sig. Process. **11**(8), 1240–1253 (2017)
4. Hannun "Sequence Modeling with CTC" (2017). https://distill.pub/2017/ctc/
5. Graves, S., Gomez, F.F., Schmidhuber, J.: Connectionist temporal classification: labelling unsegmented sequence data with recurrent neural networks. In: ICML (2006)
6. Chan, W., et al.: Listen, attend and spell. arXiv: 1508.01211 https://arxiv.org/abs/1508.01211
7. Amodei, D., et al.: Deep speech2: end-to-end speech recognition in English and Mandarin arXiv: 1512.02595 https://arxiv.org/abs/1512.02595
8. Zeghidour, N., Usunier, N., Synnaeve, G., Collobert, R., Dupoux, E.: End-to-end speech recognition from the raw waveform. arXiv:1806.07098 (2018)
9. Mamatov, N.S., Niyozmatova, N.A., Abdullaev, S.S., Samijonov, A.N., Erejepov, K.K.: Speech recognition based on transformer neural networks. In: International Conference on Information Science and Communications Technologies: Applications, Trends and Opportunities, ICISCT 2021 (2021)
10. Mamatov, N., Niyozmatova, N., Samijonov, A.: Software for preprocessing voice signals. Int. J. Appl. Sci. Eng. **18**(1) (2021). https://doi.org/10.6703/IJASE.202103_18(1).006

11. Narzillo, M., Abdurashid, S., Parakhat, N., Nilufar, N.: Automatic speaker identification by voice based on vector quantization method. Int. J. Innov. Technol. Explor. Eng. **8**(10), 2443–2445 (2019). https://doi.org/10.35940/ijitee.J9523.0881019
12. Wiedecke, B., Narzillo, M., Payazov, M., Abdurashid, S.: Acoustic signal analysis and identification. Int. J. Innov. Technol. Explor. Eng. **8**(10), 2440–2442 (2019). https://doi.org/10.35940/ijitee.J9522.0881019
13. Narzillo, M., Abdurashid, S., Parakhat, N., Nilufar, N.: Karakalpak speech recognition with CMU sphinx. Int. J. Innov. Technol. Explor. Eng. **8**(10), 2446–2448 (2019). https://doi.org/10.35940/ijitee.J9524.0881019

Equal Temperament and Just Intonation Feature Based Analysis of Indian Music

D. V. K. Vasudevan[1(✉)], Nagamani Molakatala[2], Nikil Priyatham[1,2], Ravikant Gautam[2], and M. Rajender[1,2]

[1] University Campus School, University of Hyderabad, Hyderabad, Telangana, India
violinvasu@gmail.com, m.rajender@hai-india.co.in

[2] School of Computer and Information Sciences, University of Hyderabad, Amezan Research Centre,Bangalore, India
{nagamanics,20mcmi04}@uohyd.ac.in

Abstract. It is a well-known fact that most western music is based on equal temperament, but Indian music is based on just intonation. However, the west has always been in the lead in creating electronic music processing tools like keyboards and software like Ableton Live. These instruments and programmes made by the west and intended mainly for Western music are increasingly used in Indian film music and a lot of fusion music (Equal temperament).A quantitative assessment is required to determine how severe the compromise is when using this software or instruments for Indian music production. We discovered that, despite numerous papers stating the distinction between the two types of scales (just intonation and equal temperament), there isn't a single thorough study that carries out trials and documents the honour. We are conducting this study for the first time. In this piece, we experiment with numerous Indian instrumental and vocal compositions to compare equal temperament with merely intonated scales. We discovered that the compromise is minimal when the music is mostly instrumental but significant when it is vocal. We also identified intriguing patterns in the musical notes (of Indian music) that were "closer" to the equivalent equal temperament notes than just intonated notes by examining the frequency values. For this study's sake, we evaluate plain notes exclusively; we do not examine musical notes containing the words "gamakam" or "meend." Since only plain notes are discussed in this text, Carnatic and Hindustani music can benefit from them. Finally, we make the dataset available to the general public to encourage more research in this area.

Keywords: Equal temperament · Just intonation · Carnatic music · Gamakam · Moorchana · Rhythm

1 Introduction

It is a well-known fact that western music uses an equal-tempered scale while Indian music uses merely intonated scales. Most electronic instruments (as well

H. Zaynidinov et al. (Eds.): IHCI 2022, LNCS 13741, pp. 109–120, 2023.
https://doi.org/10.1007/978-3-031-27199-1_12

as music software) used in Indian music (including film music and other genres) are created using the equal temperament design principles, which inspired us to conduct this study. We want to find out through our study if this is the right method or if these electronic instruments and software need to be modified to fit Indian music. According to the results of our study, there isn't much of a distinction between the two types of instrumental music. Therefore, adopting these tools or software based on equal temperament principles is a very minor compromise. We did discover, however, that a "master ear" can still distinguish between the two. We found that the difference is significant in the case of vocal music, and it is obvious that electronic music or software needs to be explicitly created for Indian music. We make the audio files and this work available to inspire more investigation. Additionally, the remainder of the paper is structured as follows: The octave concept, scales, and the two most common forms of scales (intonated and equi-tempered scales) are covered in Sect. 2, along with their respective differences. The associated work is described in Sect. 3.

1.1 Basic Concept of Music

Indian music underwent a lengthy, subtle transformation from simple Vedic Scales to highly developed art music. The study would show that the foundational concept for all subsequent music advancements toward becoming a fine art is the rich music tradition and the practice of reciting Vedic hymns. In a nutshell, this introductory chapter will highlight the significance of musical heritage. From the Vedic era to the present, evolution has undergone several stages. Ragas, musical forms, Talas, musical instruments, notation, and other areas have undergone extensive development. The scale is a critical distinction between Western and Indian music. Indian music uses a just intonated scale, whereas Western music uses an equal-tempered scale. Technically, the frequency difference between different notes in an octave is determined by the type of scale. Let's take some time to learn the basic musical notions before we continue to focus on the technical aspects. The terminology used in Indian music is briefly introduced in the following part. Table 2 illustrates the relationship between an introductory note, tune, and frequency.

Table 1. Notes and it's signal parameters and types Comparision Indian with Western[Kamini, 2015]

Shruti Name	Indian note	Western note	Frequency in Hz	Common notes
Teevra	Sa	C	261.63	1
Kumudvati	re		273.38	
Manda	re +(komal Re)	Db	279.07	2
Chandovati	Re-		290.7	
Dayawati	Re(Shudha Re)	D	294.33	3
Ranjani	ga		310.08	
Ratika	ga+(Komal Ga)	Eb	313.96	4
Raudri	Ga(Shudha)	E	327.03	5
Krodhi	Ga+		331.12	
Vajrika	ma(shudha ma)	F	348.84	6
Prasarini	ma+		353.20	
Preeti	Ma(Teevra Ma)	F#	367.92	7
Marjani	MA+		372.52	
KShiti	Pa	G	392.52	8
Rakta	dha		413.44	
Sandipani	dha+(komal dha)	Ab	418.61	9
Aalaapini	Dha-(shudha)	A	436.05	10
Madanti	Dha		441.50	
Rohini	ni		465.12	
Ramya	ni+(komal ni)	Bb	470.93	11
Ugra	Ni (Shudha)	B	490.56	12
Kshobhini	Ni+		496.69	
Teevra	Sa (Taar Saptak)	C	523.26	13

1.2 Note (Swara), Octave (Frame) and Scales (Frequency Range)

Music is a language of expression of emotions in encripted with notes(sware) and its frame work based on the type of Raga they are playing. Hence as like language music also have its literals.These literals are called notes or "Swara" [?] used in Indian music are Sa/ Shadaj, Re/ Rishab, Ga/Gandhar, Ma/Madhyam, Pa/Pancham, Dha/Dhaiwat and Ni/Nishad. This corresponds to the western diatonic scale. In Indian music, Sa and Pa (1st and the perfect 5th) have a fixed pitch. The 2nd, 3rd, 4th, 6th and 7th notes are variables. Seven notes make up an octave, referred to as a "Saptak." Due to its higher frequency, the eighth note is an octave higher than the first note. An octave which is frame of 8 notes is produced by a frequency ratio of 2:1. In Indian music, three primary octaves are employed. Mandra Saptak, Madhya Saptak, and Taar Saptak are the lower, middle, and higher octaves. Mandra saptak "Sa" would be 120 Hz and Taar saptak "Sa" 480 Hz if the Madhya saptak "Sa" were 240 Hz. In the Indian music system, the seven natural notes are derived according to the ratio of 3:2, 4:3, 5:3, 5:4, 9:8 and 15:8, which can be expressed in ratios as - 1, 9/8, 5/4, 4/3, 3/2, 5/3, 15/8. Indian music, however, employs more than seven notes. Indian music uses 22 microtones, or Shrutis, instead of the 12 notes in an octave used in Western music since it was once believed that there were 22 distinct notes in an octave. The 22 shruthis, however, have been roughly translated into 12 notes by musicologists. Technically speaking, an octave is a collection of frequencies in the range [x, 2x], where 'x' denotes the set's initial frequency and '2x' denotes its final frequency. "X" is known as the base note. The selection of "x" depends on several factors, including the instrument, the singer's age, gender, and the composition to be performed. Some of the example octaves are mentioned below:

Table 2. Initial and Final Frequencies of the Octave measures scale

S.NO	x(Initial frequencey)	2x(Final Frequency)
1	100 Hz	200 H
2	200 Hz	400 Hz
3	400 Hz	800 Hz
4	800 HZ	1600 Hz

A set of notes in a pitch-based arrangement is called a musical scale. A scale is a collection of related notes widely used in music as the basis for chord progressions and melodies. Most popular music is composed of notes of a single major or minor scale, though some compositions incorporate notes from multiple scales.

The Two Major Types of Scales: The entire octave is measures in two different scales based on their parameter consideration.

Equally Tempered Scale: In a method of equal tuning or musical temperament, every pair of adjacent notes has the same frequency ratio. This indicates that all neighbouring notes have a consistent frequency ratio. This note spacing is used in western music [7]. The frequency of the "nth" musical note in an octave can be calculated mathematically using the following equation.

```
notefrequency = Basenotefrequency * 2(n/12)
```

Where [0,12] is the octave and "n" is the note position. Keep in mind that any two consecutive notes' frequency ratios are fixed. Musicians could now transpose any piece of music into any key thanks to this. Humans naturally perceive music through just intonation. Several musicians believe that the shift from just intonation to equal temperament ruined harmony [8]. **Just Intonated Scale:** Pure intonation, often known as just intonation, is any musical tuning in which ratios of small whole integers relate to the frequencies of notes. This type of tuning is known as a just pure interval. Two sounds in every just break are members of the same harmonic series. The term "justly tuned" is not usually used to describe large integer frequency ratios, such as 1024:927. Ptolemy claimed that the later ancient Greek modes employed just intonation as their tuning system. Many musical cultures around the world, both ancient and modern, use this tuning system, and it was also the aesthetic ideal of Renaissance thinkers. Each note is defined as some multiple of the base note [9]. This system is heavily used in Indian classical music. Prof. Sambamoorthy describes the exact as just intonation of "ancient Indian music". They mention it again in the Table 3 for convenience. When sounds with these apparent mathematical relationships sound nice (consonant) together, the ear recognises harmony (the words "harmony" and "arithmetic" share a common root). Also, the human ear could accurately calculate exact ratios between frequencies [8]. It may also be essential to remember that just intonation was not exclusively used by Indian civilization; it was also used by many other cultures, including Chinese, Japanese, Egyptian, and even societies that existed before the Europeans. For musical convenience, modern Europeans have switched to equities pared scale. Western artists divided the octave into twelve equal intervals, but not because it sounds better that way-far from it. It has an audible thumping between sustained pitches and is a little buzzy, so they can transpose any music to any key. The compromise of equal temperament has recently come to be seen as a blunder by many composers [10]. They hold that musical logic takes precedence over the auditory sensuality of music when changing from one key to another. Figure 1 shows the general representation of Equal temperament and Just intonation notes.

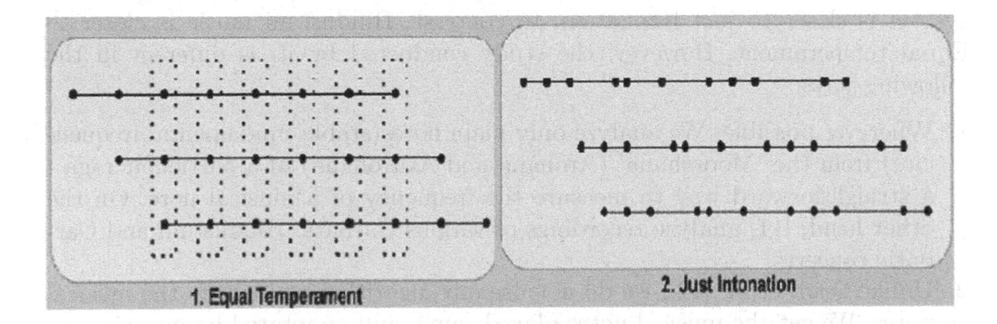

Fig. 1. Representation of Equal temperament and Just intonation notes

Analysis of Carnatic Music about Just Intonation and Equal Temperament Scales:- Let's take the frequency range: 220 Hz–440 Hz
Total notes: 12(basic)

Equal Temperament: notefrequency = Basenotefrequency*2(n/12)
'n' is the number of notes
n = 0; notefrequency = 220*2(0/12) = 0
Next frequency = 220 + 0 = 220
n = 1; notefrequency = 220*2(1/12) = 36.67
Next frequency = 220 + 36.67 = 256.67
n= 2; 220*2(2/12) = 73.33
Next frequency = 220 + 73.33 = 293.33
Like this series of Equal temperament, frequency will be: 220, 256.67, 293.33............403.33, 440

Just Intonation: According to Table 1, every consecutive note frequency has a predefined ratio.
For ratio (1/1): 220 * (1/1) = 220,
For ratio (16/15): 220*(16/15) = 234.67,
For ratio (9/8): 220*(9/8) = 247.5,
.
.
.
For ratio (2/1): 220(2/1) = 440
So the series will be: 220, 234.67, 247.5 440

2 Review of Related Literature

We want to acknowledge that there have been similar studies in the past to assess the tuning of sung Indian classical music [11]. Xerra et al. [11] have analyzed Hindustani and Carnatic music separately and found that the Carnatic music system is closer to just intonation. In contrast, Hindustani music is closer to Equal temperament. However, the study conducted by us is different in the following ways:

1. Wherever possible, We analyze only plain notes(stable fundamental frequencies) from the 'Moorchana' (Arohana and Avarohana) of a particular raga - a straightforward way to measure the frequency of a musical note. On the other hand, [11] analyze recordings of various Carnatic Hindustani and Carnatic concerts.
2. Unlike Serra et al. [12], we do not use any algorithm to identify the musical notes. We get the musical notes played, sung, and annotated by musicians of the highest repute. Hence, our note recognition is more accurate than [11].
3. Serra et al. [12] use only vocal music to conduct experiments, whereas we cover a much larger ground by conducting experiments on Violin, Vocal, and Mandolin.
4. [3] An overview of computational musicology, a branch of artificial intelligence, is provided in this work (AI). It describes the development of the area of computational musicology and the current research being done about Western and Indian music. Additionally, it contrasts Western Music with Indian

Table 3. Western and Indian notes with frequency ratio [3]

Western Notation	Indian Notation	Frequency Ratio (natural)
C	Sa	1/1
C#	Re	16/15
D	Re	9/8
Eb	Ga	6/5
E	Ga	5/4
F	Ma	4/3
F#	Ma	7/5
G	P	3/2
Ab	Dh	8/5
A	Dh	5/3
Bb	Ni	9/5
B	Ni	15/8
C'	Sa'(Next Octave)	2/1

Classical Music (ICM). It highlights their key differences to show how Western Music analysis is incompatible with Indian Classical Music, as discussed in Table 3. [1] By utilizing the structured character of Indian Carnatic Music, they suggest a model in this work produce music for Indian Classical Music, specifically Carnatic Music.

3 Experiments

The idea of a musical note expresses the pitch and duration of a sound. It shows the sound's pitch. Instead of classifying the sounds by frequency, musicians assigned them letter names. Different regions use different naming conventions for musical notes. The Solfege system is used in nations like Italy, Spain, Greece, France, Portugal, and others. Indian culture includes the Swara system. The significant cultural contrast between Indian and western music sets them apart the most. While these links are missing from Western music, spirituality and interpersonal relationships are the foundation of traditional classical Indian music. Everything is in the writing and annotations. Figure 2 contains the conducted experiment to check whether the Indian music (including Carnatic and Hindustani music) plain notes when sung or played on instruments is closer to just intonation or equal temperament scale. We cover a much larger ground by conducting experiments on Violin, Vocal, and Mandolin. All of these experiments are displayed in Tables 4, 5, and 6.

Figure 3 shows the graphical representation of frequency extracted from the violin along with the frequency of Just Intonation and Equal Temperament. Similarly, Fig. 4 and 5 shows the graphical representation of frequency extracted

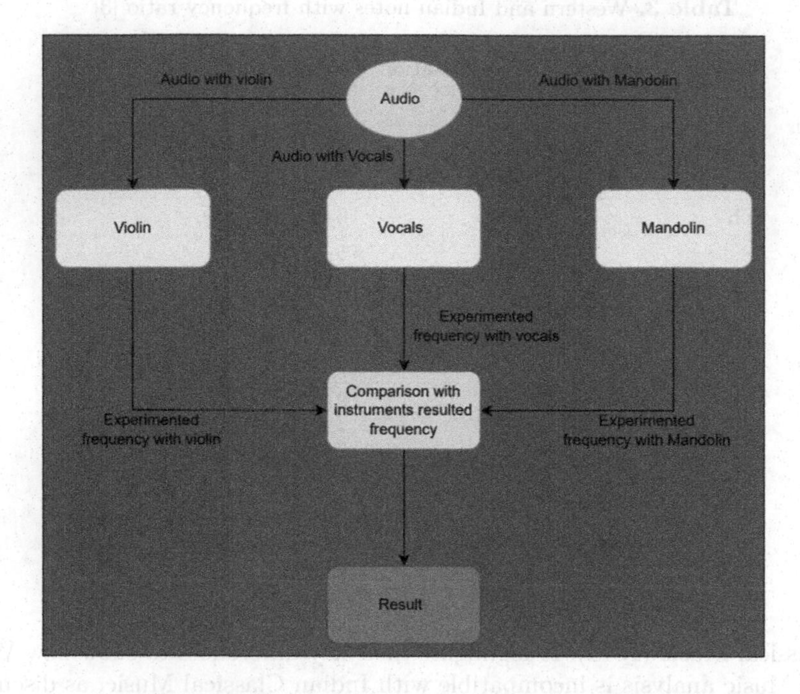

Fig. 2. Process of the empirical Data Extraction

Table 4. Violin data and their extracted JI and ET and comparing with actual values

Note	AV	JI	D	D%	ET	D	D%
S	329.6	329.6	0	0%	329.6	0	0%
R1	346	351.57	5.57	1.6%	349.19	3.19	0.9%
R2	366	370.8	4.8	1.31%	369.96	3.96	1.08%
G1	396	395.52	0.48	0.12%	391.96	4.04	1.02%
G2	412	412	0	0%	415.26	3.26	0.79%
M1	435	439.46	4.46	1.02%	439.96	4.96	1.14%
M2	4461	463.5	2.5	0.54%	466.12	5.12	1.11%
P	496	494.4	1.6	0.32%	493.84	2.16	0.43%
D1	523	527.36	4.36	0.83%	523.20	0.20	0.03%
D2	550	549.33	0.67	0.12%	554.31	4.31	0.78%
N1	577	593.28	23.28	4.4%	587.28	10.28	0.017%
N2	621	618	3	0.48%	622.20	1.20	0.19%
S	660	659.2	0.8	0.12%	659.2	0.8	0.12%

Table 5. Vocals data and their extracted JI and ET and comparing with actual values

Note	AV	JI	D	D%	ET	D	D%
S	193	193	0	0%	193	0	0%
R1	202	205.87	3.87	1.87%	204.48	2.48	1.21%
R2	217	217.13	0.13	0.06%	216.64	0.37	0.17%
G1	230	231.6	1.6	0.69%	229.52	0.48	0.21%
G2	248	241.25	6.75	2.8%	243.17	4.84	1.99%
M1	259	257.33	1.67	0.65%	257.62	1.38	0.53%
M2	278	271.41	6.59	2.43%	272.94	5.06	1.85%
P	291	289.5	1.5	0.52%	289.17	1.83	0.63%
D1	310	3.08.8	1.2	0.39%	306.37	3.63	1.19%
D2	320	321.67	1.67	0.52%	324.59	4.59	1.41%
N1	344	347.4	3.4	0.98%	343.88	0.11	0.03%
N2	369	361.88	7.12	1.97%	364.34	4.67	1.28%
S	387	386	1	0.26%	386	1	0.26%

Table 6. Mandolin data and their extracted JI and ET and comparing with actual values

Note	AV	JI	D	D%	ET	D	D%
S	271	271	0	0%	329.60	0	0%
R1	288	289	1	0.34%	287.11	0.89	0.31%
R2	305	305	0	0%	304.17	0.83	0.27%
G1	323	325.5	2.5	0.77%	322.28	3.22	0.22%
G2	342	339	3	0.88%	341.43	0.57	0.17%
M1	362	361.3	0.7	0.19%	361.74	0.26	0.07%
M2	384	381	3	0.78%	383.25	0.75	0.19%
P	407	406.5	0.5	0.12%	406	1	0.24%
D1	437	433.6	2.4	0.55%	430.18	6.62	1.53%
D2	462	451.67	10.33	2.28%	455.77	6.23	1.36%
N1	488	487.8	0.2	0.04%	482.87	5.13	1.06%
N2	514	508	6	1.18%	511.58	2.92	0.57%
S	544	542	2	0.37%	542	2	0.37%

from vocal and Mandolin along with the frequency of Just Intonation and Equal Temperament for the same song. And these graphs are also showing the resulting frequency is near to Just Intonation frequency or near to Equal Temperament frequency.

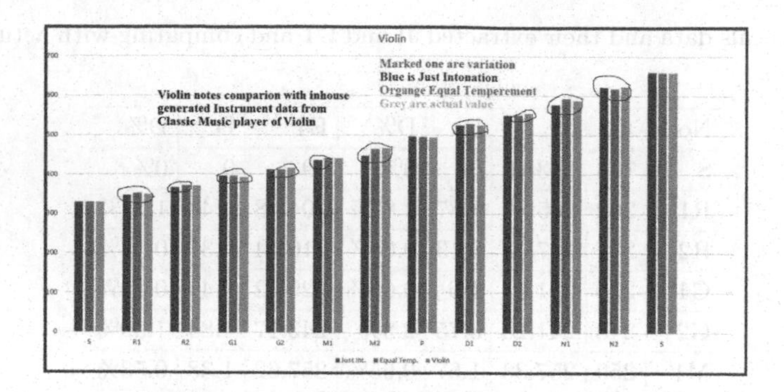

Fig. 3. Violin data set their ET and JI values comparison with real values

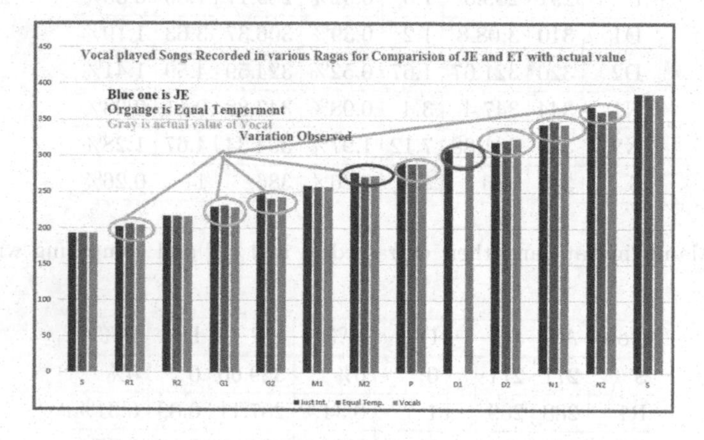

Fig. 4. Vocals

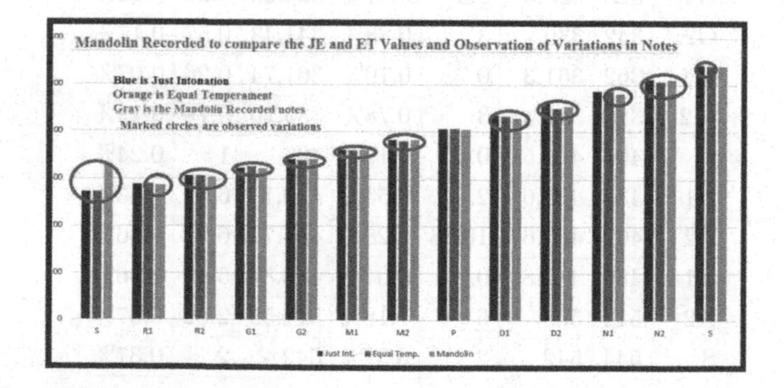

Fig. 5. Mandolin

4 Analysis and Result

The graphs above show that the fundamental frequency difference between recording devices is neither more temperamental nor closer to a just intonated

scale. This encourages us that we might be able to play Indian music using eerie frequencies! Remember that the violin's N1 significantly differed from the ideal just intonated frequency. Concerning other notes, the frequency difference (23 Hz) between the performed Swara and the just intonated frequency of N1 on the violin is too significant. To confirm this, we repeatedly ran the recording experiments for this particular case. Every time, the result was the same. An additional study (as well as tests) on the fundamental ratio of N1 is required in light of this conclusion (used in the Kharaharapriya ragam). Prof. Sambamoorthy identifies four types of "Nishadam" and, consequently, four different ratios in his thorough analysis of the 22 Swara sthana system [10,11]. Finding out which of the four forms the real Swara played closest to that would be interesting. So, based on experiments analysis, we came up with the following results:
Violin: The experimented result of the violin is 76.92 % near Equal Temperament and 38.46 % near Just Intonation.
Vocal: The experimented result of the vocal is 76.92 % near Equal Temperament and 30.76 % near Just Intonation.
Mandolin: The experimented result of the Mandolin is 53.84 % near Equal Temperament and 61.53 % near Just Intonation.

5 Conclusion and Future Scope

One of the many differences between western and Indian music that we have emphasised in this study is the use of note motions, or gamakam, which are distinctive to Indian classical music. To maintain the quality of Indian music, the digital instruments must be tuned carefully, and the necessary precautions must be taken when designing the instruments.

To understand the human cognitive parameters that are tuned using this music subject, it is important to create music data sets with experts for various Raga with vocal and instrumental accompaniment. This is future potential for musicology as one of the major research areas in human computer interactive systems.

References

1. Viramgami, G., et al.: Indian classical music synthesis. In: 5th Joint International Conference on Data Science & Management of Data (9th ACM IKDD CODS and 27th COMAD) (CODS-COMAD 2022), Bangalore, India, 8–10 January 2022, p. 2. ACM, New York (2022)
2. Marin, M., Bhattacharya, J.: Music induced emotions: some current issues and cross-modal comparisons. Music Educ., 1–38 (2011)
3. Sentiment Analysis: A Definitive Guide. MonkeyLearn (2022). https://monkeylearn.com/sentiment-analysis/
4. Sampurna, O., Shadava.: Computational musicology for raga analysis in Indian classical music: a critical review (2017)

5. Mathur, A., Vijayakumar, S.H., Chakrabarti, B., Singh, N.C.: Emotional responses to Hindustani raga music: the role of musical structure. Front Psychol. **6**, 513 (2015). PMID: 25983702; PMCID: PMC4415143. https://doi.org/10.3389/fpsyg.2015.00513
6. Bajpai, G.: Music and Personality (2017)
7. Burns, E.M., Ward, W.D.: Intervals, scales, and tuning. Psychol. Music **2**, 215–264 (1999)
8. Jourdain, R.: Music, the Brain, and Ecstasy. William Morrow and Company, New York (1997)
9. Gannon, J. W., Weyler, R.A.: Just intonation tuning. U.S. Patent No. 5,501,130 (1996)
10. Dvavimsati (22) sruti chart. https://people.rit.edu/pnveme/raga/Sruti.html, Rochester Institute of Technology
11. The classical Indian just intonation Tuning System, Early Experiments in Indian Music, Prof. P Sambamoorthy. https://www.plainsound.org/pdfs/srutis.pdf
12. Serra, J., et al.: Assessing the Tuning of Sung Indian Classical Music. In: ISMIR (2011)
13. Kaminimusic.com web article (2015). https://www.kaminimusic.com/notes-octaves-scale/
14. Gurrala, S., et al.: Multilayer tag extraction for music recommendation systems. In: Kim, J.-H., Singh, M., Khan, J., Tiwary, U.S., Sur, M., Singh, D. (eds.) IHCI 2021. LNCS, vol. 13184, pp. 248–259. Springer, Cham (2022). https://doi.org/10.1007/978-3-030-98404-5_24
15. Molakathaala, N., et al.: Melody-based hindi song retrieval using SVM. In: Satapathy, S.C., Bhateja, V., Favorskaya, M.N., Adilakshmi, T. (eds.) Smart Intelligent Computing and Applications, Volume 2. Smart Innovation, Systems and Technologies, vol 283. Springer, Singapore (2022). https://doi.org/10.1007/978-981-16-9705-0_4

Emotion Classification Through Facial Expressions Using SVM and Convolutional Neural Classifier

Varsha Singh(✉), Ravi Kumar Singh, and Uma Shanker Tiwary

Indian Institute of Information Technology, Allahabad, India
varshagaur@gmail.com, ust@iiita.ac.in

Abstract. Emotions being an influential personal state of feelings, have phenomenal importance, and facial expressions are one's instinctive reflections and photoprints of emotions. These facial expressions that count to be 55% of the total human communication cannot remain unnoticed, especially in today's expanding world of Human-Computer interaction, where the need of the hour is to train computers to recognize human emotions from facial expressions of images. Four models are developed in this work for emotion classification. This work utilizes HOG (descriptor) and SVM for the first model while employing CNN models with varying input strategies with and without down-sampling in the remaining three models to predict the given FER dataset images into either of the seven universal facial expressions. The first model extracts the histogram of oriented gradient (HOG) from the images and applies classification with a support vector machine (SVM). The second model inputs raw pixel image data for training. The third model uses a novel hybrid feature strategy that maneuvers a combination of HOG features and pixel data of images. The last model uses the same architecture as the previous two CNN models but with a balanced dataset (all classes having the same number of images). Batch normalization, dropout, and L2 regularization reduced the overfitting of models, and the GPU improved the training speed. The hybrid technique (Model-3) performed better than model-1, model-2, and model-4 in terms of accuracy and F1 score. The performance evaluation speaks about the falter arising with downsampling in model-4.

Keywords: Facial expression · Emotion classification · Convolutional neural network (CNN) · Support vector machine (SVM) · Histogram of oriented gradients (HOG)

1 Introduction

Facial emotion recognition is the process of identifying human facial emotions through the face. People widely vary with different accuracies at recognizing others' emotions. In this spread of varying accuracies, technology can prove to channel emotion recognition and develop some models to find accurate human facial

H. Zaynidinov et al. (Eds.): IHCI 2022, LNCS 13741, pp. 121–131, 2023.
https://doi.org/10.1007/978-3-031-27199-1_13

emotions. Facial emotion recognition has applications in various fields, including clinical science, behavioural science, etc. Communication mediums include body gestures, facial expressions, etc., apart from speech. Body gestures help to communicate modulations in speech, whereas facial expressions reflexively display human emotions. Detecting facial expressions is today's crucial requisite for human-machine interfaces too. Several advancements in facial expression detection have been made in the past few years, like techniques for extracting facial features and classifying expressions. However, still developing a system for automated facial expression detection is the required level. This work performs facial recognition in multiple ways, such as by using descriptor (Histogram Of Orientation Gradient) and SVM (Support Vector Machine) for the first model while varying input strategies for Convolutional Neural Network (CNN) in the other models. Then, CNN and SVM predict the label as either of the following facial emotions: neutral, happiness, fear, sadness, disgust, anger, or surprise. The motive of combining two or more techniques and preprocessing to achieve comparable results is achieved successfully.

2 Dataset

The project utilizes the FER2013 dataset from the Kaggle website that contains 35,887 (48 pixels * 48 pixels) grayscale images depicting facial expressions. The dataset has seven different classes, and each image belongs to one particular class of facial emotion. All the images have almost centred faces and occupy a similar amount of space. For example, Fig. 1 shows seven different facial emotions.

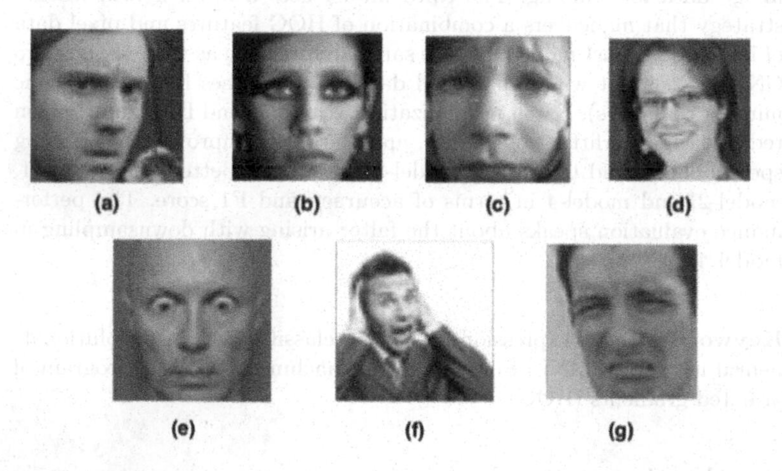

Fig. 1. Seven facial emotions are a) angry, b) neutral c) sad d) happy e) surprise f) fear g) disgust

Numbers between 0 to 6 describe the image's class number and thus facial emotion as angry, neutral, sad, happy, surprise, fear, and disgust, respectively. The size of the dataset is (35887, 48, 48, 1). The classified description of the dataset (image count and respective facial emotions) are:

- Angry has 4953 images
- Disgust has 547 images
- Fear has 5120 images
- Happy has 8988 images
- Sad has 6077 images
- Surprise has 4002 images
- Neural has 6198 images.

The dataset has been divided into three separate sets, namely the training set (to train the model), the validation set (for tuning the hyper-parameter), and the testing set (to test the model). There are 28196 images in the training set, 3546 images in the validation set, and 3545 images in the testing set. The pixel data of images is observed and then normalized.

3 Background

Previously, researchers have progressed the research in developing automatic expression classifiers [8,10]. The facial emotion recognition systems embody the classification of faces into several sets of original emotions, such as happiness, sadness, and anger [5]. The face produces individual muscle movements to produce an objective face, and the Facial Action Coding System (FACS) is the psychological framework used to describe facial movements. It is a method to classify human facial movements by appearance using Action Units (AU). They are the relaxation or contraction of one or more muscles. Many techniques are used for facial emotion recognition, such as Bayesian Networks, Artificial Neural Networks, and Hidden Markov Model (HMM) [3]. Some techniques can be combined to improve the accuracy of the System. However, these techniques have limited and poor accuracy in terms of image classification [11]. Furthermore, these techniques also require high computational resources compared to CNN [12].

4 Methods

The four models for Facial Emotion Recognition are as follows:

1. Model-1 (HOG + SVM): This model employs a histogram of oriented gradients (HOG) for feature extraction and a support vector machine (SVM having RBF kernel) for classifying the facial emotions of facial images [4]. HOG is one of the facial descriptors in machine learning and computer vision but can also be utilized for quantifying and representing both shape and texture. The popularity of HOG can be advocated for its ability to characterize the

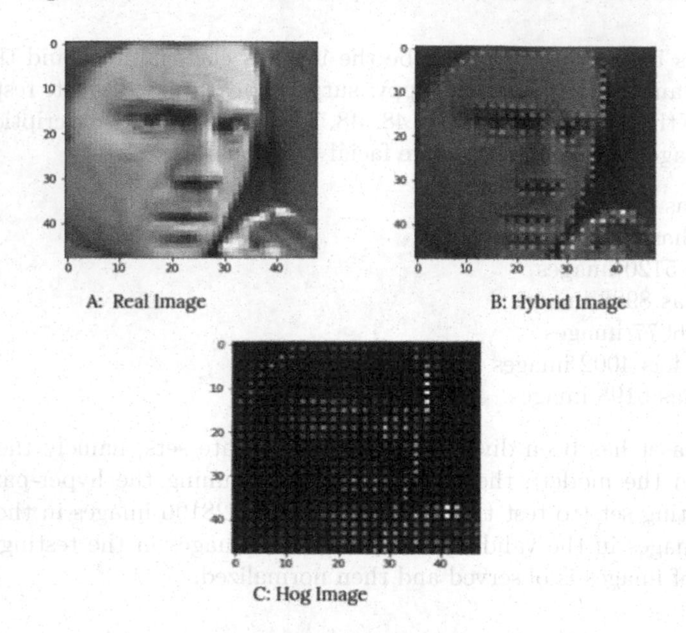

Fig. 2. Input images for models

local object's appearance and shape using the distribution of local intensity gradients (horizontal and vertical gradients) [7]. HOG descriptor is modelled to provide the dimensionalities of the real-valued feature vectors, depending upon the following parameters: orientation, pixels per cell, and cells per block. Radial Basis Function (RBF) is used in SVM as the kernel method [6].

2. Model-2 (Real Images + CNN): This model uses a convolutional neural network (CNN model) to engulf both the purposes of feature extraction and classification of facial emotion and inputs real images (grayscale images) (Fig. 2). The model is structured with 8 convolutional layers, a non-linear activation function, ReLU, and then led to 4 fully connected layers [2]. Dropout, batch normalization (BN), and max-pooling operations are used after each layer. In the end, the network has a dense layer that computes the scores and softmax loss function [1].
3. Model-3 (Hybrid Image + CNN): This model wields the CNN model for both feature extraction and classification of facial emotion and is similar to the second model in terms of the functioning of convolutional layers, non-linear activation function, the count of fully connected layers, over-fitting reduction strategy of dropout, batch normalization, max-pooling operations, the structure of the dense layer, score computation, loss function type, etc. but differs in input. It takes hybrid images (Real image + HOG image) as input (Fig. 2). Each hybrid image is obtained by combining the real image (used in model 2) and the HOG image (used in model 1).

4. Model-4 (CNN Balanced Dataset): The datasets are not always balanced in terms of the count of images lying in each class of facial emotion classification and can not be completely relied upon for accuracy. Similar is the case with the FER dataset. It has a total of 35887 images that fall in either of the 7 different emotions. The exact category counts are as follows: Angry has 4953 images, Disgust has 547 images, Fear has 5120 images, Happy has 8988 images, Sad has 6077 images, Surprise has 4002 images, Neutral has 6198 images. This model exploits the concept of downsampling to balance the dataset. Five hundred forty-seven images from each above-mentioned category are collected as "disgust" has the least image count, 547, among all the categories. The new dataset now has 3829 images [4].

5 Analysis

5.1 Preprocessing Part

In **Model-1**, the dataset (35887 images) is divided into 2 sets as the training set containing 28196 images and the testing set with 7691 images. The size of each image is 48 pixels * 48 pixels and is divided into different equi-sized blocks that contain pixels.

In **Model-2** and **Model-3**, the dataset is divided into 3 sets, namely training set, validation set and testing set. Dataset has 35,887 images, each of size 48 pixels * 48 pixels. The training set is used for training the model, which contains 28196 images, each of 48 pixels * 48 pixels. The testing set is used for testing the model and contains 3845 images, each of 48 pixels * 48 pixels. The validation set is used for tuning the hyperparameters and contains the remaining 3846 images, each of 48 pixels * 48 pixels.

In **Model-4**, the dataset is divided into 3 downsampling training sets, a validation set, and a testing set. Dataset has 3829 images, each of size 48 pixels * 48 pixels. The training set is used for training the model that contains 3008 images, each of 48 pixels * 48 pixels. A Testing set is used for testing the model and contains 410 images, each of 48 pixels * 48 pixels. The validation set is used for tuning the hyperparameters and contains the remaining 411 images, each of 48 pixels * 48 pixels.

5.2 Experiment

Model-1

1. Compute gradient magnitude/direction at each pixel using a cell area of 8*8 pixels.
2. Create a histogram of generated 64 gradient vectors (8*8).
3. Split each cell of the image into angular bins (Fig. 3) where each bin corresponds to a gradient direction from 0°C to 180°C. (20°C per bin).

4. Repeat all the above steps until the entire image is covered by convolution operation to obtain the gradient images.
5. Normalization: This is an optional step but sometimes used to improve the performance of the HOG descriptor like with:
   - Gamma/power law normalization
   - Square-root normalization
6. After finding HOG features from the training set (size of Hog features = (size of images, 900))
   Xtrain = (28196, 900),
   Xtest = (7691, 900),
   ytrain =(28196, 1),
   Ytest = (7691, 1).
7. Train the SVM (RBF kernel, gamma = 0.1) on the training set.
8. Test the SVM model on the testing data.
9. Calculate the accuracy as well as the confusion matrix.

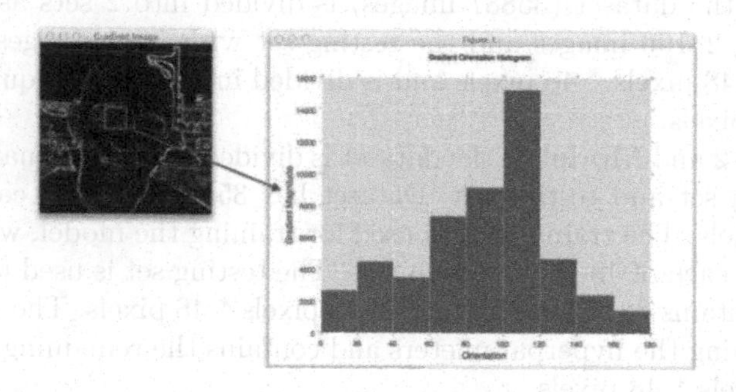

Fig. 3. Cell in an image is splitting into the angular bin. Adapted from [13]

The Accuracy of the SVM matcrializes as 52.9 (C = 100, gamma = 0.01, kernel ='rbf') (Shown in Table 1)

Table 1. Overall accuracy of all models

MODEL	MODEL-1	MODEL-2	MODEL-3	MODEL-4
ACCURACY	52.92	65.15	65.2	50.48
Weited F1-score	44.65	65	65.11	45.69

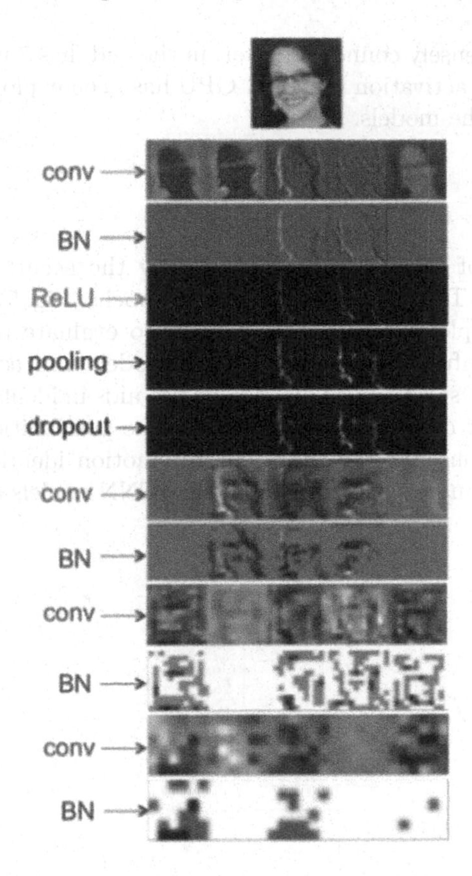

Fig. 4. Visualisation of activation maps at each level of CNN Model. Adapted from [13]

In **Model 2, Model 3** and **Model 4**, the architecture consists of eight convolutional layers and four fully connected layers (for calculation of loss and score) followed by batch normalization and dropout. The first CNN layer (Fig. 4) uses 64 filters, each sized 3 * 3, stride (size 1) and ReLU as the activation function. The second layer uses 64 filters, each sized 3 * 3, stride (size 1), batch normalization, max-pooling with 2 * 2 sized filter, dropout as 0.5, and ReLU as activation function. The third CNN layer is similar to the first layer but contains 128 filters, each sized 3 * 3. The fourth layer is similar to the second layer but with 128 filters, each 3 * 3 sized. The fifth CNN layer is similar to the third layer but uses 256 filters. The sixth layer is homogeneous to the fourth layer but with 256 filters, each of size 3*3. The seventh CNN layer is similar to the fifth CNN layer but uses 512 filters. The eighth layer is analogous to the sixth layer but with 512 filters. The first fully connected layer has 512 neurons in a hidden layer and ReLU as an activation function followed by a dropout of 0.4. The second fully connected layer has 256 neurons in a hidden layer and Relu as an activation function followed by a dropout of 0.4. The third fully connected layer has 128 neurons in a hidden layer, ReLU as an activation function followed by

0.4 dropouts. The densely connected layer, in the end, has 7 neurons in a hidden layer and a Softmax activation function. GPU has been exploited to increase the processing time of the models.

6 Result

The performances of the models are judged by the accuracies of the models (Table 1 and Fig. 8). The graphs of accuracy vs epochs (Fig. 5) and loss vs epochs (Fig. 6) have been plotted for all the models to evaluate their performances. Fig. 7 shows the confusion matrix of all the models and assists in evaluating their performances respectively. This matrix also aids in identifying the probable presence of incorrect classifications, which are the possibilities with the human emotion identifications, too, as the accurate emotion identification of faces is always difficult. It can be easily visualized that CNN models are better than the HOG-SVM model.

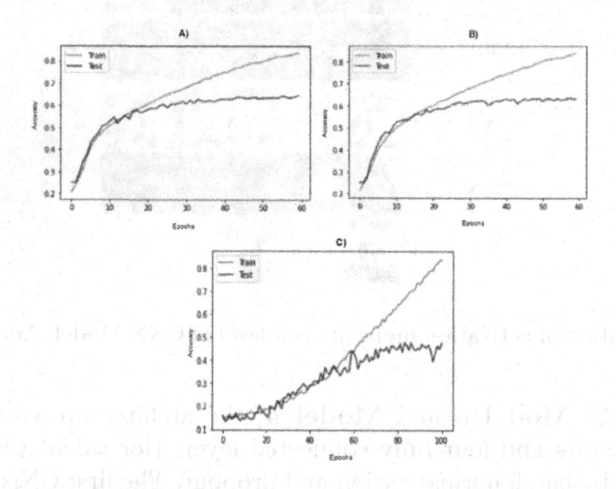

Fig. 5. Accuracy vs epochs for CNN models: A) Model2, B) Model3, C) Model4

CNN model taking real images as input and CNN model inputting hybrid images have the results approximate to each other. Model-3 is better than other models. **Model2** has trained for 60 epochs where accuracy is 65.15 and f1-score 65.00 **Model3** has trained for 60 epochs where accuracy is 65.20 and f1-score 65.0 **Model4** has trained for 101 epochs where accuracy is 50.48 and f1-score 45.69. The graphs of Accuracy vs Epochs (Fig. 5) and Loss vs Epochs (Fig. 6) have been plotted for all the CNN Models which apprise that the increase in the number of epochs decreases the loss of the models whereas increases accuracy for both the test set and the train set.

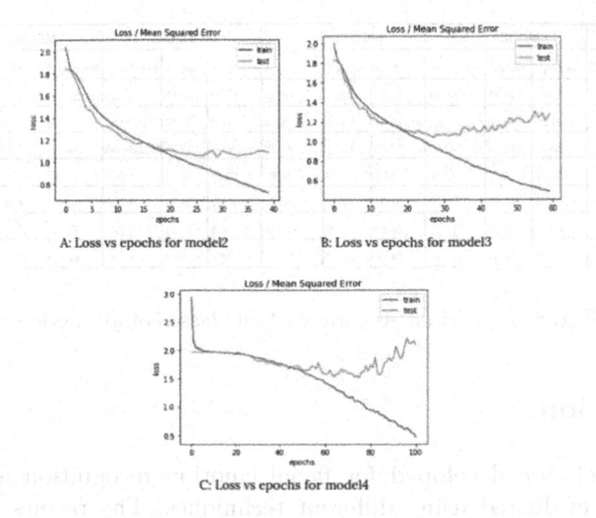

Fig. 6. Loss vs Epochs for CNN Models 2, 3, and 4

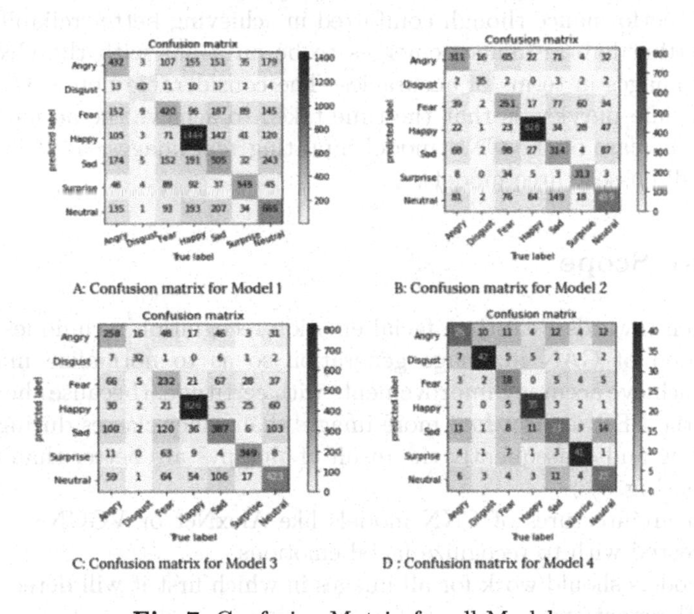

Fig. 7. Confusion Matrix for all Models

MODELS	MODEL-1			MODEL-2			MODEL-3			MODEL-4		
EMOTIONS	Precision	Recall	F-1 Score	Precision	Recall	F-1 Score	Precision	Recall	F-1 Score	Precision	Recall	F-1 Score
Angry	40.67	40.87	40.77	59.64	60.86	60.24	61.57	48.58	59.31	32.05	43.1	36.76
Disgust	51.28	70.58	59.4	76.08	60.34	67.3	65.3	55.17	59.81	65.62	71.18	68.29
Fear	38.25	44.53	41.15	52.29	45.71	48.77	50.87	42.25	46.16	48.64	31.03	37.89
Happy	74.97	66.2	70.31	87.8	85.98	86.88	82.68	85.77	84.2	72.91	59.32	65.42
Sad	38.78	40.52	39.7	52.33	48.23	50.23	51.32	59.44	55.08	32.39	38.98	35.38
Surprise	63.51	70.05	66.61	85.51	72.96	78.74	78.6	81.35	79.95	70.68	69.49	70.08
Neutral	50.07	47.46	48.73	54.06	69.12	60.67	58.6	63.7	60.95	42.54	39.65	41.07

Fig. 8. Individual Accuracy of all classes of all models

7 Conclusion

The four models are developed for facial emotion recognition and their performances are evaluated using different techniques. The results manifest that the CNN models are better than the Descriptor + SVM model. Contrasting to the first seeming lookout, the down-sampling did not help in improving the model's performance though comforted in achieving better reliability. CNN model inputting hybrid images emerges to be on a par with the CNN model taking real images in terms of accuracies. The considerable noteworthy upshot in terms of time purview is that the time taken to achieve the actual accuracy is lesser in the case of the CNN model inputting real images. Model3 is better than other discussed three models.

8 Future Scope

- Expansion towards Real-time facial emotion recognition techniques [9].
- Application of GAN in image generation so as to normalize images and thereby achieve accuracy improvements with certitude. (because the "Happy" class of the FER dataset had more images than other classes during training and testing and consequently the result of "happy" are better than the other emotions (Fig. 8))
- Different architectures of CNN models like AlexNet or VGGNet [5] can be experimented with to recognize facial emotions.
- These models should work for all images in which first it will detect face and predict the emotions.
- Attempts towards an extension for the models that can recognize emotions from the images of faces having different viewing angles.
- Development in direction of models that can input images containing multiple faces and then predict all their emotions after detecting the faces from that image.
- Discovering alternatives of multi-head self-attention mechanism for further enhancement of performance of vision transformers.

References

1. Albawi, S., Mohammed, T.A., Al-Zawi, S.: Understanding of a convolutional neural network. In 2017 International Conference on Engineering and Technology (ICET), pp. 1–6 (2017)
2. Alizadeh, S., Fazel, A.: Convolutional neural networks for facial expression recognition (2017)
3. Cohen, I., Sebe, N., Sun, Y., Lew, M.S., Huang, T.S.: Evaluation of expression recognition techniques. In: Bakker, E.M., Lew, M.S., Huang, T.S., Sebe, N., Zhou, X.S. (eds.) CIVR 2003. LNCS, vol. 2728, pp. 184–195. Springer, Heidelberg (2003). https://doi.org/10.1007/3-540-45113-7_19
4. Dalal, N., Triggs, B.: Histograms of oriented gradients for human detection, vol. 1, pp. 886–893 (2005)
5. Shichuan, D., Tao, Y., Martinez, A.M.: Compound facial expressions of emotion. Proc. Natl. Acad. Sci. **111**(15), E1454–E1462 (2014)
6. Durmuÿsoglu, A., Kahraman, Y.: Facial expression recognition using geometric features. In 2016 International Conference on Systems, Signals and Image Processing (IWSSIP), pp. 1–5 (2016)
7. Korkmaz, S. A., Akÿciÿcek, A., Bt'nol, H., Korkmaz, M.F.: Recognition of the stomach cancer images with probabilistic hog feature vector histograms by using hog features. In: 2017 IEEE 15th International Symposium on Intelligent Systems and Informatics (SISY), pp. 000339–000342 (2017)
8. Littlewort, G., Frank, M., Lainscsek, C., Fasel, I., Movellan, J.: Automatic recognition of facial actions in spontaneous expressions. J. Multimedia **1**, 09 (2006)
9. Simonyan, K., Zisserman, A.: Very deep convolutional networks for large-scale image recognition. arXiv proprint arXiv:1409.1550 (2014)
10. Tian, Y.I., Kanade, T., Cohn, J.F.: Recognizing action units for facial expression analysis. IEEE Trans. Pattern Anal. Mach. Intell. **23**(2), 97–115 (2001)
11. Maruyama, T., et al.: Comparison of medical image classification accuracy among three machine learning methods. J. X-ray Sci. Technol. **26**(6), 885–893 (2018)
12. Chaganti, S. Y., et al.: Image classification using SVM and CNN. In: 2020 International Conference on Computer Science, Engineering and Applications (ICCSEA). IEEE (2020)
13. Rosebrock, A.: Deep learning for computer vision with python: Starter bundle. PyImageSearch (2017)

On the Evaluation of Generated Stylised Lyrics Using Deep Generative Models: A Preliminary Study

Hye-Jin Hong[1], So-Hyeon Kim[2], and Jee-Hang Lee[1,2](✉)

[1] Department of Human-Centered AI, Sangmyung University, Seoul, Republic of Korea
jeehang@smu.ac.kr

[2] Department of AI and Informatics, Sangmyung University, Seoul, Republic of Korea

Abstract. Deep generative models such as a family of GPT have exhibited super-human performance in natural language generation. However, the evaluation of the generated lacks the automated solutions and mostly requires human involved manual experiments. This paper explores the possibility of a computational means to evaluate the generated contents in an automated way. We in particular conducted the experiment with stylised lyrics which requires careful consideration in the evaluation since the lyrics generation takes into account individual characteristics of artists. To this end, we first carried out the lyrics generation through fine-tuning with K-Pop songs in three different genres using the KoGPT-2 to effectively transfer the individual artists' persona and style. Afterwards we conducted the evaluation of stylised lyrics with another deep generative model, BERT, to measure the similarity between the lyrics generated and that in the training data, both within and between artists. The results showed the highest score between the generated and the original lyrics within the same artist but lower similarity than that between the artists, which the phenomena was not captured in a typical evaluation metric such as BLEU. Although this is a preliminary approach, this shows a possibility to automatically evaluate the generated contents in which individual characteristics were infused without human effort.

Keywords: Persona · Chatbot · GPT-2 · BLEU · BERT

1 Introduction

Deep generative models such as the GPT family have exhibited fascinating performance in natural language generation (NLG). A representative example is a chatbot, a software system that can interact and/or communicate with the people in a form of natural language, which is widely used in e-commerce, financial services, healthcare and messaging applications [1–4]. As it gains much attention, the NLG research has rapidly extended to generate not only more naturalistic but also highly individuated presentations to promote the quality of user experience in practical services. As an example, a field of NLG

H.-J. Hong and S.-H. Kim—These authors contributed equally.

H. Zaynidinov et al. (Eds.): IHCI 2022, LNCS 13741, pp. 132–139, 2023.
https://doi.org/10.1007/978-3-031-27199-1_14

has focused on reflecting unique personas such as gender, age or a backstory as well as having a human-like personality [5]. According to a study by Lee et al. [6], this attempt is likely to enhance the trustworthiness between NLG agents and people whilst relieving people's feeling of reluctance to use. Since the agent is able to deliver the textual contents with individuated tone and manner in relation to user profiles, it would encourage people to have more continuous interactions which in turn lead them to have a high-quality user experience and a sense of trust.

However, to build such a system with personality and style is not well explored for a couple of obstacles. The one is the persona sparsity in the dataset. Usually the dataset for the training is likely to be constructed from various dialogue corpora. Here, the persona was sparsely observed in the training dataset which incurred the difficulty to embed the personality consistently. This brings about the inconsistency of personality in the sequence of generated textual contents during the conversation [7]. The other is the evaluation. As widely accepted, the generated textual contents with personality and style is highly idiosyncratic so that the evaluation process is likely to rely much on the human involved experiments. It is not only a cost-inefficient means but also under the risk of subjective valuation caused by individual bias. However, this issue is neither well explored nor the automated computational means are suggested.

In this paper, we explore the possible resolutions for the latter, a computational means to evaluate the generated contents embedding persona and style in an automated way. In particular, we investigated it using the contrasting examples which stylised lyrics generated with individual artists' characteristics. To this end, we first carried out the lyrics generation through fine-tuning of the KoGPT-2 with K-Pop songs in three different genres to effectively transfer the individual artists' persona and style. Afterwards we conducted the evaluation of stylised lyrics with another deep generative model, BERT, to measure the similarity between the lyrics generated and original lyrics for fine-tuning, both within and between artists. We expect the comparison using BERT, a well known for natural language understanding (NLG) tasks, will be able to highlight the consistency on the lyrics within the same artist, and the disparity on the lyrics between the different artists.

We note that it seems a stylised lyrics generation would not be seen as a typical way to design a persona-infused NLG model. But it could be a good source to examine our proposal. Since the lyrics of an artist predominantly contains the artist's own persona that is already widely known thus to recognise the persona and style in the generated lyrics would be able to be a relatively easy task we assume.

The remainder of the paper is organised as follows: After the brief introduction in Sect. 1, we describe the experimental settings to build a stylised NLG model in Sect. 2. It is followed by a proposed evaluation method and its results in Sect. 3. We conclude the paper with the discussion and future works in Sect. 4.

2 Experimental Settings

Baseline Backbone. We adopted a KoGPT-2 as a baseline backbone which is the Korean version of vanilla GPT [8]. It was a pre-trained model using Open-AI's GPT-2 Small with approximately 20 GB of Korean dataset [9].

Datasets. Recalling that our main objective is to evaluate the stylised generated lyrics by investigating whether the characteristics of the generated lyrics are different for each artist, it is important to select the artists having been shown the unique set of vocabulary with distinctive way of lyrics writing. With this criteria in mind, we first chose ten candidate K-Pop artists to construct the training datasets consisting of title and full lyrics by crawling their songs. We constructed a dataset of 823 songs as a result. Afterwards, we chose the final three K-Pop artists whose number of songs and the total length of lyrics were sufficient to perform fine-tuning the baseline backbone as indicated in [11]. Here, we only used Korean in all lyrics; all Foreign words and expressions in each song, and songs released in Foreign languages such as English, Japanese or Chinese were excluded. We finally carried out the training of the model with the dataset. Table 1 shows the statistics on the dataset.

Table 1. Statistics of lyrics data by singer

	Mean (words per song)	Mean (songs)	Std. Dev
Ten candidate artists	87.8	82.3	40.60
Three final artists	154.57	130.6	8.50

Model Training. We performed fine-tuning on the KoGPT-2 model using the Korean lyrics dataset. Since the maximum sequence length offered by the baseline backbone was 1024 as a default value, we excluded the data with a longer sequence length than that of baseline. We used Adam optimizer with learning rate 3e−5, Cross Entropy for a loss function.

Generation of Stylised Lyrics. It was tokenized in units of morphemes, and in this process, KoNLPy's morpheme analyzer Kkma library was used for Korean analysis. Only nouns were extracted from the entire lyrics dataset, and 20 words with high frequency of appearance were selected. Among them, three most frequently observed words in the three artists' lyrics (e.g., "now", "love", "time") were chosen for the seed to generate lyrics. The results of the generated lyrics are attached in Appendix. We note that only Korean lyrics are presented.

3 Evaluation of Stylised Lyrics

For the precise evaluation, we measured two essential metrics: (i) style similarity between each artist's original lyrics, (ii) style similarity between each artist's generated lyrics. As a complement to these, we additionally measured the proportion of Foreign words in both original and generated lyrics of each artist. We assume that the frequency of foreign words in K-Pop could also indirectly expose the artists' writing style.

We adopted BLEU [12] and BERT [13] score to measure the degree of similarity. As widely adopted in machine translation, these scoring models based upon natural language understanding (NLG) compute the similarity between machine-translated text and human-created reference text with regards to syntactic (with BLEU) and semantics (with BERT).

In this paper, we assume the generation of stylised lyrics as a problem of machine translation - a deep generative model can learn the artist-created lyrics and translate them into the stylised lyrics. Thus, the relationship between machine-translated and human-created reference text could be equivalent to that between machine-generated lyrics and artist-created original lyrics. Within this context, we used those deep-learning based scoring models for the proposal of an automated computational evaluation framework.

To conduct the evaluation, we first generated the stylised lyrics with the aforementioned word 'time', one of most frequently observed in the artists' songs. We then computed the BLEU score using the stylised lyrics (we call it the generated later) and artist-created lyrics (we call it the reference later). For the reference lyrics, we sampled them from the entire training dataset which contain the largest number of a keyword 'time' in the lyrics. For the generated lyrics, we sampled five lyrics from the entire set of generated lyrics with the keywork 'time'. We afterwards performed the automated evaluations, BLEU and BERT score test, with the generated and the references for 25 times (5 reference lyrics x 5 generated lyrics).

For your information, we briefly introduce the genre of each artist: (i) IU is an overground singer-songwriter in the Ballad genre, (ii) Sunwoojunga (we call it SJ later) is an underground singer-songwriter actively performing an indie music, and (iii) MonstaX (we call it MX later) is an overground Korean Idol boy band performing Hiphop and R&B.

Syntactic Similarity Between Lyrics of Each Artist. BLEU score is a means for an automatic evaluation dominantly used in neural translation. It measures the degree of similarity between the translated text by generative models and the reference text translated by humans. In other words, BLEU score computed the similarity between machine-translated and human-translated text. BLEU particularly focuses on the structural similarity between translated and reference text. It compares consecutive segments of the machine translated text with those in the human-created reference text. It afterwards counts the number of matches between the two texts. The larger the number of matches is, the higher the similarity between them is. We note that BLEU does not consider the grammatical correctness of the machine translated and its accuracy with regards to the meaning [12].

We would like to capture the similarity at the level of words, specifically the degree of concurring with the successive phrases in the two using BLEU score. We assume that the higher the BLEU score is, the more frequent use of favoured words and phrases thus the more similar the style of lyrics writing is. Figure 1 shows the result, an average of BLEU scores on 25 pairs mentioned above.

As quickly reckoned, the style of IU and SJ could be closer, but MX could not. Aligned with the quick impression, the similarity between IU and SJ is higher (0.7, approximately) than that between IU and MX (0.66) and SJ-MX (0.66) in the case of the reference. The same pattern is observed in the case of generated.

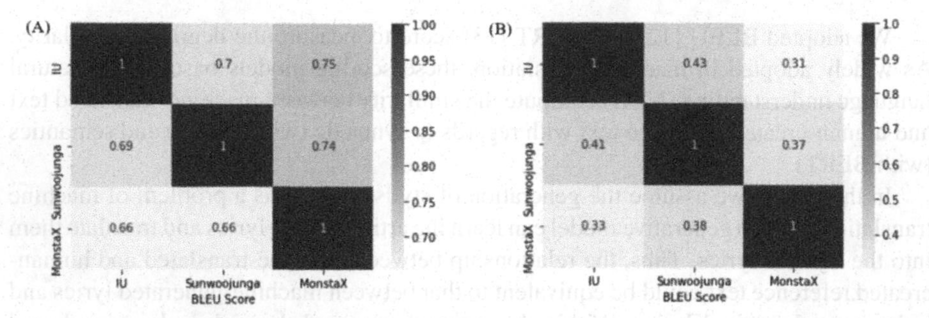

Fig. 1. Syntactic similarity (BLEU Score). (A) Syntactic similarity between the reference lyrics (in the training dataset) of each artist. (B) Syntactic similarity between the generated lyrics of each artist.

Semantic Similarity Between Lyrics of Each Artist. BERT score is an evaluation metric for the generated text. It focuses on the computation of the semantic similarity between the generated and the reference text whereas BLEU score focuses on the syntactic aspect. In other words, it does not consider the exact matching of consecutive segments (e.g., tokens, words) in the two texts, but does seek to measure the segment similarity using the contextual information. Once a pair of the generated and the reference text comes into BERT (a deep generative model specialised in NLG), it extracts the context embedding of each text and computes the cosine similarity between two embedding vectors. According to the experimental results on the machine translation tasks, BERT score correlates better with human experiment [10].

As BERT score is powerful to capture the semantic similarity between two given texts, we assume that BERT score between the generated lyrics and the reference lyrics successfully reflects the concordance with respect to the meaning behind the songs. Here, the higher BERT score between the two lyrics is, the more similar meaning the two lyrics have. BERT Score for the lyrics was measured with the same dataset as used in the BLEU score test. Figure 2 shows the result, an average of BERT scores for 25 pairs described above.

In both cases, the highest BERT score is observed when comparing the same artist's lyrics in the case of the reference lyrics. This pattern is consistent with all artists (Fig. 2-A). In the case of the generated lyrics, it is clear to see the same pattern as that in the former case (Fig. 2-B). These results confirm that the generated lyrics are semantically similar to the reference lyrics of each artist. In addition, generated lyrics implicitly possess the artist's own style and characteristics when comparing the generated lyrics between individual artists. MX is again distinctive to the rest of them, IU and SJ as shown nearly the same as that in the case of BLEU score.

Miscellanies. To check the similarity between the training dataset and the style of the generated lyrics further, we additionally measured the proportion of English words in the reference lyrics. As a result, it is clear that the lyrics of MX had the highest English ratio (MX 38.56%, SJ 23.06%, IU 15.43%). Likewise, MX showed again the highest English ratio (MX 26.68%, SJ 21.20%, IU 0.0%) in the case of the generated lyrics.

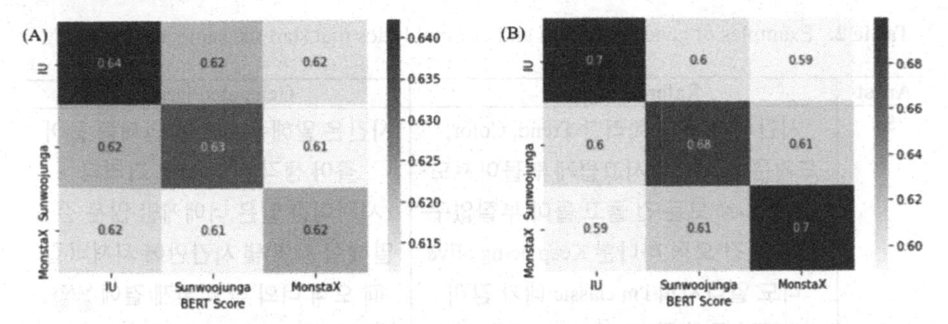

Fig. 2. Semantic similarity. (BERT Score). (A) Semantic similarity between the reference lyrics (in the training dataset) of each artist. (B) Semantic similarity between the generated lyrics of each artist.

Taken together, it is clear that the generated and the reference within the artist are highly similar while the two lyrics are less similar between the artists. These results imply that the computational means to evaluate the similarity would be able to capture the difference between stylised lyrics written by individual artists.

4 Conclusion

In this paper, we showed that it is possible to generate lyrics by taking into account the artists' style and persona, and suggest a novel method that the automated computational means for the evaluation of the stylised lyrics. Using the KoGPT 2, we generated the stylised lyrics then performed the experiments in due course. Evaluation results confirm that automatic evaluation metrics such as BLEU and BERT successfully capture the similarity between the generated and the reference of each artist. In addition, the degree of similarity of generated lyrics between individual artists are distinctive. It potentially implies that the contrasted similarity between the artists may reflect the different style and persona are embedded in the generated lyrics.

However, it is not sufficient to support the hypothesis that the evaluation of stylised lyrics only with automatic evaluation indicators such as BLEU and BERT is enough. Therefore, we plan to conduct the correlation analysis with the result of the human subjects experiments and the results we presented in the paper. Moreover, another correlation study using the result of both BLEU (Fig. 2) and BERT (Fig. 3) scores should be required to confirm the statistical significance and adequacy of the proposed evaluation framework..

Acknowledgement. This research was supported by (i) the Samsung Research Funding Center of Samsung Electronics under Project Number No. SRFC-TC1603-52, and (ii) the National Research Foundation of Korea (NRF) grant funded by the Korean government (No. 2020R1G1A1102683).

Appendix

Comparing the original lyrics with the generated lyrics, Sunwoojunga has a similar structure in which English words are inserted in the middle and the same English sentences

Table 2. Examples of generated lyrics and original lyrics that start the same word "시간".

Artist	Reference lyrics	Generated lyrics
SJ	시간이 흐르니 화려한 Trend, Color, 뜨거운 Feedback 시끄럽게 휘몰아쳐도 I don't care 모든 건 돌고 돌아 부질없구 영원한 건 오직 하나뿐 Keep being alive 너도 알다시피 I'm classic 네가 감히 오르지 못할 곳 I'm classic 누가 감히 건들지 못할 Soul	시간은 말해주는데 넌 언제쯤 울어 울어 생각은 다 하고 그러나 시간이란 말은 너에게만 많은 걸 말해줘 자면 몇 시간만에 지쳐버릴 때 오직 너와 나 이렇게 곁에 You 있어줘You're my brainshut you do look away 난 언제나 그랬어 You're my brainshut you do look away
IU	시간 좀 내요 얘기 좀 해요 한 마디 없이 꼭 다문 입술 나 그 목소리를 꼭 들을래요 다음에 보면 꼭 인사해요 또 봐요 미스터 채플린 바람에 날린 동그란 저 모자 달리기를 하나 왜 서둘러 시계톱니처럼 똑같은 모퉁일 돌아돌아 뛰네요 어디서 와서 어디로 갈까 말고도 궁금한 게 이렇게 난 많은데 잠시만 서서 얘기 좀 해요	시간의 존재도 시간과 함께 소멸되고파 기억은 사라져버려요 잠시만 눈을 떼면 세상 모든 것이 멈추죠 아프지 않은 곳으로 다시 돌아갑니다 그대 눈을 뜨면 그대 맘도 알아줄 거예요 그렇게 하루 하루를 살아요 시간이 지나도 끝나지 않은 그대 사랑도 잊혀버렸죠 이제 그만 잊어요 잊지 않을게요
MX	시간에도 내가 있었음 하고 난 또 지난 추억들을 두드려 You 처음 마주한 봄 You 따스했던 네 손 You 하나하나 모두 you Ooh 가끔 헤매여도 Ooh 빛을 잃더라도 Always be with you 너의 옆에 있을게 BEBE Monday Tuesday Wednesday Thursday Friday Saturday	시간마저 아쉬워 아쉬웠던 날들 You're so sick ofday 짧은 꿈인 줄로만 알았는데 아니던 사실 널 둘러싼 수많은 수많은 시선중 내게만 알려졌던 days 길 잃은 기분이 들어 나의 모든 걸 내려놨어 널 다시 붙잡고 싶어 Cuz I'm not sick ofday

are repeated, as shown. In the case of IU, lyrics with the same ending in '- 요' are being generated. Even in the case of Monsta X, both the original lyrics and the generated lyrics have a structure in which English words are included in the middle and the same English sentence structure is repeated. In conclusion, when the generated lyrics and original lyrics are compared, structurally similar lyrics are generated (see Table 2).

References

1. Shawar, B.A., Atwell, E.: Different measurement metrics to evaluate a chatbot system. In: Proceedings of the Workshop on Bridging the Gap: Academic and Industrial Research in Dialog Technologies, pp. 89–96 (2007)
2. Nagarhalli, T.P., Vaze, V., Rana, N.K.: A review of current trends in the development of chatbot systems. In: 2020 6th International Conference on Advanced Computing and Communication Systems (ICACCS), pp. 706–710. IEEE (2020)
3. Report of chatbot market size. https://www.grandviewresearch.com/industry-analysis/chatbot-market
4. Chandel, S., Yuying, Y., Yujie, G., Razaque, A., Yang, G.: Chatbot: efficient and utility-based platform. In: Arai, K., Kapoor, S., Bhatia, R. (eds.) Intelligent Computing, vol. 858, pp. 109–122. Springer, Cham (2019). https://doi.org/10.1007/978-3-030-01174-1_9
5. Pradhan, A., Lazar, A.: Hey Google, do you have a personality? Designing personality and personas for conversational agents. In: CUI 2021–3rd Conference on Conversational User Interfaces, pp. 1–4 (2021)
6. Zheng, Y., et al.: A pre-training based personalized dialogue generation model with persona-sparse data. In: Proceedings of the AAAI Conference on Artificial Intelligence, pp. 9693–9700 (2020)
7. Lee, S.K., Yun, J.Y.: A convergence study on chatbot persona and user experience of financial service - focused on loan service. Korean Soc. Sci. Art **37**(4), 257–267 (2019)
8. KoGPT2. https://github.com/SKT-AI/KoGPT2
9. Radford, A., et al.: Language models are unsupervised multitask learners. OpenAI Blog **1**(8), 9 (2019)
10. Devlin, J., et al.: Bert: pre-training of deep bidirectional transformers for language understanding. arXiv preprint https://arxiv.org/abs/1810.04805 (2018)
11. Hong, H.-J., Kim, S.-H., Lee, J.H.: Engineering a deep-generative model for lyric writing based upon a style transfer of song writers. In: Proceedings of the Korea Information Processing Society Conference. Korea Information Processing Society, pp. 741–744 (2021)
12. Papineni, K., et al.: Bleu: a method for automatic evaluation of machine translation. In: Proceedings of the 40th annual meeting of the Association for Computational Linguistics, pp. 311–318 (2002)
13. Zhang, T., et al.: Bertscore: evaluating text generation with bert. arXiv preprint arXiv:1904.09675 https://arxiv.org/abs/1904.09675 (2019)

GWD: Graded Word Drop Model for When Type Questions for Hindi QA

Vani, Sumit Singh, Puja Burman, Anmol Jain, and Uma Shanker Tiwary(✉)

Indian Institute of Information Technology, Allahabad, India
vanichandna@gmail.com, sumitrsch@gmail.com, pujaburman30@gmail.com, jainanmol84@gmail.com, ustiwary@gmail.com

Abstract. This paper proposes a preprocessing methodology, namely, Graded Word Drop (GWD) and its algorithm as a solution to the problem of bigger contexts in Extractive Question Answering for Hindi language, focusing mainly on "When" (कब) type questions. This paper discusses in detail the problems associated with bigger contexts, such as increased prediction times, misleading text as a part of bigger context etc. It then discusses three methodologies, viz., Boolean Model and two new proposed methodologies of Word Drop. We used cross-linguality of transformer models, mBERT, XLM-RoBERTa and MuRIL and fine-tuned them using SQuAD dataset. We used 84 Hindi-language, "When" (कब) type questions from chaii (Challenge in AI for India) dataset for evaluation. The GWD preprocessed text gave improvement over non-preprocessed results in terms of both accuracy and F1-score and achieved 53.57%, 38.09%, 55.95% accuracy, and 63.21, 68.37 and 67.09 F1-score in mBERT, XLM-RoBERTa and MuRIL respectively and improved prediction times by five fold in all these models.

Keywords: Graded Word Drop · Static Word Drop · mBERT · XLM-RoBERTa · MuRIL · SQuAD

1 Introduction

The extractive Question Answering involves a context, which is a passage, and a question, for which the answer is expected to be retrieved from the context. Such type of Question Answering is termed, Extractive Question Answering.

In supervised learning, the training dataset also involved answer text, which is the answer to the question from the passage.

Question Answering domain has shown immense progress over the last decade, with models like LSTM (Long short-term memory), RNN (Recurrent neural network), BiDAF (Bidirectional Attention Flow), transformer-based models like BERT (Bidirectional Encoder Representations from Transformers) [3] and its variants and datasets like SQuAD (Stanford Question Answering Dataset) [10,11], WikiQA, Natural Questions, XQuAD (Cross-lingual Question

H. Zaynidinov et al. (Eds.): IHCI 2022, LNCS 13741, pp. 140–153, 2023.
https://doi.org/10.1007/978-3-031-27199-1_15

Answering Dataset), MLQA (Multilingual Question Answering Dataset), TyDi (Topologically Diverse) etc. The progress has been such that it has exceeded human performance in such problems.

This progress has been, however, in the English language. As per Cochraine, "Only about 6% of the world's population are native English speakers, and 75% of people don't speak English at all" [13].

In this paper, we focus on a resource-scarce language, Hindi and propose a novel preprocessing model, GWD or Graded Word Drop Model which, when applied to the dataset, helps improve the accuracy of Hindi models. We applied the GWD Model on the evaluation dataset and gave the preprocessed data as input to the top three models of cross-language multilingual capacity, viz., mBERT [3], XLM-RoBERTa [2] and MuRIL [8]. We utilised the cross lingual capability of the models by fine-tuning the same on SQuAD v1 dataset. We shortlisted 84 "When" (कब) type questions from chaii [4] dataset for evaluation. All the three models, mBERT, XLM-RoBERTa and MuRIL, showed an improvement in both accuracy and F1-score by 23.81%, 1.19%, 3.57%, and 21.20, 6.96 and 1.17, respectively and prediction times by five fold, over non-GWD preprocessed context.

In the following sections, we discuss major Question Answering work done for Indian languages prior, the challenges related to the existing Question Answering BERT model, observations we had over fine-tuning of BERT model, the solution offered by using the proposed GWD Model and the results.

The code to reproduce the results of all the three preprocessing models is available at https://github.com/vanichandna/Word-Drop. The fine-tuned mBERT, XLM-RoBERTa and MuRIL models for the testing purpose can be from HuggingFace using "vanichandna/bert-base-multilingual-cased-finetuned-squadv1", "vanichandna/xlm-roberta-finetuned-squad" and "vanichandna/muril-finetuned-squadv1" as model and tokenizer respectively.

2 Related Work

The first attempts in Indian Question Answering models dated a decade back and started with pipeline based approaches. There are no benchmark datasets, and most attempts have been monolingual.

The initial Hindi Question Answering System [12] used a pipeline-based approach. The QA system was divided into five components: Query Preprocessing, Query Generation, Database Search, Related Document and Answer Extraction. They made their database of 60 Question Answer pairs, 15 pairs each for four different types of questions, when (कब), where (कहाँ), how many (कितने), what time (किस समय). The answer extraction algorithms were rule-based and varied according to the question type. Average accuracy of 68% was achieved. They used static data for evaluation.

Another initial attempt for an Indian language was made for Bengali Question Answering System [1]. They also followed pipelined-based approach and had incorporated three components in their system, question analysis, sentence extraction and answer extraction. As opposed to [12], they avoided rule-based approach and used similarity-based answer extraction and subsequent ranking. Due to a lack of any existing resources for the language, they built their own corpus of 184 factoid question-answer pairs using 14 Wikipedia articles. They annotated the same by employing 20 Bengali language experts.

As time passed, more advanced algorithms were attempted to solve the Question Answering problem for the Hindi language.

[6] proposed a multilingual QA system using both English and Hindi, which is a resource-scarce language and English, a resource-rich language. This was done to be able to utilise resources of the English language. This model had three components, question classification, document(s)/passage(s) extraction and appropriate answer(s) extraction. They used Convolutional Neural Networks (CNNs) and Recurrent Neural Networks (RNNs) for question classification, Lucene4 for document retrieval, term-coverage and proximity-score for answer extraction. They created 5,495 question-answer pairs with the questions and answers, both in English and Hindi. They incorporated two types of questions, factoid and short descriptive types. The answers were categorized into 6 coarse and 63 finer types from 500 curated articles in six domains from the web.

[5] is one of the end-to-end models which uses a deep neural network model for multilingual QA, in which question and answer can be either in Hindi or English or both. They also used word embeddings to create a language-independent graph-based snippet generation algorithm. They used bidirectional Gated Recurrent Units (GRU) to learn the shared representation of questions from different languages. They used MQA dataset and translated 18,454 data rows from SQuAD to Hindi for their purpose.

Transformer-based models like mBERT (multilingual Bidirectional Encoder Representations from Transformers) have lately been the focus of Question Answering problem as one of the best choices. These are multilingual cross-trained models. They can be utilized for various NLP (Natural Language Processing) based tasks by fine-tuning them on the specific tasks. They offer unified architecture over different tasks.

Other variants of BERT like XLM-RoBERTa (Cross-lingual Language Model - Robustly Optimized BERT) and MuRIL (Multilingual Representations for Indian Languages) have shown improvement over mBERT model in various tasks. mBERT and XLM-RoBERTa are pretrained on the top 104 languages of the world and 100 languages, respectively. Whereas MuRIL is pretrained on 17 languages, out of which 1 is English and the remaining 16 are native Indian languages.

One of the unique datasets is Stanford Question Answering Dataset (SQuAD). It is a reading comprehension dataset. There are two versions of SQuAD, SQuAD1.1 and SQuAD2.0. Wikipedia was used to collect contextual data. Questions and answering pairs were crowd-sourced accordingly. SQuAD 1.1, contains 107,785 question-answer pairs on 536 articles. SQuAD2.0 combines question-answering pairs of SQuAD1.1 with 53,775 unanswerable questions from the same paragraphs.

chaii (Challenge in AI for India) was another dataset created by Google Research India as a part of a competition in Kaggle. It has 1,115 question-answer pairs. Out of which, 747 are in Hindi and 368 are in Tamil.

3 Challenges

1. A lot of Question Answering Systems exist for resource-rich languages. There is a gap in dataset availability and hence, the fine-tuned models for low-resource languages like Hindi.
2. BERT is state-of-the-art architecture for multilingual extractive Question Answering problem. However, there is always some "specific" preprocessing required for each specific problem to get accurate results.
3. In this "specific" preprocessing step, certain parameters are adjusted, and the model is trained and evaluated. To further improve the results, the pre-processing parameters need to be changed again and training and evaluation process are repeated. This makes the fine-tuning step costly, using up a lot of time and compute resources and resulting in this step being non-environment friendly.
4. Even though BERT and its variants have been pre trained heavily on big corpora, they take long times to predict individual answers for extractive type QAs.

4 Observation While Fine-Tuning Bert Model for Question Answering Problem

Figure 1(a) contains a context of 776 words or 1435 tokens using Muril tokenizer. It took MuRIL model, fine-tuned on SQUAD, more than 2 s to predict the answer. The answer extracted, however, is incorrect. When the same context was reduced, keeping every other parameter the same, to one of 161 words or 286 tokens, we got the correct answer in less than half a second.

We observed two things here:

1. Bigger the context, the lower the prediction speed.
   Question Answering BERT-based models can take a big context as input but cannot process it all simultaneously. These models can process context sequentially, taking a maximum of 512 tokens in a single sequence. If the number of tokens exceeds the maximum limit, it results in overflow and truncation. Hence, bigger contexts which are truncated, are considered sequentially. In the above case, the bigger context was tokenised into 1435 tokens. BERT had to process them in ceil(1435/(512 − x − y)) sequences, where x is the length of delimiter tokens, and y is the length of question tokens. However, BERT processed them in a single sequence in the case of the smaller context, containing 286 tokens. So, the processing speed up is expected to be four fold. In practice, as we saw, the speed up is more than 4 times. This could be attributed to the overhead experienced in handling overflows.

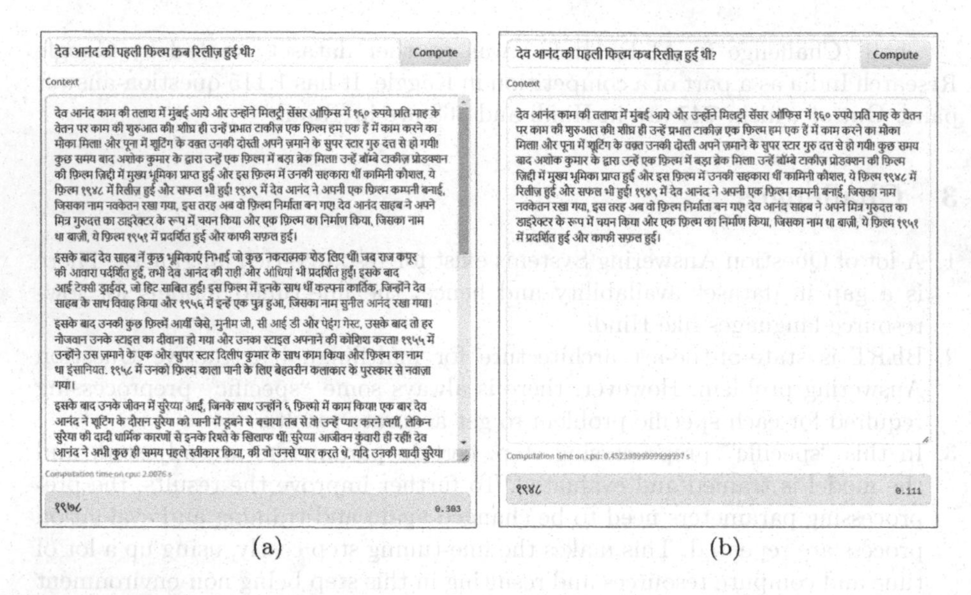

(a) (b)

Fig. 1. QA example with (a) long context and (b) short context

2. Bigger contexts result in lower accuracy than smaller contexts, provided both the contexts contain the expected answer.
   These are the two possible causes:
   (a) Best matched answer might be misleading in bigger contexts. This problem increases with an increase in the size of the context.
   (b) In the case of larger contexts, like those taken from Wikipedia, when tokenised using BERT Tokeniser, results in 4–5k tokens on average, they may range from 1.5k–25k tokens, resulting in overflows and truncation. So it might be possible that the answer gets split between two adjacent sequences. Hence, the correct answer is not matched. "Stride" might help, but its static value might not support every question.

5 Methodology

The solution to the problem of bigger contexts is to reduce the context size. We evaluated an existing model, the Boolean model, for this purpose. We proposed two variations of WD or Word Drop Model, SWD (Static Word Drop) and Graded Word Drop Model. The input to these models is 'question' and 'context'. The input 'question' needs to be processed before feeding to these models. Hence, there are three predominantly identified modules:

1. POS Extractor
2. Synonym Generator
3. Context Conciser Model

5.1 POS Extractor

POS Extractor extracts the crucial words from the question and ignores the others. For this purpose, stopwords [7] are identified and removed from the question. The output of this module is a list of POS words (Fig. 2).

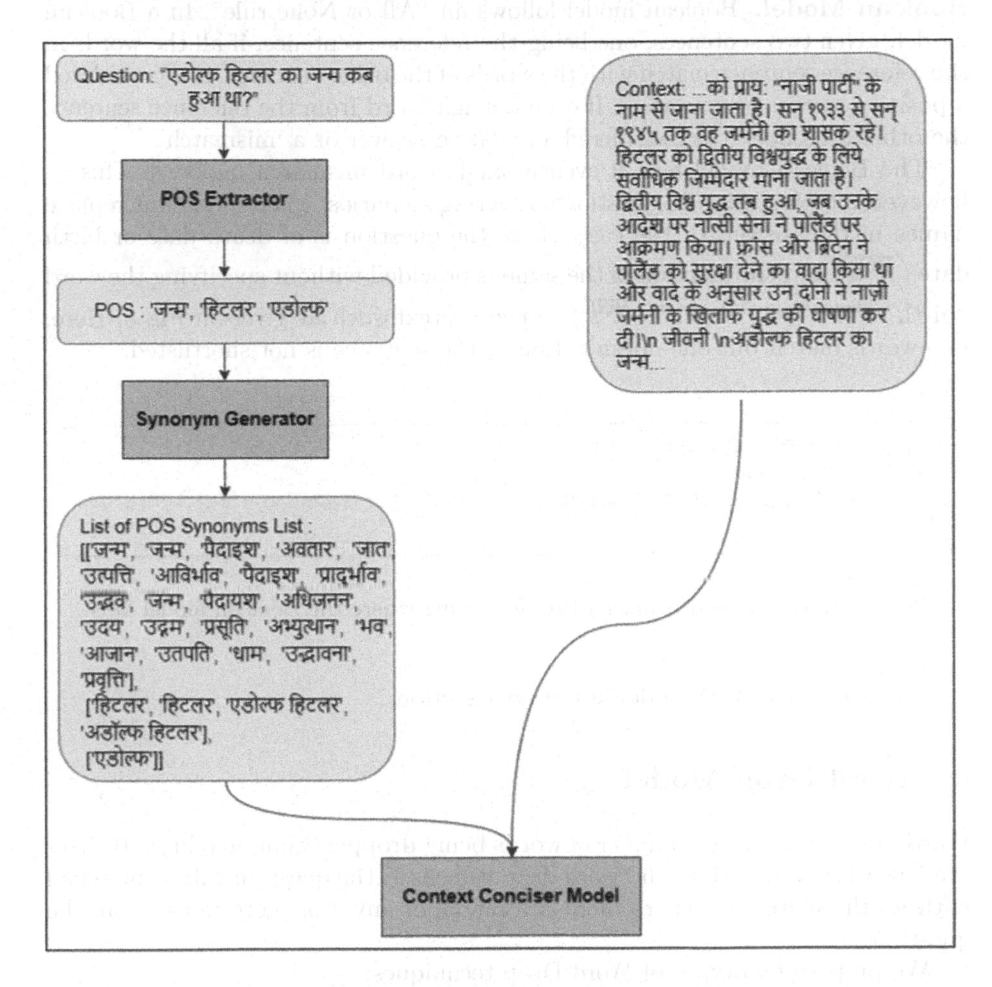

Fig. 2. Preprocessing Methodology

5.2 Synonym Generator

Synonym Generator generates synonyms of words extracted in the previous step using indo-wordnet [9]. These synonyms are stored along with the actual word in the form of lists. The input of this step is a list of unique words, and the output of this step is a list of lists, where each sublist contains one of the input words and its synonyms. This nested list is given as input to either of the three models.

5.3 Context Conciser Model

Boolean Model. Boolean model follows an "All or None rule". In a Boolean model, given two sentences, one being the reference sentence, if all the words in the reference sentence match with the words of the other sentence, it is considered a positive answer or a 'match'. If even a single word from the reference sentence the other sentence, it is considered a negative answer or a 'mismatch'.

The Boolean model fails if even a single word mismatch happens. This is, however, a typical case. In question answering scenarios, where pronouns replace names in the answer, or the case where the question is of death date or birth date ('मृत्यु') and the mention of the same is provided without specifying the word "birth" ('जन्म') or "death" ('मृत्यु') in the context such as given in Fig 3. Here, two words match but one doesn't. Hence, the sentence is not shortlisted.

Question: देव आनंद का जन्म कब हुआ था?
POS: 'देव', 'आनंद', 'जन्म'
Context: देव आनन्द उर्फ़ धरमदेव पिशोरीमल आनंद (२६ सितंबर १९२३- ३ दिसम्बर २०११) हिन्दी फ़िल्मों के एक प्रसिद्ध अभिनेता थे।

Fig. 3. An example for Boolean Pre-processing where the Boolean model fails

We discuss the WD model in the next section.

6 Word Drop Model

Word Drop refers to the number of words being dropped from matching. If there are 5 words in a question, and word drop value is 1, the question will be matched with all the sentences, where there is a match of any 4 or more words from the question.

We propose two types of Word Drop techniques:

1. Static Word Drop or SWD
2. Graded Word Drop or GWD

6.1 SWD: Static Word Drop

The number of words to be dropped in each question, to look for the answer in the context, is decided at the beginning and they will remain the same for all questions. As we increase the number of word drops, the probability of getting the correct answer in the new, more specific and shorter context increases.

But the problem is, as we increase the number of word drops, it also increases the junk in our newer context in cases where the answer is found at lower word drop. Imagine the situation for the cases where we found answers in 0-word drop, but to increase accuracy, we went up to 4-word drops.

In the example given in Fig. 4, in (a), the answer is found in context, summarized using 0 word drop. But when we increased the number of word drops to 3 in (b), the junk in the context increased. To handle this problem, we use the Graded Word Drop or GWD.

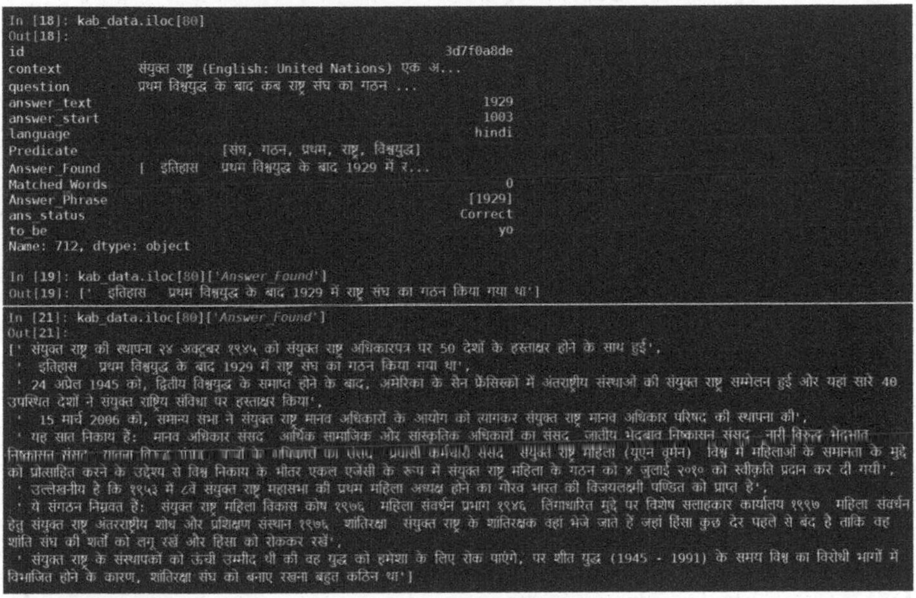

```
In [18]: kab_data.iloc[80]
Out[18]:
id                                                        3d7f0a8de
context          संयुक्त राष्ट्र (English: United Nations) एक अ...
question         प्रथम विश्वयुद्ध के बाद कब राष्ट्र संघ का गठन ...
answer_text                                                    1929
answer_start                                                   1003
language                                                      hindi
Predicate                   [संघ, गठन, प्रथम, राष्ट्र, विश्वयुद्ध]
Answer_Found     [ इतिहास  प्रथम विश्वयुद्ध के बाद 1929 में र...
Matched Words                                                     0
Answer_Phrase                                                [1929]
ans_status                                                  Correct
to_be                                                            yo
Name: 712, dtype: object

In [19]: kab_data.iloc[80]['Answer_Found']
Out[19]: ['  इतिहास  प्रथम विश्वयुद्ध के बाद 1929 में राष्ट्र संघ का गठन किया गया था']
```

```
In [21]: kab_data.iloc[80]['Answer_Found']
Out[21]:
[' संयुक्त राष्ट्र की स्थापना २४ अक्टूबर १९४५ को संयुक्त राष्ट्र अधिकारपत्र पर 50 देशों के हस्ताक्षर होने के साथ हुई',
 '  इतिहास  प्रथम विश्वयुद्ध के बाद 1929 में राष्ट्र संघ का गठन किया गया था',
 ' 24 अप्रेल 1945 को, द्वितीय विश्वयुद्ध के समाप्त होने के बाद, अमेरिका के सैन फ्रेंसिस्को में अंतराष्ट्रीय संस्थाओं की संयुक्त राष्ट्र सम्मेलन हुई और यहां सारे 40 उपस्थित देशों ने संयुक्त राष्ट्रिय संविधा पर हस्ताक्षर किया',
 '  15 मार्च 2006 को, समान्य सभा ने संयुक्त राष्ट्र मानव अधिकारों के आयोग को त्यागकर संयुक्त राष्ट्र मानव अधिकार परिषद की स्थापना की',
 ' यह सात निकाय हैं:  मानव अधिकार संसद  आर्थिक सामाजिक और सांस्कृतिक अधिकारों का संसद  जातीय भेदबाव निष्कासन संसद  नारी विरुद्ध भेदभाव निष्कासन संसद  [illegible] प्रवासी कर्मचारी संसद  संयुक्त राष्ट्र महिला (यूएन वूमेन)  विश्व में महिलाओं के समानता के मुद्दे को प्रोत्साहित करने के उद्देश्य से विश्व निकाय के भीतर एकल एजेंसी के रूप में संयुक्त राष्ट्र महिला के गठन को ४ जुलाई २०१० को स्वीकृति प्रदान कर दी गयी',
 ' उल्लेखनीय है कि १९५३ में ८वें संयुक्त राष्ट्र महासभा की प्रथम महिला अध्यक्ष होने का गौरव भारत की विजयलक्ष्मी पण्डित को प्राप्त है',
 ' ये संगठन निम्नवत हैं:  संयुक्त राष्ट्र महिला विकास कोष १९७६  महिला संवर्धन प्रभाग १९४६  लिंगाधारित मुद्दे पर विशेष सलाहकार कार्यालय १९९७  महिला संवर्धन हेतु संयुक्त राष्ट्र अंतरराष्ट्रीय शोध और प्रशिक्षण संस्थान १९७६  शांतिरक्षा  संयुक्त राष्ट्र के शांतिरक्षक वहां भेजे जाते हैं जहां हिंसा कुछ देर पहले से बंद है ताकि वह शांति संघ की शर्तों को लगू रखें और हिंसा को रोककर रखें',
 ' संयुक्त राष्ट्र के संस्थापकों को ऊंची उम्मीद थी की वह युद्ध को हमेशा के लिए रोक पाएंगे, पर शीत युद्ध (1945 - 1991) के समय विश्व का विरोधी भागों में विभाजित होने के कारण, शांतिरक्षा संघ को बनाए रखना बहुत कठिन था']
```

Fig. 4. Output after application of SWD (a)when WD = 0 and (b)when WD = 3

6.2 GWD: Graded Word Drop Model

The number of words to be dropped in each question to look for the answer in the context is decided dynamically at the runtime. In this case, the process starts with 0-word drops or Boolean model, and if there is no match of question with any of the sentences in the context, the number of word drop increases only for that particular Question-Context pair.

This is a significant improvement over Static Word Drop as accuracy with increase in word drop increases, avoiding the junk added to the "already answered questions".

Algorithm for Graded Word Drop. The inputs to the Algorithm are:

1. Context, CTXT

2. Set of POS synonyms, POS_SYN, this is given as a list of lists.
3. Number of words which need to be dropped from matching, UW. This value is fixed at 0 for the Graded word drop.
4. boolean FOUND

The context is split up line by line, and each line is compared with the question for matching. Suppose the line doesn't contain any date or time information, it is discarded. Otherwise, the number of matched words are counted. Any word in the line must match any synonym of a sublist to be considered as a single match. A line is added as a possible answer context if the sum of the number of matches and number of expected dropped words is greater than or equal to the number of POS_SYN sublists (Fig. 5).

```
Algorithm Graded_Word_Drop(Context CTXT , Set_of_Pos_synonyms
POS_SYN, unmatched_words UW)
{
    ans_lines = []

    lines = split CTXT into lines

    for each line in lines

    {

        if the line does not contain date/time value

            continue

        match = count the number of words that match atleast one
                synonym of each word in the POS_SYN list

        if match + UW >= len(POS_SYN)

            then ans_lines += line

    }

    if ans_lines is empty

        then call Graded_Word_Drop(CTXT, POS_SYN, UW + 1)

    else return ans_lines
}
```

Fig. 5. Algorithm for GWD: Graded Word Drop

After a complete run-through of all the context lines, if we don't find even a single matched line, then number of words to be dropped is incremented, and the algorithm is repeated afresh. Otherwise, the list of lines is returned.

An example of Context Reduction using GWD Model is shown in Fig. 6.

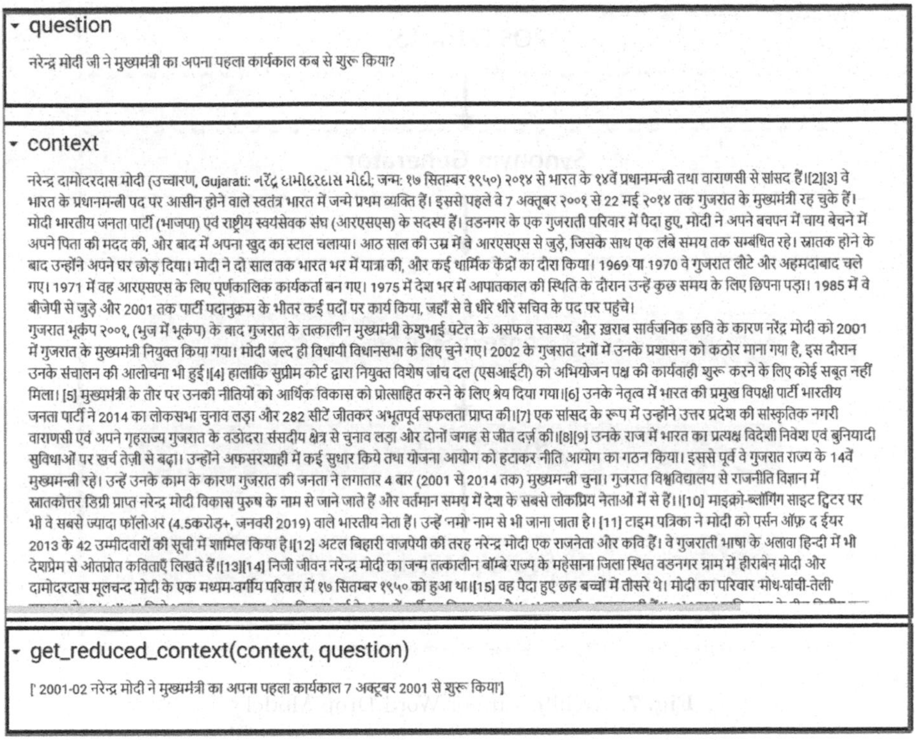

question

नरेन्द्र मोदी जी ने मुख्यमंत्री का अपना पहला कार्यकाल कब से शुरू किया?

context

नरेन्द्र दामोदरदास मोदी (उच्चारण, Gujarati: નરેંદ્ર દામોદરદાસ મોદી; जन्म: १७ सितम्बर १९५०) २०१४ से भारत के १४वें प्रधानमन्त्री तथा वाराणसी से सांसद हैं।[2][3] वे भारत के प्रधानमन्त्री पद पर आसीन होने वाले स्वतंत्र भारत में जन्मे प्रथम व्यक्ति हैं। इससे पहले वे 7 अक्तूबर २००१ से 22 मई २०१४ तक गुजरात के मुख्यमंत्री रह चुके हैं। मोदी भारतीय जनता पार्टी (भाजपा) एवं राष्ट्रीय स्वयंसेवक संघ (आरएसएस) के सदस्य हैं। वडनगर के एक गुजराती परिवार में पैदा हुए, मोदी ने अपने बचपन में चाय बेचने में अपने पिता की मदद की, और बाद में अपना खुद का स्टाल चलाया। आठ साल की उम्र में वे आरएसएस से जुड़े, जिसके साथ एक लंबे समय तक सम्बंधित रहे। स्नातक होने के बाद उन्होंने अपने घर छोड़ दिया। मोदी ने दो साल तक भारत भर में यात्रा की, और कई धार्मिक केंद्रों का दौरा किया। 1969 या 1970 वे गुजरात लौटे और अहमदाबाद चले गए। 1971 में वह आरएसएस के लिए पूर्णकालिक कार्यकर्ता बन गए। 1975 में देश भर में आपातकाल की स्थिति के दौरान उन्हें कुछ समय के लिए छिपना पड़ा। 1985 में वे बीजेपी से जुड़े और 2001 तक पार्टी पदानुक्रम के भीतर कई पदों पर कार्य किया, जहाँ से वे धीरे धीरे सचिव के पद पर पहुंचे।
गुजरात भूकंप २००१, (भुज में भूकंप) के बाद गुजरात के तत्कालीन मुख्यमंत्री केशुभाई पटेल के असफल स्वास्थ्य और ख़राब सार्वजनिक छवि के कारण नरेंद्र मोदी को 2001 में गुजरात के मुख्यमंत्री नियुक्त किया गया। मोदी जल्द ही विधायी विधानसभा के लिए चुने गए। 2002 के गुजरात दंगों में उनके प्रशासन को कठोर माना गया है, इस दौरान उनके संचालन की आलोचना भी हुई।[4] हालांकि सुप्रीम कोर्ट द्वारा नियुक्त विशेष जांच दल (एसआईटी) को अभियोजन पक्ष की कार्यवाही शुरू करने के लिए कोई सबूत नहीं मिला। [5] मुख्यमंत्री के तौर पर उनकी नीतियों को आर्थिक विकास को प्रोत्साहित करने के लिए श्रेय दिया गया।[6] उनके नेतृत्व में भारत की प्रमुख विपक्षी पार्टी भारतीय जनता पार्टी ने 2014 का लोकसभा चुनाव लड़ा और 282 सीटें जीतकर अभूतपूर्व सफलता प्राप्त की।[7] एक सांसद के रूप में उन्होंने उत्तर प्रदेश की सांस्कृतिक नगरी वाराणसी एवं अपने गृहराज्य गुजरात के वडोदरा संसदीय क्षेत्र से चुनाव लड़ा और दोनों जगह से जीत दर्ज़ की।[8][9] उनके राज में भारत का प्रत्यक्ष विदेशी निवेश एवं बुनियादी सुविधाओं पर खर्च तेज़ी से बढ़ा। उन्होंने अफसरशाही में कई सुधार किये तथा योजना आयोग को हटाकर नीति आयोग का गठन किया। इससे पूर्व वे गुजरात राज्य के 14वें मुख्यमन्त्री रहे। उन्हें उनके काम के कारण गुजरात की जनता ने लगातार 4 बार (2001 से 2014 तक) मुख्यमन्त्री चुना। गुजरात विश्वविद्यालय से राजनीति विज्ञान में स्नातकोत्तर डिग्री प्राप्त नरेन्द्र मोदी विकास पुरुष के नाम से जाने जाते हैं और वर्तमान समय में देश के सबसे लोकप्रिय नेताओं में से हैं।।[10] माइक्रो-ब्लॉगिंग साइट ट्विटर पर भी वे सबसे ज्यादा फॉलोअर (4.5करोड़+, जनवरी 2019) वाले भारतीय नेता हैं। उन्हें 'नमो' नाम से भी जाना जाता है। [11] टाइम पत्रिका ने मोदी को पर्सन ऑफ़ द ईयर 2013 के 42 उम्मीदवारों की सूची में शामिल किया है।[12] अटल बिहारी वाजपेयी की तरह नरेन्द्र मोदी एक राजनेता और कवि हैं। वे गुजराती भाषा के अलावा हिन्दी में भी देशप्रेम से ओतप्रोत कविताएँ लिखते हैं।[13][14] निजी जीवन नरेन्द्र मोदी का जन्म तत्कालीन बॉम्बे राज्य के महेसाना जिला स्थित वडनगर ग्राम में हीराबेन मोदी और दामोदरदास मूलचन्द मोदी के एक मध्यम-वर्गीय परिवार में १७ सितम्बर १९५० को हुआ था।[15] वह पैदा हुए छह बच्चों में तीसरे थे। मोदी का परिवार 'मोध-घांची-तेली'

get_reduced_context(context, question)

[' 2001-02 नरेन्द्र मोदी ने मुख्यमंत्री का अपना पहला कार्यकाल 7 अक्टूबर 2001 से शुरू किया']

Fig. 6. Context Reduction using GWD

7 Generic Architecture of Word Drop Model

Figure 6 shows the architecture of a Word Drop Model. POS Extractor and Synonym Generator are discussed in Sect. 5.1 and 5.2 respectively.

Context Splitter splits the context based on the chosen granularity and outputs text divided into sub-texts. If the granularity is decreased, the preciseness of the context decreases and accuracy increases and vice versa. This choice is, however, context-dependent. Suppose there is high lexicographic similarity between the context and the questions. An increased granularity is a wiser option.

Date Checker checks if each sub-text being considered contains any date value. This is because this is the kind of answer we expect to "When" (**कब**) type questions.

Words Matcher, finally matches the words of the question, we received from synonym generator and sub-text to find the top match (Fig. 7).

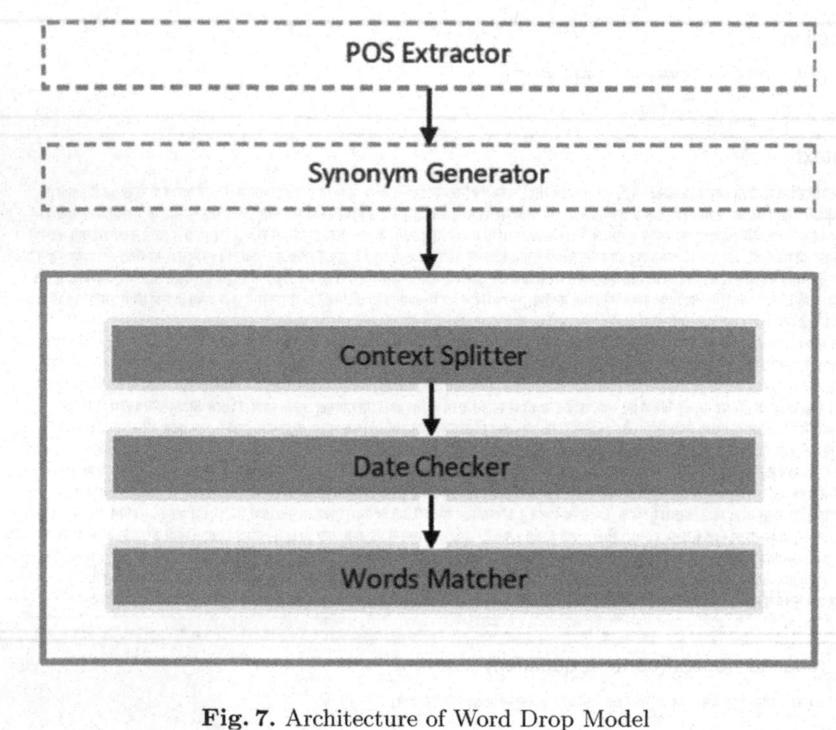

Fig. 7. Architecture of Word Drop Model

8 Experiment and Results

When 84 Question Context pairs from chaii dataset, which are of 'When' (कब) type, were fed into the SWD Model, we obtained the results shown in Table 1.

Table 1. Accuracy of Preprocessed Context when SWD is applied.

No of words dropped from matching	Accuracy	Average Number of sentences
0	26.1	1.3
1	64.2	3.6
2	80.9	7.9
3	91.6	15.9
4	95.2	21.0
5	96.4	23.1
6	97.6	24.5

We notice here that the average number of sentences or paragraphs being included in the concised context increases exponentially as the number of dropped words increase.

When 84 Question Context pairs from chaii dataset, which are of 'When' (कब) type, were fed into the GWD Model, we get an accuracy of 68/84 = 80.9% with an average number of sentences = 2.6 and an average number of words dropped from matching = 0.97. We used these 68 QA pairs obtained from GWD and applied state-of-the-art models to them to compare and see how reducing the context size affects the accuracy (in terms of Exact Match and F1-scores) and average prediction time.

Table 2. Comparison of Exact Match, F1-score and Average Prediction Time for 68 questions with and without GWD.

Context/Model	mBERT			XLM-RoBERTa			MuRIL		
	EM	F1	PT	EM	F1	PT	EM	F1	PT
Original Context	30.88	42.51	17.47	30.88	59.52	11.02	52.94	67.30	12.74
GWD Processed Context	61.76	70.24	1.16	41.17	75.87	0.83	61.76	78.31	0.95

We used three BERT based models, mBERT, XLM-RoBERTa and MuRIL. We chose these three models as they are cross-trained on various languages, including Hindi and English. Hence, we could fine-tune them using resource-rich language, English. We fine-tuned these models using SQuAD v1.0 on Google Colab Pro.

mBERT: We fine-tuned bert-base-multilingual-cased over 4 epochs, using AdamWeightDecay optimizer, learning rate 1e−05.

XLM-RoBERTa: We fine-tuned xlm-roberta-base over 3 epochs, using AdamWeightDecay optimizer, learning rate 2e−05.

MuRIL: We fine-tuned muril-base-cased over 3 epochs, using AdamWeight-Decay optimizer, learning rate 1e−05.

As in Table 2, the reduced context improves the Exact Match, F1-score as well as prediction time in all the models, viz., mBERT, XLM-RoBERTa and MuRIL. There is an improvement in Exact Match of approximately 30.88%, 10.29%, 8.82% and 27.73, 16.35, 11.01 in F1-score respectively in mBERT, XLM-RoBERTa and MuRIL. Prediction time improvement has ranged from 11 fold to 17 fold.

As mentioned earlier, we first preprocessed this data using GWD, which gave an accuracy of 80.9%. This accuracy becomes the limiting factor in an overall accuracy improvement. If we consider the complete set of 84 QA pairs and try to find out the accuracy and F1-score, we still see an improvement in mBERT in terms of accuracy and F1-score of 20.24% and 16.13 respectively and prediction time improvement has ranged from 10 fold to 13 fold over all the models (see Table 3).

Table 3. Comparison of EM and F1-score of 84 questions with and without GWD.

Context/Model	mBERT			XLM-RoBERTa			MuRIL		
	EM	F1	PT	EM	F1	PT	EM	F1	PT
Original Context	29.76	42.01	13.70	36.90	63.14	10.31	52.38	65.92	11.37
GWD Processed Context	50.00	58.14	0.90	33.33	61.41	0.76	52.00	65.31	0.85

8.1 Improvement on Graded Word Drop Model

We further tried to improve GWD Model by adding an extra step, where we dropped exactly 1 word extra after having found the results to improve accuracy of our preprocessing model, such that the average word count is 1.92 and average lines are 7.23. Now the accuracy of our preprocessing model is 76/84 = 90.47%. This improved the overall accuracy and average prediction time in case of all the models, as shown in Table 4.

Table 4. Comparison of EM and F1-score of 84 questions with and without improved GWD.

Context/Model	mBERT			XLM-RoBERTa			MuRIL		
	EM	F1	PT	EM	F1	PT	EM	F1	PT
Original Context	29.76	42.01	16.21	36.90	61.41	10.37	52.38	65.92	10.91
GWD Processed Context	53.57	63.21	2.92	38.09	68.37	1.98	55.95	67.09	2.05

9 Conclusion

In this paper, we proposed GWD or Graded Word Drop Model for pre-processing "when" (कब) type questions in Extractive Question Answering domain for pre-processing. This preprocessing part reduces the size of the context to focus on the main parts of the context where the answer might lie. Along with that, we have also seen an application of Boolean model for the same purpose and that the Graded Word Drop model performs better inherently than the former, in a way that the Boolean model has shown an accuracy of 26.1% and the Graded Word Drop has shown an accuracy of 90.47%. The preprocessing model limits the accuracy of the overall Question Answering system. Even then, an overall improvement is seen in the completed Question Answering model by the application of the GWD Model, an increase in Exact Match by 23.81%, 1.19%, 3.57% and F1-Score by 21.20, 6.96 and 1.17 in mBERT, XLM-RoBERTa and MuRIL respectively and an overall decrease in prediction time by five times. This algorithm can be explored for other languages as well as for other question types. This also opens the gateway for further improvement of the model and any new model which could concise the context more accurately and precisely.

References

1. Banerjee, S., Naskar, S.K., Bandyopadhyay, S.: BFQA: a Bengali factoid question answering system. In: Sojka, P., Horák, A., Kopeček, I., Pala, K. (eds.) TSD 2014. LNCS (LNAI), vol. 8655, pp. 217–224. Springer, Cham (2014). https://doi.org/10.1007/978-3-319-10816-2_27
2. Conneau, A., et al.: Unsupervised cross-lingual representation learning at scale. In: Proceedings of the 58th Annual Meeting of the Association for Computational Linguistics, pp. 8440–8451 (2020). https://doi.org/10.18653/v1/2020.acl-main.747
3. Devlin, J., Chang, M.-W., Lee, K., Toutanova, K.: BERT: pre-training of deep bidirectional transformers for language understanding. arXiv (2018). https://doi.org/10.48550/ARXIV.1810.04805
4. Google Research India. Chaii - Hindi and Tamil Question Answering, 11 August 2021. Kaggle. https://www.kaggle.com/competitions/chaii-hindi-and-tamil-question-answering/data
5. Gupta, D., Ekbal, A., Bhattacharyya, P.: A deep neural network framework for English hindi question answering. ACM Trans. Asian Low-Resour. Lang. Inf. Processi. **19**(2), 1–22 (2020). https://doi.org/10.1145/3359988
6. Gupta, D., Kumari, S., Ekbal, A., Bhattacharyya, P.: MMQA: a multi-domain multi-lingual question-answering framework for English and Hindi. In: Proceedings of the Eleventh International Conference on Language Resources and Evaluation, LREC 2018 (2018)
7. Jha, V., N, M., Shenoy, P.D., R, V.K.: Hindi language stop words list. Mendeley Data, 4 December 2018. https://data.mendeley.com/datasets/bsr3frvvjc/1
8. Khanuja, S., et al.: MuRIL: multilingual representations for Indian languages. arXiv (2021). https://doi.org/10.48550/ARXIV, https://arxiv.org/pdf/2103.10730.pdf
9. Panjwani, R., Kanojia, D., Bhattacharyya, P.: pyiwn: a Python based API to access Indian language WordNets. In: Proceedings of the 9th Global Wordnet Conference, pp. 378–383 (2018). https://aclanthology.org/2018.gwc-1.47
10. Rajpurkar, P., Jia, R., Liang, P.: Know what you don't know: unanswerable questions for SQuAD. In: Proceedings of the 56th Annual Meeting of the Association for Computational Linguistics (Volume 2: Short Papers), Melbourne, Australia, pp. 784–789. Association for Computational Linguistics (2018)
11. Rajpurkar, P., Zhang, J., Lopyrev, K., and Liang, P.: SQuAD: 100,000+ questions for machine comprehension of text. In: Proceedings of the 2016 Conference on Empirical Methods in Natural Language Processing, Austin, Texas, pp. 2383–2392. Association for Computational Linguistics (2016)
12. Sahu, S., Vasnik, N., Roy, D.: Prashnottar: a Hindi question answering system. Int. J. Comput. Sci. Inf. Technol. **4**, 149–158 (2012). https://doi.org/10.5121/ijcsit.2012.4213
13. The Cochrane Collaboration. Cochrane evidence in different languages. Cochrane, 24 February 2021. https://www.cochrane.org/news/cochrane-evidence-different-languages

Masked Face Recognition Model with Explainable AI

Hyeon Ah Sung[1], Seunghyun Kim[1], and Eui Chul Lee[2](✉)

[1] Department of AI & Informatics, Graduate School, Sangmyung University, Seoul, Republic of Korea
202132040@sangmyung.kr

[2] Department of Human-Centered Artificial Intelligence, Sangmyung University, Seoul, Republic of Korea
eclee@smu.ac.kr

Abstract. Due to the recent COVID-19 pandemic, people tend to wear masks indoors and outdoors. Therefore, systems with face recognition, such as FaceID, showed a tendency of decline in accuracy. Consequently, many studies and research were held to improve the accuracy of the recognition system between masked faces. Most of them targeted to enhance dataset and restrained the models to get reasonable accuracies. However, not much research was held to explain the reasons for the enhancement of the accuracy. Therefore, we focused on finding an explainable reason for the improvement of the model's accuracy. First, we could see that the accuracy has actually increased after training with a masked dataset by 12.86%. Then we applied Explainable AI (XAI) to see whether the model has really focused on the regions of interest. Our approach showed through the generated heatmaps that difference in the data of the training models make difference in range of focus.

Keywords: Face Recognition · Masked Face · Explainable AI · Heatmap · Convolutional Neural Network · Deep learning

1 Introduction

A lot of studies have been made to recognize people's faces. Face recognition system is also being used all around us in a various way. For example, Apple uses it for logging in to devices with FaceID, Meta uses it to tag others in Facebook and Instagram, and Airports use it for immigration. Moreover, it is also for finding lost children and criminals. However, challenges have occurred due to the outbreak of COVID-19. Everyone had to wear face masks everywhere, so it became challenging to recognize people's faces with existing methods. The accuracy has been compromised due to the face masks. Therefore, a lot of research focused to improve the accuracy of the model with covered face [1]. Most studies have focused on datasets. A new database was established by recruiting additional data wearing masks or covering the mask through landmark on existing data. Then the datasets were trained to existing deep learning models. Some of

H. Zaynidinov et al. (Eds.): IHCI 2022, LNCS 13741, pp. 154–159, 2023.
https://doi.org/10.1007/978-3-031-27199-1_16

them accomplished remarkable accuracy, which is more the 99% [2, 3]. However, due to the nature of the deep learning model, it is not easy to explain the result. In our research, we focus on explaining the model differences through the heatmaps.

2 Related Work

Explainable AI(XAI) is a field of study to explain deep learning models. The most famous XAI includes **class activation map (CAM)** [4] and **gradient-weighted CAM (Grad-CAM)** [5]. Both of them use heatmap to show which part of the image the convolutional neural network (CNN)-based deep learning model focuses on the most. However, in the case of face recognition, the approach is different from the existing classification model because the number of classes is too large and the number of images per class is little. **Explainable Face Recognition (XFR)** [6] is a technique that explains why the face was matched to the output of the face recognition system among XAIs. The XFR algorithm used in this paper uses Convolution Network to generate a heatmap that visualizes the image region that best describes the output of the network.

Improved Residual Networks (IResNet) [7] is an improved version of Residual Networks (ResNet), as its name implies. The three key differences in IResNet are the flow of information through the network layers, the residual building block, and the projection shortcut. It improves the performance without increasing the number of parameters and computational cost. Original ResNet interfered with the propagation of information over the network when the number of layers increased, but IResNet separated the network steps and facilitated information propagation using different building blocks depending on the location within each step. In addition, in traditional ResNet, projection shortcut is used when the dimension of the building block does not fit the dimension of the next building block, which is found in the main information propagation path and plays an important role because it can easily perturb the signal or cause information loss. Therefore, IResNet uses an improved projection shortcut with no parameters and significantly improved performance. Existing ResNet used bottleneck building blocks to control the number of parameters and computational costs as depth increased, but IResNet can include four times more spatial channels in the building block than existing ResNet while also controlling parameters and computational costs. Therefore, we have used the structure for baseline network to train and explain the effectiveness of the masked data.

As mentioned earlier, since face recognition is used and studied in various fields, research on loss function as well as deep learning models has been conducted. **Angular Margin Loss (ArcFace)** [8] is a novel loss function proposed to improve the softmax function in facial recognition. The method was proposed in 2018, but it is still a loss function that shows state-of-the-art (SOTA) performance in the field of face recognition. The main idea of this technique is that when the model is trained in the direction of reducing the value of the loss, the embedding values tend to converge in the center. However, this hinders learning the differences between classes. Therefore, it locates the embedding vector values on a super spherical surface and uses the angular margin as a loss value. Equation 1 shows the loss function of ArcFace.

$$L = -\frac{1}{N}\sum_{i=1}^{N}\log\left(\frac{e^{s(\cos(\theta_{y_i}+m))}}{e^{s(\cos(\theta_{y_i}+m))}+\sum_{j=1,j\neq y+i}^{n}e^{s\cos(\theta_j)}}\right) \quad (1)$$

Our study intends to compare whether various face recognition deep learning models poured out of the COVID-19 pandemic situation are learning the difference in data properly by applying XAI.

3 Proposed Method

3.1 Dataset and Preprocessing

Datasets for face recognition can be collected easily compared to other kinds of data, and it can be used in many ways. Famous datasets such as VGG face, CASIA-Webface, and MS-Celeb-1M are used for training the deep learning models and LFW, AgeDB-30, and CFP-FP are used for testing the model.

This study was conducted using the Korean face images constructed with the support of the Korea Intelligent Information Society Promotion Agency with the funds of the Ministry of Science and ICT. The dataset we used was downloaded from AI Hub [9].

Since artificial intelligence based facial application technology is very dependent on races included in the dataset, we split the same dataset into train and test in the ratio of 8:2. Person ID was separated in advance to prevent overfitting.

We cropped the face image with Multitask Cascaded Convolutional Networks (MTCNN) [10] model and resized it to (112, 112). After data preprocessing 269,098 images were used for training with 320 people and 80 people's images were for testing. Then the dataset was duplicated to two set, one is the original, and the other is with mask. We used face landmarks to generate masks on to the faces. Sample images could be found in Fig. 1.

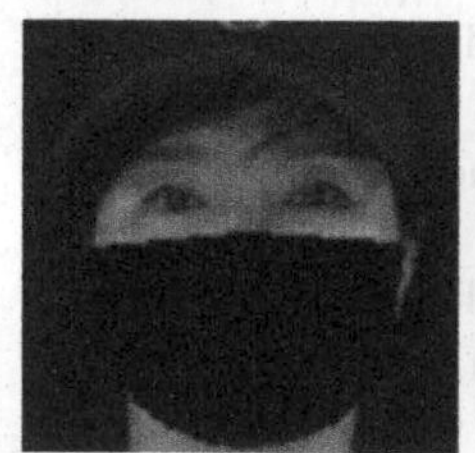 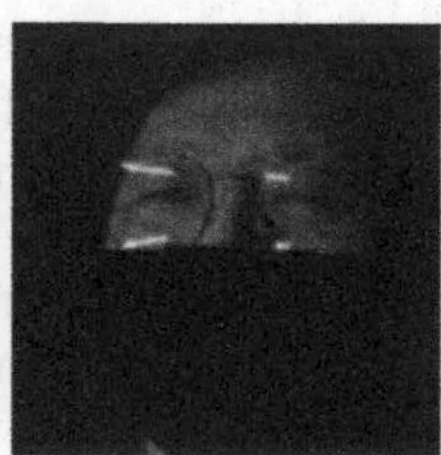 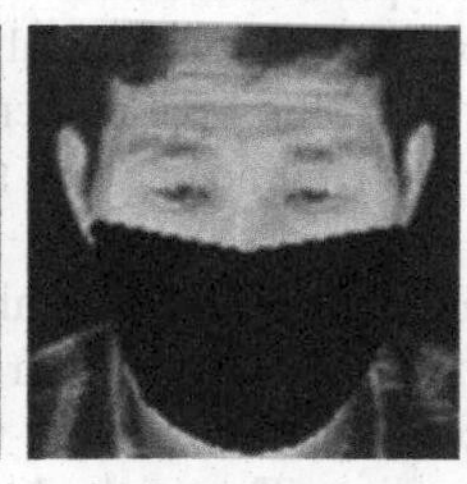 

Fig. 1. Sample images of mask generated data

3.2 Face Recognition Model

Face recognition is one of the most studied fields in artificial intelligence. Therefore, a lot of method accomplished outstanding results. Overall research, ResNet is still the promising model to be used. Therefore, we proceed with learning with a slightly better performing ResNet-based model, i.e., a basic model that learns faces from IResNet. This was a choice considering that the mask-wearing data did not perform well enough when learned through original ResNet. We used 101 layers, so our overall model's structure is IResNet-101.

3.3 Explainable AI

In order to interpret the decision-making process of the face recognition model, we created a white-box object for iResNet-101. Also, to create heatmaps that can explain the models, we used Excitation Backprop (EBP) [11] and pass images in three channels as input data. Then we normalized images and divided them into the maximum value. Only pixels smaller than the specified scale factor were saved, and we divided them once again with the scale factor. The scale factor is used to determine the maximum value of the normalized heatmap and in this case, we set it to 1.0. The jet color map was used to derive the heatmap of the extracted image, and the heatmap for the first convolutional layer is used for the output. The test image derived using the XFR system based on the face recognition model algorithm used in this paper can be viewed in the result section.

4 Experimental Results

The model trained with unmasked dataset achieved the accuracy of 71.97% and the other model trained with masked dataset achieved 84.83% which was 12.86% higher. The accuracy was derived from the test images separated in advance. Figure 2 show the cosine distances between same person's images and difference people's images. As can see, model trained with masked dataset shows a clear difference between two.

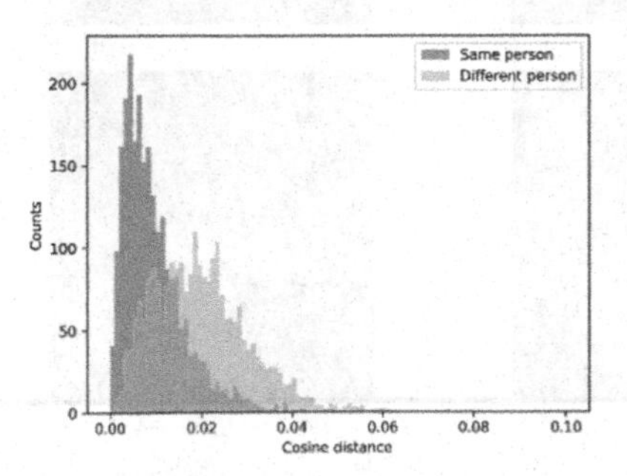

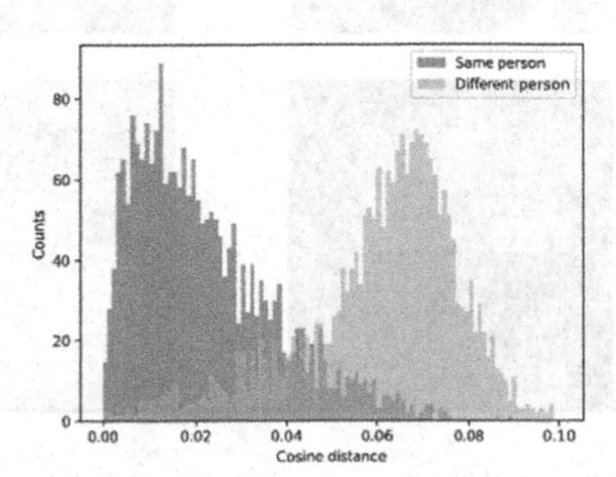

Fig. 2. Genuine and imposter graphs of each model from the test data. Model trained with original data (left) and with masked data (right).

Then we made heatmaps to show the models' focusing regions. Table 1 show some sample images for each model.

Table 1. Heatmaps of two different models. First column represents the input picture, second is the output heatmaps of model trained with original (unmasked) dataset and the last column is the output heatmaps masked dataset.

Train dataset	Trained with unmasked dataset	Trained with masked dataset
	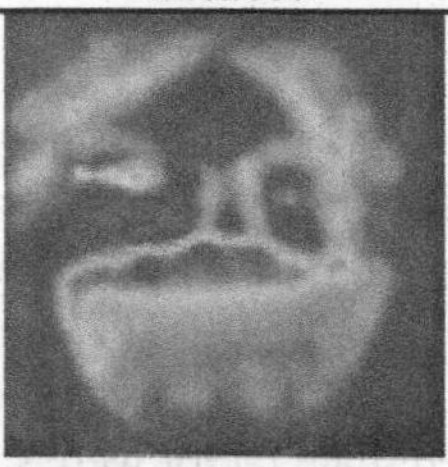	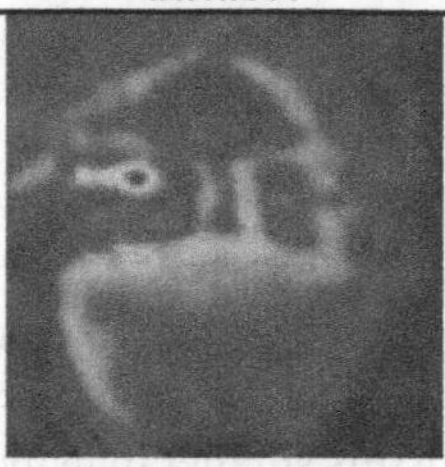
	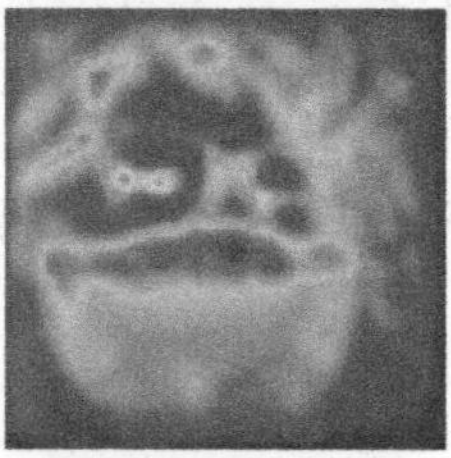	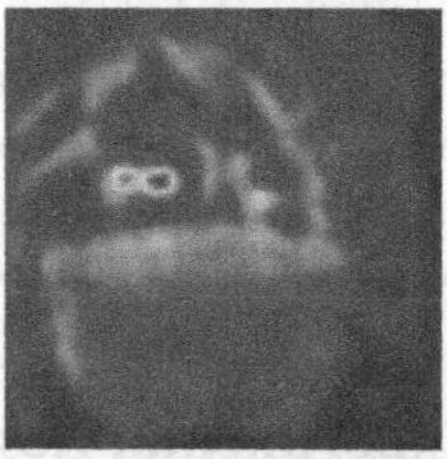
	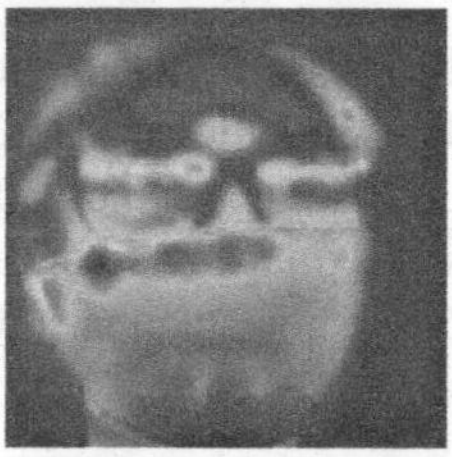	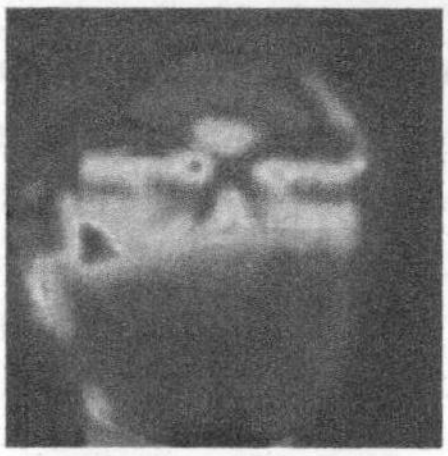
	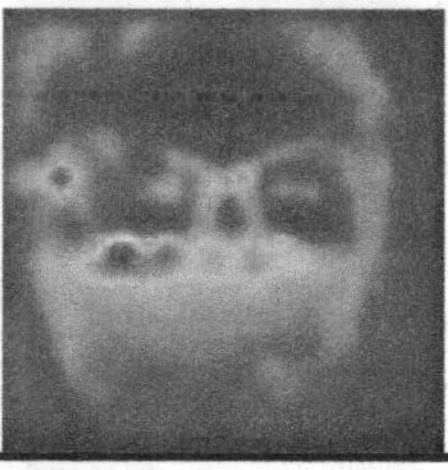	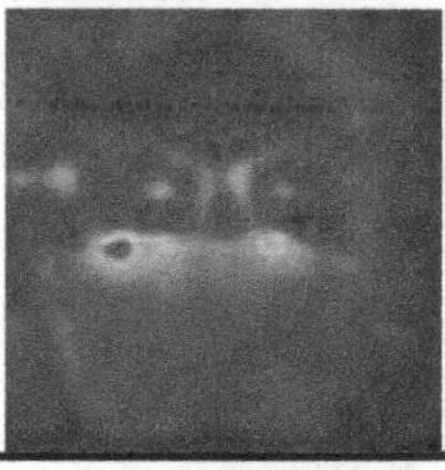

Images on the second column are generated with the model trained with original images. Which means that faces with no masks were used when training and faces with masks were used for testing. On the other hand, model trained with faces with masks' result is on the last column. As we can see, heatmaps on the second column focus on masked area which is not the region of interest when recognizing people with masked faces.

5 Conclusion

In this paper, we tried to check whether the face recognition model learns the data difference properly. As a result, we were able to confirm that in the case of models trained with mask-covered images, the test accuracy was higher in the mask-covered data than in the existing model. In addition, we demonstrate the reason why the model's accuracy is high using XAI. In future studies, in addition to the models used in this paper, various models commonly used in face recognition will be included to increase persuasion. Furthermore, we would like to examine the effect of objects that cover various face areas such as sunglasses as well as masks.

Acknowledgement. This work was supported by the Ministry of Trade, Industry & Energy (MI, Korea) (P0019323).

References

1. Alzu'bi, A., Albalas, F., Al-Hadhrami, T., Younis, L.B., Bashayreh, A.: Masked face recognition using deep learning: a review. Electronics **10**(21), 2666 (2021)
2. Pann, V., Lee, H.J.: Effective attention-based mechanism for masked face recognition. Appl. Sci. **12**, 5590 (2022)
3. Deng, H., Feng, Z., Qian, G., Lv, X., Li, H., Li, G.: MFCosface: a masked-face recognition algorithm based on large margin cosine loss. Appl. Sci. **11**, 7310 (2021)
4. Zhou, B., Khosla, A., Lapedriza, A., Oliva, A., Torralba, A.: Learning deep features for discriminative localization. In: 2016 IEEE Conference on Computer Vision and Pattern Recognition (CVPR), Las Vegas, NV, USA, pp. 2921–2929 (2016)
5. Selvaraju, R.R., Cogswell, M., Das, A., Vedantam, R., Parikh, D., Batra, D.: Grad-CAM: visual explanations from deep networks via gradient-based localization. In: 2017 IEEE International Conference on Computer Vision (ICCV), pp. 618–626 (2017)
6. Williford, J.R., May, B.B., Byrne, J.: Explainable face recognition. In: Vedaldi, A., Bischof, H., Brox, T., Frahm, J.M. (eds.) Computer Vision, vol. 12356, pp. 248–263. Springer, Cham (2020). https://doi.org/10.1007/978-3-030-58621-8_15
7. Duta, I.C., Liu, L., Zhu, F., Shao, L.: Improved residual networks for image and video recognition. In: 2020 25th International Conference on Pattern Recognition (ICPR). IEEE (2021)
8. Deng, J., Guo, J., Xue, N., Zafeiriou, S.: ArcFace: additive angular margin loss for deep face recognition. In: 2019 IEEE/CVF Conference on Computer Vision and Pattern Recognition (CVPR), pp. 4685–4694 (2019)
9. Choi, Y., et al.: K-FACE: a large-scale KIST face database in consideration with unconstrained environments (2021)
10. Xiang, J., Zhu, G.: Joint face detection and facial expression recognition with MTCNN. In: 2017 4th International Conference on Information Science and Control Engineering (ICISCE), pp. 424–427 (2017)
11. Zhang, J., Bargal, S.A., Lin, Z., Brandt, J., Shen, X., Sclaroff, S.: Top-down neural attention by excitation backprop. Int. J. Comput. Vis. **126**(10), 1084–1102 (2018). https://doi.org/10.1007/s11263-017-1059-x

UX Design Workshop for Building Relationships Between Humans and Intelligent Objects Using 'T + e = B' Toolkit

Eui-chul Jung, Younhee Cho, and Hyewon Kim(✉)

Design Department, Seoul National University, Seoul, South Korea
jenk1030@snu.ac.kr

Abstract. 'T + e = B' (Trigger plus emotion equals behavior toolkit) was developed for designers to create and evaluate design concepts using nine triggers as inspirational tools. We claim with T + e = B that emotion is the key to finding a missing link between a single decision-making process and the lasting behavior for trigger design. To prove it, we will present participants with the *Frame of AB, OT, and DB*, Trigger Cards, and Hint Cards to help them derive a behavioral strategy for their target audience. This workshop aims to think about the symbiotic relationship between intelligent objects and humans from a UX design perspective, and to consider the direction of HCI design. Therefore, in this workshop, contemplate and discuss the direction of artificial intelligence (AI) chatbots or robots that can emotionally communicate with humans based on human emotions with participants from various cultural and social backgrounds. At the workshop, participants will (1) use metaphors to select desired relationships between humans and intelligent objects to define existing behavioral patterns and obstacles, (2) ideate triggers using Hint and Trigger Cards. To this end, especially explore various ways of acquiring trust and intimate relationships between intelligent objects and humans, or creating positive user experiences.

Keywords: Trigger design · Emotion · Behavioral strategy · Desired behavior · Obstacle

1 Introduction

The emergence of various intelligent objects such as chatbots, robots, and autonomous vehicles is said to bring happiness to the future of mankind, but many people think that these objects take away human jobs and lead to misery. It can be seen that the future will depend on the relationship between the person who developed these things and the intelligent objects. This workshop aims to think about the symbiotic relationship between intelligent objects and humans from a UX design perspective, and to consider the direction of HCI design. Therefore, in this workshop, contemplate and discuss the direction of artificial intelligence (AI) chatbots or robots that can emotionally communicate with humans based on human emotions with participants from various cultural and social backgrounds. To this end, especially explore various ways of acquiring trust

H. Zaynidinov et al. (Eds.): IHCI 2022, LNCS 13741, pp. 160–165, 2023.
https://doi.org/10.1007/978-3-031-27199-1_17

and intimate relationships between intelligent objects and humans, or creating positive user experiences.

T + e = B was developed for designers to create and evaluate design concepts using nine triggers as inspirational tools. The basic concept of T + e = B Workshop recognizes the need to practice desired behaviors in everyday life that we think should or want to be done but are difficult to do (Jung et al., 2019). In the Fogg Behavior Model (2009), a trigger "causes someone to perform a target behavior." Current researches using triggers are mostly focused on changing one's instant decision-making rather than inducing constant behavior. Therefore, the purpose of this workshop is to present the trigger design with emotion as a behavioral strategy for making lasting positive relationships between humans and intelligent objects by exploring its new design possibilities.

2 Proposed Activities

T + e = B workshop will lead participants to identify the desired relationship between humans and intelligent objects and objectify and analyze their present limitations. To do this, organizers will present a set of triggers promoting emotion. To begin the workshop, participants will select a theme out of eight metaphor relationships between humans and intelligent objects (Table 1) that they have a will to create.

The workshop will be conducted as follows.

1. Introduce (30 min) and ice-breaking (10 min)
2. Activity 1: Use metaphors to organize the desired behavior, actual behavior, and obstacles (20 min)
3. Activity 2: Exploring the situations and emotions in which the behavior occurs from the user's point of view (40 min)
4. Activity 3: Establishing a design strategy and generating ideas using the given examples (50 min)
5. Activity 4: presentation by team (min)

2.1 Activity 1: Use Metaphors to Organize Desired Behavior, Actual Behavior, and Obstacles (30 min)

Use metaphors to define desired relationship between humans and intelligent objects. Metaphorically expressing the relationship between intelligent objects and humans helps participants to immerse themselves in thinking.

After selecting a theme based on their understanding of the design circle, each team should specify what GL (Goal functions/relationship) and DB (Desired Behavior; in this workshop, desired behavior means goal functions of intelligent objects or desired relationship between human and intelligent objects) they will achieve the desired relationship. And their current behavior and obstacles (current difficulties and idealistic ideas) should be defined. Through this, participants can immerse themselves in metaphorically expressing how intelligent objects and humans will relate.

On the *Frame of AB, OT, and DB* (Fig. 1), participants will put post-its written with the Actual behavior, Obstacle (current difficulties and idealistic ideas), and Desired

Table 1. Metaphor relationship between human and intelligent objects

Metaphors	Scenario & User	Focus of Exploration
(past) Windows puppy, Her (movie) #a first-time friend	Using it first-time	Making a positive first impression
Teacher – student relationship #Expert	Learning/finding new information, knowledge	Reliability, high information acceptance attitude (convincing)
Guide dog – owner relationship # Guides, guides, help (information guidance)	When you're lost in an unfamiliar situation or complex information	Reliability, convenience, comfort
A three-legged race relationship #colleague	Trust and entrust a part of my work	Reliability, convenience, high information acceptance attitude (convincing), familiarity
Glasses, wheelchairs relationship #helper, physical aid	When doing physically limited work	Reliability, convenience
Secretary #supporter	Managing my work schedule	Safety (data-related), convenience
Butler #supporter for our family	Managing my daily life and everything that happens at home	Safety (data-related), convenience, familiarity, sharing with family (other users)
Best friend #friends, pets	Who understands me/needs me	Emotional exchange, familiarity, sociality
Emergency rescue #emergency escape, emergency report button, backup	When an emergency occurs	Reliability (dependence)

Behavior (Goal functions/relationship). With these elements placed on *Frame*, it will help participants to objectify their current condition of the behavior and design triggers against obstacles they have. Ideas on new triggers against the obstacles will overwrite old triggers.

2.2 Activity 2: Exploring the Situations and Emotions in Which the Behavior Occurs from the USEr's Point of View (40 min)

The T + e = B toolkit consists of three components: (1) 'Main Trigger Cards' are categorized based on nine key emotions, along with tips to help designers develop their ideas, (2) 'Sub-trigger Cards' contains examples of sub-triggers belonging to the main trigger, (3) 'Hint Cards' contains keywords and questions that can be inspired from the

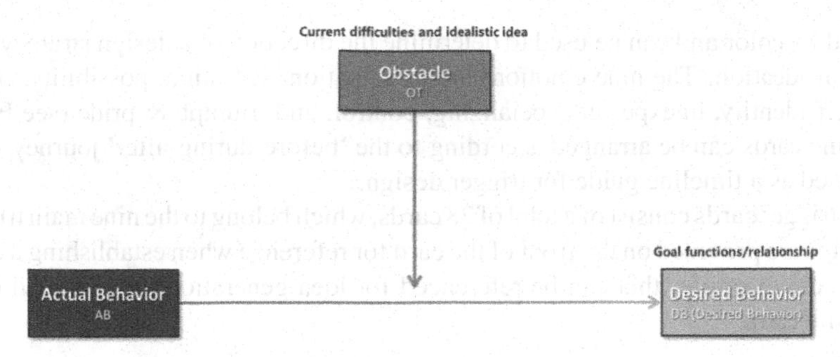

Fig. 1. Frame of AB, OT, and DB

user's point of view. These components are based on Brains, Behavior & Design Toolkit, Game theory, and 'flow' theory.

In this activity stage, select an appropriate or inspiring Hint card that helps turns an actual action into a target action. Hint cards give you something to think about so that you can approach the situation in which the action takes place from the user's point of view. A total of 49 hint cards provide inspiring questions on the back of the card so that you can establish a strategy to eliminate/reduce/overcome factors that hinder your target behavior by utilizing the main trigger and sub-trigger. Participants can get ideas from answering questions on the hint card.

On the Frame of AB, OT, and DB, participants will put Hint card that they selected. Afterward, write down answers on post-its each Hint card, and attach them to the corresponding hint card (Fig. 2).

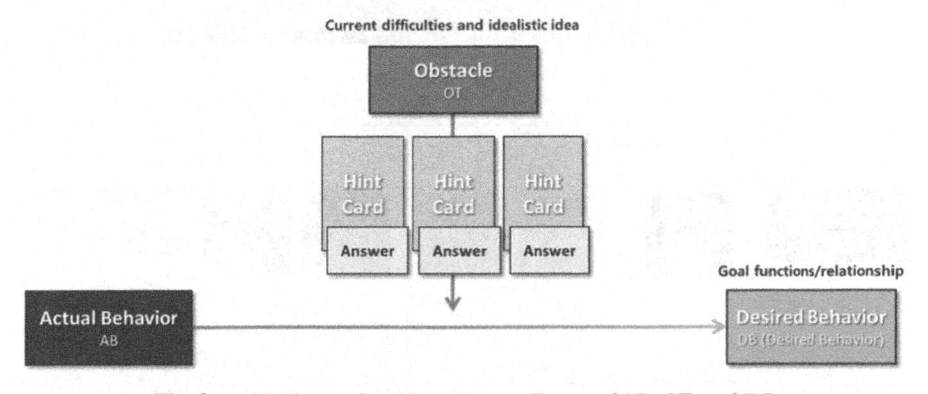

Fig. 2. Add Hint cards and answer on *Frame of AB, OT, and DB*

2.3 Activity 3: Establishing a Design Strategy and Generating Ideas Using the Given Examples (50 min)

In this activity stage, explore Trigger and Sub-Trigger cards to get inspiration and develop ideas and concepts. Nine Emotion Trigger Cards is an ideation tool for designing triggers with emotions. Main Trigger Cards are nine key emotion trigger cards that can be identified by color and can be used to determine the direction of a design strategy in the process of ideation. The nine emotions are anticipation, sensation, possibility, competition, self-identity, unexpected, socializing, control, and triumph & pride (see Fig. 3). These nine cards can be arranged according to the 'before-during-after' journey, which can be used as a timeline guide for trigger design.

Sub-trigger cards consist of a total of 28 cards, which belong to the nine main triggers. Design tips are provided on the front of the card for reference when establishing a design strategy, and examples that can be referenced for idea generation are provided on the back of the card.

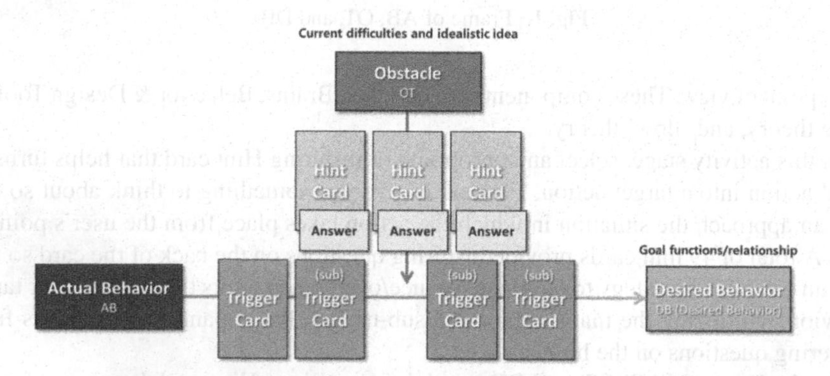

Fig. 3. Add Trigger or Sub-Trigger cards and answer on *Frame of AB, OT, and DB*

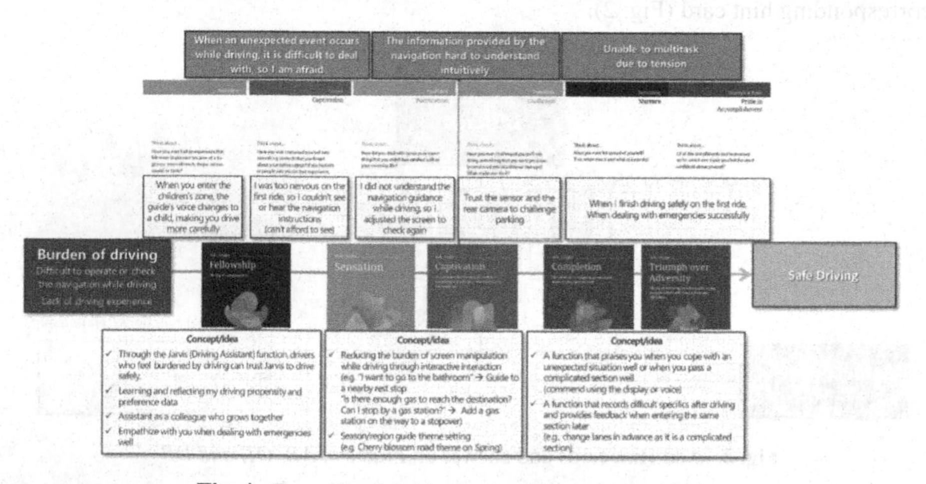

Fig. 4. Example of the final output from past workshop

Participants can get ideas from reference cases and answer questions on Trigger and Sub-Trigger cards. On the *Frame of AB, OT, and DB*, participants will put Trigger and Sub-Trigger cards that they selected (Fig. 3). Afterward write down ideas or concepts on post-its each Trigger cards, and attach them to the corresponding Trigger cards. The following Fig. 4 is an example of the final output from a past workshop.

2.4 Activity 4: Presentation by Team (20min)

Participants will present their trigger design ideas and discuss the validity of 'T + e = B'.

2.5 Workshop Result

Participants will make a behavioral design strategy for the desired relationship between human and intelligent objects on the chosen item (product, service, system, etc.). They will expect to gain insight on how to design a UX (user experience) with the pleasure triggers that provide users control over their choice of actions on workshop themes.

Acknowledgement. This work was supported by Creative-Pioneering Researchers Program (Grant 600-20210006) through Seoul National University.

References

1. Wood, W., Neal, D.T.: A new look at habits and the habit-goal interface. Psychol. Rev. **114**(4), 843–863 (2007)
2. Olander, F., Thogersen, J.: Understanding of consumer behaviour as a prerequisite for environmental protection. J. Consum. Policy **18**(4), 345–385 (1995). https://doi.org/10.1007/BF01024160
3. Fogg, B.J.: A behaviour model for persuasive design. ACM Int. Conf. Proc. Ser. **350**, 1–7 (2009)
4. Jung, E., et al.: Trigger design circles: a behavioral design strategy to practice desired behaviors. International Association of Societies of Design Research (IASDR), Manchester, UK, September 2019

A Longitudinal Study of the Emotional Content in Indian Political Speeches

Sandeep Kumar Pandey[1,2(✉)], Mohit Manohar Nirgulkar[1], and Hanumant Singh Shekhawat[1]

[1] Indian Institute of Technology Guwahati, Guwahati, Assam, India
{sandeep.pandey,mohitmanohar,h.s.shekhawat}@iitg.ac.in
[2] Samsung Research and Development Institute, Bangalore, India
sandeep.p@samsung.com

Abstract. Emotions are an impacting factor when public speaking is concerned. Mainly, politicians utilize emotion in their public speeches as a tool to generate a better connection with the audience. As such, this study proposes a longitudinal study of emotional content in speeches of Indian politicians, encompassing several primary elections over 15 years. A speech emotion dataset is proposed for the same, annotated using human annotators. We also present experimental analysis on the collected dataset using a standard approach such as Attention-based CNN+LSTM architectures and transfer learning using multiple standard emotion datasets. The model achieves a recognition accuracy of 73.18% using pre-training. A longitudinal study spanning over three Lok Sabha elections is also presented, demonstrating how the politicians modulate emotions in speech over several elections.

Keywords: Speech Emotion · CNN+LSTM · Political Speech · Deep Learning · Longitudinal

1 Introduction

Speech signal serves as the preferred choice of communication between two individuals over written or non-verbal communications. This is primarily because speech signal carries a multitude of information - intended message on the primary level and paralinguistic information such as emotion, speaker identity, gender, etc., on the secondary level [22]. Emotions play a crucial role in making the speech sound more natural and provide additional meaning to the context of the message. As such, the same utterance can convey different information under different emotions [19]. This has motivated researchers to inspect several aspects

This research was supported under the India-Korea joint program cooperation of science and technology by the National Research Foundation (NRF) Korea (2020K1A3A1A68093469), the Ministry of Science and ICT (MSIT) Korea and by the Department of Biotechnology (India) (DBT/IC-12031(22)-ICD-DBT).

H. Zaynidinov et al. (Eds.): IHCI 2022, LNCS 13741, pp. 166–176, 2023.
https://doi.org/10.1007/978-3-031-27199-1_18

of speech emotion recognition and its application in various domains such as education [9], health [1], military [24], politics [13], etc.

Emotions play a significant role in public speaking, and hence, politicians utilize them as a tool to modify the emotional content of their speeches. This, in turn, helps to make the speech more connecting and set an agenda by making a connection with the masses [23]. Several factors impact the evaluation of a candidate in elections, such as education, background, achievements during the previous tenure, etc., apart from public speeches. However, the study in [4] presents emotions as a rhetorical tool that shapes perceptions about a candidate, emphasizing emotions such as Anger and Fear to shape political discourse. Another interesting study conducted in [7] highlights the use of emotions in six million US Congress speeches over a time frame of 1858-2014. The study highlighted that the emotional content was found to be more in speeches concerned about war and patriotism. Also, it stated that the emotionality in speeches was found to be higher for democrats, women candidates, ethnic minorities, etc.

This field of study is well explored from a psychological point of view. However, due to the lack of annotated data on political speeches, signal processing and deep learning are still unexplored for this task. Recent work in [5] utilized the Canadian House of Commons audio-visual dataset, explored emotion and sentiment analysis using standard Natural Language Processing (NLP) techniques, and concluded that although text transcripts contain sentiment information, emotional arousal could not be effectively captured through text transcripts. Utilizing the speeches of Indian politicians, the study in [8] tried to identify how each politician in the dataset incorporates different proportions of the four basic emotion categories - angry, happy, neutral, and sad in their speeches.

In this work, we propose a deep learning-based Speech Emotion Recognition (SER) framework for automatic emotion identification from speech utterances of Indian Politicians. The study is divided into two parts. The first part is concerned with the dataset collection and annotation strategy used. The second part of the study employs Attention-based CNN+LSTM architecture for the task of emotion recognition. Finally, we present an analysis of the emotional content per politician over several years, such as the Lok Sabha elections of 2009, 2014, and 2019.

The remainder of the paper is organized as follows. Section 2 presents the dataset description along with the annotation strategy used. Section 3 gives a detailed description of the data pre-processing and feature extraction. The architecture components are also discussed in this section. Section 4 presents the experimental evaluation on the political speech dataset. The paper concludes with Sect. 5.

2 Dataset Description

Due to the non-availability of the necessary speech corpus to conduct a longitudinal analysis, we propose IITG Political Speech Dataset, which integrates audio components and their labels. This dataset is built upon the work IITG Politician

Dataset [8] which was not longitudinal in nature. Political speeches can be found easily on YouTube, which we use as our primary data source. Various channels (YouTube), such as news and political parties, have repositories of speeches in video format.

We selected ten politicians belonging to diverse political ideologies, both from the ruling and opposition teams. These politicians represent different views and styles of speech in the database due to an equal number of males and females. We selected politicians based on the language in which they primarily deliver speeches - Hindi in our case. The names of the politicians are anonymized for privacy reasons and are assigned random abbreviations like - NM, YA, RG, SI, etc. Speeches were delivered indoors (hall or auditorium) and outdoors (complex stadiums, courts, or stadiums), with significant background noise.

The Lok Sabha (lower house of the parliament) elections in India are held at an interval of 5 years and the political party bagging the 2/3rd of the seat forms the government. As such, the campaign for Lok Sabha elections by various politicians is very emotionally enriched and interesting. The speeches held during the Lok Sabha elections in 2009, 2014, and 2019, as well as speeches delivered throughout the election year, were chosen for this study. Figure 1 shows the distribution of speeches by politicians over the different election years. The audio component is extracted from the video of speeches using the python library pafy and then segmented into equal duration Wav files of 6 s each. The naming convention adopted for the speech segments is

$$AA\_YYYY\_VXXX\_SXXXXX.wav$$

Here, *AA* is assigned random abbreviation of the politician's name, *YYYY* is the election year, *VXXX* is the speech number of that year, and *SXXXXX* is the segment number of that audio/speech. These speech segments are divided according to four emotional states: angry, happy, neutral, and sad. For the task of annotation of the speech segments, we assigned five annotators. To make the labeling process fair, we have used a two-step process as in [8]. The first step is to determine if the audio contains any emotional information, not influenced by the text content of the speech. This is done by assigning a score to each segment depending on how confident the examiners feel on one to ten scales (1 - definition of low confidence and 10 - indicating high confidence). We then assign a categorical label to the segments, which is the second step. Because political discourses are deeply emotional, different observers may view them differently. We used the following procedure to provide a category label in this case.

- If the majority of listeners (or volunteers) agree on certain emotion, that emotion is labeled on the audio sample. The label's confidence score would be the majority's mean score.
- If two listeners favor one emotion and two prefer the other, the emotion with the higher mean confidence score is labeled. The confidence score of the label would be the same as the mean used to break the tie.
- If each listener assigns a different label, the audio clip is discarded.

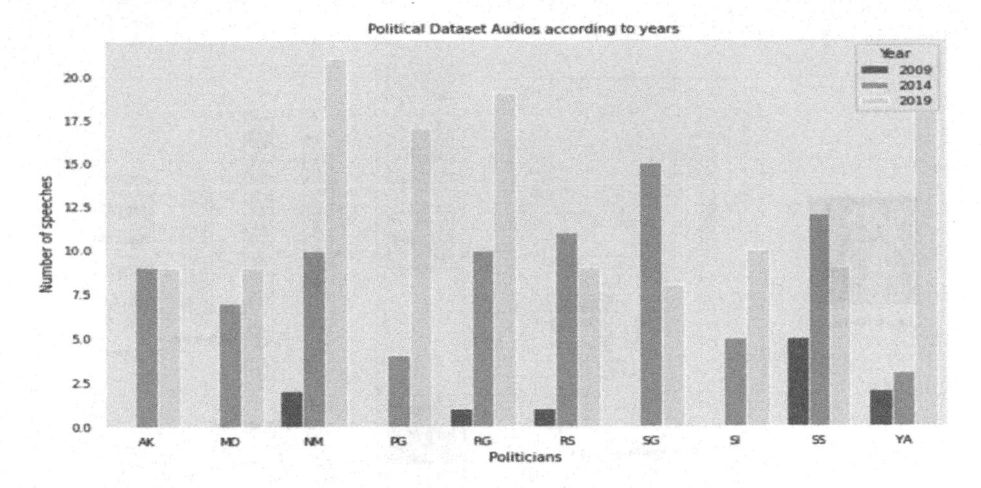

Fig. 1. Number of speeches delivered by the politicians in a given year.

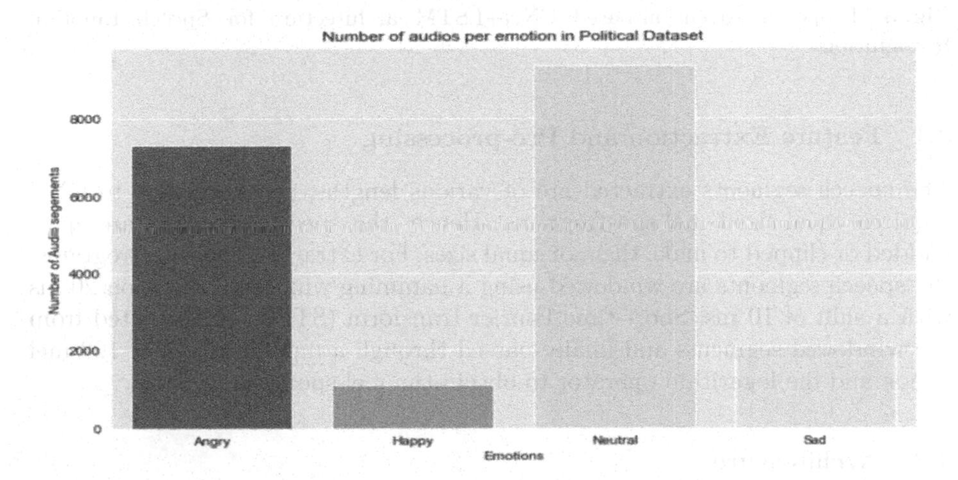

Fig. 2. Distribution of audio segments per emotion class in IITG Political Speech Dataset

3 Experiment Details

For the task of identifying emotional states underlying a speech utterance, we exploited mel-spectrograms as feature representations of speech [10]. We utilized an Attention-based CNN+LSTM architecture [14,28] to model the emotion categories, the details of which are described below.

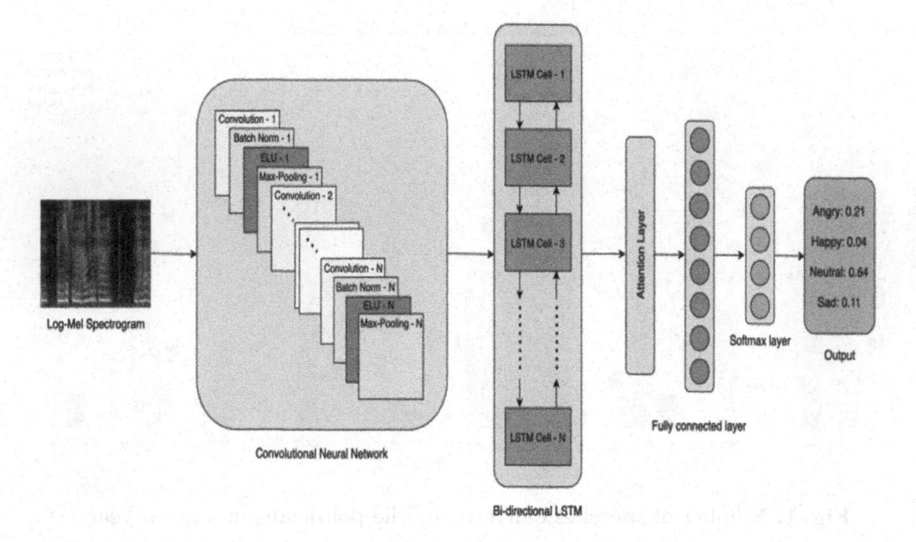

Fig. 3. Proposed Attention-based CNN+LSTM architecture for Speech Emotion Recognition

3.1 Feature Extraction and Pre-processing

The speech segments extracted are of various lengths, and the input to CNN requires equal-sized mel-spectrograms. Hence, the speech segments are zero-padded or clipped to make them of equal sizes. For extracting mel-spectrograms, the speech segments are windowed using a hamming window of duration 20 ms with a shift of 10 ms. Short-time Fourier transform (STFT) is calculated from the windowed segments and finally passed through a mel filterbank of 128 mel bands and the logarithm operator to obtain the mel-spectrogram [21].

3.2 Architecture

We utilized the CNN+LSTM architecture with attention as in [8]. CNNs account for extracting local features using filters, leveraging the image form of spectrograms [26]. LSTMs take care of the temporal dependencies in the speech signal, as emotional information is spread temporally and not concentrated at certain time instants [25]. The Attention-based CNN-LSTM model employs four local feature learning blocks (LFLBs), one bidirectional LSTM layer, an attention layer that aids in generalization, followed by a fully connected layer and a softmax layer to compute class probabilities predicting emotions.

Local Feature Learning Block (LFLB). The LFLB in this Network architecture is made up of one 2-D convolutional layer, a batch-normalization layer, an activation layer, and one max pooling layer. CNN can extract prominent features from input data without needing painstaking feature engineering using

2D spatial kernels. Because of CNN's local connections and shared weights, the convolution layer reads Log Mel Spectrograms via kernel reading. Following the convolution layer, batch normalization is applied to normalize the activations on each batch by keeping the activation rate near zero and the standard deviation close to one. The activation function known as the Exponential Linear Unit (ELU) is used, which is known to reduce the impact of the vanishing gradient. Unlike other functions, ELU has negative values, which push the mean of activations closer to zero, aiding learning and improving performance [18]. Max pooling is the process of publicly generating feature map pads and then using them to produce a sample (integrated) feature map. It introduces a minor degree of translation invariance, which implies that translating a picture has no significant effect on the value of most integrated effects. It reduces the number of parameters for the successive layers by reducing the mapping feature [17].

Attention-based Long Short-Term Memory (LSTM). The Bidirectional LSTM is comprised of two LSTMs, stacked one behind another. Bi-LSTMs effectively increase the amount of data available on the network, thereby improving the algorithm. By allowing special LSTM units to replace the hidden units in RNNs, LSTMs can solve vanishing or exploding gradient problems. These LSTM units allow the network to accumulate and forget exhausting data while maintaining a constant error flow backward in time [27]. The final LFLB output is sent to the LSTM layer, which learns the long-term context's dependencies. The sequence of high-quality presentations found in the construction of CNN + LSTM is transferred to an Attention layer. In [15,20], it is found that the Attention mechanism can extract the elements that are important to the emotion of the utterance; hence we aggregate its output and pass it to the dense layer, which feeds its output to the softmax layer for predicting emotions.

4 Experimental Evaluation

The experiments on IITG Political Speech Dataset are conducted in two phases. Firstly, the emotion label is predicted using the trained Attention-based CNN+LSTM architecture as described in 3.2. Secondly, we analyze the amount of each emotion category present in a politician's speech over different election years.

We use the same model configuration for both Pretraining based and Non-Pre-Training based scenarios on the Political speech datasets with four LFLBs with the number of output filters as 32, 32, 64, 128, kernel size for all the blocks is 3×3 with stride 1×1. The blocks also contain Batch-normalization, which normalizes the input. The dropout is used with a probability of 0.2 for all the blocks. Next is the Bi-LSTM layer with 96 units, the results of which are fed to an attention layer. Finally, the outputs of the attention layer are passed into a dense (32,4) and softmax layer for emotion predictions. For even better results, the model trains with Early Stopping, Learning Rate Scheduler, and Model

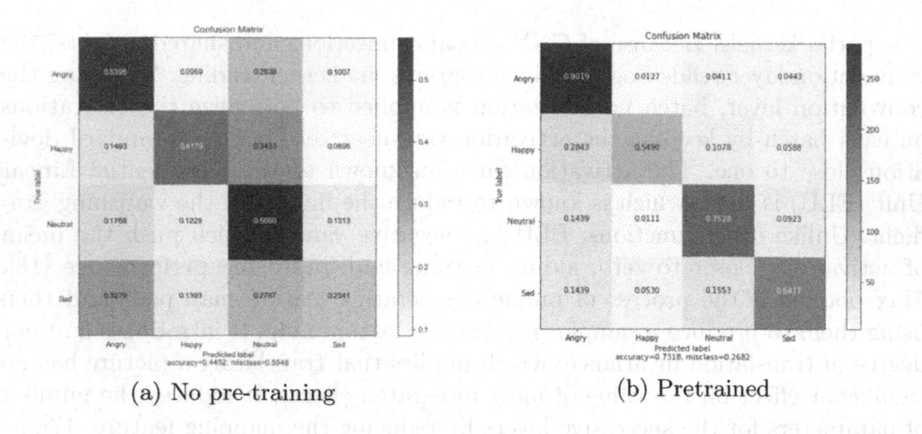

(a) No pre-training (b) Pretrained

Fig. 4. Confusion matrices for the IITG Political speech dataset using -(a) No pre-training and (b) Pretraining using multiple speech emotion datasets

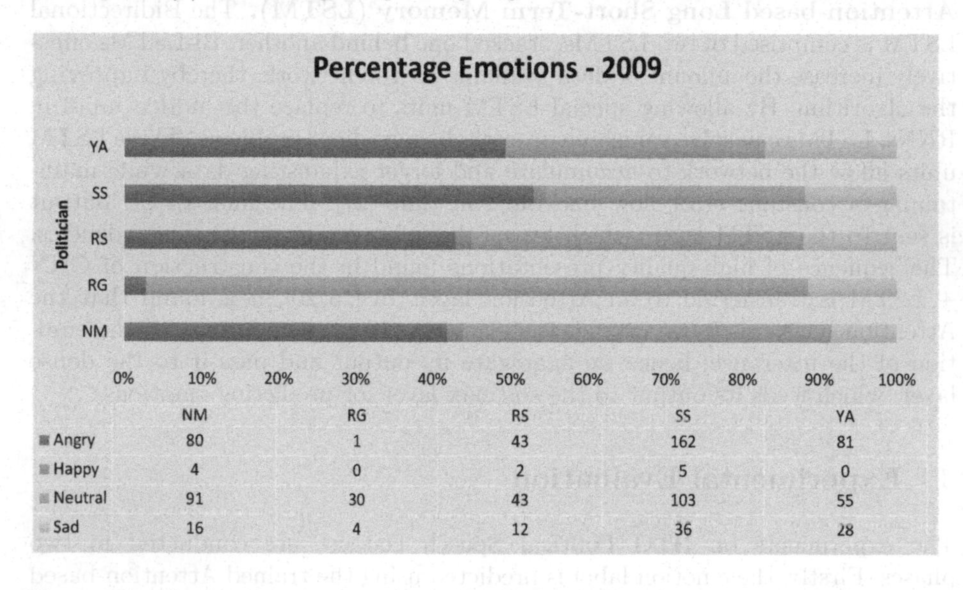

Fig. 5. Distribution of emotions in public speeches of politicians in IITG Political Speech Dataset for the election year 2009

Checkpoint callbacks, which monitor the loss, and accuracy and maintain the best model, respectively.

The model achieves a recognition performance of UAR 44.52% with the proposed architecture, when the weights are randomly initialized. This low performance can be attributed to the low amount and unbalanced data and the naturalness of the speech emotions due to the real-life speech scenarios. However, when the proposed model is pre-trained using a combination of speech emotion datasets such as Emo-DB [2], TESS [6], RAVDESS [12], Shemo-DB [16], CREMA-D [3], and SAVEE [11], the recognition accuracy increases to 73.18%.

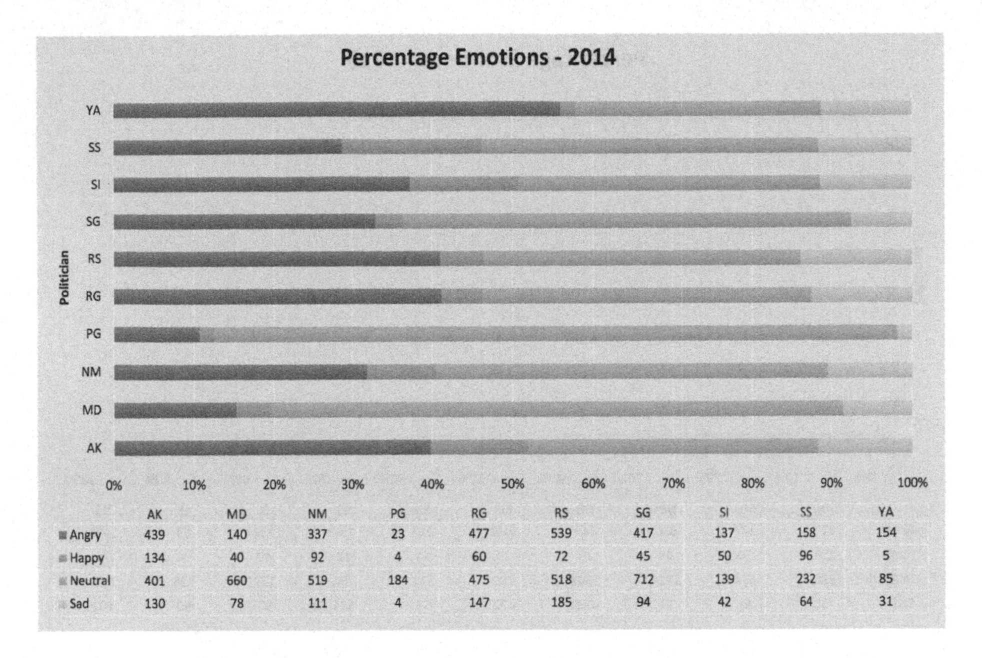

Fig. 6. Distribution of emotions in public speeches of politicians in IITG Political Speech Dataset for the election year 2014

Figure 4 presents the confusion matrix for the predicted emotions of the IITG Political Speech Dataset test set for both the Non-Pretraining and Pretraining-based models. As is evident from the confusion matrix, the sad emotion class is difficult to model as the presence of sad emotions in public speeches is very rare. Moreover, Angry and Neutral are the dominant classes in terms of the number of data samples and recognition accuracy. This clarifies the various statements in literature that Anger is used as a potent tool for shaping emotional discourse during public speeches in elections.

For analyzing the emotion content variation per politician over the election years 2009, 2014, and 2019, a histogram-based plot is presented, which demonstrates the amount of each emotion class present per politician. As can be seen from Fig. 5, the number of politicians present in the study is less due to the non-involvement of the rest other politicians in the 2009 Lok Sabha Elections. However, it can be observed that the politicians' RS and NM tried to balance emotional content across all the four emotion classes compared to the rest, who were more biased towards Anger or Neutral. Similarly, Fig. 6 incorporated more politicians, and it can be seen that AK, SI, SS, and NM tried to balance the emotional content. However, the balance was disturbed for RS compared to his 2009 speeches. Moreover, the trend in Lok Sabha 2019 elections, as seen in Fig. 7 is tilted more toward Anger emotion, giving a hint at the change in political dynamics and viewpoints of the politicians.

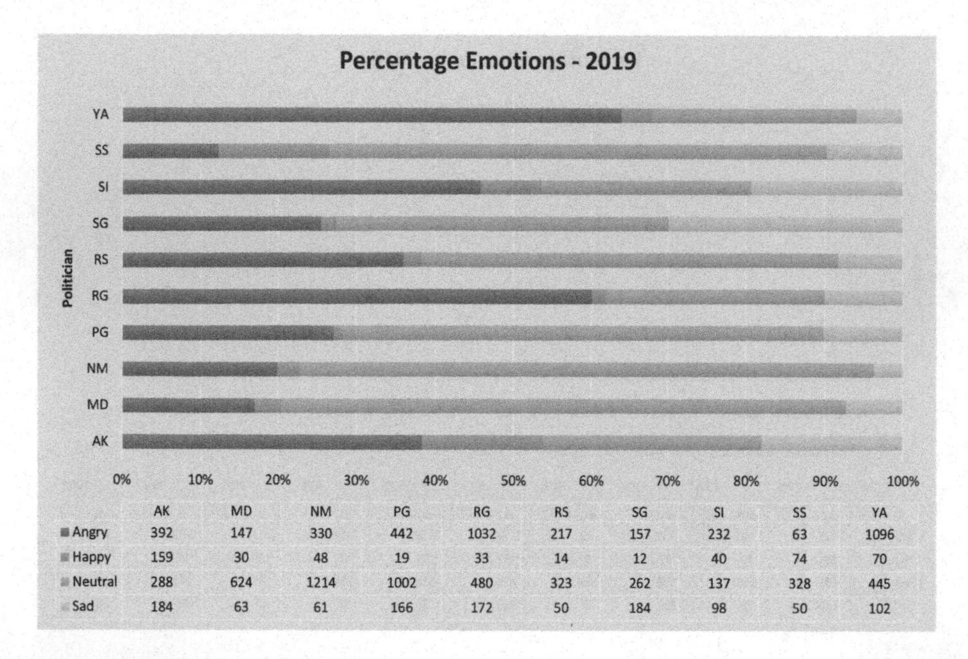

Fig. 7. Distribution of emotions in public speeches of politicians in IITG Political Speech Dataset for the election year 2019

5 Conclusion

In this study, we presented the data collection strategy for the Political Speech dataset. The speeches available on YouTube are successfully processed to obtain the audio part. The audio segments are extracted and given unique identifiers to make labeling easier. Five volunteers label the dataset, assigning one of the four basic emotions (Angry, Happy, Neutral, and Sad) to an audio chunk and assigning a confidence score. We also attempted to describe the technique of obtaining audio features; an analysis of political speech data was also provided, demonstrating how politicians use Anger in their election campaigns, along with a healthy amount of happiness and sadness. Furthermore, Speech Emotion Recognition is effectively performed by employing the Log Mel Spectrograms generated from the audios in the Multiple datasets and Political Speech datasets as input to the deep learning Attention-based CNN+LSTM architectures.

References

1. Bhaduri, S., Chakraborty, A., Ghosh, D.: Speech emotion quantification with chaos-based modified visibility graph-possible precursor of suicidal tendency. J. Neurol. Neurosci. **7**(3) (2016)
2. Burkhardt, F., Paeschke, A., Rolfes, M., Sendlmeier, W.F., Weiss, B., et al.: A database of German emotional speech. In: Interspeech, vol. 5, pp. 1517–1520 (2005)

3. Cao, H., Cooper, D.G., Keutmann, M.K., Gur, R.C., Nenkova, A., Verma, R.: Crema-d: crowd-sourced emotional multimodal actors dataset. IEEE Trans. Affect. Comput. **5**(4), 377–390 (2014)
4. Cislaru, G.: Emotions as a rhetorical tool in political discourse (2012)
5. Cochrane, C., Rheault, L., Godbout, J.F., Whyte, T., Wong, M.W.C., Borwein, S.: The automatic analysis of emotion in political speech based on transcripts. Polit. Commun. **39**(1), 98–121 (2022)
6. Dupuis, K., Pichora-Fuller, M.K.: Toronto emotional speech set (tess) (2010)
7. Gennaro, G., Ash, E.: Emotion and reason in political language. Econ. J. **132**(643), 1037–1059 (2022)
8. Goel, S., Pandey, S.K., Shekhawat, H.S.: Analysis of emotional content in Indian political speeches. In: Singh, M., Kang, D.-K., Lee, J.-H., Tiwary, U.S., Singh, D., Chung, W.-Y. (eds.) IHCI 2020. LNCS, vol. 12615, pp. 177–185. Springer, Cham (2021). https://doi.org/10.1007/978-3-030-68449-5_18
9. Gong, M., Luo, Q.: Speech emotion recognition in web based education. In: 2007 IEEE International Conference on Grey Systems and Intelligent Services, pp. 1082–1086. IEEE (2007)
10. Issa, D., Demirci, M.F., Yazici, A.: Speech emotion recognition with deep convolutional neural networks. Biomed. Signal Process. Control **59**, 101894 (2020)
11. Jackson, P., Haq, S.: Surrey audio-visual expressed emotion (savee) database. University of Surrey, Guildford, UK (2014)
12. Livingstone, S.R., Russo, F.A.: The Ryerson audio-visual database of emotional speech and song (ravdess): a dynamic, multimodal set of facial and vocal expressions in North American English. PloS one **13**(5), e0196391 (2018)
13. Lutz, C.A., Abu-Lughod, L.E.: Language and the politics of emotion. In: This book grow out of a session at the 1987 annual meeting of the American Anthropological Association called Emotion and Discourse. Editions de la Maison des Sciences de l'Homme (1990)
14. Meng, H., Yan, T., Yuan, F., Wei, H.: Speech emotion recognition from 3d log-mel spectrograms with deep learning network. IEEE Access **7**, 125868–125881 (2019)
15. Mirsamadi, S., Barsoum, E., Zhang, C.: Automatic speech emotion recognition using recurrent neural networks with local attention. In: 2017 IEEE International conference on acoustics, speech and signal processing (ICASSP), pp. 2227–2231. IEEE (2017)
16. Mohamad Nezami, O., Jamshid Lou, P., Karami, M.: ShEMO: a large-scale validated database for Persian speech emotion detection. Lang. Resour. Eval. **53**(1), 1–16 (2019)
17. Murray, N., Perronnin, F.: Generalized max pooling. In: Proceedings of the IEEE Conference on Computer Vision and Pattern Recognition, pp. 2473–2480 (2014)
18. Nwankpa, C., Ijomah, W., Gachagan, A., Marshall, S.: Activation functions: comparison of trends in practice and research for deep learning. arXiv preprint arXiv:1811.03378 (2018)
19. Pandey, S.K., Shekhawat, H.S., Prasanna, S.M.: Deep learning techniques for speech emotion recognition: a review. In: 2019 29th International Conference Radioelektronika (RADIOELEKTRONIKA), pp. 1–6. IEEE (2019)
20. Pandey, S.K., Shekhawat, H.S., Prasanna, S.: Attention gated tensor neural network architectures for speech emotion recognition. Biomed. Signal Process. Control **71**, 103173 (2022)
21. Rabiner, L., Juang, B.H.: Fundamentals of Speech Recognition. Prentice-Hall, Inc., Hoboken (1993)

22. Schuller, B., et al.: Paralinguistics in speech and language-state-of-the-art and the challenge. Comput. Speech Lang. **27**(1), 4–39 (2013)
23. Shields, S.A.: The politics of emotion in everyday life: appropriate emotion and claims on identity. Rev. Gen. Psychol. **9**(1), 3–15 (2005)
24. Tokuno, S., et al.: Usage of emotion recognition in military health care. In: 2011 Defense Science Research Conference and Expo (DSR), pp. 1–5. IEEE (2011)
25. Xie, Y., Liang, R., Liang, Z., Huang, C., Zou, C., Schuller, B.: Speech emotion classification using attention-based LSTM. IEEE/ACM Trans. Audio Speech Lang. Process. **27**(11), 1675–1685 (2019)
26. Yenigalla, P., Kumar, A., Tripathi, S., Singh, C., Kar, S., Vepa, J.: Speech emotion recognition using spectrogram & phoneme embedding. In: Interspeech, vol. 2018, pp. 3688–3692 (2018)
27. Yu, Y., Si, X., Hu, C., Zhang, J.: A review of recurrent neural networks: LSTM cells and network architectures. Neural Comput. **31**(7), 1235–1270 (2019)
28. Zhao, J., Mao, X., Chen, L.: Speech emotion recognition using deep 1D & 2D CNN LSTM networks. Biomed. Signal Process. Control **47**, 312–323 (2019)

Building a Local Classifier for Component-Based Face Recognition

Shavkat Kh. Fazilov[1(✉)], Olimjon N. Mirzaev[1,2], and Shukrullo S. Kakharov[1]

[1] Research Institute for the Development of Digital Technologies and Artificial Intelligence, 17A, Buz-2, Mirzo Ulugbek, 100125 Tashkent, Republic of Uzbekistan
sh.fazilov@mail.ru

[2] Tashkent University of Information Technologies named after Muhammad al-Khwarizmi, 108, Amir Temur Avenue, 100200 Tashkent, Republic of Uzbekistan

Abstract. The problem of constructing a supervised local classifier is solved in this article. In essence, this task is a classical recognition problem. The mathematical formulation and method of solving the problem, the results of experimental studies, as well as a detailed description of the main stages of building a local classifier, based on the assessment of the interdependence between fragments (face components) of the image of a recognizable face, are given. This article also discusses local face recognition methods based on the analysis of face components. A new approach to component-based face recognition is proposed, which (compared with existing analogs) has a number of advantages, the main of which is that when using local classifiers, it is possible to make soft decisions presented in the form of a list of candidates with weights assigned to them. The acceptance of such decisions ensures the stability of the system.

Keywords: Pattern recognition · Face recognition · Classifier · Image preprocessing · Feature extraction · Image analysis · Face image · Feature vector

1 Introduction

At present, many approaches to face recognition of high reliability and accuracy were developed. However, some issues still need to be solved due to various problems such as changes in head position, lighting conditions, and facial expressions. To solve these problems, new methods are being developed, on the basis of which reliable face recognition systems are created. However, these methods require a large amount of memory and are rather complex from a computational point of view.

This study is related to the consideration of local face recognition methods based on the analysis of face components. In this case, the biometric identification system sequentially implements the following stages of processing the original face image.

H. Zaynidinov et al. (Eds.): IHCI 2022, LNCS 13741, pp. 177–187, 2023.
https://doi.org/10.1007/978-3-031-27199-1_19

In the first stage, face detection in the original image is performed. The purpose of this stage is to determine if the input image contains a human face or not. Various methods are widely used for face image detection and localization, such as the Viola Jones detector [1], histogram of oriented gradient (HOG) [2,3], and principal component analysis [4].

In the second stage, if a face was detected in the original image, this image was pre-processed in order to improve its quality. The need for such processing is due to the fact that changes in lighting, aspect angle, and facial expressions can interfere with the correct recognition of faces and make it difficult to develop a more reliable recognition system [5–7]. In the third stage, the face image was decomposed into local areas containing its main components (eyes, nose, mouth, etc.). The main task in this stage was to extract the indicated components of the face and describe their shapes using geometric characteristics (dimensions and distances).

In the fourth stage, sets of features were extracted, which determine an informative description of the considered components of the face. The purpose of the implementation of this stage is to form an informative description of the face components identified at the previous stage by sets of identification features. In the case of redundancy of features in these sets, the problem of reducing the dimension of the feature space was solved [8,9].

In the final, fifth stage, face recognition was realized based on the comparison of the feature vector calculated for the input face image with the feature vectors of face images stored in the database. There are two main applications of face recognition, one is called identification and the other is called verification. In the case of identification, the tested face is compared with a set of faces in order to identify the most likely match, and in the case of verification, the tested face is compared with a known face from the face database in order to make a "true"/"false" decision [10].

One of the possible schemes for implementing the existing methods of component-based face recognition is shown in Fig. 1. These methods differ mainly in the choice of certain components of the face (eyebrows, eyes, nose, mouth, etc.), methods for determining the features (descriptors) of the analyzed face components, and recognition methods. According to this scheme, the corresponding feature values are determined for the selected face components. Then a generalized feature vector is formed, which is fed into the recognition block.

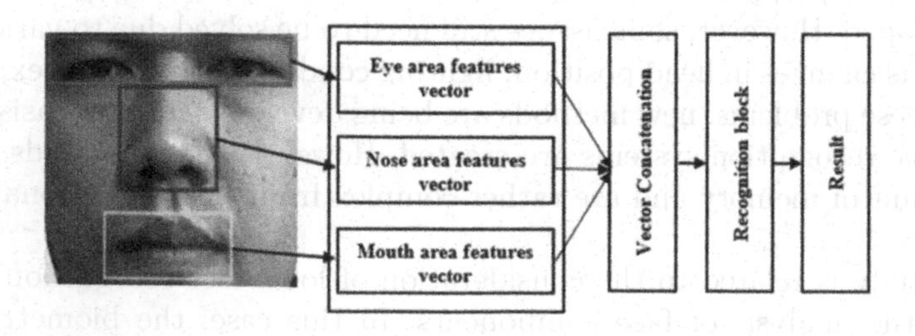

Fig. 1. Implementation scheme for component-based face recognition

In this paper, another approach to component-based face recognition is considered, implemented in the form of a scheme shown in Fig. 2. According to this scheme, as in the previous case (Fig. 1), the corresponding feature values are determined for the selected face components, which are fed to the input of local classifiers that implement the soft recognition process. Then, based on the results of the implementation of these classifiers, the integrator generates the final solution for the recognized face.

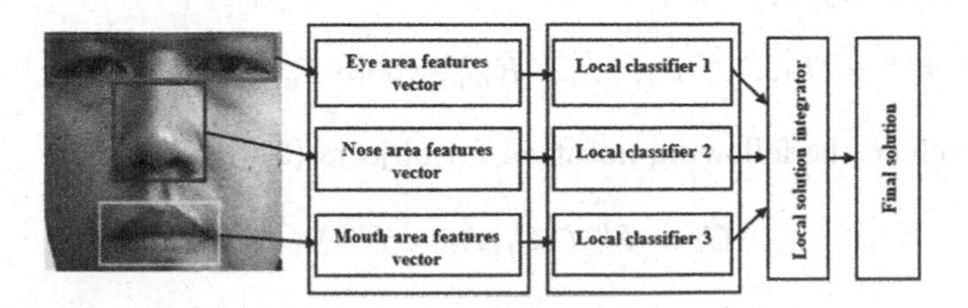

Fig. 2. The proposed scheme for the implementation of component-based face recognition

The second scheme of the implementation of component-based face recognition, compared to the first scheme, has a number of advantages, the main of which is that when using local classifiers, it is possible to make soft decisions that are presented in the form of a list of candidates with weights assigned to them. The acceptance of such decisions ensures the stability of the system operation: small changes at the input of any local classifier lead to small changes both at its output and at the output of the entire system. The final decision is generally a hard one.

In this paper, we solve the problem of constructing a supervised local classifier. This problem is a classical recognition problem, which is formulated as follows [11].

In this paper, we solve the problem of constructing a supervised local classifier. This problem is essentially a classical problem of pattern recognition, which is formulated as follows [11].

2 Problem Statement

Consider the problem of pattern recognition (problem Z) in the standard statement formulated in [12–14]. Let the set of admissible objects $\mathbb{F}$ be given, represented as face images. The initial data on each admissible object $\mathbb{F}$ are given in the form of a three-dimensional matrix (color image) X of size (where c – is the number of color channels; m and n are the number of rows and the number of columns, respectively) [5]:

$$X = (X_1, X_2, X_3), X_c = |x_{cij}|_{n \times m}$$

It is assumed that elements $\mathbb{F}$ form ℓ disjoint classes $C_1, \ldots, C_j, \ldots, C_\ell$ [13,14]:

$$\mathbb{F} = \bigcup_{j=1}^{\ell} C_j, \; C_i \cap C_j = \emptyset, i \neq j, \; i, j \in \{1, \ldots, \ell\}. \tag{1}$$

Partition (1) is not completely defined, and there is only some initial information I_0 on classes $C_1, \ldots, C_j, \ldots, C_\ell$. Let there be some sample $\widetilde{\mathfrak{F}}^m \left(\widetilde{\mathfrak{F}}^m \subset \mathfrak{F}\right)$ consisting of m objects:

$$\widetilde{\mathfrak{F}}^m = \{F_1, \ldots, F_u, \ldots, F_m\}, \text{where } F_u \in \mathbb{F}, \; u = \overline{1, m}. \tag{2}$$

Let us introduce the following notation for objects (2):

$$\widetilde{C}_j = \widetilde{\mathfrak{F}}^m \cap C_j, \widetilde{A}_j = \widetilde{\mathfrak{F}}^m \setminus \widetilde{C}_j.$$

Then the initial information I_0 about the classes can be given in the form [13]:

$$I_0 = \{F_1, \; \widetilde{\alpha}(F_1), \ldots, F_u, \; \widetilde{\alpha}(F_u), \ldots, F_m, \; \widetilde{\alpha}(F_m)\},$$

where $\widetilde{\alpha}(F_u)$ is the information vector of object F_u $(F_u \in \widetilde{\mathfrak{F}}^m)$: $\widetilde{\alpha}(F_u) = (\alpha_{u1}, \ldots, \alpha_{uj}, \ldots, \alpha_{u\ell})$. Here α_{uj} is the value of the predicate, which has the following form:

$$P_j(F_u) = \begin{cases} 0, & \text{if } F_u \in \widetilde{C}_j; \\ 1, & \text{if } F_u \notin \widetilde{C}_j. \end{cases} \tag{3}$$

The problem is to construct such an algorithm A that allows (based on the initial information I_0) calculating the values of elementary predicates (3):

$$A(F_u) = \widetilde{\alpha}(F_u).$$

3 Solution Method

The construction of a local classifier in this study is based on the solution to problem Z, in which the face components (for example, eyes, nose, mouth) act as objects to be recognized. Therefore, when describing the classifier, using the concept of "fragment", we will mean the studied component of the face.

The proposed local classifier model is based on the assessment of the interdependence between image fragments of a recognizable face and is set in the form of the following main steps.

1. Formation of Basic Fragments of Face Images. Here a set of basic fragments Σ is formed, depending on parameter $k : \Sigma = \Xi_1, ..., \Xi_k$ [15–17]. Here $k = m_H \times n_W, m_H$ and n_W are the numbers of image division into fragments in height and width, respectively.

2. Determination of a Set of Identification Features of the Face Image. In this stage, the dependency model between the basic fragments is built. Let Ξ_u and Ξ_v be the basic fragments of the considered image of face F. Then the dependence model between them is given in the following form [6,11,18]:

$$y = \mathcal{F}_q(\bar{c},\ x),\ \ y \in \Xi_u,\ \ x \in \Xi_v(q \in [1,\ k]), \tag{4}$$

where $\mathcal{F}_q$ is the function belonging to a given class of functions $\mathbb{F}$, i.e., $(\mathcal{F}_q \in \mathbb{F})$.

The calculated values of the vector of unknown parameters $\bar{c}$ $(\bar{c} = (c_0, \ldots, c_u, \ldots, c_h))$ of dependency model (4) characterize the identification features of the face image.

A second-order polynomial is taken as $\mathbb{F}$. Then model (4) is given in the form

$$y = c_0 + c_1 x + c_2 x^2$$

where c_0, c_1, c_2 are the parameters that are determined based on the least squares criterion [19,20].

3. Extraction of Subsets of Closely Coupled Features of the Face Image. In this stage, n' "independent" subsets of identification features of the face image are determined [11,18,21].

4. Formation of a Set of Representative Features of the Face Image. In this stage, a set of representative features of the face image consisting of n' features is formed [18,22]:

$$\mathfrak{X}' = (\mathfrak{x}_{i_1}, \ldots, \mathfrak{x}_{i_q}, \ldots, \mathfrak{x}_{i_{n'}}).$$

As a result of this stage,$n\prime$ representative features for each face image.

5. Determination of the Difference Function $d(F_u, F)$ Between Face Images F_u and F. In this stage, the difference function is determined, which characterizes the difference measure between the images F_u and F and is given as [11,17,23,24]:

$$d(F_u, F) = \sum_{q=1}^{n} \lambda_q \left(a_{ui_q} - a_{i_q}\right)^2,$$

where λ_q is the unknown coefficient that characterizes the importance of the corresponding feature of the face images.

6. Determination of the Generalized Difference Function Between the Face Image F and a Class K_j. Here, a generalized difference function is defined; it characterizes the difference between the face image F and class K_j. The generalized difference function can be given in the following form [21–25]

$$\mathcal{D}(K_j, F) = \sum_{\mathfrak{F}_u \in \tilde{K}_j} \gamma_u d(F_u, F), \tag{5}$$

where γ_u is the parameter characterizing the importance of the image F_u in the training sample.

7. Determination of the Proximity Function $R(K_j, F)$ Between the Face Image F and the Class K_j. Here, the proximity function between the class K_j and the face image F is defined using radial functions [22,23,25]:

$$R(K_j, F) = \frac{1}{1 + \tau \mathcal{D}(K_j, F)}, \tag{6}$$

where $\mathcal{D}(K_j, F)$ is determined by formula (5).

8. Determination of the Generalized Proximity Function $B(K_j, F)$ Between a Face Image F and a Class K_j. At this stage, a generalized proximity function is determined, which characterizes the similarity of point F with the points belonging to class K_j. Each subset (class) of points is characterized by its representative points [22–25].

Assume that using radial functions (6) membership estimates for point F were calculated: $R(K_1, F), \ldots, R(K_j, F), \ldots, R(K_l, F)$. Then the function of the following form is taken as the generalized proximity function F to class K_j [24]

$$B(K_j, F) = \mathscr{b}'_j R(K_j, F) - \sum_{u \neq j}^{l} c'_u R(K_u, F), \tag{7}$$

where $R(K_j, F)$ is determined by formula (6).

Thus, the stages listed above, determine the classifier model based on the assessment of the relationship between the fragments of the image of the recognizable face.

4 Experimental Studies

To study the effectiveness of various classifier models, experiments on face recognition on a component-based approach were conducted. The experiments involved the local classifier proposed in this study, based on the assessment of the interconnectedness between image fragments (face components), designated as A1, the classifier model that implements the k-nearest neighbors' method, designated as A2, and the classifier model based on the calculation of estimates, designated as A3.

The experiment was conducted using the following databases of frontal face images:

– FERET [26], containing 1187 face images. Image size 256 × 384 pixels;

- LAB DPS [10], consisting of 9806 images. Image size 210 × 250 pixels.

The effectiveness of the tested classifiers was evaluated based on the accuracy and time of face component recognition. The distribution of the initial data into the training and control samples was 80% and 20%, respectively.

Table 1 shows the results of face component recognition based on the "FERET" image database.

Table 1. Recognition results for the face image database "FERET"

Algorithm	Recognition accuracy				Time spent on the analysis of one image (ms)
	Right eye	Left eye	Nose area	Mouth area	
A1	0.951	0.952	0.898	0.822	3.2
A2	0.917	0.901	0.764	0.766	10.4
A3	0.931	0.922	0.807	0.781	8.2

Table 2 shows the recognition results from the face image database "LAB DPS"

Table 2. The results of recognition by the gallery of face images "LAB DPS"

Algorithm	Recognition accuracy				Time spent on the analysis of one image (ms)
	Right eye	Left eye	Nose area	Mouth area	
A1	0.916	0.925	0.805	0.811	3.1
A2	0.835	0.856	0.703	0.702	9.7
A3	0.868	0.883	0.732	0.723	8.6

The higher accuracy of face component recognition in the FERET database is explained by the fact that the quality of face images in this database is somewhat higher than in the "LAB DPS" database.

The data given in Tables 1 and 2 illustrate the higher recognition quality of the proposed classifier A1 compared to classifiers A2 and A3. This is due to the fact that when training the classifier A1, models of the interdependence between the considered face fragments are formed (see stage 2 of the implementation of this classifier), which provide a more complete extraction of useful information from the training sample.

In addition, the data from Tables 1 and 2 show a significant gain in face component recognition time using the A1 classifier compared to the A2 and A3 classifiers. This is explained by the fact that in the learning process using the A1 classifier, there is a significant reduction in the dimension of the feature vector representing the recognizable component of the face. This gain in recognition time is also due to the fact that classifiers A2 and A3 use all objects of the training sample to make a decision in the recognition process.

Figure 3 shows the dependence of the quality of the tested classifiers on the size of the training sample with the same test data for the eye area. At a small size of the training sample, the proposed classifier model (A1) works better in comparison with the analogs. The advantage of classifier A1 with a small size of the training sample, compared to classifiers A2 and A3, is due to the fact that classifier A1 in the learning process (see stage 4 of the implementation of this classifier) forms a set of representative features, the number of which (n') is much less than the number of image identification features faces (n) used by classifiers A2 and A3, i.e. $n' << n$. This result confirms the classical statement in pattern recognition that an increase in the number of features leads to the need to increase the size of the training sample to ensure correct recognition.

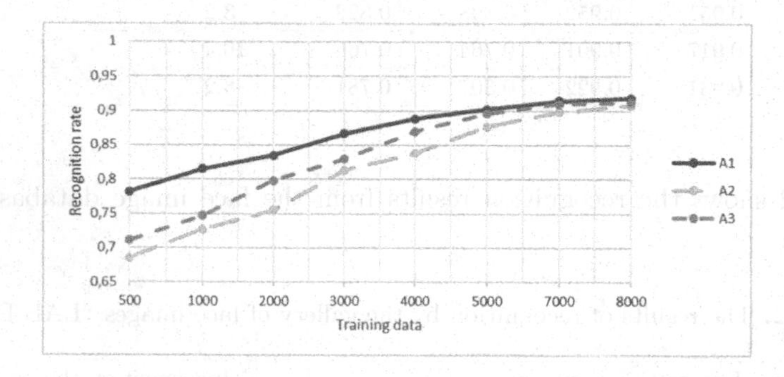

Fig. 3. The quality of local classifiers depending on the size of the training sample

Table 3. Results of integration of local classifiers solutions

Integrator	Recognition rate		
	Region 1 (eyes + nose)	Region 2 (nose + mouth)	Region 3 (eyes + nose + mouth)
Integrator 1	0,941	0,895	0,973
Integrator 2	0,913	0,876	0,945
Integrator 3	0,928	0,884	0,952

Table 3 shows the results of integrating the solutions of the local classifier A1, obtained by recognizing combinations of the main components of the face (eyes, nose, mouth). In accordance with the face recognition scheme shown in Fig. 2. In this scheme, the same classifier A1 was used as local classifiers 1, 2, 3. The neural network (Integrator 1), Bayesian (Integrator 2), and log-linear (Integrator 3) considered in [27] were used as integrators. The A1 classifier showed the best face recognition result based on all considered face components (Region 3). The recognition quality of Region 1 is better than Region 2 as the upper part of the face is more informative than its lower part. Sufficiently close values of the quality of recognition of Region 1 and Region 2 indicate that the local classifier A1 retains sufficient performance with partial occlusions:

- eyes are covered, for example, with dark glasses (Region 2 holds);
- the mouth area is covered, for example, by a thick beard, medical mask or scarf (Region 1 holds).

5 Conclusion

This study considers an approach to biometric identification of a person's personality using component-based face recognition, taking into account the interdependence between image fragments, which makes it possible to speed up the recognition process by implementing parallel recognition of face components, followed by the formation of a final solution based on the integration of local solutions.

To obtain local solutions, a local classifier was proposed, based on the assessment of the interdependence between image fragments of a recognizable face.

The use of this classifier provides higher accuracy of face recognition, in comparison with known analogs, with a small amount of training sample (Fig. 3), as well as a significant gain in time when recognizing face components (Tables 1 and 2).

In addition, the proposed local classifier, based on the assessment of the interdependence between image fragments of a recognized face, retains sufficient performance in case of partial occlusions.

References

1. Viola, P., Jones, M.J.: Robust real-time face detection. Int. J. Comput. Vis. **57**, 137–154 (2004)
2. Rettkowski, J., Boutros, A., Göhringer, D.: HW/SW co-design of the HOG algorithm on a Xilinx Zynq SoC. J. Parallel Distrib. Comput. **109**, 50–62 (2017)
3. Annalakshmi, M., Roomi, S.M.M., Naveedh, A.S.: A hybrid technique for gender classification with SLBP and HOG features. Clust. Comput. **22**, 11–20 (2019)
4. Kortli, Y., Jridi, M., Al Falou, A., Atri, M.: Face recognition systems: a survey. Sensors **20**(342) (2020)
5. Kukharev, G., et al.: Methods for face images processing and recognizing in biometrics. Polytechnika, St. Petersburg (2013)
6. Radjabov, S.S., Mirzaeva, G.R.: Preprocessing of face images for person identification. South-Siberian Sci. Bull. Biysk **2**(22), 50–55 (2018). https://doi.org/10.25699/SSSB.2018.22.15459
7. Fazilov, S., Urinov, E., Kakharov, S., Khashimov, A.: Improving image contrast: challenges and solutions. In: 2021 International Conference on Information Science and Communications Technologies (ICISCT), Tashkent, pp. 1–5. IEEE (2021). https://doi.org/10.1109/ICISCT52966.2021.9670106

8. Fazilov, S., Mirzaev, N., Mirzaev, O., Mirzaeva, G., Ibragimova, S., Rustamov, B.: Feature extraction model in systems of face images for person identification. In: Proceedings of the 9th International Conference Advanced computer information technologies, ACIT 2019, Ceske Budejovice, Czech Republic, 5–7 June 2019 (2019). https://doi.org/10.1109/ACITT.2019.8780089
9. Fazilov, S., Radjabov, S., Atakhanov, M., Khashimov, A.: Access control system based on facial image analysis. In: AIP Conference Proceedings, vol. 2432, p. 060007 (2022). https://doi.org/10.1063/5.0089655
10. Narzillo, M., Bakhtiyor, A., Shukrullo, K., Bakhodirjon, O., Gulbahor, A.: Peculiarities of face detection and recognition. In: 2021 International Conference on Information Science and Communications Technologies (ICISCT), Tashkent, pp. 1–5. IEEE (2021). https://doi.org/10.1109/ICISCT52966.2021.9670086
11. Kamilov, M.M., Fazylov, S.H., Mirzaev, N.M., Radzhabov, S.S.: Modeli algoritmov raspoznavaniya, osnovannyh na ocenke vzaimosvyazannosti priznakov. Nauka i tehnologiya, Tashkent (2020)
12. Zhuravlev, Y.I.: Selected scientific works. Master, Moscow (1998)
13. Fazilov, S., Mirzaev, O., Saliev, E., Khaydarova, M., Ibragimova, S., Mirzaev, N.: Model of recognition algorithms for objects specified as images. In: Proceedings of the 9th International Conference Advanced Computer Information Technologies, ACIT 2019, Ceske Budejovice, Czech Republic, 5–7 June 2019, pp. 479–482. IEEE (2019). https://doi.org/10.1109/ACITT.2019.8779943
14. Mirzaev, N., Saliev, E.: Feature extraction model in systems of diagnostics of plant diseases by the leaf images. In: Proceedings of the III International Forum "Instrumentation Engineering, Electronics and Telecommunications - 2017", Izhevsk, Russia, 22–24 November 2017, pp. 20–27. Publishing House of Kalashnikov ISTU, Izhevsk (2018). https://doi.org/10.22213/2658-3658-2017-20-27
15. Mirzaev, N., Saliev, E.: Recognition algorithms based on radial functions. In: 3rd Russian-Pacific Conference on Computer Technology and Applications, FEFU, Vladivostok, Russia (2018). https://doi.org/10.1109/RPC.2018.8482213
16. Darlington, R.B., Hayes, A.F.: Regression Analysis and Linear Models: Concepts, Applications, and Implementation. Guilford Press, New York (2017)
17. Olive, D.J.: Linear Regression. Springer, New York (2017)
18. Fazilov, Sh.Kh., Mirzaev, N.M., Radjabov, S.S., Mirzaev, O.N.: Determining of parameters in the construction of recognition operators in conditions of features correlations. In: Proceedings of 7th International Conference on Optimization Problems and Their Applications, Omsk, Russia, 8–14 July 2018, pp. 118–133 (2018). https://ceur-ws.org. Accessed 01 Aug 2022
19. Fazilov, S.K., Mirzaev, N.M., Mirzaeva, G.R., Tashmetov, S.E.: Construction of recognition algorithms based on the two-dimensional functions. In: Santosh, K.C., Hegadi, R.S. (eds.) RTIP2R 2018. CCIS, vol. 1035, pp. 474–483. Springer, Singapore (2019). https://doi.org/10.1007/978-981-13-9181-1_42
20. Li, J., Lu, B.-L.: An adaptive image Euclidean distance. Pattern Recogn. **42**, 349–357 (2009). https://doi.org/10.1016/j.patcog.2008.07.017
21. Hamdamov, R.H., Mirzaev, O.N., Fazilova, M.M., Mirzaeva, G.R., Ibragimova, S.N.: Model' algoritmov raspoznavaniya, postroennyh v prostranstve predpochtitel'nyh priznakov. Problemy vychislitel'noy i prikladnoy matematiki **5**, 75–86 (2020)
22. Mirzaev, O.N., Radjabov, S.S., Mirzaeva, G.R., Usmanov, K.T., Mirzaev, N.M.: Recognition algorithms based on the selection of 2D representative pseudo-objects. In: Jordan, V., Tarasov, I., Faerman, V. (eds.) HPCST 2021. CCIS, vol. 1526, pp. 186–196. Springer, Cham (2022). https://doi.org/10.1007/978-3-030-94141-3_15

23. Ibragimova, S.N., Radjabov, S.S., Mirzaev, O.N., Tavboyev, S.A., Mirzaeva, G.R.: Recognition algorithm models based on the selection of two-dimensional preference threshold functions. In: Djeddi, C., Siddiqi, I., Jamil, A., Ali Hameed, A., Kucuk, İ (eds.) MedPRAI 2021. CCIS, vol. 1543, pp. 354–366. Springer, Cham (2022). https://doi.org/10.1007/978-3-031-04112-9_27
24. Fazilov, S., Khamdamov, R., Mirzaeva, G., Gulyamova, D., Mirzaev, N.: Models of recognition algorithms based on linear threshold functions. In: Journal of Physics: Conference Series, vol. 1441, p. 012138 (2020). https://doi.org/10.1088/1742-6596/1441/1/012138
25. Russell, J., Cohn, R.: Radial Basis Function. Book on Demand, New York (2012)
26. Color FERET Database/NIST. https://www.nist.gov/itl/products-and-services/color-feret-database. Accessed 03 Aug 2022
27. Gorskij, N.: Raspoznavanie rukopisnogo teksta. Ot teorii k praktike. Polytechnika, St. Petersburg (1997)

Dominance Submissiveness Predisposition Scale (DSPS): Development and Validation

Ankita Shah(✉) and Uma Shanker Tiwary

Indian Institute of Information Technology, Allahabad, Prayagraj 211012, India
ashah2013.as@gmail.com, ust@iiita.ac.in

Abstract. Personality of a person is governed by his emotions. Dominance, Valence and Arousal are the three dimensions of Emotions and are analysed to know individual differences. There have been few attempts to uncover the factors of Dominance-Submissiveness, and little is known about how it varies among persons. In order to overcome both of these concerns, this study is used to develop the Dominance-Submissiveness predisposition Questionnaire (DSPS), an estimate of individual differences in the importance of several qualities. Exploratory Factor Analysis (EFA) was performed on first set of 189 items (N = 42). Following this study, Confirmatory Factor Analysis (CFA) was performed using a refined 48-item scale on a second sample (N = 43). A five-factor structure was analysed using this technique, and the results showed that the model fit was good (CFI = 0.75, RSMEA = 0.068). The five sub-scales of dominance-submissiveness predisposition represented by the factors are as follows: Emotional and Working on The Front Line; Defensiveness; Initiative and Conformity to Societal Conventions; Believe on Self Abilities; and Control Over Self & Emotions. All the sub-scales indicated high internal consistency and reliability. Stating construct validity, each sub-scale also showed a distinctive correlation with other dimensions assessing personality characteristics, emotions, and objectives. Overall, the results indicate that the DSPS is a viable and trustworthy tool for evaluating individual differences in terms of Dominance-Submissiveness.

Keywords: Dominance · Submissiveness · Scale · Emotions · Analysis · Individual Differences

1 Introduction

Emotions play an important role in human decision-making and interpersonal communication [1]. Furthermore, different emotional states may have an internal impact on human communication and the ability to memorise relevant information. These emotions are expressed by vocal or nonverbal signs such as voice inflection, facial expressions, and gesticulations [2]. From person to person, the various emotional states can be evaluated and recognised. Computational routines face a

H. Zaynidinov et al. (Eds.): IHCI 2022, LNCS 13741, pp. 188–200, 2023.
https://doi.org/10.1007/978-3-031-27199-1_20

challenge as a result of this. Various researchers have tried to evaluate emotions based on Basic Emotion Model, Dimensional Emotion Model, Appraisal Based Model. All emotions are considered distinct emotions in the Basic Emotion Model and distinguishable in human face expressions. In Dimensional Emotion Model, the affective states are described by the one or more dimensions. An appraisal based model sorts and extracts emotions, by continuously and cognitively appraising human emotions to an occurrence.

A fundamental framework for personality description is provided by the PAD Temperament Model, which includes of the traits of Pleasure-Displeasure, Arousability, and Dominance-Submissiveness. It is a psychological model for describing and measuring emotional states developed by A. Mehrabian and James A. Russell (1974) [3]. To represent all emotions, PAD uses three quantitative dimensions: Pleasure, Arousal, and Dominance. Its first application was in an environmental psychology theory, with the central idea being that physical settings have an emotional impact on humans. Later, Peter Lang and others employed it to construct a physiological philosophy of emotion. James A. Russell utilised it to create a hypothesis on emotional episodes (emotionally compelling events that lasted only briefly). In PAD, PA developed as a circumplex framework of emotion and titled "core effect". As part of the review process, the D component of PAD was reformulated (a rational evaluation of the circumstance that caused the emotion).

The PA component of PAD was transformed into a circumplex model of emotion experience, and those two components were labeled "core affect." The D part of PAD was re-conceptualized as part of the evaluation process (a cold cognitive assessment of the situation eliciting the emotion). The psychological construction theory of emotion is a more refined variation of this concept.

According to APA (American Psychological Association) Dictionary of Psychology, Dominance-Submission, a crucial aspect of interpersonal interaction, was discovered using factor analysis. It is a continuum that spans from strong dominance (active, chatty, extroverted, forceful, domineering, authoritative) to strong subordination (quiet, reserved, obedient, and feeble). Also called ascendance-submission.

Current state-of-the-art system is The PAD (Pleasure, Arousal, Dominance) model was employed to do research in consumer marketing (extra time, unplanned spending), nonverbal communication (studying body language in psychology), virtual emotion characters (the creation of animated creatures to express emotions), job recruitment and many more. However, in the emotion literature, dominance has not been studied as thoroughly as valence and arousal [4].

2 Material and Methods

Questionnaire Development. Theoretical and empirical literature acknowledging Dominance-Submissiveness constructs were analysed as a preliminary step for item collection. To find a variety of probable factors. The following concepts, principles and tools were examined: Dominance-Submissiveness, Pleasure Arousability Dominance Scales [5], California Psychological Inventory Dominance Scale [6], Personality Research Form - Dominance scale [7], Multidimensional Personality Questionnaire - Social Potency scale [8], Submissive Behavior Scale [9], an adolescent Social Goals Questionnaire [10], IPIP-NEO [11], IPIP-NEO-120 [12].

Following this process, 46 potential rewards of Dominance-Submissiveness disposition were identified: timidness, decision-making, recognition in public interactions, tendency to receive care from others, leadership and creativity at work, role played in conversations, defence of personal views, courage to say no, willingness to be vital, ease in speaking in public, problem - solving skills on one's own, self - control, sensitivity to intimidation by higher authority, ability to voice an opinion, confidence in one's abilities, need for moral support, reliance on experts, conforming to peer standards, Dominance-Submissiveness over others at workplace, susceptibility to fads, adherence to own convictions, conformity to society's customs, authority over one's own circumstances, reliance on authority, belief in one's own ideas, control over others, leadership, willingness to follow instructions, perspectives on laws and rules, patience, concern for others' opinions, assurance in social circumstances, insistence on one's own privileges, avoidance of conflict, courtesy, putting oneself first, capacity to impart wisdom, consciousness, ability to focus, tenacity, achievement, vociferousness and competency. It's crucial to keep in mind that the goal of this step was to create a wide variety of Dominance-Submissiveness Question Items without assuming that the different questions would match the factor structure of a Dominance-Submissiveness Questionnaire.

To capture the content of this broad range of Dominance-Submissiveness traits, questionnaire items were developed and guarantee that each factor's hedonic value was assessed, mostly questions have prefix as "I enjoy" [13]. For instance, the Dominance-submissiveness trait competence was evaluated using the statement "I enjoy playing games less when it is just for fun and there is no clear winner." Total of 204 items were produced during this phase (two to twenty items for each proposed type of Dominance-submissiveness trait) and order was randomised that was implemented using Fisher-Yates's algorithm and participant id as seed. Cues were added to the Question Items.

Word2Vec similarity analysis was done using Spacy. Item-pairs under same factor having similarity greater than 0.95 were excluded. Factor: Decision making, Item 7 (I do not like to be the one in a group making decisions.) and Item 10 (I like to be the one in a group making decision.): 0.98. Thus, Item 7 removed. Incase, only three items are in the factor, instead of removing the item, we rephrased it. In total, 6 question Items were removed.

Pilot questionnaire was presented to Graduate-level, Post Graduate and Research students. For the purpose of sensitively capturing individual indifference, a five-point scale (1 = Strongly disagree - 5 = Strongly agree) was used. "Consider the following statements as to what you like to do when you engage with others, or rate how much each item applies to you," the instructions said.

The statements apply to everyone throughout your life. Rank each statement according to how much it applies to you, from 1 to 5. NOTE: Consider whether you would value something if you had never done it.

Sample 1: Exploratory Factor Analysis (EFA)

Candidates. Candidates were asked to fill up the questionnaire in return for Assignment credit. For reaching experimental subjects and performing thorough surveys of broad population samples, Centre for Cognitive Computing is employed. Participants in the research signed up through CCC and were then told to fill the questionnaire and profiling. Participants received credits and a certificate of participation upon completing roughly 90 min answering the questionnaire.

Pilot questionnaire having 198-item was finished 37 times. It was a long questionnaire, thus divided DSPS scale to multiple parts. So that participants can fill it in multiple sitting without fear of losing the form. Due to the lack of variation in their responses, two participants were excluded (e.g., a participant answered "Neither agree nor disagree" to most of the question items). Thereby only 35 participants in Pilot 1. In Dominance-Submissiveness predisposition scale, Items having frequency(response) greater than 50% or equal to be excluded. The Question items having less variance on the participant's scores were removed. Procedure produced a total of 190 questions, and the main questionnaire was subsequently made by randomly ordering them.

Proceeded with the Main Study. Participants (14 females; 71 males;) were aged 20–39 years old (mean = 24.3, SD = 3.7). The highest completed education level of the sample was as follows: 62.3% Bachelor's degree, 28.2% Master's degree and 9.4% Postgraduate degree.

Data Analysis Procedure. To investigate the Dominance-Submissiveness item set's underlying structure, a range of EFAs ran utilising IBM SPSS Statistics for Windows, Version 21.0 [14]. Item-total correlation, reliability analysis, was carried out on the perceived Dominance scale comprising 189 items. Cronbach's alpha revealed that the questionnaire had acceptable reliability, = 0.911. Most things seemed to be worthwhile retaining, deleting them would cause the alpha to drop. The only exception was item 9, that would raise the alpha to $\alpha = 0.914$. As a result, removing this item should be considered.

Large number of Question items were detected to be highly collinear. Several Items having collinearity near zero were removed using Stata [15]. In total, 104 items was removed.

Principal Axis Factor Analysis was done in accordance with varimax rotation to investigate the fundamental structure for the 84 items of the Dominance Questionnaire. According to Keiser Criterion, Scree plot, Parallel Analysis and MLP Analysis number of factors retained were Five.

EFA Results. Since the questionnaire were framed in a manner they needed to be answered, thus no missing data was there. EFA detected five components having eigen values more than four, implying a five-factor structure. The Question-items having factor loadings ≥0.4 were removed. The first factor, which appears to assess Emotional and Working on The Front Line, had high factor loadings for initial 12 items. 2 had low indexes and -ve loadings. The 2nd factor, appeared to represent Defensiveness, has high loadings for later 9 items. The third factor, Initiative and Conformity to Societal Conventions, had high loadings on the next nine items. The fourth factor, Believe on Self Abilities, had high loadings on the next eleven items. The fifth factor, Control Over Self & Emotions, loaded heavily on the table's final seven items. "People advise me to keep check on my emotions." had its highest loading from the Control Over Self & Emotions factor, but it too had a significant loading in Defensiveness factor. These five factors were defined as follows: Emotional and Working on The Front Line, Defensiveness, Initiative and Conformity to Societal Conventions, Believe on Self Abilities and Control Over Self & Emotions. (see Table A at https://github.com/Ashah2013/DSPS).

Item Reduction. Many further steps were used to lessen the Question-Item set. The EFA results were used to make all decisions. To begin, items not loading strongly on either of the five factors (36 items; all loaded < 0.40 each factor) were removed. Second, item that had loadings greater than 0.40 loaded on at least two factors was removed (1 item was removed). Finally, only the best Question item from each factor was retained to create a concise scale. As a result, a 48-item scale with five subscales was created.

CFA was done on the selected 48 items with Sample 1 to investigate the effectiveness of the suggested 48-item prior proceeding to main data collection.

Sample 2: Confirmatory Factor Analysis (CFA). In the second phase, sample 2 was gathered to confirm the structure, validity and reliability for DSPS having 48 items.

Initially, CFA was performed over 48-item Dominance-Submissiveness scale. Secondly, Sample 2 participants also filled all standard proposed questionnaires forms related to individual difference, mental stability, and social desirability to establish the validity of the Dominance-Submissiveness predisposition. All correlational studies were done with the help of IBM SPSS Amos 26.0 for Windows [16].

Candidates. Candidates were again asked to fill up the questionnaire in return for Assignment credit as done in case of Exploratory Factor Analysis. Finally, a part of Sample 2's participants (N = 43) filled the Dominance-Submissiveness predisposition.

Measures. In addition to the Dominance-submissiveness questionnaire, participants filled the required below questions for construct validity purpose:

DSM-5 Level 1 Test [17]. This is a 23-item measure with thirteen subscales that each assess a different psychiatric domain. It assesses the person's mental stability, actually used as a screening tool. It works exactly like a Likert Scale, which ranges from 0–4. 0 denotes never and 4 denotes every day.

A participant is assumed to be mentally fit if he clears the DSM-5 Level 1 Test and then allowed to proceed further filling of questionnaire.

Arousal Predisposition Scale [18]. This is a 12-item questionnaire that measure individual differences in arousability. Likert scales are used to provide response for each item ranging 0–4 scale (0 = No, definitely not, 4 = Yes, definitely).

We hypothesized that Arousal and Dominance are correlated.

Valence Predisposition Scale. This is a thirty-six-item scale in development. We developed a Valence Predisposition Scale based on the two major categories "Positive Valence" and "Negative Valence" and the specific emotions it characterizes viz. anger, fear, joy etc. Likert scales are used to provide responses.

We hypothesized that Valence Predisposition would be somehow correlated with all other subscales.

3 Results

No missing data was there. The five-factor model generated from Sample 1 got an acceptable model value using the replication sample data, Sample 2 [2(215) = 628.139, p = .017; CFI = 0.75; RMSEA = 0.068, 90% CI = 0.07–0.08]. The question-items factor loading was generated to be acceptable as they were coming in 0.4–0.68 and mean was found to be 0.35, with SD (standard deviation) as 0.36) and are shown in Table B at https://github.com/Ashah2013/ DSPS.

3.1 Reliability

Mean inter-item correlations (MICs), Correlations and Cronbach alphas according to factor scores are presented in Table 1. Cronbach alphas were acceptable for all factors, (mean = 0.82, SD = 0.04; range = 0.77–0.87). The MICs of factors evaluating comparatively little construct (Clark and Watson, 1995) of reasonable values with mean as 0.56, SD (standard deviation) as 0.05 and range as 0.51–0.65. This implies that the Question items under each factor are having only one dimension according to their corresponding factor. Asterisks shows that the correlations are significant.

Table 1. Sample 2(N = 43): Correlations descriptive

	1	2	3	4	5	Mean+(SD)	MIC
1. F1	0.55					0.44(0.32)	0.16
2. F2	0.16	0.57				0.50(0.04)	0.25
3. F3	−0.04	**0.46***	0.55			0.53(0.06)	0.2
4. F4	**0.30†**	−0.35	0.12	0.59		0.43(0.01)	0.53
5. F5	**0.33***	−0.23	0.13	**0.58***	0.75	0.53(0.06)	0.35

Significance of Correlations:
† $p < 0.100$
* $p < 0.050$
** $p < 0.010$
*** $p < 0.001$

3.2 Validity

The amount to which a test measures what it is designed to measure is referred to as its Validity. CR (Composite Reliability), maximal reliability, omega threshold is 0.7. Removed Question Item 131(f1j), 119(f3g) since CR < 0.7. Each factor indicated CR ≥ 0.7. Thus, the construct reliability is established. For the Convergent Validity, AVE ≥ 0.3. For the Discriminant Validity, according to criterion of Fornell-Larcker, the MSV i.e., Maximum Shared Squared Variance value must be usually less as compared to the AVE i.e., Average Variance Extracted. Also, the values in bold should be greater than the all the values in the same column and same row given in previous table. Thence, factors are Discriminant Valid.

The relationships between the three Emotion scales and other was investigated using Pearson correlational analysis (Table 2).

Table 2. Validation of values based on factors in Sample2

	CR	AVE	MSV	MaxR(H)
F1	0.74	0.30	0.11	0.82
F2	0.72	0.33	0.22	0.81
F3	0.74	0.30	0.22	0.80
F4	0.73	0.36	0.34	0.75
F5	0.72	0.56	0.34	0.79

Abbreviations: CR, Construct Reliability; AVE, Average Variance Extracted; MSV, Maximum Shared Variance; MaxR(H), Maximum Reliability

The Dominance-submissiveness scale demonstrated expected relationships with the external scales, indicating relatively unique individual differences (Table 3).

Table 3. Pearson correlations and significance among DSPS and other scales

Dimension1	Dimension2	Correlation	Significance	Significant?
VPS	DSPS	0.03	93%	n
DSPS	APS	0.50	0%	y
VPS	APS	−0.31	1%	y

Valence Predisposition Scale was negligibly correlated with Dominance - Submissiveness Predisposition Scale. Dominance - Submissiveness Predisposition Scale was highly positive correlated with Valence Predisposition Scale. Valence Predisposition Scale was negatively moderate correlated with Arousal Predisposition Scale.

For the Significance test, since Likert scale is ordinal and test of correlation needs to be done, Spearman's rho test was chosen. We assumed a H0: Null Hypothesis: There is no correlation between data sets (Valence, Arousal and Dominance) and a H1: Alternate Hypothesis: There is some correlation between data sets. Significance between Dominance-Arousal Predisposition Scale show very strong evidence for rejecting Null Hypothesis (H0).

Significance between Valence and Arousal Predisposition Scale show strong evidence for rejecting Null Hypothesis (H0). Significance between Dominance and Valence Predisposition Scale cannot reject Null Hypothesis (H0). Insufficient evidence to believe H1 (Table 4).

Table 4. Pearson correlations between DSPS factors and external measures

	DSPS				
	F1	**F2**	**F3**	**F4**	**F5**
VPS	0.49**	0.38**	0.19	0.27*	0.31**
APS	−0.36**	−0.02	0.41**	−0.07	−0.53**

Significance of Correlations:* $p < 0.050$** $p < 0.010$

DSPS F1 Emotional and Working on The Front Line was significantly positively associated with valence. DSPS F2 Defensiveness was positively associated with valence and significant. DSPS F3 Initiative and Conformity to Societal Conventions was positively associated with both Valence and Arousal. DSPS F4 Believe on Self Abilities was positively associated with Valence. DSPS F5 Control Over Self & Emotions was significantly associated with both Valence and Arousal (Table 5).

Table 5. Significant difference among factors of Dominance-submissiveness on the basis of different aspect of participants

Scale	Factor	P-value (sig)	Significant?
F1	Gender	0.009	y
	Course	0.374	–
F2	Gender	0.724	–
	Course	0.720	–
F3	Gender	0.456	–
	Course	0.150	–
F4	Gender	0.850	–
	Course	0.002	y
F5	Gender	0.884	–
	Course	0.630	–

in F4-Believe on Self abilities, Male mean score +.17Female mean score −.86

Table 6. Significant difference based on Course pursued by participants

Test	P value (sig)	Significant?
B.tech vs M.tech	0.023	y
B.tech vs Ph.D	0.032	y
M.tech vs Ph.D	1	–

B.tech mean score 4.19, M.tech mean score 4.54, Ph.D mean score 4.51

Significance test within scales based on Factors was done. Individual samples t-test was carried out for the gender significance (MvsF). It showed significant difference for the Factor F4- Believe on Self Abilities, others were insignificant. For the courses (Btech vs Mtech vs PhD), analysis through ANOVA was done. Factor F1- Emotional and Working on The Front Line was significantly different, thus post-Hoc test for significance detected (Games-Howell Test) was done.

The results indicated a significant difference in mean score among three different courses pursued, F (2,82) = 4.93, p = 0.009. Pairwise comparison of the means using Games-Howell Test revealed significant difference between B.tech and other 2 courses pursued ($p < 0.05$). More specifically, in the B.tech (M = −.25, SD = .97, p = 0.023) the Dominance mean scores of Factor1 were significantly lower than the M.tech (M = .41, SD = 1) and Ph.D (M = .41, SD = .55, p = 0.032). Other than this, non-important point of interest found on the other courses pursued by participants ($p < 0.05$; see Table 6).

4 Discussion

The 48-item DSPS is a detailed evaluation of individual difference might to succumbed to external factors or not. We found five sub-scales utilizing EFA and CFA for DSPS that correspond to five major factors: Emotional and Working on The Front Line, Defensiveness, Initiative and Conformity to Societal Conventions, Believe on Self Abilities, Control Over Self & Emotions. The findings resulted DSPS having a distinct model structure and significant features that assesses Individual differences.

DSPS F1 Emotional and Working on The Front Line was positively correlated with Valence and negatively with Arousal, the results are significant, as concluded by "subjective emotional valence and arousal dynamics have specific physiological correlates" [19]. DSPS F2 Defensiveness was positively associated with valence and significant, suggesting this subscale not at all related with Arousal. "The role of a single valence dimension for behaviour prediction is limited. Although positive affect has been associated with a generative behavioural orientation (exploring, achieving positive outcomes, risk taking, little loss aversion) and negative affect with a defensive behavioural orientation" [20]). DSPS F3 Initiative and Conformity to Societal Conventions was positively correlated with both Valence and Arousal, but association with Arousal is significant. It revealed an inverse pattern of associations to DSPS F2 Defensiveness Although the relationship between them is somewhat significant having r as 0.4 and p value less than 0.05, but that does not mean they are measuring the same thing. DSPS F4 Believe on Self Abilities was positively associated with Valence and significant. "Beliefs about personal ability to influence a particular situation may initiate and sustain coping behaviour, increasing the likelihood of experiencing positively valanced feelings of power" [21]. Finally, DSPS F5 Control Over Self & Emotions showed the expected high positive correlations with valence and moderate negative correlation with arousal. Positive valence resulted in enhanced control of vehicle [22–24]. High arousal resulted in a quicker response in hazard diagnosis [24–26].

Significant difference was found between the Gender (Male vs Female) was found when it comes to Dominance factor F1 Emotional and Working on The Front Line. It has different implications for the different gender. Men and women are likely to be believed more agentic and communal respectively. "Dominant and submissive acts were shown to be bidimensional, equally desirable for men and women, and gender stereotypic" [27]. It is a common paradox, that men are born to be leaders and women followers. But the world is changing now, the women leadership is pervasive. In 2021, women account for 26% of all CEOs and managing directors worldwide, whereas only 15% in 2019. "Institutional requirements for gender equity and inclusion can be transformational in shaping male and female preferences and female access to leadership" [28].

Another significant result was found among the participants based on the Courses they are enrolled in. Dominance factor F4 Believe on Self abilities has different implications for different age group people. Pairwise comparison of the means using Games-Howell Test revealed significant difference between B.tech and other 2 courses pursued ($p < 0.05$) M.Tech and Ph.D. The findings are in agreement with previous research indicating that older adults are less confident in their abilities than younger adults [29,30].

It is critical to recognise limitations of the DSPS. To begin, dominance-submissiveness is a complicated construct; as a questionnaire, the DSPS will imply some simplification, which may mask minute concepts. Second, it is one of the initial studies to analyse the fundamental model of dominance-submissiveness empirically. Future research should reproduce the factor structure in other datasets, as well as the test-retest reliability in bigger samples. Finally, there may be other dominance-submissiveness factors that are not investigated by the DSPS and have yet to be accurately identified in the available research. The DSPS, on the other hand, provides a potential foundation for further experimentally assessing individual differences in characteristics.

5 Conclusion

Dominance-Submissiveness Predisposition Questionnaire and factor analysis is one of the initial studies to establish and assess the Individual Difference. Using EFA and CFA, five factors are discovered in this study: Emotional and Working on The Front Line, Defensiveness, Initiative and Conformity to Societal Conventions, Believe on Self Abilities, Control Over Self and Emotions. These five factors of Dominance-Submissiveness were discovered to be strongly and differentially related to a wide range of self-reported character traits, beliefs, and objectives. It is established that the DSPS is a trustworthy and precise assessment with application in the experiment of Dominance-Submissiveness.

References

1. Ekman, P.: Basic emotions. In: Handbook of Cognition & Emotion, pp. 301–320. Wiley, New York (1999)
2. Liu, Y., Sourina, O.: Real-time subject-dependent EEG-based emotion recognition algorithm. In: Gavrilova, M.L., Tan, C.J.K., Mao, X., Hong, L. (eds.) Transactions on Computational Science XXIII. LNCS, vol. 8490, pp. 199–223. Springer, Heidelberg (2014). https://doi.org/10.1007/978-3-662-43790-2_11
3. Mehrabian, A., Russell, J.A.: An Approach to Environmental Psychology. The MIT Press, Cambridge (1974)
4. Osgood, C.E., Suci, G.J., Tannenbaum, P.H.: The Measurement of Meaning. University of Illinois Press, Urbana (1957)
5. Mehrabian, A.: Pleasure-Arousal-Dominance: a general framework for describing & measuring individual differences in temperament. Curr. Psychol. **14**, 261–292 (1996). https://doi.org/10.1007/BF02686918
6. Gough, H., Bradley, P.: CPI Manual. Consulting Psychologists Press, Palo Alto (1996)
7. Jackson, D.N.: Personality Research Form. Sigma Assessment Systems Inc., Port Huron (1999)

8. Tellegen, A., Waller, N.G. (eds.): Exploring Personality Through Test Construction: Development of the Multidimensional Personality Questionnaire the Sage Handbook of Personality Theory & Assessment, vol. II. Sage, London (2008)
9. Allan, S., Gilbert, P.: Submissive behaviour & psychopathology. Br. J. Clin. Psychol. **36**, 467–488 (1997)
10. Jarvinen, D.W., Nicholls, J.G.: Adolescents' social goals, beliefs about the causes of social success, & satisfaction in peer relations. Dev. Psychol. **32**(3), 435–441 (1996). https://doi.org/10.1037/0012-1649.32.3.435
11. Goldberg, L.R.: A broad-bandwidth, public domain, personality inventory measuring the lower-level facets of several five-factor models. Pers. Psychol. Europe **7**, 7–28 (1999)
12. Johnson, J.A.: Measuring thirty facets of the Five Factor Model with a 120-item public domain inventory: development of the IPIP-NEO-120. J. Res. Pers. **51**, 78–89 (2014)
13. Snaith, R.P., Hamilton, M., Morley, S., Humayan, A., Hargreaves, D., Trigwell, P.: A scale for the assessment of hedonic tone the Snaith-Hamilton Pleasure Scale. Br. J. Psychiatry **167**(1), 99–103 (1995)
14. IBM Corp. Released 2012. IBM SPSS Statistics for Windows, Version 21.0. IBM Corp, Armonk
15. StataCorp.: Stata Statistical Softwrae: Release 17. StataCorp LLC, College Station, TX (2021)
16. Arbuckle, J.L.: Amos (Version 26.0) [Computer Program]. IBM SPSS, Chicago (2019)
17. Web resource: DSM-5 Self-Rated Level 1 Cross-Cutting Symptom Measure-Adult
18. Coren, S.: The arousal predisposition scale: normative data. Bull. Psychon. Soc. **28**, 551–552 (1990). https://doi.org/10.3758/BF03334078
19. Sato, W., Kochiyama, T., Yoshikawa, S.: Physiological correlates of subjective emotional valence & arousal dynamics while viewing films. Biol. Psychol. **157**, 107974 (2020). ISSN 0301-0511
20. Seo, M.-G., Bartunek, J.M., Barrett, L.F.: The role of affective experience in work motivation: test of a conceptual model. J. Organ. Behav. **31**, 951–968 (2010)
21. Vera, S., David, S., Klaus, S.: Levels of valence. Front. Psychol. (2013)
22. Chan, M., Singhal, A.: The emotional side of cognitive distraction: implications for road safety. Accid. Anal. Prev. **50**, 147–154 (2013)
23. Hancock, G., Hancock, P., Janelle, C.: The impact of emotions & predominant emotion regulation technique on driving performance. Work **41**(Suppl. 1), 3608–3611 (2012)
24. Trick, L.M., Brandigampola, S., Enns, J.T.: How fleeting emotions affect hazard perception & steering while driving: the impact of image arousal & valence. Accid. Anal. Prev. **45**, 222–229 (2012)
25. Navarro, J., Osiurak, F., Reynaud, E.: Does the tempo of music impact 25 human behavior behind the wheel? Hum. Factors **60**(4), 556–574 (2018)
26. Ünal, A.B., de Waard, D., Epstude, K., Steg, L.: Driving with music: effects on arousal & performance. Transp. Res. Part F: Traffic Psychol. Behav. **21**, 52–65 (2013). https://doi.org/10.1016/j.trf.2013.09.004
27. McCreary, D.R., Rhodes, N.D.: On the gender-typed nature of dominant & submissive acts. Sex Roles **44**(5/6), 339–350 (2001)
28. Smith, J.E., von Rueden, C.R., van Vugt, M., Fichtel, C., Kappeler, P.M.: An evolutionary explanation for the female leadership paradox. Front. Ecol. Evol. **30** (2021). https://doi.org/10.3389/fevo.2021.676805

29. Berry, J.: Memory self-efficacy in its social cognitive context. In: Hess, T.M., Blanchard-Fields, F. (eds.) Social Cognition & Aging, pp. 70–96. Academic Press, New York (1999)
30. Miller, L.M.S., Lachman, M.E.: The sense of control & cognitive aging: toward a model of mediational processes. In: Hess, T.M., Blanchard-Fields, F. (eds.) Social Cognition & Aging, pp. 17–41. Academic Press, New York (1999)

Design of a Mixed Reality-Based Immersive Virtual Environment System for Social Interaction and Behavioral Studies

Sophia Matar[1], Alfred Shaker[1], Saifuddin Mahmud[1], Jong-Hoon Kim[1](✉), and Jan-Willem van't Klooster[2]

[1] Advanced Telerobotics Research Lab, Computer Science, Kent State University, Kent, OH, USA
jkim72@kent.edu

[2] The BMS Lab, University of Twente, Enschede, The Netherlands
http://www.atr.cs.kent.edu, https://bmslab.utwente.nl/

Abstract. The advancements in immersive technologies allow us to create more sophisticated environments designed to help engage users by merging the physical world with a digital or simulated reality. These can range from completely immersive virtual environments to mixed reality immersive environments, where the virtual world and the real world collide. Virtual reality is a completely immersive environment where the users' reality is replaced with a simulated environment, and the hardware works to convince the user that they are in a different world. In contrast, augmented reality is a mixed type of reality, combining both the virtual and the natural world by augmenting the real world with digital assets and components. While both types of experiences contribute to creating rich collaborative environments, a limitation, and sometimes inconvenience, is present with the requirement of wearing a head-mounted device (HMD), creating restriction that prevents users from having physical interactions with others. Rather than interacting in the virtual space, we propose a concept that provides the structure for a physical space where users can interact with the shared mixed reality environment, an environment projected to help create the collaborative aspect in this project without any wearable devices. This paper will present the developed system and implemented four-dimensional interactions and demonstrate the feasibility of the structured experience we have created.

Keywords: Immersive Environments · Mixed Reality Environment · Augmented Reality · Virtual Reality

1 Introduction

Immersive technology helps create environments where the line between the physical world and the virtual world is blurred, producing a sense of immersion.

H. Zaynidinov et al. (Eds.): IHCI 2022, LNCS 13741, pp. 201–212, 2023.
https://doi.org/10.1007/978-3-031-27199-1_21

The technology is able to trick your senses into thinking you are experiencing something different. Examples of these technologies are virtual reality and augmented reality. Virtual reality is a technology that generates an environment replacing the physical world with a simulated environment with the use of various head-mounted devices (HMDs). In comparison, Augmented reality is a technology that enhances the physical world with interactable digital elements that are augmented on top of the real physical world, either using markers or depth-mapping. There are many facilities and labs where the purpose is to create immersive environments using all types of immersive technology. An example of these labs is the Curved Corner Room, a room-sized immersive projection display that provides an interactive high, present virtual world. The corner area of a room is used as a screen, and a wide-angle image is projected onto the curved surface. Using this system, one PC and projector system can generate an immersive virtual environment covering the user's view [9]. But immersive technology has been able to advance and become more impressive and sophisticated, expanding from a simple set-up to having multiple projectors as well as having a single synced screen wrapping around three walls of the room. Allowing bystanders to not only view the virtual environment but also interact with it in a more immersive way than before [16]. There are various studies where this technology has been applied and iterated on [7,12], and in this study, we show our approach to it using the Blank Lab facility.

1.1 Blank Lab

The Blank Lab serves as a "black box" environment for researching group activities that cross the boundaries of immersive technology, such as augmented, virtual, and extended reality tools. The area is adaptable, so projects incorporating these technologies can be developed and demonstrated. Whether users are experimenting with projections to create "cave-like" experiences or using AR/VR headsets, the objective is to support group experiences for up to 15 people in all circumstances. The main area of the Blank Lab includes a suspended grid ceiling that enables the installation of a hung curtain wall and the ability to reconfigure the support for cameras, projectors, etc. The curtain wall makes it possible to install speakers, wiring, and other supporting technologies behind the wall without the public being aware. The ability to create, modify, and support the presentation of these technologies in the main area is provided by servers and workstations housed in a control room next to the main space. To allow immersive experiences, the entire environment is sound and light isolated (not sound or light "proof"). [3] The Blank Lab is a multipurpose space many use to have their projects displayed. Since not every project is done with the needed setup and configuration, the lab has special software to ensure every project can be projected in the correct format. Allowing the Blank Lab to be a shared space has its pros and cons, pro in regards to being a space so many can use for a wide range of projects. Con regarding the ability to not configure the lab specifically for individual projects, since the cost of maintenance is relatively high for the upkeep of the lab so, a single configuration is set, and using the different software in the labs' computers, the various projects can be manipulated as needed, using scalable displays for the syncing of the nine projects, and Pro-tools for the syncing for the 13 speakers in the lab as depicted at Fig. 1. This is one of the critical challenges in this project that the proposed system should perform all intractable features as well as immersive virtual environments under restrictions of *Salable display and Pro-tools driving software.*

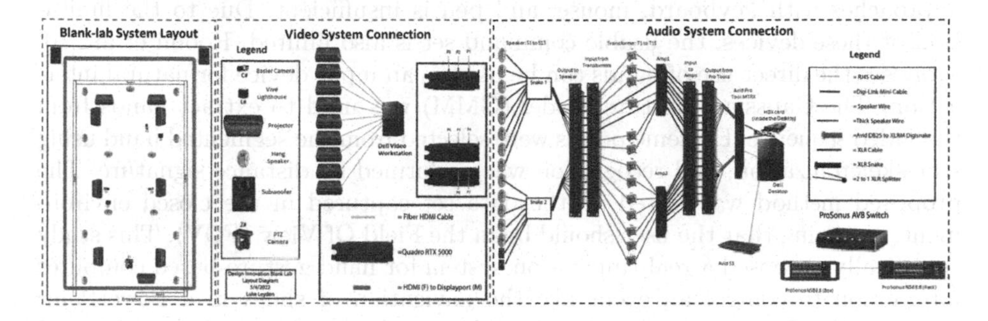

Fig. 1. The Blank-Lab system layout and configuration

This paper is structured as follows: Section Two describes similar projection room projects as related works. In Section Three, we have detailed the system overview behind the ATR Immersive Virtual Environment. Section Four, the implementation on the ATR Immersive Virtual Environment system. Section Five contains the discussion points to validate our system. Finally, Section Six depicts the future research plan and conclusion.

2 Related Work

Virtual reality allows users to experience an immersive virtual environment (IVE) in a multi-sensory fashion. While most previous works focused on providing the VR user with a realistic and highly immersive VR experience, bystanders in the physical environment are often not involved. Zenner et al. [16] has introduced a projection-based system to include bystanders in the virtual experience by projecting the virtual environment onto the registered, physical counterpart. Using controllers, the system allows bystanders to interact with the IVE and perceive the virtual environment semi-immersively. Similar to this paper, this research plans to create an environment where multiple bystanders can interact with the virtual environment, except this environment will not use any controllers. Instead, the participants can use hand gestures to manipulate the environment.

A large-scale display system with immersive human-computer interaction (HCI) is an essential solution for virtual reality (VR) systems. In contrast to the traditional human-computer interactive VR system that requires the user to wear heavy VR headsets for visualization and data gloves for HCI. Wang et al. [15] proposes a method that utilizes a large-scale display screen, with or without 3D glasses, to visualize the virtual environment, and bare-handed gestures are used to manipulate the virtual environment. Similar to this paper, gesture recognition is used to have users interact with the virtual environment. In contrast, this research utilizes the 270-degree view it has, using the screen which stretches across three of the labs' walls.

With the development of ubiquitous computing, current user interaction approaches with keyboard, mouse, and pen is insufficient. Due to the limitation of these devices, the usable command set is also limited. Roomi et al. [10] proposes the direct use of hands can be used as an input device for natural interaction. The Gaussian Mixture Model (GMM) was used to extract hands from the video sequence. Extreme points were selected from the segmented hand using star skeletonization, and recognition was performed by distance signature. The proposed method was tested on the data set captured in the closed environment, assuming that the user should be in the Field Of View (FOV). This study specifically proposed a real-time vision system for hand gesture-based computer interaction to control an event like the navigation of slides in a PowerPoint Presentation. In this paper, a hand tracking code using python OpenCV and CVZone is written. With the same goal as the GMM, this script helps detect the main segments of the hand, and using that data it can implement the hand model in the unity 3D environment. Similarly, a real-time hand tracking system will be used in the ATR immersive virtual reality system to have the users manipulate the virtual environment projected around them.

Rompay et al. [14], inspired by research that demonstrated the positive effects of nature-based imagery on well-being and cognitive performance, this study aims to research to what extent nature imagery can also enhance creative performance. Imagery presenting green settings varying in predictability and spaciousness was displayed before and during a creative drawing task in a high school classroom. After finishing the task, the students were given a questionnaire to complete, comprised of self-report measures for perceived creativity and positive affect. Unpredictability and spaciousness enhanced creative performance, with images combining these factors being particularly inspiring. This study helped provide the necessary findings to demonstrate that nature imagery can increase creativity in individuals and warrant follow-up studies that may further clarify the role of spaciousness, unpredictability, and other creativity-enhancing features of nature imagery. Using the virtual environment that the nature imagery came from, the implementation of the virtual nature in the Blank Lab, with the addition of the interactive elements, will help immerse the participant when using the virtual environment. Instead of only showing the imagery of the virtual nature, the user will be able to interact and manipulate the environment using simple hand gestures.

Lin et al. [8] proposes a Virtual Reality system to reduce anxiety and stress using music therapy called Virtual Harmony. The system allows users to listen to music and play different instruments in a 3D environment with real-world panoramic video as background scenes. This system avoids using a traditional joystick or keyboard; the participants can use their hands to play the virtual instruments creating a more realistic feeling. Similarly, in this paper, the participants can use their hands to control the virtual environment, helping create an immersed environment for the user. The user will also be able to choose between two virtual environment types, the first being an environment with a real-world panoramic video as background scenes and the second being a completely digitized environment as the background scene.

Houwelingen-Snippe et al. [13] proposed that digital nature can substitute for real nature for those with limited access to green space or confined to their homes. In a large-scale online survey, participants watched videos of digital nature, varying in nature and spaciousness. Results show a significant increase in connectedness to the community after watching digital nature. Instead of watching a video of the virtual nature, participants will be able to interact and manipulate the virtual environment projected around them. Without the need for any hardware, participants can use different gestures to help control the environment.

Among all these VR technologies applications, virtual tours allow people to be immersed in remote environments, as seen in studies by Shaker et al. [11] and Hendricks et al. [5]. In Shaker's work, the study involved using a virtual reality tour to quantify the effects VR has on individuals with developmental disabilities and see if VR can be used as a form of therapy to improve their mental state. The study had users interact with the scene using gaze and controller interfaces and was created using 360 photographs of an actual location, which made the experience more immersive. Their study showed that this kind of technology positively impacts individuals with developmental disabilities and that VR has potential in the therapy field for such individuals and others as well. Further, Hendrick et al. took the technology of the virtual tour a step further by creating a virtual tour that was dynamic in nature by creating a content management system (CMS) for admins that allowed them to edit and add new content to the virtual tour. The client-side application for the tour comprised the main interface and a skeleton scene populated by requesting its data from the server, which allowed for the previously mentioned CMS to function well and provide a seamless, uninterrupted experience for the end-users.

Both virtual reality (VR) and immersive projection display systems have a wide range of use, from Gaming to simulation training. As proposed by Rory Clifford et al. [4], Virtual reality has been utilized for training in all disciplines. Virtual Situation Awareness training is used effectively as a training tool, proposed to be less hazardous, lower in cost, and could create better learning outcomes than performing field tasks alone. In this study, Clifford proposes to verify the suitability and effectiveness of VR Head-Mounted Displays (HMDs) and immersive projection display systems to discover if they afford any advantages over traditional displays in terms of usability, task performance, and ability to acquire situation awareness in wildfire aerial firefighting scenarios. The same training simulation is used for all three methods tested in this research. The application adapted to each device's configuration, starting with High Definition TV (HDTV), then a 270-degree cylindrical simulator projection display

(SimPit), and finally an Oculus Rift. The results showed an improvement in situation awareness in the immersive systems over the non-immersive HDTV in each Situation Awareness. Indicated by the results the SimPit is the most suitable device for situation awareness training due to several factors: Being able to see your body, having access to tools such as notepads, etc., and having an entire unrestricted field of view. The SimPit could also be used for a longer duration due to lower simulator sickness. Significant simulator sickness is observed in the HMD over the SimPit. Similarly, in our research, both the BMS Virtual Nature Healing Environment [1] and the Virtual Music Therapy: Virtual Harmony [8] applications, respectively, have the ability to adapt to both the Oculus Rift and the Blank Labs configuration, creating an immersed environment.

Horan's et al. research proposes the design of a reconfigurable VR System [6]. CAVE Automated Virtual Environments (CAVE) is an alternative to HMDs; CAVE is a reconfigurable VR system that incorporates 6-degree-of-freedom haptic interactions and 3D ambisonic audio. The CAVE VR system provides the fundamental system designers use to develop applications to help optimize the interaction and audio technologies available in the system. Three main configurations are presented with experiments in this research, each tested with three VR technologies from different disciplines; Micro-robotic Cell Injection, Virtual Dashboard for Bicycle, and Architectural Visualization. All applications were tested on a desktop as well as the Reconfigurable VR System. The Architectural Visualization application was also tested with a HMD; similarly to the ATR Immersive Virtual Environment system, the application is transferable and configurable between multiple devices.

The last two studies indicate that immersive experience was enhanced when participants use the applications under an immersive projection display system. By using this indication, we have designed the system architecture for the ATR Immersive Virtual Environment which provides immersive virtual environments, creates an immersive space where users are able to share social interacts, and manipulates a shared mixed reality environment with hand gestures and sounds. In the following section, the proposed system architecture is explained.

3 System Overview

The ATR Immersive Virtual Environment System focuses on creating a mixed reality-based immersive environment that helps provide the sense of being able to interact with the virtual environment. In this research, one of the main goals is to create an application where users can interact with the virtual environment without needing any hardware. The ATR Immersive Virtual Environment System is a system that combines both the audio and video systems in the Blank lab, which is one of the key challenges mentioned at Sect. 1.1. The ATR immersive virtual environment severe helps create an immersive environment where users can interact with the virtual environments without the constraint of needing to wear a headset or the limitation of having to experience the space alone. Figure 2 shows the system architecture for the ATR Immersive Virtual Environment System. Starting with the ATR immersive virtual environment severe, the

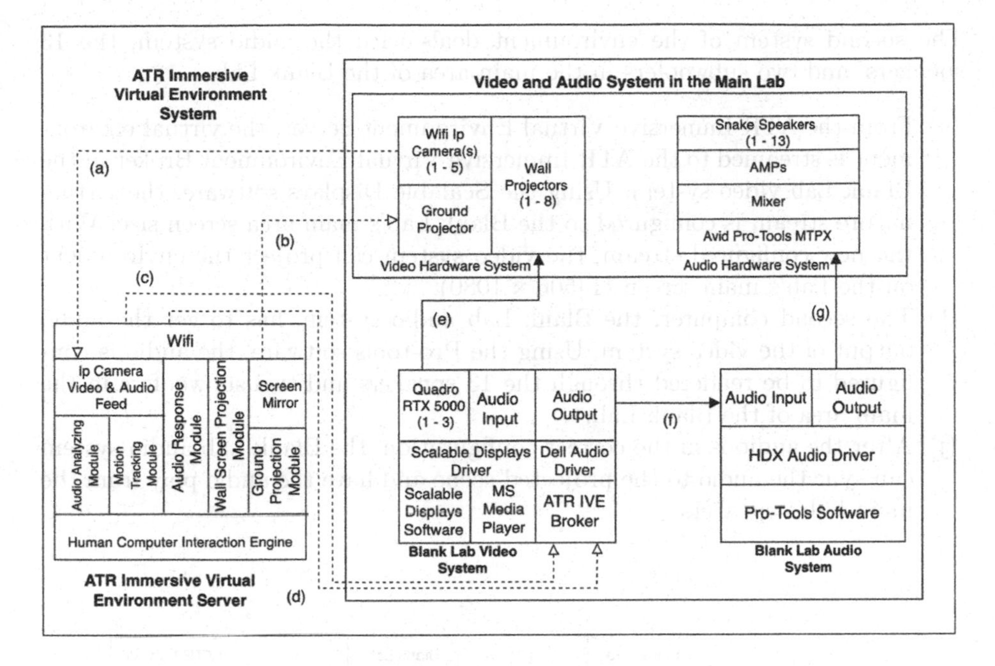

Fig. 2. Immersive Virtual Environment System Architecture

server will be processing all the input from the camera in the blank lab to ensure that the projected scene and the ground projections are fully updated with the correct scenes.

(a) The video feed from the IP camera(s) will be used to ensure the environment is constantly up to date, fitting how the user interacts with it. The camera will stream its feed into the ATR immersive Virtual Environment server. In the server, the motion tracking module will be able to track the user's position and hand gestures, which are used to control the virtual environment. With the motion tracking module, the Human-Computer Interaction Engine (Fig. 3) will also be used to update the surround screen and the ground projection with the respective scene and the speakers with the required audio. There will be two ways to project the virtual environment in the Blank Lab.
(b) The first projection is for the ground scenes. Since an external projector will be installed in the Lab, the scene generator will mirror the scene fitting for the ground of the Blank Lab.
(c) The audio module, audio that gets picked up from the Ip camera(s), is used along with the user's hand gesture to manipulate the virtual environment. From a voice command to help open the menu to having a specific sound cause a graphical element change in the background.
d) The Blank Labs' wall projectors. The Blank Lab has two separate computers, one for the video system, concentrating on the nine projectors in the Lab.

The second system of the environment deals with the audio system, the 13 speakers, and two subwoofers in the main area of the Blank Lab.

(e) From the ATR Immersive Virtual Environment Server, the virtual environment is streamed to the ATR Immersive Virtual Environment Broker in the Blank Lab video system. Using the Scalable Displays software, the virtual nature stream is configured to the Blank Lab's main area screen size. With the new configured stream, the video system can project the environment on the Lab's main screen (14506×1080).
(f) The second computer, the Blank Lab audio system, has to get the audio output of the video system. Using the Pro-tools software, the audio is configured to be rendered through the 13 speakers and two subwoofers in the main area of the Blank Lab.
(j) After the audio is in the correct configuration, the Blank Lab audio system can sync the audio to the projected scene and have the audio play from the main Lab's speakers.

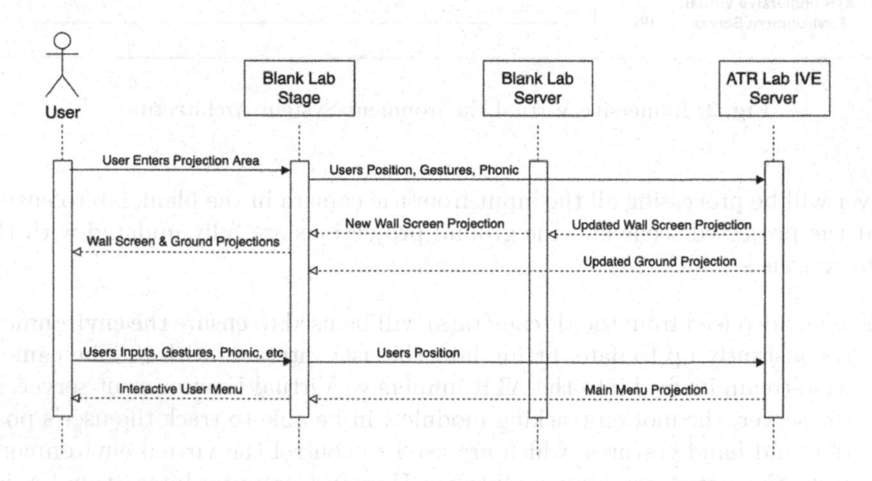

Fig. 3. User - Immersive Virtual Environment Interaction Diagram

The Human-Computer Interaction Engine (Fig. 3) will be used to update the surround screen and the ground projection with the respective scene. The Human-Computer Interaction Engine uses the feed from the motion tracking module and the audio detection module to generate the needed audio and the updated scene for either the ground projection or the wall screen projections.

4 Implementation

In this paper, we focus on the validity of key concepts and verification of essential functionalities for the proposed system under restrictions of the current blank lab environment as described at Sect. 1.1. These included the ability to collect and process the data from the users' actions to create a seamless application where users can interact with the virtual environment. The user's position will be detected using the live video feed from the Wifi IP Cameras. The position of the user affects two main aspects of the project. The first is the 'activation' of the application. When the user is in the projected area, the main menu appears, and the user can start using the application. The participant will be able to manipulate the virtual environment by using different hand gestures. The motion tracking module and the human-computer interaction engine will recognize the various gestures and the different commands corresponding to each gesture. The second element is the interactive ground projection. Depending on which scene the user chooses as their background, the corresponding ground projection will also be displayed based on the scene the participant chooses to experience. The ground projection will also have interactive elements controlled based on the participant's position (Fig. 4).

Fig. 4. Wall Projections in the Blank Lab. (left to right); Main Menu, Virtual Nature Healing Environment, Virtual Music Therapy

Main Menu: The application's main screen introduces the options the user has for what type of scene they would like to experience. Both examples provided are beneficial for one's mental health, the only difference being that one focuses on the healing qualities of nature and the other through music.

4.1 Virtual Nature Healing Environment

This BMS software was created with the means of healing using nature. The same healing qualities were present in the virtual nature environment as in the real-world nature environments. With the Virtual healing environment, the user has the ability to design and experience the virtual environment of their choosing. In this research, we reconstructed the user interface to allow users to interact with the application through gestures, from looking around the environment to editing the space around them.

4.2 Virtual Music Therapy; Virtual Harmony

This virtual therapy system developed to reduce anxiety and stress by using music therapy is another possible experience the user can choose. The system allows users to listen to music and play different instruments in a 3D environment with real-world video as their background. In the Blank Lab, the user will be able to experience these environments in the isolated lab, helping to create an environment where the user can be immersed in whichever application they select to experience. The user can control the application with their hands, from choosing the real-world video they set as their panoramic background to choosing the instrument they choose to play. All of the applications can be manipulated using the users' gestures.

Using the ATR Immersive Virtual Environment System, participants can implement any virtual environment to create a shared, immersed experience. The virtual environment isn't limited to a nature environment. This paper proposes nature examples, such as the BMS Virtual Nature Healing Environment [2]; since this paper is developing the BMS lab's research [1], the paper uses the same application as the previous research. The other example, Virtual Music Therapy: Virtual Harmony [8], is used because it is an application developed by the ATR Lab, similar to the software proposed in this paper. The ATR Immersive Virtual Environment System helps provide an essential foundation for users to experience any environment they choose in an immersed manner.

5 Discussion

The concept introduced in this paper is a development of the previous research completed by Houwelingen-Snippe [13], and Rompay [14]. Their study using the Virtual Nature Healing Environment [2], a software created by the BMS Lab [1]. Present in their research papers, a replica of the natural environment has been designed and used to test how virtual environments affect social aspirations. Also, Virtual Harmony [8], a virtual therapy system developed to reduce anxiety and stress by using music therapy, is another possible experience the user can choose. The system allows users to listen to music and play different instruments in a 3D environment with real-world video as their background. The main objective of this research was the development of interactive software with the inclusion of the BMS lab's virtual environment, Virtual Harmony, and the ability to incorporate the software in a space where an interactive application could be constructed. The Blank Lab has a fixed setup, so the lab could not be customized to the specifics of this project. With the limitations present, the Blank Lab is set up to configure a multitude of projects. For the correct configurations, the Blank Labs' workstation has software installed to help sync all the projectors creating a seamless digital display; Scalable Displays, a software that automatically wraps and blends multiple projectors. For the proper audio play, Pro-tools, a digital audio workstation, is used to help distribute the audio for a more ambient sound. We were able to achieve our desired outcome by working around these limitations and with the facilities provided. In this research, we created an application that combined two different virtual environment projects and configured them to work in the Blank Lab to create an immersive experience that eliminates the need for users to handle any hardware while using the application. The screens in the blank lab surround the user from three directions, so the user will experience the environment as if they have stepped into the application; we have also added a ground projected stage to help immerse the users even more.

6 Conclusion and Future Work

This paper proposes a concept that provides the structure for a physical space where users can interact with the shared mixed reality environment. The environment is projected to help create the collaborative aspect of this project without any wearable devices. The developed system and implemented interactions demonstrate the feasibility of the structured experience that has been made. As per our past research, the Virtual Nature Healing Environment has been used to collect data and find the Awe Score of the users who experience the virtual space. Similarly, in future research, the application will be tried and configured for different devices, i.e., laptop and phone screens and a VR headset, in hopes of discovering how increasing immersion in the virtual environment could affect human aspirations through various test surroundings with a diverse demographic. Bio-metric testing will also be necessary to help test how much the ATR immersive virtual environment system affects one's Awe Score compared to experiencing the application on different devices. A future step for enhancing the project is improving the ground projection module. The module has been introduced briefly in the paper. Although it hasn't been thoroughly executed, we believe having that extra element will help immerse the user. Another possible addition to the project is an audio response module, which will be the next step for this application to help further the research. Users will be able to create a sound, i.e., clapping, and yelling, which triggers the respected response from the software. We believe that introducing phonic commands will help create a more notably immersive environment. Developing a software that can configure any virtual environment, helping it build an utterly immersive space, is where this project is advancing towards.

Acknowledgement. This research was partially supported by the KSU Blank lab and the ATR lab. We thank our colleagues Prof. J.R. Campbell and Shannon Hines from the KSU Blank Lab, as well as Lucia Rabago Mayer from the BMS lab, who provided insight and expertise that greatly assisted the research, although they may not agree with all of the interpretations/conclusions of this paper.

References

1. BMS Lab, a lab of the social sciences faculty at the University of Twente. https://bmslab.utwente.nl/
2. BMS Lab Virtual Nature Healing Environment. https://bmslab.utwente.nl/virtual-nature-healing-environment/
3. Kent State University Design Innovation Hub Blank Lab. https://www.kent.edu/designinnovation/blanklab

4. Clifford, R.M.S., Khan, H., Hoermann, S., Billinghurst, M., Lindeman, R.W.: The effect of immersive displays on situation awareness in virtual environments for aerial firefighting air attack supervisor training. In: 2018 IEEE Conference on Virtual Reality and 3D User Interfaces (VR), pp. 1–2. IEEE (2018)
5. Hendricks, S., Shaker, A., Kim, J.-H.: Design of a VR-based campus tour platform with a user-friendly scene asset management system. In: Kim, J.-H., Singh, M., Khan, J., Tiwary, U.S., Sur, M., Singh, D. (eds.) IHCI 2021. LNCS, vol. 13184, pp. 337–348. Springer, Cham (2022). https://doi.org/10.1007/978-3-030-98404-5_32
6. Horan, B., Seyedmahmoudian, M., Mortimer, M., Thirunavukkarasu, G.S., Smilevski, S., Stojcevski, A.: Feeling your way around a cave-like reconfigurable VR system. In: 2018 11th International Conference on Human System Interaction (HSI), pp. 21–27. IEEE (2018)
7. Huang, H.-M., Liaw, S.-S., Lai, C.-M.: Exploring learner acceptance of the use of virtual reality in medical education: a case study of desktop and projection-based display systems. Interact. Learn. Environ. **24**(1), 3–19 (2016)
8. Lin, X., et al.: Virtual reality-based musical therapy for mental health management. In: 2020 10th Annual Computing and Communication Workshop and Conference (CCWC), pp. 0948–0952. IEEE (2020)
9. Ogi, T., Hayashi, M., Sakai, M.: Room-sized immersive projection display for tele-immersion environment. In: 17th International Conference on Artificial Reality and Telexistence (ICAT 2007), pp. 79–86. IEEE (2007)
10. Mohamed Mansoor Roomi, S., Jyothi Priya, R., Jayalakshmi, H.: Hand gesture recognition for human-computer interaction. J. Comput. Sci. **6**(9), 1002–1007 (2010)
11. Shaker, A., Lin, X., Kim, D.Y., Kim, J.-H., Sharma, G., Devine, M.A.: Design of a virtual reality tour system for people with intellectual and developmental disabilities: a case study. Comput. Sci. Eng. **22**(3), 7–17 (2020)
12. Takatori, H., Hiraiwa, M., Yano, H., Iwata, H.: Large-scale projection-based immersive display: the design and implementation of largespace. In: 2019 IEEE Conference on Virtual Reality and 3D User Interfaces (VR), pp. 557–565 (2019)
13. van Houwelingen-Snippe, J., van Rompay, T.J.L., Allouch, S.B.: Feeling connected after experiencing digital nature: a survey study. Int. J. Environ. Res. Public Health **17**(18), 6879 (2020)
14. van Rompay, T.J.L., Jol, T.: Wild and free: unpredictability and spaciousness as predictors of creative performance. J. Environ. Psychol. **48**, 140–148 (2016)
15. Wang, X., Yan, K.: Immersive human-computer interactive virtual environment using large-scale display system. Futur. Gener. Comput. Syst. **96**, 649–659 (2019)
16. Zenner, A., Kosmalla, F., Speicher, M., Daiber, F., Krüger, A.: A projection-based interface to involve semi-immersed users in substitutional realities. In: 2018 IEEE 4th Workshop on Everyday Virtual Reality (WEVR) (2018)

Privacy-Preserving Digital Intervention for Mental Health Using Federated Learning

Ankit Kumar Singh, Ajit Kumar(✉), and Bong Jun Choi

Soongsil University, Seoul, South Korea
aks.bihta@gmail.com, ajitkumar.pu@gmail.com, davidchoi@soongsil.ac.kr

Abstract. Digital health care is emerging along with the growth and acceptability of modern ICT technologies such as IoT and Artificial Intelligence. Today's many health services are being served by computer-related technologies like telemedicine, teleconsultation, and remote diagnosis. However, mental health still lacks ICT technology integration due to privacy and highly sensitive data sharing. Recently, privacy-preserving technologies have been researched and applied to various privacy-required domains. To harness the benefits of these technologies and address the bottleneck of existing solutions in mental health-related solutions, we have proposed a federated learning-based solution for classifying human emotion into seven classes adoration, amusement, anxiety, disgust, empathic pain, and fear and surprise. Our proposed solution preserves data privacy while providing classification accuracy nearly equal to the traditional centralized machine learning solutions.

Keywords: Privacy-Preserving · Machine Learning · Mental Health · Depression · Emotion classification · Federated Learning

1 Introduction

Today, affording health care and its availability is quite challenging due to the cost and complexity involved in health-related services. In recent years, digital health care has paved the way to realizing affordable and available health care to all. The popular acquisition of OneMedical (telehealth services) by Amazon for $3.49 billion is a testimonial of the future growth of digital health care[1].

[1] https://www.reuters.com/markets/deals/amazon-buy-one-medical-35-billion-deal-2022-07-21/.

This research was supported by the Ministry of Science and ICT (MSIT) Korea under the National Research Foundation (NRF) Korea (NRF-2022R1A2C4001270), by the MSIT Korea Korea under the India-Korea Joint Programme of Cooperation in Science & Technology (NRF-2020K1A3A1A68093469), and by the ITRC (Information Technology Research Center) support program (IITP-2020-2020-0-01602) supervised by the IITP (Institute for Information & Communications Technology Planning & Evaluation).

H. Zaynidinov et al. (Eds.): IHCI 2022, LNCS 13741, pp. 213–224, 2023.
https://doi.org/10.1007/978-3-031-27199-1_22

In comparison to general health, mental health has different challenges like the privacy of patients and availability of clinical psychologist, medical adherence (person with depression diagnosis and the similar problem tends to avoid medication). Digital technology-based intervention is an excellent supporting tool to address these challenges of diagnosis, detection, and support for mental health-related problems.

With the advancement and adoption of machine learning (ML) techniques in various domains, many recent research and applications in mental health care are also applying ML techniques for tasks like emotion classification and depression detection from video, text, or audio data. However, traditionally these solutions use *centralized machine learning* in which data from sources (patients, doctors, or devices) are collected at a central server for training and building a machine learning model. Such a centralized approach has privacy risks, making applying ML for mental health applications infeasible. Users with symptoms or mental illness are sensitive and may not want to share this information with anyone other than trusted doctors.

Recently, *federated learning* (FL) has been adopted in digital health systems, and many recent works show various applications of federated learning in the health domain [26,27]. An FL system can be implemented using support from existing technologies like Internet-of-Things (IoT), edge computing, and cloud computing [12,30]. FL Driven Internet of Medical Things (FLDIoMT) is an example of an emerging framework for digital health care [10]. Similar to general digital health care, it is also being used for the mental health domain; for example, Lee et al. [16] provides a performance assessment of FL on two clinical benchmark datasets, i.e., MIMIC-III for in-hospital mortality prediction and PhysioNet/CinC ECG dataset for a multi-class (atrial fibrillation, normal sinus rhythm, alternative rhythm, and noisy) ECG classification. However, there are some challenges and bottlenecks in federated learning, such as different attacks (model and data poising, label or attribute inference) and other inherent problems mentioned in [33]. Feng et al. [11] have demonstrated attribute inference attack on emotion recognition system trained in FL setting; while Chang et al. [7] have proposed defenses against adversarial attacks using adversarial training and randomized inference. Although we have not considered attacks on the FL system in our current work, this will be a future direction to explore due to the high-risk domain of health and mental health. Recent works have claimed that FL performance is nearly equal to the centralized approach (4–6% accuracy loss [2]) and better than the model trained on individual user's data [18]. In this proposed work, we want to explore the application of FL for mental health-related tasks, such as depression detection from emotional states. We have trained multiple emotion classifiers (seven classes) on four individual and one combined audio dataset along with augmented samples (noise and pitch) in a federated learning and centralized setup. Our proposed solution preserves data privacy and provides classification accuracy nearly equal to traditional centralized machine learning solutions.

The proposed work has made the following contributions:

- We provide a performance comparison of various models trained using federated learning on four datasets that aim to detect mental health through emotion classification of speech.
- We also tested and presented the result of feature augmentation like noise and pitch on model performance.
- Finally, we compared the performance of federated and centralized learning on individual and combined datasets.

2 Background and Related Work

Federated learning is the latest technique and is being used in various domains to preserve the privacy of data to be used for model training. Few recent research works have also used FL for mental health-related tasks. Borger et al. [4] have applied federated learning (cross-silo) for violence incident prediction using clinical notes as data. Authors have used a natural language processing approach in an FL setting. Kerkouche et al. [15] have claimed to enhance the privacy of traditional federated learning by infusing differential privacy and used the proposed method to build a mortality prediction using hospital data comprising patient in-and-out records. Dang et al. [9] have provided an evaluation of exiting FL algorithms trained on the EHR dataset (collected at multiple centers) for in-hospital mortality prediction and acute kidney injury (AKI) prediction. Similarly, in a recent work, heart rate and CT image data are used to create a multi-modal dataset and train and test vertical FL for personnel health detection. The authors created a lightweight feature extraction sub-model for addressing the device computation capability and a Fast and Secure (FS) module to reduce the data size to be communicated between device and server [14]. FedHealth [8] apply transfer learning in a federated learning approach to provide a privacy-preserving and personalized model. The authors demonstrated the performance of the proposed method for activity recognition and real Parkinson's disease diagnosis.

The authors in Fed-ReMECS have extended their previous work ReMECS for federated learning. This work uses multi-modal data to build a classification model for emotion state classification. The model's performance is tested using the DEAP dataset [23]. Liu et al. [19] have used multiple mobile data (keyboard pattern, online communication, and app usage) to build a mood prediction system. Authors have compared unimodal and multi-modal centralized approaches in privacy (using the Selective-Additive Learning framework) and non-privacy settings. In FedMood [31,32], authors have applied federated learning for mood detection using data (key press duration and time for alphanumeric and special characters, and accelerometer value) collected using mobile devices and labeled by doctors using Hamilton Depression Rating Scale (HDRS) and Young Man Mania Scale (YMRS) assessment questionnaire. DTbot uses real-time federated learning on voices and video and analyzes and predicts emotions [20]. In [28], authors have adopted a semi-supervised technique along with federated learning for speech emotion recognition to address the client's unlabelled and low data issue. In a federated learning setting, heart activity signals from smartwatches have been used to build a stress detection system [5] By processing health data (authored by patients using the PHQ-9 questionnaire) and creating an attention-based embedding for federated learning, authors have created mental health classification (depression or No-depression) system [1]. Similar to previous work, Suhas BN and Saeed Abdullah [2] have used transfer learning with Federated Averaging (FedAvg) [22] and Federated Matching Averaging (FedMA) [29] for depression detection using DAIC-WOZ dataset. Authors claimed that FL methods performed better than the best models from prior work. Li et al. [17] have used federated learning for depression detection from social media text. Authors have used data from *Weibo* and proposed CNN Asynchronous Federated optimization (CAFed) that improves communication cost and speeds up training convergence. Table 1 provides a summary of related works that have applied federated learning to the health or mental health domain.

3 Digital Intervention for Mental Health Using Centralized and Federated Machine Learning

The use of digital technologies (including ICT) for diagnosis, improvement, and prevention of mental health is inspired by the principles and scopes mentioned under affective computing [24]. In recent years, with machine learning in various domains, many traditional digital solutions for mental health are also being replaced by ML models. Like other domains, the data is also the key to the success of machine learning-based approaches to mental health. However, data is more sensitive in health, and even more, privacy requirement is cited for mental health applications. Traditional centralized machine learning collects data from various sources (patient or doctor end) at a centralized server. Then ML models are trained, and this creates privacy concerns for users. Figure 1 highlights the steps carried out in a centralized approach (the fine-grained steps may vary in different approaches).

Recently, many privacy-preserving solutions have been developed considering the privacy requirement and user concerns, including the need for machine learning-based digital interventions for mental health. In this proposed work, we have experimented with federated learning for building an emotions classification model from audio data. Figure 2 shows the steps of federated learning for the building ML model. Federated learning is a distributed machine learning method; in this approach, data is not collected in any centralized location. A server is used to moderate the model training among various selected participants in multiple rounds. Each participant gets a global model from the server and performs training locally on its data. At each round, each participant sends model updates to the server, and the server keeps aggregating all updates to improve the initial model at the end of each training round. In simple terms, data never leaves the generation source (participant), and only update, i.e., the difference between locally trained model and global update, is exchanged. So,

Table 1. Summary of previous works related to the application of federated learning in mental health and related domains

Works	Aggregation algorithm/s	Dataset	FL type	Task
Arijit et al. [23], 2022	FedAvg	DEAP	Cross-Device	Emotion classification
Borger et al. [4], 2022	FedAvg	Dutch medical text	Cross-Silo	Violence risk assessment
Xu et al. [32], 2021	Local TrainingCDS, FedAVG IIL, CIIL	BiAffect	Cross-Silo	Depression detection
Feng et al. [11], 2022	FedSGD, FedAvg	IEMOCAP, CREMA-D, MSP-Improv	Cross-Device	Attribute inference attack on speech emotion recognition
Yoo et al. [33], 2021	FedAvg	IEMOCAP	Cross-Device	Speech emotion recognition
Chang et al. [7], 2022	Average	DEMoS	Cross-Device	Adversarial attacks for speech emotion recognition
Can et al. [5], 2021	FedAvg	Heart rate	Cross-Device	Stress-level monitoring
Kerkouche et al. [15], 2021	FedAvg	EHR	Cross-Silo	Mortality prediction
Dang et al. [9], 2022	FedAvg, FedProx, FedAvgM, FedAdagrad, FedAdam, FedYogi	eICU	Cross-Silo	Mortality and AKI prediction
Ji et al. [14], 2022	Average	RSNA-ICH-Det, PTB Diagnostic ECG dataset	Cross-Device	Person status detection
Ahmed [1], 2022	FedAvg	UCI Smartphone, Parkinson's Disease Dataset	Cross-Device	Activity classification

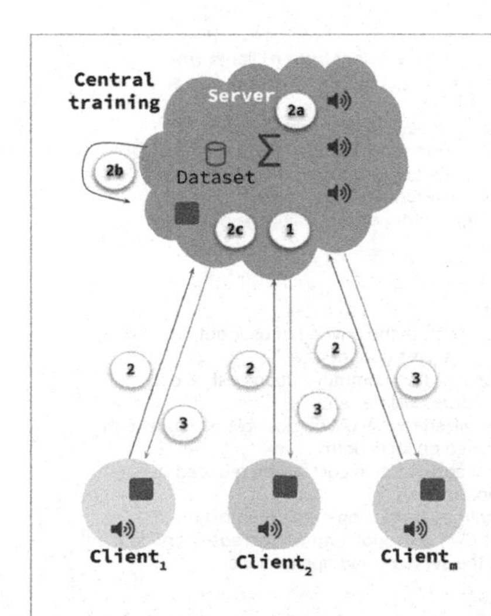

1. Server select *m* client from *n* clients. *(m < n)*
2. Selected *m* clients send data to central server.
   - 2a. Server aggregate data received from clients.
   - 2b. Train common model θ_t on aggregated data, produce θ_{t+1}.
   - 2c. Send θ_{t+1} to the clients.
3. Receive θ_{t+1} from server for local inference.

Challenges:
1. Privacy of user data
2. Data communication cost
3. Personalized model not possible
4. High Computation cost
5. Discourage Mental health patient to use service and application

Fig. 1. Illustration of Data collection, Model Training, and Distribution in the Centralized Machine Learning.

federated learning inherently promises privacy by design for training ML models. In sum, the proposed work aims to study the usability of FL and its performance against the centralized approach. The experimental setup and results are presented and discussed in the further section.

4 Experiments and Results

4.1 Experimental Setup

Hardware: We have used a computer having Intel(R) Core(TM) i9-10980XE CPU @ 3.00 GHz processor, 251 GB RAM, and 2xNVIDIA GeForce RTX 3090 (each has 24 GB memory) for training ML models in centralized and federated learning settings.

Software: All experiments are performed in a system running Ubuntu 18.04 64-bit OS, and all the scripts are written in Python 3.9 using different modules and frameworks such as TensorFlow2, and Keras3. For implementing Federated Learning, we have used TensorFlow Federated (TFF) [3].

4.2 Dataset and Feature Engineering

In federated learning, we aim to train a global model based on data from multiple clients or participants. Currently, the data related to mental health research, such as depression detection, mood or emotional detection, etc., are collected for centralized learning. A deep learning model can be trained with good accuracy in

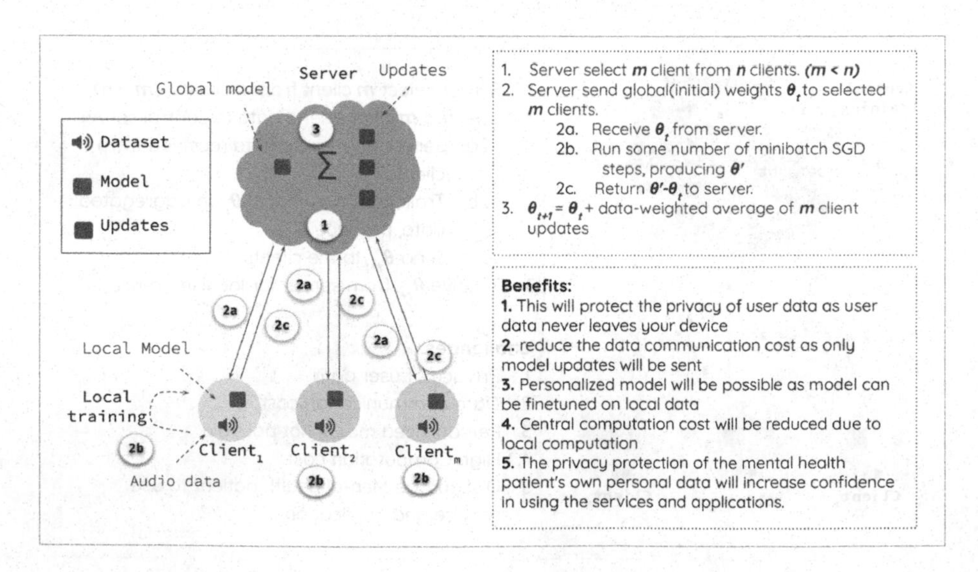

Fig. 2. Illustration of Data, Model Training, and Distribution in the Federated Learning.

such an approach due to the centralized data source (sufficient data for training a model). However, to simulate federated learning, we split the dataset into multiple parts (each part represents one client). In cross-device FL, where there are more participants, each end up having a very less number of samples. In the proposed work, we have identified four audio datasets for emotion classification (which have a similar type of 6–7 class) and merged them to create a larger dataset sufficient for FL. We have used Ravdess [21], Crema [6], TEES [25], and Savee [13] dataset. Table 2 shows the total sample of each class in each dataset.

Table 2. Description of dataset and class distribution

Dataset	Neutral	Fear	Disgust	Angry	Sad	Surprise	Happy	Total
Ravdess [21]	288	192	192	192	192	192	192	1440
Crema [6]	1087	1271	1271	1271	1271	NA	1271	7442
TESS [25]	400	400	400	400	400	400	400	2800
Savee [13]	120	60	60	60	60	60	60	480
Combined	1895	1923	1923	1923	1923	652	1923	12162

Ravdess [21] dataset is collected from 24 professionals (60 trials from each) in north American accent in two emotional intensities (normal and strong) for seven classes of emotions. Savee [13]. Crema [6] dataset is the recordings of 91 actors (male (48) and female (43)) simulated emotional expressions (happy, sad, anger, fear, disgust, and neutral). TEES [25] is mainly based on 200 spoken words in seven emotions from two actresses representing different ages (26 and 64 years). Savee [13] has recordings from students and researchers from the University of Surrey. In addition to the actual data files, we have also created augmented samples by applying noise, pitch, and a combination of both. Finally, we used the feature extractor on each raw data and its three augmented variants to extract three features: Zero Crossing Rate (ZCR), Root Mean Square energy (RMS energy), and Mel-frequency Cepstral Coefficient (MFCC). Figure 3 depicts the feature engineering process from raw data to feature vector.

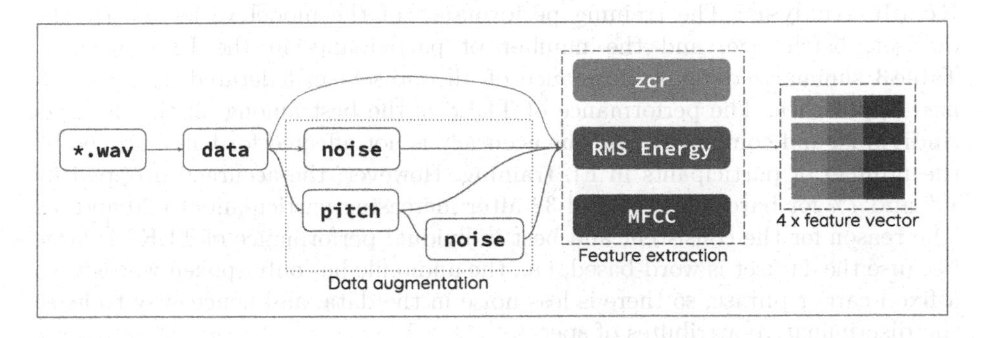

Fig. 3. Feature Engineering: Dataset merging, augmentation, feature extraction and stacking.

4.3 Results and Analysis

Model architecture: The deep learning model has five 1D Convolution blocks, each containing a Conv1D layer followed by a BatchNormalization and MaxPooling1D layers. The model takes a feature vector of size 1620, and the output layer gives a vector of size 7, i.e., each value is a probability of each of the seven classes. Figure 4 show the input to output layers of the model.

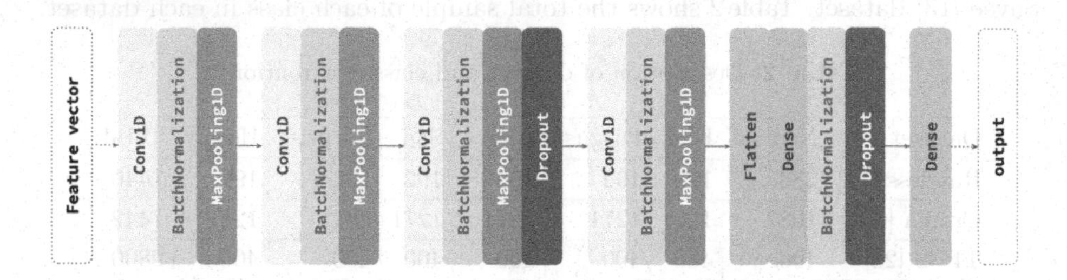

Fig. 4. Model Architecture of Deep Learning Model.

Federated Learning: Training and Testing for the training model in federated learning, we have used the TensorFlow Federated (TFF) [3] framework and performed three training experiments that simulate three sets of 10, 20, and 30 participants. Each client's local model updates were aggregated using FedAvg [22]. Further, each experiment was carried out with different batch sizes, i.e., 16, 32, and 64. The model performance of each training (total of 9 for each dataset) is shown in Table 3. The table also shows the total round used for each training, and they are not equal because each model training varies due to size and differences in data distribution.

Centralized Learning: Training and Testing for the training model in a centralized setting, we have used the TensorFlow (Keras) framework and performed training experiments with different batch sizes, i.e., 16, 32, and 64. The model performance of each training (total of 9 for each dataset) is shown in Table 3.

Result Analysis: The training performance of the model varies as per the dataset, batch size, and the number of participants in the FL approach. Table 3 summarized the performance of all datasets in federated and centralized approaches. The performance of TEES is the best among all the datasets (individual and combined), and the accuracy is not affected for batch size 64 by the number of participants in FL training. However, the accuracy dropped by 1% and 7% for batch sizes 16 and 32 after increasing participants to 20 and 30. The reason for the consistent and best individual performance of TEES is large because the dataset is word-based, i.e., the audio file has only spoken words with a fixed carrier phrase, so there is less noise in the data, and hence easy to learn the discriminative attributes of speech.

In the federated learning approach, the best performance of the combined dataset is 86% with a batch size of 64 and a total of ten participants. The model

Table 3. Training and Validation performance of Federated and Centralized Learning

Federated Learning										**Centralized Learning**		
Client	**10**			**20**			**30**					
Batch Size	**16**	**32**	**64**	**16**	**32**	**64**	**16**	**32**	**64**	**16**	**32**	**64**
Crema-D												
Accuracy (train)	0.83	0.83	0.82	0.54	0.43	0.73	0.52	0.41	0.65	0.9002	0.9147	0.925
Round/val	299	299	299	121	30	237	120	23	299	0.8895	0.9002	0.9137
Ravdess												
Accuracy (train)	0.9	0.98	0.89	0.93	0.96	0.97	0.69	0.95	0.94	0.9924	0.993	0.997
Round/val	127	299	40	299	299	171	84	299	180	0.9392	0.9361	0.9358
Savee												
Accuracy (train)	0.97	0.98	0.99	0.95	0.95	99	0.93	0.98	0.97	0.9876	0.9993	0.9922
Round/val	299	251	190	299	102	299	219	267	299	0.9583	0.9583	0.9375
TESS												
Accuracy (train)	0.99	0.98	0.99	0.98	0.92	0.99	0.97	0.92	0.99	0.9984	0.9979	0.9996
Round/val	299	104	299	299	34	299	247	47	213	0.9982	0.9973	0.9937
Combined												
Accuracy (train)	0.83	0.84	0.86	0.72	0.73	0.73	0.53	0.62	0.5	0.934	0.9318	0.9354
Round/val	299	299	299	299	263	244	86	161	30	0.9309	0.9316	0.919

performance keeps reducing with increasing participants, and the accuracy is reduced to only 50% (batch size 64) for a total number of 30 participants. The best FL performance is 7% lower than the centralized approach, which got the best performance of 93% (irrespective of batch size). However, it is interesting to observe that, except for Crema-D, the performance of the other three individual dataset is nearly equal for best from federated learning and centralized learning. The reason would be that these datasets are collected in a controlled environment with a very low number of participants, so there is less diversity in the dataset, and both learning approaches achieved similar performance. In comparison, Crema-D has more samples and more variation, which gets wider in federated learning due to the participation of data. Hence, performance is lower than the centralized approach.

5 Conclusion and Future Works

The proposed work aimed to train and explore the model training for mental health-related classifiers in federated learning and compare its performance with centralized learning. Based on various training and testing on four individual datasets and their combined version, we observed that FL performance is slightly lower (7%) than the centralized approach. We observed that the result can be improved if we have less diversity in the dataset.

The proposed work has focused on audio data. So, learning from other modalities individually and in a multi-model setup will be explored in future work.

Current work has also assumed and worked with an already labeled dataset which is not a valid assumption for real-life implementation of federated learning. Techniques for learning from unlabelled or small-scale labeled data will be researched and adopted in future work. So, the dataset from more natural environments can be used for the model training, unlike the dataset collected in the controlled and using few participants.

One of the extensions of the proposed work is to deploy federated learning with wearable devices to learn a global model for depression and anxiety detection in a privacy-preserving and non-intrusive manner.

Acknowledgement. This research was supported by the MSIT Korea under the NRF Korea (NRF-2022R1A2C4001270), by the MSIT Korea Korea under the India-Korea Joint Programme of Cooperation in Science & Technology (NRF-2020K1A3A1A68093469), and by the ITRC support program (IITP-2020-2020-0-01602) supervised by the IITP.

References

1. Ahmed, U., Lin, J.C.W., Srivastava, G.: Hyper-graph attention based federated learning method for mental health detection. IEEE J. Biomed. Health Inform. (2022)
2. Bn, S., Abdullah, S.: Privacy sensitive speech analysis using federated learning to assess depression. In: 2022 IEEE International Conference on Acoustics, Speech and Signal Processing (ICASSP), ICASSP 2022, pp. 6272–6276. IEEE (2022)
3. Bonawitz, K., Eichner, H., Grieskamp, W., et al.: TensorFlow federated: machine learning on decentralized data (2020)
4. Borger, T., et al.: Federated learning for violence incident prediction in a simulated cross-institutional psychiatric setting. Expert Syst. Appl. **199**, 116720 (2022)
5. Can, Y.S., Ersoy, C.: Privacy-preserving federated deep learning for wearable IoT-based biomedical monitoring. ACM Trans. Internet Technol. (TOIT) **21**(1), 1–17 (2021)
6. Cao, H., Cooper, D.G., Keutmann, M.K., Gur, R.C., Nenkova, A., Verma, R.: CREMA-D: crowd-sourced emotional multimodal actors dataset. IEEE Trans. Affect. Comput. **5**(4), 377–390 (2014)
7. Chang, Y., Laridi, S., Ren, Z., Palmer, G., Schuller, B.W., Fisichella, M.: Robust federated learning against adversarial attacks for speech emotion recognition. arXiv preprint arXiv:2203.04696 (2022)
8. Chen, Y., Qin, X., Wang, J., Yu, C., Gao, W.: Fedhealth: a federated transfer learning framework for wearable healthcare. IEEE Intell. Syst. **35**(4), 83–93 (2020)
9. Dang, T.K., Lan, X., Weng, J., Feng, M.: Federated learning for electronic health records. ACM Trans. Intell. Syst. Technol. (TIST) (2022)
10. Fan, J., Wang, X., Guo, Y., Hu, X., Hu, B.: Federated learning driven secure internet of medical things. IEEE Wirel. Commun. **29**(2), 68–75 (2022)
11. Feng, T., Hashemi, H., Hebbar, R., Annavaram, M., Narayanan, S.S.: Attribute inference attack of speech emotion recognition in federated learning settings. arXiv preprint arXiv:2112.13416 (2021)
12. Hakak, S., Ray, S., Khan, W.Z., Scheme, E.: A framework for edge-assisted healthcare data analytics using federated learning. In: 2020 IEEE International Conference on Big Data (Big Data), pp. 3423–3427. IEEE (2020)

13. Haq, S., Jackson, P.J., Edge, J.: Audio-visual feature selection and reduction for emotion classification. In: Proceedings of the International Conference on Auditory-Visual Speech Processing (AVSP 2008), Tangalooma, Australia (2008)
14. Ji, J., Yan, D., Mu, Z.: Personnel status detection model suitable for vertical federated learning structure. In: 2022 the 6th International Conference on Machine Learning and Soft Computing, pp. 98–104 (2022)
15. Kerkouche, R., Acs, G., Castelluccia, C., Genevès, P.: Privacy-preserving and bandwidth-efficient federated learning: an application to in-hospital mortality prediction. In: Proceedings of the Conference on Health, Inference, and Learning, pp. 25–35 (2021)
16. Lee, G.H., Shin, S.Y.: Federated learning on clinical benchmark data: performance assessment. J. Med. Internet Res. **22**(10), e20891 (2020)
17. Li, J., Jiang, M., Qin, Y., Zhang, R., Ling, S.H.: Intelligent depression detection with asynchronous federated optimization. Complex Intell. Syst. **9**, 115–131 (2022). https://doi.org/10.1007/s40747-022-00729-2
18. Liu, J.C., Goetz, J., Sen, S., Tewari, A.: Learning from others without sacrificing privacy: simulation comparing centralized and federated machine learning on mobile health data. JMIR Mhealth Uhealth **9**(3), e23728 (2021)
19. Liu, T., et al.: Multimodal privacy-preserving mood prediction from mobile data: a preliminary study. arXiv preprint arXiv:2012.02359 (2020)
20. Liu, Y., Yang, R.: Federated learning application on depression treatment robots (DTbot). In: 2021 IEEE 13th International Conference on Computer Research and Development (ICCRD), pp. 121–124. IEEE (2021)
21. Livingstone, S.R., Russo, F.A.: The Ryerson audio-visual database of emotional speech and song (RAVDESS): a dynamic, multimodal set of facial and vocal expressions in North American English. PLoS ONE **13**(5), e0196391 (2018)
22. McMahan, B., Moore, E., Ramage, D., Hampson, S., Arcas, B.A.: Communication-efficient learning of deep networks from decentralized data. In: Artificial Intelligence and Statistics, pp. 1273–1282. PMLR (2017)
23. Nandi, A., Xhafa, F.: A federated learning method for real-time emotion state classification from multi-modal streaming. Methods (2022)
24. Picard, R.W.: Affective Computing. MIT Press, Cambridge (2000)
25. Pichora-Fuller, M.K., Dupuis, K.: Toronto emotional speech set (TESS). Scholars Portal Dataverse **1**, 2020 (2020)
26. Shyu, C.R., et al.: A systematic review of federated learning in the healthcare area: from the perspective of data properties and applications. Appl. Sci. **11**(23), 11191 (2021)
27. Smith, A.: A study on federated learning systems in healthcare. Ph.D. thesis (2021)
28. Tsouvalas, V., Ozcelebi, T., Meratnia, N.: Privacy-preserving speech emotion recognition through semi-supervised federated learning. In: 2022 IEEE International Conference on Pervasive Computing and Communications Workshops and other Affiliated Events (PerCom Workshops), pp. 359–364. IEEE (2022)
29. Wang, H., Yurochkin, M., Sun, Y., Papailiopoulos, D., Khazaeni, Y.: Federated learning with matched averaging. arXiv preprint arXiv:2002.06440 (2020)
30. Wu, Q., Chen, X., Zhou, Z., Zhang, J.: Fedhome: cloud-edge based personalized federated learning for in-home health monitoring. IEEE Trans. Mob. Comput. (2020)
31. Xu, X., et al.: Privacy-preserving federated depression detection from multisource mobile health data. IEEE Trans. Ind. Inf. **18**(7), 4788–4797 (2021)

32. Xu, X., Peng, H., Sun, L., Bhuiyan, M.Z.A., Liu, L., He, L.: FedMood: federated learning on mobile health data for mood detection. arXiv preprint arXiv:2102.09342 (2021)
33. Yoo, J.H., Jeong, H., Lee, J., Chung, T.-M.: Federated learning: issues in medical application. In: Dang, T.K., Küng, J., Chung, T.M., Takizawa, M. (eds.) FDSE 2021. LNCS, vol. 13076, pp. 3–22. Springer, Cham (2021). https://doi.org/10.1007/978-3-030-91387-8_1

How Can Humans and Robots Live Together?: The 5 Types of Human-Robot Relationship

Karam Park and Eui-Chul Jung(✉)

Department of Design, Seoul National University, Seoul 08826, South Korea
{karam,jech}@snu.ac.kr

Abstract. Users purchase home service robots to replace their household chores. However, several previous research showed some people have a new relationship type with home service robots, regardless of the main reason for purchase. Using the findings of previous studies on the user's usage behavior of the robotic vacuum cleaner, one of the home service robots, this research analyzed the relationship between human and robotic vacuum cleaners through the 5 types of human-robot relationship framework. The human-robot vacuum cleaner relationship was shown to have four relationships: 1) Hierarchical power, 2) Reciprocity, 3) Coalitions and 4) Attachment. In the relationship 1) Hierarchical power and 2) Reciprocity types, people regard robot cleaners as 'tools', desiring the equipment to do their cleaning efficiently, and robots obey human commands. On the other hand, 3) Coalitions and 4) Attachment types correspond to robot relationships with people who recognize robot cleaners as specific individuals and form new relationships. Through understanding human-robot relationship types, it is necessary to design with an appropriate UX concept that considers the relationship types, experience, and technology for developing and designing future robot services. Considering the types of human-robot relationship, it will be easy to understand what the priorities are when designing several kinds of robots.

Keywords: Human-Robot Interaction Design · The type of human-robot relationship · User experience · Service design

1 Introduction

In several industries, robots are taking the place of workers. In particular, expectations are high for home service robots that live with humans and carry out household chores that people do not want to do, as demand has risen due to the recent growth in single-person and elderly families. According to a survey conducted by Dautenhahn et al. [2], 80 percent of respondents anticipated robots in their private space to work as assistants in the future, and most respondents expected them to clean their houses. The reason why demand and anticipation for domestic robots are high is that people think that the use of home service robots helps to decrease physical tiredness and allows users to enjoy their own time without worrying about housework. Companies like iRobot, Samsung, LG, and Xiaomi have recently been actively introducing robot cleaners, a sort of home service robots, to the market in an effort to match this expectation.

H. Zaynidinov et al. (Eds.): IHCI 2022, LNCS 13741, pp. 225–229, 2023.
https://doi.org/10.1007/978-3-031-27199-1_23

Home service robots differ from industrial robots in that they live together in the user's residence space and help the user with household chores. In addition, several previous studies had revealed that human and robot form a new relationship, unlike the purpose that users purchase, as a tool to replace housework. The aim of this study is to find out what role robots currently have and to categorize human-robot social relationships by types. Through the categorization of human-robot relationships, this study would like to find out how to develop the user's robot experience in the future robot service development.

2 Method

This study aims to investigate the relationship between humans and robots through the analysis of previous research on the user experience of robotic vacuums, which are the most commonly used among home service robots used to assist with housework. Using the five human-human domains of social life of Bugental [1], the relationship between humans and robot cleaners was divided into five types, and user behaviors and experiences found in previous studies were classified by five types.

2.1 The 5 Types of Human-Robot Relationship

Bugental [1] classified human-human social relations into five types: Hierarchical Power, Reciprocity, Coalition, Attachment and Mating. Figure 1 explains the social relationship between human and robot using the framework of Bugental [1]. The dotted line in Fig. 1 shows the behavior or specific goals of humans and robots, and the arrows express negotiations and interactions between the two. The difference in the size of the circle reflects the degree of power and equivalence, and the overlapping part represents the degree of psychological attachment.

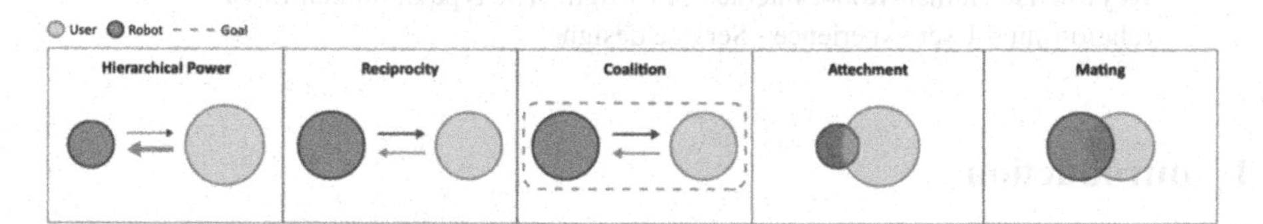

Fig. 1. The five types of humans-robots relationship

1) Hierarchical Power is a type of relationship in which humans recognize robots as 'tools' and robots act according to human instructions. 2) Reciprocity is a type of interaction where humans recognize a robot as 'tools' and give orders, and the robot responds by performing a new task that has been personalized by learning about the user on its own sensors. 3) Coalitions is also a type of relationship in which humans perceive robots as 'tools' but also as 'specific individuals'. Furthermore, humans and robots perform tasks together for specific purposes. 4) Attachment is a type of relationship in which humans perceive a robot as a 'specific individuals' and form an attachment to the robot. 5) Mating includes the type of recognition of robots as an individual that shares

sexual and mental communication. The types of relationship between human-robot identifying robots as tools and issuing commands are stated as 1) Hierarchical power and 2) Reciprocal. On the contrary, 3) Coalitions, 4) Attachment, and 5) Mating types can be grouped into relationships where human perceive robot as particular individuals and interact with emotions. However, 5) Mating is about the relationship of sexual and mental communication, so it is difficult to apply to robot vacuum cleaners due to this different perspective. That is why this study explains human-robotic vacuum only with types 1), 2), 3), and 4).

3 Result

It was discovered through the findings of earlier studies that there are four types of relationships when analyzing the behaviors and feelings that humans encounter when using robot vacuum cleaners: 1) Hierarchical power, 2) Reciprocity, 3) Coalition, 4) Attachment. The user's behavior in the human-robot cleaner relationship types shown in previous research are as follows.

3.1 Hierarchical Power

In this relationship type, users set the function of the robot cleaner with their smartphone application, give instructions to the robot cleaner, and the robot cleaner cleans according to the instructions. As revealed in the study of Dautenhahn. K. et al. [2], it showed the same pattern as people expected home service robots in the home to serve as assistants and machines.

3.2 Reciprocity

Robotic vacuums have higher intelligence than before, such as artificial intelligence, so they perform personalized cleaning tasks according to user behavior patterns and living environments without requiring the user's special orders. For example, Samsung's Bespoke Jetbot AI model monitors pets in real time when a dog owner is away from home. In addition, that robot cleaner provides customized functions such as sending an alarm to the user if the dog barks or does not move for a long time [5]. This example is included in 2) Reciprocity because the robotic vacuum provides personalized functions with AI sensors.

3.3 Coalition

The robotic vacuum also allows users to carry out new housework to maintain, such as emptying dust cans and cleaning filters inside robots like other appliances and electronic devices [6]. It is interesting to note that the long-term usage of robot vacuum cleaners encourages new housework, such as motivating users to clean other rooms or clean up objects on the floor [8]. Therefore, in this type, it can be seen that the user perceives robot cleaners as a cooperative and stimulating being rather than a tool.

3.4 Attachment

In this relationship type, previous studies had discovered that users had been developing a sense of familiarity and relationships. Users, for instance, gave robotic vacuum cleaners names and decorated their appearance [3, 7]. In addition, users inferred the robotic vacuum's different personalities depending on the movement, sound, and appearance of the robot vacuum cleaner [4]. Users build social relationships with robot cleaners, unlike the core purpose of domestic robots, which is interesting in that robots are not designed to exchange emotions with people and form social relationships. For example, even if a robotic vacuum does not respond to user's talks, the user had tried a conversation with the robotic vacuum was a typical example of building a one-sided relationship [8].

4 Discussion

Analyzing the case of robot cleaner, which is one of the sorts of home service robots, human and robot cleaner were forming four types of social relationships: 1) Hierarchy Power, 2) Reciprocity, 3) Coalition, and 4) Attachment. In the 1) Hierarchy Power and 2) Reciprocity relationship types between human-robot vacuum cleaners, human view robot vacuum cleaners as 'cleaning tools', and they want that tools to perform cleaning orders well. On the other hand, 3) Coalition and 4) Attachment relationship types showed that humans perceive robot vacuum cleaners as specific individuals and form new relationships. Although it is hard to find in the case of robot vacuum cleaners, robot from the movie 'Her', which showed possible mental communication with virtual object with intelligence and sexual satisfaction, can be thought of as 5) Mating relationship type.

In human life, robots are forming various types of relationships with people. Therefore, it is important to find out what role the robot shares with the user when developing and designing the robot in the future. Taking the development and design of robot vacuum cleaners as an example, in relation types 1) Hierarchy Power and 2) Reciprocity, robot cleaners need to be developed to perform cleaning well as 'a tool', focusing on the functional aspect of robot vacuum cleaners. The relationship types of 3) Coalition and 4) Attachment need to focus not only on the experience of cleaning, but also on the part of new relationships and experiences living in the same space that robot developers hard to expect. Therefore, when updating and developing robotic vacuum cleaners, it is necessary to identify the types of relationships that robot users are most likely to show and present an appropriate UX concept through consideration of personas or user scenarios.

5 Conclusion

This study analyzed how robots currently form social relationships with humans and what roles they are positioned as by dividing them into five relationship types through the results of previous studies on robotic vacuums. Since humans have various types of relationships with robots, robot developers and designers need to not only clearly identify the type of relationship and provide technology in developing future robot services, but also design with UX concepts that consider appropriate user-robot relationships and experiences. Considering the relationship types, it will be easy to understand what needs to be considered as the top priority when designing various types of robots. This research has divided five relationship types through previous studies, but it is necessary to prove the usability of the framework through future studies. In addition, it needs to be verified through extra experiments whether other kinds of robots can also explain the relationship between humans and robots through this framework.

References

1. Bugental, D.B.: Acquisition of the algorithms of social life: a domain-based approach. Am. Psychol. Assoc. **126**(2), 187–219 (2000)
2. Dautenhahn, K., Woods, S., Kaouri, C., Walters, M., Koay, K.L., Werry, I.: What is a robot companion-friends, assistant or butler?. In: 2005 IEEE/RSJ International Conference on Intelligent Robots and Systems, pp.1192–1197. IEEE, New York (2005)
3. Forlizzi, J., DiSalvo, C.: Service robots in the domestic environment: a study of the roomba vacuum in the home. In: HRI'06: Proceedings of the 1st ACM SIGCHI/SIGART Conference on Human-Robot Interaction, pp.258–265. ACM, New York (2006)
4. Hendriks, B., Meerbeek, B., Boess, S., Pauws, S., Sonneveld, M.: Robot vacuum cleaner personality and behavior. Int. J. Soc. Robot. **3**(2), 187–195 (2011)
5. Samsung Newroom Homepage. https://news.samsung.com/global/bespoke-jet-bot-aiplus-the-smart-choice-for-floor-cleaning-and-pet-monitoring-even-when-youre-away-from-home. Accessed 17 Sept 2022
6. Smarr, C.-A., et al.: Domestic robots for older adults: attitudes, preferences, and potential. Int. J. Soc. Robot. **6**(2), 229–247 (2013). https://doi.org/10.1007/s12369-013-0220-0
7. Sung, J.Y., Guo, L., Gringter, R.E., Christensen, H.I.: " My roomba is rambo" : intimate home appliances. In: Krumm, J., Abowd, G.D., Seneviratne, A., Strang, T. (eds.) UbiComp 2007: Ubiquitous Computing. Lecture Notes in Computer Science, vol. 4717, pp. 145–162. Springer, Heidelberg (2007)
8. Sung, J.Y., Grinter, R.E., Christensen, H.I.: Domestic Robot Ecology. Int. J. Soc. Robot. **2**(4), 417–429 (2010)

Effectiveness of Deep Learning Based Filtering Algorithm in Separation of Human Objects from Images

S. P. Khalilov[2(✉)], I. Yusupov[1], M. G. Mannapova[1], N. B. Nasrullayev[2], and F. Botirov[1]

[1] Tashkent University of Information Technologies Named After Muhammad al Khwarizmi, Tashkent, Uzbekistan
fayzullo@tuit.uz

[2] Nurafshan Branch of Tashkent University of Information Technologies Named After Muhammad Al Khwarizmi, Tashkent, Uzbekistan
kh.surajiddin@gmail.com, n.nasrullayev@nbtuit.uz

Abstract. This article focuses on the use of filtering algorithms based on noise deep learning in high-efficiency detection of moving human objects in real-time video frames. The frame separation method developed for the detection of moving human objects has been modified to provide real-time image processing. In the first method, an efficient frame splitting method was used to detect the moving object from the original image. In this method, it is possible to detect moving objects in the image at a high speed, but the disadvantage is that the efficiency index is lower due to the fact that the image noise filtering tools are not used. To increase the efficiency of the algorithm, the use of the LBF-algorithm - (Machine Learning Approach for Filtering), which reduces the additional quality of the noise, was proposed.

Keywords: Machine Learning Approach for Filtering · Convolutional Neural Networks · Models · Computer · Image · Segmentation · Frame · Object · Artificial neural network · Haara · Layer · Database · Algoritm · Formula · Video

1 Introduction

In recent years, along with the development of information technologies, it is observed that the volume of human data sets has increased significantly. An example of this is the proliferation of new surveillance cameras, smart-phone gadgets, and the acquisition of information from sensor cells in this society.

Most of the advances in artificial intelligence today are related to artificial neural networks. As we all know, in humans, neural network consists of interconnected neurons of the brain. By sharing information with each other, neurons enable thought processes, allowing us to make decisions, recognize objects, and do many other things. With the development of technology, scientists have transferred the basic principles of neural networks to the level of computer models [1, 4].

H. Zaynidinov et al. (Eds.): IHCI 2022, LNCS 13741, pp. 230–238, 2023.
https://doi.org/10.1007/978-3-031-27199-1_24

Artificial Neural Networks Convolutional Neural Networks (CNN) are used to process visual data in images and videos. If the layers of the neural network are effectively organized to achieve high-level results, the efficiency of the expected result will be high.

Based on Internet data, Machine Learning and CNN have been found to be the most effective in visual reanalysis of images based on neural networks. They have been highlighted for several years for their ability to perform operations on large volumes of data. This requires a computer based on fast graphics processors (GPU). CNN applications include high-functioning artificial intelligence systems. For example, robots working on the basis of Sunni intellect, self-driving cars based on virtual assistants can be used as an example.

To better understand the working principle of CNN, it is necessary to compare it with human vision. When a person sees a car, he can distinguish it based on its wheels, doors, headlights, color and shape. In the same way, the whole appearance of a person can be distinguished by the parts of his arms and legs, the color of his clothes, the structure of his face, and the color of his body. In the above two situations, a person makes a decision and distinguishes which is a machine and which is a person. A CNN is a multilayer network that contains different types of layers. There are three main layers that define a CNN, which are the switch layer, the pooling layer, and the softmax layer. The figure below shows an example of a CNN architecture consisting of two vertex layers, two concatenation layers, and a softmax layer [14].

2 Related Works

Swertky layer is the main layer of CNN construction. Each channel of the layer has its own filters, the core of the layer processes the original image fragments (summarizes the results according to the characteristics of each object). The peculiarity of the Svertki layer is that it improves the analysis based on a small number of parameters in the image during the processing. This layer uses input and output vectors of the same size [15]. The elements of the input vector can take any real values, but the sum of the output vector itself takes the values that are real in (1.1).

$$(x_i) = \frac{e^{x_i}}{\int_{j=1}^{n} e^{x_j}} \tag{1.1}$$

In this case $x = (x_1, \ldots, n_n)$ are the input vectors of the layer and the number of classes of objects in the n-image.

It is necessary to have a large amount of data to learn human objects from images and to solve processing problems. Today, such large volumes of data are widely used in scientific research in developed countries. Input images use three channels based on the RGB color model to represent each pixel. This color model determines the R (red) G (green) B (blue) pixel values of the image.

2.1 Importance of Computer Vision to Identify Human Objects from Images

Scientific research in the field of digital processing of images focused on the study of the most important features of images, including the identification of human objects from images. Most of these studies are based on image segmentation, background removal, object edge detection, and other approaches. However, on the basis of the above-mentioned methods, it was aimed only at identifying human objects in a certain defined area. In it, other researchers have used Haara substitutions to learn and collect (pixel) features in images.

The disadvantages of the above methods are that they require special knowledge to extract and study the features of human objects in the image. In the detection of human objects from the image, traditional methods of dividing into frames, defining object borders, separating the object from the background, and the movement of the camera in the observation area were used.

2.2 Human Object Detection Using Deep Learning-Based Approaches

In recent years, deep learning-based hardware has made it possible to process large amounts of data with the help of software. In addition, existing problems in several areas have been solved with the help of deep learning. In particular, effective results have been achieved in the fields of speech signal processing, network data filtering, medical and vision. Computers have reached a level where they can match humans when it comes to solving vision problems. In order to improve the reliability and efficiency of deep learning-based methods, several studies have been conducted on human object detection and accounting from images.

Learning methods for human object recognition from images differ from computer vision-based methods. First, a deep learning system for human discrimination in trained neural networks automatically extracts features from large volumes of data to identify and train distinct parameters. Another advantage of multilayer neural networks is that they can be quickly and efficiently adapted to different domains, and the time spent on extracting parameters can be spent on setting robust parameters to build a proportional network.

Database Collection

The lack of reliable data hinders effective resolution of identification and segregation problems. Data sets studied in scientific research were analyzed to find information relevant to scientific research. For example, according to the site qanalysts.com, collecting information about a person is the most important factor (face, body part and his actions). In this regard, several video images of moving human images were extracted using image segmentation and foreground and background subtraction algorithms to create a dataset in scientific research. Table 1 below provides information on the number of human images in motion.

The images collected in the database are in RGB format with a size of 300*300 pixels. Below are examples from the collected images (Fig. 1).

Classification of Objects

Image object classification is a collection of predefined derived classes for searching and recognizing objects from images. Based on these sets, it is possible to identify objects in the image and distinguish each object from another. Includes several steps to extract objects from an image. First, object detection is the process of categorizing several

Fig. 1. Database collection

different objects in an image and demarcating them with bounding boxes (frames). Second, object segmentation is the process of dividing a digital image into several pieces. The purpose of segmentation is to easily and conveniently analyze the image or divide the image into segments. More specifically, segmentation is the process of assigning each pixel in an image to visualize pixels with similar characteristics. Thirdly, pattern-based segmentation allows to distinguish several objects belonging to the same class (Fig. 2).

Fig. 2. Classification of objects

3 Our Methodology

3.1 Framing

A video is a set of consecutively collected cards, that is, if there is a moving object in the video image, then it is possible to observe the change of the pixels and coordinates of the image in each successive frame. If we divide the movement of objects in the video image into frames A and B, it is possible to distinguish the presence of a moving object in the image by comparing the initial state of the frame A and the state of the frame B [3]. That is, the following formula is used to calculate the sequence of video images obtained from surveillance cameras.

$$S(a,b) = \begin{cases} 1, |f_k(a,b) - f_{k-1}(a,b)| \geq Z \\ 0, |f_k(a,b) - f_{k-1}(a,b)| < Z \end{cases} \tag{1.2}$$

where a, b are the frames marked by the invention, f is the total number of frames in the video image Z is the border of the video image. In Fig. 3 below, it can be observed that the video image is split into a series of frames (Fig. 4).

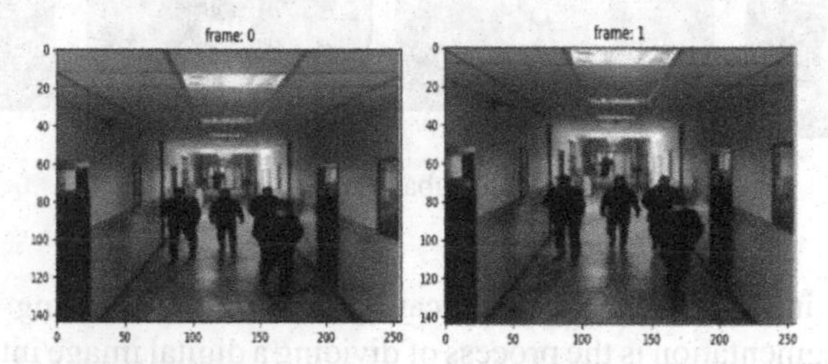

Fig. 3. Compare the image by dividing it into frames

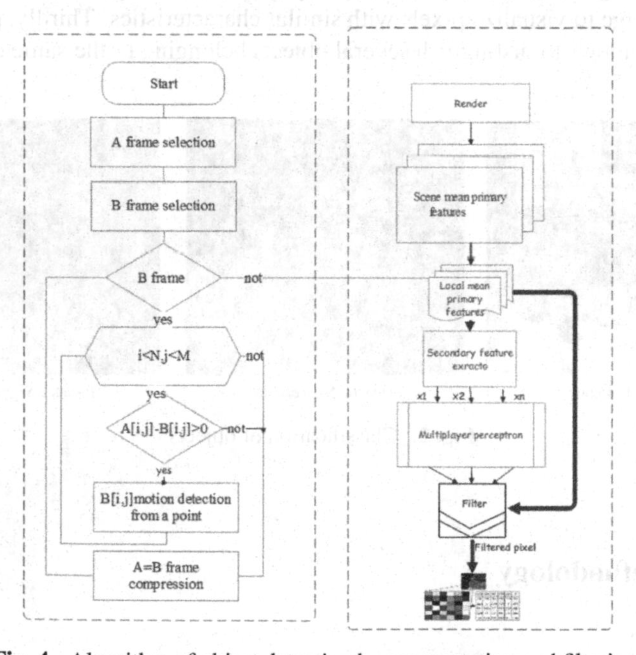

Fig. 4. Algorithm of object detection by segmentation and filtering.

3.2 The Proposed Algorithm from Figure 1 Consists of Two Steps

In the first step, the algorithm uses mathematical statistical operators to sample the moving objects in the video, from the first A frame of the image, since the image in the first card is fixed. These frames are taken as a background model of the image using random values from a normal distribution of pixel brightness in the time plane and as a preliminary analysis. At least from all pixels in N frames, it is necessary to recognize values with constant brightness values. These pixels belong to the background, because these values do not change, and the background is the smallest set of matrix elements. Determination of average sample values of brightness for image pixels with deviations of coordinates [i, j, k] is carried out based on the following formula.

$$\delta_{i,j,0} = \frac{\sum_{k=1}^{n} y_{i,j,k}}{n}, \tag{1.3}$$

where $\delta_{i,j,0}$- mathematical expectation (average brightness value) i, j - pixels of the initial frame, $y_{i,j,k}$ – k-brightness value of i, j -pixels in k card (in bits), n - total number of frames selected for analysis.

Based on the application of the formula (1.3) given above to each row of image pixels in the video sequence, it is possible to determine the brightness distribution of the image pixels.

$$\sigma_{i,j,0}^{2} = \frac{\sum_{k=1}^{n} \left(y_{i,j,k} - \delta_{i,j,0}\right)^{2}}{n-1} \tag{1.4}$$

where $\sigma_{i,j,0}^{2}$– is the luminance variance of the i, j– th pixel in the background matrix.

Formulas (1.3) and (1.4) make it possible to determine $\mu_{i,j,k}$–probability that the object belongs to the background depending on the variability of the values of the pixels whose coordinates are $[i, j]$- in the k -frame.

$$\mu_{i,j,k}\left(y_{i,j,k} \,|\, \delta_{i,j,0}, \sigma_{i,j,0}^{2}\right) = \frac{exp^{\frac{\left(y_{i,j,k}-\delta_{i,j,0}\right)^{2}}{2\sigma_{i,j,0}^{2}}}}{\delta_{i,j,0}\sqrt{2\pi}}, \tag{1.5}$$

For motion detection, based on the formula (1.4), it is possible to determine the successive deviations for the corresponding values of the image pixels by determining the background model of the image.

$$\frac{\left|y_{i,j,k} - \delta_{i,j,0}\right|}{\sigma_{i,j,0}^{2}} \le e \tag{1.6}$$

The background matrix resulting from determining whether a certain pixel belongs to a certain frame is made according to the criteria that determine the range of optimal values for the brightness of the pixel.

Based on the formula (1.6), it is determined whether the elements of the image pixel background matrix belong to the background or foreground.

$$F\left(y_{i,j,k}\right) = \begin{cases} 1, & \text{if result() satisfies the condition} \\ 0, & \end{cases} \tag{1.7}$$

If the change of brightness values of pixels in the amount of 1….*n* satisfies the condition of the formula (1.6), the pixel belongs to the background, that is, appropriate conclusions are accepted on the selection of statistical data of voxels [13].

The above algorithm is adapted to solve the problem of noise and noise transmitted from the camera. In addition, a clear advantage of the algorithm is that it uses a normal model distribution in three-dimensional RGB format for color images. If is difficult to deal with the problem of noise added by the important off-camera environment and image pre-processing algorithms. To solve these problems, it is desirable to develop special algorithms for noise reduction and removal.

The advantage of the first stage of the algorithm proposed above is that it is possible to detect moving objects in the image at a high speed, but the disadvantage is that the efficiency index is lower due to the fact that no noise filtering tools are used in the image. To increase the efficiency of the algorithm, the additional quality of the LBF-algorithm reduces noise - (Machine Learning Approach for Filtering) [13] is suggested to use. This algorithm uses a multi-layer neural network model based on regression, because this algorithm is an efficient system that uses a simple method to identify complex relationships between non-linear input and output images. In addition, the algorithm is adapted for direct parallel operation using the GPU, and the efficiency indicator is high (Fig. 5).

Fig. 5. Algorithm results.

For the results of the experiment, data consisting of 600 images containing human objects from 35 video clips was collected for human detection. Among the collected image data, 7% of moving human objects, i.e. 41 moving objects, were incorrectly detected.

The noise tolerance of the proposed method and the estimation of detection probability were calculated based on the sequence given in the table. For this, first of all, the location coordinates of the moving object should be determined. The noise detected for the video frames is compared with the original [9] pixel coordinates of the image.

$$\begin{cases} p_t = 1, i = jk = 1 \\ p_t = 0, i \neq j || k = 1 \end{cases} \tag{1.8}$$

this is the average of the p value obtained after performing N operations.

$$P = \sum_{y=1}^{N} p_y / N \tag{1.9}$$

the result is the average probability of correctly detecting a moving object from a noisy video image. N-number of frames in a video sequence. If the algorithm detects a moving object in the video, $p_t = 1$, otherwise $p_t = 0$.

Table 1. Probability of correct detections

	Noise/signal = 0.1	**Noise/signal = 0.3**	**Noise/signal = 0.4**
1	2	3	4
A method of separation from the background	0.8	0.8	0.8
GMM background separation	0.9	0.9	0.9
Method of Lucas-Kanade	0.8	0.5	0.3
The proposed method	0.95	0.95	0.95

From the table above, it can be observed that the proposed method and algorithms using image background subtraction and Gaussian distribution mixture are robust to different levels of noise. But the disadvantage of the background removal algorithm is that when the level of noise increases, the efficiency of object detection decreases and false detections increase. The main disadvantage of the Lucas-Kanade method is its variability with respect to the shock. As the signal-to-noise ratio increases, the probability of correct detections decreases.

4 Conclusion

The results of the conducted scientific research revealed that the problem of high-accuracy detection of moving objects in the monitored area with various types of noise in different video images was solved. In order to maintain a balance between accuracy and speed, framing methods have been analyzed and modified. The initial structure of the proposed object detection, based on the results of the experiment, can be concluded that the efficiency of the method of comparing frames to separate moving objects from video images was 91%.

References

1. Roth, P.M., Winter, M.: Technical Report ICG-TR-01/08, Institute for Computer Graphics and Vision, p. 68. Graz University of Technology, Austria, January (2008)
2. Murphy, K.P.: Models for generic visual object detection. Technical report, Department of Computer Science, p. 8 University of British Columbia, Vancouver, Canada (2005)
3. Fomin Ya, A.: Recognition of images: theory and practice, p. 429 (2012)
4. Khamdamov, U., Mirzayev, A., Khalilov, S.: Use of spline models in the analysis of signals collected during the measurement of technological processes. Int. J. Adv. Trends Comput. Sci. Eng. **9**(4), 4451–4456 (2020)
5. Khamdamov, U., Abdullayev, A., Mukhiddinov, M., Xalilov, S.: Algorithms of multidimensional signals processing based on cubic basis splines for information systems and processes. J. Appl. Sci. Eng. **24**(2), 141–150 (2021)
6. Redmon J., Farhadi, A.: YOLO9000: better, faster, stronger. In: Proceedings of the IEEE Conference on Computer Vision and Pattern Recognition, pp. 7263–7271 (2017)
7. Schwegmann, C.P., et al.: Very deep learning for ship discrimination in synthetic aperture radar imagery. In: 2016 IEEE International Geoscience and Remote Sensing Symposium (IGARSS), pp. 104–107. IEEE (2016)
8. Wijlings, J.R., Van De Sande, K.E., Gevers, T., Meulders, A.W.: Int. J. Comput. Vis. **104**(2), 154–171 (2013)
9. Chang, Y.L., Anagaw, A., Chang, L., Wang, Y.C., Hsiao, C.Y., Lee, W.H.: Ship detection based on YOLOv2 for SAR imagery. Remote Sens. 11(7), 786 (2019)
10. Mirzayev, A., Khalilov, S.: Basic principles of wavelet filtering. In: 2021 International Conference on Information Science and Communications Technologies (ICISCT)
11. Ferrari, V., Jurie, F., Schmid, C.: From images to shape models for object detection. Intl. J. Comput. Vis. **87**(3), 284–303 (2010)
12. Fergus, R., Perona, P., Zisserman, A.: Object class recognition by unsupervised scale-invariant learning. In: Proceedings of Computer Vision and Pattern Recognition, vol. 2, Sec. II, pp. 264–271 (2003)
13. Viola, P., Jones, M.: Robust real-time object detection. Intl. J. Comput. Vis. **57**(2), 137–154 (2004)
14. LeCun, Y., Huang, F.-J., Bottou, L.: Learning methods for generic object recognition with invariance to pose and lighting. In: Proceedings of Computer Vision and Pattern Recognition. vol. 2, pp. 97–104 (2004)
15. Umarov, M., Muradov, F., Azamov, T.: Traffic sign recognition method based on simplified Gabor wavelets and CNNs. Int. Conf. Inf. Sci. Commun. Technol. (ICISCT) **2021**, 1–5 (2021). https://doi.org/10.1109/ICISCT52966.2021.9670118
16. Djumayozov, U.Z., Makhamova, D.A., Umarov, M.A.: Numerical methods for solving the two-dimensional boundary value problem of the elasticity theory. Int. Conf. Inf. Sci. Commun. Technol. (ICISCT) **2020**, 1–5 (2020). https://doi.org/10.1109/ICISCT50599.2020.9351436
17. Rakhimov, M., Elov, J., Khamdamov, U., Aminov, S., Javliev, S.: Parallel implementation of real-time object detection using OpenMP. In: International Conference on Information Science and Communications Technologies: Applications, Trends and Opportunities, ICISCT 2021 (2021)
18. Musaev, M., Rakhimov, M.: Accelerated training for convolutional neural networks. In: 2020 International Conference on Information Science and Communications Technologies, ICISCT 2020 (2020)

Building the Groundwork for a Natural Search, to Make Accurate and Trustworthy Filtered Searches: The Case of a New Educational Platform with a Global Heat Map to Geolocate Innovations in Renewable Energy

Seongyun Ku[1], Sunghwan Kim[1], Minji You[2], and Mark D. Whitaker[1](✉)

[1] Department of Technology and Society, Stony Brook University, State University of New York, Korea (SUNY Korea), Songdo, Incheon, South Korea
{seongyun.ku,sunghwan.kim}@stonybrook.edu,
mark.whitaker@sunykorea.ac.kr

[2] Department of Bioscience Engineering, Ghent University, Ghent University Global Campus (GUGC), Songdo, Incheon, South Korea
minji.you@ghent.ac.kr

Abstract. This is a technical and theoretical paper conceptualizing problems to be solved and a few solutions in order to have groundwork for a more 'natural search' to make effective educational use of the Internet on any later filtered search, in this case, in renewable energy innovation. Additionally, this platform for geolocating energy innovations would generate interest in energy resource development trends and encourage industrial progress in the energy field. To do this, our interest is in erecting a strong foundation in natural search to build a better filtered search in other words. However, pragmatically, how can this be done if popular search engines return different kinds of filtered biased searches from the start? We analyze a current prototype platform designed by our team to be a fresh educational platform with a global heat map for geolocation of energy innovations. This platform attempts to "hack" (in a positive way, as a creative work around) and to approximate a natural search using various methods. In short, we discuss a doubly pressing issue in education and sustainability: first, finding a natural search, and second, filtering it upon energy innovations and sharing that information effectively. We list many problems and some solutions toward a more unbiased natural search in discussing our solutions for sampling innately low-'page ranked' energy innovations around the world. We aim to create a platform to better to visualize and to plot trends and locations of energy innovation across the world for expanding that innovation, yet it depends on having a more accurate and trustworthy database to start with in a natural search.

Keywords: Renewable · Energy · Innovation · Platform · Geolocation · Heat map · Python · React · Search engine · Natural search · Problem · Solution

H. Zaynidinov et al. (Eds.): IHCI 2022, LNCS 13741, pp. 239–250, 2023.
https://doi.org/10.1007/978-3-031-27199-1_25

1 Introduction

1.1 A World of Search Biases: Conceptualizing Four Web Search Biases

In the 21st century, to maintain ourselves and to develop, energy sources must be retained above a certain level. However, with growing depletion of and pollution from many non-renewable energy sources, our world is entertaining seriously a desire for an energy transition toward more renewable energies—even if a lot of ideas of current transitions are hardly optimal. Therefore, it would accelerate a positive energy transition if the world could have an online platform to judge, find, and debate the fuller options of more ideal renewable innovations and the locations associated with them.

However, one problem is all current online search engines are a biased resource unable to do this for us. They filter searches in at least four biased ways, many of them used simultaneously: (1) current popularity bias; (2) geolocation bias giving different regions different results from the same searches; (3) personalization biases that are historical to a particular user's computer(s) or accounts based on previous search strings, histories, and clicks; (4) and a company's own editorial filtering biases that 'boost or deboost' (propagandize or censor) particular results from primed or blacklisted websites biased toward maximizing advertising payments or ideologically important narratives and minimizing various controversies (of the former). Therefore, many search biases are '(un)naturally' shown at present. It is hardly a joke: most people think that they are programming a search result from a search engine, when really the search result is programming them, in a biased way, to believe a warped view of the Internet and the world itself. [1] This state of affairs is dangerous both civically and educationally. It is dangerous civically because it leads to centralized misinformation that over time creates an uninformed and propagandized populace, and this undermines representative republican/democratic principles of government of the people, by the people, and for the people. It is dangerous educationally since such information bias frustrates renewable energy innovation and fair judgment or awareness of different technologies that get censored by definition as something always against the status quo politically, economically, or culturally.

So, how do you design a platform that would "hack" (in a good creative sense) biased search engine data mostly based on current popularity, when innovations are rarely popular by definition? Then, how can you show such energy innovations and their locations to people for their global education, without a bias itself? The first point is solved by asking what would be the best "hack" to get around the four filtering biases above (only some of which are algorithmic), instead of simply choosing a different filtered algorithm? As Reynolds writes, "...hacking has been associated with illegal activity. There is a positive [and earlier] definition to hacking. A 'hack' can mean a clever solution that solves problems. Hacks are unorthodox solutions that extend the capability of an application beyond its conventional or intended use." [2] The second point is solved by a global heat map for visualization of information. A heat map can avoid linear ranking of information on any criteria.

The easiest hack in order to have a more 'natural search' for energy innovation is to look through collected articles across multiple search engines. By analyzing the articles about energy resources and plotting these on a heat map, each country's specifically-developed energy sources and their regional and topographical characteristics can be seen as data trends. However, in thinking about hacking solutions toward this particular kind of filtered search, we were forced to consider the bigger idea first: how can we build groundwork for a more 'natural search' in the first place? Moreover, does merely searching across multiple four-biased search engines really result in a more natural unbiased search, or just compounds some biases while solving others? So, how can we solve more than our platform's problems, though could solve the biased civic and educational problems of using current search engines to (mis)inform ourselves. We fail to have all the answers, though we can begin by listing the problems we found that have to be solved to get us to a more 'natural search' in the first place, and we invite others to share their solutions as we mention our solutions in building this prototype educational platform for featuring an accurate global review of renewable energy innovations. To foreshadow, we find at least one search engine, Google, shows its ability to do unbiased natural search, and we are pleased that Google even uses our solution of heat maps for representing unbiased searches, but Google's current business model based on the four biases above relegates these more utopian search ideas to the backpages.

In short, we are suggesting a list of problems to be solved to get unbiased natural searches, and we are attempting technically to solve them in various APIs and database querying strategies. We share our prototype solutions to this double task of approximating a natural search upon which we make our filtered search. We invite others to share their solutions to help us on this project.

1.2 Literature Review of Definitions of Different Kinds of Searches: Natural, Ideal, and Balanced Typologies Between Redundancy and Variety

As a definition of 'natural search' we mean a process that avoids the four biases above at least. This view of natural search is different than others' versions of 'ideal search' algorithms because these mostly only try to address the first bias of current popularity and monopoly power combined. This first bias has been called the "rich get richer" bias in how highly viewed links in search results get the most clicks, further biasing the search and our cultures by the effect of monopoly power on a search itself. Meanwhile, low ranked links are rarely clicked on at all and that biases our cultures as well. [3, 4] Cho writes: "are we somehow penalizing newly created pages that are not very well known yet? Are popular pages getting even more popular and new pages completely ignored?" [3] Their data said yes. This is what Pasquale has called the problem of a "convergence culture" [5] clogging up the search as well, which he says was foreseen by media theorists in the 1980s as an information problem due to the growing convergence of a singular information source across all different media. [6] Others try to address the second and third bias against a natural search by employing levels of randomness and by merging multiple search engine results, to demote some bias. [7] However, the unresolved issue that is particularly a problem in searching for renewable energy innovation is that to really get a natural search the fourth bias of filling search results with intentional propaganda and/or censorship is key to the current reigning world of search engines. This censorship even is aided by states that mandate search engines censor online search results as well. That poses a problem for our platform interested in more renewable energy innovations against a status quo that already relies on intentional *technological* suppression deeply, now extended further than ever in our digital convergence culture into the fourth search bias of intentional *information* suppression. [8–15].

We like other's formulation of a typology of all search engine strategies as ranging between 'redundancy' and 'variety.' [16] This frames a greater 'natural search' result as one with both strategies instead of biased only to one:

> This is a [hard] conceptual task…to find a common variable operating behind all possible research discovery tools. [We can do] so by drawing from the information-theoretical distinction of redundancy and variety. Redundancy means that knowing one piece of information allows one to draw inferences about other information; the revelation of the other datum would then seem expected, predicted and superfluous. Variety, however, means that the presence of one (known) piece of information does not disclose other (as yet unknown) information. The revelation of the other datum then seems surprising, unexpected and novel. To briefly summarize the results of an application of the redundancy/variety-distinction…one obtains a conceptual typology based on a dimension ranging from redundancy-reproducing to variety-enhancing research discovery tools. The redundancy-reproducing tools can be (1a) citation-based or (1b) query-based, while the variety-enhancing ones may be (2a) category-based or (2b) randomness-based. [16].

Our solutions for a 'natural search' followed this view well by having a bit of both: *redundancy reporting* that is both citation-based and query-based (on fifteen discrete energy innovations), yet *variety-enhancing* that is category-based (on two main categorical searches of 'renewable' and 'non-renewable') as well as somewhat randomness-based (in our plural database querying strategies). Nishikawa-Pacher argues "[a] well-functioning scientific system…would combine both variety and redundancy," and we would agree. However, the problem still is many search engines which we use as data sources about the world give us a filtered and redundant Internet search based only on current popularity. Such a one-sided bias in search results toward redundancy and toward only the present fails to allow us easily to develop or catalogue the variety around the world over the years in mentions of energy innovations, particularly in the world's peripheries and particularly if less links were in evidence for the innovations if governments were suppressing technology or the search results.

2 Prototype Description

2.1 Motivation for Focus on Non-renewable and Renewable Energies

We are creating a prototype website to find out how much people are interested in energy based on the number of articles in each search and based on geo-located information taken from such articles. Both kinds of information of scale and location will be plotted on a global 'heat map' of energy topics. Like all big ideas, this can be described simply. However, the execution of the idea deals with two major issues: building a more 'natural search' itself as accurate and trustworthy, and upon that whether our filtering choices are appropriate. We address both natural and filtered search issues after discussing our motivation for this focus to build an online educational resource featuring energy innovations on a global heat map.

Energy is the focus as it was judged as the most essential component of almost all modern and industrial economic activities. Its flows join literally industrial production, transportation, communication, commerce, and household economies. For evidence, seven of the ten largest industrial sectors in the world in 2021 relied on energy flows directly: telecommunication, automobiles, oil and gas exploration and production, food industry, information technology, e-commerce, and construction. [17] Furthermore, eight of the ten largest industrial firms themselves are oil (6) or automobile (2) firms. The remaining two (Apple in telecommunications and Walmart in food/apparel distribution) both rely on energy for their scale to work. In 2015, "oil account [ed] for less than 10% of world GDP, but much of the world's capital stock is designed to use oil; when oil becomes more expensive, that capital becomes less productive." [18] Later, in 2019, oil was judged as only 3.8% of total world GNP [18, 19], yet its energy flows remain the lifeblood and bottleneck of any growing national or global economy. Clearly, if supply and demand of any energy source is not smooth, the ripple effect on global and domestic economies increase, and the economic foundation of a specific region may collapse. This is exemplified by how Sri Lanka lost the ability to pay for oil imports by Summer 2022 and soon after had a revolution. Their President resigned and escaped overseas in July 2022. It is exemplified by how much of Europe is at risk of deindustrializing itself by Autumn 2022 via its sanctions against Russian energy imports. Many European governments are falling already as of September 2022. When the price of oil increases, or if certain energies or their additives cannot be had exemplified by how diesel fuels and diesel fuel additives worldwide are expected to be unable to be bought by August 2022, the related diesel energy scarcity will be a problem that spreads to all economic activities, shrinking industrial activities and resulting in a decrease in the economic growth rate. [20] Thus, it is important to popularize energy innovations for economic stability and for smoother energy transitions.

The amount and cost of energy consumed in industrial societies is large. Modern useful sources of energy depend mainly on mining coal or drilling for crude oil, with only small yet rising percentages in various renewable energies. Since many fuels are non-renewable, it is important to conserve resources to save energy and to develop new energy sources. Non-renewable resources may not be depleted ever to zero, though it is possible that long before that point non-renewable materials may be economically more costly than other renewable options. This would push renewable energy transitions since scarcity historically is the "mother of invention" [21] as past materials get more costly. This kind of price structure encourages material transitions.

We argue that our lives will not be normal unless alternative renewable energies are developed as older non-renewable energy prices rise. Recently, based on this trend, the need for new and renewable energy is increasing. In addition, the electric energy we use is supplied by hydro, geothermal, solar, and wind power generation in increasing amounts compared to oil- and coal-derived electricity, yet the scale of renewables so far is insufficient compared to the larger demand, and renewables currently are regularly far more costly than non-renewables. Moreover, environmental pollution is becoming more serious due to the excessive use of select fuels worldwide. These many problems can be solved through improving people's awareness at the same time as innovating science and technology. These are the interactive motivations for the prototype platform.

2.2 Problems Found and Methods Chosen for Solutions So Far in the Prototype for a Natural Search to a Filtered Search

There is a dual problem to solve pragmatically if we are going to use current online tools to try to hack our way to a natural search and then employ a filtered search on renewable energy innovations. Such energy innovations innately 'swim upstream' against the flow of the algorithms of search engines that give high rank only to current popularity and redundancy while downranking novelty and variety. [3, 16, 22] Below, we mention how we plan to solve the four biases against natural search. We sometimes criticize how others attempt to solve the same problems.

Open Source API for Multiple Search Engines. One idea for removing one search engine's many biases is to use multiple search engines. This has been suggested before [7], and we are doing this. However, this still leaves all four biases from multiple search engines in tandem, so by itself this is hardly really a good solution. The suggestion of a level of partial randomization [7] is involved in this further solution in this by some. Currently, we do not attempt randomization though we will utilize multiple search engines. To get around aggregate bias, instead, we will use an intentional selection of multiple geopolitically-competing countries' different search engines (like sampling both Russia's Yandex and US's Google for instance) that is some protection against the fourth bias of a singular political economic pressure of convergence culture or of censorship on the aggregate search results. However, this may still pollute the global heat map with multiple countries' own fads, trends, and censorship in its searches or publications about the world at large. Thus, we will further filter by sampling smaller and larger countries in filtered searches, as well as sampling richer and poorer countries in filtered searches in the search engines of geopolitical adversaries. In the prototype, we begin with open-source APIs that have been found that can handle over ten search engines, though we are only sampling seven at present: (US Microsoft) Bing, (US) Google, (US) Yahoo, (Russian) Yandex, (Chinese) Baidu, and (Korean) Naver and Daum. In the future, some less-used Indian search engines may be included as well, like Qmamu or Epic Search, just for geopolitical variety. Historically, countries like India were non-aligned during the Cold War between the USA and the USSR, and this means geopolitically they have encouraged their own localized technological paths and innovative solutions. This plurality of national cultural search engines is some guard against a single large country polluting the global heat map with its own country's fads, trends, and censorship in its searches about the world's energy innovations. However, multiple languages are a programming hassle, mentioned below.

Plus, our API for multiple search engines can block the filtering from near-monopoly Google. Thus we will have results of multiple articles without filtering, and our validation can get rid of this. We are not going to take the validation process from Google though will be getting the raw data, and we can validate it instead. This is a far more accurate plan for a natural search.

Global Heat Map, Using Google Maps API as the Base Map. A heat map is a thermal distribution chart that displays two-dimensional numerical data in color and appears on the website based on the number of energy-related articles in a specific region. Our team is creating an unbiased heat map to provide information on energy innovation and even friend-finding on common energy innovations around the world. We could find out where the action is, and multiple different groups worldwide could find each other more readily and coordinate better, expanding innovation as well. The advantage of using a heat map as a visualization method is that it provides intuitive insight into big data by representing it as two-dimensional data and coloring figures of the aggregate ranked data all at once, without the bias of a list that is visually ranked from top to bottom. To create such a platform, Python will be used for the back-end's article crawling and React is used for the front-end visually to show it. Heat maps shift from yellow to red as the number of articles related to a particular energy innovation increases in a specific area. If the number of related articles is small, it is in green. Other ideas for the heat map are different color gradations for different energies (Fig. 1).

Fig. 1. Example of Global Heat Map for Energy News (Origin: team prototype)

It is a shame that Google's API of their main search engine does not offer a more 'natural search' replete with heat maps. Ironically, Google does have three locations where they show a natural search even organized with our solution of heat maps. Google's main search function maintains the four biases mentioned above, yet Google does do natural search heat maps though only in data analysis of people's aggregate search terms and their geo-location accurately, instead of showing search results themselves accurately. If they did the latter, Google would go against their own 'upranked' search engine business model of taking advertising money to create a directed, inflated, personalized, and suppressed search. Amazingly, multiple Google API's showing more natural search occur three times on their platform—with little comment there of how different this natural search is compared to their main biased search. The more natural searches at Google are seen in Google Ngram Viewer, in the heat maps of Google Trends, and in the Google 'Year in Trends.'

Additional Map Overlay of Global Ecoregions. Ecoregions are specific ecological regions of our world that are full of their own unique ecological relationships between plants, animals, soil types, and microclimate. There are estimated to be around 846 to 867 ecoregions. [23, 24] Our hypothesis is that similar ecoregions within or across the world's 14 biomes (defined as much larger geographic zones mapped as having unique life and climate due to common temperature, precipitation, altitude, or latitude) may have common energy innovations. In other words, it would be a useful public service to map energy innovations upon fine-grained maps of ecological relations like ecoregions or biomes. This would be useful for noting environmental impact assessment conditions that catalyze similar or different energy innovations. The ecoregional map equally is useful for thinking about the lives of particular environmental underclasses in any ecologically and socially marginalized sections of nations that feel the brunt of environmental toxicities and thus who would have the most interest in energy innovation as a consequence. [25] Ecoregions are used in parallel with the global heat map (likely with a fader selection) instead of only abstract geopolitical country borderlines. In short, this is in order to show how different ecological zones and different marginalized peoples worldwide in various nation states could (or should?) have an interest in different energy developments. It would be worth mapping for this cross-inspiration between marginalized groups.

> The 2001 map of the terrestrial ecoregions of the world (Olson et al., 2001) facilitated the design of representative networks of protected areas. It has also been used to depict species distributions, to model the ecological impacts of climate change, to develop landscape-scale conservation plans, and to report on progress toward international targets. The revised map, named Ecoregions2017©Resolve, that is the basis for this scheme is unchanged for large sections of the seven biogeographical realms but differs from the original map in four regions: the Arabian Peninsula, some of the desert and drier ecoregions of the African continent, Antarctica, and the southeastern United States….[For the world, this makes]…14 biomes and their constituent 846 ecoregions. [24] (Fig. 2).

Crawling the Web for Fifteen Specific Energies. Instead of searching for abstract terms like 'fossil fuels' or 'renewables' alone, we avoid crawling the web for euphemistic terms or only big categories. Instead, we are both categorical and very specific to the actual energy choice to make accurate global heat maps that show different energies. This is a strategy approved of by others to merge different principles of redundancy and variety in searching toward a more accurate natural search. [16] For example, instead of a web-crawl search for 'fossil fuels,' two main variety-enhancing categorical search terms were used along with more redundancy-enhancing discrete energies within such categories: Non-Renewable (8) [oil/petroleum, gasoline, propane, methane, LPG, natural gas, coal, nuclear energy] and Renewable (7) [solar heat energy, solar light energy, bioenergy, wind energy, geothermal energy, hydrogen energy, hydropower]. Search can be expanded later to other terms easily.

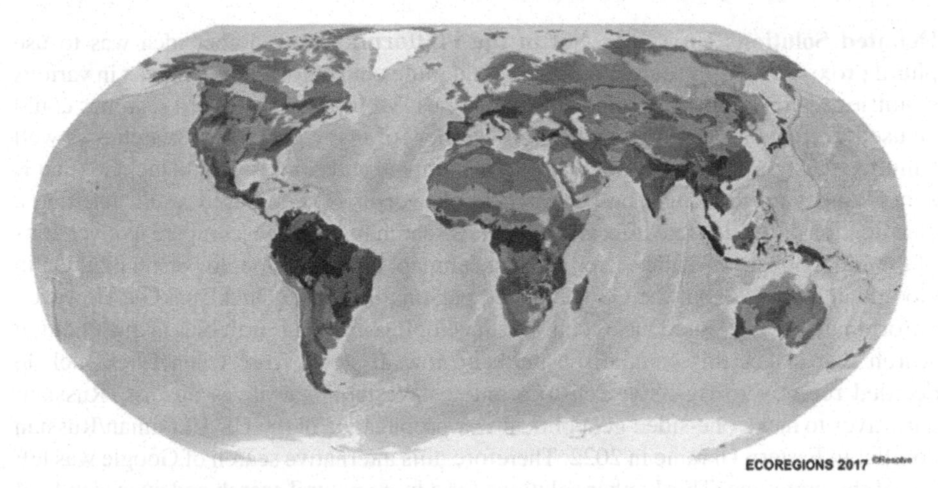

Fig. 2. Global Heat Map Overlay: 846 Terrestrial Ecoregions of the World ("TEOW") based on Olson et al.'s 867 ecoregions [23], updated to Dinerstein et al.'s 846 ecoregions [24].

Crawling the Web for Only Journalism and News Subcategories. We attempt some level of currency and trustworthiness of data by sampling toward only journalistic news writing, since journalistic training typically includes the basic facts and locations of all stories with a city byline. We rely on sampling journalism as a potentially less biased source than simply raw text search results over all web pages of sound or unsound information with uneven provenance. Thus, for better or for worse, we bias our filtered search by further filtering for professional journalism even if these sources are judged as much of a source of good information as it is 'fake news' and omission or suppression of facts sometimes. Since the heat map is not judging content of the article and only referencing major words for energies, this is a more trustworthy presentation because the heat map will be avoiding any analysis of the media spin or attitude of an article, since the heat map is only showing aggregate media activity and aggregate locations of stories.

US City Names in 50 States and 2 Territories as Prototype Base; World Later. The end goal is to search for journalistic news stories about energy linked to all major cities, provinces, and states around the world. However, to test the programming, the prototype of the energy heat map will start with searching and classifying news stories on energy on the global heat map from only major cities in the 50 US states and 2 US territories. It was thought that searching for only states or provinces would hardly be fine-grained enough to plot on a global heat map accurately—particularly a heat map with a high resolution layer for demarcations like the ecoregional map.

English in the Prototype Only; Multiple Languages Sourced Later. There are problems to rebuild code for each language of the search, if we are serious about a natural search as the basis for our later filtered search. Therefore, for the prototype, only English will be used to test the ideas even if this is surely going to miss out on many regional energy innovations. We will get serious about multiple-language coding later when simultaneously coding for world cities.

Debated Solutions Currently Not in the Platform. First, another idea was to use plural proxy servers for searching. This would mute out geolocation or blocks in various countries search engines, to mute out such issues. Multiple proxy-based searches could be used to create the 'poor or rich countries' view of many geolocated searches as well simultaneously. Second, another idea was to compare and contrast DuckDuckGo (that is a more anonymized Google-based search result) versus Google-based results itself, as a clarification of the discrete effects of Google's search biases if we compare two versions of the monopolistic Google search on the heat map, one with Google personalization in Google and one without the Google-added personalization in DuckDuckGo. However, unfortunately, even DuckDuckGo now is developing its own fourth bias against natural search that makes this comparison unlikely now. It stems from when DuckDuckGo decided recently to rig its search to promote 'Western' narratives against 'Russian' narratives to make one-sided geopolitical war propaganda in the US-Ukrainian/Russian conflict in Eastern Ukraine in 2022. Therefore, this alternative search of Google was left out of the prototype. Third, other solutions for a more natural search and even a style of heat map presentation exist on the Commodity Ecology platform. This platform makes a more natural search on its platform by turning the platform itself into a global civic posting platform for all ecoregions and the world instead of scraping search engines. Its version of heat maps are in the Commodity Ecology 'commodity wheel' prototypes that will show aggregate user interest per 130 commodity categories of innovation. Thus, searching through a global posting structure discussing only innovative materials and material news innately represents a more natural civil interest in energy (and all other materials) and avoids filtering the heat maps via biased search engines or journalism per se. [26, 27].

2.3 Plan of Procedural Testing of the Platform

We follow a four step model to assure that we vary only one issue in the programming at a time to aid troubleshooting. We will only do the 'web crawl' in this order: 1. on at least one search engine and at least one search term. This is the first hurdle to make it functional. 2. Then, we will expand to multiple search terms on the same search engine, and then, 3. Expand the web crawl to a singular term on multiple search engines, and then finally, 4. The goal of crawling for multiple terms on multiple search engines with the implementation of our unique data validation process. In this way, we know discretely what programming problems are as we make it more complex. If we tried to do all four levels at once, it could be hard to troubleshoot.

3 Conclusion

In this paper, a method was proposed for comparing the level of interest in non-renewable and renewable energy options by country through a global heat map website based on the number of energy-related articles by region judged from multiple search engines. Toward a groundwork of a more natural search for our filtered search by crawling multiple search engines and other tactics mentioned above, we avoid list-based ranked searches on energy with our proposed solution to visualize the filtered search results via geolocation on a global heat map based on the number of articles of each energy news story. In this visual way, it will be possible to verify more intuitively the world of energy innovations unlike other biased search methods or simply a ranked list. We hope the solution of a global heat map for energy innovation will increase people's interest in energy choices and at the same time contribute to greater innovation in the energy industry. Plus, the platform will compare various countries, regions, energies, and search engines over time through trend graphs and various filters as well.

References

1. Pasquale, F.: The Black Box Society: The Secret Algorithms That Control Money and Information. Harvard University Press, Cambridge (2015)
2. Reynolds, R. P., The Librarian as Hacker, Getting More from Google (2007). https://scholarworks.sfasu.edu/cgi/viewcontent.cgi?article=1011&context=libfacpub
3. Cho, J., Roy, S.: Impact of search engines on page popularity. In: Proceedings of the 13th International Conference on World Wide Web (WWW 2004). Association for Computing Machinery, New York, NY, USA, pp. 20–29 (2004). https://doi.org/10.1145/988672.988676
4. Lewandowski, D.: Is Google responsible for providing fair and unbiased results? In: Taddeo, M., Floridi, L. (eds.) The Responsibilities of Online Service Providers. LGTS, vol. 31, pp. 61–77. Springer, Cham (2017). https://doi.org/10.1007/978-3-319-47852-4_4
5. Pasquale, F.: Internet Nondiscrimination Principles: Commercial Ethics for Carriers and Search Engines, University of Chicago Legal Forum, 2008, Article 6 (2008). https://chicagounbound.uchicago.edu/uclf/vol2008/iss1/6
6. de Sola Pool, I.: Technologies of Freedom: On Free Speech in an Electronic Age. Harvard University Press, Cambridge (1983)
7. Pandey, S., Roy, S., Olston, C., Cho, J., Chakrabarti, S.: Shuffling a stacked deck: the case for partially randomized ranking of search engine results. In: 31st VLDB Conference (2005). https://www.cs.cmu.edu/~spandey/publications/randomRanking-vldb.pdf
8. Bearden, T.: Energy from the Vacuum: Concepts and Principles. Cheniere Press, Tokyo (2001)
9. Eisen, J.: Suppressed Inventions and Other Discoveries. Avery Publishing Group, New York (2001)
10. Lindemann, P.: The Free Energy Secrets of Cold Electricity. Clear Tech Inc, Washington, Metaline Falls (2001)
11. Cook, N.: The hunt for zero point: inside the classified world of antigravity technology (2003)
12. Valone, T.: Practical Conversion of Zero-Point Energy, Revised Integrity Research Institute, Washington DC (2005)
13. Bearden, T., Bedini, J.: Free Energy Generation—Circuits and Schematics: 20 Bedini-Bearden Years, 2nd edn. Cheniere Press, Tokyo (2006)
14. Black, E.: Internal combustion: how corporations and governments addicted the world to oil and derailed the alternatives (2007)
15. Manning, J., Garbon, J.: Breakthrough Power: How Quantum-leap New Energy Inventions Can Transform Our World, Second Edition. Malloy, Michigan (2011)
16. Nishikawa-Pacher, A.: A typology of research discovery tools. Journal of Information Science (2021). https://journals.sagepub.com/doi/full/10.1177/01655515211040654
17. Novicio, T.: 10 Biggest Industries in the World in 2021 (2021). https://finance.yahoo.com/news/10-biggest-industries-world-2021-150703784.html

18. Mueller, J.S., Brown, S.P.A.: Oil and the Global Economy, Resilience.org (originally at The Energy Xchange, October 9, 2015) (2015). https://www.resilience.org/stories/2015-10-09/oil-and-the-global-economy/
19. Jegede, A.: Top 10 Largest Industries in the World, Trendrr.net (2022). https://www.trendrr.net/8487/largest-industries-world-us-revenue-profitable/
20. Adams, M.: The Diesel Engine Oil Crisis Explained, Brighteon, Entire U.S. supply of diesel engine oil may be wiped out in 8 weeks, no more oil until 2023 due to 'Force Majeure' additive chemical shortages (2022). https://www.brighteon.com/4eab4bd0-db72-4afa-b633-6efc782473b8, https://www.nytimes.com/2006/08/10/opinion/10sass.html?_r=0
21. Sass, S.L.: Scarcity, Mother of Invention, New York Times Editorial, Aug. 10 (2006). http://www.nytimes.com/2006/08/10/opinion/10sass.html?_r=0
22. Cho, J., Roy, S., Adams, R.E.: Page quality: in search of an unbiased web ranking. In: SIGMOD 2005: Proceedings of the 2005 ACM SIGMOD International Conference on Management of Data, pp. 551–562 (2005). https://doi.org/10.1145/1066157.1066220
23. Olson, D., Dinerstein, E., et al.: Terrestrial Ecoregions of the World: A New Map of Life on Earth. BioScience **51**(11), 933–938 (2001) https://doi.org/10.1641/0006-3568(2001)051[0933:TEOTWA]2.0.CO;2
24. Dinerstein, E., et al.: An ecoregion-based approach to protecting half the terrestrial realm. BioScience **67**(6), 534–545 (2017)
25. Pellow, D.N.: Environmental racism: inequality in a toxic world, In: Romero, M., Margolis, E. (eds.) The Blackwell Companion to Social Inequalities, pp. 147–164. Wiley-Blackwell (2008)
26. Whitaker, M.D. Shin, G.: A circular economy for the world: the commodity ecology model for achieving all sustainable development goals, J. APEC Stud. (Asian Pac. Econ. Cooperation), **12**(2), 77–99 (2020)
27. Whitaker, M.D.: The commodity ecology mobile (CEM) platform illustrates ten design points for achieving a deep deliberation in sustainable development goal #12. In: Singh, M., Kang, D.-K., Lee, J.-H., Tiwary, U.S., Singh, D., Chung, W.-Y. (eds.) IHCI 2020. LNCS, vol. 12615, pp. 414–430. Springer, Cham (2021). https://doi.org/10.1007/978-3-030-68449-5_41

Implementation of Virtual Sea Environment with 3D Whale Animation

Uipil Chong and Shokhzod Alimardanov(✉)

Graduate School of Industry, University of Ulsan, Ulsan, Republic of Korea
upchong@ulsan.ac.kr

Abstract. Global marine tourism is emerging as a new blue ocean, and VR/AR technology is coming to our real world. This research concerns virtual sea environment development to make marine tourism more accessible and safer. The importance of our research is that virtual tourism is more cost efficient and accessible while being more immersive and entertaining than traditional TV or Radios. We have researched different wave spectra models to simulate the sea environment in Ulsan, Korea. Wind speed is used as a main disturbing force of the sea and gravity constant as a restoring force. The processing workload on CPU were reduced by utilizing GPU for heavy calculations. The results of this marine environment study were applied to the development of contents for whale tourism. We developed the 3D whale animation using the Unity. The scene was presented to people on the whale watching tour boat. The survey data showed interest and support from participants. We could render sea surface in real-time and simulate Ulsan sea environment with various wind speeds.

Keywords: Unity · Wave spectra · Marine Tourism · Whale Animation · Sea Surface Rendering

1 Introduction

The purpose of this research is to render a sea environment depending on various wind speeds to simulate several weather conditions. Once achieved, the virtual space can be consumed as a cost efficient and environmentally safer alternative for experiencing marine tourism.

Current study utilizes a wave spectra model to create complex wave heights in frequency domain. Then, the result is converted from frequency domain to spatial domain using Inverse Fourier Transform (IFFT). Additionally, the examples of marine life, such as whales were animated.

A virtual environment may be displayed on a head-mounted display, a computer monitor, or a large projection screen. Head and hand tracking systems are employed to enable the user to observe, and manipulate the virtual environment [1].

The main difference between VR systems and traditional media (such as radio, television) lies in three dimensionality of Virtual Reality structure. Immersion, presence, and interactivity are peculiar features of Virtual reality that draw it away from other

H. Zaynidinov et al. (Eds.): IHCI 2022, LNCS 13741, pp. 251–256, 2023.
https://doi.org/10.1007/978-3-031-27199-1_26

representational technologies. Virtual reality does not imitate real reality, nor does it have a representational function. Human beings have inability to distinguish between perception, hallucination, and illusions [1].

We assume that 3D whale animations in a virtual reality could produce the satisfaction of watching whale at the surface level of ocean. It is especially helpful for those who have limited access to a marine tourism. Animated whales would entertain people while introducing their life.

2 Background

2.1 Height Field Generation

Ocean simulation is largely about finding vertical displacement of ocean waves. We reproduce ocean waves as a sum of many sinusoidal waves. The sum of the sinusoidal waves is calculated using Fast Fourier Transform. Equation (1) gives the height of our ocean in spatial domain [2]:

$$\eta(x, t) = \sum_k h(k, t)e^{ikx} \tag{1}$$

where η is the height of the wave, k–wavevector, t – time, x – horizontal direction of the area, h – height of our water wave. In later paragraphs, the Eq. (1) will be broken down and discussed briefly.

2.2 Time Dependent Height Function

Equation (1) represents the general algorithm of wave height in the spatial domain. By breaking down the Eq. (1) we get a time dependent height function. Jerry Tessendorf [2] guarantees that the result of the Eq. (2) will always be real number.

$$h(\mathbf{k}, t) = h_0(\mathbf{k})e^{i\omega(k)t} + h_0(-\mathbf{k})e^{-i\omega(k)t} \tag{2}$$

where ω is angular frequency.

2.3 Stationary Spectrum

Breaking down Eq. (2) we get our stationary spectrum for height calculations [2]:

$$h_0(k) = \frac{1}{\sqrt{2}}(\xi_1 + i\xi_2)\sqrt{2S(\omega)D(\theta, \omega)\frac{d\omega(k)}{dk}\frac{1}{k}\Delta k_x \Delta k_z} \tag{3}$$

where ξ_r and ξ_i are random scalars drawn from the Standard Normal Distribution. $S(\omega)$ is the non-directional wave spectrum, $D(\theta, \omega)$ is directional spreading function.

2.4 Wave Spectrum and Directional Spread

Wave Spectrum. The wave spectrum indicates the height of the water wave at a certain frequency. There are many wave spectrums that generate different kinds of wave heights depending on the weather conditions. Current study utilizes TMA spectrum model to produce random heights. TMA spectrum combines JONSWAP spectrum with the Kitaigorodskii Depth Attenuation Function [3].

The JONSWAP spectrum is a one-dimensional frequency spectrum which has been developed by Hasselman (1973) based on data collected during the Joint North Sea Wave Project [4].

JONSWAP spectrum alone cannot produce visually realistic ocean waves. In addition to said spectrum, we use Kitaigorodskii Depth Attenuation Function which produces smooth looking ocean waves [5, 6].

Directional Spread. The study of how wave energy disperses directionally, relative to the primary wind direction, is called directional spread. Its function is written $D(\theta, \omega)$ where θ is the angle of wave relative to wind direction. See reference [7] for thorough discussions.

2.5 Unity

Unity3D is a cross-platform game creation system developed by Unity Technologies containing a game engine and an integrated development environment (IDE). It is used to develop games for websites, desktops, handheld devices, and consoles. Previously released on Mac OS in 2005, it was then expanded to target more than fifteen platforms. It includes an asset server and Nvidia PhysX physics engine (Cross-platform Application Development using Unity Game Engine) [8].

This research benefits from Unity by using its powerful animation technology. The animation features provided out of the box include full control of animation, retargetable animations, sophisticated state machines hierarchies and transitions, event calling from the animation playback, etc. While 3D models of whales were designed using other software, such as, Blender, and 3DS Max, Unity allowed to import them with one click, and start animating.

3 Implementation

Finally, we implement the spectra and start simulation in Unity. Compute shaders, C# scripts and surface shaders are used for the rendering of the ocean surface (see Fig. 1).

Compute shaders are programs that run on the graphics card, outside of the normal rendering pipeline. They can be used for massively parallel GPGPU algorithms, or to accelerate parts of game rendering. Usually, compute shader files are written in HLSL (High Level Shader Language) and compiled or translated into all necessary platforms automatically. Thus, we use terms "Compute shader" and "HLSL" interchangeably. Compute shaders are responsible for calculation of the spectrum. In addition, we exploited an IFFT [9] implementation in compute shaders to delegate heaviest processing away from CPU.

At first, compute shaders input initial data, such as wind speeds, and water depth. Secondly, it calculates the wave spectrum, displacements, and slopes. Up until this point everything calculated is in the frequency domain. At the third step, we convert the data from frequency domain to discrete time domain using IFFT.

A Unity C# script is used to control the data flow. For example, it gets input from an artist (unity developer) and sends it to the HLSL program. When HLSL finishes its process, C# script retrieves the output. Meanwhile, we produce the ocean geometry in C#. Since the spectrum gives the wave heights not the water geometry, we need geometrical body to apply those values.

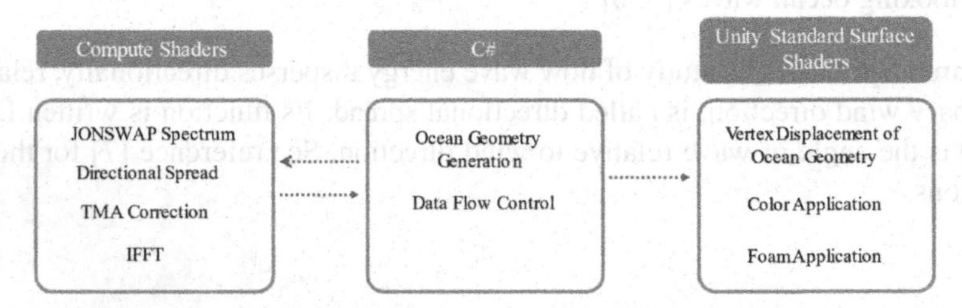

Fig. 1. Unity Ocean Simulation Workflow.

Next, C# sends the values of the spectrum as well as the surface geometry to a unity surface shader. Surface shaders, as the name implies, define the physical characteristics of materials. They are responsible for both calculating the final color of each pixel within a material and performing the light calculations that define the shading of each pixel on the surface. Most surface shaders in Unity are extensions of the default Standard Surface Shader, which makes the creation process more intuitive and allows artists to define the look of their surfaces more freely.

The data that unity surface shader inputs contains displacements (water heights), and derivatives of the displacements. In the shader, the displacements are used to modify the vertices of the ocean geometry. The derivatives of the displacements are used to calculate normal vectors. Normal vectors are parameters that help the surface shader to apply colors and shadows.

Foam effects are also applied in the unity surface shader. However, the detection of the foam points is done using the determinant of Jacobian matrices [5]. When the Jacobian determinant has a negative value, the shader applies the foam texture.

3.1 Experimental Environment

In Ulsan, Republic of Korea, whale watching tour ships gather people every weekend, especially in summer. The probability of visitors seeing whales or dolphins at surface level is very low. We intend to solve this problem by giving an alternative option to experience sea environment with the mammals virtually. Several animated whales and dolphins were added in the rendered sea environment.

4 Results

Figure 2 is the resulting sea surface rendering. To achieve more realistic virtual water surface, we need to tweak several parameters, such as, color, foam intensity, wind speed and more. Fortunately, many parameters need to be set only once and they match the other weather conditions seamlessly. The only two things that require frequent tweaks are wind speed and foam intensity. Figure 2 demonstrates the resulting. At the top (a) is an agitated sea level where wind blows at the speed of 10 m/s. The foam intensity is also modified to match the behavior of waves. The image in the middle (b) is a rendering of a sea level with wind speed going up to 4 m/s as well as an increased foam intensity. But the waves still have relatively high picks and stay slightly agitated. At the bottom (c) is an image of a very calm sea level where wind speed is 0.1 m/s and the foam intensity set to zero.

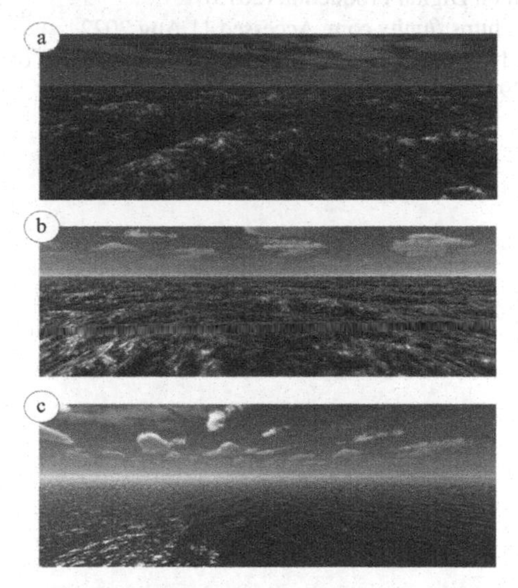

Fig. 2. Simulation results. a) An agitated sea level (10 m/s wind speed); b) A medium sea level (4 m/s wind speed); c) A calm sea level (0.1 m/s wind speed).

5 Conclusion

In conclusion, we were able to render sea surface with different weather conditions in real-time. Since our approach is based on processing calculations on the GPU, the better performance was achieved. The resulting scene is utilized for virtual sea life simulation. People acknowledged that the whale watching tourism would be more enjoyable if this kind of virtual experiences was well-developed.

Acknowledgments. This research is supported by the 2022 National Research Foundation (No. 2017R1D1A3B05030815) of the Korean Government and the University of Ulsan.

References

1. Mandal, S.: Brief introduction of virtual reality and its challenges. Int. J. Sci. Eng. Res. **4**, 304–309 (2013)
2. Tessendorf, F.: Simulating ocean water. In: SIGGRAPH Course Notes. ACM (1999)
3. Lee, N., Baek, N., Ryu, K.W.: A real-time method for ocean surface simulation using the TMA model. Int. J. Comput. Inf. Syst. Ind. Manage. Appl. (IJCISIM), **1**, 25–26 (2008)
4. Hasselmann, K., Barnett, T.P., Bouws, E., Carlson, D.E., et al.: Measurements of wind-wave growth and swell decay during the joint north sea wave project (JONSWAP). Deutsche Hydrographische Zeitschrift (1973)
5. Gamper, T.: Ocean Surface Generation and Rendering. Master's thesis (2018)
6. Kitaigorodskii, S., Krasitskii, V., Zaslavskii, M.: On phillips' theory of equilibrium range in the spectra of wind-generated gravity waves. Phys. Oceanogr. **5**, 410–420 (1975)
7. Horvath, Ch. J.: Empirical directional wave spectra for computer graphics. In: Proceedings of the 2015 Symposium on Digital Production (2015)
8. Unity Game Engine . https://unity.com. Accessed 11 Aug 2022
9. Cooley, J.W., Tukey, J.W.: An algorithm for the machine calculation of complex fourier series. Math. Comput. **19**, 297–301 (1965)

A Constant-Factor Approximation Algorithm for Online Coverage Path Planning with Energy Constraint

Ayan Dutta[1] and Gokarna Sharma[2(✉)]

[1] School of Computing, University of North Florida, Jacksonville, FL 32224, USA
a.dutta@unf.edu

[2] Department of Computer Science, Kent State University, Kent, OH 44242, USA
gsharma2@kent.edu

Abstract. We study the problem of coverage planning by a mobile robot with a limited energy budget. The objective of the robot is to cover every point in the environment while minimizing the traveled path length. The environment is initially unknown to the robot. Therefore, it needs to avoid the obstacles in the environment on-the-fly during the exploration. As the robot has a specific energy budget, it might not be able to cover the complete environment in one traversal. Instead, it will need to visit a static charging station periodically to recharge its energy. To solve the stated problem, we propose a budgeted depth-first search (DFS)-based exploration strategy that helps the robot to cover an unknown planar environment while bounding the maximum path length to a constant-factor of the shortest-possible path length. This guarantee advances the state-of-the-art of log-approximation for this problem. Simulation results show that our proposed algorithm outperforms the current state-of-the-art algorithm both in terms of the traveled path length and run time in all the tested environments with concave and convex obstacles.

1 Introduction

Coverage planning is the task of finding a path or a set of paths to cover all the points in an environment [4]. In robotics, this problem has many potential real-world applications including autonomous sweeping, vacuum cleaning, and lawn mowing. In an *online* version of the problem, the area of interest is initially unknown to the robot. Therefore, it needs to discover and avoid the unknown obstacles in the environment while covering all the points in the free space by traveling as minimum distance as possible [4].

Traditionally, this problem has been studied assuming that the robot has an unlimited energy budget, where given a robot, a single path can be planned to cover the given environment. The *offline* version of the problem where the robot(s) have a priori knowledge of the environment including obstacles has been well studied [3,7–9,11].

In practice, however, the robots do not have unlimited energy available. Therefore, even covering a standard-size environment (e.g., a farm) while simultaneously using on-board sensors (e.g., camera) becomes prohibitive with a single charge. A battery-powered robot needs to return to the charging station to recharge its battery before it runs out. Due to practical relevance, in the recent years, there has been a significant

H. Zaynidinov et al. (Eds.): IHCI 2022, LNCS 13741, pp. 257–270, 2023.
https://doi.org/10.1007/978-3-031-27199-1_27

volume of work on the *energy-constrained* coverage planning problem [12,13,15–18]. The offline version of the problem, which we denote as OFFLINECPP, is studied in [15,17,18] and the online version, which we denote as ONLINECPP, is studied in [13, 15]. The state-of-the-art algorithm for OFFLINECPP is due to [17], which provides an $O(1)$-approximation. For ONLINECPP, the state-of-the-art algorithm is due to [13], which provides $O(\log(B/L))$-approximation, where B is the energy budget and L is the size of the robot (assuming a $L \times L$ square robot). Our goal in this paper is to provide a better approximation algorithm for ONLINECPP, given an unknown but static polygonal environment P and a charging station S within P.

Model Overview and Contributions. Initially, the robot with battery budget B is at the charging station S that is inside the unknown environment P. The environment P is a polygonal area possibly containing obstacles within it; nothing is known about obstacles including the numbers of them except that the obstacles are static (i.e., they do not move during coverage process). Moreover, the environment is fixed in the sense that it does not change during the coverage process. The robot is equipped with an obstacle detection sensor (e.g., laser rangefinder) as well as position sensor (e.g., GPS). The robot is represented as a square $L \times L$ that moves rectilinearly in P (the environment is discretized into cells of size $L \times L$ forming a 4-connected grid). Initially, the robot is at the charging station S that is inside P. The robot has sufficient on-board memory to store information necessary to facilitate the coverage process. The robot has to go back to the charging station S to get its battery recharged before it runs out of it. It is assumed that the size of P is such that the robot cannot fully cover P in a single traversal with budget B. The energy consumption of the robot is assumed to be proportional to the distance travelled, i.e., the energy budget of B allows the robot to move B units distance (that is, total $\lfloor B/L \rfloor$ cells). the goal of ONLINECPP is to find a set of paths $\mathcal{Q} = \{\mathcal{Q}_1, \ldots, \mathcal{Q}_k\}$ for the robot such that

- **Condition (a):** Each path $\mathcal{Q}_i$ starts and ends at S.
- **Condition (b):** Each path $\mathcal{Q}_i$ has length $l(\mathcal{Q}_i) \leq B$.
- **Condition (c):** The paths in $\mathcal{Q}$ collectively cover the unknown environment P, i.e., $\cup_{i=1}^{n} \mathcal{Q}_i = P$,

and the following two performance metrics[1] are optimized:

- **Performance metric 1:** The *number of paths* $|\mathcal{Q}|$ in $\mathcal{Q}$ is minimized, and
- **Performance metric 2:** The *total lengths of the paths* in $\mathcal{Q}$, denoted as $l(\mathcal{Q}) = \sum_{i=1}^{n} l(\mathcal{Q}_i)$, is minimized.

We establish the following main theorem for ONLINECPP.

Theorem 1 (Main Result). *Given an unknown planar environment P possibly containing (static) obstacles and a battery-constrained robot of size $L \times L$ consisting of position and obstacle detection sensors initially situated at a charging station S inside P with energy budget B, there is an algorithm that correctly solves* ONLINECPP *and guarantees* 10*-approximation to both performance metrics compared to the optimal algorithm that has complete knowledge about P.*

[1] We do not consider the charging time of the battery since this delay does not impact the performance metrics we consider.

Theorem 1 clearly advances the state-of-the-art as it improves upon the $O(\log(B/L))$-approximation in [13] to $O(1)$.

The proposed algorithm covers an unknown environment P through a *depth first search* (DFS) traversal approach tailored for the limited energy budget B. The battery-constrained robot performs the DFS traversal while building a tree map of the environment on-the-fly. It returns to the charging station to get its battery fully recharged (stopping the DFS traversal) when the path length of the traversal becomes at most the energy budget B. Simulation results show that the proposed algorithm is up to 5.80 times faster and up to 2.53 times less costlier (in terms of traversed path length) than the current state-of-the-art algorithm [13].

Remark. Shnaps and Rimon [15] established a lower bound of $\Omega(\log(B/4L))$-approximation for ONLINECPP. In their lower bound proof, they considered an environment P that can change while robot is under the coverage process. Their lower bound construction proceeds in iterations. In the first iteration, there is a single block in the environment. In each iteration, one block is added to the environment. The algorithm's behavior is observed in each iteration and the new block is added such that it incurs more cost to the algorithm. The algorithm's cost in the environment modified in all iterations is then compared to the cost of the environment if everything is known about the environment a priori. Thus P is dynamic and changes constantly during the execution of the algorithm in the lower bound proof of Shnaps and Rimon [15], making the robot to work more to adjust to the change. The previous works on the upper bound [13,17] and this work consider that P is fixed (not changing in the coverage process) and obstacles are unknown but static. This is the reason why the lower bound of [15] does not apply to a fixed unknown environment P and only to a dynamic unknown environment P. It would be interesting to develop an algorithm for ONLINECPP matching the lower bound of Shnaps and Rimon [15] for a dynamic unknown environment P.

2 Related Work

The most closely related works to ours are [13,15,17,18]. Shnaps and Rimon [15] proposed an $1/(1-\rho)$-approximation algorithm for OFFLINECPP, where $\rho \leq B/2$ is the ratio between the furthest distance between any two cells in the environment [15]. For ONLINECPP, they proposed an $O(B/L)$-approximation algorithm. Wei and Isler [18] presented an $O(\log(B/L))$-approximation algorithm for OFFLINECPP, which has been improved to an $O(1)$-approximation in Wei and Isler [17]. Recently, Sharma *et al.* [13] provided an $O(\log(B/L))$-approximation algorithm for ONLINECPP. In this paper, we improve upon [13] and provide an $O(1)$-approximation to ONLINECPP.

A closely related area is the graph coverage. The goal is to design paths to visit every vertex of the given graph. Without energy constraints, it becomes the well-known *Traveling Salesperson Problem* (TSP) [1] and a DFS traversal provides a constant-approximation of the TSP. With energy constraints, this coverage problem becomes the *Vehicle Routing Problem* (VRP) [10]. One version of VRP is the *Distance* Vehicle Routing Problem (DVRP), which models the energy consumption proportional to the distance travelled. Most of these papers studied the offline version so that preprocessing of the environment can be done prior to exploration. This is also the case in the algorithm of [17,18] for OFFLINECPP. Coverage with multiple robots has also received a lot of attention (e.g., see [2,6]). Here, we consider coverage planning with a single robot under battery constraint.

3 Problem Setup

We use the same model as in [13,15,18].

Environment. The environment P is a planar polygon containing a single charging station S inside it. P may possibly contain polygonal obstacles. The obstacles (the environment P) are assumed to be static meaning that they do not move (does not change) during the coverage process. This assumption on the obstacles and the environment is also made in the existing algorithms for ONLINECPP. See left of Fig. 1 for an illustration of P with an obstacle O_1. P is discretized into cells forming a 4-connected grid.

Robot. We consider the robot r to be initially positioned at the charging station S. r has size $L \times L$ that it fits within a grid-cell in P. The robot r moves rectilinearly in P, i.e., it may move to any of the four neighbor cells (if the cell is not occupied by an obstacle) from its current cell. We also assume that r has the knowledge of the global coordinate system through a compass on-board, that means it knows left (West), right (East), up (North), and down (South) cells consistently from its current cell. Robot r is equipped with a position sensor (e.g., GPS) and an obstacle-detection sensor (e.g., laser rangefinder). We assume that with the laser rangefinder, the robot can detect obstacles in any of its neighbor cells. The robot has sufficient on-board memory to store information necessary to facilitate the coverage process. Moreover, we assume that initially r does not have any knowledge about P, i.e., P is an unknown environment. For the feasibility of covering all cells of P, it is assumed that P is as big as a circle of radius $\lfloor B/2 \rfloor$ with center S. The energy consumption of the robot is proportional to the distance travelled, i.e., the energy budget of B allows the robot to move B units distance.

Path. A path (route) $\mathcal{Q}_i$ is a list of grid-cells that r visits starting and ending with S. Notice that if there are some obstacles within P located in such a way that they divide P into two sub-polygons P_1 and P_2 with P_1 and P_2 sharing no common boundary, then r cannot fully cover P. Therefore, it is considered that there is no such cell c in P, i.e., there is (at least) a route from S to any obstacle-free cell of P. We call a cell *free* if it is not occupied by an obstacle. We call a cell *reachable* if it satisfies the definition below.

Definition 1 (Reachable Cell). *Any cell c in P is called* reachable *by the robot r, if and only if (a) it is a free cell, (b) it is within distance $\lfloor B/2 \rfloor$ from S, and (c) there must be at least a route of consecutive free cells from S to c.*

Problem ONLINECPP. The problem is formally defined as follows.

Definition 2. *Given an unknown planar polygonal environment P possibly containing obstacles with a robot r having battery budget of B initially positioned at a charging station S inside P,* ONLINECPP *is for r to visit all the reachable cells of P through a set of paths so that*

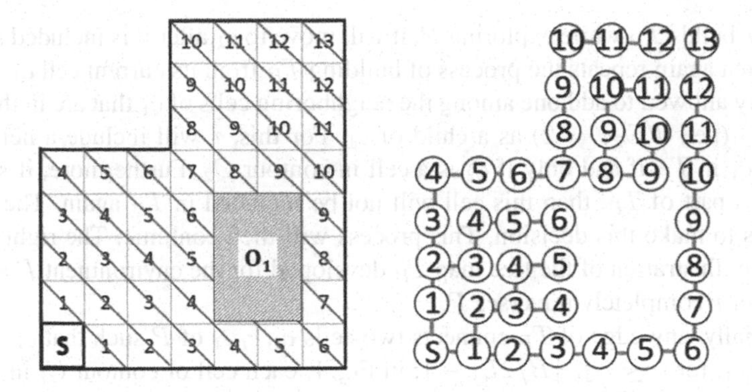

Fig. 1. **(left)** An environment P with an obstacle O_1 and a charging station S. P is shown decomposed as cells of size $L \times L$ same as the robot; **(right)** An example tree map T_P (right) constructed for the environment P on the left. The cells at any contour C_d are at depth d in T_P.

- *Conditions (a)–(c) are satisfied, and*
- *Performance metrics (1) and (2) are minimized.*

Following [13,17,18], we measure the efficiency of any algorithm for ONLINECPP in terms of *approximation ratio* which is the worst-case ratio of the cost of the designed algorithm for an environment P over the cost of the optimal, offline algorithm for the same environment P.

4 Processing Unknown Environment

Decomposition of the Environment. We decompose the environment P into square cells of size $L \times L$, which is the size of the robot itself. An *equi-distance contour* is a poly-line where the cells on it has the same distance to/from the charging station S (the left of Fig. 1). The cells on a contour can be ordered from one side to the other since the global coordinate system is known to the robot.

Let c be a cell and C be a contour. Let $d(c)$ denote the distance to S from c and let $d(C)$ denote the distance to S from C. If $d(C_j) = d(C_i) + 1$, we say that contour C_j is contour C_i's next contour. The contour C with $d(C) = 1$ is called the first contour. Each cell in a contour has at most 4 reachable cells from S that are its neighbors.

Constructing a Tree Map. Initially, the robot r is placed at the fixed charging station S. In this case, the tree, denoted by T_P, has only one node S, which we call the *root* of T_P. If there is no obstacle in P, each cell except the boundary cells of P will have exactly four neighbor grid cells.

Robot r picks the first reachable cell c_1 according to a clockwise ordering of its neighbors starting from the west and ending in the south neighbor cell. r then inserts it into T_P as a *child* of S. If the cell labeled West is a reachable cell, then r picks that cell. Otherwise, it goes in order of North, East, and South until it finds the first cell that is reachable. We now have two nodes in $T_P = \{S, c_1\}$, with c_1 as a child node of S. Furthermore, c_1 is a cell in the first contour C_1.

Since r builds T_P while exploring P, it will move to c_1 after it is included as a child in T_P. r then again repeats the process of building T_P from its current cell c_1. While at c_1, r is only allowed to add one among the neighboring cells of c_1 that are in the second contour C_2 (i.e., $d(C_2) = 2$) as a child of c_1. For this, r will include a neighboring cell c_2 of c_1 in T_P if and only if c_2 is a cell in contour C_2. Furthermore, if some cell is already a part of T_P, then this cell will not be included in T_P again. The memory at r allows to make this decision. This process will then continue. The right of Fig. 1 provides an illustration of the tree map T_P developed for the environment P shown on the left after r completely traverses P.

Essentially, any edge of T_P connects two cells c_i, c_{i+1} of P such that $c_i \in C_i$ and $c_{i+1} \in C_{i+1}$, for $0 \leq i \leq \lfloor B/2L \rfloor - 1$; in Fig. 1, each cell of contour C_i in P on the left are at depth i in T_P shown on the right. Therefore, all the cells in the first contour C_1 will be children of S (the root of T_P), all the cells in the second contour C_2 will be the children of the nodes of T_P at depth 1 from S.

5 Algorithm

In the description of the algorithm and its analysis, we assume that $L = 1$. A simple adaptation will work when $L > 1$. The pseudocode is given in Algorithm 1 and illustration of the working principle of the algorithm is given in Fig. 2.

5.1 A Naive Approach Without Energy Constraint

The main idea behind our algorithm is to let r incrementally explore the environment P while simultaneously constructing a tree map T_P to keep track of the new frontiers that still need to be visited by it.

For simplicity, suppose T_P (or P) is known a priori and r has no energy constraint (i.e., $B = \infty$). Let $\mathcal{Q}_\infty(r)$ be a route in T_P that visits all the nodes of T_P, obtained performing a *Depth First Search* (DFS) traversal of T_P. Since T_P is a tree, it is known that all the nodes of T_P can be covered by the DFS traversal by visiting each edge of T_P at most twice. Therefore, if there are n nodes in T_P, then there are $n - 1$ edges in T_P and the length of the route $l(\mathcal{Q}(r)) \leq 2n$. Moreover, any algorithm employed by r to traverse all n nodes of T_P must have length $l(Q_{OPT}(r)) \geq n$, since r can only visit a node of T_P at a time. Therefore, without any energy constraint ($B = \infty$), we have a 2-approximation algorithm. The 2-approximation can also be guaranteed for ONLINECPP when $B = \infty$ since with the knowledge of the global coordinate system, r can visit all the nodes of T_P as if T_P is known a priori, satisfying the length of the route $l(\mathcal{Q}(r)) \leq 2n$. Notice that r is building T_P and traversing it at the same time.

5.2 Incorporating the Energy Constraint

Now suppose that r has energy budget $B < 2n$. The aforementioned algorithm does not work anymore since each route of r can be at most of length B. Therefore, r needs to return to S to get recharged before the length of the robot's path exceeds B. The proposed algorithm here is to use the same idea of constructing a T_P and performing a

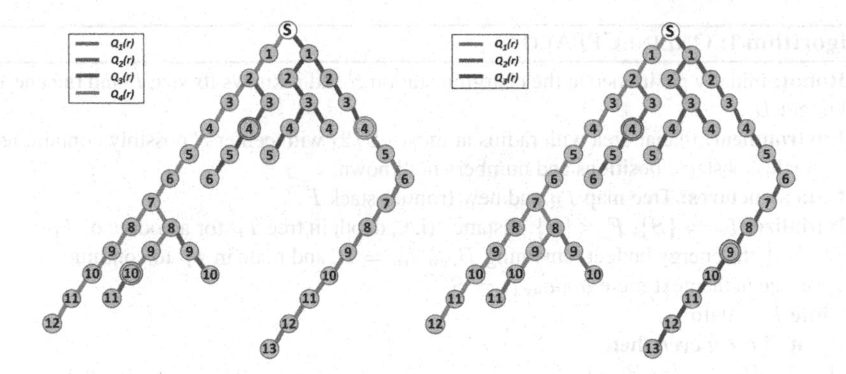

Fig. 2. An illustration of the budgeted DFS traversal by using Algorithm 1 when energy budget B is 30 (**left**) and 40 (**left**) for the environment P shown in Fig. 1. For $B = 30$, r needs 4 paths whereas only 3 paths are needed when $B = 40$. The nodes of T_P where $\mathcal{Q}_{i-1}(r)$ stops the DFS traversal are marked by double circles; the next path $\mathcal{Q}_i(r)$ continues the DFS traversal from these nodes.

DFS traversal of it as described in the previous subsection with the constraint of stopping the DFS traversal process before the route length of r exceeds B. Let $\mathcal{Q}_\infty(r) = \{S, v_1, v_2, \ldots, v_l\}$ be the route with respective nodes visited by r while running the DFS traversal assuming $B = \infty$. Let $\mathcal{Q}_i(r)$ denote a route of r visiting the nodes of $\mathcal{Q}_\infty(r)$ when $B < 2n$. The goal is to obtain $\mathcal{Q}'(r) = (\mathcal{Q}_1(r), \mathcal{Q}_2(r), \ldots, \mathcal{Q}_k(r))$, such that the three conditions listed in Sect. 1 are satisfied.

The challenge is to plan each route $\mathcal{Q}_i(r)$ in an online fashion satisfying all three criteria while minimizing both the number of paths $|\mathcal{Q}'(r)|$ and the total length of the paths $l(\mathcal{Q}'(r))$. We use the following approach: $\mathcal{Q}_1(r)$ starts from S and visits the nodes of $\mathcal{Q}_\infty(r)$ in a sequence. As soon as $\mathcal{Q}_1(r)$ reaches to a node $v_i \in \mathcal{Q}_\infty(r)$ such that $\text{dist}(v_i, S) \leq B_{remain}$, it terminates the DFS traversal and returns to S, where B_{remain} is the energy remained after each move (i.e., after every traversal of an edge of T_P, $B_{remain} = B_{remain} - 1$ with initially $B_{remain} = B$). In route $\mathcal{Q}_2(r)$, r moves to v_i (where it stopped the DFS traversal in $\mathcal{Q}_1(r)$) from S and continues the DFS traversal until it reaches to a node $v_j \in \mathcal{Q}_\infty(r)$ from which $\text{dist}(v_j, S) \leq B_{remain}$. Like last route, r then returns to S. Note that the return path to S is of length B_{remain} since T_P is a tree. This process then continues until the last node $v_l \in \mathcal{Q}_\infty(r)$ is visited in some route $\mathcal{Q}_k(r)$.

We call our algorithm OnlineCPPAlg (shown in Algorithm 1). Initially, the robot r is at S with the energy budget $B < 2n$. This is a special situation where $v = S$ and $B_{remain} = B$. Robot r then includes the child nodes of S (in contour C_1) in T_P making them child nodes of root S in T_P and moves to the leftmost child node, say $v_{1,left}$. The node $v_{1,left}$ is marked visited. B_{remain} is decreased by 1, which is now $B - 1$ (line 21). Robot r then moves from $v_{1,left}$ to the leftmost child node $v_{2,left}$ (in contour C_2). The node $v_{2,left}$ is marked visited and B_{remain} is again decreased by 1 (which becomes $B - 2$). This process continues until at some node v_i (in some contour i), $B_{remain} = D_{v_i}$, where D_{v_i} is the distance from v_i to S in the tree map T_P. Robot r then returns to S following the path in T_P (line 23). After getting fully charged at S, r follows a path in T_P to reach v_i to continue the DFS traversal.

Algorithm 1: ONLINECPPALG

1 **Robot:** Initially positioned at the charging station S and it knows its size L and the energy budget B.
2 **Environment:** Planar area with radius at most $\lfloor B/2 \rfloor$ with center S possibly containing obstacles; obstacle positions and numbers not known.
3 **Data structures:** Tree map T_P and new frontier stack F.
4 **Initialize:** $T_P = \{S\}$, $F = \{S\}$, distance (i.e., depth in tree T_P for a node v of T_P) $D_v = 0$, the energy budget remaining $B_{remain} = B$, and node in T_P to continue coverage in the next route $node_{next} = S$.
5 **while** $F \neq \emptyset$ **do**
6   **if** *B is not even* **then**
7     $B_{remain} = B - 1$;
8   Move to $node_{next}$ from S using the shortest path in T_P;
9   $D_v \leftarrow$ the distance from S to $node_{next}$ in T_P;
10   $B_{remain} \leftarrow B_{remain} - D_v$;
11   **while** $B_{remain} > D_v$ **do**
12   **if** *$node_{next}$ has unvisited child nodes in T_P or the cell on top of F is the child node of $node_{next}$ in T_P* **then**
13   **if** *the child nodes of $node_{next}$ are not already included in F and T_P* **then**
14   Include all the child nodes of $node_{next}$ in T_P and F (ordered clockwise from left to right starting from the leftmost child node and insert to F from right to left);
15       $v \leftarrow$ the node on the top of F or the leftmost in T_P (v will be the node pushed into F last);
16   Robot r removes v from F, moves to v, marks v visited in T_P;
17   **else**
18       $v \leftarrow$ the parent node of $node_{next}$ in T_P;
19   Robot r moves to v;
20     $node_{next} \leftarrow v$;
21     $B_{remain} \leftarrow B_{remain} - 1$;
22     $D_v \leftarrow$ the distance from S to $node_{next}$;
23   Robot r goes to S following a path in T_P;
24   $B_{remain} \leftarrow B$ (after r is fully changed) after reaching S;

At any grid cell of P (i.e., any node of T_P), if it has a unvisited neighbor cell in the next contour (child node in T_P), then r moves to that cell. Otherwise, r backtracks to the parent cell of its current cell in T_P (lines 18–19).

6 Analysis of the Algorithm

Correctness. We start with the following lemma.

Lemma 1. *Let $\{S, v_1, v_2, \ldots, v_l\}$ be the order of the nodes of T_P (i.e., cells in P) visited by the DFS traversal $\mathcal{Q}_\infty(r)$ for the unconstrained energy budget $B = \infty$. Let $\mathcal{Q}'(r) = \{\mathcal{Q}_1(r), \mathcal{Q}_2(r), \ldots, \mathcal{Q}_k(r)\}$ be the routes of r that collectively visit the nodes* *of T_P when $B < 2n$ using Algorithm 1. The not-yet-visited nodes of T_P (or cells of P) are visited in $\mathcal{Q}'(r)$ by Algorithm 1 in the same order as in $\mathcal{Q}_\infty(r)$.*

Proof. Consider the paths in $\mathcal{Q}'(r)$ when $B = \infty$. In this case, instead of stopping the DFS traversal at some node α and making a round trip to S from α, each subsequent paths continue their traversal without this stoppage. This simulates essentially the behavior of r when $B = \infty$ giving $\mathcal{Q}_\infty(r)$ (except the nodes visited in the round trip to S and back) and hence the not-yet-visited nodes of T_P (or cells in P) are visited in the same order in both. □

Theorem 2 (Correctness). *Algorithm 1 completely covers the environment P.*

Proof. When P is known a priori, r can visit all the reachable cells in P with an unlimited budget. If P is unknown but $B = \infty$, it is also known that through a DFS traversal, each reachable cell of P is guaranteed to be visited where the traversal path is represented by $\mathcal{Q}_\infty(r) = \{S, v_1, v_2, \ldots, v_l\}$. We have proved in Lemma 1 that when P is unknown but $B < 2n$, the not-yet-visited nodes of T_P are visited in the same order as in $\mathcal{Q}_\infty(r)$. This immediately provides the guarantee that all reachable cells of P will be visited by r. □

Approximation Ratio. We prove the following theorem.

Theorem 3 (Approximation). *Algorithm 1 achieves* 10*-approximation for both the performance metrics – the number of paths and the total length of the paths.*

Proof. Let T be a tree of depth at most $\lfloor B/2 \rfloor$. Let r be a robot with energy budget at least B. After r starts from S to visit the nodes of T, due to the limited energy budget, r may need to stop the coverage of T and visit the charging station S again before rest of the nodes in T can be covered.

Let OPT be the DFS exploration strategy for r that consists of the minimum number of routes, i.e., the minimum number of times r needs to visit S before T is completely covered. Let ALG be the DFS exploration strategy that visits the nodes of T (starting from S) using a DFS traversal where length of each route is bounded by B. As soon as the battery is fully charged at S, in the next route, r directly goes to the node of T where it stopped the DFS traversal in the last route and continues covering the unvisited nodes of T. For any tree T of depth $\lfloor B/2 \rfloor$ and any robot r of energy budget at least B initially at the root of T, we have the a recent result from Das *et al.* [5] on the number of routes $|\mathcal{Q}_{ALG}(r)|$, of the strategy ALG, compared to the number of routes $|\mathcal{Q}_{OPT}(r)|$, of strategy OPT that $|\mathcal{Q}_{ALG}(r)| \leq 10 \cdot |\mathcal{Q}_{OPT}(r)|$. Moreover, let $l(\mathcal{Q}_{ALG}(r))$ be the total length traversed by r while using the strategy ALG. Let $l(OPT(r))$ be the optimal length traversed by r. Again from Das *et al.* [5], we have that $l(\mathcal{Q}_{ALG}(r)) \leq 10 \cdot l(OPT(r))$. These results are interesting meaning that the bounds hold for any arbitrary DFS traversal $\mathcal{Q}(r)$ of T by r starting from the root of T. That means that the whole DFS traversal $\mathcal{Q}(r)$ does not need to be known to robot r beforehand (i.e., can be computed online not knowing T in advance). Moreover, each route can be constructed without any knowledge on the not-yet-visited part of T.

We now discuss how these results prove the same bounds for Algorithm 1. Consider a DFS traversal $\mathcal{Q}_\infty(r)$ of P (or equivalently T_P) by r when $B = \infty$. We have from

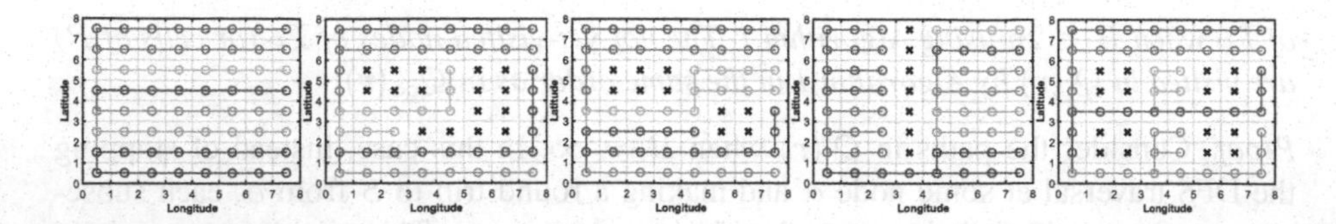

Fig. 3. An illustration of five different configurations (Conf1 to Conf5) and the paths followed by r using Algorithm 1. The obstacles are represented with 'x'. S is in the bottom-left corner cell. When plotted, later paths have been given higher priority and they are shown in the foreground.

Lemma 1 that the routes in $\mathcal{Q}'(r)$ visit the not-yet-visited nodes of P in the order same as in $\mathcal{Q}(r)$. Moreover, the tree map T_P of P formed during the exploration is of depth at most $\lfloor B/2 \rfloor$. Therefore, $|\mathcal{Q}'(r)| \leq 10 \cdot |\mathcal{Q}_{OPT}(r)|$ and $l(\mathcal{Q}'(r)) \leq 10 \cdot l(OPT(r))$. The theorem follows. □

Proof of Theorem 1: Theorems 2 and 3 prove Theorem 1 for $L = 1$. Since the cells are decomposed proportional to the robot size $L \times L$, a simple adaptation of the analysis again gives 10-approximation for Algorithm 1 for $L > 1$. The correctness analysis stays the same. □

7 Evaluation

Setting. We have implemented the proposed algorithm ONLINECPPALG using Java programming language on a desktop computer with an Intel i7-7700 CPU and 16 GB RAM. The robot size L is set to 1. We have created five different environments with both convex and concave obstacles in them. The test environments are of the same dimension – 8×8 ($l = 8$). The budget B is set to $4l$ (unless otherwise mentioned), i.e., four times the size of each side of the environment. Note that this is the lowest possible budget to completely cover the environment. The charging station S is placed at the left-bottom corner in every test environment. These environments are shown in Fig. 3 (Conf1 to Conf5). Empirically, we have mainly focused on three metrics to evaluate the quality of the proposed algorithm: 1) time to cover the environment, 2) total path length traversed by the robot, and 3) approximation ratio. We also compare our results against the state-of-the-art algorithm that solves ONLINECPP under energy constraint [13].

Results. First we empirically verify the theoretically-proved constant-factor approximation bound. Let n denote the number of reachable cells in the environment. Then $MIN = \frac{2n}{B}$ will indicate the *absolute minimum* number of paths required by the robot to completely cover the environment [13,18]. No optimal DFS strategy (OPT) can guarantee a better approximation bound than MIN. Here, we compare our experimental result against MIN as a comparison against any OPT cannot make our empirical approximation bound worse. The result is shown in Fig. 4(a). The state-of-the-art bound of $\log(B)$ (when $L = 1$) is also plotted for reference. The figure shows that in practice, the approximation bound is well below the 10-factor theoretical worst-case bound. Also, in all of the test environments, our proposed algorithm outperforms the state-of-the-art $\log(B)$-approximation algorithm [13].

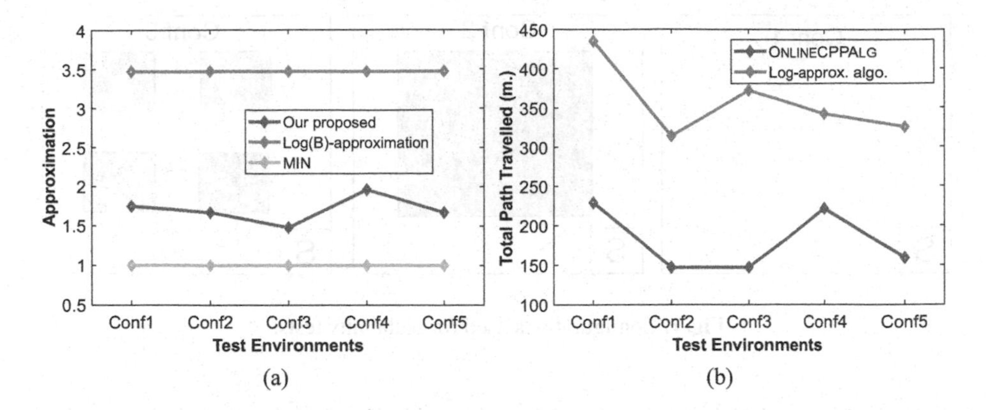

Fig. 4. a) Empirical validation of the constant-factor approximation on different environment configurations; b) Comparison of path lengths travelled by the robot using our algorithm and the state-of-the-art algorithm in [13].

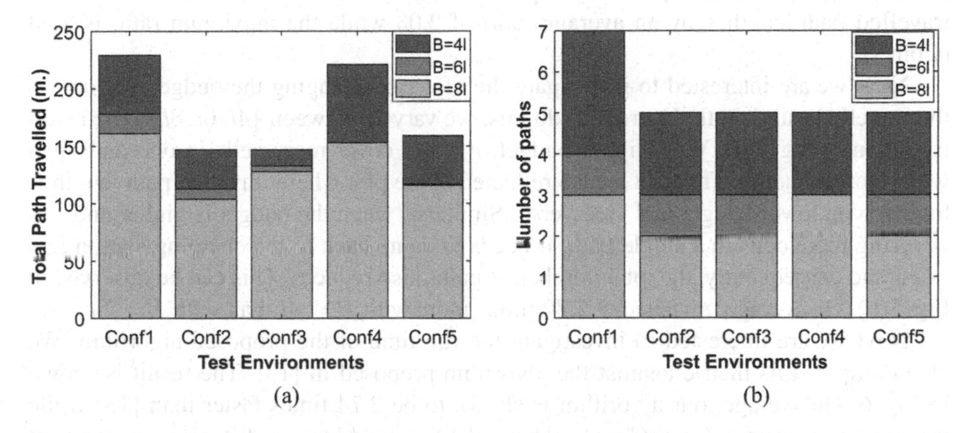

Fig. 5. Varying the budget: a) Comparison of travelled path lengths; b) Comparison of total number of paths (i.e., number of visits to S).

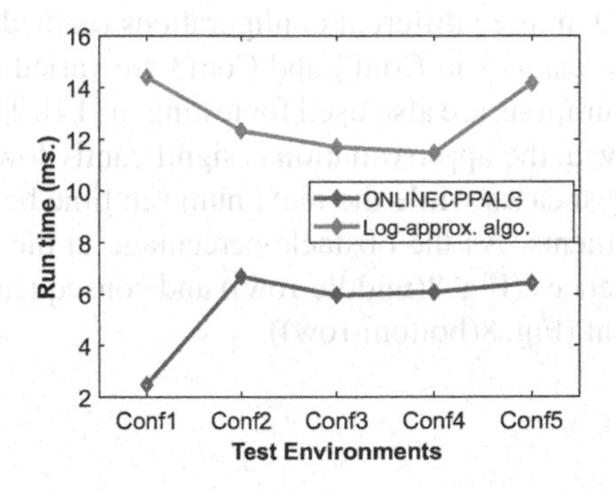

Fig. 6. Runtime comparison of Algorithm 1 (ONLINECPPALG) against the state-of-the-art $O(\log(B/L))$-approximation algorithm in [13].

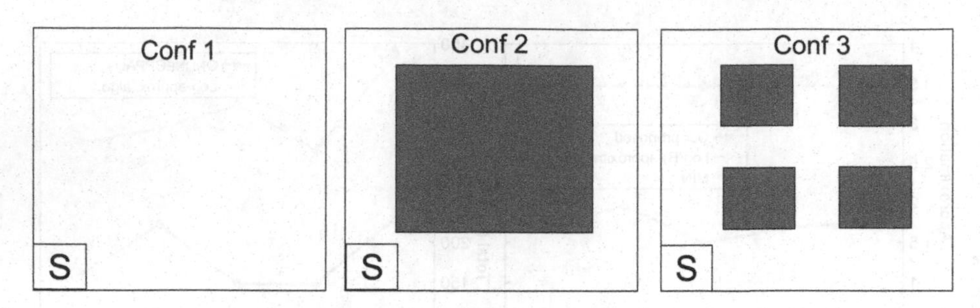

Fig. 7. Configurations used for scalability tests.

As we are minimizing the total path length travelled by the robot to completely cover the environment, we are interested to compare this metric for our algorithm against the algorithm in [13]. The result is shown in Fig. 4(b). It can be clearly observed from this plot, that our proposed algorithm outperforms the algorithm in [13] in terms of the travelled path length – by an average ratio of 2.03 while the maximum ratio is 2.53 (Conf3).

Next we are interested to investigate the effect of changing the budget amount on the travelled path length. In order to do this, we vary B between $\{4l, 6l, 8l\}$. The result is shown in Fig. 5(a). With higher budget, r could cover more cells in one path than with a lower budget. This fact is also reflected in the plot where travelled path length is higher with lower budget and vice-versa. Similarly, when the budget is higher and r is covering more cells in a single path, it needs to come back to the charging station less often and consequently, the total number of paths also reduces. This can be observed in Fig. 5(b). On average, r visited S 2.30 times more with $B = 4l$ than with $B = 8l$.

Next we are interested to investigate the run time of the proposed algorithm. We also compare this metric against the algorithm proposed in [13]. The result is shown in Fig. 6. On average, our algorithm is shown to be 2.74 times faster than [13] while the maximum ratio is 5.80 (Conf1). The paths followed by r in different environment configurations are shown in Fig. 3.

Finally, to test the scalability of our approach, we vary our environment size from 100×100 to 500×500 in three different configurations (named Conf1-3) as shown in Fig. 7. The obstacle percentages in Conf2 and Conf3 are varied between $[10, 30]$. B is set to $6l$. Similar configurations are also used for testing in [14]. The results are shown in Fig. 8. As can be observed, the approximation is significantly lower than the theoretical bound of 10 in all the test cases while the maximum run time being a negligible 69 ms for 500×500 environments. As the obstacle percentage in the environment grows, r needs to cover less distances (Fig. 8(middle-row)) and consequently it takes lesser time to cover the environment (Fig. 8(bottom-row)).

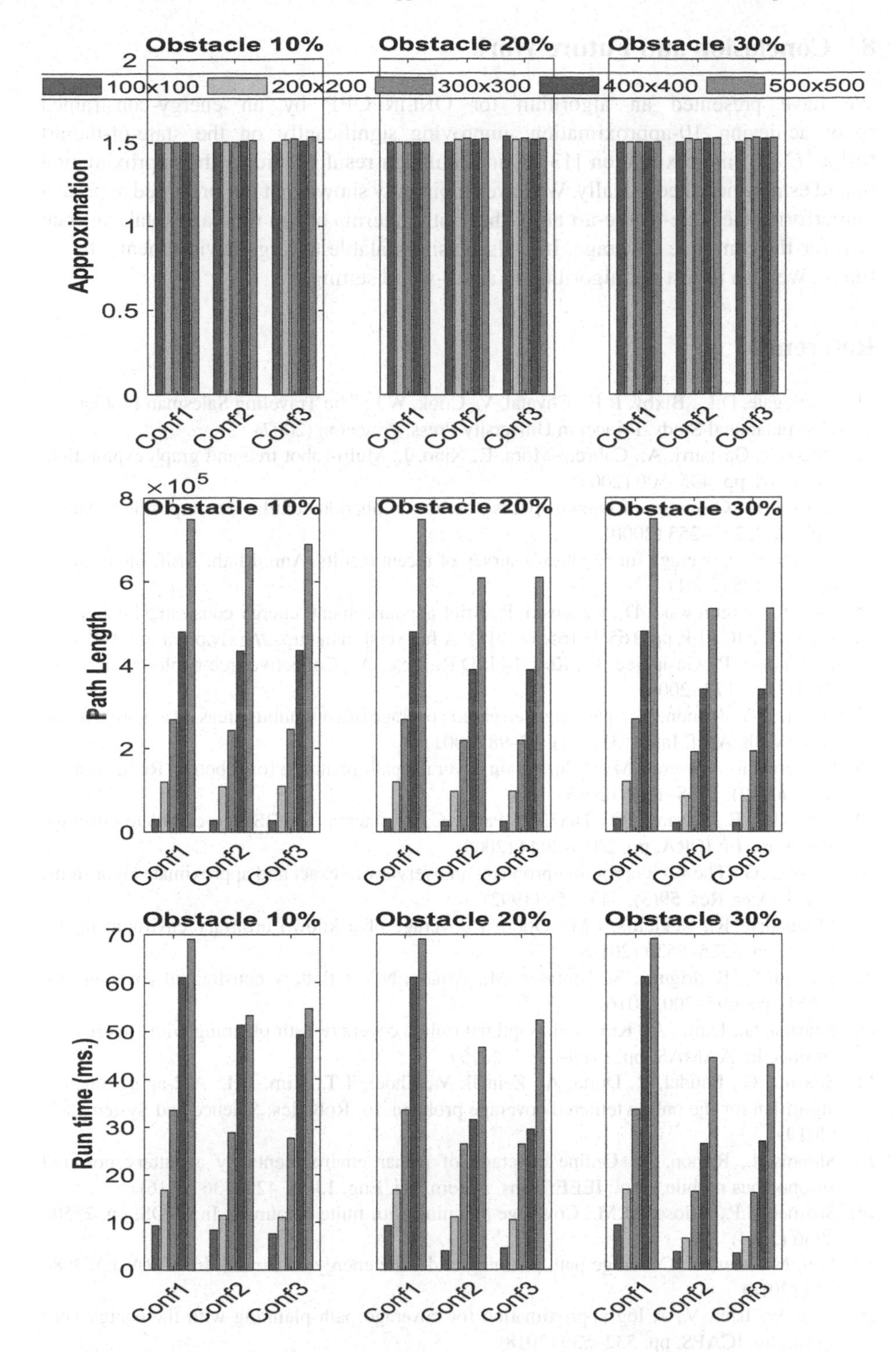

Fig. 8. Approximation ratios, path lengths, and run times of our proposed approach in environments of sizes from 100×100 to 500×500.

8 Conclusion and Future Work

We have presented an algorithm for ONLINECPP by an energy-constrained robot achieving 10-approximation, improving significantly on the state-of-the-art $O(\log(B/L))$-approximation [13]. Our simulation results validate the approximation bound established theoretically. We have empirically shown that our proposed approach outperforms the state-of-the-art algorithm both in terms of run time and total traversal cost for the complete coverage. It is also easily scalable to large environments. In the future, we plan to test our algorithm in a real-world setting.

References

1. Applegate, D.L., Bixby, R.E., Chvatal, V., Cook, W.J.: The Traveling Salesman Problem: A Computational Study. Princeton University Press, Princeton (2007)
2. Brass, P., Gasparri, A., Cabrera-Mora, F., Xiao, J.: Multi-robot tree and graph exploration. In: ICRA, pp. 495–500 (2009)
3. Choset, H.: Coverage of known spaces: the boustrophedon cellular decomposition. Auton. Rob. **9**(3), 247–253 (2000)
4. Choset, H.: Coverage for robotics-a survey of recent results. Ann. Math. Artif. Intell. **31**(1–4), 113–126 (2001)
5. Das, S., Dereniowski, D., Uznanski, P.: Brief announcement: energy constrained depth first search. In: ICALP, pp. 165:1–165:5 (2018). A full version in https://arxiv.org/abs/170910146
6. Fraigniaud, P., Gąsieniec, L., Kowalski, D.R., Pelc, A.: Collective tree exploration. Netw. **48**(3), 166–177 (2006)
7. Gabriely, Y., Rimon, E.: Spanning-tree based coverage of continuous areas by a mobile robot. Ann. Math. Artif. Intell. **31**(1–4), 77–98 (2001)
8. Galceran, E., Carreras, M.: A survey on coverage path planning for robotics. Robot. Auton. Syst. **61**(12), 1258–1276 (2013)
9. González, E., Álvarez, O., Díaz, Y., Parra, C., Bustacara, C.: BSA: a complete coverage algorithm. In: ICRA, pp. 2040–2044 (2005)
10. Laporte, G.: The vehicle routing problem: an overview of exact and approximate algorithms. Eur. J. Oper. Res. **59**(3), 345–358 (1992)
11. Mannadiar, R., Rekleitis, I.M.: Optimal coverage of a known arbitrary environment. In: ICRA, pp. 5525–5530 (2010)
12. Mishra, S., Rodríguez, S., Morales, M., Amato, N.M.: Battery-constrained coverage. In: CASE, pp. 695–700 (2016)
13. Sharma, G., Dutta, A., Kim, J.H.: Optimal online coverage path planning with energy constraints. In: AAMAS, pp. 1189–1197 (2019)
14. Sharma, G., Poudel, P., Dutta, A., Zeinali, V., Khoei, T.T., Kim, J.H.: A 2-approximation algorithm for the online tethered coverage problem. In: Robotics: Science and Systems XV (2019)
15. Shnaps, I., Rimon, E.: Online coverage of planar environments by a battery powered autonomous mobile robot. IEEE Trans. Autom. Sci. Eng. **13**(2), 425–436 (2016)
16. Strimel, G.P., Veloso, M.M.: Coverage planning with finite resources. In: IROS, pp. 2950–2956 (2014)
17. Wei, M., Isler, V.: Coverage path planning under the energy constraint. In: ICRA, pp. 368–373 (2018)
18. Wei, M., Isler, V.: A log-approximation for coverage path planning with the energy constraint. In: ICAPS, pp. 532–539 (2018)

Real-Time Image Based Plant Phenotyping Using Tiny-YOLOv4

Sonal Jain[1(✉)], Dwarikanath Mahapatra[2], and Mukesh Saini[1]

[1] Indian Institute of Technology Ropar, Rupnagar, India
{sonal.21csz0010,mukesh}@iitrpr.ac.in

[2] Inception Institute of Artificial Intelligence, Abu Dhabi, UAE
dwarikanath.mahapatra@inceptioniai.org

Abstract. Image-based plant phenotyping is getting considerable attention with the advancement in computer vision technologies. In the past few years, the use of deep neural networks (DNNs) is well-known for segmentation and detection tasks. However, most DNN-based methods require high computational resources, thus making them unsuitable for real-time decision-making. This study presents a real-time plant phenotyping system using leaf counting and tracking individual leaf growth. For leaf localization and counting, a Tiny-YOLOv4 network is utilized, which provides faster processing, and is easily deployable on low-end hardware. Leaf growth tracking is performed by active contour segmentation of leaf localized using the Tiny-YOLOv4 network. The proposed system is implemented for top-view RGB images of the *Arabidopsis thaliana'* plants. And its performance for leaf counting is evaluated against Tiny-YOLOv3 and Faster R-CNN using the difference in count (DiC), accuracy, and F1-score measures. The model achieves an improved accuracy of 90%, absolute DiC of 0.42, F1-score of 96%, and inference time of 15 milliseconds. Further, the segmentation accuracy measures using Dice and Jaccard scores are 0.91 and 0.86, with a computing time of 0.96 s. These obtained results depict the effectiveness of the proposed system for real-time plant phenotyping.

Keywords: Object detection · Leaf counting · Plant phenotyping · YOLO · Tiny-YOLOv4 · Leaf segmentation

1 Introduction

Plant phenotyping is essential to identify the behavior of different cultivars of the same plant under diverse environmental conditions, which helps farmers in selecting a suitable cultivar. The plant phenotyping activities can be classified into three categories [6]: structural, physiological, and temporal or event-based. Structural phenotyping is done for either whole plant or plant components such as leaf, fruit, flower, etc. In this, several traits such as plant height, plant area, leaf length, and fruit volume are computed [11]. Physiological phenotyping is related to measuring traits that affect plant processes such as plant growth,

H. Zaynidinov et al. (Eds.): IHCI 2022, LNCS 13741, pp. 271–283, 2023.
https://doi.org/10.1007/978-3-031-27199-1_28

stress levels of leaves, and metabolism. While, temporal phenotyping is computed from a sequence of images to define activities such as plant growth rate, time of flowering, time of occurring new leaves, etc. [3,5]. Manual plant phenotyping is costly, time-consuming, and destructive in many scenarios. Also, it can be performed only for a small group of plants. However, large-scale plant phenotyping is required for consistent and accurate analysis of plant traits.

A real-time image-based plant phenotyping is crucial in precision agriculture for identifying various plant traits such as plant growth stage, fruit growth stage [24], architecture [12], yield potential, flowering time, etc. This can be achieved using several imaging tasks like plant segmentation, leaf counting, leaf localization, and leaf tracking. However, performing these tasks is challenging due to rapid change in leaf size, leaf occlusion, illumination problems, change in leaf direction as to the sun, complex background, change in camera distance, etc. [16]. In the last few years, DNNs based methods have become widespread for performing computer vision tasks due to their accuracy and ability to work in complex scenarios. DNN-based leaf counting methods, such as R-CNN, despite being accurate, is relatively slow for real-time decision making. Hence, in this study, we jointly leverage Tiny-YOLOv4 and active contours for efficient phenotyping.

2 Related Work

Most of the study in image-based plant phenotyping is conducted on the publicly available Arabidopsis plants dataset [15]. A review of various works on image-based plant phenotyping is presented in [14,17]. The literature survey on leaf counting can be classified into three categories: Counting via regression, counting via detection, and counting via instance segmentation.

Counting via Direct Image-to-Count Regression Model: Regression-based methods takes an input image, compute its features, and converts it into leaf count. One such model is presented by Andrei et al. [6] that uses rosette-shaped plant data from multiple sources and trains a modified ResNet-50 model to increase the robustness of leaf counting. Similarly, convolutional and deconvolution-based networks are used in [1] for leaf counting on the CVPPP 2017 challenge dataset [15]. This method works in two passes initially, SegNet architecture is utilized for segmenting the whole plant, then the segmented and original RGB image is taken input to another VGG-based network architecture for leaf counting. This method provides increased accuracy of leaf counting. However, extensive training time is required to train both networks. Similarly, a DNN-based model named Pheno-Deep is proposed in [10] that combines features from different imaging sources (visible light, near-infrared, and fluorescence) to improve leaf counting accuracy. The images from different sources are passed to the ResNet-50 model to calculate features. These obtained features are fused and passed to a fully connected layer to predict the leaf count.

Counting via Object Detection: Such approaches work in single or multiple passes for object classification and localization. The object detection method

Table 1. Summary of existing approaches for plant phenotyping

Related Work	Dataset	Method	Real-time	Phenotyping Traits	Low- end support
[8]	LCC dataset of CVPPP-2017, Banana dataset	Regression, Detection (Resnet50+FPN)	Yes	Leaf counting, Leaf center detection	No
[2]	Arabidopsis plant	Detection (Tiny-YOLOv3)	Yes	Leaf counting	Yes
[25]	CVPPP A1 data, Cauliflower dataset	Detection Improved YOLOv3	Yes	Leaf counting	No
[1]	LSC dataset, LCC dataset of CVPPP-2017	Regression (SegNet + VGG)	No	Plant segmentation, Leaf counting	No
[28]	LSC dataset of CVPPP-2017	Instance segmentation (Mask-RCNN)	No	Leaf counting, Leaf segmentation	No
[7]	CVPPP-2017 A1, A2, A3, A4 dataset	Regression (Modified ResNet50)	Yes	Leaf counting	No
[27]	Field dataset of sugar beets	Detection (FPN+ CenterNet)	Yes	Plant detection, Leaf counting	No

Note: LCC - Leaf counting challange; LSC - Leaf segmentation challange

often results in more accuracy than the regression-based method due to identifying each leaf individually. Two such popular methods are you only look once (YOLO) [21] and region-based convolution neural network (R-CNN) [22]. YOLO has been utilized recently for fire detection [18] and real-time object detection in smart glass systems [19]. YOLO is a faster method that performs object localization and classification in a single phase but is unable to detect tightly coupled objects. While RCNN, despite being more accurate suffers from the slower computation. A real-time leaf counting method based on Tiny-YOLOv3 is presented in [2]. This model is trained on a self-generated dataset of the Arabidopsis plant.

Counting via Instance Segmentation: The most recent works in leaf counting are based on instance segmentation, which counts the number of leaves by segmenting each leaf individually. The leaf detection and segmentation method by adopting mask-RCNN is presented in [28]. Similarly, a recurrent instance segmentation method is introduced in [23] that segments one instance at a time using an attention-based mechanism. Such methods are more accurate but computationally expensive to implement on low-end hardware.

Table 1 demonstrates the summary of existing work in plant phenotyping. This depicts most of the existing methods were not suitable for real-time computing on a low-end device. Hence, in this study, we propose a real-time plant phenotyping system suitable for a low-end device.

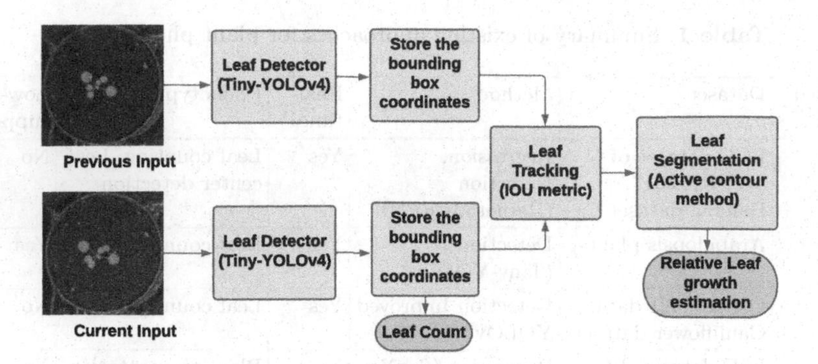

Fig. 1. Schematic diagram of proposed real-time plant phenotyping system

3 Proposed Methodology

The flow diagram of the proposed plant phenotyping system is depicted in Fig. 1. Both current and previous images are fed to the system to measure plant growth. The first step is to detect and localize leaves in the images. In the proposed work, we have utilized a Tiny-YOLOv4 network [26] for leaf localization and counting. The output of YOLO is a set of bounding box coordinates. These bounding box coordinates are processed further to track individual leaf areas. Finally, the bounding box areas are segmented to identify leaf regions.

The implementation of a proposed plant phenotyping system consists of two tasks for each plant image:

1. Count the number of leaves to identify the new leaf (Leaf counting).
2. Identify the growth in old leaf correspond to previous frame (Leaf alignment or tracking).

3.1 Leaf Counting

Leaf counting is achieved through leaf detection using Tiny-YOLOv4. The Tiny-YOLOv4 is a compressed version of YOLOv4 designed for low-end devices. It has fewer parameters and simpler network architecture that can execute on a device with low computational power. It provides faster training and has a high FPS value, which makes it a suitable choice for real-time object detection. For leaf counting, the Tiny-YOLOv4 network is trained using Darknet [20] deep learning framework. This framework speeds up the execution and can be deployed to a low-end hardware device. The output of the YOLOv4 network detects a leaf by creating a bounding box around the leaf area along with class probability and bounding box coordinates. These bounding box coordinates are stored in a CSV file. Counting the number of rows in this CSV file gives the leaf count. The sample output using a YOLOv4 network is shown in Fig. 2. The leaf count of the current and previous images is compared to find the number of new leaves, which indicates the plant growth.

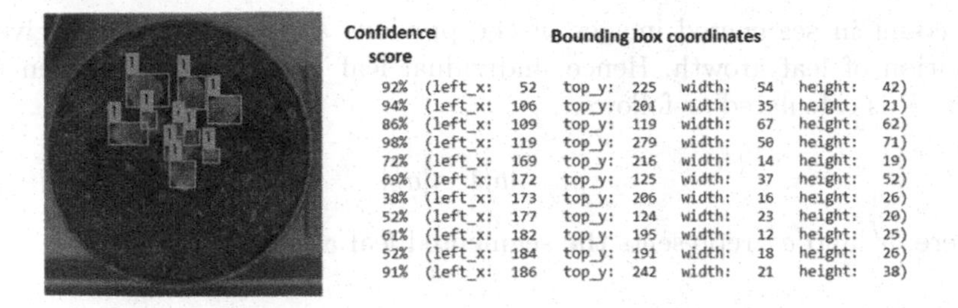

Fig. 2. Output Sample from YOLOv4.

3.2 Leaf Growth Tracking

For leaf growth tracking, the first task is to locate the individual leaf across subsequent images of the plant. This is achieved by comparing the Intersection over Union (IoU) metric between current and previous bounding box coordinates. Let $\mathcal{A} = \{a_1,\ a_2 ... a_m\}$ be the set of bounding boxes detected in the previous frame, and $\mathcal{B} = \{b_1,\ b_2 ... b_n\}$ represent the set of bounding boxes detected in the current frame. Here, a and b represent the area of the corresponding bounding box. Let O_{ij} represent IoU between a_i and b_j:

$$O_{ij} = \frac{a_i \cap b_j}{a_i \cup b_j} \tag{1}$$

Our end goal is to compare the growth of each leaf in the current frame with respect to the previous frame. As the images are captured from a fixed camera, it is assumed that the movement of leaves across subsequent images is very slight. Hence, in every subsequent image of the plant, the bounding box having the highest IoU value with the previous image leaf bounding box locates the individual leaf area. Therefore, for each bounding box b_i in $\mathcal{B}$, we find a matching a_k in $\mathcal{A}$ as follows:

$$k = \underset{j}{\operatorname{argmax}}\ O_{ij}. \tag{2}$$

Let $\mathcal{P}' = \{(i,k)|b_i \in \mathcal{B}, a_k \in \mathcal{A}\}$ be the set of tuples obtained using Eq. (2). Not all tuples represent valid correspondence. For example, a new leaf does not have any corresponding leaf in the previous frame. Therefore, we find the set of valid correspondences as follows:

$$\mathcal{P} = \{(i,k)|b_i \in \mathcal{B}, a_k \in \mathcal{A}, O_{ij} > \theta\} \tag{3}$$

Here θ is found to be 0.75 for the given dataset. The value of θ should be chosen based on the leaf growth rate and the time interval between subsequent images.

After locating the individual leaf across different frames, active contour-based image segmentation is performed to recognize its growth. The image segmentation task outputs the leaf area part white and masks the rest. Comparing the pixel count in segmented images of the previous and current frames gives an estimation of leaf growth. Hence, individual leaf growth g_i for a given tuple $(i, j) \in \mathcal{P}$ is calculated as follows:

$$g_i = |b_i^f| - |a_k^f| \tag{4}$$

Here, b_i^f and a_k^f represents the segmented leaf area.

4 Experimental Results and Discussion

4.1 Dataset

The dataset used in this work consists of top view RGB images of the 'Arabidopsis thaliana' plant. The dataset is taken from two different sources named A1, and A2 [4,15]. The sample images of plants from these two datasets are depicted in Fig. 3. The A1 dataset consists of images taken from two different setups denoted *Ara*2012 and *Ara*2013. Figure 3(a) represents the wild variety of Arabidopsis plants, while Fig. 3(b) represents sample images of different cultivars of Arabidopsis plants. Figure 3(c) depicts sample plant images from the A2 dataset. For leaf counting, the model is trained on the A1 dataset, and its performance is tested on both the A1 and A2 datasets to ensure the generalization capacity of the model. From the A1 dataset, 350, 50, and 50 images are used for training, validation, and testing, respectively. The bounding box annotation of leaves provided in the available dataset includes the area of the leaf stalk along with the leaf, which may reduce the accuracy of the object detection framework. Hence, the plant dataset is annotated manually in the Roboflow platform (https://app.roboflow.com/) by creating a bounding box around the leaf area. The available data is resized into 416 × 416 and augmented with 90° rotation (clockwise and anti-clockwise), variation in brightness, and saturation varied from −25% to +25%. Hence, for each image, three output samples are created, which makes the final training dataset of 1050 data samples. From the A2 dataset, 100 samples are utilized as an evaluation dataset.

4.2 Evaluation Metrics

To evaluate the performance of the leaf counting approach using Tiny-YOLOv4 standard evaluation metrics such as difference in count, mean square error, absolute difference in count, and percentage agreement are used. The evaluation metrics are defined as follows:

- Difference in count (DiC) = $\frac{1}{N}\sum_{i=1}^{N}(y - y')$
- Mean square error (MSE) = $\frac{1}{N}\sum_{i=1}^{N}(y - y')^2$
- Absolute difference in count (|DiC|) = $\frac{1}{N}\sum_{i=1}^{N}|y - y'|$
- Agreement (%) = $\frac{1}{N}\sum_{i=1}^{N}\begin{cases}1 & y - y' = 0\\0 & y - y' \neq 0\end{cases}$

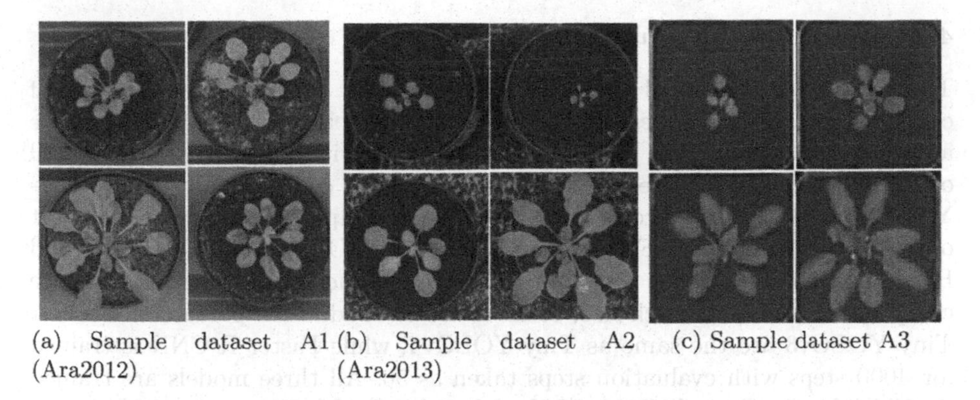

(a) Sample dataset A1 (Ara2012) (b) Sample dataset A2 (Ara2013) (c) Sample dataset A3

Fig. 3. Sample images of 'Arabidopsis thailnana' plant dataset

Here, y and y' denote actual and predicted leaf count, and N represents the total number of samples. These evaluation metrics are helpful to quantify leaf counting accuracy. However, the output bounding box may not represent the actual leaf or count more than one leaf for a single. To quantify such scenarios following cases are considered: True-positive (TP) when the bounding box denotes the actual leaf, False-positive (FP) when the bounding box denotes the leaf falsely or predicts multiple counts for the same leaf, and False-negative (FN) when the model fails to detect the actual leaf. To identify the correlation between actual and predicted bounding boxes, the intersection over union (IoU) metric is utilized with a threshold value of 0.5. An IoU value above 0.5 is considered true detection, and below 0.5 is considered false detection.

Table 2. Network evaluation metrics on A1 and A2 datasets

Metric	A1 Dataset			A2 Dataset		
	Tiny-YOLOv4	Tiny-YOLOv3	Faster R-CNN	Tiny-YOLOv4	Tiny-YOLOv3	Faster R-CNN
DiC	0.0 (0.65)	0.30 (0.93)	0.45 (1.53)	0.35 (0.99)	−1.09 (1.67)	2.16 (2.25)
|DiC|	0.42 (0.49)	0.60 (0.77)	1.30 (1.80)	0.63 (0.84)	1.49 (1.32)	2.56 (1.78)
MSE	0.42	0.96	2.54	1.11	3.99	5.76
% Agreement	57.57	54.54	22.12	52.38	20.95	19.52
Accuracy	0.90	0.87	0.79	0.89	0.82	0.71
Precision	0.98	0.95	0.92	0.97	0.97	0.87
F1-Score	0.96	0.93	0.89	0.94	0.90	0.84
mean AP(@.5)	97.79%	96.16%	90.06%	95.97%	93.26%	84.34%
Inference time(s)	0.015	0.011	1.120	0.015	0.011	1.120

4.3 Experimental Setup and Results

The Tiny-YOLOv4 model is trained in the python Jupyter notebook for leaf counting and further processing of results. The model training parameters have a max batch size of 4000 and a batch size of 32 with a subdivision of 8. All other hyperparameters are the same as the default configuration file of Tiny-YOLOv4. The performance of Tiny-YOLOv4 is compared with state-of-the-art object detection methods Faster R-CNN and Tiny-YOLOv3. Tiny-YOLOv3 and Faster R-CNN model has been trained on the same dataset for performance comparison. The max batch size, batch size, and subdivision parameter used for Tiny-YOLOv3 are the same as Tiny-YOLOv4, while Faster R-CNN is trained for 4000 steps with evaluation steps taken as 50. All three models are trained on NVIDIA GeoForce MX330 GPU with 8 GB RAM. The evaluation metrics obtained using these methods are shown in Table 2. Here, the value in the brackets depicts the standard deviation.

From the results, it can be noted that Tiny-YOLOv4 provides overall better performance in leaf counting and leaf localization on both A1 and A2 datasets. While the accuracy and F1-score reduced significantly for Faster R-CNN and Tiny-YOLOv3 on the A2 dataset. Moreover, the standard deviation of DiC and |DiC| metrics is high in Tiny-YOLOv3 and Faster R-CNN as compared to Tiny-YOLOv4, which depicts high variation in actual and predicted leaf count. Also, the Tiny-YOLOv4 has the advantage of lower inference time, which makes it a suitable choice for real-time plant phenotyping. The actual and predicted leaf counting results using Tiny-YOLOv4 on the A2 dataset are shown in Fig. 4, depicting the predicted leaf count using Tiny-YOLOv4 matches the actual leaf count very closely. Moreover, the leaf count graph for a single plant using Tiny-YOLOv4 over 17 consecutive days is shown in Fig. 5. It can be noted from the

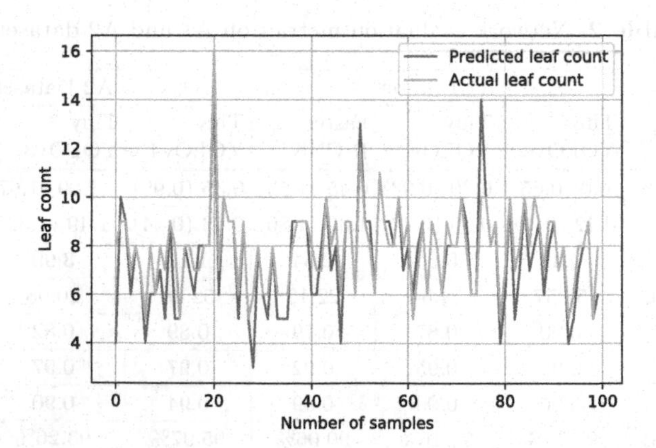

Fig. 4. Predicted leaf count on dataset A2 using Tiny-YOLOv4

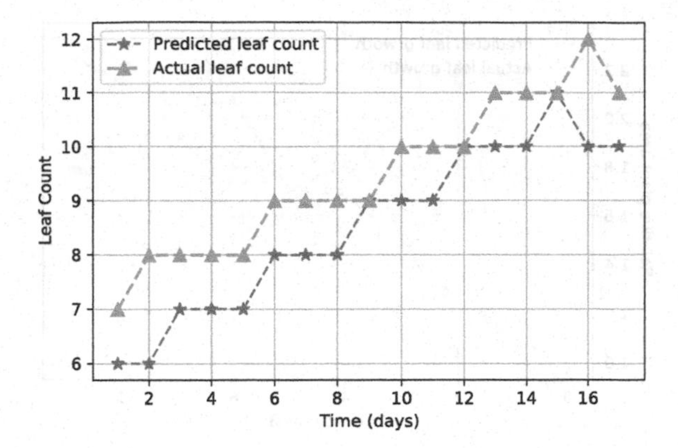

Fig. 5. Leaf count of a plant over seventeen consecutive days

graph that there is a difference of one count at the time of emergence of a new leaf due to the leaf size being very small on the initial days. Further, a decrease in leaf count is observed on the 17th day due to the complete occlusion of the leaf with time as the number of leaves increases.

Further, the performance of leaf growth tracking is evaluated using active contour segmentation of leaf identified using the proposed leaf tracking approach. The sample leaf tracking output of a single leaf for ten consecutive frames is depicted in Fig. 6. The growth is defined as relative plant area computed from segmented leaf area normalized by the actual leaf segmented area on the first day. The leaf growth graph for ground truth and predicted segmented leaf is depicted in Fig. 7, which shows the proposed method can capture the leaf growth with sufficient accuracy.

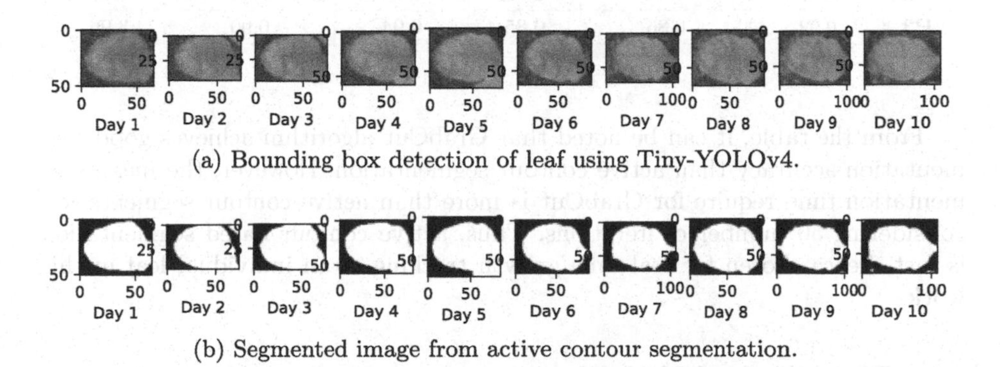

(a) Bounding box detection of leaf using Tiny-YOLOv4.

(b) Segmented image from active contour segmentation.

Fig. 6. Leaf image of plant over ten consecutive days

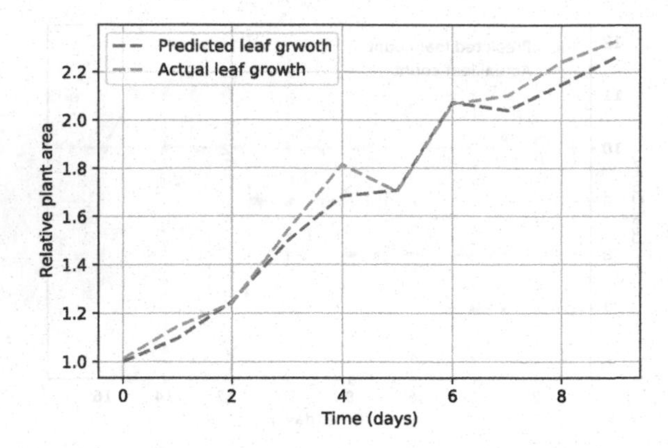

Fig. 7. Leaf growth curve over ten consecutive days

To evaluate the segmentation performance, the accuracy of leaf segmentation using the active contour method is compared with the GrabCut-based segmentation approach. The segmentation results are evaluated using the Dice score and Jaccard index. These metrics are calculated for ten sequential images of three plants named P1, P2, and P3 from the A1 dataset. The evaluation results are shown in Table 3.

Table 3. Evaluation metrics for leaf segmentation

Data	Active Contour			GrabCut		
	Dice Score	Jaccard Index	Time (s)	Dice Score	Jaccard Index	Time (s)
P1	0.86	0.81	1.05	0.89	0.85	3.97
P2	0.96	0.93	0.99	0.98	0.94	4.05
P3	0.92	0.86	0.85	0.94	0.90	3.96

From the table, it can be noted that GrabCut algorithm achieves good segmentation accuracy than active contour segmentation. However, the mean segmentation time require for GrabCut is more than active contour segmentation considering 50 number of iterations. Thus, active contour based segmentation is fast, hence chosen for real-time growth tracking of an individual leaf in this work.

4.4 Discussion and Limitation

Here, we present a real-time plant phenotyping approach by leaf counting and growth estimation of individual leaves in the plant. The proposed system works well for different varieties of the 'Arabidopsis thaliana' plant. The leaf counting approach using the YOLO object detection network has been specified in the literature. While we extend this work by leaf alignment and tracking using the IoU metric followed by active contour segmentation for leaf growth estimation.

The existing solutions of leaf alignment and tracking use a complex optimization method [29] to track the growth of multiple leaves in a video. Such a method applies only to a specific plant species due to utilizing the shape attribute of the leaf. However, this paper represents the utilization of Tiny-YOLOv4 bounding box output along with the IoU metric for leaf alignment and tracking on images taken on a subsequent day. Further, the proposed method does not use a shape-based model for leaf segmentation and tracking, hence applicable for different plant species with well-separated leaf structures. The experimental results depict the applicability of this method for leaf growth estimation in plants. For segmentation, the active contour-based segmentation approach is used. The proposed system can be extended to high throughput imaging systems (containing several plants in a single image) by adding one more step of plant detection before leaf counting [13].

This approach works well in a controlled environment due to limited variation in illumination conditions throughout the plant's growing period. However, a complex segmentation approach based on a deep neural network (DNN) is needed in a field environment, where illumination changes with time. The proposed system applies to all the rosette-shaped plants with well-separated leaves captured in a top-down manner in a controlled environment. However, for leaf counting in plants with overlapped leaf scenarios and plants that grow in height, the image captured from a single viewpoint is not sufficient. In such scenarios, the 3D imaging technique [9] or multimodel imaging will be beneficial. The possible future research direction is to design standard phenotyping traits of different plant species that should be identified through image analysis, thus allowing global comparison of obtained results.

5 Conclusion

This paper presents a real-time plant phenotyping system through leaf counting, alignment, and individual leaf growth tracking. The Tiny-YOLOv4 network trained using the Darknet framework is utilized for leaf localization and counting. Further, The trained performance is evaluated on two different publicly available datasets. The model achieved a leaf counting accuracy of 90% and |DiC| of 0.42 on the test dataset, which denotes the generalization capacity of the model. Further, the growth analysis of individual leaves is performed by leaf tracking using the IoU metric and active contour segmentation of the localized leaf. The proposed system can be easily deployed on low-end hardware, thus showing its applicability to real-time growth estimation of any rosette-shaped plant.

Acknowledgements. This work is supported by the grant received from DST, Govt. of India for the Technology Innovation Hub at the IIT Ropar in the framework of the National Mission on Interdisciplinary Cyber-Physical Systems.

References

1. Aich, S., Stavness, I.: Leaf counting with deep convolutional and deconvolutional networks. In: Proceedings of the IEEE International Conference on Computer Vision Workshops, pp. 2080–2089 (2017). https://doi.org/10.1109/ICCVW.2017.244
2. Buzzy, M., Thesma, V., Davoodi, M., Mohammadpour Velni, J.: Real-time plant leaf counting using deep object detection networks. Sensors **20**(23), 6896 (2020). https://doi.org/10.3390/s20236896
3. Choudhury, S.D., Stoerger, V., Samal, A., Schnable, J.C., Liang, Z., Yu, J.G.: Automated vegetative stage phenotyping analysis of maize plants using visible light images. In: KDD Workshop on Data Science for Food, Energy and Water, San Francisco, California, USA (2016)
4. Cruz, J.A., et al.: Multi-modality imagery database for plant phenotyping. Mach. Vis. Appl. **27**(5), 735–749 (2015). https://doi.org/10.1007/s00138-015-0734-6
5. Das Choudhury, S., Bashyam, S., Qiu, Y., Samal, A., Awada, T.: Holistic and component plant phenotyping using temporal image sequence. Plant Methods **14**(1), 1–21 (2018). https://doi.org/10.1186/s13007-018-0303-x
6. Das Choudhury, S., Samal, A., Awada, T.: Leveraging image analysis for high-throughput plant phenotyping. Front. Plant Sci. **10**, 508 (2019). https://doi.org/10.3389/fpls.2019.00508
7. Dobrescu, A., Valerio Giuffrida, M., Tsaftaris, S.A.: Leveraging multiple datasets for deep leaf counting. In: Proceedings of the IEEE International Conference on Computer Vision Workshops, pp. 2072–2079 (2017). https://doi.org/10.1109/ICCVW.2017.243
8. Farjon, G., Itzhaky, Y., Khoroshevsky, F., Bar-Hillel, A.: Leaf counting: fusing network components for improved accuracy. Front. Plant Sci. **12**, 575751 (2021). https://doi.org/10.3389/fpls.2021.575751
9. Gibbs, J.A., Pound, M.P., French, A.P., Wells, D.M., Murchie, E.H., Pridmore, T.P.: Active vision and surface reconstruction for 3d plant shoot modelling. IEEE/ACM Trans. Comput. Biol. Bioinf. **17**(6), 1907–1917 (2019). https://doi.org/10.1109/TCBB.2019.2896908
10. Giuffrida, M.V., Doerner, P., Tsaftaris, S.A.: Pheno-deep counter: a unified and versatile deep learning architecture for leaf counting. Plant J. **96**(4), 880–890 (2018). https://doi.org/10.1111/tpj.14064
11. He, J.Q., Harrison, R.J., Li, B.: A novel 3D imaging system for strawberry phenotyping. Plant Methods **13**(1), 1–8 (2017). https://doi.org/10.1186/s13007-017-0243-x
12. Koornneef, M., Hanhart, C., van Loenen-Martinet, P., Blankestijn de Vries, H.: The effect of daylength on the transition to flowering in phytochrome-deficient, late-flowering and double mutants of arabidopsis thaliana. Physiol. Plantarum **95**(2), 260–266 (1995). https://doi.org/10.1111/j.1399-3054.1995.tb00836.x
13. Lee, U., Chang, S., Putra, G.A., Kim, H., Kim, D.H.: An automated, high-throughput plant phenotyping system using machine learning-based plant segmentation and image analysis. PLoS One **13**(4), e0196615 (2018). https://doi.org/10.1371/journal.pone.0196615
14. Li, Z., Guo, R., Li, M., Chen, Y., Li, G.: A review of computer vision technologies for plant phenotyping. Comput. Electron. Agric. **176**, 105672 (2020). https://doi.org/10.1016/j.compag.2020.105672

15. Minervini, M., Fischbach, A., Scharr, H., Tsaftaris, S.A.: Finely-grained annotated datasets for image-based plant phenotyping. Pattern Recogn. Lett. **81**, 80–89 (2016). https://doi.org/10.1016/j.patrec.2015.10.013
16. Minervini, M., Scharr, H., Tsaftaris, S.A.: Image analysis: the new bottleneck in plant phenotyping [applications corner]. IEEE Signal Process. Mag. **32**(4), 126–131 (2015). https://doi.org/10.1109/MSP.2015.2405111
17. Mochida, K., et al.: Computer vision-based phenotyping for improvement of plant productivity: a machine learning perspective. GigaScience **8**(1), giy153 (2019). https://doi.org/10.1093/gigascience/giy153
18. Mukhiddinov, M., Abdusalomov, A.B., Cho, J.: Automatic fire detection and notification system based on improved YOLOv4 for the blind and visually impaired. Sensors **22**(9), 3307 (2022). https://doi.org/10.3390/s22093307
19. Mukhiddinov, M., Cho, J.: Smart glass system using deep learning for the blind and visually impaired. Electronics **10**(22), 2756 (2021). https://doi.org/10.3390/electronics10222756
20. Redmon, J.: Darknet: open source neural networks in C. https://pjreddie.com/darknet/ (2021)
21. Redmon, J., Divvala, S., Girshick, R., Farhadi, A.: You only look once: unified, real-time object detection. In: Proceedings of the IEEE Conference on Computer Vision and Pattern Recognition, pp. 779–788 (2016). https://doi.org/10.1109/CVPR.2016.91
22. Ren, S., He, K., Girshick, R., Sun, J.: Faster R-CNN: towards real-time object detection with region proposal networks. In: Advances in Neural Information Processing Systems, vol. 28 (2015). https://doi.org/10.5555/2969239.2969250
23. Romera-Paredes, B., Torr, P.H.S.: Recurrent instance segmentation. In: Leibe, B., Matas, J., Sebe, N., Welling, M. (eds.) ECCV 2016. LNCS, vol. 9910, pp. 312–329. Springer, Cham (2016). https://doi.org/10.1007/978-3-319-46466-4_19
24. Roy, A.M., Bhaduri, J.: Real-time growth stage detection model for high degree of occultation using densenet-fused YOLOv4. Comput. Electron. Agric. **193**, 106694 (2022). https://doi.org/10.1016/j.compag.2022.106694
25. Tu, Y.-L., Lin, W.-Y., Lin, Y.-C.: Toward automatic plant phenotyping: starting from leaf counting. Multimed. Tools Appl. **81**(9), 11865–11879 (2022). https://doi.org/10.1007/s11042-021-11886-w
26. Wang, C.Y., Bochkovskiy, A., Liao, H.Y.M.: Scaled-YOLOv4: scaling cross stage partial network. In: Proceedings of the IEEE/CVF Conference on Computer Vision and Pattern Recognition, pp. 13029–13038 (2021). https://doi.org/10.1109/CVPR46437.2021.01283
27. Weyler, J., Milioto, A., Falck, T., Behley, J., Stachniss, C.: Joint plant instance detection and leaf count estimation for in-field plant phenotyping. IEEE Robot. Autom. Lett. **6**(2), 3599–3606 (2021). https://doi.org/10.1109/LRA.2021.3060712
28. Xu, L., Li, Y., Sun, Y., Song, L., Jin, S.: Leaf instance segmentation and counting based on deep object detection and segmentation networks. In: 2018 Joint 10th International Conference on Soft Computing and Intelligent Systems (SCIS) and 19th International Symposium on Advanced Intelligent Systems (ISIS), pp. 180–185. IEEE (2018). https://doi.org/10.1109/SCIS-ISIS.2018.00038
29. Yin, X., Liu, X., Chen, J., Kramer, D.M.: Joint multi-leaf segmentation, alignment, and tracking for fluorescence plant videos. IEEE Trans. Pattern Anal. Mach. Intell. **40**(6), 1411–1423 (2017). https://doi.org/10.1109/TPAMI.2017.2728065

Influence of Packet Switching and Routing Methods on the Reliability of the Data Transmission Network and the Application of Artificial Neural Networks

D. Davronbekov(✉), J. Aripov, Sh. Jabbarov, R. Djuraev, and D. Matkurbonov

Tashkent University of Information Technologies Named After Muhammad Al-Khwarizmi, Tashkent, Uzbekistan
d.davronbeckov@gmail.com

Abstract. In this article, a study was made of the influence of packet switching and routing methods on the reliability of a packet-switched data transmission network. Methods for routing and calculating the reliability of data transmission networks are considered. Analytical expressions for the readiness factor of communication channels between the nodes of the simulated network in virtual mode are given, as well as in the presence of redundancy, analytical expressions for the readiness factor of communication channels between the nodes of the simulated network in the datagram mode. Calculations of the reliability of data transmission networks with packet switching in the virtual mode and in the datagram mode are presented.

The features of the use of artificial neural networks in predicting the reliability of data transmission networks are considered. The scope of artificial neural networks is constantly expanding. In telecommunication systems, they are used in solving the following important problems: switching control, adaptive routing, traffic control, optimal distribution of network channel load, information encoding/decoding, etc. An algorithm for managing a data transmission network based on artificial neural networks is shown. The use of this algorithm for predicting the reliability indicators of telecommunication networks is proposed.

Keywords: data transmission network · switching · packet · virtual mode · datagram mode · readiness factor · node · channel · prediction · artificial neural network

1 Introduction

Modern data transmission networks (DTN) are built using packet switching technology with further packet routing. Since the routing of each packet occurs independently, the solution of routing problems becomes of paramount importance. Thus, packet routing is the cornerstone of the entire IP technology [1–7].

The development of packet-switched data networks is inextricably linked with the complication of their topology, which leads to the ever-increasing importance of the routing problem, since the quality of the solution to this problem directly affects the performance, efficiency of use, and reliability of the network as a whole [8–11].

H. Zaynidinov et al. (Eds.): IHCI 2022, LNCS 13741, pp. 284–296, 2023.
https://doi.org/10.1007/978-3-031-27199-1_29

2 Methods for Routing and Calculating the Reliability of Data Transmission Networks

Routing is the choice of the most rational way of transmitting information, and its solution is one of the main tasks of the network layer.

Modern packet-switched data networks are built using more switches and routers and other network devices that have become very complex [1–3]. Therefore, in packet-switched data networks, network status monitoring and management of network devices and the network as a whole become extremely complex. The router includes two main components:

- determination of optimal routing paths (routes) (problem of route construction);
- transportation of packets through the network (routing task).

Routers independently determine traffic routes based on information received via routing protocols (BGP, OSPF, etc.) from neighboring nodes [4–6].

Packet-switched data networks include packet routing technology. Each packet includes control information identifying the destination station (computer, terminal, host). This means that each packet of one information flow is routed independently of the others. Thus, routing in IP technology becomes the main factor affecting the performance and efficiency of the network as a whole [4, 5, 7].

Routing methods can be divided into two large classes: routing with virtual channels, datagram (dynamic) routing [2, 3].

A switched virtual circuit is established when a special connection request packet is sent to the network. This packet passes through the switches and lays a virtual circuit. This means that the switches remember the route for this connection and when subsequent packets of this connection arrive, they always send them along the laid route.

If a switch or link fails along the path of a virtual link, the connection is broken and the virtual link must be re-laid. At the same time, it will naturally bypass the failed sections of the network [2, 3, 12].

The datagram method does not require a connection to be established beforehand and therefore works without delay before transmitting data. This is especially advantageous for small data transfers where the connection setup time can be comparable to the data transfer time. In addition, the datagram method quickly adapts to changes in the network [2, 3, 13].

When designing data networks, it is necessary to take into account many criteria and factors in order to build a reliable network infrastructure, taking into account the topology, communication channels, active equipment, and routing protocols. The latter play a key role in the distribution of information flows in communication channels. With short-term failures of some network segments, the use of static routing is not reasonable, due to the lack of operational flexibility.

One of the important factors characterizing the data transmission network is its reliability indicator [14, 18–20]. The most important complex indicator of the reliability of packet-switched data transmission networks is its readiness factor (K_G), which determines the probability that the network will be in a working state at an arbitrary point in time [8–10].

The readiness factor in packet-switched data transmission networks is influenced by the readiness factors of its nodes and channels. Each of the data network devices is a device with certain characteristics related to their reliability.

Since K_G is a probabilistic quantity, then to calculate K_G of a data transmission network, one should use the mathematical apparatus of probability theory [8–11].

One of the methods for calculating K_G of a complex topology data transmission network is the method of its sequential decomposition into a set of networks with linear topology. The decomposition process is carried out until the remaining structures become parallel-serial. K_G of series-connected devices according to the theorem on the product of the probabilities of independent events is calculated by the following formula [23]:

$$K_{G_{s-c}} = K_{G1} \times K_{G2} \times \ldots\ldots \times K_{G_n}, \quad (1)$$

where $K_{G1} \ldots\ldots K_{G_n}$ is the readiness factor of series-connected elements.

The readiness factor of parallel series-connected elements is calculated according to the following relation [24]:

$$K_{G_{parallel}} = 1 - \left(1 - K_{G1}\right) \times \left(1 - K_{G2}\right) \times \ldots\ldots \times \left(1 - K_{G_n}\right). \quad (2)$$

Thus, the readiness factor of the formalized topology of the data transmission network can be represented as a function of K_G network nodes and K_G network channels. The form of the function K_{GDTN} will depend on its topology and be determined by (1), (2):

$$K_{G\text{DTN}} = f\left(K_{G_p}; K_{G_{node}}\right). \quad (3)$$

There are two packet switching methods: the virtual connection method and the datagram method [4–6].

3 Calculation of the Reliability of Data Transmission Networks with Packet Switching in Virtual Mode

In packet switching technology, using virtual channels, a logical connection is established, organizing a virtual channel, between the end devices of subscribers, which is used during the entire session. In networks with virtual channels, local addresses of packets are used when making a forwarding decision (Fig. 1) [2, 12].

When transmitting in virtual mode, a temporary virtual connection and a permanent virtual channel are distinguished. In the temporary virtual connection mode, the connection is established only for the duration of the message transmission. The formation of a virtual channel in this case is performed at the request of the user and is similar to the procedure for establishing a connection in a circuit-switched network. A permanent virtual channel between two users is organized for a certain time, not related to the duration of the communication session.

As the model under consideration, a linear topology network of five nodes is used (Fig. 2), in which the readiness factor of the communication line between nodes 1 and 5 is calculated. Internet segments are used as communication lines. Statistical information was processed using the Internet as an information transmission medium. To perform

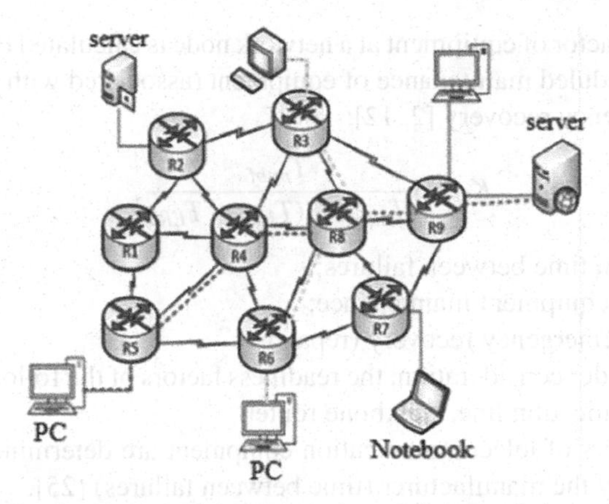

Fig. 1. The principle of operation of the virtual channel

calculations, the network topology is formalized to a diagram (Fig. 2), in which, depending on the composition of the equipment, a router of the Cisco type can act as nodes [25].

The readiness factor of communication channels between 1 and 5 nodes of the simulated network is calculated by the formula [2, 12]:

$$K_{G1-5} = K_{G1} \times K_{G1-2} \times K_{G2} \times K_{G2-3} \times K_{G3} \times K_{G3-4} \times K_{G4} \times K_{G4-5} \times K_{G5}. \tag{4}$$

where K_{G1}, K_{G2}, K_{G3}, K_{G4}, K_{G5} are the readiness factors of the corresponding nodes; $K_{G1\text{-}2}$, $K_{G2\text{-}3}$, $K_{G3\text{-}4}$, $K_{G4\text{-}5}$ – readiness factors of communication lines between nodes.

All considered topologies of the data transmission network are reduced to formalized ones, consisting of the same nodes and the same communication channels. Readiness factors of all considered nodes (K_{Gnode}) and all considered channels (K_{Gch}) are taken equal to each other.

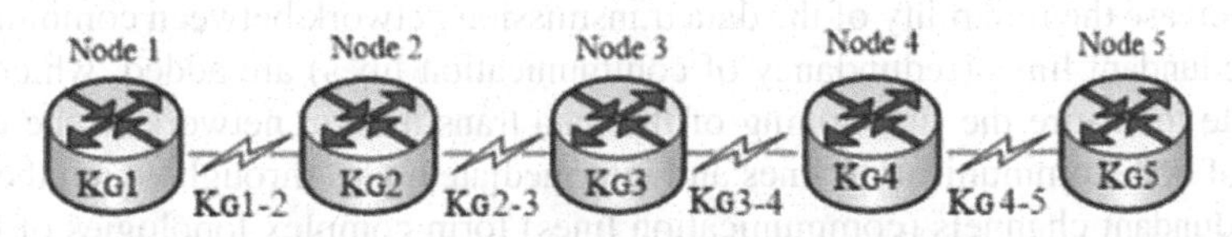

Fig. 2. Formalized topology of the simulated network

Taking $K_{G1} = K_{G2} = K_{G3} = K_{G4} = K_{G5} = K_{Gnode}$, and $K_{G1\text{-}2} = K_{G2\text{-}3} = K_{G3\text{-}4} = K_{G4\text{-}5} = K_{G\text{-}i}$, the formula for the readiness factor of the formalized network topology can be written as [2, 12]:

$$K_{G1-5} = K_{G-node}^{5} \times K_{G-i}^{4}. \tag{5}$$

The readiness factor of equipment at a network node is calculated taking into account the time for unscheduled maintenance of equipment (associated with its shutdown) and the time for emergency recovery [2, 12]:

$$K_G = \frac{T_{mtbf}}{T_{mtbf} + (T_{EM} + T_{ER})}, \tag{6}$$

where T_{mtbf} – mean time between failures;

T_{EM} - time for equipment maintenance;

T_{ER} - time for emergency recovery (repair).

In the model under consideration, the readiness factors of the following elements are considered communication line, backbone router.

Readiness factors of telecommunication equipment are determined on the basis of the data declared by the manufacturer (time between failures) [25].

The readiness factor of the communication line between the network nodes is assumed to be $K_{G\text{-}i} = 0{,}999$.

The Cisco router readiness factor is calculated based on the following data [25]:

- MTBF for platform and interface cards is $T_{mtbf-r} = 7$ years or 61320 h;
- time for annual maintenance (shutdown of the equipment is required) $T_{EM\text{-}r} = 4$ h × 7 years = 28 h;
- time for emergency recovery in case of an unforeseen failure (according to statistics, 1–2 times during the service life, we accept 2 times) $T_{ER\text{-}r} = 16$ h × 2 = 32 h.

By (6) we get:

$$K_{G-r} = \frac{61320}{61320 + (28 + 32)} = 0{,}99902. \tag{7}$$

The main equipment of communication nodes - the router according to (5) we get:

$$K_{G1-5} = 0{,}99902^5 \times 0{,}999^4 = 0{,}991135. \tag{8}$$

Let us consider the influence of redundancy of communication channels on the efficiency of the data transmission network.

To increase the reliability of the data transmission network between communication nodes, redundant links (redundancy of communication lines) are added, which makes it possible to ensure the functioning of the data transmission network in the event of failures of both communication lines and intermediate nodes through which these lines pass. Redundant channels (communication lines) form complex topologies of the data transmission network, in which the influence of nodes and communication lines on the reliability characteristics of the data transmission network as a whole is expressed in different ways. In accordance with (1) and (2), the readiness factor of a data transmission network built according to a linear topology with redundant communication lines (Fig. 3) for the path between nodes 1 and 5 can be written as (6). Simplifying (6), we obtain [2, 12]:

$$\begin{aligned} K_{G1-5}^{with\ reserve\ (2)} &= K_{G1} \times (1 - (1 - K_{G1-2\backslash 1}) \times (1 - K_{G1-2\backslash 2}) \times K_{G2} \\ &\times (1 - (1 - K_{G2-3\backslash 1}) \times (1 - K_{G2-3\backslash 2}) \times K_{G3} \times (1 - (1 - K_{G3-4\backslash 1}) \\ &\times (1 - K_{G3-4\backslash 2}) \times K_{G4} \times (1 - (1 - K_{G4-5\backslash 1}) \times (1 - K_{G4-5\backslash 2}) \times K_{G5}. \end{aligned} \tag{9}$$

Simplifying (9), we get:

$$K_{G1-5}^{withreserve(2)} = K_{Gnode}^{5} \times \left(1 - (1 - K_{Gch})^{2}\right)^{4}. \quad (10)$$

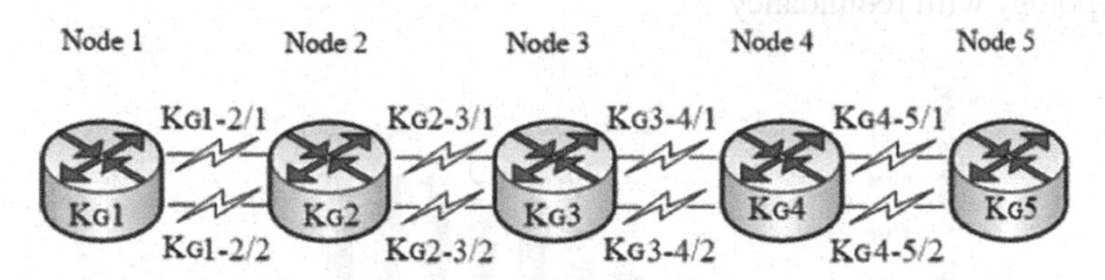

Fig. 3. Formalized redundant linear network topology communication channels

To assess the influence of K_G network elements on K_G of the data transmission network as a whole, we construct graphs of dependence of K_G of the data transmission network for each considered topology.

Let's build a dependency graph K_G of formalized topologies of each type with stabilized K_G network channels. For analytical calculation and plotting, we will accept K_G of the communication center (K_{Gnode}) varying in the range from 0,99 to 0,9999 with a step of 0,00099 in accordance with Table 1.

The readiness factor of the network channels (K_{Gch}) will be taken as stabilized at a value of 0,999. The results of the calculation according to (6), (9), (10) will be entered in the corresponding columns of Table 1 and plotted (Fig. 4).

Table 1. Calculation of *KG* formalized topologies depending on *KGnode*

K_{Gnode}	K_G formalized topologies	
	linear without reserve	line with reserve
0,99	0,947191	0,950986
0,99099	0,951937	0,955750
0,99198	0,956701	0,960534
0,99297	0,961485	0,965336
0,99396	0,966287	0,970158
0,99495	0,971109	0,974999
0,99594	0,975950	0,979860
0,99693	0,980810	0,984740
0,99792	0,985690	0,989639
0,99891	0,990589	0,994557
0,9999	0,995508	0,999496

Figure 4 clearly demonstrates that at low values of K_{Gnode} with a stabilized value of K_{Gch}, the readiness factor of a linear topology with all types of redundancy is almost identical.

At the same time, under conditions of low values of K_{Gnode}, the highest value of the readiness factor of the data transmission network segment is achieved when using a linear topology with redundancy.

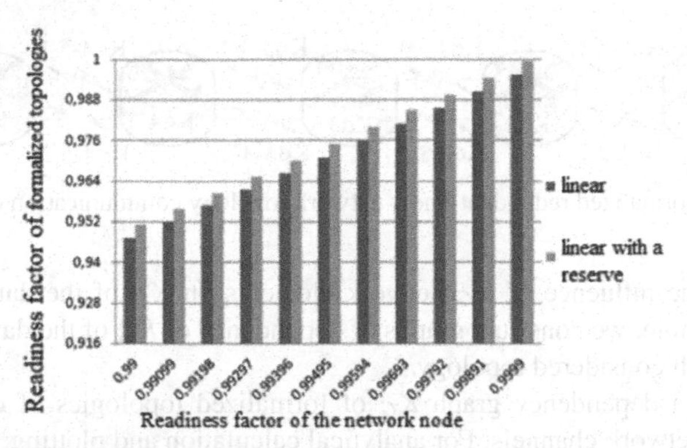

Fig. 4. Dependence of the readiness factor of formalized data transmission network topologies on the readiness factor of the network node

4 Calculation of the Reliability of Data Transmission Networks with Packet Switching in the Datagram Mode

A packet-switched network in datagram mode is a multi-connected network in which each node communicates not with two neighboring nodes, but with many more of them. The network design is based on the following principle: at least three communication lines should come from each network node, since to ensure the high survivability of the system (network), it is necessary to have a large number of nodes. Each packet is addressed separately and interpreted as an independent entity with its own control commands [13].

Switching devices route each packet (datagram) independently, directing it through the network, and intermediate nodes determine the next route segment of the next packet. As a result, they arrive at the addressee randomly. A sequence number is included in the packet header, and the receiving device uses it to reassemble the packets and recreate the original message.

The decision on which node to send the incoming packet to is made on the basis of a table containing a set of destination addresses and address information that uniquely identifies the next (transit or end) node. Such tables have different names - for example, for Ethernet networks they are usually called a forwarding table, and for network protocols such as IP and IPX, they are called a routing table (routingtable) (Fig. 5).

As the model under consideration, a linear topology network of five nodes is used (Fig. 6), in which the readiness factor of the communication line between nodes 1 and 5 is calculated [2, 13].

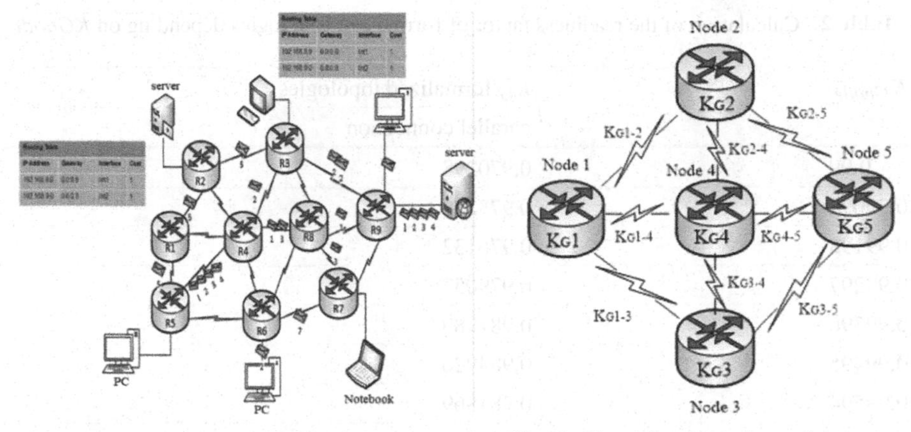

Fig. 5. Datagram principle of packet transmission

Fig. 6. Formalized topology of the simulated network

The readiness factor of the communication line between 1 and 5 nodes of the simulated network is calculated by the formula [2, 13]:

$$\begin{gathered} K_{G1-5} = K_{G1} \times (1 - (1 - K_{G1-2} \times K_{G2} \times K_{G2-5}) \times (1 - K_{G1-4} \times K_{G4} \times K_{G4-5}) \\ \times (1 - K_{G1-3} \times K_{G3} \times K_{G3-5}) \times (1 - K_{G1-3} \times K_{G3} \times K_{G3-4} \times K_{G4} \times K_{G4-5}) \\ \times (1 - K_{G1-3} \times K_{G3} \times K_{G3-4} \times K_{G4} \times K_{G2-4} \times K_{G2} \times K_{G2-5}) \times (1 - K_{G1-2} \\ \times K_{G2} \times K_{G2-4} \times K_{G4} \times K_{G4-5}) \times (1 - K_{G1-2} \times K_{G2} \times K_{G2-4} \times K_{G4} \times K_{G3-4} \\ \times K_{G3} \times K_{G3-5}) \times (1 - K_{G1-4} \times K_{G4} \times K_{G2-4} \times K_{G2} \times K_{G2-5}) \times (1 - K_{G1-4} \\ \times K_{G4} \times K_{G3-4} \times K_{G3} \times K_{G3-5})) \times K_{G5}. \end{gathered} \tag{11}$$

Simplifying (11), we obtain [2, 13]:

$$K_{G1-5} = K^2_{G-y} \times (1 - \left(1 - K_{G-y} \times K^2_{k-i}\right)^3 \times \left(1 - K^2_{k-y} K^3_{k-i}\right)^4 \times \left(1 - K^3_{k-y} K^4_{k-i}\right)^2). \tag{12}$$

By (12) we get [2, 13]:

$$\begin{gathered} K_{G1-5} = 0,99902^3 \times (1 - \left(1 - 0,99902 \times 0,999^2\right)^3 \\ \times \left(1 - 0,99902^2 \times 0,999^3\right)^4 \times \left(1 - 0,99902^3 \times 0,999^4\right)^2) = 0,99706. \end{gathered} \tag{13}$$

Let's build a dependency graph of the readiness factor of formalized topologies of each type with a stabilized readiness factor of the network channels. For analytical calculation and plotting, we will accept the readiness factor of the network node varying in the range from 0,99 to 0,9999 with a step of 0,00099.

The readiness factor of the network channels (K_{Gch}) will be taken as stabilized at a value of 0,999. The results of the calculation by (11), (12) will be entered in the corresponding columns of Table 2 and plotted (Fig. 7).

The performed analysis showed the possibility of using the reliability theory apparatus for analyzing the structural reliability of packet-switched data transmission networks.

Table 2. Calculation of the readiness factor of formalized topologies depending on *KGnode*

K_{Gnode}	K_G formalized topologies
	parallel connection
0,99	0,970298
0,99099	0,973212
0,99198	0,976132
0,99297	0,979057
0,99396	0,981989
0,99495	0,984926
0,99594	0,987869
0,99693	0,990818
0,99792	0,993772
0,99891	0,996733
0,9999	0,9997

As an indicator of the effectiveness of the functioning of the data transmission network and their elements, the readiness factor was chosen as a normalized indicator of the reliability of the data transmission network and their elements.

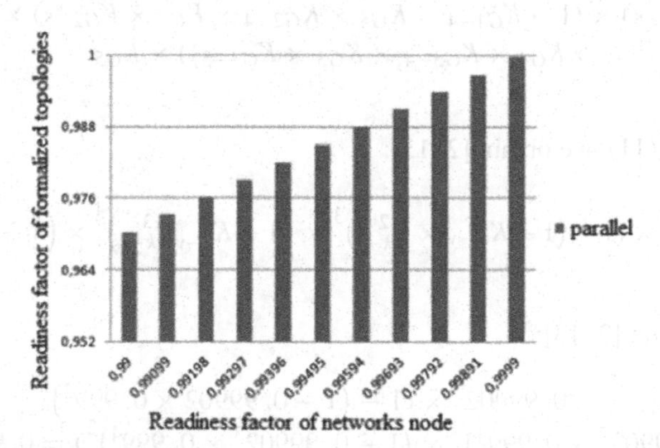

Fig. 7. Dependence of the readiness factor of formalized data transmission network topologies on the readiness factor of networks node

5 Features of the Use of Artificial Neural Networks in Predicting the Reliability of Data Transmission Networks

Modern data transmission networks are networks with a complex architecture, which includes a large number of equipment of various types and purposes, interconnected by various types of communication lines, as well as complex software, protocols, etc. [21, 22]. Each of these components affects the reliability of the data network. During the operation of the data transmission network, various kinds of failures, failures and errors occur, which worsen the reliability of the network [15–19].

Therefore, to manage a modern data transmission network, there is a need to use effective methods based on the use of a certain tool for predicting traffic based on previous data values [28], which will make it possible to predict the reliability of a data transmission network in the future.

The analysis [26, 28] showed that artificial neural networks (ANNs) are the most effective tool for forecasting. A neural network is a highly parallel dynamic system with a directed graph topology that can receive output information through the response of its state to input actions [2–9].

The scope of artificial neural networks is constantly expanding. In telecommunication systems, they find application in solving the following important problems [29]: switching control, adaptive routing, traffic control, optimal load distribution of network channels, information encoding/decoding, etc.

Control networks are subject to the requirements of low inertia, the ability to independently solve tasks, and have sufficient memory [29].

In [28], the following data transmission network control algorithm based on artificial neural networks was proposed (Fig. 8).

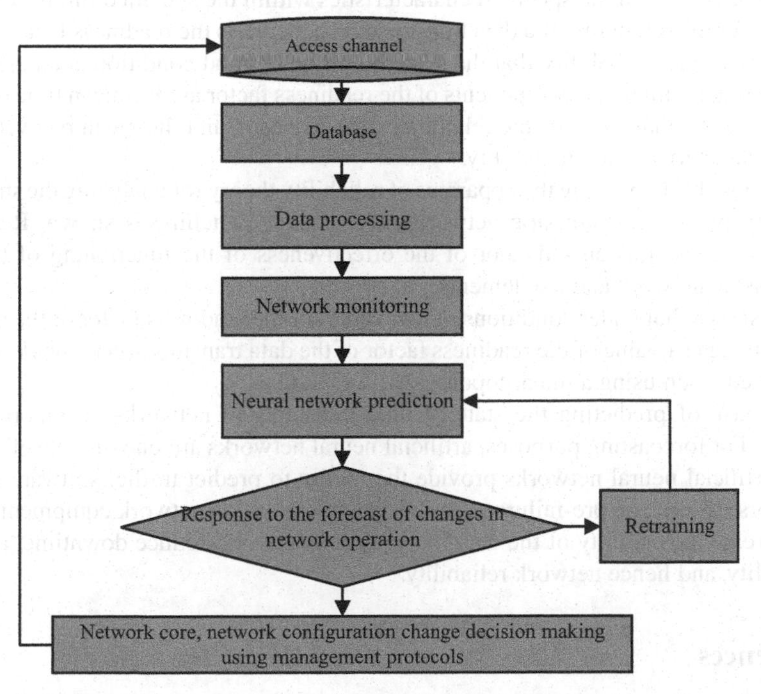

Fig. 8. Data transmission network control algorithm based on artificial neural networks [28]

For the prediction process, statistical information about traffic parameters is collected; data volumes; failures, failures, errors in the software and equipment of the data transmission network, etc. Based on the collected data, the artificial neural network is trained.

Based on the predicted values and priorities of traffic, failures, errors and failures of equipment and software, routes are formed and the results are analyzed, routes with the largest specific weight are selected, and a decision is made on choosing a route. Also, when making a routing decision, feedback is provided taking into account an unsuccessful decision. In this case, a re-request is made to train the neural network, which makes it possible to more accurately approximate the following prediction results [28].

Using an artificial neural network, based on checking the fulfillment of conditions for the residual component (forecast error) for independence (lack of autocorrelation), randomness, and normality of distribution, the adequacy of the short-term forecasting model is checked for adequacy [28].

6 Conclusions

One of the important requirements for a data transmission network is the requirement to ensure its reliability, which means that the network must ensure long-term operation while maintaining all the specified characteristics within the specified limits. The main criterion for the reliability of a data transmission network is the readiness factor, which characterizes the probability that the system will be in good condition at an arbitrarily chosen point in time. The components of the readiness factor are the mean time between failures, which characterizes the reliability of the system, and the mean recovery time, which characterizes maintainability.

The possibility of using the apparatus of reliability theory for analyzing the structural reliability of data transmission networks with packet switching is shown. Readiness factor was chosen as an indicator of the effectiveness of the functioning of the data transmission network and its elements.

It is shown that under conditions of low values of the readiness factor of the network node, the highest value of the readiness factor of the data transmission network segment is achieved when using a linear topology with redundancy.

The task of predicting the state of data transmission networks is important and relevant. For forecasting purposes, artificial neural networks are one of the well-trained ones. Artificial neural networks provide the ability to predict traffic, software failures and errors, failure and pre-failure states of data transmission network equipment, which will increase the stability of the data transmission network, reduce downtime, increase availability, and hence network reliability.

References

1. Gordienko, V.N., Krukhmalev V.V., Mochenkov O.D., Sharafutdinov R.M. Optical telecommunication systems. Hotline - Telecom, p. 368 (2011)

2. Ringenblum, P.G.: Improving the method of taking into account the influence of information security threats on the efficiency of the functioning of a corporate telecommunications network. Dissertation for the degree of candidate of technical sciences, Novosibirsk, 164 p. (2016)
3. Goldstein, B.S., Sokolov, N.A., Yanovsky G.G.: Communication networks.-St.Petersburg: "BHV - Petersburg", p. 400 (2014)
4. Olifer, V.G., Olifer N.A.: Computer networks. Principles, technologies, protocols. Textbook for high schools. Ed .- in Peter (2010)
5. Tanenbaum, A.S.: Computer Networks. Wetherall - 5th Edition, p. 962
6. Juraev, R.Kh., Dzhabbarov, Sh.Y., Umirzakov, B.M.: Data transfer technologies. Educational allowance (2008)
7. Stallings, W.: Data and Computer Communications, p. 901. Pearson Prentice Hall, New Jersey (2007)
8. Juraev, R.X., Djabbarov, S.Y., Baltaev, J.B.: Systems technical service and exploitation networks telecommunications. Textbook: Aloqachi, p. 234 (2019)
9. Polovko, A.M., Gurov, S.V.: Fundamentals of the theory of reliability. - 2nd ed., revised and additional - St. Petersburg: BHV-Petersburg, p. 704 (2006)
10. Yu, V., Shishmarev, M.: Reliability of technical systems: a textbook for students. Higher textbook institutions. Publishing Center Academy, p. 304 (2010)
11. Lisienko, V.G., Trofimova, O.G., Trofimov, S.P., Druzhinina, N.G., Dyugai, P.A.: Modeling of complex probabilistic systems: Textbook. Allowance, Ekaterinburg: URFU, p. 200 (2011)
12. Juraev, R.Kh., Aripov, J.A.: Features of calculating the reliability of a data transmission network with packet switching in virtual mode. In: The importance of information communication technologies in the innovation development of economic sectors. Republican scientific and technical conference. Collection of lectures. II-part. - Tashkent, pp. 22–24 (2022)
13. Aripov, J.A.. Reliability of a data transmission network with packet switching in datagram mode. In: The importance of information communication technologies in the innovation development of economic sectors. Republican scientific and technical conference. Collection of lectures. II-part. - Tashkent, pp. 20–22 (2022)
14. Davronbekov, D.A.: Analysis of quantitative indicators of reliability of elements and components of complex radio engineering systems. J. Muhammad al-Xorazmiy avlodlari. **2**(8), 89–92 (2019)
15. Davronbekov, D., Kamalov, Y., Aripov, J.: Mathematical model for improving the reliability of the data transmission network. J. Muhammad al-Khwarizmiy avlodlari **2**(16), 88–90 (2021)
16. Davronbekov, D.A., Matyokubov, U.K., Aripov, J.A.: Analysis of reliability issues in telecommunications networks and power grids. Scientific Collection "InterConf", Proceedings of the 4th International Scientific and Practical Conference "Recent Scientific Investigation", vol. 91, pp. 434–438, Norway (2021)
17. Davronbekov, D., Kamalov, Yu., Aripov, J.: Mathematical model of the control system of the priority data transmission network. Int. J. Trend Sci. Res. Develop. (IJTSRD), **6**(2), 993–997 (2022)
18. Davronbekov, D.A., Matyokubov, U.K.: Influence of communication lines on reliability in mobile communication systems. In: 2021 International Conference on Information Science and Communications Technologies: Applications, Trends and Opportunities, ICISCT (2021). https://doi.org/10.1109/ICISCT52966.2021.9670377
19. Davronbekov, D.A., Matyokubov, U.K.: Algorithms for calculating the structural reliability of a mobile communication system. In: 2021 International Conference on Information Science and Communications Technologies: Applications, Trends and Opportunities, ICISCT (2021). https://doi.org/10.1109/ICISCT52966.2021.9670315

20. Matyokubov, U.K., Davronbekov, D.A.: The impact of mobile communication power supply systems on communication reliability and viability and their solutions. Int. J. Adv. Sci. Technol. **29**(5), 3374–3385 (2020)
21. Hakimov, Z.T., Davronbekov, D.A.: Equalization of spectral characterist of optical signals by acousto-optic filters. In: 2007 3rd IEEE/IFIP International Conference in Central Asia on Internet, ICI (2007). https://doi.org/10.1109/CANET.2007.4401704
22. Davronbekov, D.A., Khakimov, Z.T.: Joint application of a runiing wave amplifier and acoustic- optical configurable filter for linearization of the passage spectral characteristics of FOCLEs. In: 2020 International Conference on Information Science and Communications Technologies, ICISCT (2020). https://doi.org/10.1109/ICISCT50599.2020.9351490
23. Mitrokhin, V.E.: Structural reliability of ring telecommunication networks and their elements. In: Theoretical and applied issues of modern information technologies: Proceedings of the All-Russian Scientific and Technical Conference (Ulan-Ude, 18–22 September 2001), pp. 132–135. Ulan-Ude: VSGU Publishing House (2001)
24. Krainov, A.Y., Meshcheryakov, R.V., Shelupanov, A.A.: Reliability model of information transmission in a secure distributed telecommunication network. Bull. Tomsk Polytech. Univ. **313**(5), 60–63 (2008)
25. Odom, S., Nottingham, H.: CISCO routers. Kudits -Obraz, p. 528 (2003)
26. Komashinsky, V.I., Smirnov, D.A.: Neural networks and their application in control and communication systems. Hotline-Telecom, p. 94 (2003)
27. Galushkin, A.M.: Neural networks: fundamentals of theory. Burning tea line-Telecom, p. 496 (2010)
28. Semeikin, V.D.: Data transmission network management using artificial neural networks. T-Comm. **7**, 118–121 (2013)
29. Fedotov, V.V.: Applications of neural networks in telecommunication networks. Izvestiya vuzov. North Caucasian region. Nat. Sci. Appl. **5**, 90–94 (2004)

Do Users' Values Influence Trust in Automation?

Liang Tang(✉), Priscilla Ferronato, and Masooda Bashir(✉)

School of Information Science, University of Illinois at Urbana-Champaign, Urbana, IL, USA
{ltang29,mnb}@illinois.edu

Abstract. As automation continues to play an increasing role in our daily life, it is essential to understand human-automation trust and the factors that influence that trust. Human values related research has become increasingly crucial to the HCI community over the past decade, but there has not been a uniform standard measure to frame and define human values, and there is a lack of clarity regarding the value-centered approach. Motivated by Schwartz's theory of fundamental human values, in this study, we investigate the relationship and the influence of human values on the formation of human trust towards automation. Specifically, how does HAI design prioritize values, which values should be emphasized during the design process, and which values will be diminished? This study, therefore, aims to provide an empirical examination of how human values can be incorporated into the design and development of automation. As a result of this study, we reveal that humans' complacency towards trust and values are more closely aligned. In addition, values such as "openness to change" seem to contribute the most to maintaining human-automation trust. We believe our study provides the first empirical evidence for the importance of considering human values when studying human-automaton trust and provides guidance for future human-centered design.

Keywords: Human Automation Interaction · Interaction Design · Human Trust

1 Introduction

Autonomous systems (AS) are the essential mediums that are being used to mediate the interaction between human-human or human-technology and are defined as "actively selects data, transforms information, makes decisions, or controls processes" [2]. Since the advent of automation, society has been dependent on automating a variety of tasks and endeavors [1]. A collaborative intelligence approach to automation can enable humans and automation to complement each other's strengths. When both types of intelligence are used in conjunction, we can produce a better result than can be obtained by either alone, thereby making human trust in automation an essential component of human-automation interaction [2].

Automated Systems (AS) are being built and designed to possess higher levels of intelligence and decision-making authority, thus enabling scaling into unpredictable, unplanned situations. Acceptance of these expansions will be determined partly by how much the users trust the AS. However, the calibration of trust can be affected by various

H. Zaynidinov et al. (Eds.): IHCI 2022, LNCS 13741, pp. 297–311, 2023.
https://doi.org/10.1007/978-3-031-27199-1_30

factors, including cultural, environmental, and systemic factors [3], making it difficult to operationalize the concept of trust comprehensively.

Several research studies within and outside of psychology have examined trust as a central theme within human-centered automation design. One theory we will use for this paper is proposed by Bashir & Hoff, where the interaction is described as dynamic and involves different layers of trust. We believe this theoretical framework provides a detailed interpretation of human trust in automation, conceptualizing trust from three broad perspectives, each with its research topics and methodological paradigms: (a) Integrity or characteristic of a trustor [4]. (b) The relevant factors of the trustor-trustee relationship so that the trustee's actions correspond with the trustee's goals and motivations [3]; (c) the quality or characteristic of a trustee [5]. While the extant literature has offered much information on how trust can be developed and sustained from the automation perspective (trustor), comparatively little attention has been paid to the human side (trustee), explicitly dealing with value as an influencing factor. The concept of value gives us cues on what we want to achieve and how to behave ourselves. The value system used in this paper serves as a set of guidelines in peoples' lives [6] and thus has a natural and robust relationship to behavior. As we interact with the subject, this instinct lets us quickly construct a sense of a person or automation. There are several approaches to assessing one's trust toward automation, and one of the essential measures of trust and influence is the degree of complacency. Complacency in automation can be defined as a "self-satisfaction that may result in non-vigilance based on an unjustified assumption of satisfactory system state." It may cause insufficient operator response, leading to automation failures and tragic outcomes [35]. The present research, therefore, aims to explain this missing link and apply statistical analysis to determine the relative importance of these factors in predicting trust levels and investigate broader implications for human-automation collaboration design.

2 Research Background

2.1 Development of Trust Toward Automation

In the same way that trust is crucial in interpersonal relationships, it also determines the willingness of humans to respond to automated systems in the circumstances characterized by uncertainty [4]. The trust toward automation has been widely studied in the past decade, which can be defined as "the attitude that an agent will help achieve an individual's goals in a situation characterized by uncertainty and vulnerability and an essential mediator of performance in work environments" [2]. The foundation of automation trust refers to the fundamental theory of interpersonal trust [7], and the integrated model of human trust in automation proposed by Muir [8] showed that models of interpersonal trust capture some of the key characteristics of the dynamic and nature of human-automation trust.

When people confront uncertainties in a relationship, both types of trust describe "situational attitudes". Trust in automation is driven by humans' innate cognitive, emotional, and social predispositions and context [2], then composed and developed by three informational bases: performance, process, and purpose of automation. However, the development of interpersonal trust is in line with the other three components of perceived trustworthiness: ability, benevolence, and integrity (the extent to which the trustee follows a set of moral rules or guidelines) [9]. Although humans and automation can socially interact with each other [10], humans still make important distinctions when engaging with automation compared to other humans. Two of the distinctions are (1) automation lacks anthropomorphism and associated societal expectations [11]; (2) human actions can be intentional, while automation cannot. Automation primarily consists of programs and algorithms that are only designed to perform specific tasks. [12] As such, automation cannot develop its intentions in the same manner as humans. If we wish to construct human-automation trust based on what we know about interpersonal trust, understanding the antecedents of trust formation in automation becomes increasingly essential for developing and designing autonomous systems.

2.2 Variability in Human Trust

The level of trust in automation can be influenced by various factors that are not necessarily related to the characteristics of automation. Individual differences between trustors are one of the factors that influence the variety of responses to the technology [2]. Evidence suggests that personality traits and social constructs such as empathy can naturally emerge and become significant determinants of interpersonal trust during human-to-human interaction. In this regard, some of the attributes commonly associated with human morality may be more advantageously integrated into automaton design from a human-centered perspective [13]. Many dimensions of human trust have been shown to be closely related to the degree of dependence on automation and machine. Trust, a social psychological concept, has a significant influence on the use, misuse, abuse, and disuse of automation as a key factor in the successful deployment of automation [11]. To put it in different words, a human's trust needs to be calibrated and adjusted to meet their needs and avoid disastrous consequences [14]. By such means, a vital issue to address is how to assist users in determining the appropriate level of trust users can place in automation. Thus, it is paramount to gain deep insight into the factors responsible for the progression of trust, improve trust calibrations, and develop a new human-centered design of automation.

Human automation trust evolves in a complex context; therefore, appropriate trust toward automation relies on a deeper understanding of individual differences. When automation is conceived as co-workers with humans, individual differences that affect and assist this coordination will be a significant factor in this coordination [15]. Therefore, in the context of HAI, considering the effects of these individual differences derived from the coordination of teams may be just as important as in interpersonal relations.

Human-automation trust is shaped by a complex set of factors. Hoff and Bashir propose a three-layered trust model that distinguishes different variability in human-automation trust and highlights the distinct categories of human-automation trust reflected in Marsh and Dibben [16]: dispositional trust, situational trust, and learned trust (Fig. 1).

Within this model, individual differences have been highlighted as important variables which influence the disposition of people to trust automated systems. One important aspect here is that dispositional trust is influenced greatly by biological and environmental factors. Researchers have found that the level of dispositional trust of automation can

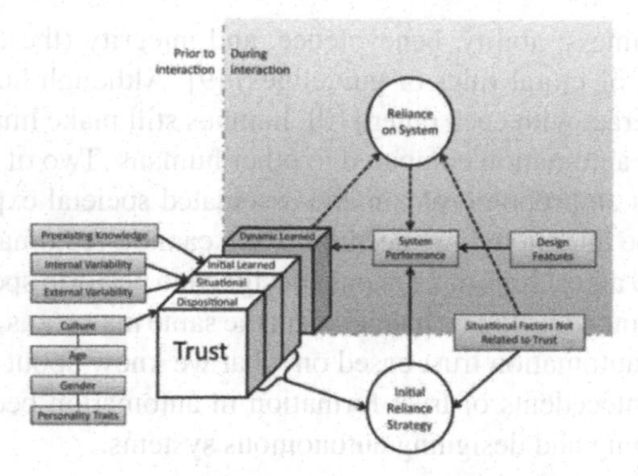

Fig. 1. Three layer - trust model proposed by Hoff & Bashir P. 421

vary across four primary sources of variability in this most basic layer as culture, age, gender, and personality, and that learned trust is influenced by the adequacy of its performance, purpose, and process [4]. However, how these personal characteristics relate to human-automation trust is still vague and not commonly studied even though it is of great value. Some researchers have conducted studies to explore whether trust in automation systems is enhanced if the moral judgment of the system aligns with the participant's preferred values by embedding a value centric-model. Despite the vast body of research surrounding this matter, an adequate study for assessing the role of value in dispositional trust is still lacking. As part of our research, we aim to fill this gap by identifying how various levels of value held by individuals contribute to differing trust levels in regard to automation exclusively, which has not been investigated widely.

3 Traits and Value

Among the studies that focus on the relationship between individual differences and trust, personality is one of the most frequently discussed characteristics. Researchers have found that when certain aspects of automation remain constant, personality moderates certain degrees of human perceptions of automation and becomes a significant determinant of human behavior [17]. Hoff and Bashir's three-layer trust model also recognizes the operator's personality as a final component of dispositional trust that reflects an individual's tendency to be trusting of others throughout his or her life. Human trust researchers acknowledged the fact that the Big Five Inventory (BFI) traits show all positive relationships on trust behavior except high neuroticism perceived trust to be low [18–20].

Personality describes actions presumed to flow from "what a person is like" regardless of their intention, whereas value refers to the individual's motivating goal and conceptions of the desirable that guide behavior [6, 21]. Similar to the role of personality on trust formation, a person's values are expected to influence trust formation when making decisions in interpersonal or organizational settings. Trust researchers have long asserted that some attributes normally associated with moral values can be helpfully applied to the computer-driven system [22, 23]. When automation becomes more intelligent and responsive to the needs of human users, the degree to which trust is built upon the integrity of a trustee might not be determined by how well the trustee is performing but rather by how closely their actions are aligned with shared values. Over the decades, values and traits have been viewed in a wide variety of ways, but the prevailing view is that they are two distinct constructs that interact to influence both attitude and behavior strongly. While personality can influence the development of the value system as descriptive variables [24], values guide individual judgments about appropriate trusting behavior in themselves and others [25] and become reliable cross-situational predictors of both attitudes and behavior as motivational variables [26]. Human trust researchers have acknowledged the fact that the Big Five Inventory (BFI) traits show all positive relationships on trust behavior (except high neuroticism where perceived trust is low) and suggest that operators may demonstrate higher trust levels when they possess extraversion, emotional stability, and intuitive rather than sensing personalities [18–20]. Research has also drawn parallels between Schwartz's values and the Big Five personality traits [27–29] as a result shown in Fig. 2.

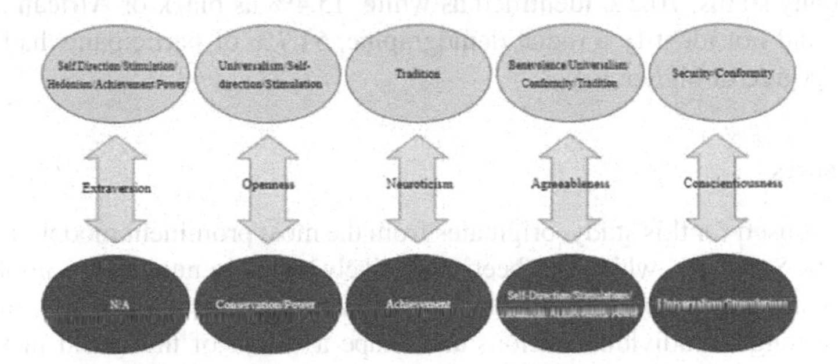

Fig. 2. Correlation chart between BFI traits and Schwartz's Value

Notes: This figure provides the association between the Big Five personality and Schwartz's value. The upper arrow indicates a positive correlation between variables, down arrow indicates a negative correlation.

Value is also often used as the best representative of the individual's culture [26] since culture is a group-level structure that may not result in meaningful consequences related to automation adoption at the individual level. Although cultures tend to vary in terms of trait-value relationships, values often have a substantial impact on the worldview of people and their behavior.

All of this leads up to our belief that values have a similar effect as personality traits on human-automation trust to a certain extent and suggests a complicated relationship between moral values and human-automation trust.

We make several specific hypotheses about the relationship between values and trust. The main hypothesis is that a user's value system will have a significant influence on their trust in automation, from which three specific predictions were derived:

H1: Participants' values and the values' related dimensions will significantly influence participants' complacency to trust in automation.
H2: Participants' values and their propensity to trust in automation are not significantly related.

H3: Participants' values, namely openness will have a significant and positive correlation with trust level in automation.

4 Method

4.1 Participants

The goal of this research is to identify the relationship between humans' values and trust in the automation system. Two-hundred and seventy-three participants were recruited through the Amazon MTurk platform within the United States to complete an online survey questionnaire via the Qualtrics platform. Each participant received compensation of $1.00. Among the data, one hundred and seventy participants reported their gender as female, one hundred and two as male. Most participants were adults (Age > 18), 50.5% of reported participants were married, 33.7% were single, and the rest reported another family status. 70.2% identified as white, 15.4% as black or African American and 14.4% did not identify a racial demographic. 54.7% of participants had earned a bachelor's degree or higher.

4.2 Measures

Value, as it is used for this study, originates from the most prominent model of 10 values developed by Schwartz, which has been extensively tested in numerous samples across over 70 countries [6]. Schwartz makes a significant point in his research that values provide meaning to individual actions and shape a vision of the world in which the individual thinks and behaves. His model identifies 10 basic value types that can be found across an extensive range of cultures and differ in their underlying motivational base. The 10 values initially proposed can be further grouped into four higher-order dimensions within this motivational continuum [6], as shown in Fig. 3.

Fig. 3. Schwartz's Value Chart

Within the circle, the further two values are from each other, the more dissimilar their underlying motivations are [6]. One of the most central and distinctive aspects of this value system is that value attributes are distinguished from one another by their goals. A person may have a variety of values simultaneously, and their significance may vary [6, 31]. When two values conflict, we are likely to act in accordance with the more important value.

(1) In this project, value systems were evaluated using Schwartz's PVQ-RR questionnaire, which was previously intensively discussed in human-human trust. Every value reveals its subject's goals, aspirations, and wishes, all of which convey the significance of a specific value. The PVQ-RR scale consists of 21 context-bounded items that assess participants' goals, aspirations, and wishes. The PVQ-RR asks participants to check one of six labeled boxes from "Very much like me" to "Not like me at all." An example is, "Thinking up new ideas and being creative is important to him/her. He/she likes to do things in his/her own original way." This reflects one's value of self-direction.
(2) A trustor's propensity to trust is a general willingness and tendency or disposition to trust another [32]. Propensity to trust may be an essential predictor of trust intentions, but not trust actions or behavior. Indeed, the propensity to trust has been related to the self-report of trust intentions [17] and trust beliefs [33] in ambiguous situations. Under these circumstances, the eight-item propensity to trust scale used by Mayer and Davis [9] for assessing trust beliefs and intention, where each participant's propensity to trust was evaluated by summing up the scores of the eight questions, where a higher score indicated the higher propensity. The propensity to trust scale consists of eight-item that measure propensity and specify the level of agreement to each statement in five points. An example is, "One should be very cautious with strangers."
(3) In the initial examination, complacency indicated a particular failure or "mindless" behavior in that operators invest unjustified trust and receive an insufficient response from the automation, which includes errors of omission and commission followed by delayed response [34]. In this case, complacency is complacent behavior—as reflected in poor monitoring of automation or inappropriate overreliance on automation. The automation-induced Complacency Potential Rating Scale (CPRS) from Singh, Molloy, and Parasuraman [35] was employed in this research to evaluate "an individual's overall tendency to trust automation, independent of context or a specific system" [4] but more from the behavioral and dynamic aspect of trust. The CPRS scale measures certain positive attitudes toward automation, and the items are on a five-point scale ranging from 1 (strongly disagree) to 5 (strongly agree). The focus of the CPRS seems to be on trust in various forms of automation across the four factors, including assessments of confidence, reliance, trust, and safety-related complacency. For example, "I feel safer depositing my money at an ATM than with a human teller" [35]. The adapted CPRS scale in this study was narrowed to a pool of 12 of which ten are original items from the CPRS, while the other two questions have been revised to reflect the current socio-technical context better. Participants' CPRS scores were calculated by adding the points derived from all 12

questions, with higher CPRS scores indicating higher levels of trust in automation and a greater tendency to exhibit over-trust behaviors.

5 Analysis

5.1 Correlation Analysis

This study first used Pearson correlations to examine the direction and strength of bivariate relationships between variables. Second, dominance analysis was used to determine the relative importance of each predictor to the trust in automation. If a predictor contributes more to the subsets of predictors than any other predictor, it is considered dominant or more important. We performed all statistical analyses, including dominance analyses, using STATA version 13.

To examine the relationship between propensity to trust and values, the regression analysis was used; as shown in Table 1, the stimulation (B = 0.715) and self-direction (B = −0.755) are statistically significant at the 0.05 level (P = 0.025), and p = 0.044 which indicates that higher self-direction is related to lower propensity to trust, and higher stimulation is related to higher propensity to trust.

Table 1. Linear regression of value and propensity to trust

Source	SS	df	MS			
Model	496.028778	10	49.6028778	Number of obs	=	273
Residual	6196.96756	262	23.6525479	F(10, 262)	=	2.10
Total	6692.99634	272	24.6066042	Prob> F	=	0.0250
				R-squared	=	0.0741
				Adj R-squared	=	0.0388
				Root MSE	=	4.8634

Propensity	Coefficient	Std. err.	t	P>|t|	[95% conf.	interval]
Benevolence	.2803579	.3829317	0.73	0.465	-.4736576	1.034373
Universalism	.4325798	.4077141	1.06	0.290	-.3702337	1.235393
Achievement	-.1824182	.3027968	-0.60	0.547	-.7786432	.4138068
Power	.4068409	.4405363	0.92	0.357	-.4606014	1.274283
Hedonism	-.2351552	.3402674	-0.69	0.490	-.9051621	.4348517
Self-Direction	-.755775	.3737827	-2.02	0.044	-1.491776	-.0197745
Stimulation	.7150624	.3177662	2.25	0.025	.0893618	1.340763
Security	-.4134652	.3650918	-1.13	0.258	-1.132353	.3054223
Conformity	.2577169	.3313274	0.78	0.437	-.3946865	.9101203
Tradition	.5083519	.3864775	1.32	0.190	-.2526454	1.269349
cons	19.93566	2.112676	9.44	0.000	15.77567	24.09564

Table 2 where examines the relationship between value and complacency. Similar results were observed but appear contrarily. The Universalism (B = 1.9749) and Self-direction (B = 1.19) propensities show statistical significance at the 0.05 level (P = 0.001) and (P = 0.032) and are positively related to complacency to trust. The stimulation propensity (B = −2.68) is closer to significance, with a P value of 0.061, but is not statistically significant at the 0.05 level. These findings suggest that individuals prioritizing universalism and self-direction values are likely pursuing a higher complacency toward automation.

The regression model with complacency in trust accounts for 13.03% (adj. R2 = 0.1303) of the variance, while the one with the propensity to trust accounts for 3.8%.

Table 2. Linear regression of value and complacency to trust

Source	SS	df	MS		
				Number of obs	= 273
				F(10, 262)	= 5.07
Model	2634.01486	10	263.401486	Prob> F	= 0.0000
Residual	13599.6188	262	51.9069421	R-squared	= 0.1623
				Adj R-squared	= 0.1303
Total	16233.6337	272	59.6824768	Root MSE	= 7.2046

Complacency	Coefficient	Std. err.	t	P>|t|	[95% conf.	interval]
Benevolence	-.5326656	.5672769	-0.94	0.349	-1.649668	.5843365
Universalism	1.974962	.6039897	3.27	0.001	.78567	3.164254
Achievement	.4527404	.4485647	1.01	0.314	-.4305102	1.335991
Power	.8731856	.6526127	1.34	0.182	-.4118477	2.158219
Hedonism	.488427	.5040738	0.97	0.333	-.5041245	1.480978
Self-Direction	1.190461	.5537235	2.15	0.032	.100146	2.280775
Stimulation	-.8873743	.4707403	-1.89	0.061	-1.81429	.0395416
Security	-.0214858	.5408487	-0.04	0.968	-1.086449	1.043478
Conformity	-.4681992	.49083	-0.95	0.341	-1.434673	.4982743
Tradition	.519627	.5725296	0.91	0.365	-.6077181	1.646972
cons	31.56427	3.129729	10.09	0.000	25.40165	37.72689

(adj. R2 = 0.038). It is surprisingly found that value within the openness to change dimension, including stimulation, and self-direction, have both significant effects on the direction of propensity and complacency. We replicated similar results from prior personality studies and observed parallels between personality traits and personal values which we found that value accounted for 3%–13% of the total variance in trust beliefs (propensity to trust) and trust behaviors (complacency to trust) while personality was founded to be accounted for 5–18% of the variance in trust behavior and intentions [36]. The finding of this result suggests that values and traits have similar effects statistically in explaining human-automation trust, and value has enough accountability to explain and moderate certain perceptions of human trust toward automation.

5.2 Dominance Analysis

Previous research has shown that values are often too abstract. For determining or assigning one single behavior or even one behavior to a specific value since there could be varying interpretations of which behavior represents a value depending on social or cultural contexts [37]. People's values are largely shaped by environmental factors, which include culture, education, parental upbringing, and life events [38]. In order to fully understand the value-trust relationship, it is necessary to present a more detailed analysis that relates value to the priorities given to a larger pool of human factors. To further investigate the robustness of our results, we ran a dominance analysis and ensemble each possible combination of the independent variables for ranking importance determinations by aggregating fit metrics across multiple models. In our model, we entered the value, age, ethnicity, gender, education, marital status, income, and region to examine the effects of all possible dispositional trust determinants on propensity/complacency to trust and look at how value combines with other enduring traits to guide the trust formation process.

After introducing different dispositional sets in this linear regression, as shown in Table 3. All Subsets Fit Stat. = 0.1789 result represents the amount of the McFadden pseudo-R-square that explained 17.89% of variance. Among the top 5 factors that are most important determinations, the demographics characterize (ethnics, age) explain individual variance better than value, accounting for 17.76% and 30.49% of the predicted variance, followed by the two openness-dimension value-stimulation, self-direction, and the power (11.45%, 5%, 7.58%).

Table 3. Linear Regression and the importance of predictors on the propensity to trust

General dominance statistics: Linear regression
Number of obs = 273
Overall Fit Statistic = 0.1789

Propensity	Dominance Stat.	Standardized Domin. Stat.	Ranking
Benevolence	0.0075	0.0418	8
Universalism	0.0033	0.0183	11
Achievement	0.0032	0.0180	12
Power	0.0136	0.0758	4
Hedonism	0.0022	0.0124	14
Self-direction	0.0090	0.0500	5
Stimulation	0.0205	0.1145	3
Security	0.0049	0.0276	10
Conformity	0.0084	0.0468	7
Tradition	0.0085	0.0475	6
Age	0.0318	0.1776	2
Ethnics	0.0546	0.3049	1
Gender	0.0070	0.0393	9
Education	0.0002	0.0011	17
Marital	0.0028	0.0155	13
Income	0.0012	0.0068	15
Region	0.0004	0.0020	16

Compared to the linear regress-based data analysis reported in Table 3, the relative importance ranking/fit stat within the complacency test has changed (Table 4). The second data analysis of the value revealed that the value of universalism, self-direction, and power explain 52.4% of the predicted variance (28.29%, 21.78%, and 8.78%, respectively). The first results, while surprising, are consistent with research showing that one's propensity to trust has demonstrated a weaker correlation with value (adj. R2 = 0.038/0.1789). It is conditioned by demographical/contextual factors other than the value which support H2, and the higher. Variance of value(adj. R2 = 0.1303/0.2262) within the complacency model has provided grounded evidence for H1. Our results also suggest that trust towards automation is more related to individuals' relatively enduring and evolution-based values, such as self-direction and stimulation within the openness to change dimension and universalism. In this case, we will accept H3.

Table 4. Linear Regression and the importance of predictors on the complacency to trust

General dominace statistics: Linear regression
Number of obs = 273
Overall Fit Statistic = 0.2262

Complacency	Dominance Stat.	Standardized Domin. Stat.	Ranking
Benevolence	0.0043	0.0191	10
Universalism	0.0640	0.2829	**1**
Achievement	0.0049	0.0218	9
Power	0.0199	0.0878	**4**
Hedonism	0.0034	0.0152	12
Self-direction	0.0493	0.2178	**2**
Stimulation	0.0053	0.0233	8
Security	0.0066	0.0290	6
Conformity	0.0027	0.0119	13
Tradition	0.0058	0.0259	7
Age	0.0173	0.0764	5
Ethnics	0.0355	0.1570	**3**
Gender	0.0001	0.0006	17
Education	0.0020	0.0086	14
Marital	0.0038	0.0168	11
Income	0.0008	0.0037	15
Region	0.0005	0.0022	16

6 Limitation and Future Research

The implications of the current study involve guiding the future researcher to consider human values as design criteria in the same way we view traditional standards regarding reliability, usability, efficiency, and explainability. In this study, we hypothesize that humans do not have enough empirical knowledge to guide these interactions when they initially interact with complex AS. Therefore, our study measures initial trust on the basis of current interactions rather than aggregated or long-term trust. Although the dominance analyses might be beyond the scope of the present value study, its result potentially illustrates the broad implications of individual differences on dispositional trust. A fuller understanding of the value-trust relationship requires us to relate value to the priorities given to the fuller set of factors and expand the differing perceptions of value variables under vulnerability and uncertainty conditions. In this study, we examine only the influence of value on the dispositional trust process, as it is the first step to investigating the relationship between value and trust. Several limitations were also possibly associated with this study:

(1) Trust is a multidimensional concept, and its formation is a dynamic process within a variety of contexts that this study does not fully capture. (2) Trust is highly complex and situational, the current study relies on self-report questionnaires, which trust is not insightful and meaningful enough to assess value, and trust might result in common-method bias. It is best to evaluate trust based on behavioral measures related to trust and other physiological signals for future research. (3) Specific human values are universally held but may play out differently in different cultures. So, future research needs to consider the value of an individual's cultural orientation in evaluating its importance. (4) In this article, we examine human trust in automation as a whole, and future research could be subdivided to explore trust in different types of automation. Accordingly, future

studies need to manipulate more dispositional factors to assess trust value more accurately toward automation and avoid an overly specific value-centered approach that only focuses on one dimension of human value.

7 Conclusion

Nowadays, automation is increasingly equipped to receive/process information and interact with people, thus transforming them into more team members than simply a tool as more potent forms of collaboration emerge. As a result, there will be an ever-increasing need to understand why and how people place and calibrate their trust in automation, as well as investigate the factors that influence the degree of human-automation trust in real-life settings.

Trust has distinct moral properties when deciding to trust or gain trust from an individual [39]. Therefore, this study proposes a list of values, particularly those with a moral component, that are essential to designing trustworthy autonomous systems. Our research findings confirm that even though there is a weak relationship between propensity and value, humans' complacency towards trust and values is more closely aligned, as trust propensity may be a more significant predictor of trust intentions than trust behavior. Furthermore, the study indicates that some values were more predominant than others, with the value construct "openness to change" contributing the most to maintaining human-automation trust. One possible explanation as to why people with more open-minded presented higher trust in AS can be related to the main characteristic of openness, which is important to avoid misunderstanding and mistrust and to align goals and expectations of other team members to achieve a common and mutual understanding of the task. Thus, the people with higher openness might consider adapting adequate approaches to develop a shared understanding and reduces mistrust and conflict of interest. However, this rule cannot be applied across all long-term situations because openness might only improve initial trust formation processes; therefore, further analysis is necessary. At the same time, it is considerable for automation embedded with high openness to initiate and engage in trusting behaviors to elicit cooperation with less open people because trust is a self-reinforcing phenomenon [41]. Our findings also leave open the question of how value-centered design can maintain a strong and sensitive balance of openness, as too much openness to shared intentions, such as absolute altruism, will introduce other risks, thereby reducing trust. In summary, we examine the role of value as a dispositional trust measure, discover the link between a user's value system and their trust toward automation, and find that the value system might serve as a critical lubricant for the cooperation of humans and autonomous systems for facilitating the trust factor in designing the automation. While a clear picture of human-automation trust has not yet been posted to the world, more research needs to be done to test the feasibility of embedding human value into automation. That is, research needs to empirically study whether our finding is indeed suitable as a methodology for future automation design.

References

1. Cummings, M.L., How, J.P., Toupet, O.: The impact of human–automation collaboration in decentralized multiple unmanned vehicle control. https://ieeexplore.ieee.org/document/6099676/. Accessed 6 June 2022
2. Lee, J.D., See, K.A.: Trust in automation: designing for appropriate reliance. Hum. Factors J. Hum. Factors Ergon. Soc. **46**(1), 50–80 (2004). https://doi.org/10.1518/hfes.46.1.50.30392
3. Hancock, P.A., Billings, D.R., Schaefer, K.E., Chen, J.Y., de Visser, E.J., Parasuraman, R.: A meta-analysis of factors affecting trust in human-robot interaction. Hum. Factors J. Hum. Factors Ergon. Soc. **53**(5), 517–527 (2011). https://doi.org/10.1177/0018720811417254
4. Hoff, K.A., Bashir, M.: Trust in automation. Hum. Factors J. Hum. Factors Ergon. Soc. **57**(3), 407–434 (2014). https://doi.org/10.1177/0018720814547570
5. Merritt, S.M., Ilgen, D.R.: Not all trust is created equal: dispositional and history-based trust in human-automation interactions. Hum. Factors J. Hum. Factors Ergon. Soc. **50**(2), 194–210 (2008). https://doi.org/10.1518/001872008x288574
6. Schwartz, S.H.: Universals in the content and structure of values: theoretical advances and empirical tests in 20 countries. Adv. Exp. Soc. Psychol. **25**, 1–65 (1992). https://doi.org/10.1016/s0065-2601(08)60281-6
7. Rempel, J.K., Holmes, J.G., Zanna, M.P.: Trust in close relationships. J. Pers. Soc. Psychol. **49**(1), 95–112 (1985). https://doi.org/10.1037/0022-3514.49.1.95
8. Muir, B.: Trust in automation: Part I. Theoretical issues in the study of trust and human intervention in automated systems. Ergonomics **37**(11), 1905–1922 (1994). https://doi.org/10.1080/00140139408964957
9. Mayer, R.C., Davis, J.H., Schoorman, F.D.: An integrative model of organizational trust. Acad. Manag. Rev. **20**(3), 709–734 (1995). https://doi.org/10.5465/amr.1995.9508080335
10. Anon: The media equation: how people treat computers, television, & new media like real people &places. Comput. Math. Appl. **33**(5), 128 (1997). https://doi.org/10.1016/s0898-1221(97)82929-x
11. Parasuraman, R., Riley, V.: Humans and automation: use, misuse, disuse, abuse. Hum. Factors J. Hum. Factors Ergon. Soc. **39**(2), 230–253 (1997). https://doi.org/10.1518/001872097778543886
12. Hopko, S.K., Mehta, R.K.: Neural correlates of trust in automation: considerations and generalizability between technology domains. Front. Neuroergonomics **2**, 26 (2021). https://doi.org/10.3389/fnrgo.2021.731327
13. Thagard, P.: What is trust? https://www.psychologytoday.com/us/blog/hot-thought/201810/what-is-trust. Accessed 6 June 2022
14. Pereira, A., Leite, I., Mascarenhas, S., Martinho, C., Paiva, A.: Using empathy to improve human-robot relationships. In: Lamers, M.H., Verbeek, F.J. (eds.) Human-Robot Personal Relationships. LNICSSITE, vol. 59, pp. 130–138. Springer, Heidelberg (2011). https://doi.org/10.1007/978-3-642-19385-9_17
15. Ross, J.M., Szalma, J.L., Hancock, P.A., Barnett, J.S., Taylor, G.: The effect of automation reliability on user automation trust and reliance in a search-and-rescue scenario. In: Proceedings of the Human Factors and Ergonomics Society Annual Meeting, vol. 52, no. 19, pp. 1340–1344 (2008). https://doi.org/10.1177/154193120805201908
16. Sheridan, T.B.: Individual differences in attributes of trust in automation: measurement and application to system design. Front. Psychol. **10**, 1117 (2019). https://doi.org/10.3389/fpsyg.2019.01117
17. Marsh, S., Dibben, M.R.: The role of trust in information science and technology. Ann. Rev. Inf. Sci. Technol. **37**(1), 465–498 (2005). https://doi.org/10.1002/aris.1440370111

18. Colquitt, J.A., Scott, B.A., LePine, J.A.: Trust, trustworthiness, and trust propensity: a meta-analytic test of their unique relationships with risk taking and job performance. J. Appl. Psychol. **92**(4), 909–927 (2007). https://doi.org/10.1037/0021-9010.92.4.909
19. Sharan, N.N., Romano, D.M.: The effects of personality and locus of control on trust in humans versus artificial intelligence. Heliyon **6**(8), e04572 (2020). https://doi.org/10.1016/j.heliyon.2020.e04572
20. Szalma, J.L., Taylor, G.S.: Individual differences in response to automation: the five factor model of personality. J. Exp. Psychol. Appl. **17**(2), 71–96 (2011). https://doi.org/10.1037/a0024170
21. Li, X., Hess, T.J., Valacich, J.S.: Why do we trust new technology? A study of initial trust formation with organizational information systems. J. Strateg. Inf. Syst. **17**(1), 39–71 (2008). https://doi.org/10.1016/j.jsis.2008.01.001
22. Tufts, J.H.: Ueber Werthaltung und Wert. Psychol. Rev. **3**(3), 352–353 (1896). https://doi.org/10.1037/h0064352
23. Boddington, P.: AI and moral thinking: how can we live well with machines to enhance our moral agency? AI Ethics **1**(2), 109–111 (2020). https://doi.org/10.1007/s43681-020-00017-0
24. Han, S., Kelly, E., Nikou, S., Svee, E.-O.: Aligning artificial intelligence with human values: reflections from a phenomenological perspective. AI Soc. **37**, 1383–1395 (2021). https://doi.org/10.1007/s00146-021-01247-4
25. Anon: Values and value-orientations in the theory of action: an exploration in definition and classification. In: Toward a General Theory of Action, pp. 388–433 (1951). https://doi.org/10.4159/harvard.9780674863507.c8
26. Parks-Leduc, L., Feldman, G., Bardi, A.: Personality traits and personal values. Pers. Soc. Psychol. Rev. **19**(1), 3–29 (2014). https://doi.org/10.1177/1088868314538548
27. Ponizovskiy, V., Grigoryan, L., Kühnen, U., Boehnke, K.: Social construction of the value-behavior relation. Front. Psychol. **10**, 934 (2019). https://doi.org/10.3389/fpsyg.2019.00934
28. Roccas, S., Sagiv, L., Schwartz, S.H., Knafo, A.: The big five personality factors and personal values. Pers. Soc. Psychol. Bull. **28**(6), 789–801 (2002). https://doi.org/10.1177/0146167202289008
29. Fischer, R., Boer, D.: Motivational basis of personality traits: a meta-analysis of value-personality correlations. J. Pers. **83**(5), 491–510 (2014). https://doi.org/10.1111/jopy.12125
30. Mehta, A., Morris, N.P., Swinnerton, B., Homer, M.: The influence of values on e-learning adoption. Comput. Educ. **141**, 103617 (2019). https://doi.org/10.1016/j.compedu.2019.103617
31. Bardi, A., Schwartz, S.H.: Values and behavior: strength and structure of relations. Pers. Soc. Psychol. Bull. **29**(10), 1207–1220 (2003)
32. Mayer, R.C., Davis, J.H.: The effect of the performance appraisal system on trust for management: a field quasi-experiment. J. Appl. Psychol. **84**(1), 123–136 (1999). https://doi.org/10.1037/0021-9010.84.1.123
33. McKnight, D.H., Cummings, L.L., Chervany, N.L.: Initial trust formation in new organizational relationships. Acad. Manag. Rev. **23**(3), 473–490 (1998). https://doi.org/10.5465/amr.1998.926622
34. Parasuraman, R., Manzey, D.H.: Complacency and bias in human use of automation: an attentional integration. Hum. Factors J. Hum. Factors Ergon. Soc. **52**(3), 381–410 (2010). https://doi.org/10.1177/0018720810376055
35. Parasuraman, R., Molloy, R., Singh, I.L.: Performance consequences of automation-induced "complacency." Int. J. Aviat. Psychol. **3**(1), 1–23 (1993). https://doi.org/10.1207/s15327108ijap0301_1
36. Alarcon, G.M., Capiola, A., Pfahler, M.D.: The role of human personality on trust in human-robot interaction. In: Trust in Human-Robot Interaction, pp. 159–178 (2021).https://doi.org/10.1016/b978-0-12-819472-0.00007-1

37. Hanel, P.H.P., Vione, K.C., Hahn, U., Maio, G.R.: Value instantiations: the missing link between values and behavior? In: Roccas, S., Sagiv, L. (eds.) values and Behavior, pp. 175–190. Springer, Cham (2017). https://doi.org/10.1007/978-3-319-56352-7_8
38. Rokeach, M.: The Nature of Human Values. Free Press
39. Anon: The moral value of trust. In: Trust in Medicine, pp.162–170 (2019). https://doi.org/10.1017/9781108763479.011

Automation of Calibration Procedure for Milk Non Automatic Weighing Instrument (NAWI) Process Using AI Methods

Nagamani Molakatala[1], Vimal Babu Undru[2(✉)], Shalem Raju Tambala[3(✉)], M. Tejaswini[4(✉)], M. Teja Kiran[5(✉)], M. Tejo Seshadri[6(✉)], and Venkateswara Sagar Juturi[7(✉)]

[1] University of Hyderabd, Hyderabd, India
nagamanics@uohyd.ac.in
[2] Legal Metrology Telangana, Hyderabd, India
uvbabu43@gmail.com
[3] Legal Metrology A.P., Guntur, India
shalemraju.tambala@gmail.com
[4] Osmaniya University, Hyderabd, India
mtejaswini12@gmail.com
[5] Geetam University, Visakhapatnam, India
cheekuchinna1@gmail.com
[6] Asian Institute of Technology Thailand, Khlong Luang, Thailand
tejosheshadri24@gmail.com
[7] American International Group Inc., New York, USA
sagarJuturi@gmail.com

Abstract. Automation is the Trend of Industrial 4.0 for rational process and prevent fraudulent activities. It is the simplification of human task assigned to machine and also brings the transparency in the system so that any fraudulent that causes human health and wealth can be traced and eliminated. The present work focusing on ca liberation correction as Milk is collected at procurement centers on a weight basis, using the Non-Automatic Weighing Instrument (NAWI) of Class III - Medium Accuracy Weighing Instruments. The specification of the gravity of milk as per standards and actual have a discrepancy which is causing the losses for the milk farmers and seriously affecting their economy with general NAWI system used for the process. In corp orating the AI based techniques on cloud based collected information storage to bring the automation methods for integrating all dairy form in a place to standardise the system and eliminate the discrepancy in calculation. Daily the farmers sell lakhs of Milk to the Milk purchasing units. The densities of the Milk brought by farmers are noted, and the data is stored in the cloud system for calibration verification through AI methods. The daily transactions are recorded stored in meta file. The cloud data of transaction in purchasing of Milk (with the incorrect calibration of the Weighing Instrument put to use by the officials) can be arrived at precisely with the help of computational methods using ML and DL techniques. These methods

H. Zaynidinov et al. (Eds.): IHCI 2022, LNCS 13741, pp. 312–322, 2023.
https://doi.org/10.1007/978-3-031-27199-1_31

will be useful for the farmers at large to find out the loss precisely and the gain to the purchaser. Integration of existing legacy method with proposed automation procedures of AI and ML/DL method is main focus of this work to bring fraud free environment in metro-logical system.

Keywords: NAWI · Computational Techniques · Specific Gravity · Milk · Weighing · Farmers

1 Introduction

The caliberation and correction of the metrology is duty of the metro logical department who care the correction and for the benefit of the public as part of the Good Governance. The present work on the Milk procurement correction in measurement for the benefit of the farmers study explored. The milk is procured from the farmers at different locations and is processed to obtain the dairy products. The Milk is procured from the farmers on volume basis. It is pertinent to mention here that the Weighing Machine is calibrated to the One Kilogram which is equivalent to the one liter of water. The value of the Milk is paid in terms of weight of Milk. As the specific gravity of the Milk differ with the Water the farmer shall incur loss for the milk sold. The difference of milk and water weights and its gravity difference is one of the reason for caliberation discrepancy. One liter of water and milk have same volume but the weight comparison milk is heavier than the milk. Gravity of milk varies from 1.028 to 1.032, for 1 l of milk.

To use a Non-Automatic Weighing Instrument **(NAWI)** for specific gravity products, more than one calibration change is needed to get the correct value to the farmer. Calibration of general NAWI differs from milk-related NAWI because of more specific gravity. When the weight of 1 kg is converted concerning the water, it comes to 1.030 kg for all practical purposes. Therefore the machine shall be calibrated to **970 g**. The departmental officials are erroneously calibrating devices to **950 g instead of 970 g** to measure the one liter of milk. Thereby the farmer is losing **20 ml** for every one liter of milk. The purchaser gains the cost of 20 ml for each liter of milk. Daily, lakhs of transactions will occur in a Dairy processing unit. The justification of the farmer can be done by measurement correction with the automation process is main focus of this research work.

Standardization across the globe is another important parameter. The standards used at different countries for collection of milk, usage in commercial transactions is gram, Ounce, Liter, pound etc. The quantity of 1 mL Milk weighs 0.958614 g. The one gram of Milk is equivalent to 1.043173 mL. One liter of butter weighs 911 g. One stick of butter weighs about 113.4 g (118.3 mL). The international standard denotation of weight is 'gram (g), or gramme'.

1 g = 0.211643 teaspoons (1 tsp, 1 t, 1 ts, or 1 tspn) or one teaspoon is equivalent to 4.7g. One tablespoon is equal to 0.5 (oz.) ounces. 1 tablespoons (1 tbsp, 1 T, 1 Tbls, or 1 Tb) = 14.786765 mL. One tablespoon of butter is equal to 1/8 of a stick or 1/2 ounce. 32 tbsp = 1 lb (Pound). The Butter in the United States is sold by the pound, with 4 sticks, or 16 ounces.

2 Necessary Study and Information Review

The author presented the basic requirements to be marked on the food packages. The labeling on the package enable the consumer to know the food product details and their shelf life safety, ingredients, transport, nutrient values [1]. The author presented that the flavored evaporative compound like esters, aldehydes, ketones, alcohols, sulphur, acids, terpenes and are decreased over the period when kept stored [2]. In this paper the author presented the cost effective method mixed linear integer programming model in transporting the milk to an extent of 10% [3].

Butter is one of the most popular fat-rich dairy products. Butter consists of butterfat, water, and milk proteins and is most frequently manufactured from cow or buffalo milk. It is generally used as a spread and a condiment. It is also used in cooking applications. Manufacture of creamery butter has been confined to the colder regions of the world, where gravity creaming has been successful. Initially, butter was manufactured by using wooden hand churns. But now almost all the butter industries are employing continuous butter making technology [4]. The economics of quality and quantity of milk collected at different locations of farms and transport cost are evaluated using Operations Research [5].

The animal Butter, Ghee are most commonly used in the making sweets, used in households in the India and it is used as ingredient in prepared food in middle east and African Countries. The main purpose of products of Milk i.e., AMF, Anhydrous Milk Fat and other are to increase the shelf life at atmospheric condition. The removal of water from milk, to avoid oxidation of milk fat, proper packing, storage will extend the quality and life [6].

With the help of weighing scale the weight of each Leg of Cow, Milking, frequency of kicks are analyzed to monitor the activity [7]. The Milk collection problem (MCP) is presented with computational results (VNS) in the paper. An integrated mathematical model is presented keeping distance, total network costs, routing, incompatibility, and loading constraints [8]. The cow milking is monitored before and after milking for better management. The Milk production can be monitored with the image processing technique [9]. With the help of IoT the pH (6.45 to 6.67) and electrical conductivity (4.65 mS/cm to 5.26 mS/cm) of milk is determined for checking the adulteration [10].

The contribution of the Buffalo Milk and the yield of butter when compared with the Cow, by products uses are depicted in the paper. In India most of the Milk is obtained from the Buffalo. The temperature effects on the products like Cheese, fermented products are presented [11]. The butter can be extracted from the Camel with the churning method and their properties are assessed [12]. The hydrogenated oils (PHOs) contain trans-fat in margarine. The samples are detected with the help of FTIR-ATR spectrometer [13].

Milk is collected at procurement centers on weight basis using Non Automatic Weighing Instrument (NAWI) of Class III - Medium Accuracy Weighing Instruments. Specific gravity of Milk in India varies from 1.028 to 1.032 where as Specific gravity of water is 1.

One liter of Milk and One liter of water have same volume, but one liter of milk is heavier than one liter of water because of more specific gravity of milk. In order to use Non Automatic Weighing Instrument (NAWI) for products of Specific gravity more than 1 calibration changes are needed.

3 Calibration of Milk Related Non Automatic Weighing Instrument (NAWI)

Calibration of General NAWI differs from milk related Non Automatic Weighing Instrument (NAWI) because of more specific gravity. In general NAWI, 10 kg weight is calibrated as 10 kg weight only. But in milk related Non Automatic Weighing Instrument (NAWI) 10 kg weight of milk when converted w.r.t. water it becomes 10.3 kg (Milks specific gravity is treated as 1.030 for all practical purposes) and it is calibrated correspondingly at 9.7 kg. We put 10 kg weight on balance, it has to show 9.7 kg instead of 10 kg. 1 L of milk when weighed on this Non Automatic Weighing Instrument (NAWI) it will a show a reading of 1 kg after calibration. The Figure shows the idea about the system in Fig. 1 shown below.

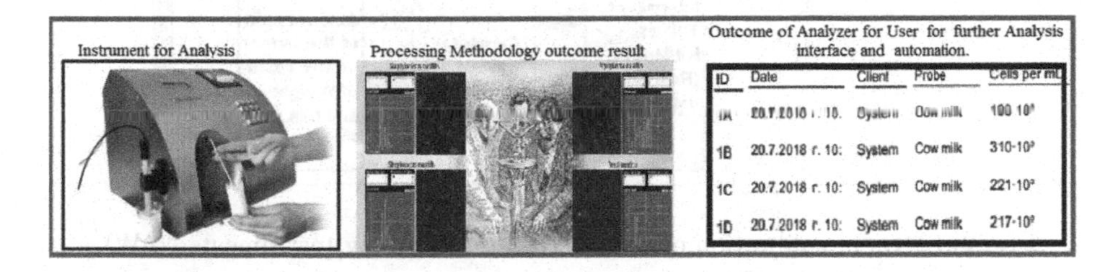

Fig. 1. General idea of NAWI system milk calibration process.

3.1 Malpractices Associated with Milk Non Automatic Weighing Instrument (NAWI)

Ironically, some of the Non Automatic Weighing Instrument (NAWI) at collection centers are wrongly calibrated below 9.7 kg for 10 kg generally at 9.5 kg, i.e., for every 1 L 20 ml more milk is collected from farmers.

3.2 Measures for Calibration of Milk:

One of the leading dairy is procuring 1 lakh liters per day in Coastal Districts of Andhra Pradesh. Suppose 50% of Non-Automatic Weighing Instrument (NAWI) are calibrated at 9.5 kg for 10 kg instead of 9.7 kg. Then they are collecting 20 ml more per one liter. This Faulty calibration when applied to 50% of the Product i.e., 50,000 l the resultant loss to the farmers is alarming as shown below.

(50,000 × 20 mL)/1000 mL=1000 L

Approximate Cost of Milk per liter is 60 Indian Rupees,

a. The cost for 1000 liters = 1000 × 60 = 60,000 rupees. (This is for 1 day)

b. For 1 month: 30 × 60,000 = Rs: 18, 00,000 (18 Lakhs) c. For 1 year: 12 × 18, 00,000 = Rs: 2, 16, 00, 000 (2 Crore 16 Lakhs)

A simple wrong calibration is causing the farmers this much of loss. Then the role of Legal Metrology Officer in calibration of Non Automatic Weighing Instrument (NAWI) plays a vital role in protecting the farmer's interests (Fig. 2).

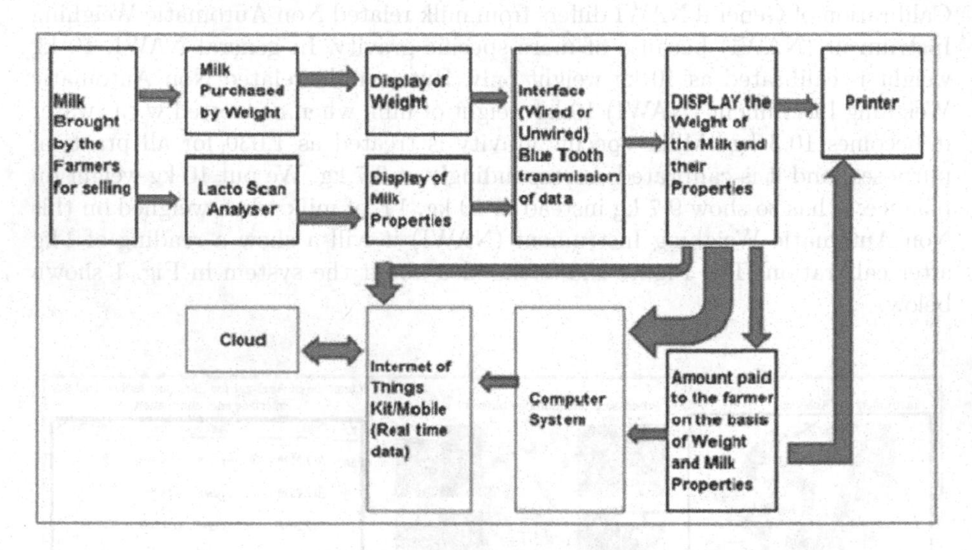

Fig. 2. LMOs in calibration of Non Automatic Weighing Instruments NAWI.

3.3 Proposed Method of Automation in Milk Calibration

Price of Milk is determined based on Milk Fat, Quantity of Milk obtained from Lacto Scan Analyser and Non Automatic Weighing Instrument respectively. Lacto scan analysers are interfaced with weighing instrument and printer. Calculate the Fat Content of Milk from reference Lacto Scan Analyser or Gerber Method. Calculate the Fat Content of Same Milk with Lacto Scan Analyser available at Milk Procurement Centre. Compare the results. The Ultrasonic Lacto scan readings can be secured with cloud calibration. The data can be shared with computer system; Bluetooth enabled applications including the Mobile applications to have real time values. The ultrasonic technology is better technology compared with the Infrared technology without intervention of the man (Fig. 3).

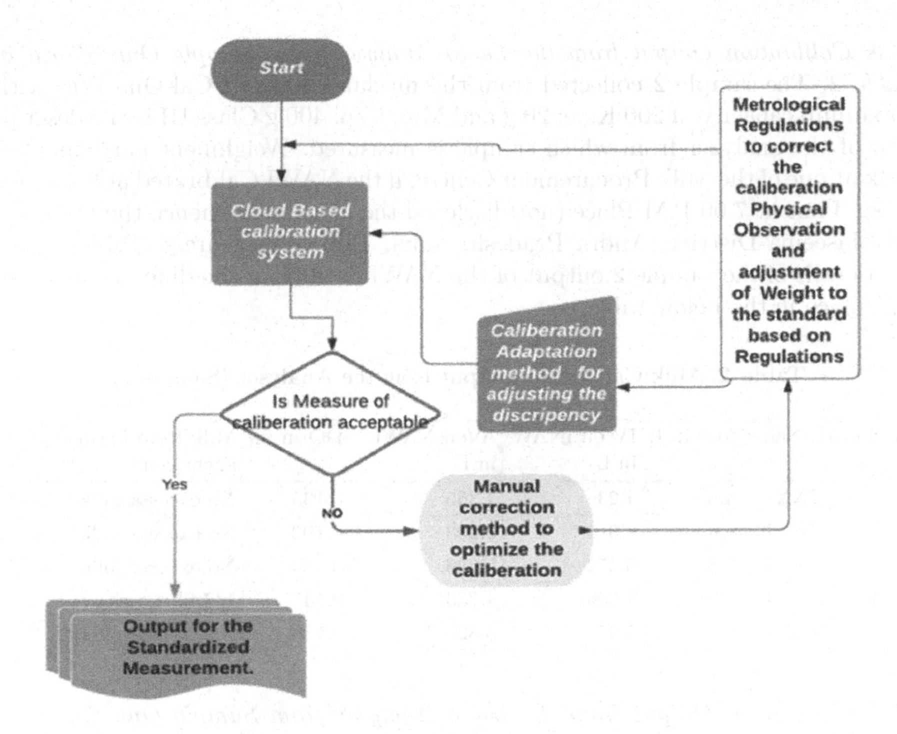

Fig. 3. The Infrared technology when there is no intervention of the man.

4 Preliminary Experiments and Results

This section gives the sample data and its manual analysis to check the discrepancy in the measurement of

Milk Calibration Output from the Lacto Analyser from Sample One Shown in Table 1. The sample 1 collected from the machine make of Swan Star with maximum capacity of 150 Kg.e: 20 g and Min. Cap:400 g Class III is the description of the analyzer from which sample is measured. Weighment particulars of Milk at one of the Milk Procurement Centre, if the NAWI Calibrated at 9.5 kg for 10 kg Time at 6.30 AM Place:(not disclosed the place name hence the masked), Coastal District, Andra Pradesh. Table 1 measures. Table 1 gives the details of the sample1 output of the NAWI data in the below table.

Table 1. Milk Calibration output from the Analyser (Sample 1).

S.no	FN	MD in L	IV on INAW inL	AV on NAWI in ltrs	DO in ml	Milk from Farmer agent centre
1	XXX1	4.5	4.275	4.365	0.09	collected 90 ml excess
2	XXX2	6.4	6.08	6.208	0.128	collected 120 ml excess
3	XXX3	5.8	5.5	5.626	0.116	collected 116 ml excess
4	XXX4	4.9	4.655	4.753	0.98	collected 98 ml excess
5	XXX5	6	5.7	5.82	0.12	collected 120 ml excess

Milk Calibration Output from the Lacto Analyser from Sample One Shown in Table 2. The sample 2 collected from the machine make of Cal-One Plus with maximum capacity of 200 Kg.e: 20 g and Min. Cap: 400 g Class III is the description of the analyzer from which sample is measured. Weighment particulars of Milk at one of the Milk Procurement Centre, if the NAWI Calibrated at 9.4 kg for 10 kg Time at 7.00 P.M Place:(not disclosed the place name hence the masked), Rayalaseema District, Andra Pradesh, India. Table 2 measuress. Table 3 gives the details of the sample 2 output of the NAWI data from the different machine and place in the below table.

Table 2. Milk Calibration output from the Analyser (Sample 2).

S.no	F Name	MD in L	IV on INAW in L	AVon NAWI in L	DO in ml	Milk from Farmer agent centre
1	XXX1	4.5	4.23	4.365	0.135	No execess milk
2	XXX2	6.4	6.016	6.208	0.192	No execess milk
3	XXX3	5.8	5.452	5.626	0.174	No execess milk
4	XXX4	4.9	4.606	4.753	0.147	147ml excess
5	XXX5	6	5.64	5.82	0.180	No execess milk

Milk Calibration Output from the Lacto Analyser from Sample One Shown in Table 2 The sample 3 collected from the machine make of Cal-One Plus with maximum capacity of 200 Kg.e: 20 g and Min. Cap: 400 g Class III is the description of the analyzer from which sample is measured. Weighment particulars of Milk at one of the Milk Procurement Centre, if the NAWI Calibrated at 9.7 kg for 10 kg Time at 7.00 P.M Place:(not disclosed the place name hence the masked), Ideal District, Andra Pradesh, India. Table 3 measuress. Table 3 gives the details of the sample 2 output of the NAWI data from the different machine and place in the below table.

Table 3. Milk Calibration output from the Analyser (Sample 3).

S.no	F Name	MD in L	IV on INAW in L	AV on NAWI in L	DO in ml	Milk from Farmer agent centre
1	XXX1	4.5	4.365	4.365	0	collected 0ml excess
2	XXX2	6.4	6.016	6.208	0.192	collected 0 excess
3	XXX3	5.6	5.626	5.626	0	collected 174ml excess
4	XXX4	4.9	4.753	4.753	0	collected 147ml excess
5	XXX5	6	5.82	5.82	0	collected 180ml excess

4.1 Analysis of the the Caliberation Results

The three case studies of field work empirical analysis of the data and its observation made through the visualization process gives the loss for the farmers due to the minute variation in measurement of the weight discrepancy caused by instrument. This needs always the knowledge about the concepts otherwise the difficulty to manually correct in all the instrument testing in the field and do corrections. De-centralized monitoring system and lack of the human intervention knowledge about the system may cause serious public loss. The centralization through technology and make the data available for ca liberating authority to identify and make generalization of caliberation corrections. The Table 1, 2 and 3 have the total 25 row items in original which are displayed in the visualization shown in Fig. 4 & 5.

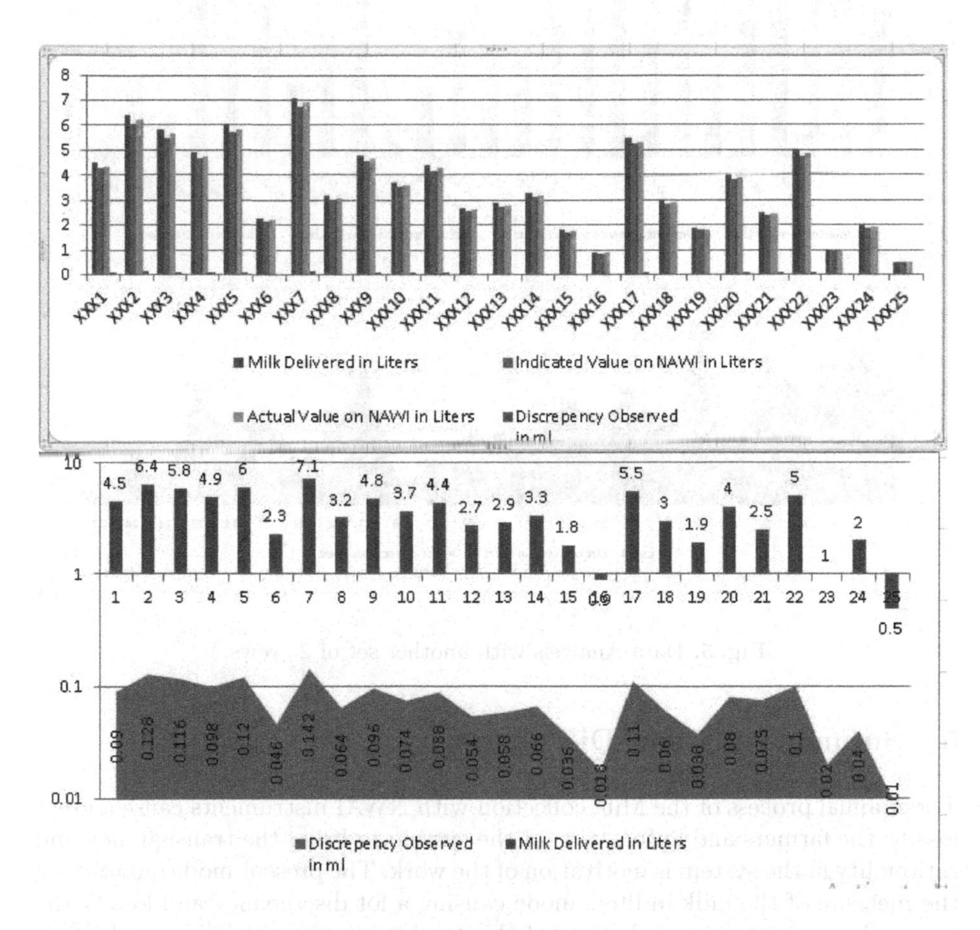

Fig. 4. Data Analysis for 25 rows.

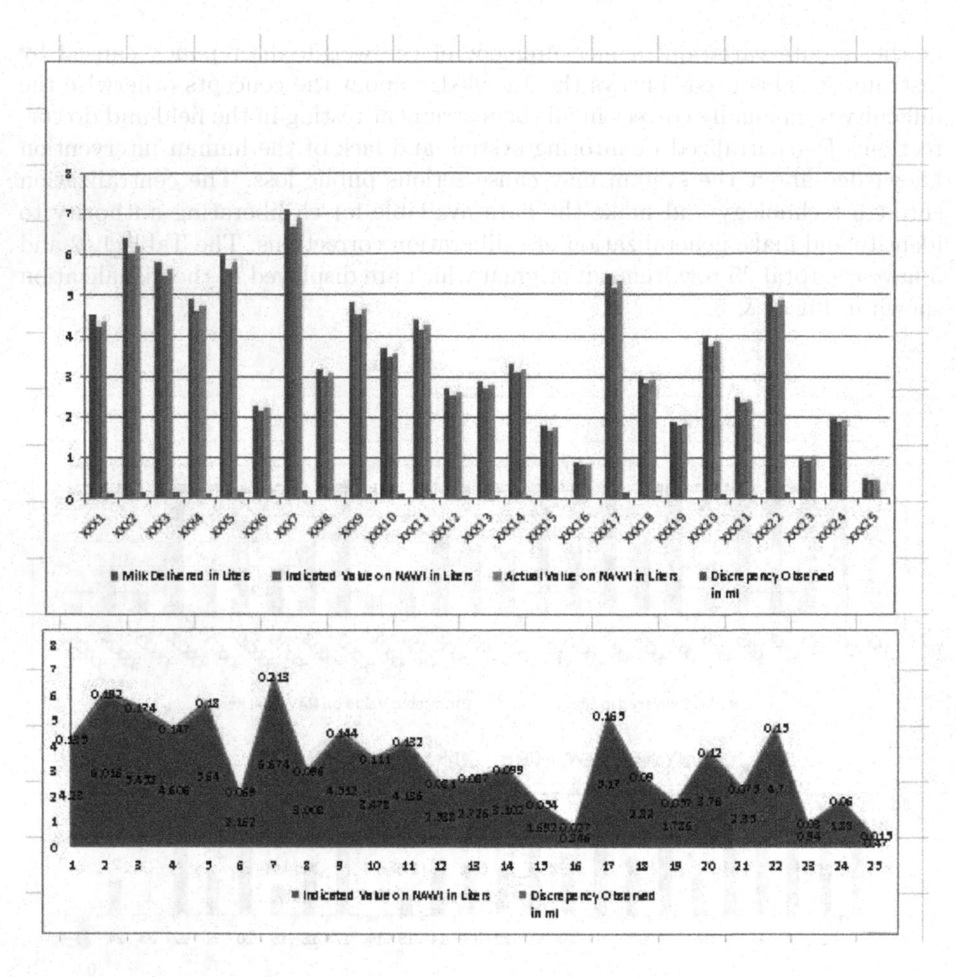

Fig. 5. Data Analysis with another set of 25 rows.

5 Summarizing and Discussion

The manual process of the Milk collection with NWAI instruments cause a great loss to the farmers and automation of the system to bring the transparency and rationality in the system is motivation of the work. The present mode quantifying the measure of the milk in liters mode causing a lot discrepancy and loss to the farmer hence with proposed study of this work recommend to discard the litre based measurement. The quantity of milk discrepancy can be presented properly while measuring Specific gravity into consideration for milk, calibrate with 10 kg @ 9.7 kg i.e. hence t 10 kg on NAWI must be displayed as 9.7 kg only facility need to incorporating through software method. The display more digits on the screen properly so that 2nd and 3rd digit blocking presently can be eliminated for properly caliberation correction. The first two table content shows the execess amount of the milk per litre collected at the field which is causing loss to the farmer and third table shows the no loss case study.

6 Conclusion and Future Scope

The process of the automation need the concepts of Human Machine Interactive methods incorporated into the existing legacy system to make more secure calibrations to eliminates any fraudulent process in the system. The proposed system of incorporation of cloud computing and automatic correction process and feedback for proper calibration adjustment according to the standards of Metro-logical analysis of the product from farmers to bring the correctness in the system. The Milk from the weighing machine shall be procured based on the specific gravity of the Milk and duly calibrated based thereon in the interest of the farmers. These process of checking with human interventions are proposed to automate with ML and DL method in possible existing legacy model for function fraud free environment.

6.1 Future Scope

Manual intervention of observation is completely automatized by adopting the AI based methods to incorporate rationality and transparency to the milk selling farmers. Also more data analysis by acquiring the real time large data set to incorporating real time environment and check for better accuracy in caliberation and to eliminate biased human interventions. This brings the quality of the food production in the proposed plan into the system.

Acknowledgement. The authors are acknowledging the milk purchasing centers and Both Telangana and Andra Pradesh state Legal Metro logy department for information and collaborative contributions.

References

1. McCrea, D.: Food safety assurance systems: labeling and information for consumers. Reference Module in Food Science (2023). https://doi.org/10.1016/B978-0-12-822521-9.00012-5
2. Tahmas-Kahyaoğlu, D., Cakmakci, S.: Changes during storage in volatile compounds of butter produced using cow, sheep or goat's milk. Small Ruminant Res. **211**, 106691 (2022). https://doi.org/10.1016/j.smallrumres.2022.106691. Department of Food Engineering, Kastamonu University, Kastamonu, Turkey, Department of Food Engineering, Atatürk University, Erzurum, Turkey, Department of Food Engineering, İnönü University, Malatya, Turkey
3. Paredes-Belmar, G., Montero, E., Leonardini, O.: A milk transportation problem with milk collection centers and vehicle routing. ISA Trans. **122**, 294–311 (2022). https://doi.org/10.1016/j.isatra.2021.04.020

4. Deosarkar, S.S., Khedkar, C.D., Kalyankar, S.D.: Encyclopedia of Food and Health. Butter: Manufacture, pp. 529–534 (2016). https://doi.org/10.1016/B978-0-12-384947-2.00094-5
5. Paredes-Belmar, G., Montero, E., Lüer-Villagra, A., Marianov, V., Araya-Sassi, C.: Innovative Applications of O.R. vehicle routing for milk collection with gradual blending: a case arising in Chile. Eur. J. Oper. Res. (2022). https://doi.org/10.1016/j.ejor.2022.03.050
6. Amamcharla, J.K., Singh, R.: Encyclopedia of Dairy Sciences, 3rd edn., pp. 695–706. Butter Oil and Ghee (2022). https://doi.org/10.1016/B978-0-12-818766-1.00381-0
7. Pastell, M., et al.: Assessing cows' welfare: weighing the cow in a milking robot. Biosyst. Eng. **93**(1), 81–87 (2006). https://doi.org/10.1016/j.biosystemseng.2005.09.009
8. Polat, O., Kalayci, C.B., Topaloğlu, D.: Modelling and solving the milk collection problem with realistic constraints. Comput. Oper. Res. **142**, 105759 (2022). https://doi.org/10.1016/j.cor.2022.105759
9. Shorten, P.R.: Original papers computer vision and weigh scale-based prediction of milk yield and udder traits for individual cows. Comput. Electron. Agric. **188**, 106364 (2021). https://doi.org/10.1016/j.compag.2021.106364
10. Lal, P.P., et al.: IoT integrated fuzzy classification analysis for detecting adulterants in cow milk. Sens. Bio-Sens. Res. **36**, 100486 (2022). https://doi.org/10.1016/j.sbsr.2022.100486
11. Arora, S., Sindhu, J.S.,Khetra, Y .: Encyclopedia of Dairy Sciences, 3rd edn., pp. 784–796. Buffalo Milk (2022). https://doi.org/10.1016/B978-0-12-818766-1.00125-2
12. Berhe, T., Seifu, E., Kurtu, M.Y.: Physicochemical properties of butter made from camel milk. Int. Dairy J. **31**(2), 51–54 (2013). https://doi.org/10.1016/j.idairyj.2013.02.008
13. Salas-Valerio, W.F., et al.: In-field screening of trans-fat levels using mid- and near-infrared spectrometers for butters and margarines commercialized in the Peruvian market. LWT **157**, 113074 (2022). https://doi.org/10.1016/j.lwt.2022.113074

Low-Cost Entry-Level Educational Drone with Associated K-12 Education Strategy

Bailey Wimer, Justin Dannemiller, Saifuddin Mahmud, and Jong-Hoon Kim(✉)

Advanced Telerobotic Research Lab, Computer Science, Kent State University, Kent, OH, USA
jkim72@kent.edu
https://www.atr.cs.kent.edu/

Abstract. Recent advancements in robotic technologies have boosted new robot applications to perform many tasks that used to be limited to humans. Given this trend towards the ubiquity of robotics in day-to-day life, demand for professionals with expertise in robot development and maintenance is highly expected to increase. It is essential that such talents be fostered in future generations at an early age in order to meet this demand. However, several challenges arise in trying to accomplish this mission. First of all, it might be difficult to hold children's attention when teaching them a new or potentially challenging subject, especially when the subject is not attractive for them. In addition, lack of mathematical understanding, which can make it challenging to comprehend key principles, is another problem with robotics teaching at young ages. In this paper, we propose a highly-accessible drone design and an accompanying education strategy, which, together, can alleviate the aforementioned problems. The major goals of the developed drone are to remain affordable as well as to enhance students' attention and motivation while maintaining a high level of functionality and safety. In the proposed educational strategy, students will be divided into four different levels; each level will have its own procedure to enable educators to integrate robotics into existing curricula. At each level, moreover, students will learn increasingly more complex robotics subjects by interacting with the drone through a user-friendly visual coding interface.

Keywords: STEM Education · Drones · Robotics Education · Computational Thinking · Visual Coding · Embedded Systems

1 Introduction

Robots are quickly becoming a vital part of modern society. Studies suggest that the use of robotics in industry is expected to see massive growth in the near future [10,14]. The future generation will be expected to develop, modify, and maintain robots which will potentially be used in all facets of everyday life. Therefore, it is crucial to start educating students early so that they have time to develop a deep understanding of robotics. Furthermore, there are many other benefits for bringing robots into the classroom.

H. Zaynidinov et al. (Eds.): IHCI 2022, LNCS 13741, pp. 323–335, 2023.
https://doi.org/10.1007/978-3-031-27199-1_32

One such benefit is that it allows educators to implement a hands-on strategy to education. Several studies have shown that hands-on education practices, when applied in STEM education, have yielded better performances from students [8,9]. Robots are one method through which this process can implemented. Benitti and Spolaôr found, in their review of over 60 publications, that robots are incredibly flexible and beneficial for STEM education [2]. Moreover, this strategy is of massive importance to robotics education, as it means that students will be more likely to engage with the educational content and develop a deeper understanding of pertinent robotics concepts.

Recently, in an attempt to bring robotics into classrooms, educators and researchers alike have begun to examine drones due to their complex nature. For example, drones must maintain proper balance to fly, and this is often achieved using a complex control structure known as a PID controller [3,13]. Many students are also very intrigued by drones, which can help educators to keep hold of the students' attention. In a study of teachers' predictions about students, Tournaki and Podell found that students who do not pay attention are more often treated negatively by teachers [15]. Unfortunately, not all educators have the ability to build drones for their classes. However, there are several commercial options that exist for this purpose.

Currently, there are many commercially-available drones which have proven to be successful for education. However, the use of these drones in classrooms is limited by a lack of fundamental features, such as safety. In a study on the subject, it was found that the DJI S1000+, the largest drone tested, was significantly more likely to cause injury than the other drones due to its size and weight [7]. Another feature that many drones are missing is a way for students without programming experience to learn from them. One drone which aims to solve these problems is the DJI Tello Drone which is both compact and lightweight, minimizing injuries. However, the Tello Drone does not allow students to explore robotics in depth. Moreover, it is moderately expensive, and likely out of the price range of the schools which could benefit from it the most.

Overall, while there is a promising outlook for the future of drones in education, the commercial technology currently available is lacking. There are many problems surrounding drones in classrooms, including safety concerns, price, and low-level control. Any drone that elegantly addresses all of these issues will be wildly beneficial to the future of robotics education for K-12 students.

2 State of the Arts

Many researchers in the field of robotics have examined ways that drones can be improved for use in the classroom. In 2018, Brand et al. proposed the PiDrone, an open-source educational drone built using a Raspberry Pi [5]. The PiDrone project saw much success, with 24 of 25 students successfully building a drone and completing all five lessons. However, the class, which was intended as an introduction to robotics, made no accommodations for younger students or students without programming experience. Additionally, the PiDrone's price point could pose a barrier for some educators. Finally, the drone design did little to handle safety beyond teaching students to exercise caution.

Often times, a crash into an object will result in a damaged drone, potentially beyond repair. To combat this problem, Breuch and Fislake examined the Airblock Drone, designed by Makeblock [6]. The Airblock drone is made up of seven distinct segments, which, upon impact, will break apart to mitigate damage to any of the core components. The study found that this feature helped motivate students due to the lack of consequences in the case of a crash. Moreover, this drone used a visual programming language known as Scratch in order to enable students as young as eight years old to program the drone. However, the Airblock drone does not enable direct control of the drone or have the ability to examine more complex robotics topics. The lessons designed with the Airblock drone target younger children, and no experimentation was done with designing a curriculum for other age groups. It is crucial that educational drones simultaneously appeal to students of varying technical backgrounds.

Researchers have found that an abstracted method of drone control works very well at motivating students in different age groups to learn about robotics [4]. Of the students in the study, 84% would participate in a similar challenge in the future, further demonstrating the success of drones in education. Despite the success, however, further improvements could be made. They have students control a simulated drone using a program known as Gazebo, which led to several issues. They suggest that the study could be more successful by using a physical drone. Additionally, 72% of students did not find it easy to familiarize themselves with the development environment (Matlab).

Another drone design was proposed by Jeong et al. in 2019 [11]. Concerned with safety, the researchers also proposed a 3D printed gyroscopic fixture that limits the drone's motion to rotation about the gyroscopes axes. In doing so, the researchers reduced the risk of injury while maintaining a large range of motion for the drone. They found much success in their design, however their drone was targeted and tested for college-level education. As such, the drone lacked accessibility to inexperienced programmers. Additionally, having a bottom-up approach to drone education, the study intended that the students would complete everything on their own, from designing the circuit board to writing the control algorithms. Once again, that would pose several issues for integrating with education below the college level.

There are a multitude of problems that arise when using drones for education in K-12 classrooms. In the next section, we propose a drone system architecture that intends to solve these problems and a corresponding education strategy describing how to best incorporate drone education with existing curricula.

3 System Overview

This system overview provides a look into the design of the overall system as well as the protocols in major components and educational strategies designed to enable the system to be used in an educational environment. First, we examine the design of the software architecture for the system. Next, we describe our custom-built communication protocol that is used by the software. Finally, the educational strategy that enables the ATR Drone to be utilized with students of any age is explained in the final subsection.

3.1 Software Architecture

The proposed ATR Educational Drone system has three major components which interact to comprise the overall system, as shown in Fig. 1. The first component, Scratch, is an open-source visual programming interface developed and maintained by MIT [12]. The second component is the ATR Linker, a middleware program written in Python. The first two components are executed on a student's computer, while the final one, the ATR Drone, is executed on the microcontroller of the drone itself. The purpose and components of each section will be explained in the coming paragraphs.

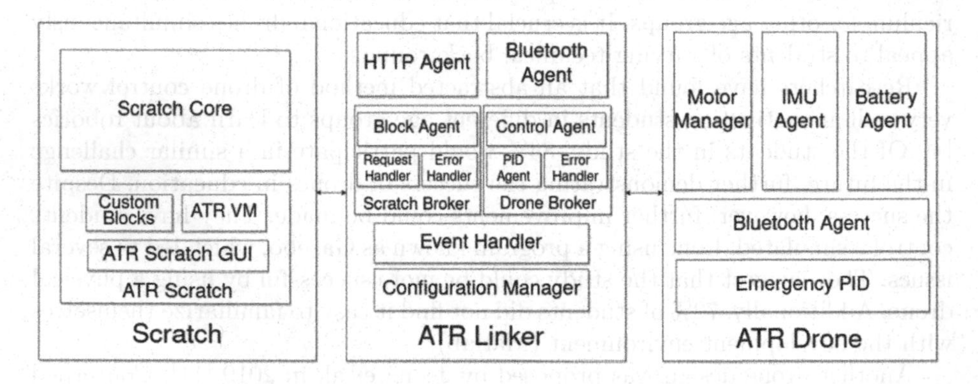

Fig. 1. Software architecture diagram

3.1.1 Scratch

The first component, Scratch, serves as a user interface through which students can code and control the ATR Drone. ATR Scratch is a custom modification added to Scratch to enable communication with the ATR Linker and Drone. A student operates Scratch through a drag-and-drop interface that allows program creation through the use of predefined "blocks" of code. These blocks are written in plain English, making it easier for a person unfamiliar with programming to utilize them. ATR Scratch utilizes Custom Blocks, a suite of customized Scratch blocks designed to interface seamlessly with the ATR Linker and Drone. The ATR Scratch GUI sub-component is responsible for modifying the existing Scratch GUI to display and execute these custom blocks. To execute a block, the ATR VM identifies and sends a corresponding HTTP request to the ATR Linker. Scratch and the customized blocks designed for this project will be discussed in further detail in the "Implementation" section of this paper.

3.1.2 ATR Linker

The ATR Linker is comprised of six separate sub-components, the most important of which being the Scratch Broker and the Drone Broker. Within the Scratch

Broker, the Block Agent maintains a knowledge of all blocks and the current state. Its purpose is to validate requests whenever a block is executed. If the request is found the be invalid, it is passed to the Error Handler. Otherwise, the Request Handler determines the proper actions.

The Drone Broker is responsible for managing the ATR Drone component. The actions which the drone must take are determined by the Control Agent. In order to maintain stability in the drone's movement, the Control Agent utilizes the PID Agent, which houses the PID Controllers for the drone. This ensures that the drone not only remains stable while hovering and moving. In the case of any errors with the drone or communication occurring, the Error Handler determines the proper steps necessary to mitigate damages.

All communication between sub-components of the ATR Linker is handled by the Event Handler. Communication between the ATR Linker and other components is handled by two sub-components: the HTTP Agent and the Bluetooth Agent. The HTTP Agent hosts the HTTP Server through which is responsible for the two-way communication between the Scratch Broker and the Scratch Component. Likewise, the Bluetooth Agent handles the Bluetooth connection, and subsequent two-way communication, between the Drone Broker and the ATR Drone component. Finally, the Configuration Manager handles all possible configuration options, such as the port of the HTTP server or the connection timeout duration for the Bluetooth connection.

3.1.3 ATR Drone

The ATR drone component of the architecture encapsulates all programs responsible for the drone's operations. As shown in Fig. 1, this component is broken into four subsystems. The first subsystem is the Motor Manager, which controls the motors based on information recieved from the ATR Linker. The second subsystem, the IMU Agent, calculates the ATR Drone's current orientation. This is achieved through the use of the drone's internal measurement unit, or IMU, which is present on the circuit board. The third subsystem is the Battery Agent, which reports important data on the state of the battery. Information from each subsystem is sent ot the ATR Linker for processing. This, and all other communication with the ATR Linker, is handled by the Bluetooth Agent. The final subsystem is the Emergency PID, which is a PID controller that takes over in the case of a loss of connection to safely land the drone.

3.2 ATR Protocol

Due to the structure of the software being split into three separate components, there exist two communication channels: one between Scratch and the ATR Linker and the other between the ATR Linker and the ATR Drone. Scratch and the ATR Linker communicate through the use of the Hypertext Transfer Protocol (HTTP), using custom-designed requests. Likewise, the ATR Linker and ATR Drone communicate across a Bluetooth connection using the ATR Bluetooth Protocol described below.

Table 1. ATR Bluetooth Protocol Frame.

Name	Header 0	Header 1	Identifier	Payload
Size	1 byte	1 byte	1 byte	n bytes
Description	$0 \times CC$	$0 \times FF$	0×01 OR 0×02	Depends on Identifier

3.2.1 ATR Bluetooth Protocol

The ATR Bluetooth Protocol is a Bluetooth messaging protocol used to transmit data between the ATR Drone and the ATR Linker while maintaining a small frame size to minimize latency. The transmitted messages consist of two parts: the header and the payload. The header indicates important characteristics of the message, and the payload contains the data being sent.

Table 2. Linker Control Payload (Identifier 0×01).

Name	M1 Power	M2 Power	M3 Power	M4 Power
Size	1 byte	1 byte	1 byte	1 byte
Description	0–100	0–100	0–100	0–100

As shown in Table 1 below, every message using this protocol begins with two header bytes which are used to indicate the start of a new message. The message then includes an identifier byte, which marks the size and structure of the payload. The identifier is either 0×01 or 0×02 if the message is coming from the ATR Linker or ATR drone, respectively. The format and contents of each payload is shown below in Tables 2 and 3.

Table 3. Drone Sensor Payload (Identifier 0×02).

Name	IMU X	IMU Y	IMU Z	Battery Percentage
Size	2 bytes	2 bytes	2 bytes	1 byte
Description	0–360	0–360	0–360	0–100

3.2.2 ATR HTTP Protocol

All communication between Scratch and the ATR Linker is handled through HTTP POST and GET requests. For POST requests, the desired data is first encoded into JSON and then sent by Scratch to the ATR Linker. For a GET request, Scratch requests data, such as that indicating the drone's current orientation, from the ATR Linker, which, in turn, returns a JSON-encoded response.

3.3 Education Strategy

The education strategy incorporates four disparate levels of abstraction to allow students, regardless of their age or knowledge of the associated topics, to better understand robotics. As shown in Table 4, the assignment of students into these four levels is performed on the basis of student age range or grade. For each level, the objectives are split into two categories: robotics objectives and general computer science objectives, as shown in Table 5. Robotics objectives are topics and ideas related specifically to the field of robotics. Computer science objectives are concepts and topics related to computer science and programming as a whole.

At the L1 level, students will have access to basic controls of the drone, such as controlling the movement of the drone over a path of their creation. The lower-level control and balance of the drone will be entirely handled by the linker. At this level, as shown in Table 5, students' robotics objectives will revolve around basic control of the drone and closed loop control [16]. Their computer science objectives will comprise entirely of learning how the visual programming interface works and what can be accomplished using it.

Table 4. Education levels with associated school grade ranges and age ranges.

Level name	School grade range	Age range
L1	Kindergarten – Grade 5	5 years old - 10 years old
L2	Grade 6 - Grade 8	11 years old - 13 years old
L3	Grade 9 - Grade 10	14 years old - 15 years old
L4	Grade 11 - Grade 12 and beyond	16 years old - 17+ years old

Beginning in L2, students will have access to lower level controls of the drone. This affords the students greater freedom in regards to the drone's actions and mechanics. Additionally, at this level, students are introduced to their first tuning [1] and system error through the K_p coefficient. Students' robotics objectives at this level are open-loop control [16] and, as mentioned before, basic tuning and system error. The computer science lessons at this level center around basic programming concepts such as variables and loops.

Table 5. Education levels with associated objectives.

Level name	Robotics objectives	Computer science objectives
L1	Basic drone control, closed loop control	Visual programming
L2	Open loop control, basic system errors	Basic programming principles
L3	PID controller basics, advanced tuning	Functions and Re-usability
L4	Build a PID controller	Basic text-based programming

The third level, L3, consists of students acquiring an even deeper level of control of the drone. At L2, a student is able to specify a number of degrees to rotate. This fine-tune control over the drone's orientation is not afforded to L3 students. Moreover, students at this level will have access to all three coefficients of the PID controller. As such, their robotics objectives revolve around how different aspects of a PID controller work and how to more efficiently tune it. The students' computer science education at this level will focus on how functions work and how to design (within the visual programming interface) code that is reusable.

At the final level, L4, students are given the lowest level of control over the drone. Students are no longer able to rely on the drone to manage its balance and throttle as it flies in a desired direction. The students must directly oversee these aspects through careful control of each motor and its power. As such, they will need to specify their own functions to be able to efficiently complete the tasks that, in previous levels, only take one block of code. Their robotics objective at this level, moreover, is to be able to build a PID controller completely from scratch. This will be done using the blocks provided to them, such as variables and multiplication. The resulting PID controller will be used to control the drone and allow it to fly while maintaining its balance.

4 Implementation

This section focuses on the implementation of each aspect of the ATR Drone. For this paper, we focus on functional implementation of each system. These systems include the printed circuit board, the software architecture, the message passing protocol, and the education strategy. The methodologies behind each of these systems' implementation is described in the following section.

Table 6. Hardware Components in Fig. 2

Identifier	Description
1	3D Printed Gyroscope
2	3D Printed Connector
3	7×20 mm micro DC motors
4	ATMEGA328P-AU Microcontroller
5	3.7V LiPo Battery Connection
6	STC4054GR Battery Charging Chip
7	USB-C Connection Port
8	Serial Converter

4.1 Hardware

All of the hardware necessary to make the ATR Drone function is built into a custom-designed printed circuit board (PCB). The PCB enables necessary functions such as Bluetooth connectivity, motor control, orientation sensing, and the ability to program using a serial USB connection. The components responsible for these functions are shown in Fig. 2 below. The first, and most important of those components is the microcontroller, an ATMEGA328P-AU. This chip acts as the brains of the drone, managing and controlling all other parts of the PCB. The PCB also contains an IMU Integrated Circuit (IC), namely the MPU-6050, responsible for determining the orientation of the drone. Additionally, the board boasts an HM-BT4502, an IC which is responsible for the drone's Bluetooth connections.

The PCB also contains several components dedicated to managing the board's power. Firstly, the board features a USB type C port, which, via the serial converter, provides power to the board and a serial connection to a computer. The battery connector allows a 3.7 v lithium-polymer battery to provide power to the drone during flight. This battery can be recharged by the board which employs the STC4054GR chip for this purpose. The remaining, lower-level components of the board are responsible for circuit protection and power management. There are also four MOSFETs on the board, which are responsible for driving the motors.

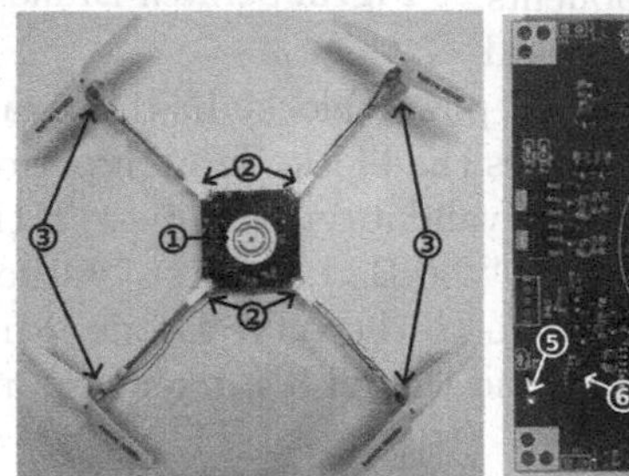

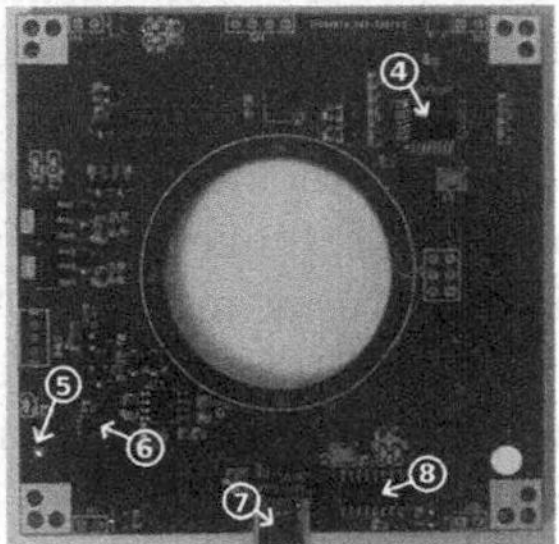

Fig. 2. ATR Drone and its control board

The motors are 7 mm by 20 mm micro DC motors which are paired with 135 mm diameter rotors. They were selected over other similar motors for their cheap price, small form factor, and high rotations per minute. Each motor is then connected to the PCB using a combination of four millimeter hollow aluminum tubes and 3D printed parts.

Another fundamental element of the ATR Drone can be seen in the shape of the PCB. The PCB has a large hole in the center, with the purpose of attaching a 3D printed gyroscopic device, similar to that used in our previous study by Jeong et al. [11]. In a fashion similar to that study, the ATR drone will be attached, through its gyroscope, to a standing aluminum fixture to restrict the drone's horizontal movement during testing. In doing so, the drone can be tested in an entirely safe manner to ensure that stable flight is achieved before risking injury that may come from a poorly-tuned drone. Moreover, this structure, along with the ATR Drone, cost a combined $42 USD at the time of this study.

4.2 Software

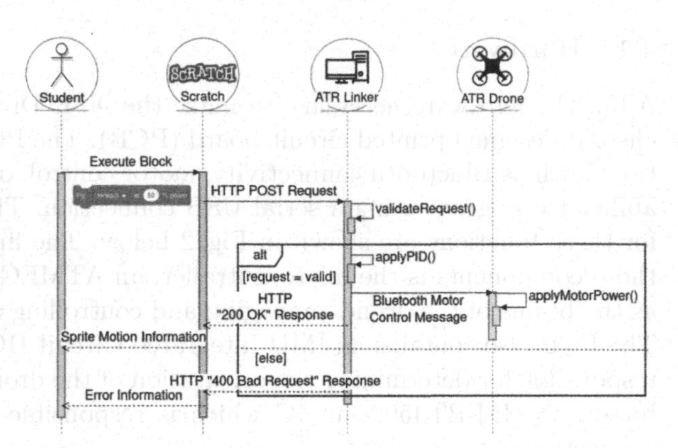

Fig. 3. Scratch-Drone Interaction Sequence diagram.

As shown before, the software is broken into three components: Scratch, the ATR Linker, and the ATR Drone. The three components are implemented using a combination of JavaScript, Python, and C++. To better understand the roles of and interactions between these three software components, suppose a student pressed the "Fly forward at 50% power block. In this example, as illustrated in Fig. 3, when the student executes the block, Scratch performs an HTTP POST request to the ATR Linker which validates the request. If the request is invalid, the ATR Linker returns an HTTP "400 Bad Request." Otherwise, it applies the PID controller and sends the data to the ATR Drone. Finally, the ATR Drone applies the new motor powers, and the ATR Linker responds to Scratch with an HTTP "200 OK" response.

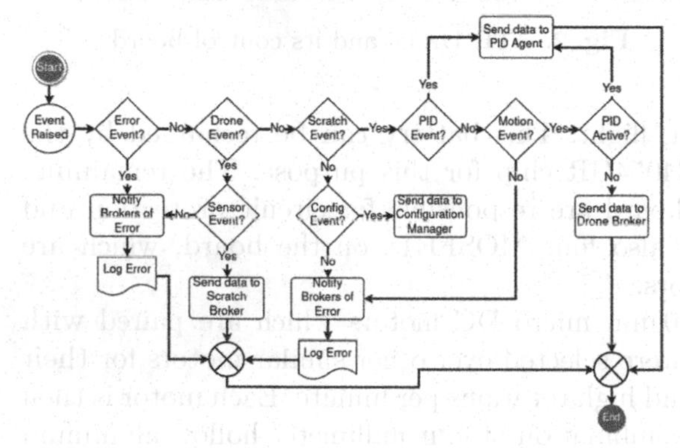

Fig. 4. Event Handling Flow of the ATR_Linker

The Event Handler, part of the ATR Linker, is responsible for handling communication between the Drone Broker and the Scratch Broker. Either broker can raise an event to signify that it needs to pass data to some other element within the ATR Linker. Whenever an event is raised, the Event Handler must decide how that event is handled as described in a flow chart in Fig. 4.

4.3 ATR Protocol

Figure 5, shown below, represents the flow of information from the time that a Scratch block is executed to the ATR Drone taking action. For this example, the "Turn right" block is used. When the block is executed, the first step occurs within the Scratch GUI, where the block is mapped to its corresponding function which then executes on the Scratch VM with proper parameters. The execution of this function converts the data into a POST request in accordance with aforementioned protocol. That POST request, which includes the direction to turn, is then sent to the HTTP Agent for further processing.

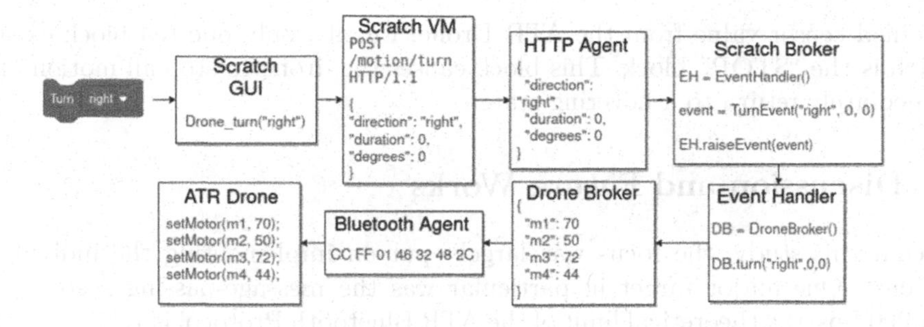

Fig. 5. Information flow diagram.

The HTTP Agent converts the POST request into usable data for the Scratch Broker. When the Scratch Broker receives that data, a turn event is raised, containing all necessary data. This turn event is then handled by the Event Handler, which extracts the relevant data from the event and forwards it to the Drone Broker. The Drone Broker converts this data from a turn request into the correct motor powers that must be sent to the drone. The Bluetooth Agent then takes that data and formats it based on the aforementioned ATR Bluetooth Protocol. Finally, when the ATR Drone receives this information, it updates its motor power to match what is required and complete the turn request.

4.4 Scratch Blocks

As shown in Fig. 6, students are provided with a different set of Scratch blocks at each of the aforementioned levels. Each set of blocks enables students to have a different level of control over the drone, which will allow them to accomplish the goals at each corresponding education level. Moreover, the blocks are also divided into four categories, based on what their use entails: "Motion," "Modify Values," "Read Values," and "Stop." The dark blue blocks belong to the "Motion" category, meaning that they are responsible for controlling the drone's motion and positioning. The "Modify Values" category, colored yellow, handles variables, such as the PID coefficients. The light blue blocks, signifying the "Read Values" category, read and return a

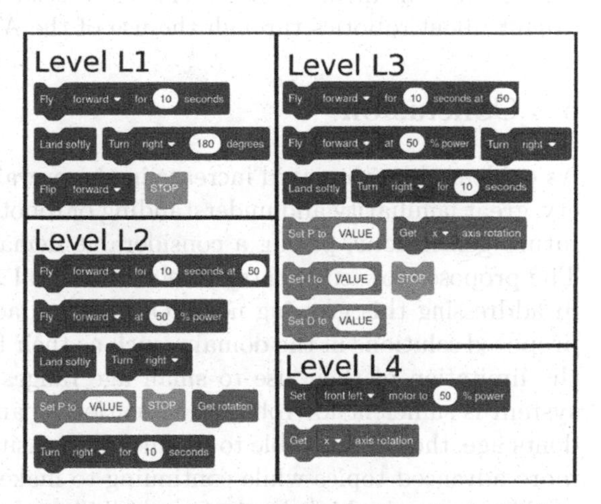

Fig. 6. Scratch "Blocks" split into levels L1-L4

specified sensor value from the ATR Drone. Finally, only one red block exists, which is the "STOP" block. This block causes the drone to stop all motion and immediately return to a hovering state.

5 Discussion and Future Works

Within this study, the focus was largely put on implementing the individual systems. One major target in particular was the message-passing system. At 115,200 bps, the theoretical limit of the ATR Bluetooth Protocol is 1,440 message frames per second. Experimentally, this proved to more than fast enough to support our desired flow of data. Additionally, we verified that the ATR HTTP Protocol easily handles over 1,000 messages per second in our architecture. We were able to successfully use these protocols to send a message from Scratch, through the ATR Linker, and to an Arduino, standing in for the drone.

This proof of concept will pave the way for further development and research relating to the ATR Drone. In the future, the entire system will be tested together to ensure that it will function properly. Moreover, once the remainder of the system is confirmed to work as intended, the ATR Linker will be expanded to work with an entire suite of ATR educational robots. This system will be used as a tool for robotics education at the K-12 level, through which students and educators may deeply explore the domain of robotics.

Additionally, future work will include putting the education strategy into practice through a cross-sectional study of subjects ages six through eighteen. This study will involve several classes of students within that age range being taught about robotics through the use of the ATR Drone.

6 Conclusion

As evidenced by the rapid increase in the prevalence of robotics in modern society, great familiarity and understanding of robotics systems will be required from future generations, posing a considerable demand for robotics education tools. The proposed low-cost, entry-level educational ATR Drone will be instrumental in addressing this growing need. Our system addresses the limitations of other proposed solutions in the domain, such as their financial accessibility, safety, and the limitation of their use to small age ranges. In particular, our ATR Drone system is sufficient for robotics education at any age. This means that, as students age, they will be able to seamlessly transition their robotics education into more advanced topics while continuing to make use of the same hardware.

Moreover, the high level of accessibility of this drone will lower the barrier to entry for the field of robotics. Especially within low-income homes or schools, this drone will be of large importance due to the low cost. This may enable students who otherwise would not have had access to the tools to partake in robotics education. Additionally, the open-source nature of this project means that students with a higher level understanding of robotics can learn the intricacies of the system to delve even further into the topic.

Finally, the message-passing and related protocols described in this paper have uses well beyond just the ATR Drone. This system allows communication between Scratch and any other HTTP server. Furthermore, the ATR Bluetooth Protocol, being a very lightweight protocol, is already being used in several of our other projects.

References

1. Bansal, H.O., Sharma, R., Shreeraman, P.: PID controller tuning techniques: a review. J. Control Eng. Technol. **2**(4) (2012)
2. Benitti, F.B.V., Spolaôr, N.: How have robots supported stem teaching? Robotics in STEM Education, pp. 103–129 (2017)
3. Bennett, S.: Development of the PID controller. IEEE Control Syst. Mag. **13**(6), 58–62 (1993)
4. Bermúdez, A., Casado, R., Fernández, G., Guijarro, M., Olivas, P.: Drone challenge: a platform for promoting programming and robotics skills in k-12 education. Int. J. Adv. Robot. Syst. **16**(1) (2019)
5. Brand, I., Roy, J., Ray, A., Oberlin, J., Oberlix, S.: Pidrone: an autonomous educational drone using raspberry pi and python. In: 2018 IEEE/RSJ International Conference on Intelligent Robots and Systems (IROS), pp. 1–7. IEEE (2018)
6. Breuch, B., Fislake, M.: First steps in teaching robotics with drones. In: Merdan, M., Lepuschitz, W., Koppensteiner, G., Balogh, R., Obdržálek, D. (eds.) RiE 2019. AISC, vol. 1023, pp. 138–144. Springer, Cham (2020). https://doi.org/10.1007/978-3-030-26945-6_13
7. Campolettano, E.T., et al.: Ranges of injury risk associated with impact from unmanned aircraft systems. Ann. Biomed. Eng. **45**(12), 2733–2741 (2017)
8. DeCoito, I., Myszkal, P.: Connecting science instruction and teachers' self-efficacy and beliefs in stem education. J. Sci. Teach. Educ. **29**(6), 485–503 (2018)
9. Ekwueme, C.O., Ekon, E.E., Ezenwa-Nebife, D.C.: The impact of hands-on-approach on student academic performance in basic science and mathematics. High. Educ. Stud. **5**(6), 47–51 (2015)
10. Eric, E., Geuna, A., Guerzoni, M., Nuccio, M.: Mapping the evolution of the robotics industry: a cross country comparison (2018)
11. Jeong, H., et al.: Design of a safe, low cost, open-source, and open-hardware educational drone (2019)
12. Nikiforos, S., Kontomaris, C., Chorianopoulos, K.: MIT scratch: a powerful tool for improving teaching of programming. In: Conference on Informatics in Education, pp. 1–5 (2013)
13. Rivera, D.E., Morari, M., Skogestad, S.: Internal model control: PID controller design. Ind. Eng. Chem. Process. Des. Dev. **25**(1), 252–265 (1986)
14. Tantawi, K.H., Sokolov, A., Tantawi, O.: Advances in industrial robotics: from industry 3.0 automation to industry 4.0 collaboration. In: Proceedings of the 4th International Conference on Technology Innovation Management and Engineering Science, pp. 1–4 (2019)
15. Tournaki, N., Podell, D.M.: The impact of student characteristics and teacher efficacy on teachers' predictions of student success. Teach. Teach. Educ. **21**(3), 299–314 (2005)
16. de Wit, C.C., Siciliano, B., Bastin, G. (eds.): Theory of Robot Control. CCE, Springer, London (1996). https://doi.org/10.1007/978-1-4471-1501-4

Creating a Modular and Decentralized Smart Mailbox System Using LoRaWan Networks

Prakash Shekhar[1(✉)], Abdolla Hegazy[1], Ajay Gupta[2], and Ammar Kamel[3]

[1] Portage Public Schools, Portage, MI 49024, USA
{prakash.shekhar,abdolla.hegazy}@portageps.org
[2] Wireless Sensornets HPC Laboratory, Western Michigan University, Kalamazoo, MI 49008, USA
ajay.gupta@wmich.edu
[3] M.S., Mustansriya University, Baghdad, Iraq
ammar@uomustansiriyah.edu.iq

Abstract. As more "Big-Tech" companies are developing IoT devices for the consumer, many users and professionals alike, have qualms about the direction that IoT devices are taking. Issues such as how IoT devices can protect user's privacy while optimizing for efficiency and modularity have arose. This paper addresses these worries by developing a Smart Mailbox system to address the shortcoming of current IoT technologies. This paper also shows that numerous functionalities can be easily added to existing devices to ease our daily lives. For mailboxes, these enhancements could be to identify whether the mailbox is open or closed, whether a small package or a letter was delivered in the mail, whether the mail inside the box is subjected to high heat, etc. Our proposed design uses alternative networks, such as LoRaWan, which provides benefits and assurances to the end-user such as decentralization and security. While we specifically focused on creating a Smart Mailbox for the average consumer, the technologies used in this paper can be very quickly utilized for other IoT products and specialized for enterprises' needs. When producing our design, we experimented with different types of distance-detecting sensors to determine how to produce the most accurate results that would be resistant to external factors such as temperature. Our results indicated that for the case of a Smart Mailbox, a combination of a temperature sensor and an ultrasonic sensor created the most consistent results. Finally, we found that the most optimal placement for the sensor would be on the mailbox's door itself.

Keywords: Internet of Things · LoRaWan · Helium · HITL

1 Introduction

The IoT concept promises to revolutionize the current service model and make appliances do more for customers. As more and more objects surrounding us are

H. Zaynidinov et al. (Eds.): IHCI 2022, LNCS 13741, pp. 336–347, 2023.
https://doi.org/10.1007/978-3-031-27199-1_33

embedded with electronics, it allows for the data these devices generate to be integrated together to enhance their collective performance and utility. Wireless sensor networks that are energy efficient and secure are crucial in advancing this idea of IoT in an environmentally and ethically sustainable way. Embedding sensors in appliances means that there are eyes and ears everywhere, including private spaces. This raises serious ethical issues about how to use this data. It is essential to limit this information to protect the owners of those devices and not share it in ways that could potentially be abused. Many existing technologies, such as Ring and Nest, store data obtained from service apps and use them for undisclosed purposes ranging from machine learning training to spying on the end-user. A better option would be to have these devices connected by a decentralized network that would allow these devices to process data locally and choose what data to share with outside its network. The LoRaWan [5] framework provides a low-cost design that addresses these issues.

This Smart Mailbox offers many benefits that traditional mailboxes and previously existing Smart Mailbox programs do not have. The most obvious benefit is that our design offers real-time mail detection to the user. Our design's connectivity to the LoRaWan network offers more security to the user's mail data, as it is a fully decentralized network. Opposed to something like Amazon's Ring mailbox detection sensor [6], which uses LTE, a technology that is relatively easier to breach than a decentralized network like LoRaWan. Another great difference is the modularity of our board, allowing for users to further expand on this with additional accessories like weight sensors.

A Smart Mailbox has a variety of applications including being a key component in delivery infrastructure. Combined with Human-in-the-Loop technology, our Smart Mailbox design could be a powerful tool that optimizes convenience and security for the end user.

This project aims to create a Smart Mailbox system, a compact solution that demonstrates the advantages of the LoRaWan wireless network over existing technologies [1] that provide similar capabilities. In addition, we explore a few different design configurations and report their performance.

2 Addressing Issues with Existing Smart Mail Solutions

2.1 Comparison to Preexisting Prototypes

While various smart mailbox prototypes have been developed in the past, they contain flaws that our system addresses. One solution, as described in ADDSMART, uses an Arduino Yun as the baseboard. With the described prototype, a few issues are evident, notably the limited modularity and security of the system. While the design offers modularity, it is limited to an SD card and a single USB connection. The RakWireless 5005-O, in comparison, features four slots that can host a spectrum of features ranging from touch ID to solar panels. As for system security, the ADDSMART system uses RFID technology to communicate. End-to-end encryption can be implemented in RFID with tools such as SAMS [8]; however, there was no mention of any security tool being used in ADDSMART.

In our smart mailbox solution, the board utilizes LoRaWan to communicate, adding difficulty for eavesdroppers [4] attempting to access private data transmitted between devices. The authors of ADDSMART also mention that RFID's range is limited on high frequencies (HF). In their system, they use a passive HF tag, which has a maximum radius of about three feet [9]; Our system operates on the Helium network, which has a 10-mile radius removing this obstacle.

Another smart mailbox system, presented in Face Recognition Based Multifunction IoT Smart Mailbox, uses an Arduino Uno baseboard alongside a Raspberry Pi 2 Model B. The proposed solution communicates to the user using LTE and SMS, which is less optimal for a smart mailbox solution as a LoRaWan network like Helium. Helium is in general more secure than LTE as it is a decentralized network. In addition, SMS requires the user to have a carrier plan with unlimited texts or pay a fine of approximately one cent per text. Helium charges $.00001 per every 24 bytes leading to greater savings for the end user. Another issue regarding LTE is the assumption that the user will be in the same country as the mailbox is system. If the user is in a different country than the mailbox system, which is the case for many corporate users, they could potentially either not receive updates from the system altogether, or face a severe fine from their carrier for each message. Our solution can be accessed from anywhere in the world with an internet connection.

2.2 Privacy and Security

With Big-Tech companies entering the IoT sphere, many issues arise with the technologies we have allowed to monitor our homes and other private spaces. Two significant concerns arise from both users and security experts when IoT devices are involved. First and foremost, the issue that occurs when talking about any type of device collecting private data on consumers: privacy. As seen from a 2018 survey on user privacy in IoT-based applications [3], 84% of users are concerned with privacy issues when it comes to activity monitoring. It bothers them that their data will be visible to others [3]. The LoRaWan technology uses symmetric cryptography to send and receive data from the devices, ensuring that it is nearly impossible for eavesdroppers to get their hands on the data being transmitted between devices.

2.3 Low-Cost Devices

Devices such as Mailbox sensors (especially in non-urban homes) are too far from the house and, therefore, cannot use the home's internet connection to communicate in the way that other Smart Home devices do. To solve this problem, current wireless smart mail systems such as Amazon's Ring Mailbox Sensor utilize cellular technologies, which send all the data from the device to an external server processing. Our Smart Mailbox system does all the calculations and processing locally on an inexpensive and easily programmable board. In addition, it only sends a signal to the LoRaWan network to indicate the mailbox's status (open, closed, empty, not-empty, etc.). This significantly reduces both the cost of the device as well as the cost of sending data while allowing for the devices to be easily reprogrammed for different purposes.

2.4 Low-Cost Infrastructure

A Smart Mailbox is likely to be more useful in rural areas where the mailboxes are not located close to homes. Usually, these areas do not have stable internet infrastructure or a high density of cellular towers (for the LTE data that Amazon's Ring Mailbox relies on). With the LoRaWan technology, users can install a relatively cheap hotspot using a LoRaWan Gateway (under $200) in their houses and have a 10-mile radius of coverage. The LoRaWan Gateway could also connect to other IoT devices, lowering the cost of expansion of other IoT devices. Now, users do not have to subscribe to a provider for their wireless devices to function or deal with the instability that comes with relying on centralized networks. Instead, our board interprets the information on the board itself. This information is then sent to Helium [2], a decentralized and open-source LoRaWan framework, where the data can then be relayed to clients. If the user or organization who uses the device so chooses, they can opt to store the data in a database such as Google's Firebase or Amazon's AWS before being sent to clients, which is a simple process. This aggregate data can be used to develop different services such as mailbox theft prevention and machine learning applications centered around user behavior with mailbox systems. Alternatively, the device can be configured to directly send data to a client app through the DDS protocol, another decentralized service.

2.5 Extensibility

When manufacturers such as Google and Amazon create a product to be put onto the market, they have no incentive to support technology that talks to devices from other manufacturers. This approach often leads to reduced modularity or the user's inability to alter the product to fit their needs. If Amazon decides to stop supporting the development of a product, it is practically defunct. This is an annoyance for users who want additional functionality without buying a completely new system due to being locked into a company's ecosystem. Our design uses a base board with multiple slots, where additional sensors could be added without purchasing a new device, based on open design and architecture. This design also benefits enterprise users with niche needs, as they can customize the design to fit their needs. Along with the board's extensibility, the data collected from the board could be used for other purposes such as a package management infrastructure or, when coupled with Human-in-the-Loop technology, a system for reporting suspicious events.

3 A Smart Mailbox Design

When designing a Smart Mailbox, we aimed to address all four of these concerns: privacy, efficiency, modularity, and decentralization. Firstly, we will briefly discuss the LoRaWan networks and why they are beneficial to our specific project. Unlike any other home appliance, Mailboxes are usually farther away from the user's source of connection. For example, in many suburban and rural properties, the mailbox is far away from the router, making it impossible for devices to get a stable connection. Current technologies such as Amazon's Ring Mailbox device utilize LTE to connect to the internet and pass data. However, as shown before, this is not an effective method, as it is centralized, inefficient, and expensive. A LoRaWan solution is perfect for the task, as it is both low power and long-range.

We specifically chose Helium, one of the most popular and decentralized LoRaWan networks, to ensure there was maximum accessibility. The decentralized nature of the block-chain that the Helium Network operates on and our decision to do all of the computations on the board itself ensures that there is no user data that goes on a public network. Data that is being transferred is in the form of a simple flag that indicates what action was just taken. This addresses privacy concerns that users may have with IoT devices placed next to sensitive information (Fig. 1).

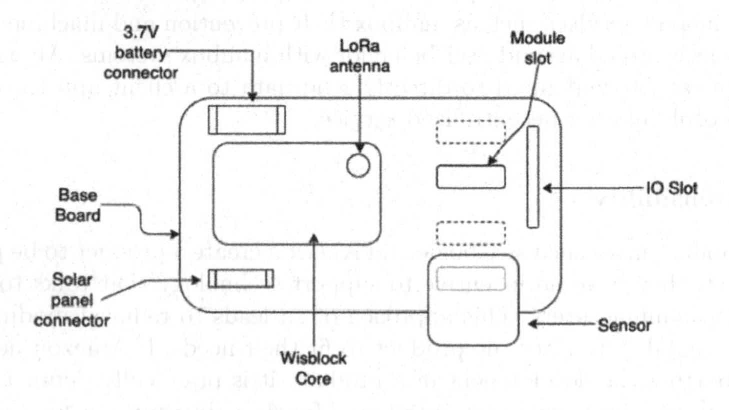

Fig. 1. Board schematic

Our prototype board design is simple and small (around 35×65 mm including the installed modules), optimal for placing in a consumer mailbox, which may be a smaller space, such as in an apartment mailbox. Firstly, the underlying board that we are utilizing comes with several functionalities. The board can be supplied with power in three ways: USB, a Li-Po battery, or a solar panel connector. We opted to use the USB connection and Li-Po battery as the solar panel would not be consistent or easily shielded from the weather. The board cannot directly connect to the LoRaWan or Helium network, so it is necessary to install a connector; we decided to use RAKwireless's Wisblock Core module to connect to the network with a 900–1200 mHz antenna. There are four module slots on the board itself, two on the front and two on the back; however, adding more modules in the IO Slot is also possible.

Another benefit of using the board that we did, is that it comes with four connectors that are directly on it. It also comes with an IO connector so that there can be additional off-board connections[1]. This allows for the user to attach different kinds of sensors to the board which can be used for collecting certain data and adding functionality to the system. This also allows for the Smart Mailbox to be expanded from just a consumer item to a fully-fledged enterprise system that can be used by corporations. For example, our design for a prototype only included a sensor to detect an event: whether the mailbox opened or closed. It would only take a few lines of code on the board and the client app, as well as a load sensor to integrate measuring the weight that is in the mailbox. This would be useful for consumers as they could identify the type of package they are receiving; the lighter the load is, the more likely it is to be paper mail, and a heavier object would be a package.

4 Findings

To determine if mail was placed in the mailbox or not, we had the option of using two sensors. Firstly, there was the ultrasonic sensor, which had two nodes, one of which would send a physical wave to the closest object with a transmitter, and the other one which would receive the wave and use a formula to determine the distance to the object. The other sensor we used was a Time-of-Flight sensor that would perform the same function as an ultrasonic one, utilizing IR light instead of a physical wave.

When measured at room temperatures, which were 20–23° C, the ultrasonic sensor did not vary far from the real measurement of the Mailbox, deviating only about a couple of millimeters. This trend followed in the IR sensor, where at room temperature, the measurements were virtually the same. The distances measured begins to separate between the two sensors as the temperature decreases as graphed in Fig. 2. As the temperature decreases to 10° C, the distance measured by the ultrasonic sensor increases to 845 mm, already far exceeding the length of the whole mailbox. Whereas, if we examine the Time-of-Flight sensor, it remains much closer to the actual distance of the sensor. The difference between these sensors only becomes more severe as the temperature continues to reduce; at 0° C the distance measured by the ultrasonic reaches 1000 mm, nearly double that of the real measurement. Again, the Time-Of-Flight sensor maintains its pattern, staying within a few millimeters of the real measurements. Our coldest measurement was at −23° C, where the ultrasonic sensor measured at around 2000 mm, almost quadruple that of the Time-Of-Flight and real measurement.

[1] Note: form factor can be significantly reduced via ASIC design. For the current project, our goal was to assess the feasibility of the system using off-the-shelf components.

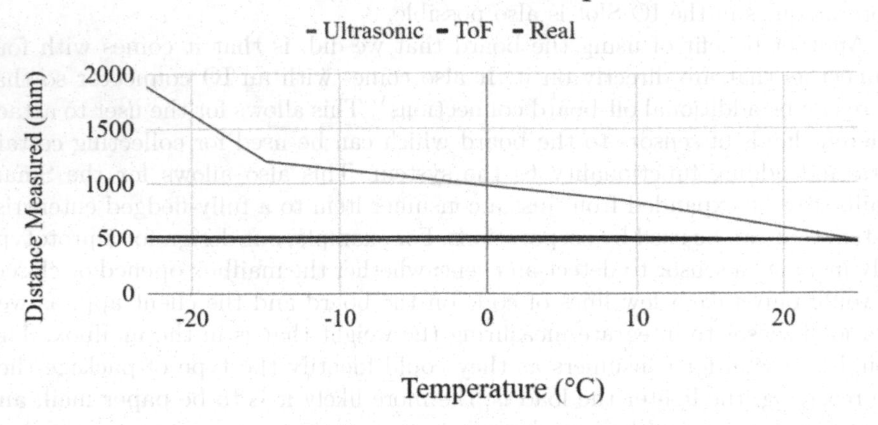

Fig. 2. Sensors in colder temperatures

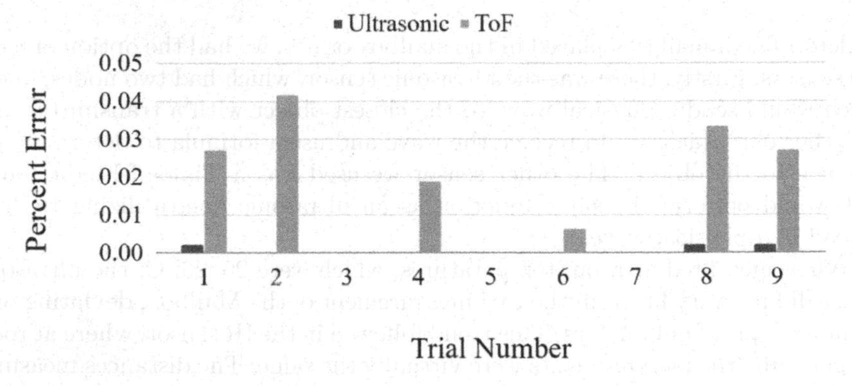

Fig. 3. Error in readings of both sensors

While the Time-of-Flight reported the distance more consistently throughout the different temperatures; when at the same temperature, it is less consistent than the ultrasonic. Both sensors were placed inside of a mailbox, which measured at 480 mm; we ran a total of 9 trials on the mailbox, measuring the accuracy of the sensors each time. The amount varied in the ultrasonic did not seem to follow a pattern, however, The Time-of-Flight sensor's % error always varied more than the ultrasonic sensor; the distances it measured had a consistently higher error than the ultrasonic sensor's readings. The ultrasonic sensor only varied from the real measurements in 3 out of the 9 trials; when it did vary during the 1st, 8th, and 9th trials, the error was around 0.002% every time. The Time-Of-Flight sensor was more volatile in its measurements as seen in Fig. 3.

The standard deviation of the ToF sensor's measurements was around 0.020, almost 20 times larger than the ultrasonic sensor's standard deviation of 0.001 (Fig. 5).

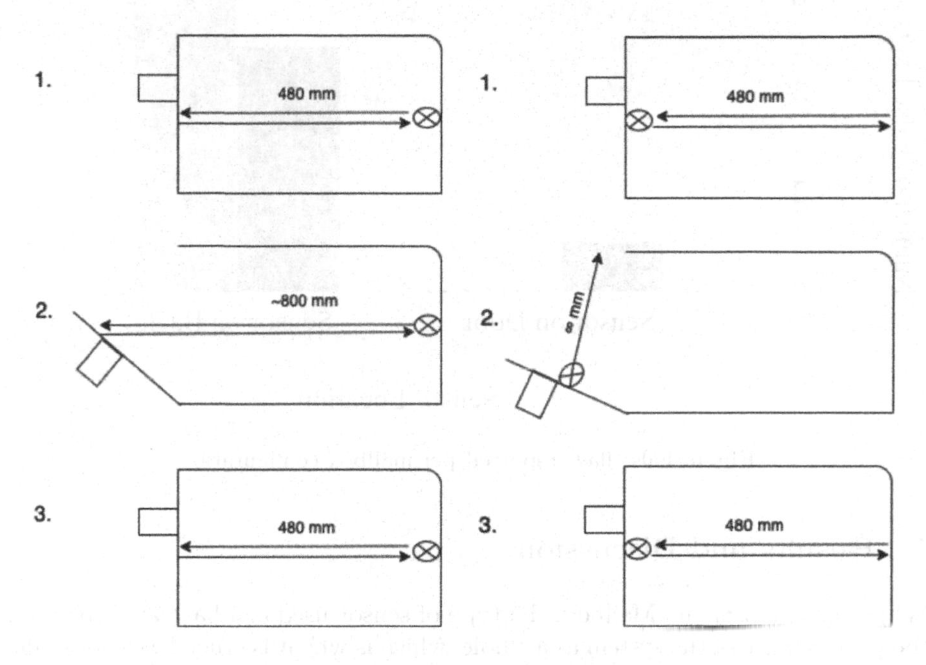

Fig. 4. Sensor placement configurations

There were four mailbox configurations tested in our experiment, for which we ran a total of 40 trials. A configuration of having the sensor placed on the door of the mailbox and facing the back (Fig. 4, right) was tested a total of 20 times, 10 times using the ultrasonic sensor, and 10 times using the Time-of-Flight sensor. Another configuration tested was having the sensor attached to the backside of the mailbox and facing the door (Fig. 4, left). This configuration was also run a total of 20 times, with 10 trials of the ultrasonic sensor, and 10 trials of the Time-of-Flight sensor. In order to ensure that our results were dependent on the configuration and not external factors, we used the same amount and type of mail within every configuration; a stack of 150 mm tall mail, which reached up to half of the mailbox's height. This stack contained an assortment of letters and magazines which one would find in a typical mailbox. In configuration one, where the ultrasonic sensor was facing the backside of the mailbox, out of 10 total trials, there was only one false flag returned by the sensor. In configuration two, where the Time-of-Flight sensor was on the door and facing the backside of the mailbox, there were no false flags shown by the sensor. Configuration 3, where the ultrasonic was attached to the back of the mailbox returned 5 false flags. During the testing of Configuration 4, which was comprised of the Time-of-Flight sensor facing the front side of the mailbox, there were 6 false flags.

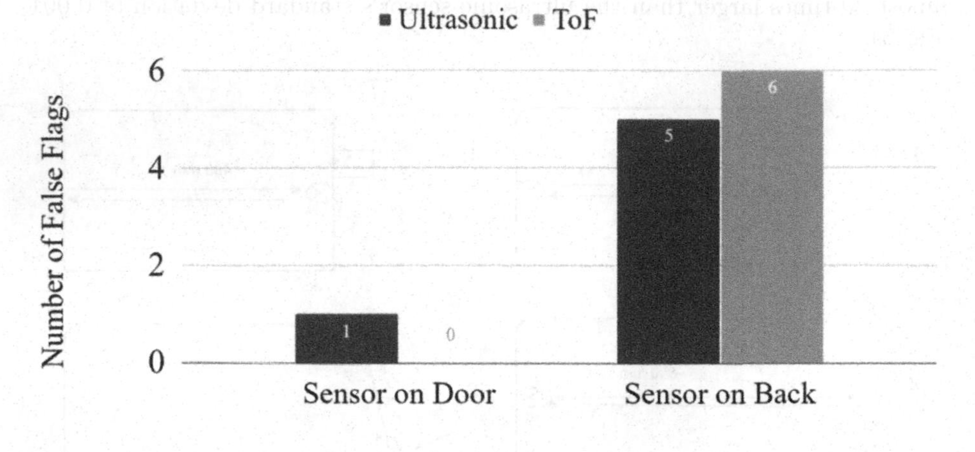

Fig. 5. False flags reported per mailbox configuration

5 Results and Discussion

When making a Smart Mailbox, the type of sensor used can have an impact on the performance of the system as a whole, which is why it is crucial we choose the most optimal sensor for our specific needs. As our experiments have shown, the two sensors both have drawbacks, with the ultrasonic sensor being sensitive to temperature and the Time-of-Flight sensor being sensitive to distance; however only one of these problems is solvable given our constraints.

$$\text{Formula for ultrasonic distance measurement} \quad D = (V_{sound} * t)/2 \tag{1}$$

$$\text{Formula for velocity of Sound} \quad V_{sound} = V_0 + 0.606T \tag{2}$$

As shown in Fig. 3, the difference between the Time-Of-Flight and the ultrasonic sensor's measurements were trivial at room temperatures, only varying by a few millimeters at most. This situation drastically changes as the temperature gets colder for the ultrasonic sensor with the distance jumping to insane measurements such as 1884 mm, almost quadruple the size of the actual mailbox. This is due to how the ultrasonic sensor measures distance to begin with. As seen by the equation in Eq. 1, the ultrasonic sensor measures the distance D in millimeters, utilizing the velocity of sound and the amount of time, t, the sensor measures in between the signal being sent and received from the nodes. The issue that arises is that the velocity of sound is not a constant; it is based on several other factors. As seen in Eq. 2, the velocity of sound uses the constant speed of sound at 0° C and the temperature of the wave that it is traveling through, T. This becomes an issue as the temperature rapidly increasing would trigger a function that monitors the distance and falsely alerts users that the mailbox has been opened.

It would be nearly impossible for us to accommodate for every single environment that the device could be placed in; however, there are solutions that exist. We have stressed how vital the system's modularity is; the benefits can be seen in this scenario, where we placed a temperature sensor next to the ultrasonic sensor on the board. This allows us to continually monitor the temperature of the environment and adjust the formula for calculating the distance in the ultrasonic sensor; thus always having an accurate formula for calculating distance. The Time-of-Flight sensor does not return consistent data values, even when the sensor is the same distance from an object. The Time-of-Flight sensor would frequently return different data values ranging from 460 mm to 480 mm, even when its distance was kept the same. The sensor's accuracy would get more inaccurate the closer an object was to it. This was proven in Depth Errors Analysis and Correction for Time-of-Flight (ToF) Cameras [7], where the researchers showed that as the object got closer to the sensor, it became more prone to error. While the paper specifically discussed ToF sensors in cameras and their depth perception, the same principles can be applied here, as the ToF technology is the same. The ultrasonic sensor returned more accurate measurement readings compared to the Time-of-Flight sensor. While both of the sensors' measurement readings were similar, the ultrasonic sensor always returned more consistent readings. With the Time-of-Flight's readings only being about 20 mm off at most, this could be detrimental to a Smart Mailbox system as the size of the system increases. For instance, if the mailbox was much bigger, the range of data would be much bigger, eventually returning a false distance so great that the sensor would think there was an action taken, when in reality nothing happened, causing a false alarm. After determining the optimal sensor, we wanted to see where the best place to put the sensor was. Originally, we tested the accuracy of each sensor by placing it on the back of the mailbox; however, we recognize that this is not the most optimal design, as if there is actual mail, the sensor in the back may not be able to see the front of the mailbox to alert the user when it opens. Another option we tested was to secure the sensor on the top part of the inside of the mailbox's door, facing the back of the mailbox (Fig. 4, right). With the sensor on the mailbox handle, the device could say with more certainty that the door had opened and that it was not a false alarm. Placing the sensor on the front of the mailbox is effective because when the door is opened, the sensor will point towards the sky and not be able to receive the signal back; which would more clearly indicate to the board that there has been a change. After performing the tests on each possible model, the most effective design becomes clear. When the sensor was placed on the back of the mailbox, there were 11 out of 20 false flags. This was most likely due to the ultrasonic's ability to get an accurate signal through the mail that was placed. It is evident that the model such that the sensor is placed on the door provided the most accuracy and least amount of false alarms for the user; it also provided the most accurate measuring when the door had been opened or closed, as the distance read would fluctuate vastly allowing for the program to accurately determine when the mailbox had opened and closed.

6 Conclusion

This project addresses complaints that consumers have about current IoT solutions by creating a modular and decentralized MailBox system. Firstly, we pointed out the problems with current Smart Mailbox Systems. Then we discussed and measured the optimal design for the Smart Mailbox. We have shown that for the specific case of Smart Mailboxes, it is better to have an Ultrasonic sensor than a Time-of-Flight, even though it is sensitive to the temperature of the environment, due to the modularity of our solution, there is an easy solution to be found. We have also demonstrated the optimal placement for the sensor within the mailbox.

The applications for a Smart Mailbox system are immense: from a simple mobile app that informs users when their packages arrive to a building block in a complex package management infrastructure. Our proposed Smart Mailbox combined with Human-in-the-Loop technology could provide an optimal combination of convenience and security in delivery systems. Human-in-the-Loop involves non-static human-inputted parameters to a system. In the case of the Smart Mailbox, the parameters are the types of mail and the frequency of Mailbox opening. For example, a Human-in-the-Loop system integrated into a Smart Mailbox could report discrepancies in user behavior which could be used to prevent mail theft.

For further improvement of a Smart Mailbox system, there should be improvements of the testing environment to correctly deem which sensor and positioning are truly the best. For instance, as the data has shown, the response time readings were affected by the temperature of the environment it was placed in. One way to overcome this would be to design a thermal-proof case for the sensor and board. That way, whatever internal temperature the sensor already is, it would not be impacted by things like the surrounding heat or cold temperatures.

Acknowledgements. We thank anonymous reviewers - the review and comments gave us insights on improving the manuscript.

This work was supported in part by the National Science Foundation under Grant OAC-2017289, National Institute of Health under Grant 1R15GM120820-01A1 and WMU FRACAA 2021-22.

References

1. Lavric, A., Petrariu, A.I.: Lorawan communication protocol: the new era of IOT. In: 2018 International Conference on Development and Application Systems (DAS), pp. 74–77 (2018)

2. Helium - Introducing the People's Network (2022). Helium.com, https://www.helium.com/. Accessed 13 July 2022
3. Psychoula, I., Singh, D., Chen, L., Chen, F., Holzinger, A., Ning, H.: Users' privacy concerns in IOT based applications. In: 2018 IEEE SmartWorld, Ubiquitous Intelligence & Computing, Advanced & Trusted Computing, Scalable Computing & Communications. Cloud & Big Data Computing, Internet of People and Smart City Innovation (SmartWorldSCALCOMUICATCCBDComIOPSCI) (2018)
4. Lorawan® is secure (but implementation matters), LoRa Alliance®, 17 November 2020. https://lora-alliance.org/resource_hub/lorawan-is-secure-but-implementation-matters/. Accessed 27 May 2022
5. What Are LoRa and LoRaWAN? The Things Network, The Things Network, 12 December 2021. https://www.thethingsnetwork.org/docs/lorawan/what-is-lorawan/. Accessed 12 July 2022
6. Ring Mailbox Sensor. Ring (2022). https://ring.com/products/ring-mailbox-sensor. Accessed 12 July 2022
7. He, Y., Liang, B., Zou, Y., He, J., Yang, J.: Depth errors analysis and correction for time-of-flight (TOF) cameras. Sensors **17**(1), 92 (2017)
8. Vasishta, K.: Security considerations for embedded system RFID readers. RFID J. (2020). elatec-rfid.com, https://www.elatec-rfid.com/fileadmin/Documents/Press-Release/ELATEC-In-format-Security-Considerations-for-RFID.pdf. Accessed 20 Sep 2022
9. Understanding Choosing RFID Tag Based on the Tag Frequency (2013). Rfid4u.com, https://rfid4u.com/rfid-frequency/. Accessed 20 Sep 2022

A Converting Model 3D Gaze Direction to 2D Gaze Position

Chaewon Lee[1], Seunghyun Kim[1], and Eui Chul Lee[2(✉)]

[1] Department of AI and Informatics, Graduate School, Sangmyung University, Seoul, Republic of Korea

[2] Department of Human-Centered Artificial Intelligence, Sangmyung University, Seoul, Republic of Korea

eclee@smu.ac.kr

Abstract. As gaze estimation research continues, the study can be classified into a 3D-based approach that estimates gaze orientation as an angle and a 2D-based approach that calculates the gaze position on a plane. That is, the 3D-based approach yields yaw and pitch, while the 2D-based approach yields the X and Y positions on the plane. Although research has been conducted for a long time, there is no way to compare gaze information calculated through each method. In order to calculate the 2D plane gaze position through yaw and pitch, the user's three-dimensional geometrical position information of face, camera, and the 2D plane should be considered in complexity. However, for the two approaches with the common purpose of gaze estimation, we believed that there would be a way to compare each other without complex three-dimensional considerations and complex calibrations. In this paper, we propose a method to accurately estimate the gaze position on a 2D plane through 2D geometric feature descriptor of face images for gaze orientation with machine learning. Experimental results on data in a laptop environment, the mean average error on X-axis was around 1.91 cm, and 0.71 cm on Y-axis, respectively.

Keywords: Gaze Estimation · Yaw and Pitch · 3D gaze orientation · 2D gaze position · Machine learning

1 Introduction

From the late 1900s to the present, research on gaze estimation has been actively conducted [1–3]. With the development of technology, more sophisticated gaze estimation is implemented through 3D coordinates, unlike the past where only the 2D coordinates system was used [4]. With the use of 3D coordinates, the concepts of pitch, roll, and yaw were also introduced. **Pitch** is rotation about the X-axis, **roll** is rotation about the Y-axis and **yaw** is rotation about the Z-axis [5]. Accordingly, the number of studies in which the result values of gaze estimation models are those are increasing [4, 6]. As the methods of deriving the results from the 2D and 3D coordinate systems were naturally divided, the performance was compared in each area. However, we believe that the field of gaze estimation will advance further if we use the similarity of the two approaches

H. Zaynidinov et al. (Eds.): IHCI 2022, LNCS 13741, pp. 348–353, 2023.
https://doi.org/10.1007/978-3-031-27199-1_34

to compare performance across area rather than within each area. We used an approach of converting results from a 3D coordinate system into 2D to compare how we derive results from a 2D coordinate system with those from a 3D coordinate system. That is because there is a lot of information required to convert from 2D to 3D, but when switching from 3D to 2D, the information is already obtained, so it is relatively simple. A simple linear regression model was used as a method of converting results from a 3D coordinate system into 2D. In the past, to change yaw and pitch to 2D coordinates, the formula method required a lot of information. Therefore, the conversion could not be done without the necessary information. In addition, when conducting research related to gaze estimation, accessible open datasets are used, and the formula method is even more difficult to be used because each dataset has fewer common features. Therefore, we propose a simple method that can be converted using only few information. In our proposed method, we estimated the X, Y position with yaw, pitch, and other given features. We tested the result with all the features and selected features which showed high correlation with the model's target.

2 Method

The dataset used to implement the proposed method is MPII Face Gaze [7]. The dataset is an open access public one which contains face images of people staring at a guided point with some features within as label values. In order to derive the best performance in terms of accuracy, we considered various gaze position inference networks. However, complex network structures increase inference time while improving accuracy. Therefore, we pursued a lightweight network structure, and it was confirmed that the simple linear regression model exhibited the most accurate and fast performance. The analysis results showing that some of the 26 feature descripts have an overwhelming correlation with the 2D gaze position supports our decision. We built our model in Python 3.8 using Scikit-learn 1.1.1.

2.1 Dataset

The MPII Face Gaze dataset [8] is used for implementation of the proposed method. The reason for using the MPII Face Gaze is that the dataset in which both 2D and 3D output values can be used. MPII Face Gaze is consisted of 213,659 pictures collected from 15 participants. The data collection is conducted using a laptop application so that volunteers look at a designated point and take photos of themselves. Therefore, the user's picture, target coordinates, and various parameters are collected. We used the 3D gaze coordinates that can infer the user's distance and the angle from the camera in addition to the corresponding four information to infer the X and Y coordinate values through yaw and pitch from the corresponding dataset.

2.2 Model

Before learning the model, we first checked the correlation between used features and the target X, Y coordinates. The features that have been used for correlation analysis are yaw, pitch, and all features provided as labels by MPII Face Gaze. Among the provided features, for 'which eye is used', one-hot encoding was used. Brief information of the used features is shown in Table 1.

Table 1. Feature description [8]

Feature index	Feature description
0~11	(X, Y) position for the six facial landmarks, which are four eye corners and two mouth corners
12~17	The estimated 3D head pose in the camera coordinate system based on 6 points-based 3D face model, rotation and translation
18~20	Face center in the camera coordinate system, which is averaged 3D location of the 6 focal landmarks face model. Not it is slightly different with the head translation due to the different centers of head and face
21~23	The 3D gaze target location in the camera coordinate system
24	Which eye (left or right) is used. (Left: 0, right: 1)
25, 26	Yaw, pitch

Figure 1 shows the correlation the features in Table 1 with the X and Y position. The number on the X-axis is mapped to the feature index in Table 1. In Fig. 1, the upper graph shows the correlation with the X point, and the lower graph shows the correlation with the Y point. As shown in Fig. 1 below, it is confirmed that the target gaze as 3D coordinates (X, Y, Z) and pitch has the greatest correlation compared to other features. Because our proposed method is to convert the model resulting in yaw and pitch into 2D coordinates, we trained the model using a total of five features including yaw even though the yaw value has a relatively low correlation as shown in Fig. 2.

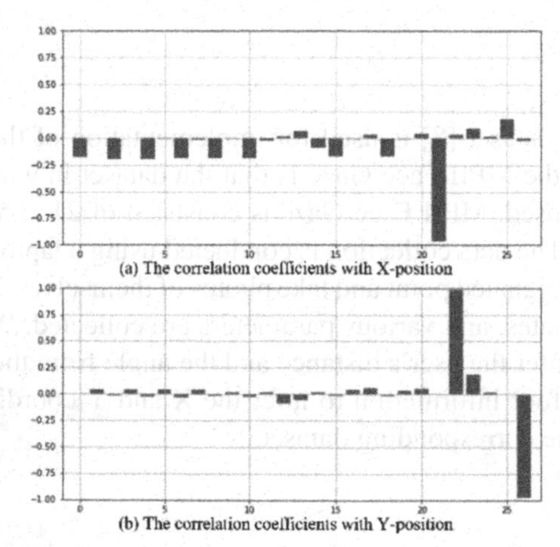

(a) The correlation coefficients with X-position
(b) The correlation coefficients with Y-position

Fig. 1. Correlation coefficient of the features with X, Y position. (a) and (b) graphs are the coefficients with X and Y position respectively

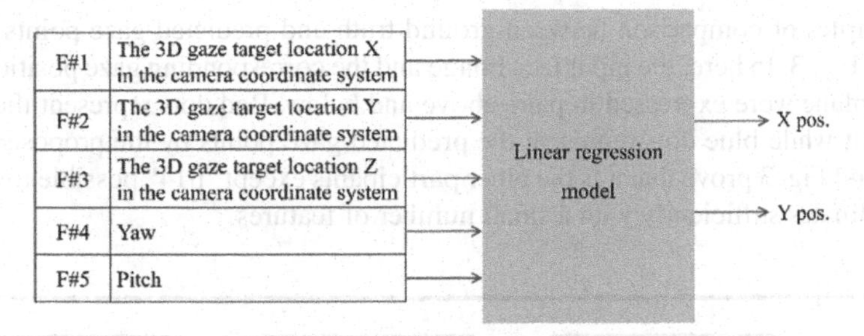

Fig. 2. The proposed 2D gaze position estimation model using five features

3 Result

Table 2 summarizes the results when all features are included and when only five selected features are used according to the correlation analysis. All participants had the same physical monitor size of 28.65 cm in width(W) and 17.9 cm in height(H), and participant number 6 was excluded because the physical monitor size was different. The result is an error in *cm*, rounded the number to three decimal places. Also, we are displayed results in two ways. First is simple error calculated as MAE (Mean Absolute Error). Second is the ratio of monitor size to error (X-position error/Width or Y-position error/Height).

Table 2. Mean error in X-axis and Y-axis per participant

Person	All feature error				Selected feature error			
	X (*cm*)	Y (*cm*)	Ratio (X/W)	Ratio (Y/H)	X (*cm*)	Y (*cm*)	Ratio (X/W)	Ratio (Y/H)
p00	4.022	1.527	0.140	0.085	3.065	1.350	0.107	0.075
p01	0.809	0.590	0.028	0.033	0.534	0.524	0.019	0.029
p02	1.398	1.075	0.049	0.060	1.289	1.297	0.045	0.072
p03	1.129	0.320	0.039	0.018	1.012	0.309	0.035	0.017
p04	0.904	0.774	0.032	0.043	0.380	0.595	0.013	0.033
p05	3.216	0.274	0.112	0.015	2.685	0.229	0.094	0.013
p07	1.049	0.421	0.037	0.024	0.619	0.300	0.022	0.017
p08	1.764	1.128	0.062	0.063	2.122	1.160	0.074	0.065
p09	1.172	0.350	0.041	0.020	1.231	0.257	0.043	0.014
p10	1.248	0.285	0.044	0.016	0.943	0.240	0.033	0.013
p11	0.781	1.210	0.027	0.068	0.691	0.882	0.024	0.049
p12	1.145	0.380	0.040	0.021	0.634	0.251	0.022	0.014
p13	2.542	0.999	0.089	0.056	2.659	1.107	0.093	0.062
p14	9.347	2.343	0.326	0.131	7.841	1.772	0.274	0.099
Average	2.180	0.834	**0.071**	**0.043**	1.836	0.734	**0.060**	**0.038**

Examples of comparison between ground truth and predicted gaze points are presented in Fig. 3. In here, the input face image and the corresponding gaze position on the monitor plane were expressed in pairs above and below. Red dots represent the ground truth point while blue dots represent the predicted gaze points by the proposed model. Table 2 and Fig. 3 prove that it is the other participants except "p14" possible to estimate 2D coordinates sufficiently with a small number of features.

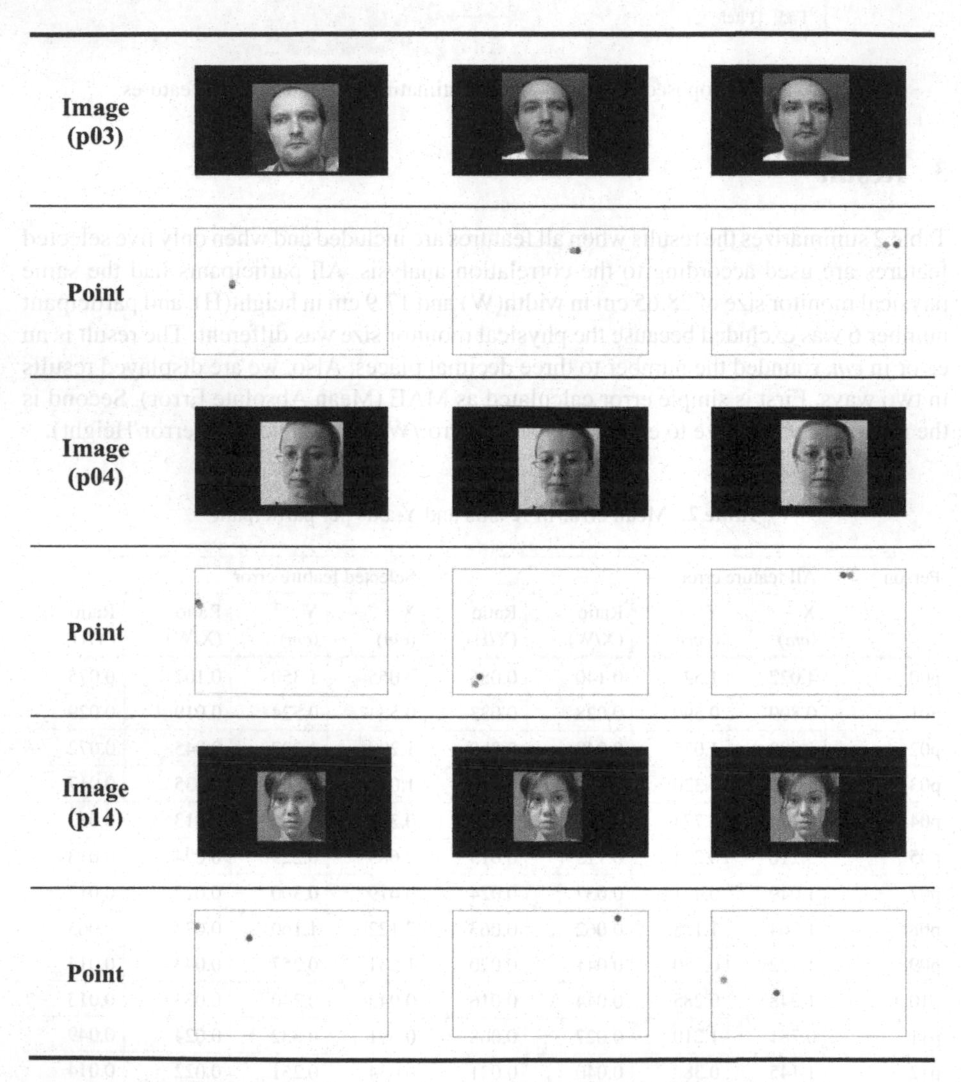

Fig. 3. Ground truth and predicted X, Y coordinates with input images.

4 Conclusion

Gaze estimation can be performed by calculating the gaze position on 2D plane or 3D orientation. Each method may have advantages and disadvantages depending on the purpose of use. However, both methods use sane sources in that gaze estimation is performed from images including eyes and faces. This means that a trained model that performs correctly for one purpose should be able to convert for the other. Therefore, we proposed a method how to project the results predicted by the 3D model on a 2D screen. As a result of the experiment, it was confirmed that errors of 1.91 cm and 0.71 cm were shown in the X and Y axis directions, respectively, on the 2D monitor plane with dimensions of 28.65 cm in width and 17.9 cm in height. Through this result, it was confirmed that fairly accurate gaze estimation is possible without wearing equipment or infrared cameras for eye imaging. In future research, we plan to develop application software for actual use considering the real-time and optimization of the continuous motion path of the eye position estimated in frame units.

Acknowledgement. This research was supported by the ICT R&D innovation voucher project (No. 2022-0-00722) of the Ministry of Science & ICT of the Republic of Korea.

References

1. Shehu, I.S., Wang, Y., Athuman, A.M., Fu, X.: Remote eye gaze tracking research: a comparative evaluation on past and recent progress. Electronics **10**(24), 3165 (2021)
2. Pathirana, P., Senarath, S., Meedeniya, D., Jayarathna, S.: Eye gaze estimation: a survey on deep learning-based approaches. Expert Syst. Appl. **199**, 116894 (2022)
3. Liu, J., Chi, J., Yang, H., Yin, X.: In the eye of the beholder: a survey of gaze tracking techniques. Pattern Recogn. 108944 (2022)
4. Cheng, Y., Wang, H., Bao, Y., Lu, F.: Appearance-based gaze estimation with deep learning: a review and benchmark. ArXiv. (2021)
5. Roll, Pitch, and Yaw Edu. https://howthingsfly.si.edu/flight-dynamics/roll-pitch-and-yaw. Accessed 25 July 2022
6. Abdelrahman, A.A., Hempel, T., Khalifa, A., Al-Hamadi, A.: L2CS-Net: fine-grained gaze estimation in unconstrained environments. arXiv preprint arXiv:2203.03339 (2022)
7. Zhang, X., Sugano, Y., Fritz, M., Bulling, A.: MPIIGaze: real-world dataset and deep appearance-based gaze estimation. IEEE Trans. Pattern Anal. Mach. Intell. **41**(01), 162–175 (2019)
8. MPII Face Gaze Dataset. https://www.mpi-inf.mpg.de/departments/computer-vision-and-machine-learning/research/gaze-based-human-computer-interaction/its-written-all-over-your-face-full-face-appearance-based-gaze-estimation. Accessed 25 July 2022

Application of Fiber Optic Sensors in Aircraft Fuel Management System

Azizbek Umarov[1], Oripjon Zaripov[1], and Ruslan Zakirov[2](✉)

[1] Tashkent State Technical University, 2 University str., Tashkent, Uzbekistan
king1995.au@gmail.com, o.zaripov@edu.uz

[2] National University of Uzbekistan Named After Mirzo Ulugbek, 2 University str., Tashkent, Uzbekistan
zrg1980@mail.ru

Abstract. Fuel management system of modern aircraft requires a large number of fuel parameters such as fuel temperature, fuel level, fuel density e.t.c.. These parameters are measured by sensors located inside the fuel tank. The resistance and capacitance type electrical transducers are recently used as aircraft fuel parameters sensors. Due to the fact that the fuel tank volume is high level explosive risk area there are increased requirements are applied to the explosion safety of the fuel parameters sensors. From this point of view fiber optic sensors are much safer than electrical sensors due to the absence of electrical contacts which can create a spark and cause ignition in the fuel tank. The hypothesis about the possibility of using of fiber-optic sensors in the aircraft fuel tanks is considered in this paper. Applicability of a system of fiber-optic sensors with different operating principles in aircraft fuel tanks for the integrated measurement of such fuel parameters as temperature, pressure at the inlet and outlet of the fuel pump, and the fuel level in the tank is analyzed. The possibility of creating of multichannel system based on fuel parameters fiber-optic sensors is considered. Various methods of channel separation for signals transmitting from fiber-optic sensors to the central controller of fuel management system is represented.

Keywords: Aircraft · Fuel system · Sensor · Fiber-optic · Bragg grating

1 Introduction

Electrical sensors are widely used in aircraft fuel management systems. Due to internal cavity of the fuel tank is high level explosion risk area, there several aviation accident took place. The most resonant of these incidents is the crash of a Boeing-747 which performed TWA 800 flight from New York to Rome aircraft on July 17, 1996. After this tragedy many aviation safety experts focused their attention on prevention of the possibility of an explosion in the fuel tank. Ways to solve this problem have been proposed by the United States Aviation Administration, as well as aviation authorities in other countries of the world [1].

The simplest solution of the ensuring of fuel tanks explosive safety at that time was the installation of an inert gas system on the aircraft. However, since 1998 it has been

H. Zaynidinov et al. (Eds.): IHCI 2022, LNCS 13741, pp. 354–360, 2023.
https://doi.org/10.1007/978-3-031-27199-1_35

found that installation of inert gas system onboard of the aircraft to fill fuel tanks with nitrogen is too expensive and impractical.

In 2001, a working group organized by the United States Federal Aviation Administration found that refilling of inert gas systems is possible only by deep modification of the airports infrastructure [2]. But final alternative solutions of the problem have not been found until present time. Due to mentioned finding of the solution for aircraft fuel tank explosive safety is important scientific task present days.

To solve the problem in this paper using of fiber-optic sensors to control fuel parameters in the aircraft tank is proposed in this paper.

The advantages of the fiber-optic sensors in comparison with the electrical sensors, where electric current is used as a signal, are as following [3]:

- Possibility of application in an explosive environment due to the absolute spark and explosion safety;
- High mechanical strength, small dimensions, simple design and high level of reliability;
- Resistance to very high temperatures, mechanical shocks, vibrations and other environmental influences;
- Possibility to perform remote measurements without electrical contacts in explosive areas;
- Existing element base allow simple multiplexing of the signals coming from the fiber-optic sensors.

A mentioned advantage of fiber optic sensors allows us to propose hypothesis about using of the sensors instead of electrical in explosive areas of aircraft fuel tank. Using of fiber optic sensors for measurement of fuel temperature, pressure and fuel level described below.

2 Application of Fiber Optic Sensors for Aircraft Fuel Parameters Monitoring

2.1 Fuel Temperature Fiber Optic Sensor

Thermocouples with explosive risk electrical contacts are currently used for fuel temperature measuring in aircraft fuel tank [4]. To solve the problem of safe measuring of fuel temperature, a fiber-optic temperature sensor with amplitude modulation and relay-type conversion characteristic proposed in this paper.

The functional diagram of such a sensor is shown on Fig. 1. The principle of operation of the proposed sensor is as following. The sensor housing is placed in the fuel tank of the aircraft where control of fuel temperature is required. Light source radiation propagates along the input optical fiber and directed to the fuel tank after directional splitter passing.

Then the radiation is reflected from the mirror and by passing of a splitter again directed into output optical fiber. A medium formed in the sensor housing between the optical fiber and the mirror filled with a polycrystalline working substance that changes the phase state at a temperature equal to the required threshold temperature of the sensor.

At a medium temperature below the threshold temperature of the sensor, the working substance is in a solid state and intensively dissipates radiation. At a medium temperature greater or equal to the threshold of the sensor, a phase transition of the first kind occurs in the working substance. As a result the light transmission capacity increases and the level of the input signal registered by the light receiver increases.

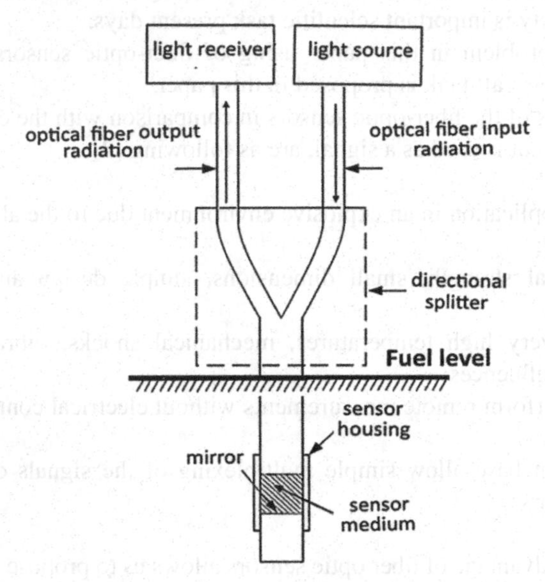

Fig. 1. A functional diagram of fiber-optic fuel temperature sensor.

Using of sensor medium substances with different melting temperatures allows us to design a sensor with the required response temperature. In the sample of the proposed fuel temperature sensor, 1, 4-dibromobenzene with a melting point of 87 °C is used as a working substance. The melting point temperature of the sensor medium substance corresponds to the fuel overheating temperature.

2.2 Fuel Pressure Fiber Optic Sensor

Presently membrane type electrical pressure sensors are used for aircraft fuel pressure measurement [5]. In this paper proposed Bragg grating type fiber optic pressure sensor.

A fiber grating with different refractive indices is used to measure fuel pressure. The grating is a section of a single-mode optical fiber. In the core of the fiber a periodic structure is induced. The structure has a certain spatial distribution, schematically shown in Fig. 2. A grating is formed in the photosensitive core of the optical fiber while the refractive index of the quartz cladding remains unchanged. Such a structure has unique spectral characteristics that make it possible to use it as a fuel pressure sensor.

Fuel pressure P is calculated from the following equation:

$$P = \frac{\lambda_2 - \lambda_1}{k_P} = \frac{\Delta\lambda}{k_P}$$

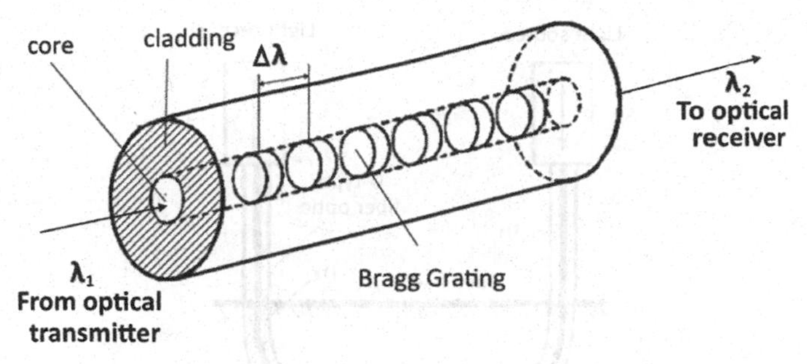

Fig. 2. Optical fiber with Bragg Grating

where λ_1 is optical transmitter signal wavelength, λ_2 is optical receiver signal wavelength, k_P is a Bragg Grating sensitivity index.

2.3 Fuel Level Fiber Optic Sensor

Fuel level control on board of aircraft is one of the most important operations to provide fuel system functionality. The main requirements for a fuel level control system are safety and high measurement accuracy. Two types of fuel level sensors are currently used to measure the fuel level. These sensors types are as follow:

1) float-inductive
2) capacitive

These types of fuel level sensors use elements of electrical circuits. In this regard, there is a possibility of a spark of an electric discharge [6]. To solve his problem further improvement of aircraft fuel level sensor is necessary. Therefore development of fuel level sensors for the aircraft fuel system without any electrical circuits is an urgent task.

Fiber-optic level sensors use an optical fiber that is immersed in fuel (Fig. 3).

The role of the optical fiber cladding is played by a tube with holes filled with fuel during immersion. The light beam created by the radiation source enters the input of the fiber optic cable. TLS001-635 (Thorlabs, Newton, NJ, USA), with a wavelength of 635 nm and a launched power of 1mW, was used as a laser beam source.

A photodiode S120 (Thorlabs, Newton, NJ, USA), with a responsivity of 0.41 A/W at 635 nm and a resolution of 1 nW, was used to detect light signals at the fiber-optic cable output.

In case of absence of fuel in the tank, all the light beam energy will be returned due to the total internal reflection of the fiber, while the intensity of the light at the output of the fiber optic cable will be the same as that of the emitted light. As the fuel level in the tank increases, the optical fiber will sink into the fuel. Due to this refractive index of the optical fiber cladding will be changed. In this case, part of the light will go into the fuel and will lead to a decrease in the intensity of the light flux at the output.

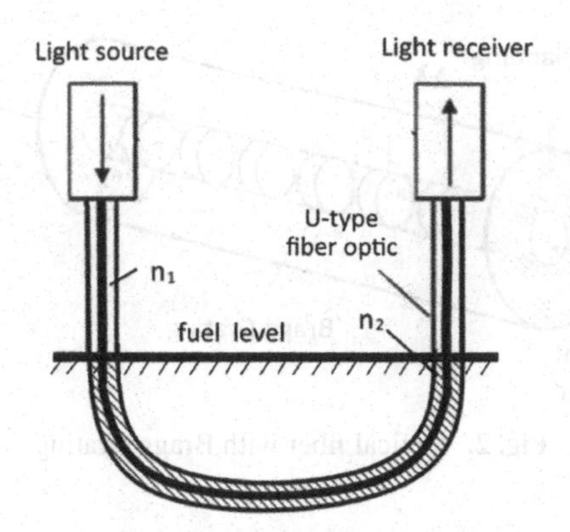

Fig. 3. Fuel level fiber-optic sensor schematic

2.4 Multi-channel Signal Processing System for Fiber Optic Sensors

The multichannel signal processing system for fiber optic sensors of the fuel system proposed in this paper uses the phenomena of transformation of spatially modulated optical signals. Information processing of fiber optic sensors includes the transformation, analysis and synthesis of multidimensional functions that describe the properties and state of the sensors.

The proposed system operation is as following. Located in the aircraft fuel tank temperature, level and pressure fiber optic sensors information send optical signals to the optical processor. The processor is an analog optoelectronic device that changes the parameters of a spatially modulated optical signal. Software for the optical processor operation is stored in the memory. Signal output unit correlates optical processor signal parameters with aircraft fuel system parameters.

Proposed in this paper system of optical information processing is an integral part of a complex optoelectronic information processing system. The system includes both electronic and optical components. The general block diagram of the proposed multi-channel (three channel is represented) information processing system for fiber-optic sensors of the aircraft fuel system is shown on Fig. 4.

The main advantages of optical information processing systems:

1) Large information capacity;
2) Multi-channel (a large number of channels processed in parallel);
3) High performance;
4) Multi-functionality

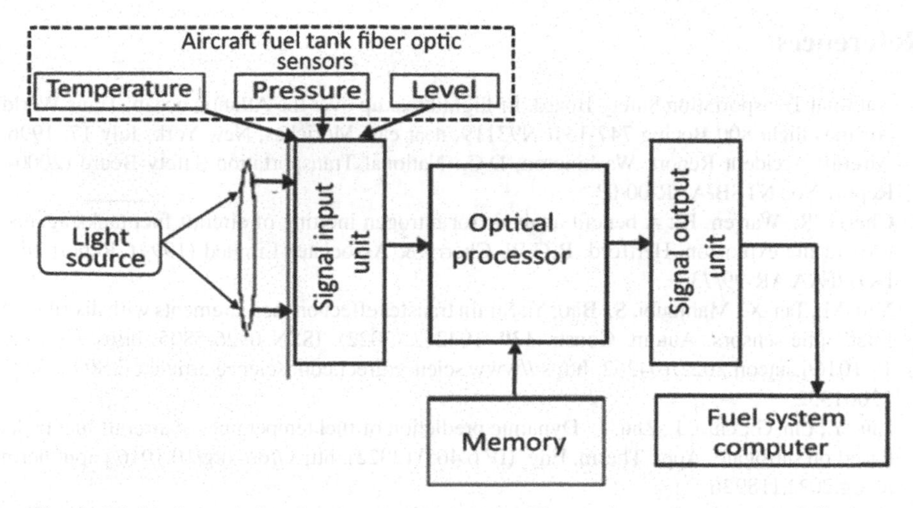

Fig. 4. Multi-channel signal processing system for fiber optic sensors functional diagram

3 Conclusion

Proposed in the paper aircraft fuel tank temperature, pressure and fuel level fiber-optic sensors proposed allows replacing of electrical sensors (thermocouples for temperature measurement, thermo resistors for pressure measurement and capacitors for fuel level conversion). Replacing electrical sensors with fiber optics will achieve the required level of fire and explosion safety in the aircraft fuel tank.

There are a lot of additional opportunities of fiber optic sensors, such as simplicity of design, the absence of mechanical parts and electrical connections, higher level of reliability. In this case the probability of the fiber-optic sensor failure is reduced to a level that does not require expensive maintenance. Fiber-optic sensors have a smaller weight and dimensions compared to electrical ones, which is an important factor in the design of on-board equipment of aircraft.

The following main factors prevent the widespread use of fiber-optic sensors for fuel parameters in aircraft tanks monitoring:

1. Currently optical fibers that are specifically designed for the manufacture of various sensors are not widely represented by the market. The existing fibers are intended for use as an optical signal propagation medium, and are not entirely suitable for their use as sensors;
2. In aviation fiber-optic communication has begun to be used relatively recently. Due to this processors for processing information from optical sensors have not yet been introduced.

Thus, research into the use of fiber optic sensors for monitoring fuel parameters should be continued.

References

1. National Transportation Safety Board. In-flight breakup over the Atlantic ocean, Trans World Airlines flight 800 Boeing 747-131, N93119, near east Moriches, New York, July 17, 1996. Aircraft Accident Report. Washington, D.C.: National Transportation Safety Board (2000). Report No.: NTSB/AAR-00/03
2. Cherry, R., Warren, K.: A benefit analysis for nitrogen inerting of aircraft fuel tanks against ground fire explosion. Hertford: R.G.W. Cherry & Associates Limited (1999). Report No.: DOT/FAA/AR-99/73.
3. Yan, M., Tan, X., Mahjoubi, S., Bao, Y.: Strain transfer effect on measurements with distributed fiber optic sensors. Autom. Constr. **139**, 104262 (2022). ISSN 0926-5805. https://doi.org/10.1016/j.autcon.2022.104262. https://www.sciencedirect.com/science/article/pii/S0926580522001352
4. Liu, Y., Lin, G., Guo, J., Zhu, J.: Dynamic prediction of fuel temperature in aircraft fuel tanks based on surrogate. Appl. Therm. Eng. (IF 6.465) (2022). https://doi.org/10.1016/j.applthermaleng.2022.118926
5. Reeves, J., Remenyte-Prescott, R., Andrews, J., Thorley, P.: A sensor selection method using a performance metric for phased missions of aircraft fuel systems. Reliab. Eng. Syst. Saf. **180**, 416–424 (2018)
6. Rayhan, S.B., Pu, X., Huilong, X.: Modeling of fuel in crashworthiness study of aircraft with auxiliary fuel tank. Int. J. Impact Eng. **161**, 104076 (2022). ISSN 0734-743X. https://doi.org/10.1016/j.ijimpeng.2021.104076. https://www.sciencedirect.com/science/article/pii/S0734743X21002633

Intelligent Multi-tariff Payment Collection System for Inter-Municipal Buses in the Department of Atlántico – Colombia

Paola-Patricia Ariza-Colpas[1(✉)], Guillermo Hernandez-Sánchez[2], Guillermo Serrano-Torné[2], Marlon Alberto Piñeres-Melo[3], Shariq Butt-Aziz[4], and Roberto-Cesar Morales-Ortega[1]

[1] Department of Computer Science and Electronics, Universidad de la Costa CUC, 080002 Barranquilla, Colombia
{pariza1,rmorales1}@cuc.edu.co

[2] Extreme Technologies, 080002 Barranquilla, Colombia
{ghernandez,gserrano}@extreme.com.co

[3] Department of Systems Engineering, Universidad del Norte, 081001 Barranquilla, Colombia
pineresm@uninorte.edu.co

[4] Department of Computer Science and IT, University of Lahore, Lahore 44000, Pakistan

Abstract. In the department of Atlántico-Colombia, inter-municipal transport companies operate that mobilize 325,000 people daily. The nature of inter-municipal transport makes it very difficult for companies and vehicle owners to have real control of the income of each bus because, unlike urban transport, the value of the ticket depends on the place of getting on and off each bus. One of the main motivations of this research is to help solve the problems associated with the management of drivers who usually hire assistants who manually and visually control each passenger's entry and exit points and, according to this criterion, calculate the amount to be paid. Charge individually. Since there is no certainty of the actual monetary income from the buses, companies and owners charge drivers a fixed daily value (fee). The objective of the platform described in this article is to manage the analysis of economic resources generated in public transport activity. This form of work affects the formality of the transport sector and generates a loss of competitiveness in the department Atlántico – Colombia.

Keywords: Multi-tariff Payment Collection System · Buses' IOT · Monitoring Buses System

1 Introduction

The Ministry of Transportation of Colombia issues decrees and resolutions to establish the cost of Intermunicipal transportation (within the department), a cost that is as-sessed according to the Origin and Destination (O-D) of the user-passenger. Despite the existence of tariff freedom, the Ministry establishes minimum rates for the provision of the Public Service of Automotive Land Transportation of Passengers by Road (SPTTAPC),

H. Zaynidinov et al. (Eds.): IHCI 2022, LNCS 13741, pp. 361–372, 2023.
https://doi.org/10.1007/978-3-031-27199-1_36

having as a starting point the Capital city and destinations the intermediate municipalities of the department, creating what is called: Routes of operation. For the specific case of this application, the routes currently authorized by the EXCOLCAR company were used: Barranquilla - Puerto Colombia, Barranquilla - Juan de Acosta, Barranquilla – Piójo, Barranquilla - Santa Catalina, Barranquilla - Tubará (By two different routes, via the sea and cotton), Barranquilla – Galerazamba and Barranquilla - La Peña. To these routes, by decree; The minimum price for the user is set, the estimated costs of which are based on studies and analysis of the sector, carried out by the Ministry of Transportation. The transporters, together with the owners of the buses, have agreed on a mechanism for charging the vehicle's product, which consists of making a single fixed daily charge to the driver, regardless of the number of users transported from an origin to a destination; Since the vehicles lack an autonomous system that allows knowing at what point a passenger approaches (origin) and at what point a passenger gets off (destination), on any of its routes, it is not possible to determine the actual output of a vehicle based on the origin user destination. Users get on the vehicle that makes this route in the municipal capitals or at any point on the road between the municipalities of Juan de Acosta and Barranquilla, paying the assistant according to the point of origin and destination. Each combination of routes has a different cost, which makes collection logistics a very complex procedure. For this reason, it is necessary to have an autonomous and intelligent mechanism that allows calculating and determining with a high degree of precision the daily production by Users-Routes or sections transported (Tariff Zones; see Fig. 1).

For this purpose, an innovation is proposed in the current collection process, introducing a hybrid system that allows users of the transport service to make payments in cash or using smart cards through an account-based system. This hybrid platform will allow the coexistence of the two means of payment, which is essential in an environment of geographical dispersion such as the Atlantic department, in which users must have the possibility of boarding buses in places far from the municipal capitals (such as farms in the middle of the road, hamlets or sidewalks) making it impossible to place smart card recharge points in said locations. It is important to specify that the information system is aligned with the implementation of algorithms based on hidden Markov chains, which allow identifying the frequency of entry of personnel into the vehicle and identifying fraud situations.

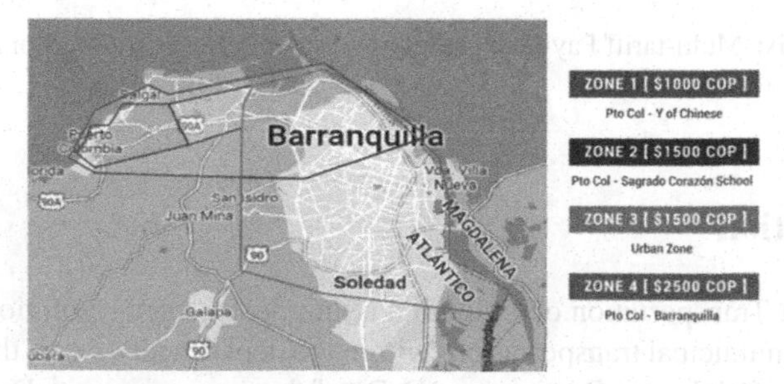

Fig. 1. Ticket prices, according to Origin-Destination on the Juan de Acosta - Barranquilla route.

2 Application of Monitoring Buses System

Through the process of a systematic review of the literature, advances can be identified by different authors regarding the implementation and use of the Internet of Things in the logistics sector.

Ref	Author	Description
[1]	Geetha, S., & Cicilia	The authors detail the implementation of a public transport system based on IOT that shows relevant information of the public transport system such as the number of passengers, both arrival and departure times, making use of cloud technology to be delivered to the user. End, using technologies such as RFID, GPS and a Wi-Fi controller.
[2]	Zambada, J., Quintero, R., Isijara, R., Galeana, R., & Santillan, L	Since school transport is one of the most widely used means in the world, the authors of this implementation have developed a system based on IOT that allows the different control entities of parents, guardians and the government in general to know the fulfillment of the speed limits and monitoring of the children's route in real time.
[3]	Zeman Masek, Krejci, Ometov, Hosek, Andreev, & Kröpfl.	The authors use different developments in order to analyze different factors associated with the process of monitoring the bus route in the city of origin of the authors, allowing to strengthen the automation of processes and services.
[4]	Krishnan, R. S., Kannan, A., Manikandan, G., KB, S. S., Sankar, V. K., & Narayanan, K. L.	Considering the effect of the pandemic on the training processes of university students, this implementation based on IOT to be able to determine the time in which students should leave to attend classes and additionally be able to determine the contagion variables of the COVID- 19 in India.
[5]	Raj, J. T., & Sankar, J.	The authors developed an information system for schoolchildren that adds to the previous works the census of passengers in addition to the variables associated with the bus and allows sending information in real time about the location of the children.

(continued)

(continued)

Ref	Author	Description
[6]	Kang, L., Poslad, S., Wang, W., Li, X., Zhang, Y., & Wang, C.	The authors have developed in China, an information system based on IOT with the purpose of guiding and showing the different atmospheric variables that can affect the proper functioning of the public transport route in China, with cloud connection.
[7]	Jisha, R. C., Jyothindranath, A., & Kumary, L. S.	The authors have developed a support system for transporting children that allows, in addition to other solutions, the calculation of the arrival time of children at their respective homes.
[8]	Boshita, T., Suzuki, H., & Matsumoto, Y.	T An information system based on IOT developed with optimization of the economic resource was developed, which allows the real-time location of the buses and thus allows passengers to be guided in the use of the transport system. LORa/Wan and 3G/LTE.
[9]	Kamal, M., Atif, M., Mujahid, H., Shanableh, T., Al-Ali, A. R., & Al Nabulsi, A.	In the United Arab Emirates, the authors have developed a station for measuring atmospheric variables at bus stops that allow decision-making in disaster areas and know the information of the buses that arrive at the different stations well in advance.
[10]	Sridevi, K., Jeevitha, A., Kavitha, K., Sathya, K., & Narmadha, K.	The authors develop a system that allows visualizing a route map showing where the bus is and being able to provide users with real-time guidance about where the buses they need to take are located.

3 Proposed Solution

The nature of inter-municipal transport makes it difficult to implement a collection system based solely on electronic means, which is why a dual system (cards and cash payment) is required because inter-municipal routes pass through roads, farms, hamlets, and small sidewalks where it would be costly (and in some cases unfeasible) to place electronic card recharge points. In general, the system must allow users to board a bus in any part of the route, even in places far from the recharging points, making use of the possibility of paying in cash.

The innovation of the proposed solution is emphasized in the union of the internet of things with artificial intelligence to carry out schedule prediction processes in which buses can be dispatched so that users can have an efficient service, which manages to connect in the cloud to be used by different users on different development platforms. The differentiation of our solution versus existing solutions is the use of hidden markov chains to perform correct path prediction processes.

Use of Contactless Card Technology (TICS): The electronic card collection systems that exist in Colombia use contactless card technology (TICS) [11], which stores the balance of the respective user on the card. This type of system has the following disadvantages: Cost: The card validation and balance management hardware that must be installed on the buses is costly both in its implementation and operation as well as in its maintenance. Security: These systems must implement distributed security methods to prevent fraud. Distributed security systems are complex and expensive. Functionality: This type of system is not very flexible in terms of functionalities such as the transfer of balances between users, payments through other channels or alter-native means, or the determination of subsidies to certain groups of the population.

In this context, collection systems based on user accounts appear, which allow a centralized administration of the corresponding balances (User/Balance) that allow lower hardware and security costs, while enabling advanced functionalities such as transfers of the balances between users or the identification of population groups for the allocation of subsidies. Although these systems are being imposed on a global level, there is no implementation of this type in Colombia, applied to the collection of inter-municipal passenger transport. This project proposes the development of a platform for electronic collection based on user accounts using Blockchain technology to guarantee the centralized security of the data. The collection platform will allow users to make use of various means of payment (cards or mobile devices) as well as a variety of recharge channels (website, B2B, SMS Messaging, mobile app, fixed points, etc.) appropriate to the socioeconomic reality of the users of the Atlantic department and given the geographic dispersion of the users in front of the municipal capitals.

To allow cash collection, the solution will also incorporate an electronic passenger counting platform in the multi-tariff environment (charges according to the origin and destination of the passenger) typical of inter-municipal transport. Electronic passenger counting systems allow counting the number of people getting on and off on each journey, but no implementation totals revenues according to the origin-destination of each passenger transported. To solve this problem, the platform will incorporate an advanced statistical calculation and data analysis system that allows estimating the total collection based on the information collected by the passenger counter and with an error rate of less than 6%. In short, the platform will integrate the current payment in cash, the modern means of payment of transport tickets, such as the use of user identification cards, payment via mobile devices, or what is now known as Open Payment, systems of open payments in public transport.

Use of Technology Based on Accounts, but Without Cash Payment: Although this may be a viable alternative in terms of cost and functionality, it has the disadvantage that it would restrict the use of the system for unbanked users and for those who live far from the municipal capitals where the placement of fixed charging points is not feasible. Applying this option, we would have a non-inclusive solution for the less favored classes of the department. This option has been deeply studied insofar as it fulfills the purpose of reducing evasion at a reasonable cost, however, it has the disadvantage that 100% of the cash is handled by the driver and one of the main problems of the department of the Atlantic is the high incidence of robberies on buses, also does not allow a centralized collection scheme. For these reasons, a hybrid platform is proposed with the following functional characteristics.

In Fig. 4, general diagram of the platform to be implemented, each of the components and procedures of the initiative to be developed is identified: Users who pay in cash, users who pay by card, ascent/descent counter equipment, card reader, wireless transmission of user identification and user up/down counts, computer at the client company's facilities, servers for CoreBank (CB) application and financial balance processes and Server with statistical Markov method implementation software (Figs. 2 and 3).

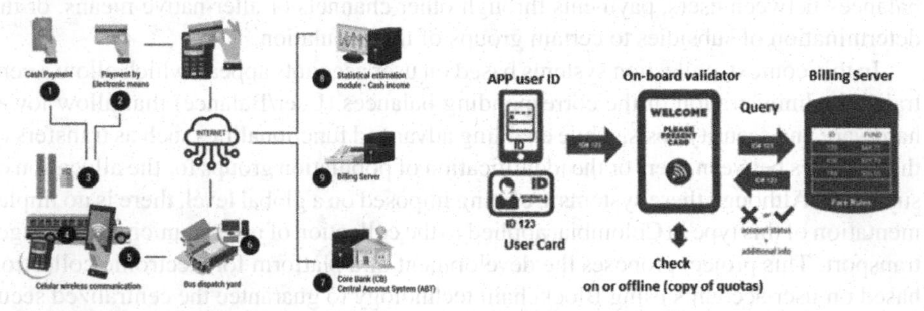

Fig. 2. Hybrid Platform Architecture. **Fig. 3.** Hybrid platform architecture.

4 Platform Operation

Each of the users will have a personalized account in the system (CoreBanking). The individual account may be recharged through the different recharge channels (M-Ticketing) available (SMS, APP, Payment button, Fixed recharge points, audio response systems, etc.). When entering the bus, the user must identify himself to the validators installed in the vehicles (readers) using any of the various electronic means that may be available (identification card, SMS message, RFID), and the value will be deducted from his account. total journey to the endpoint. If the user gets off the bus at an intermediate point along the way, he or she may present his identification means to the validator so that the system will pay the fraction of the journey not made to his account, see Fig. 5.

Example: Barranquilla - Juan de Acosta route total value of the journey COP 6,000. If the user approaches in Barranquilla (Origin) when passing his means of identification through the reader, the system deducts from his account the total value of the journey, if the user gets off in Puerto Colombia, he will be able to identify himself before the reader Again and the Billing Service will refund COP 3,500 to your account since the Barranquilla - Puerto Colombia route costs COP 2,500. A study will be carried out to determine the number of cards required to be distributed within the population of users that make use of the selected test route.

On-board validators will maintain an encrypted local copy, with Blockchain technology, to perform balance validations in disconnected or offline mode, allowing passenger boarding, in areas where there is momentarily no cellular coverage. Once it finds connectivity, it will update the transaction carried out, allowing the central balance to be updated. Account-based systems present advanced functionalities compared to systems based on contactless cards (TICS Cards) such as the possibility of transferring balances between users, online recharges, and environmental implementations, see Fig. 4.

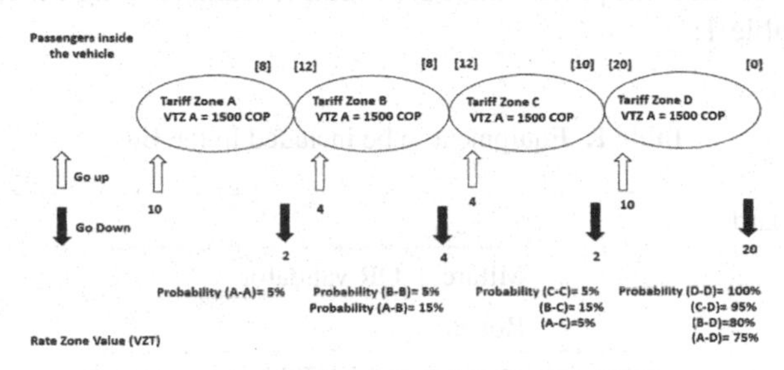

Fig. 4. Example of cash payment in 4 Tariff Zones.

Account-based systems such as the one proposed to eliminate the obligation to recharge means of payment only at fixed points, allowing online recharge through an internet connection (with PC or APP) or an audio response system, however. These online payment methods are only available to banked persons. The reality of the Atlantic department is that many of its inhabitants (especially in the south of the department) are unbanked, live far from the municipal capitals and in many cases do not have an individual Internet connection, for this reason, the system must be hybrid and allow payment of the ticket in cash.

To allow payment in cash, the buses will be equipped with electronic passenger counters that allow counting the total of the passengers getting on and off in each of the fare zones. Due to the multi-tariff nature of inter-municipal transport, the information on total passenger arrivals and departures is not enough to calculate the value collected because the collection is based on an individual count (that is, each passenger pays a different value depending on where it goes up and where it goes down). To solve this problem, the platform will implement a system based on Markovian and Bayesian algorithms and models that perform statistical approximations based on origin-destination matrices of the users who board or get off the bus. These algorithms are expected to make it possible to approximate the cash collection of the inter-municipal transport bus (based on the passenger count in the respective fare zones) with a maximum error percentage of 6%.

5 Solution Components

Within the electronic collection project, several components were developed, among which the following stand out: **UMR mobile collection unit**: includes all the devices that must be installed on buses for passenger transport and synchronized using a data plan with the central collection and fleet control system. **XTicketing**: It is the electronic collection system, in its rates, transfers, discounts, collections with the different means of payment, settlements, and collections are configured and executed. **CityApp:** Mobile App that allows you to know the services provided on the platform, available routes to move around the city, recharge, payments, and transfers.

5.1 Use of Contactless Card Technology (TICS)

Within this component of the project there are multiple teams, among which the following stand out, see Table 1:

Table 1. Equipment to be included in the Bus

Equipment on Board	
1	Mifare + QR validator
2	Router
3	Driver Console (Tablet)
4	Passenger count sensor bar
5	Panic button
6	Cameras
7	Button for cash payments
8	Passimeter

Of these teams, the Mifare + QR validator stands out, which is what allows interacting with passengers and making payments, all its logic is developed in Python that communicates through web services, sees Fig. 7.

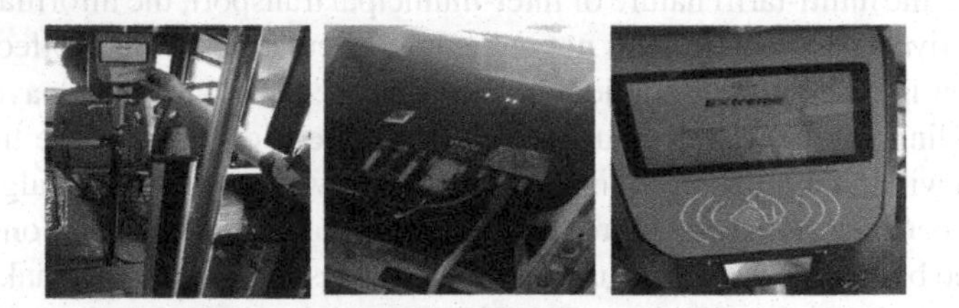

Fig. 5. Location of devices within buses.

5.2 Xticketing

For the solution, different components oriented to microservices were defined, which intend for the architecture to be scalable, fault-tolerant, and secure. The diagrams illustrate the system both at the component level and how it communicates. To create the microservices, Sprint web flux was used because it has an optimal use of memory, thus reducing the hardware costs of the ap-plication. The platform is developed in Angular which is a framework for web applications developed in TypeScript, open-source, maintained by Google, which is used to create and maintain single-page web applications. Its goal is to augment browser-based applications with Model View Controller (MVC) capability, to make development and testing easier, see Fig. 6.

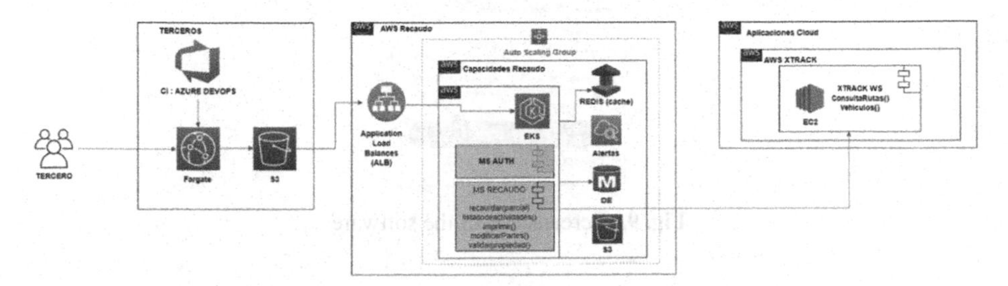

Fig. 6. Xticketing collection architecture.

The library reads the HTML that contains additional custom tag attributes, then obeys the custom attribute directives, and binds the input or output pieces of the page to a pattern represented by standard JavaScript variables. About the database, Postgresql was used, a powerful open-source object-relational database system with a solid reputation for reliability, robustness of functions, and performance.

5.3 Deployment CityApp

To deploy the application, AWS S3 will be used with CloudFront Distribution. Below is an illustration very close to the implementation of the application, this only applies to web applications, see Fig. 9. For the construction of the microservices, functions in AWS Lambda were used together with Amazon API Gateway, which interacts with the database machine in the AWS RDS service and with the other microservices of the solution, see Fig. 9.

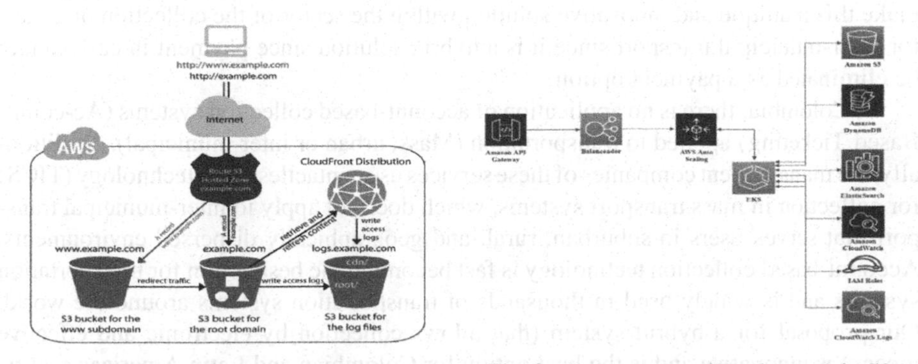

Fig. 7. Deployment in S3

Fig. 8. Components of Deployment in S3

Fig. 9. Screenshots of the software

For the development of the mobile app, Ionic was used, an open-source framework used in the development of hybrid mobile applications, which combines HTML5, CSS, and JavaScript that can be used on both Android and IO platforms. For its part, the Backend is developed in Heroku, see Fig. 8.

6 Results and Conclusions

The platform presented is an innovative platform that will allow the application of new technologies for centralized passenger collection, using population estimation tools in fare zones for users who use cash and electronic ticket collection through prepaid technology with an M-Ticketing system, which allows the possibility of using various means of recharging for the card, relying on recharge channels such as SMS messages through the balance of the cell phone, payment button on the website, transfer of balances between users, recharge using systems automatic audio response, recharging via mobile APP, recharging via fixed point network through commercial alliances in stores and electronic recharging companies, all of the above thanks to the centralization of user accounts (Ac-count-Based Ticketing). The above advantages and functionalities of the platform make this a unique and innovative solution within the sector of the collection of tickets for inter-municipal transport since it is a hybrid solution since payment in cash cannot be eliminated as a payment option.

In Colombia, there is no application of account-based collection systems (Ac-count-Based Ticketing) applied to transportation (Mass, urban or inter-municipal). Traditionally, the management companies of these services use contactless card technology (TICS) for collection in mass transport systems, which does not apply to inter-municipal transport that serves users in suburban, rural, and geographically dispersed environments. Account-based collection technology is fast becoming the best option for transportation systems and is widely used in thousands of transportation systems around the world. Our proposal for a hybrid system (that allows collection by electronic and effective means) is innovative and is the best option for Colombian and Latin American reality.

There is no hybrid collection system in Colombia applied to passenger transport. If users are restricted to using only electronic means of payment (card), it would be indirectly promoting informal transport because due to the weather conditions on the Colombian Caribbean coast, many people will prefer to take a motorcycle taxi than walk several blocks to recharge the means. electronic.

Generation of open knowledge among all members of the Alliance, since the implementation of the proposal, requires coordination between the transport company with its historical data, the Universidad de la Costa Corporation, which will provide knowledge and methodology in carrying out the project, and the companies. co-creators who will contribute their technical experience and knowledge of the sector. The proposed initiative will require, for its implementation, the hiring of new personnel, who will participate in the different stages of its creation, such as data collection personnel in the field, installation technicians and sensor supervision, etc. And once it is adopted by the Atlantic inter-municipal transport companies, it will allow the creation of new business units within the co-creators. With the proposed initiative, the department's inter-municipal transportation companies will be able to market the prepaid sale of their tickets, a system that does not exist today, and would mark a national milestone in the intermunicipal transportation ticket marketing scheme.

The implementation of statistical and mathematical methods will open a field of research for the adjustment of the models, through other computational techniques such as Neural Networks, requiring active participation of the academy and research groups thanks to the initiative presented. The initiative, with its various support technologies, will allow to carry out an intense technology transfer between companies in the public road transport sector and software companies in the department of Atlántico, as it will promote other technological proposals based on the solution proposed here for the problem to be solved, new developments and approaches can be validated and adjusted to maximize the productivity of this demanding sector. • The companies in the sector will be able to implement a unified model of centralized collection, allowing the generation of economies of scale to operate optimally, generating benefits for users and greater competitiveness for the department.

References

1. Geetha, S., Cicilia, D.: IoT enabled intelligent bus transportation system. In: 2017 2nd International Conference on Communication and Electronics Systems (ICCES), pp. 7–11. IEEE, October 2017
2. Zambada, J., Quintero, R., Isijara, R., Galeana, R., Santillan, L.: An IoT based scholar bus monitoring system. In: 2015 IEEE First International Smart Cities Conference (ISC2), pp. 1–6. IEEE, October 2015
3. Zeman, K., et al.: Wireless m-bus in industrial IoT: technology overview and prototype implementation. In: European Wireless 2017; 23th European Wireless Conference, pp. 1–6. VDE, May 2017
4. Krishnan, R.S., Kannan, A., Manikandan, G., KB, S.S., Sankar, V.K., Narayanan, K.L.: Secured college bus management system using IoT for Covid-19 pandemic situation. In: 2021 Third International Conference on Intelligent Communication Technologies and Virtual Mobile Networks (ICICV), pp. 376–382. IEEE, February 2021

5. Raj, J.T., Sankar, J.: IoT based smart school bus monitoring and notification system. In: 2017 IEEE Region 10 Humanitarian Technology Conference (R10-HTC), pp. 89–92. IEEE, December 2017
6. Kang, L., Poslad, S., Wang, W., Li, X., Zhang, Y., Wang, C.: A public transport bus as a flexible mobile smart environment sensing platform for IoT. In: 2016 12th International Conference on Intelligent Environments (IE), pp. 1–8. IEEE, September 2016
7. Jisha, R.C., Jyothindranath, A., Kumary, L.S.: IoT based school bus tracking and arrival time prediction. In: 2017 International Conference on Advances in Computing, Communications and Informatics (ICACCI), pp. 509–514. IEEE, September 2017
8. Boshita, T., Suzuki, H., Matsumoto, Y.: IoT-based bus location system using LoRaWAN. In: 2018 21st International Conference on Intelligent Transportation Systems (ITSC), pp. 933–938. IEEE, November 2018
9. Kamal, M., Atif, M., Mujahid, H., Shanableh, T., Al-Ali, A.R., Al Nabulsi, A.: IoT based smart city bus stops. Future Internet **11**(11), 227 (2019)
10. Sridevi, K., Jeevitha, A., Kavitha, K., Sathya, K., Narmadha, K.: Smart bus tracking and management system using IoT. Asian J. Appl. Sci. Technol. (AJAST) **1** (2017)
11. Ariza-Colpas, P.P., et al.: SISME, estuarine monitoring system based on IoT and machine learning for the detection of salt wedge in aquifers: case study of the Magdalena river estuary. Sensors **21**(7), 2374 (2021)

Agent-Based Modelling and Simulation of Public Transport to Identify Effects of Network Changes on Passenger Flows

Sophie Ensing and Chintan Amrit(✉)

Amsterdam Business School, University of Amsterdam,
P.O. Box 15953, 1001 Amsterdam, The Netherlands
c.amrit@uva.nl

Abstract. This research looks at the application of an agent-based model to assess the effects of changes in the public transport on passenger flows. Data from public transport company GVB was provided by the municipality of Amsterdam. We used domain knowledge, literature and data to create a simulation of a baseline scenario and a scenario with a malfunctioning tram line using Agent-Based Modelling. Our simulations show that the malfunctions result in different occupancy levels of several lines and cause an increase in waiting time for a lot of people. Future research should be aimed at implementing the model for the complete transport network in a city like Amsterdam and expanding the behaviour of the different agent types.

Keywords: public transport · route choice behaviour · agent-based simulation

1 Introduction

In a big city, malfunctions, delays or construction can often lead to problems in public transport. In these situations, passengers might have to take different routes or wait longer, which can result in greater congestion than usual. Assessing the effects of possible malfunctions or changes in public transport would enable the municipality to take preemptive action. The effects of malfunctions need to be compared to a similar scenario without malfunctions to draw any conclusions on the kind of action required. This calls for a model which is representative of real public transport flows in a particular city. Such a model could serve as a baseline for regular travel behaviour. When a scenario with malfunctioning transport lines is simulated, this can be compared to such a baseline model to better understand the effect the malfunction has on passenger flows. However, existing research does not guide how such a baseline model can be constructed to understand the functioning of public transport in a city [4].

In this research, we describe how such a baseline simulation model can be created for a given city. We demonstrate our method by creating a simulation

H. Zaynidinov et al. (Eds.): IHCI 2022, LNCS 13741, pp. 373–385, 2023.
https://doi.org/10.1007/978-3-031-27199-1_37

of public transport flows in the city of Amsterdam. For this, we obtained a data set from the public transport company of Amsterdam (called GVB). This data set contained travel information on the bus, tram and metro lines in the city of Amsterdam. The baseline simulation was created using Agent-Based Modelling (ABM), through which complex behaviour could be accounted for and high explainability could be ensured. In the context of public transport, ABM makes it possible to model and simulate route choice behaviour. This behaviour produces the flows through the public transport network in a city. If route choice behaviour can be simulated correctly, it will also be possible to analyse the effects of any malfunctions or delays in traffic. The simulation can also have benefits for decision-making processes in public transport companies because an understanding of route choice behaviour can help in designing new platforms or transport lines. The challenge of creating a model for such research is the sensitive nature of the data. Due to privacy concerns and GDPR[1], we did not have access to data on travel preferences or route choice behaviour specifically for Amsterdam. Therefore, we intend to understand the extent to which such a baseline model can be created with limited data. We use a combination of domain knowledge at the transportation company (GVB), sample transportation data and related work to create the sample population and behaviour for our baseline simulation model. We then evaluate the model with data from the transportation company. Furthermore, we use our baseline model to simulate the effects of malfunctions. We compare the baseline scenario and a scenario with malfunctions to assess the effect of changes in public transport on passenger flows.

The rest of the paper is structured as follows, Sect. 2 deals with the background literature on Route choice behaviour and Agent-based modelling (ABM), Sect. 3 gives an overview of our case domain, Sect. 4 describes our research methodology, Sect. 5 describes our results, and finally, Sect. 6 discusses the results and concludes the paper.

2 Related Work

2.1 Route Choice Behaviour

Although there exists research on route choice behaviour and traveller preferences [4], because of big differences in transport networks in different cities, it is not possible to take these findings and apply them directly to any given city. This is because comfort, prices and speed of certain types of transport vary across different cities and countries. A survey on satisfaction with public transport in different cities in Europe showed that there are some overlapping factors for all cities, but local conditions also need to be taken into consideration [3]. Some cities show exceptions that need to be researched further to find the underlying reason. Research has shown, however, that some general features influence route choice behaviour.

[1] https://gdpr-info.eu/.

Several important factors determine route choice behaviour, namely, on-board travel time, cost, frequency, waiting time and the number of transfers [[2,5,7]]. However, the parameter cost is not interesting for this research, because the difference in cost for the modes of transport in Amsterdam is very minimal. The buses, trams and metros in Amsterdam all have the same boarding rate and the rest of the costs are very close as well. If trains and regional transport are included as well, it becomes a bigger factor. Earlier research often uses the shortest path approach, because it is assumed this is also the fastest and cheapest route [6]. When there are common lines between two stations, other factors also become important. Common lines imply that there are two lines between a certain origin-destination pair. When this is the case, waiting time becomes more important [6]. The problem here is not necessarily having to wait, but the uncertainty regarding the arrival of the next transport option [2]. Modes of transport with a high frequency and reliability have a higher guarantee of a fast transfer, which is why a lot of people have a preference for this type of transport [1]. The importance of different variables largely depends on the type of passenger [7]. Regular commuters, like work commuters or students, have a slight preference for the fastest route. Elderly people usually prefer trips without any transfers and dislike waiting time relatively more than on-board travel time compared to other groups.

2.2 Agent-Based Modelling

In agent-based modelling (ABM), the model represents a system with many individual agents who act in a specific environment. The agents can have properties from position and speed to age and wealth. ABM is a good method to discover why people, in public transport, for example, make certain decisions and what happens in the system based on these decisions [10]. The agents can interact with the environment they are in and it is also possible to model the interaction between agents.
In the context of route choice behaviour, agents are a good way to represent travellers and their behaviour. An agent as a traveller in the environment of public transport can choose several modes of transport based on their current situation. For example, someone travelling to work might prefer the fastest route. This can be formulated then as a rule for this particular agent. The environment here is the whole public transport network, in which other agents act as well. Agents can see the behaviour of the agents around them. So, if a certain mode of transport is very busy this might influence an agent's behaviour as well. The goal for every agent is to get from points A to B concerning their rules. To decide on how to get to B from point A, several options might be compared to their rules. The agent is autonomous in this network, meaning that the agent makes its own decisions. Finally, the agent can adapt and learn. If a certain mode of transport is always delayed, the agent might try another mode of transport and see if this is an improvement.

When analysing travel behaviour, a distinction is made between static (pre-planning) and dynamic (within a trip planning) route choice behaviour [10].

Concerning route choice behaviour, most people travelling in public transport will have planned their trip before entering the public transport network. Malfunctions and delays will either be known before arriving at the origin of their trip or after arriving. This means the route choice has to be altered. Even though the route is changed, it is still considered static as the decision is being made before the start of travelling. Dynamic planning might occur when an accident occurs during a trip or if there are delays, but this is generally less common.

Since GVB has no clear behaviour model, the most useful information is the type of travel membership a passenger has. There is some domain knowledge about the different types of membership and the behaviour that is expected from such passengers. Travellers with a business membership probably travel during rush hours and prefer the shortest route. For this group of people, the trip duration is very important. For other groups, like elderly people, waiting time and the number of transfers is prioritised. Duration, waiting time and the number of transfers will be used as variables in our model. These three variables can be used to create a variety of route choice decisions and it is also possible to represent different types of people. All variables are supported by research and GVB to have an impact on route choice decisions and they can be calculated without additional data. Additional data would be needed for example to include variables like personal preference, reliability and comfort. Choosing a small set of parameters also ensures high explainability and provides a good starting point for the model. The behaviour can easily be increased in complexity or altered when more information is available.

3 Case Study at the Amsterdam Travel Company (GVB)

Every year, GVB creates around two thousand new schedules. This ranges from a completely new schedule every December to small schedule changes for a short period. There are situations like construction work, in which GVB is informed about the issue and can prepare an alternative schedule in time. Some situations like delays, accidents and malfunctions in the infrastructure are not planned. In these cases, passengers can be informed directly by civil servants on site or via digital boards at the stops. If the inconvenience is severe, alternative transport options are provided to the passengers. In most cases, however, it is just a matter of communicating the situation as soon as possible. Often, no further action is taken because doing so would take longer than waiting for the next bus/tram/metro to arrive. This does, however, result in additional congestion at certain stops or in lines that provide an alternative. Another reason to change the schedule might be patterns in the occupancy of certain lines. The occupancy of every line is measured every thirty minutes. If this occupancy is above a certain threshold that is based on the actual capacity of the vehicles, the data will be analysed in more detail by GVB. Sometimes the high occupancy can be a sign that people are missing a certain connection.

GVB has identified different passenger groups, but these groups are used mostly for marketing purposes. In the context of route choice behaviour, there are no predefined groups. To accommodate the needs of different types of people, the most important source of data are the memberships and products that are linked to a person's transport chip card. Even though the memberships can not be linked to the routes a person takes, it is possible to make some assumptions about the type of traveller. For example, people with a business member who travels mostly during rush hours are probably going to work. A few people at GVB with a lot of experience and domain knowledge use this information to get a picture of the different passenger groups. There are certain places, like hospitals and retirement homes, where the stops and lines will not be altered (or as little as possible) due to the passenger groups. People at hospitals and retirement homes that take public transport need to be able to get to a near stop. These groups also prefer little waiting time and no transfers. A difficult passenger group to track are tourists. A lot of tourists take routes that might not be the obvious choice. This might be because tourists generally prefer to take trams as they can look outside and go straight through the city. Tourists are also hard to track in the data because most tourists use single-use tickets in public transport.

4 Methodology

To speed up the simulation process and ensure high explainability, only a subset of the public transport network was used. The lines for the sub-network needed some overlapping stations and good coverage of the city. This makes it possible to simulate routes in different parts of the city and also simulate different types of routes for every origin-destination pair. The lines that were selected for the simulation were tram lines 12 and 24 and metro lines 50, 52 and 53. This resulted in a network with 67 different stations. To further reduce the computation, all simulations were made for a weekday (aggregation of Monday through Thursday) and a weekend day (Saturday) for only a few hours of the day. The hours that were simulated were 08:00–09:00, 11:00–12:00, 14:00–15:00, 17:00–18:00 and 20:00–21:00. These hours have diverse distributions over the day and include both morning and evening rush hours. By doing this, the effects of malfunctions can still be analysed in different situations in an acceptable time frame.

4.1 Data Exploration and Preparation

The data of GVB contains information about complete trips and sub-trips that people have taken. A sub-trip is a part of the complete trip in one mode of transport without any transfers. A trip can consist of several sub-trips. An example of a trip can be someone travelling from metro station *Noord* to tram stop *Muntplein*. To complete this trip, a transfer is needed between the metro and tram. This trip is therefore made up of a sub-trip from *Noord* to *Centraal Station* in metro line 52 and a sub-trip from *Centraal Station* to *Muntplein* in tram 24.

In the GVB data, only the origin and destination of the sub-trips are matched. There are strict privacy regulations that make it impossible to match the complete trip. The risk of matching all data is that in some situations it might be able to deduce to whom this travel data belongs.

As mentioned, the complete routes were not available in the data. To generate routes, the passenger distribution per hour and the probabilities of all origins and destinations per hour were analysed. Figure 1 shows the passenger distribution for all days of the week. This shows that the weekdays are quite similar, but the weekends differ quite a bit. All weekdays show clear peaks during the rush hours 08:00–09:00 and 17:00–18:00. Wednesday and Friday seem to have slightly fewer passengers, which could suggest these are popular part-time working days. Saturday and Sunday follow a completely different pattern from weekdays. There is no morning or evening rush hour and the busiest hours are midday, Saturdays being busier than Sundays. The average number of passengers per hour was used as a base for the simulation.

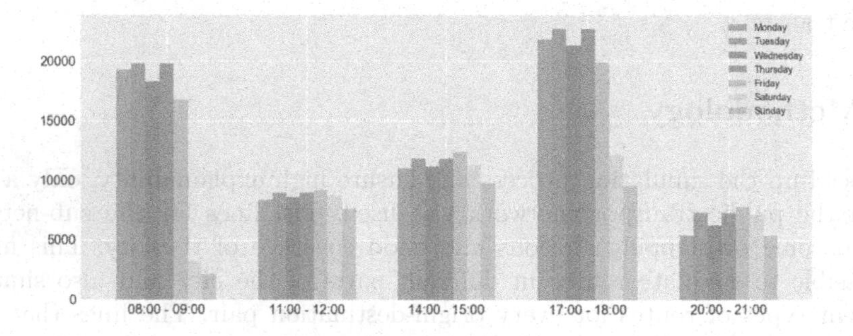

Fig. 1. Passenger distribution

For all people in the simulation, a route has to be calculated. Even though complete route information is unavailable, the GVB data was used as a base. For every hour, the probability was calculated of departing from and arriving at a particular station. These probability distributions were also compared for all days of the week. Since the probability distributions of Monday through Thursday were relatively similar, these days were aggregated. By aggregating the noise, like past malfunctions, the delays will have less impact on these distributions.

4.2 Simulation Setup

For our simulation the *Transport Network Analysis*[2] package from TU Delft was used as the base code for this research. The model provided was written in Python, using SimPy[3], NetworkX[4] and the custom package of TU Delft to simulate agent-based behaviour.

[2] https://github.com/TUDelft-CITG/Transport-Network-Analysis.
[3] https://simpy.readthedocs.io/en/latest/.
[4] https://networkx.org/.

The sub-network of the transport lines in the city was created in NetworkX. Each node represents a station and each edge is a connection between two stations. Each edge has a duration weight in minutes and a line attribute which is the name and direction of the transport line. Figure 2 shows the network that was created and the real transport network on a map. The network in NetworkX does not completely follow the same shape, because all edges are straight lines, but the locations of the stops are matched correctly.

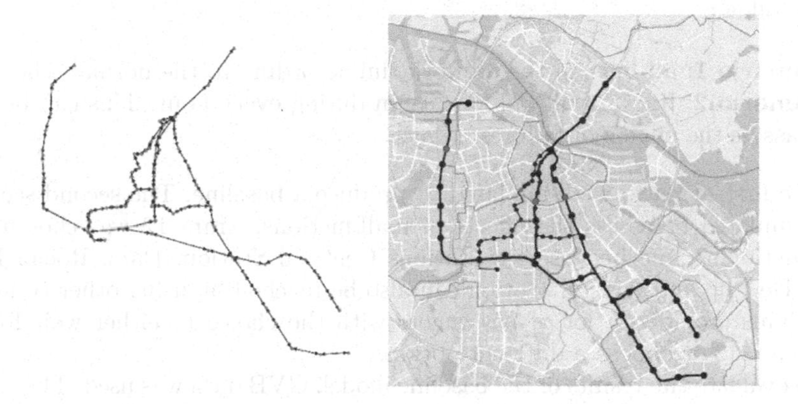

Fig. 2. Transport lines in networkx (left) and on maps.amsterdam.nl (right)

The agents in the simulation are the vehicles and the passengers. The vehicles have a name, route and start node. The *Transport Network Analysis* package provides classes to make the vehicles move over the network and pick up passengers. At every node, the vehicle will check for passengers. If there are passengers that want to take this mode of transport, they will be loaded into the vehicle. At every node, it is also checked if any passengers want to get off the vehicle and they are removed from the vehicle. After the (un-)loading process, the vehicle drives over the network edge to the next node in its route. The duration in minutes between two stops is stored as an attribute for that edge. The generation of the vehicles is based on the scheduling of every line. The frequency of metro 52 is six minutes, so every six minutes a new vehicle with the right attributes for this line is generated at both ends of the line.

The passengers were generated based on the average number of passengers for that hour. An arrival rate is calculated based on this number, to evenly distribute all passengers over the whole hour. A probability distribution is used to generate an origin and destination for each passenger. Between these two stations, all possible routes are generated with NetworkX. To reduce the number of unrealistic routes as a result of using a sub-network, only routes with less than three transfers were considered. There are three different classes: (i) the preferred trip is the one with the shortest duration, (ii) the preferred trip is the one with the least amount of transfers, and (iii) the preferred trip is the one with the least amount of waiting time.

Therefore, one of the parameters of the simulation was the probability distribution for the three different classes. Each simulation can be executed with a different distribution of people, and it can be altered for every hour if needed. All simulations were run for a weekday (aggregation of Monday–Thursday) and a weekend (Saturday). As mentioned before, five hours are selected for every day to reduce computation. All simulations were run with an even probability distribution for all classes (all 0.33) because the real population is unknown. To make sure the results converge, each simulation was run ten times. The scenarios are as follows:

- **Scenario 1**: all lines in the network run according to the normal schedule.
- **Scenario 2**: line 12 will fail for 30 min during every hour. This can be used to assess the effects of failures/delays.

The first scenario was simulated to produce a baseline. The second scenario was simulated to assess the effects of malfunctions. Tram 12 was chosen as a malfunctioning line because the stations Centraal Station, Dam, Roelof Hartplein, De Pijp and Amstel Station can also be reached by using other transport lines. This provided a lot of passengers with the choice to either wait for the next tram 12 or choose a different option.

To evaluate the results of the baseline model, GVB data was used. The results that were possible to evaluate with the data were the origin, destination and subtrips from all passengers. To compare simulation data with measurement data, we use the *goodness of fit* [8] through the mean average error (MAE) metric, as it is easier to interpret and more understandable than the RMSE metric [9].

5 Results

This section will discuss the simulation results about origin stations, destination stations, sub-trips, trip data and vehicle occupancy.

Table 1 shows the MAE for both scenarios 1 and 2 for weekdays and Saturdays when the simulation was run for different Origins. The MAE represents the average difference between the number of people with a certain origin in the simulation and in the GVB data. The results show that during the rush hours on weekdays the MAE is highest, where the predictions are between 20 and 23 people off, on average.

Table 2 shows the MAE for both scenarios 1 and 2 for weekdays and Saturdays when the simulation was run for different destinations. The MAE is higher in every case than the MAE for the origin stations, with the highest value of 45 people.

Table 1. Origin Weekday (columns 2 and 3)and Saturday (columns 4 and 5)

Hour group	MAE S1 W	MAE S2 W	MAE S1 S	MAE S2 S
08:00–09:00	22.59	22.72	3.01	3.07
11:00–12:00	4.67	5.57	6.33	5.80
14:00–15:00	6.13	6.68	6.99	8.20
17:00–18:00	19.57	20.55	7.70	7.17
20:00–21:00	3.94	3.65	3.90	4.45

Table 2. Destination Weekday(columns 2 and 3) and Saturday (columns 4 and 5)

Hour group	MAE S1 W	MAE S2 W	MAE S1 S	MAE S2 S
08:00–09:00	25.75	25.80	5.29	5.23
11:00–12:00	5.92	6.10	7.98	7.63
14:00–15:00	17.76	17.79	9.05	8.82
17:00–18:00	44.98	45.17	14.10	14.11
20:00–21:00	6.93	7.23	7.09	6.59

For all simulations the average waiting time, the average amount of transfers and average duration were calculated. This was done for scenarios 1 and 2 for weekdays and Saturdays. Figure 3 shows a box plot of the results from 08:00–09:00 (the other hour groups showed similar results). The plots show the spread of the waiting time is larger in scenario 2. The mean duration and mean waiting time do not seem to be affected. Figure 4 shows the counts of the number of transfers. There seems to be no difference between the two scenarios in the number of transfers.

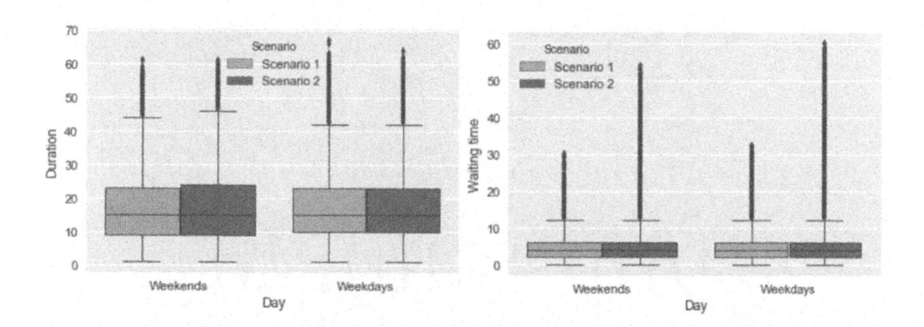

Fig. 3. Box plot of duration(left) & waiting time (right)

Tram 12

In scenario 2, tram 12 had a malfunction problem for 30 min. The first trams to run again depart at 08:35 and 08:33. The results clearly show a much higher

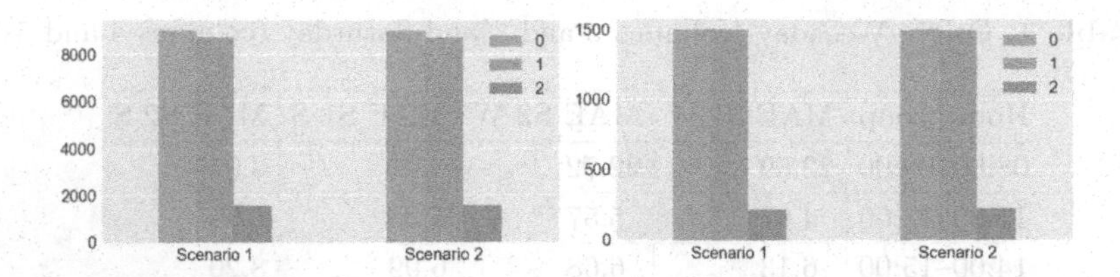

Fig. 4. Weekdays (left) and Saturdays (right) frequency count of transfers

number for the occupancy of this tram in scenario 2 than in scenario 1. The number corresponding to a station on the x-axis represents the number of people in the vehicle to this particular station. For example, if this value is 50 at station De Pijp this means that 50 people were in the vehicle while driving to station De Pijp.

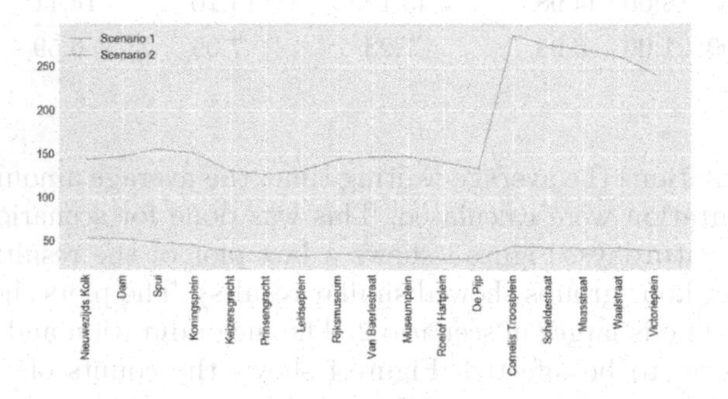

Fig. 5. Tram 12 to Amstelstation (08:35)

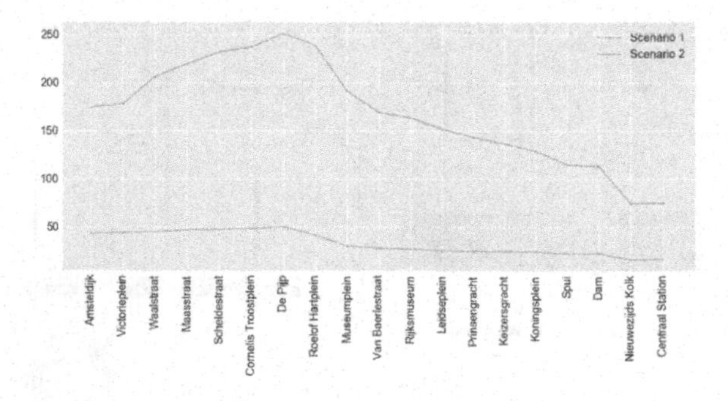

Fig. 6. Tram 12 to Centraal Station (08:33)

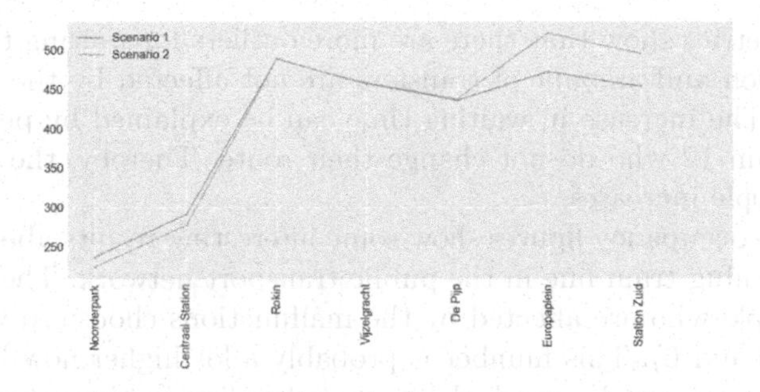

Fig. 7. Metro 52 to Zuid (08:36)

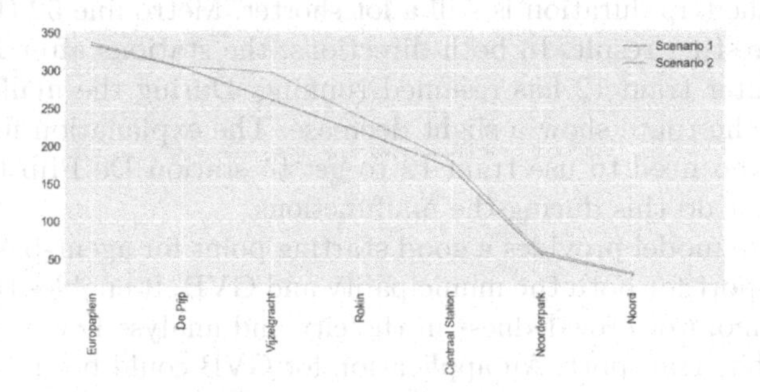

Fig. 8. Metro 52 to Noord (08:41)

6 Discussion and Conclusion

Agent-based modelling provides a good methodology to simulate passenger flows in a public transport network. With limited data about passenger behaviour, research and domain knowledge have provided a basis for our model. Three different classes were used in a subset of the public transport network in Amsterdam to create our baseline model. The results show that our model can generate passengers and routes that are somewhat similar to reality. To accurately assess the model, it should be further tested on the complete public transport network. After this, the behaviour of the passenger groups can be altered as well to further improve the model.

Overall the busy hours show a larger MAE. Since there are a lot of passengers during these hours, it might take more iterations of the simulation for the results to converge and thereby decrease the MAE. The higher errors for the destination stations can be assigned to the fact that the probability distributions are derived from a single hour. The problem with this is that a lot of destination stations within an hour are destinations from a trip that was started in the previous hour and a lot of actual destinations will fall within the next hour. Since the GVB data was aggregated by hour, it is not possible to correctly assign the destinations to the exact hour the trip started.

The trip metrics show that there are more outliers for waiting time and that the trip duration and amount of transfers are not affected by the malfunctions in scenario 2. The increase in waiting time can be explained by people who are waiting for tram 12 who do not change their route. Thereby, the waiting time for a lot of people increases.

The vehicle occupancy figures show some interesting results about the effects of a malfunctioning tram line in the public transport network. The results show that most people who are affected by the malfunctions choose to wait for tram line 12 (Figs. 5 and 6). This number is probably a lot higher now because some stations simply can not be reached by any other line in the network. In other cases, it might still be the preferable route if this means there is no need to transfer or if the trip duration is still a lot shorter. Metro line 52 (Figs. 7 and 8) shows an interesting result. In both directions, the stations after De Pijp show a clear peak after tram 12 has resumed running. During the malfunctions, the same parts in the route show a slight decrease. The explanation for this is that some people who need to use tram 12 to get to station De Pijp to transfer to metro 52 cannot do this during the malfunctions.

Our baseline model provides a good starting point for agent-based modelling of public transport for both the municipality and GVB. It enables the municipality to further monitor crowdedness in the city and analyse how it is affected by changes in public transport. An application for GVB could be the identification of critical lines in the network. The results show that tram 12 was a critical line in the network as its failure resulted in a lot of people having to wait at their stop. Our simulation could also be used to see what influence the new schedules might have on the passenger flows in the network, by editing the network structure in the model. Even though this research is focused on Amsterdam, it provides a methodology that can be applied to other cities as well. We have shown that even though there is little data available about passenger behaviour, the baseline model can still produce interesting insights.

References

1. Anderson, M.K., Nielsen, O.A., Prato, C.G.: Multimodal route choice models of public transport passengers in the greater Copenhagen area. EURO J. Transp. Logistics **6**(3), 221–245 (2017)
2. Beirão, G., Cabral, J.S.: Understanding attitudes towards public transport and private car: a qualitative study. Transp. Policy **14**(6), 478–489 (2007)
3. Fellesson, M., Friman, M.: Perceived satisfaction with public transport service in nine European cities. J. Transp. Res. Forum **47**, 93–105 (2012)
4. Kuo, Y.H., Leung, J.M., Yan, Y.: Public transport for smart cities: recent innovations and future challenges. Eur. J. Oper. Res. **306**, 1001–1026 (2022)
5. Lai, X., Fu, H., Li, J., Sha, Z.: Understanding drivers' route choice behaviours in the urban network with machine learning models. IET Intell. Transp. Syst. **13**, 427–434 (2018)
6. Liu, Y., Bunker, J., Ferreira, L.: Transit users' route-choice modelling in transit assignment: a review. Transp. Rev. **30**(6), 753–769 (2010)

7. Schmöcker, J.D., Shimamoto, H., Kurauchi, F.: Generation and calibration of transit hyperpaths. Procedia Soc. Behav. Sci. **80**, 211–230 (2013)
8. Truong, M.T., Amblard, F., Gaudou, B., Sibertin-Blanc, C.: To calibrate & validate an agent-based simulation model-an application of the combination framework of bi solution & multi-agent platform. In: 6th International Conference on Agents and Artificial Intelligence (ICAART 2014), pp. 172 (2014)
9. van der Does de Willebois, J.: Assessing the impact of quay-wall renovations on the nautical traffic in Amsterdam. Master's thesis, TU Delft (2019)
10. Zheng, H., et al.: A primer for agent-based simulation and modeling in transportation applications. Technical report, United States. Federal Highway Administration (2013)

Exploiting Security and Privacy Issues in Human-IoT Interaction Through the Virtual Assistant Technology in Amazon Alexa

Amrth Ashok Shenava[1], Saifuddin Mahmud[1], Jong-Hoon Kim[1(✉)], and Gokarna Sharma[2]

[1] Advanced Telerobotics Research Lab, Computer Science, Kent State University, Kent, OH, USA
jkim72@kent.edu
[2] SCALE Lab, Computer Science, Kent State University, Kent, OH, USA

Abstract. Smart connected devices, called Smart IoTs are becoming a heavily adopted product around the world. They have been integrated into their service platform and are spreading rapidly in households as common products. For example, people use them to control lights, get weather updates, play music, set a timer and do a plethora of tasks through various human interfaces in their platforms. Especially, voiced controlled interfaces and virtual assistant technology, such as Google Assistant and Amazon Alexa are quickly embraced in major companies due to user convenience. In addition, they allow third party developers to build Smart IoTs and applications for adopting these voice controlled interfaces in their platform so that they can be controlled by voice. While the technology has created value by helping people perform certain household chores through voice, it is also imperative to acknowledge the security vulnerabilities that lie within these voice controlled devices. In this paper, we demonstrate the exploitation of security and privacy issues in these smart IoTs which arc integrated with virtual assistant, especially Amazon Alexa in a custom endpoint and provide possible solutions to the platform provider as redesigning human-IoT interaction processes.

Keywords: Human-IoT Interaction · Virtual assistant technology · Amazon Alexa · security and privacy vulnerability

1 Introduction

Smart devices are becoming an integral part of modern household in the 21st century [1]. It is believed that 10% of the world consumers own smart virtual assistants devices(SVAD) [2] like alexa echo or google home. SVAD are becoming widespread, with 147 million units sold in 2019 [3], of which 26.2% were Amazon Echo/Alexa units and 20% Google Home/Assistant, which are the two that dominate the market by a large margin over the rest. There are a number

H. Zaynidinov et al. (Eds.): IHCI 2022, LNCS 13741, pp. 386–395, 2023.
https://doi.org/10.1007/978-3-031-27199-1_38

of features that contribute to the popularity of smart virtual assistants devices. Due to the emergence of technologies associated with AI such as Machine Learning (ML), voice recognition, and natural language processing, new generations of SVAD have emerged. SVAD are quite different from early voice-activated technologies that could only work with small inbuilt commands and responses. Instead, SVAD use Internet services and benefits from recent advances in Natural Language Processing (NLP), which allow them to handle a wide range of commands and questions. They enable a playful interaction, making their use more engaging [4]. They are assigned a name and a gender, which encourages users to personify them and therefore interact with them in a human-like manner. They are used to maintain shopping and to-dos lists, purchase goods, and food, play audio-books, play games, stream music, radio and news, set timers, alarms and reminders [5], get recipe ideas, control large appliances [6], send messages, make calls and many more depending on their usage context. Through these devices, users have the ability to perform a plethora of tasks just by speaking commands to the device. For example, users of voice controlled devices will be able to switch on or switch off lights, set a timer, play music and perform many more tasks. In addition Smart devices are powered by smart virtual assistants which also provide a platform for third party developers to utilize the underlying technology for hosting their application allowing users to interact with the bot through these smart voice controlled devices. The companies that develop these virtual assistants provide developer console through which third party developers can build their apps and deploy it. While the developer console encourages developers to use the pre defined template through which one can build directly using the console, it also provides the option of using custom endpoints where developers can build and write code on their own terms. If a developer opts for custom endpoints, they will have more freedom and flexibility to use any tools and API for making the bot. By default, bot development platforms provide pre-defined function which the developer can then modify. However, using these custom endpoints, it is possible to write custom functions and replace the pre-defined functions. Using custom functions, one can make external API calls [7] and manipulate information that comes from the user. Besides SVAD collects or can upload a variety of personal data during the use of various skills, including credit card numbers and daily schedules, which could pose problems for the owner in the event that the data falls into the wrong hands.

2 State of Arts

A complicated architecture underlies SVAD but all of them execute comparable functions and share some common features, despite the fact that individual SVAD across different suppliers have a few distinguishing qualities. In particular, SVAD's architectures often comprise the following in addition to other architectural components like cloud-based processing and communication with other smart devices: i) a voice-based intelligent personal assistant like Alexa from Amazon, Google Assistant from Google, Apple Siri from Apple, and Cortana from Microsoft [8]; and ii) a smart speaker like the Amazon Echo series,

Microsoft Home Speaker, Google Home Speaker, and Apple HomePod. While we concentrate on SVAD as a single full-fledged instance and ecosystem built on voice-based personal assistants.

Many IoT devices used in residential settings rely heavily on SVAD, making those systems vulnerable to vocal attacks. Lei et al. [9] evaluated whether home digital voice assistants have adequate security to defend against vocal attacks as part of their research. They came to the conclusion that these gadgets do not provide adequate defense against voice instructions delivered by someone other than the owner, and attackers may take advantage of this. They showed how equipping the Echo with a sensor may keep intruders from utilizing sound attacks to attack the device. To determine if someone was physically there in the room when the Echo issued the command or if it was only a voice, they suggested developing a virtual security button. The button detects motion of people based on Wi-Fi signals. Given that the Echo currently uses Wi-Fi and since the majority of homes are Wi-Fi-equipped, Amazon only has to make minimum upgrades. In confined home contexts and lab settings, their trial demonstrating the effectiveness of their virtual security button for properly triggering the Echo with human motion was successful. Security is one of relevant emerging challenges for the SVAD.They are particularly vulnerable due to their limited memory and their constant communication via the Internet to cloud servers. Security attacks in daily life [4] raises user concerns about the technology maturity. IoT devices are particularly vulnerable due to their limited memory and their constant communication via the Internet to cloud servers. Deogirikar and Vidhate (2017). The demand for resiliency against attacks faced by SVAD reveals resource limitations. We studied several papers published on the domain of finding security vulnerabilities in SVAD and found similarities and areas where it can be improved.

A Research was conducted by Das et. al. [10] which shows how using Alexa, Amazon's voice-activated assistant customers can use variety of web services using third-party apps and vulnerabilities associated with these operations. While such apps make it easier for users to engage with smart devices and provide a variety of extra services, they also pose security and privacy concerns because they function in a personal context. The goal of this study is to examine the Alexa skill ecosystem in detail. Anupam and his colleagues conducted the first large-scale analysis of Alexa skills, which included 90,194 unique talents from seven separate skill marketplaces. The existing skill vetting procedure has significant flaws, according to their findings. A rogue user can not only publish a skill under any random developer/company name, but they can also make backend code changes after approval to entice users into disclosing undesired information, according to the researchers. After that, they formalize the various skill-squatting strategies and assess their efficacy. They discovered that, while certain ways are more beneficial than others, talent squatting is not widely used in the real world. Finally, they look at the presence of privacy rules across different skill categories, as well as the policy text of skills that leverage Alexa permissions to access sensitive user data. Around 23.3% of these talents do not properly reveal the data types linked with the permissions asked, according to their findings. However, while Das, A. et all has brought up various statistics about how a skill is not safe, it lacks a demonstration how the access of sensitive information can be exploited and breached. In this paper, we extend on that research by demonstrating a code through which certain triggers take place when an interaction takes place between a user and a bot.

Man in the middle (MITM) attack continue to be a problem for SVAD, as Li et. al. [11], Novak showed. The use of SVAD, they claim, MITM is a disaster for both the network service providers and their customers" because it is presumed that the OpenFlow channel is already secured by the SVAD security, leaving devices using them open to MITM attacks. They claim that MITM attacks are still prevalent despite this. Chung et. al. [12] list incidents in which Intelligent SVAD has been misused. To communicate with Amazon Alexa or Google Home, they used an example of television sounds, such as commercials. They describe how Man-in-the-Middle and DDoS assaults can be used by attackers to take advantage of these gadgets. They also demonstrate how malevolent and unintended orders might hurt the device's user.

Chalhoub, G. and Flechais, I. [13] in their paper studied about how the interaction between a user and the bot has a effect on the overall security capabilities of the paper . While smart speakers are convenient and useful, but they also pose a number of security and privacy risks. To investigate the impact of user experience (UX) aspects on security and privacy, the study team conducted thirteen interviews with smart speaker users. The research team used Grounded Theory to examine the data and a qualitative meta-synthesis to confirm our findings. The researchers discovered that smart speaker users have no privacy concerns, prompting them to compromise their privacy for convenience. Various trigger points, such as bad experiences, however, elicit security and privacy requirements. The research here has brought up many important statistics and has also conducted user interviews giving us an understanding from a user perspective about privacy concerns however the paper focuses more on UX and lacks a implementation of which demonstrates an exploit which stems from the UX.

Another research conducted by Alhadlaq, A. et al. [14] focuses more on the overall ecosystem of Alexa including the availability of Alexa on iPhone and Android. Amazon's speech service, Alexa, powers Echo, a new sort of device installed in people's homes. Amazon introduced Alexa Skills Kit in 2015, a set of APIs that allow developers to create voice-driven features for Alexa called Skills. The research team examines Amazon's privacy-related decisions when developing the Skills ecosystem, as well as the developers' existing privacy policies inside it. They look at the privacy policies of all Alexa Skills and find that 75% of them don't have one. Furthermore, even among Skills that are required by Amazon's policy to have a privacy policy, 3.5% do not have one and 70% are not personalized to Alexa. The researchers go on to outline and debate Amazon's design choices for exposing user data to Skills developers. Our goal is to raise awareness of the current design choices and facilitate discussion that could help set meaningful privacy standards for this new domain of growing importance, as the ecosystem is very new and the privacy challenges it poses may be distinct from those in Apple's App Store and Android's Play Store.

3 Proposed Approach

To better understand this vulnerability, it is imperative to demonstrate using code how it can be exploited. While the papers mentioned in the State of Arts did cover important aspects of how privacy is being breached using smart devices, it did not demonstrate in detail how custom endpoints and external API can be used to not only breach privacy but to also manipulate that data. For this research, the proposed approach is to build an Alexa skill which will be hosted using custom endpoint rather than the traditional Alexa developer console. Alexa provides a pre-defined function called intent handlers through developers can manipulate and code how the skill interact based on the intent and utterance. Here, instead of modifying the code block, we will write our own custom function which will then be called inside the code block. In the custom function, the code will be modified to call certain API functions to inform the attacker of what and how the victim is interacting with the Alexa skill.

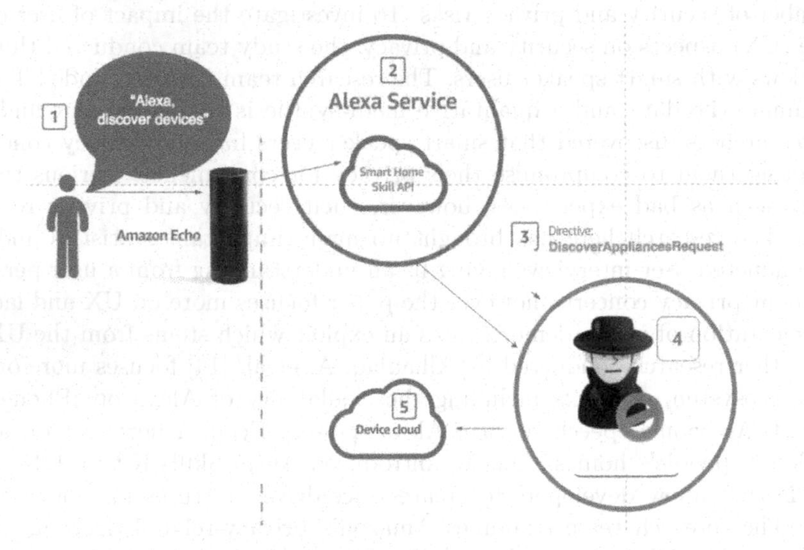

Fig. 1. Description of Scenario

Proposed Scenario: In order to demonstrate this vulnerability, a scenario is set up where there is a attacker and there is a victim. There is an assumption here that the victim has a voice controlled device and for our research, the specific assumption is the the victim has a Alexa device. The attacker has a Alexa app deployed with the intention of monitoring every interaction of between the user and the bot. Since the victim has a Alexa device, he or she can activate the bot by invoking it. After invoking the bot, the victim can utter specific commands which the bot executes. In Fig. 1, there are two actors, the attacker and the victim. The attacker published a bot to the Alexa platform which the victim can use on their Alexa device. The bot uses a custom endpoint and has custom modified functions where it manipulates the intent handlers to extract information as well as to publish details of the interaction on other platorms making it available on the public domain.

4 Implementation

For the implementation of this project, Python was used as it allows developers to build projects using a custom endpoint and libraries such as Twilio, Flask, Pyrebase, Tweepy and Alexa SDK were used as well. In order to demonstrate the security vulnerability here, a script was written in Python and Flask was used to generate a custom endpoint for the Alexa developer console. Twilio was used for making phone calls. Pyrebase is the python library for Firebase, which is a database provided by Google. Tweepy is a python library through which Twitter API can be accessed through Python. After API keys were correctly figured, certain custom functions were defined to modify the code.

4.1 Technology

For the implementation, the primary programming language used is Python and the project is hosted on the Alexa Developer Console through a custom endpoint. The custom endpoint is set up using the Flask library of python. The API used in this project include Twilio, which is a library through which programs can trigger a phone call or send a SMS text, Pyrebase, which is the python library that allows python programs to interact with Firebase, a platform that provides database and authentication services, Tweepy, which is a library through python programs can access the Twitter API.

4.2 Custom Functions

To demonstrate the vulnerability, custom functions were written that would execute based on certain commands and intents. In Fig. 2, there is a picture of the function written to steal information. When the user triggers a intent, instead of invoking the intent handler function, it will execute this function where information is updated into other platforms about

As we can see above, fsteal is a custom function that notifies the attacker if the victim happens to have a interaction about making payments.

```
def fsteal(handler_input, speak_output):
  db = qxsecutils.firebase.database()
  data = {
    "utternace" : "payment is being made"
  }
  account_sid = "<ACCOUNT_SID>"
  auth_token = "<AUTH_TOKEN>"
  client = Client(account_sid, auth_token)

  call = client.calls.create(
                          twiml='<Response><Say>Amrit is
going to make a payment using his Alexa. If there are any
more updates, I will call you.</Say></Response>',
                          to='+13306763377',
                          from_='+14154888661'
                      )
  twitter_response =
qxsecutils.client.create_tweet(text="Amrit(@0xastro98) is
going to make a payment.")

  print(twitter_response)
  print(call.sid)
  db.push(data)
  return
handler_input.response_builder.speak(speak_output)
.ask(speak_output).response
```

Fig. 2. The custom trigger function

5 Experiments and Results

For the experimental setup, an Amazon account was given access to test the bot as the bot is under development. After running the program and mentioning the various commands, Alexa responded with the output it was intended to. However, while the victim sees the standard response for the particular intent from the front, the custom function executes and relays the message of the user activity on various platforms. For this research project, the attacker would receive a call and a twitter bot would post updates about the users interaction with the bot. After every interaction. based on the intent of the interaction, the code would execute the function. Here, the victim wants to make a payment. This information will then be relayed to the attacker through various mediums. The entire proceudure is shown by the video[1] and the flowing pictures (Fig. 3).

[1] https://youtu.be/WOg9tKYcOy4.

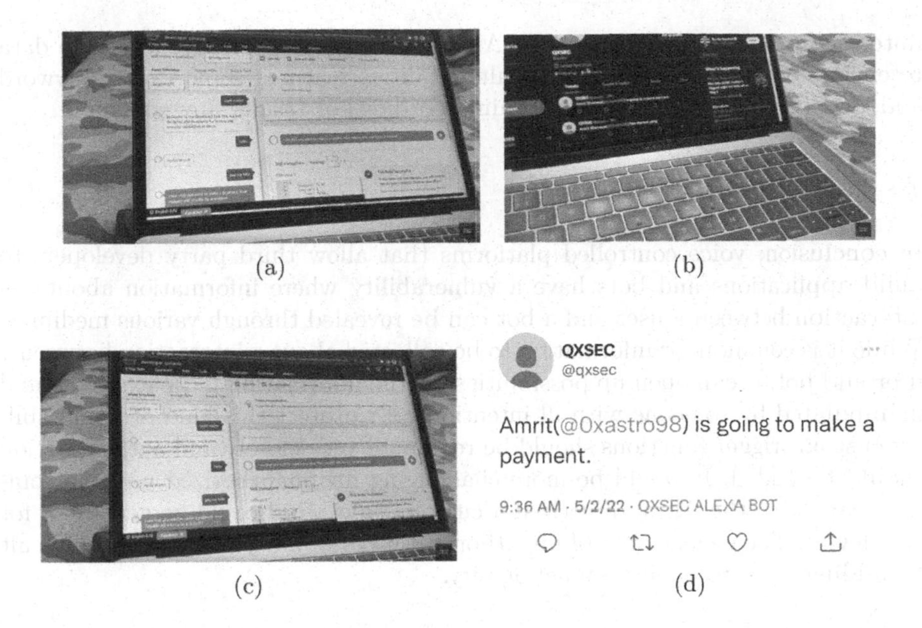

Fig. 3. Step by step of the experimental process. a) Getting access of Alexa b) Initial tweet after access c) Payment made by victim d) Information relayed to the attacker through Tweet

6 Discussion

This research was done with the intent of better understanding the implications of how these vulnerabilities can be exploited and the affect it has on its users. Privacy is all about the protection of the information of the users. Here, using a custom endpoint, the details of a user's interaction was published to platforms where it will remain on the public domain for everyone to see. The interaction is meant to be between the device and the user. The privacy of the user is being breached and violated. Bots developed by third party store interactions for research purposes however by using custom endpoint, there are endless possibilities about what one can do with the interaction and user information. In order to combat this, it is imperative to implement preventive measures through which users will have control over their data and not have a third party monitor them.

On the other side, every Alexa virtual assistant instantly sends all recorded information back to Amazon servers. The company conserves storage space by keeping certain voice recordings and erasing others whenever necessary. For the purpose of enhancing the service and assessing how well Alexa comprehends requests, Amazon personnel regularly listen to recordings. The user's initial name and account number are used to link recordings.

Users can choose to remove their interactions with Alexa on Amazon, but they cannot choose to stop Amazon from keeping some voice recordings. Indefinite record retention suggests that Amazon's servers don't have a private data retention policy. To avoid those vulnerabilities features some extra Keyword and/or PIN should be added to retain recordings or trigger any command.

7 Conclusion

In conclusion, voice controlled platforms that allow third party developers to build applications and bots have a vulnerability where information about the interaction between a user and a bot can be revealed through various mediums. While it is common for information to be collected about a interaction between a user and bot, it can open up possibilities for the information to be exploited and manipulated by someone with ill intentions. To protect this kind of vulnerability custom trigger functions should be regulated very closely and authentication should be added. It would be more challenging for unauthorized users to communicate with the Alexa directly if a custom wake word could be generated for the device. The engagement of unauthorized users would be made more difficult by adding voice recognition functionality.

References

1. https://www.oecd.org/futures/35391210.pdf
2. Ovum 2017 virtual digital assistants to overtake world population by 2021 (2017). https://ovum.informa.com/resources/product-content/virtual-digital-assistants-to-overtake-world-population-by-2021
3. Kitchenham, B., Brereton, O.P., Budgen, D., Turner, M., Bailey, J., Linkman, S.: Systematic literature reviews in software engineering - a systematic literature review. In: Information and Software Technology, vol. 51(1), pp. 7–15, (2009). special Section - Most Cited Articles in 2002 and Regular Research Papers. https://www.sciencedirect.com/science/article/pii/S0950584908001390
4. Luger, E., Sellen, A.: like having a really bad pa: The gulf between user expectation and experience of conversational agents. In: Proceedings of the 2016 CHI Conference on Human Factors in Computing Systems, ser. CHI 2016. New York, NY, USA: Association for Computing Machinery, pp. 5286–5297 (2016). https://doi.org/10.1145/2858036.2858288
5. Singleton, M.: Alexa can now set reminders for you (2017). https://www.theverge.com/circuitbreaker/2017/6/1/15724474/alexa-echo-amazon-reminders-named-timers. Accessed 21 Dec 2018
6. Martin, T.: 12 Reasons to use Alexa in the kitchen (2018). https://www.cnet.com/how-to/how-to-use-alexa-in-thekitchen/. Accessed 17 Dec 2018
7. https://developer.amazon.com/en-US/blogs/alexa/alexa-skills-kit/2018/04/alexa-skill-recipe-making-http-requests-to-get-data-from-an-external-api
8. Statista: worldwide intelligent/digital assistant market share in 2017 and 2020, by product (2018). https://www.statista.com/statistics/789633/worldwide-digital-assistant-market-share/. Accessed 21 Dec 2018
9. Lei, X., et al.: The insecurity of home digital voice assistants-vulnerabilities, attacks and countermeasures. In: Paper Presented at the 2018 IEEE Conference on Communications and Network Security (CNS)

10. Lentzsch, C., Degeling, M., Das, A., Enck, W.: Hey Alexa, is this skill safe?: Taking a closer look at the Alexa skill ecosystem. In: Network and Distributed Systems Security Symposium, New York, National Science Foundation, NY, USA (2021)
11. Li, N.E.C., Qin, Z.: TI securing SDN infrastructure of IoT-fog networks from MITM attacks **4**(5), 1156–1164
12. Chung, H., Iorga, M., Voas, J., Lee, S.: Alexa, can i trust you? Computer **50**(9), 100–104 (2017)
13. Chalhoub, G. Flechais, I.: Alexa, are you spying on me?: exploring the effect of user experience on the security and privacy of smart speaker users. ser. HCI for Cybersecurity, Privacy and Trust
14. Alhadlaq, A., Tang, A., Almaymoni, A., Korolova, A.: Privacy in the amazon Alexa skills ecosystem. ser. Privacy Enhancing Technology Symposium. Los Angeles, CA, USA: University of Southern California

User Experience in Virtual Reality Using Threshold Space in Between Different Physical Laws

Lori Minyoung Kim, Jung-Ryun Kwon, and Eui-Chul Jung(✉)

Department of Design, Seoul National University, Seoul 08826, South Korea
{ryuniss,jech}@snu.ac.kr

Abstract. The purpose of this research is to understand the Threshold Space to solve the user experience (UX) problems in virtual reality (VR) caused by the difference in physical laws. It is often misunderstood that transitioning from the physical world into a virtual space is an easy task. However, VR experiences can create a sense of heterogeneity in usability due to the difference in virtual physical laws. To overcome this problem, the researcher investigated the 5 characteristics of the differences in physical laws and the influence exerted by the presence of the Threshold Space. Most of the Threshold Space in existing VR content carried the design methods similar to 2D web/mobile-based UI/UX. Thus, it was not seen as a device to alleviate the sense of heterogeneity due to the difference in physical laws. In VR, the Threshold Space should be designed to give recognition to the VR user so that the transition or connection point is felt when entering or exiting the VR space. This can reduce experiencing confusion. Future research will study the design device suitable for the following Threshold Space for each of the different physical laws.

Keywords: Virtual Reality · User Experience · User Interface · Physical Laws · Threshold Space · Perception

1 Introduction

Everyone has experienced being half-asleep and half-awake right after waking up from a dream. In this stage, it takes time to return to reality. The same goes for Virtual Reality (VR) experiences. VR users may experience a chaotic half-awake state when entering or exiting a VR space, due to the differences between the physical world and virtual reality, which may affect the user's perception.

For the current VR seamless experience, users can quickly enter the virtual space without time and space restrictions by simply putting on an HMD (Head Mounted Device) [1]. Ironically, humans tend to perceive their surroundings based on their past experiences [2]. Providing a completely new experience – mainly, new physical laws – in VR may be difficult to accept from the user's point of view.

Also, humans tend to remember and judge experiences based on how said experience made them feel at the climax and end, rather than the average or sum of all experiences

H. Zaynidinov et al. (Eds.): IHCI 2022, LNCS 13741, pp. 396–405, 2023.
https://doi.org/10.1007/978-3-031-27199-1_39

[3, 4]. Considering this theory, the VR experience is likely to be thought of in a negative light for the following reasons: abruptly experiencing stimulating content due to a rapid transition between physical and virtual space; a sense of alienation to the user due to the cognitive confusion between the physical space in which the body actually exists and the visually perceived virtual space. One of the causes of this sense of heterogeneity is the discrepancy between the perceived space and the body due to the difference in physical laws.

How can design tools close the gap between real physical and virtual space to give users an optimal experience? This study proposes a Threshold Space to act a portal of sorts that connects the start and end points of reality and virtual spaces. According to Lexico dictionary, Threshold refers to "the magnitude or intensity that must be exceeded for a certain phenomenon to occur" [5]. It is a concept that refers to a transition stage that exists between spaces [6]. Threshold Space can take the form of UX/UI within VR environments. Therefore, the Threshold Space can be utilized to mitigate the negative effects of heterogeneous VR experiences.

This study categorizes VR content with different physical laws from reality and aims to investigate whether a UX design – Threshold Space – is provided in current VR case studies. Through this study, the design direction of the Threshold Space will be found in hopes of eventually achieving the optimal VR experience.

2 Confusion in VR Experience Due to Differences in Physical Laws

VR content usually does not follow the nature of physical laws. It follows an infinite realm with no physical limits, creating its own new physical laws. While this may very well provide a brand-new experience to users, a collision can occur between the user's visually perceived experience and their body in the physical world.

According to Merleau-Ponty's phenomenology, the human body and psychological phenomena are not strictly separate distinct domains, but rather are two different layers of 'existence' [7]. The human body corresponds to the lower layer of the existential state, and the psychological phenomenon corresponds to the upper layer of the same state. Because these two exist on a continuum, human beings have a vivid, immersive experience. However, in the VR experience, when the HMD is worn, the user's visual sense is completely blocked from the external environment and immersed in the virtual space. The user can thus experience cognitive confusion by sensing the two different layers of existence simultaneously: the physical space where the user's body is located and the visual space perceived by the user's visual information.

When brain recognition 1 and 2 are felt at the same time, the two types of information are separated, and conflict occurs in the phenomenal cognitive action between the two senses (see Fig. 1). Also, when transitioning from the virtual space to the physical space, it may be difficult to immediately recognize that the entire body is in the physical world. Such negative experiences are proportional to the degree to which the physical laws of virtual space differ from those of physical space. The five main different types of physical laws, typically used in VR contents, are classified as follows.

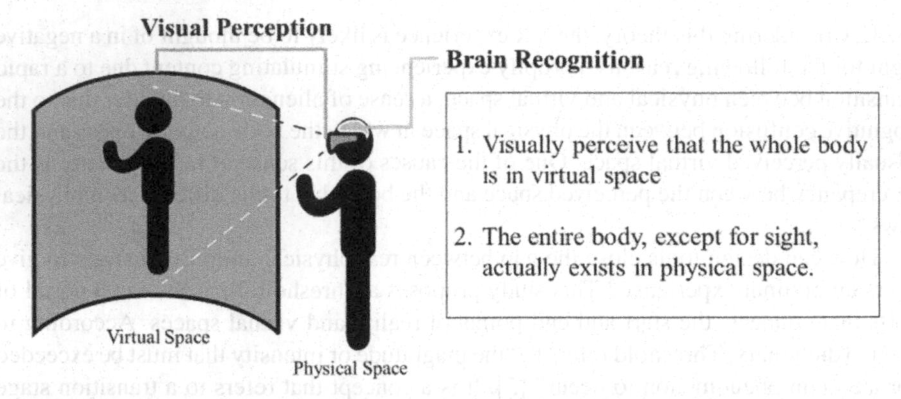

Fig. 1. Perception and Recognition in VR

2.1 Gravity

In virtual space, unlike the gravity felt in the physical realm, it follows very different laws of gravity, enabling impossible experiences such as flying, floating in the air, walking on water, hanging upside down, etc. (Fig. 2).

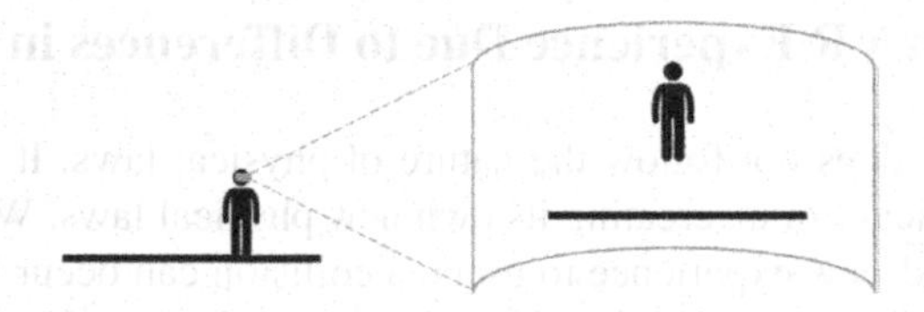

Fig. 2. Gravity Differences in VR

2.2 Scale

A phenomenon occurs in which the scale of one's or another's body can change, and the surrounding environment and the objects within it such as buildings, mountains, and trees can have differing sizes to their counterparts in the physical world (Fig. 3).

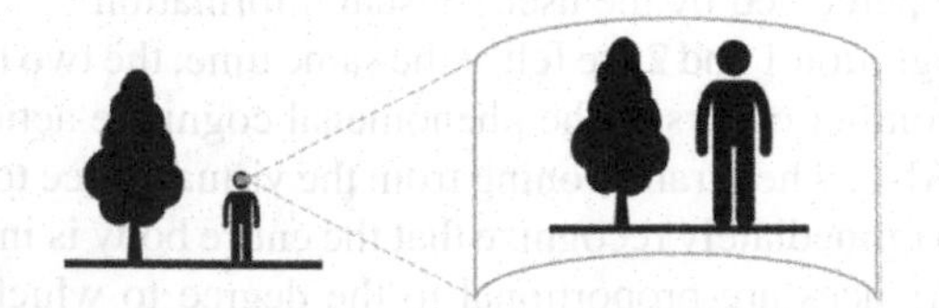

Fig. 3. Scale Differences in VR

2.3 Object and Shape

In virtual space, non-existent virtual characters such as avatars appear. Unlike the other types of physical laws, which are created to mimic the physical world, virtual spaces often create and explore non-existent things and live forms (Fig. 4).

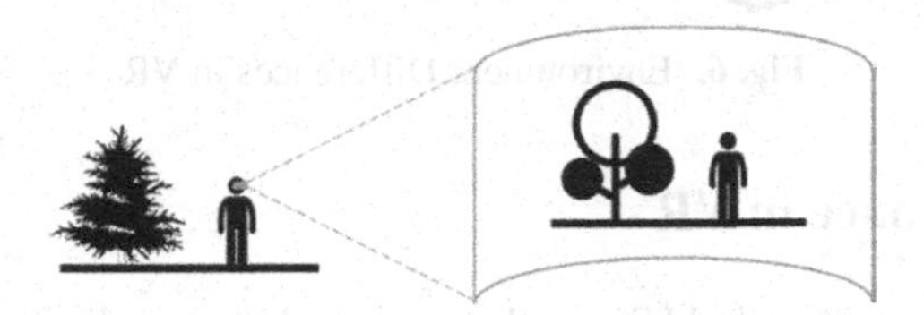

Fig. 4. Object and Shape Differences in VR

2.4 Time

VR contents can often be designed not to follow the flow and duration of time as we know it and is typically sped up drastically. For example, one day in the physical world could be months or even years in the virtual world. Users can also often change their virtual age as they please, becoming younger or older than their actual age (Fig. 5).

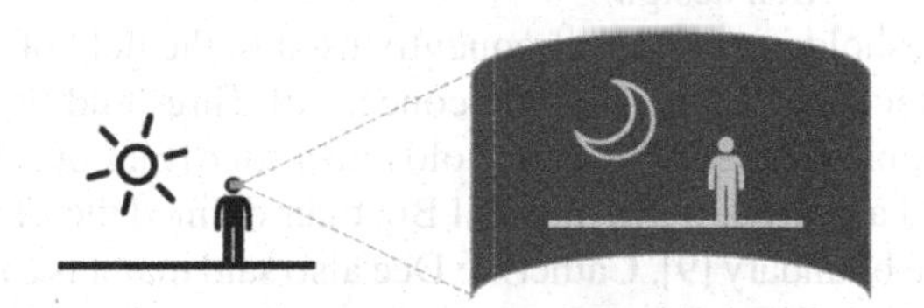

Fig. 5. Time Differences in VR

2.5 Environment

VR contents can also allow users to explore what it might be like to reside in a different environment to that we are accustomed to in the physical world, such as Mars, deep underwater, or underground (Fig. 6).

From perceiving the five representative types of differences in physical laws, users may have difficulty in usability when moving from the physical realm to virtual realm and vice versa. Furthermore, users may feel a sense of heterogeneity, disconnection, and disparity in the VR experience. It is necessary to pay attention to the negative effects that accompany difficulties in the VR experience. Creating a design device, a cushion of sorts, to minimize the possibility of said negative experience should be considered when creating VR contents.

Fig. 6. Environment Differences in VR

3 Threshold Space in VR

This study selected the Threshold Space that appears between dimensions when moving from one space to the other, to allow VR users to easily recognize and perceive the VR user experience. When using VR, the Threshold Space device can be used to overcome the heterogeneity caused by the difference in the physical laws of each space. Also, by facilitating the UX in the VR space and encouraging active participation, it is possible to increase the likelihood that users will view the VR experience in a positive light.

3.1 Threshold Space in Architecture

While VR content corresponds to interactive media design, users can experience VR in a three-dimensional space. Therefore, it is necessary to analyze the VR environment in terms of spatial composition design.

Originally, Threshold is a concept frequently used in the field of architecture. Many theorists and architects have expanded the concept of Threshold by combining it with space. Walter Benjamin extended the Threshold into a zone [6], Louis I. Kahn understood it as an entrance and a courtyard [8], and Till Boettger defined the Threshold in space as making a hole in the boundary [9]. Catherine Dee also said that it is a space that connects space and media [10], and Teyssot understood the window as the typical Threshold [11].

The Threshold Space in which various interpretations exist can be applied to various media. In Thru The Mirror (1936), Mickey Mouse passes through a hole in the space, as defined by Till Boettger, to go to the space beyond the mirror. In Doctor Strange in the Multiverse of Madness (2022), Doctor Strange enters a bizarre-shaped entrance, like the threshold interpreted by Louis I. Kahn, and moves into the multiverse (Fig. 7).

3.2 User Experience in VR Using Threshold Space

In order to design a Threshold Space with a suitable UX/UI, it is required to have a connecting device that allows users to perceive and recognize the turning point of each space with its unique physical laws. In phenomenology terms, the lower layers of the body and the upper layers of psychological phenomena should not be confused. By implementing a Threshold Space that allows users to adapt to a new environment, an optimal experience should be provided so that it can be accepted without psychological resistance. Otherwise, if only one-sided guidance is provided, not only may the user not properly understand what it is that they are experiencing, but it may also have the opposite effect of inducing a negative reaction (Fig. 8).

Fig. 7. Thru The Mirror, Mickey Mouse complete (left), Doctor Strange (right)

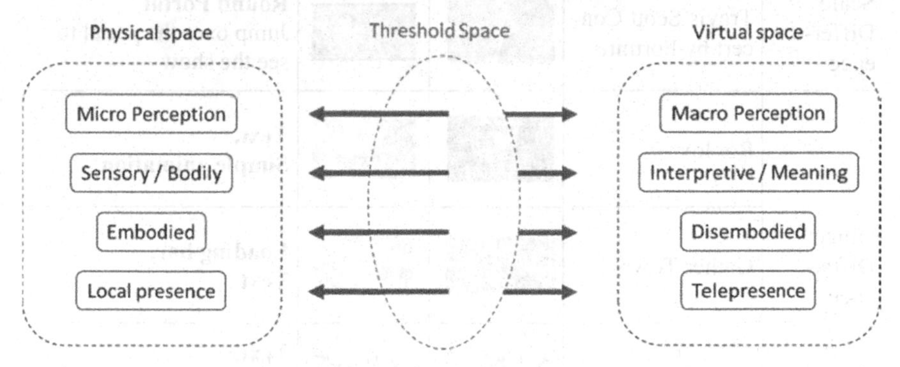

Fig. 8. Diagram of the Threshold Space existing in between the physical and virtual realm.

4 Threshold Space in Between the Different Physical Laws

Threshold Space case studies were investigated from existing VR and other digital content and classified into five different types of physical laws: gravity, scale, object & shape, time, and environment. Threshold Space – a design device for overcoming differences in physical laws – was analyzed to find the characteristics of UX. Both the good and bad examples have been divided into two subsections, describing the UX design characteristics of the Threshold Space.

4.1 Bad Examples

The case studies (Table 1) below show that lots of VR contents uses simplified UI/UX design in the Threshold Space. The design elements frequently used are listed below:

- Fade in/out
- Loading bar
- Text
- Simple animation
- Round Portal

Table 1. Examples of Threshold Space in between the Different Physical Laws in VR

Physical Laws	Example	Threshold Space	Virtual Space	Design Device For Overcoming Differences
Gravity Difference	Second White			**Text, Simple animation** Link to text before walking on water
	Horizon Worlds			**Text, Simple animation, Loading bar**
Scale Difference	Travis Scott Concert by Fortnite			**Round Portal** Jump over the portal to see the show
Object Difference	Roblox			**Text, Simple animation**
	Gather Town			**Loading bar, Text**
	Zepeto			**Text, Simple animation, Loading bar**
Time Difference	Gucci Garden By Roblox			**Text, Simple Animation** Only Daytime
	VR Human Documentary			**Fade in/out, Portal Ball**
Environment Difference	Fortnite			**Fade-Out** Starting from in-water
	BBC Home - A VR Spacewalk			**Text, Fade-Out** Starting from Space
	Journey into the Deep Sea, National Geographic			**Fade in/out** Starting from in-water

These design elements are identical to the existing 2D web/mobile-based UX/UI design language. User experience design of a web/mobile-based content coordinates technologies to drive simplicity. However, VR is a 3D content and completely separates the user's body and sensory space. Therefore, it is necessary to use more complex design language that is different from the existing web/mobile-based content. The objective of VR UX design should be to harness complexity to deliver ease, slow and progressive familiarization, and visual clues should all be used to help the user.

4.2 Good Examples

The examples below show that the UX/UI design of the Threshold Space is well designed, as they can give users a greater sense of immersion when viewing digital content. Most of the successful examples were found in 2D digital content that could in some way be adapted to VR.

In each of the cases below, there is a clear Threshold Space between easily perceived spaces and spaces with different physical laws, and the Threshold Space is designed to

Table 2. Good Examples of Threshold Space in between the Different Physical Laws

Physical Laws	Example	Threshold Space	Virtual Space	Design Device For Overcoming Differences
Gravity Difference	The Matrix (Movie)			The use a public phone to travel through space. An image is produced as if the entire space is composed of coding.
Scale Difference	Alice in Wonderland (Movie)			The main character passes through a long tunnel (rabbit hole) with different gravity and arrives in a world of a different scale.
Object Difference	Doctor Strange (Movie)			Whenever the main character moves to each multiverse, he moves through passages (doors, etc.).
Time Difference	Ready Player One (Movie)			When the characters wear an HMD, it is directed to enter a tunnel. Once through the tunnel, the characters enter a world without the constructs of time.
Environment Difference	ifland (Metaverse Platform)			Entering the space beyond the door, the visual design makes it seem as though it is a portal, and one is traveling with the speed of light to another space.

recognize the turning point whenever there is either a slow or sudden change in one or more physical law (Table 2).

American phenomenologist Don Ihde said, "Naked perception and perception via artifacts are never completely identical. Mediated perception – meaning to see the world through technology – is indissolubly linked with a transformation of perception. Mediated perception always strengthens certain specific aspects of the reality perceived and weakens others" [12]. Much like Ihde's phenomenology, when using VR, you are entering a new world that is rich in sensory experiences, and therefore requires intense immersion. Immersive UX/UI is required to enter a new world that requires strong immersion. Therefore, in order to smoothly connect the two different spaces and to alleviate the difference between the physical laws of virtual and real, it is necessary to categorize the method of entering the virtual space by designing a unique Threshold Space that is suitable for said virtual space.

5 Conclusion

The poster researched the Threshold Space to solve the user experience (UX) problem in virtual reality (VR) caused by the difference in physical laws. Current VR and other digital content were categorized into five representative types of physical laws. And the Threshold Space or UX design in the chosen content was investigated. The researcher conceptualized the foundations of Threshold Space between the virtual and physical spaces and made an overview of both the good and bad examples. Threshold space well designed cases feature: 1) a direction in which physical laws change gradually or 2) a direction that recognizes in advance that a rapid change in space occurs. Good practices of Threshold Space used a different design process than the process for designing a 2D web/mobile based content (e.g., fade in/out, text, loading bar, simple animation, round portal). Designing Threshold Space for VR experiences should not simply mean transferring 2D process to 3D, but finding a new paradigm.

The limitation of this research is evaluating case studies with transition stages from the starting point of the physical space to the virtual space. In the future, research should be conducted to demonstrate Threshold Space effect in other steps including the midpoint and the endpoint transition stage. Furthermore, we would also like to deepen our knowledge for Threshold Space, by finding and researching more of good examples of Threshold Space, designing Threshold Space parameters in different physical laws and see how these might affect user experience, in the best-case scenario, giving an optimal experience to the VR users.

References

1. Oculus VR Headsets, Games & Equipment - Meta Quest. https://www.oculus.com/blog/introducing-the-new-oculus-quest-system-experience/. Accessed 28 July 2022
2. Lee, N.: (Husserl and Merleau-Ponty) Phenomenology of Perception (2013)
3. Acampado, A.G.: Understanding experience: Dewey's philosophy. Int. J. Educ. Res. Stud. **1**(1), 01–06 (2019)
4. Yablonski, J.: Laws of UX: Using Psychology to Design Better Products & Services, 1st edn. Jon Yablonski (2020)

5. Lexico Dictionaries. https://www.lexico.com/synonyms/threshold. Accessed 28 July 2022
6. Benjamin, W., Tiedemann, R.: The Arcades Project, p. 494. Belknap Press, Cambridge (1999)
7. Merleau-Ponty, M.: Phénoménologie De La Perception/M. Merleau-Ponty (1972)
8. Wurman, R.S.: What Will Be Has Always Been the Worlds of Louis I. Kahn, p. 194. Rizzoli (1986)
9. Boettger, T.: Threshold Spaces: Transitions in Architecture: Analysis and Design Tools/Till Boettger. (2014)
10. Dee, C.: Form and Fabric in Architecture. Spon Press, London (2001)
11. Teyssot, G.: Mapping the Threshold: A Theory of Design and Interface, pp. 3–12. AA Files 57 (2008)
12. Ihde, D.: Technology and the Lifeworld: From Garden to Earth, pp.119–146 (1990)

Monitoring Pollination by Honeybee Using Computer Vision

Vinit Kujur[1(✉)], Anterpreet Kaur Bedi[2], and Mukesh Saini[1]

[1] Indian Institute of Technology Ropar, Rupnagar, India
{2021csm1010,mukesh}@iitrpr.ac.in
[2] Thapar Institute of Engineering and Technology, Patiala, India
anterpreet.bedi@thapar.edu

Abstract. Honey bees are critical in pollination worldwide and are essential for crop productivity and ecological management. Knowing more about honeybees and their interaction with plants is urgently needed given the current global pollination crisis. For monitoring pollination, non-invasive approaches are recommended because they diminish the possibility of interfering with pollinator behaviour. Traditional techniques for manually recording pollinator activity in the field can be expensive and time-consuming. In this paper, we have developed a system for pollination monitoring by honeybees using computer vision technique. However, monitoring honeybees is challenging because of their tiny size, swift speed and complex outdoor environments. To detect honeybees in a frame, we have used YOLOv7 as our deep learning model. We have fine-tuned the model on a custom-created dataset for better detection and accuracy. The dataset contains snapshots of YouTube videos in which honeybees were pollinating flowers in different environments. We examined a specific setting where the honeybee pollinated the flowers at several intervals during the video using the detector. We have generated a heatmap and pollination activity graph based on the data from the detector. This information will help with better pollination management, which will increase crop production quality and yield.

Keywords: Honey Bee · Pollination · YOLOv7 · Video Monitoring · Computer Vision

1 Introduction

Pollinators are critical to global food production and ecological management. Insect pollinators affect 35% of the worldwide agricultural area [1], supporting over 87 food crops [2]. Despite the fact that biotic pollination is vital for many crops [3], it is rarely researched.

The creation of a system to monitor flower visits and determine visitation rates, that optimise agricultural output (tonnes per hectare) from a pollination standpoint, would be beneficial to farmers globally, enhancing both their revenue and pollinator health. Crop flower visitation is influenced by both inherent

H. Zaynidinov et al. (Eds.): IHCI 2022, LNCS 13741, pp. 406–416, 2023.
https://doi.org/10.1007/978-3-031-27199-1_40

properties of the crop as well as external factors, such as the attractiveness of neighbouring vegetation and interactions with wild pollinators [4]. Furthermore, pollination management for pollinator-dependent crops should be based on direct assessments of pollinator activity, which may be performed by monitoring flower visitation rates [5].

Insects are the most significant worldwide agricultural pollinators that play a vital role in the sustainability of natural ecosystems. Out of all the insects, honeybee species are the most common pollinators [6]. Monitoring and managing honeybee pollination are critical for increasing agricultural output and food security.

Monitoring animals in their natural environment is critical for the progress of animal behavioural research, particularly pollination studies. Sensor technology advancements have facilitated low-cost Internet of Things (IoT) devices for pollination monitoring, such as cameras and microscopic insect-mounted sensors [7]. Non-invasive approaches are favoured for these goals because they decrease the possibility of study equipment interfering with behaviour. Insect-mounted sensors track the movement of tagged insects across broad regions [7]. However, the technology is unsuited for agriculture since tagging is time-consuming and may cause insect stress or change in their behaviour. Also, it is impractical on a large enough scale to be helpful in this context. Camera-based pollination monitoring can overcome these shortcomings using computer vision and deep learning [8].

Motivated by the reasons stated above we have tried to develop a system that will monitor the pollination activity by the honeybees using the computer vision technique.

However, the complexity of monitoring pollination activity by a bee in an uncontrolled environment is challenging owing to their tiny size and quick speed while functioning in complex 3D surroundings, thus making their identification difficult.

Recent research in this field has used in-situ insect monitoring algorithms [9,10]. However, they were limited in spatiotemporal resolution for efficient pollination monitoring. In [8] authors have presented a Hybrid Detection and Tracking (HyDaT) algorithm and a Polytrack algorithm [11] which detects and tracks insects in uncontrolled conditions. It uses a hybrid model consisting of a deep learning-based detection model (YOLOv4) [12] and a segmentation-based detection model [13]. However, they are limited to one location and one species. In [14] authors introduced another model which detects and tracks four insects.

The main downside of the above models is that they are limited to a specific location and environment. This is due to the fact that the dataset used to train the model was from that specific location only. Hence using these models on another flower and environment would fail as the model is not generalised for various environmental conditions. Moreover, in the above work camera is fixed at a specific distance from the plant. This makes the size of the honeybee similar in frames. Varying the distances could lead to failure of the model to detect the insect because the training set does not have samples of varying sizes.

In this study, we have used YOLOv7 [15] as our deep learning model to detect honeybees. To overcome the abovementioned issues, we have created a custom dataset containing data samples of honeybees of varying sizes and in various environments. This dataset is then used to fine-tune the YOLOv7 model for honeybee detection. We use the detections by the model to create a heatmap like visualisation of the pollination activity.

2 Methodology

2.1 Honeybee Detection

The overall setup to monitor the pollination activity is shown in Fig. 1 below. The video camera will be used to capture the video of the plant, which is under observation. Any RGB camera capable of recording HD video can be used. An edge device like Raspberry Pi is preferred because of its wide availability and low cost, which makes it easier to deploy in large regions [16]. The camera should be positioned at an adequate distance from the plant so that the deep learning model can detect the honeybee. The camera position could be at varying distances from the plant, but if the camera is too far, the bee will appear like a dark blob in the frame, which can lead to the failure of the model to detect the honeybee.

Fig. 1. Overview of Method

The video taken from the camera will then be used to analyse the areas of higher pollination activity(visitation rate of the honeybee in flower) and the time interval in which the foraging activity(visitation by honeybee) was more prominent. The deep learning model will detect the honeybee in each video frame. The detected information will then be used to generate heatmaps and activity graphs. For practical implementation of this method, the edge devices could capture the video, which will be deployed in various areas to monitor the pollination. These edge devices will then send the video data to the central server for further analysis.

To detect honey bees we have used YOLOv7 [15] model. YOLOv7 works on the trainable bag-of-freebies method to enhance object detection accuracy. It has surpassed all current object detection models and received state-of-the-art results [15]. YOLOv7 can detect and identify animals in a frame regardless of the surroundings. The detection rate and accuracy are determined by the amount and quality of training data. So, to ensure a good detection rate and accuracy, we have created a custom dataset to train the model.

Fig. 2. Image frame divided into partitions. (Color figure online)

2.2 Pollination Monitoring

The fine-tuned YOLOv7 is used as the DeepLearning model to monitor pollination activity as described in Fig. 1. The detection from the model will then be used to generate heatmaps and pollination activity graphs. The detector will give the bounding box of the detected bees in each video frame. To generate the heatmap, we have used the centroid of the bounding box. First, the image frame is divided into small partitions of equal sizes (see in the Fig. 2) and initialised with zero. The value of partition increases by one if the centroid falls on the area of the partition. The partition size depends on the distance between the camera and the plant; if the distance is large, the partition size should be small. From the Fig. 2 it can be seen that the blue partitions represent the value zero and the colour from light to red represents an increased value. For better representation of the resulting heatmap, it can be normalised to values between 0 and 1, and a smoothing filter(like a blur) can be applied.

An activity graph with respect to time can also be generated where the y-axis denotes the number of bees visiting to pollinate the flower. The number of bees in the Field of view (FOV) of the camera for each second is calculated and summed for each time interval(one minute). This value is then normalised for ten seconds, as we have assumed that the honeybee will spend around ten seconds pollinating the flower.

3 Dataset

YOLOv7 model must be trained on labelled data to learn classes of objects in that data. For the generalisation of the model to have a high detection accuracy, the dataset should be diverse, consisting of many different breeds of bees and different environments, i.e. different plants and flowers. So, two steps are involved in creating a dataset: collection of images and annotation of the collected images.

Manually collecting the dataset by going to different plant types is laborious. So instead of manual collection, we have used the Youtube videos of the honeybee pollination activity. On Youtube, there are thousands of videos where there are honey bees doing pollination of flowers. We picked a few short videos [17–22] from the sites and took snapshots at random intervals. The snapshots collected from the video are the image dataset. Picking different videos under different environments gives a more diverse dataset which will help the model to generalise better.

The second task was to annotate the honeybee to create ground truth for the model to learn from. For the same, we have used Roboflow, which is a simple web-based tool for managing and labelling images and exporting them in YOLOv7's annotation format. The different data instances can be shown in Fig. 3.

The dataset contains 500 images, with 20 images having null annotation. There are 1030 annotations with an average of 2.1 per image across 1 class(bee). The median image resolution is 1280 × 720.

4 Experimental Setup and Training

The dataset is then used to fine-tune the YOLOv7 model. We used the Google Colaboratory (Colab) environment for the training and analysis, which provides free access to a powerful GPU for inferring deep learning models. The available GPUs could be varied for each session, but for our experiment, the NVIDIA Tesla T4 with a VRAM of 16 GB and 7.5 computation capability was used for all sessions. The median image size in the dataset was 1280 × 720, so the input image size to train the model was set to "1280". Moreover, this will help to detect the bees in the HD(1280 × 720) videos. 15% of the data is used for validation. The model was trained for 250 epochs and we got max mAP of 0.895.

Fig. 3. Various data instances and the labels

Various results of the training are shown in Fig. 4. It can be seen that the training goes smoothly throughout the epochs, with both the training loss and the validation loss decreasing consistently.

The main metric to measure the performance of detector model is the mean Average Precision(mAP) and F1-score. The formula to calculate it is as follows.

$$\begin{aligned} P &= \frac{TP}{TP+FP} \qquad & R &= \frac{TP}{TP+FN} \\ AP &= \int_0^1 P(R)\,dR \qquad & F1\text{ - }score &= \frac{2 \times P \times R}{P+R} \end{aligned} \tag{1}$$

Here P is precision, R is recall, TP is the number of correctly predicted positive class, FP is the the number of incorrectly predicted positive class, FN is the number of incorrectly predicted negative class and AP is the average precision. For our case, as there is only one class, mAP will be equal to AP. Precision and recall generally have a negative correlation, where one rises and the other falls (see Fig. 5a). The value of AP is the area under the P-R curve, and a larger value means better model performance. The F1-score, which takes into consideration both the precision and recall, is the harmonic mean of precision and recall.

Figure 4b shows mAP/0.5 which is the mean average precision with the confidence(IoU threshold) of 0.5 for the training data. It can be seen that the mAP/0.5 increases considerably after a few epochs and then saturates around 200 epochs.

Figure 5 shows various results on the validation data, Fig. 5a shows Precision-Recall curve for fine-tuned detection model for bee class, where IoU threshold is 0.5 and Fig. 5b shows F1 score curve for fine-tuned YOLOv7 detection model.

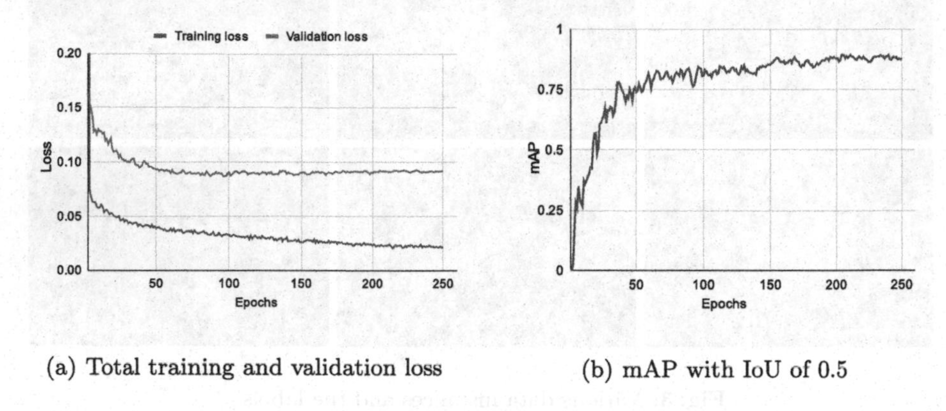

(a) Total training and validation loss (b) mAP with IoU of 0.5

Fig. 4. YOLOv7 training results on custom dataset

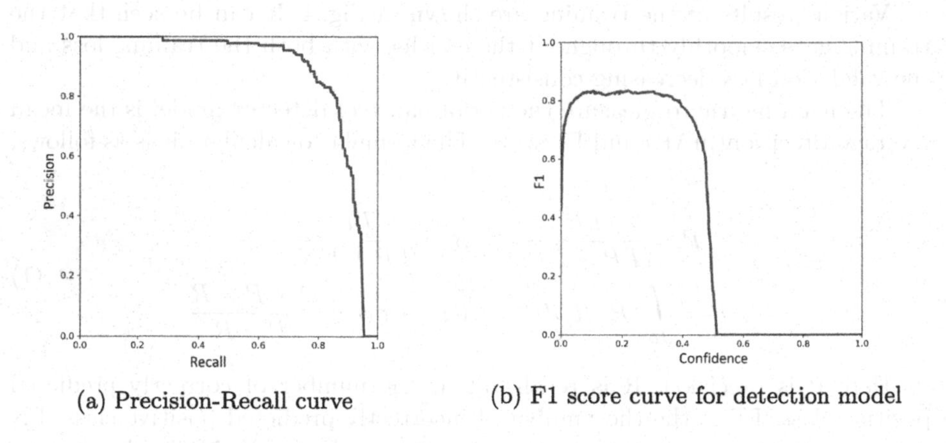

(a) Precision-Recall curve (b) F1 score curve for detection model

Fig. 5. YOLOv7 model results on validation data

5 Analysis and Results

The model was tested on a video [23,24] of duration 20 min and 35 min to analyse and generate the heatmaps. In the video, the camera is fixed in the environment. The snapshot of the video is shown in Fig. 6a and 6b. The video contains the apple blossom flower in which the honeybees pollinate at various intervals. The

(a) Snapshot of the environment [23] and the pollination activity heatmap

(b) Snapshot of the environment [24] and the pollination activity heatmap

(c) Snapshot of the environment in dataset[14] and the pollination activity heatmap

Fig. 6. Results on the test video

confidence threshold is set to 0.4, and the IoU threshold is set to 0.5 for the model. For this experiment, the image-frame size is 1280×720, and the partition size is taken as 15(i.e. 15×15 square).

The resulting heatmap of the pollination activity is shown in Fig. 6a and 6b. For representation purposes, we have blurred the resulting heatmap with a kernel size of 31×31. The light to red area denotes the location where the honeybee pollination activity(visitation rate) was significant throughout the video. These areas can be considered the flower area as the bee will spend more time pollinating the flower.

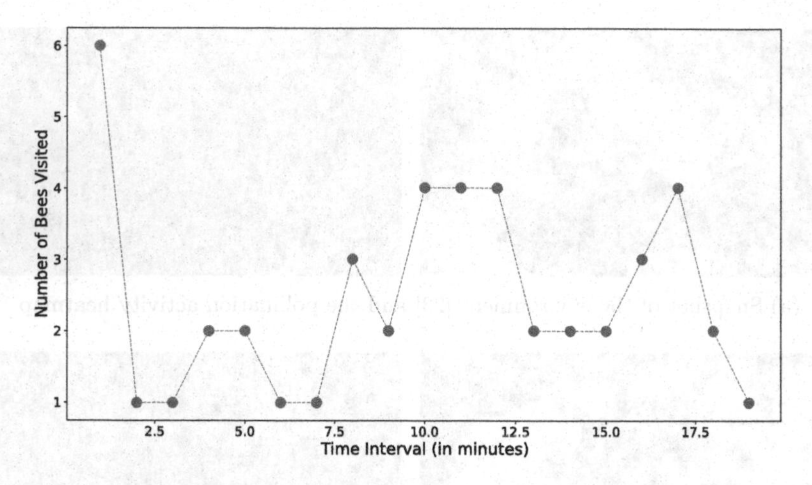

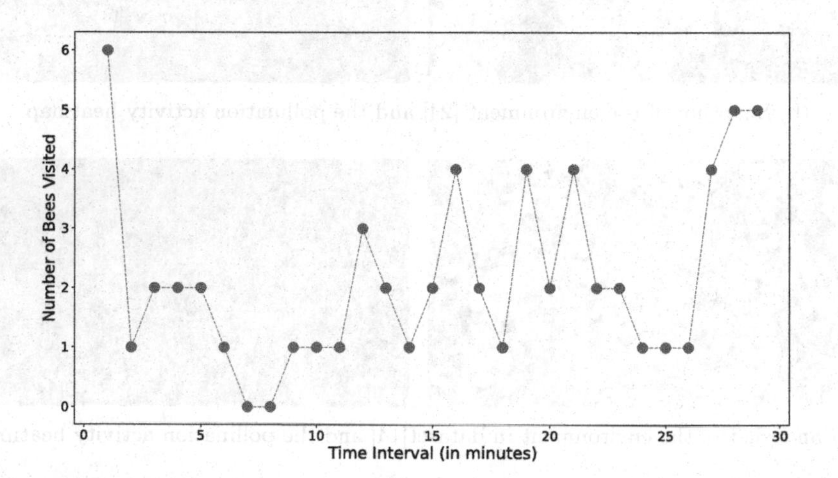

Fig. 7. Activity graph with respect to time throughout the video [23,24]

Figure 7 shows the activity graph with respect to the time period(in minutes) throughout the video. The y-axis denotes the number of bees visited to do pollination. From the graph, it can be seen that number of visits by honeybees was more initially and in the time between 10–12 min in the test video. It can also be verified by looking at the video at 10–12 min that more bees will appear in the FOV of the camera for a longer duration.

We have repeated the same procedure for a test video of duration 10 min given on the dataset by [14] in which the camera is placed on a strawberry farm. The video was resized to 1280×720 from 1920×1080 and from the Fig. 6c, it can be seen that the camera is far away compared to the previous test videos, so we have decreased the partition size to 11×11. The kernel size of blur filter was taken as 17×17 for smoothing of the heatmap. The resulting heatmap shows that the honeybee spends most of the time in the area where the strawberry flower is present.

The test video was of 20 min only, but for a monitoring system, the video duration could be of hours depending on the observation times. From the activity graph with respect to the time, we can know in which time interval the honeybee prefers to visit the flower for pollination(whether in the morning, afternoon, evening or in any particular time interval). This information will help in crop management and make better decisions to increase the output of the produce.

6 Conclusion and Future Work

In this paper, we have developed a system that will monitor the pollination activity of the honeybees using the computer vision technique. The YOLOv7 is used to detect honeybee in each frame. For better detection rate and accuracy of the detector, we have fine-tuned the model on a custom-created dataset. Dataset was created using snapshots of videos from YouTube. Different videos under different environments gave a more diverse dataset which helped the model to generalise better under different environments. The training results showed that the model trained well from the dataset.

We used this detector to analyse a particular environment where the honeybee pollinated the flowers at different time intervals in the video. All this information will help better understand the honeybee's pollination behaviour for a particular flower.

The future work will be to make this model work in low computing edge devices like JetsonNano and RaspberryPi with a good FPS. For this to work, we can do model compression via distillation and quantization. It will increase the performance of the number of inferences per second of the detector while maintaining good accuracy.

Acknowledgements. This work is supported by the grant received from DST, Govt. of India for the Technology Innovation Hub at the IIT Ropar in the framework of the National Mission on Interdisciplinary Cyber-Physical Systems.

References

1. FAO. Why bees matter; the importance of bees and other pollinators for food and agriculture (2018)
2. Aizen, M.A., Garibaldi, L.A., Cunningham, S.A., Klein, A.M.: How much does agriculture depend on pollinators? lessons from long-term trends in crop production. Ann. Botany **103**(9), 1579–1588 (2009). https://doi.org/10.1093/aob/mcp076
3. Potts, S., Imperatriz-Fonseca, V., Ngo, H., et al.: Safeguarding pollinators and their values to human well-being. Nature **540**, 220–229 (2016). https://doi.org/10.1038/nature20588
4. Rollin, O., Garibaldi, L.A.: Impacts of honeybee density on crop yield: a meta-analysis. J. Appl. Ecol. **56**(5), 1152–63 (2019). https://doi.org/10.1111/1365-2664.13355
5. Garibaldi, L.A., Sáez, A., Aizen, M.A., Fijen, T., Bartomeus, I.: Crop pollination management needs flower-visitor monitoring and target values. J. Appl. Ecol. **57**(4), 664–70 (2020). https://doi.org/10.1111/1365-2664.13574

6. Garibaldi, L.A., Requier, F., Rollin, O., Andersson, G.K.: Towards an integrated species and habitat management of crop pollination. Curr. Opin. Insect Sci. **1**(21), 105–14 (2017). https://doi.org/10.1016/j.cois.2017.05.016
7. Abdel-Raziq, H.M., Palmer, D.M., Koenig, P.A., et al.: System design for inferring colony-level pollination activity through miniature bee-mounted sensors. Sci. Rep. **11**, 4239 (2021). https://doi.org/10.1038/s41598-021-82537-1
8. Ratnayake, M.N., Dyer, A.G., Dorin, A.: Towards computer vision and deep learning facilitated pollination monitoring for agriculture. In: Proceedings of the IEEE/CVF Conference on Computer Vision and Pattern Recognition, pp. 2921–2930 (2021)
9. Bjerge, K., Mann, H.M., Høye, T.T.: Real-time insect tracking and monitoring with computer vision and deep learning. Remote Sens. Ecol. Cons. (2021). https://doi.org/10.1002/rse2.245
10. Bjerge, K., Nielsen, J.B., Sepstrup, M.V., Helsing-Nielsen, F., Høye, T.T.: An automated light trap to monitor moths (lepidoptera) using computer vision-based tracking and deep learning. Sensors **21**, 343 (2021). https://doi.org/10.3390/s21020343
11. Ratnayake, M.N., Dyer, A.G., Dorin, A.: Tracking individual honeybees among wildflower clusters with computer vision-facilitated pollinator monitoring. PLoS ONE **16**(2), e0239504 (2021). https://doi.org/10.1371/journal.pone.0239504
12. Bochkovskiy, A., Wang, C.Y., Liao, H.Y.: Yolov4: optimal speed and accuracy of object detection. arXiv preprint arXiv:2004.10934. Accessed 23 Apr 2020
13. Zivkovic, Z., Van Der Heijden, F.: Efficient adaptive density estimation per image pixel for the task of background subtraction. Pattern Recogn. Lett. **27**(7), 773–80 (2006). https://doi.org/10.1016/j.patrec.2005.11.005
14. Ratnayake, M.N., Amarathunga, D.C., Zaman, A., Dyer, A.G., Dorin, A.: Spatial Monitoring and Insect Behavioural Analysis Using Computer Vision for Precision Pollination. arXiv preprint arXiv:2205.04675. Accessed 10 May 2022
15. Wang, C.Y., Bochkovskiy, A., Liao, H.Y.: YOLOv7: trainable bag-of-freebies sets new state-of-the-art for real-time object detectors. arXiv preprint arXiv:2207.02696. Accessed 6 July 2022
16. Jolles, J.W.: Broad-scale applications of the raspberry pi: a review and guide for biologists. Methods Ecol. Evol. **12**(9), 1562–79 (2021). https://doi.org/10.1111/2041-210X.13652
17. Plant Mama Tatiana: First Signs of Spring. https://www.youtube.com/watch?v=lSr9smZJK-E, Accessed 19 July 2022
18. Painkra, G.P.: Giant bee foraging on chhuimui flower. https://www.youtube.com/watch?v=-iFtx1cmoEw, Accessed 19 July 2022
19. 98 Honey Bees: Honey Bees Foraging on Dandelions. https://www.youtube.com/watch?v=oroi3bvj_JE, Accessed 19 July 2022
20. Bisi Bisi Coffee: Busy bees pollinating lemon flowers. https://www.youtube.com/watch?v=ty6zB75Lk7c, Accessed 19 July 2022
21. Media Space: Honeybees pollinating coconut flower. https://www.youtube.com/watch?v=kcduLTz2Zoo, Accessed 19 July 2022
22. Prithvi Media Creations: Pollination Process Sunflower and Bee. https://www.youtube.com/watch?v=klEQG76OG0w, Accessed 19 July 2022
23. Ystwyth Valley Apple Breeders: Bees pollinating apple blossom: Beauty of Bath. https://www.youtube.com/watch?v=le0PAWrilZI, Accessed 19 July 2022
24. Ystwyth Valley Apple Breeders: Bees pollinating apple blossom: Lord Lambourne. https://www.youtube.com/watch?v=RMSFHlHf1tc, Accessed 19 July 2022

Modeling the Problem of Integral Geometry on the Family of Broken Lines Based on Tikhonov Regularization

N. U. Uteuliev[1], G. M. Djaykov[1,2](✉), and A. O. Pirimbetov[1,3]

[1] Nukus branch of Tashkent University of Information Technologies named after Muhammad al-Khwarizmi, Nukus, Uzbekistan
gafur_djaykov@mail.ru

[2] Tashkent University of Information Technologies named after Muhammad al-Khwarizmi, Tashkent, Uzbekistan

[3] National University of Uzbekistan named after Mirzo Ulugbek, Tashkent, Uzbekistan

Abstract. We consider the modeling of an integral geometry problem on a family of broken lines with a given weight function. One of the main problems in solving the problem of integral geometry is to construct an analytical formula that is expressed in terms of given integral data. In the general case, this process requires the creation of special computational algorithms based on the general theory of ill-posed problems. In this regard, it is advisable to use the regularization method to build stable algorithms for solving the problem.

In this paper, we obtained an analytic representation of the solution of the considered problem of integral geometry in the class of smooth finite functions. Considering when noisy integral data are always present during measurements, a stable algorithm is constructed based on the idea of Tikhonov's regularization for the numerical solution of the problem of integral geometry on a family of broken lines. The conducted numerical experiment shows that the developed algorithm effectively restores the image of the internal structure of the studied objects with sufficient accuracy.

Keywords: Problems of integral geometry · Tikhonov regularization · inversion formula · ill-posed problem

1 Introduction

Integral geometry is one of the major directions in the theory of ill-posed problems of mathematical physics and analysis [1,2]. This topical and rapidly developing area of modern mathematics is closely related to the theory of differential equations and mathematical physics, geometric analysis and has numerous applications in the mathematical study of seismic exploration problems, interpretation of geophysical and aerospace observation data, and in solving inverse problems of astrophysics and hydroacoustics [2–4]. Of particular interest to the

H. Zaynidinov et al. (Eds.): IHCI 2022, LNCS 13741, pp. 417–427, 2023.
https://doi.org/10.1007/978-3-031-27199-1_41

problems of integral geometry was the discovery of their importance for solving problems of computed tomography. Cormac's work [5] marked the beginning of the application of computed tomography methods in medicine. Based on the results obtained, computationally efficient algorithms for solving the corresponding tomography problems were developed (see [6]).

Let's consider the following problem of computed tomography.

Let a thin beam of radiation with intensity I_0 hit a layer of matter with linear absorption coefficient (attenuation) distribution $c(x)$ along the beam propagation.

The stationary equation of radiation transfer in a purely absorbing inhomogeneous medium, describing the process of radiation in matter, is a balance of particles or energy and has the form

$$\frac{dI(x)}{dx} = -c(x)\,I(x) \tag{1}$$

where $I = I(x)$ –is the intensity of the radiation, $c(x)\,\Delta x$– is the number of rays absorbed at the gap Δx.

Let us integrate the left and right parts of (1):

$$\int_{I_0}^{I_1} \frac{dI(x)}{I} dx = -\int_{L} c(x)\,dx$$

By denoting $-\ln \frac{I_0}{I_1} = q$, we get:

$$\int_{L} c(x)\,dx = q$$

So the scan yields linear integrals of the function $c(x)$ from the line L.

Thus, in X-ray computed tomography, projection data are obtained using an X-ray tube, X-ray beams of intensity I_0 are emitted, which, having passed through the substance, are recorded by detectors.

Further, we will consider this integral for a function of two variables in a different form:

$$\int_{L_k(l,\beta)} c(x,y)\,ds = q(l,\beta) \tag{2}$$

where L_k– is the family of lines where the integration is carried out and l– is the distance that the radiation travels through the substance.

The formula for reversal was first obtained in an paper by Johann Radon published in 1917 in the Proceedings of the Saxon Academy of Sciences [7]. However, Eq. (2) has the following features.

First, Eq. (2) as an integral equation has a non-standard form, namely, it does not have a kernel in an explicit form. Although the right-hand side $q(l,\beta)$ and the desired function $c(x,y)$ are two-dimensional, the integral is one-dimensional: the integral does not explicitly have lower and upper integration limits.

Second, the problem of solving Eq. (2) is ill-posed.

Thirdly, Radon obtained one of the solutions to Eq. (2), which is written as

$$c(x, y) = -\frac{1}{2\pi^2}\int_0^{\pi} d\beta \int_{-\infty}^{+\infty} \frac{\partial q(l, \beta)}{\partial l} \frac{dl}{l - (x\cos\beta + y\sin\beta)}. \quad (3)$$

But this solution is unstable due to the need to numerically calculate the derivative $\frac{\partial q(l,\beta)}{\partial l}$, moreover, the second integral in (3) is singular, since the denominator $l - (x\cos\beta + y\sin\beta)$ can vanish.

In other words, the solution of the problem is reduced to finding an explicit inversion formula or to finding the inverse Radon transform, which does not always exist and is unstable to small data changes [8].

The work [9] is devoted to methods of obtaining the internal structure of an object in a thin layer using a set of projection data measured from many angles. The work represents the first fundamental monograph, which summarizes the results of the initial, most rapid development stage of reconstructive tomography.

The works [10,11] outline the methodological foundations for constructing computational algorithms for solving mathematical problems of computed tomography diagnostics and show how these algorithms can be used in practice and create appropriate mathematical support for them.

The problem of recovery of a function by known integrals from it on a family of cones in the case of a space of even dimension was studied in the paper [12]. The singularity theorem was proved and a representation of the solution was constructed, stability estimates of the solution in Sobolev spaces were obtained and thereby weak uncorrectness of the problem was shown.

In [13] the Radon transform defined on circular cones, called the conic transform, is studied. Such transforms appear in various mathematical models arising in medical imaging, nuclear industry. This paper contains new results about the inversion of the conic Radon transform with a fixed scattering angle and derives simple explicit formulas for the inversion. Numerical simulations have been performed to demonstrate the effectiveness of the proposed algorithm in R^2.

In [14] the problems of integral geometry of the Volterra type on a family of broken lines in a strip are studied. Analytic representations of the solution in the class of smooth finite functions are obtained. Estimates of the solution of the problem in Sobolev spaces are presented, from which its weak uncorrectness follows.

In [15–17] linear problems of integral geometry in a strip with a given weight function were considered. An analytic representation of the solution in the class of smooth finite functions was obtained, and the singularity theorems of the solution of the problem were proved. An estimate of the stability of the solution in Sobolev spaces is given, from which it follows that the solution is weakly uncorrelated.

In paper [18] considers a new application of the TV-transform in primary radiation transmission imaging, more precicely as a coupled transmission-reflection tomography. In this new tomography, X-radiation emitted from the source falls on the mirror under an incidence angle and registered later by a detector. In this way when the object is scanned by all possible V-lines, then the standard rotational motion can be avoided.

In this paper, we consider the problem of recovering a function from noisy integral data on a family of broken lines with a given weight function. An explicit inversion formula is obtained. A stable algorithm for the numerical solution of the integral problem on a family of broken lines with an approximately given right-hand side is constructed based on the idea of Tikhonov's regularization.

Let us introduce the notations that we will use below:

$$(x,\, y) \in R^2, \quad (\xi,\, \eta) \in R^2, \quad \lambda \in R^1, \quad \mu \in R^1,$$
$$L_H = \{(x,\, y):\ x \in R^1,\, y \in [0,\, H],\ H < \infty\}$$

2 Inversion Formula

In the strip L_H consider families of broken lines, which are defined by the relations

$$\Upsilon(x,y) = \left\{(\xi,\eta):\ |x-\xi| = (y-\eta)\,tg\theta,\ 0 \le y \le H,\ \theta \in \left(0,\frac{\pi}{2}\right)\right\}$$

Consider the operator equations with respect to the function $u\,(x,y)$:

$$\int\limits_{\Upsilon(x,y)} g\,(x,\xi,y,\eta)\, u\,(\xi,\eta)\, ds = f\,(x,y) \tag{4}$$

(4) is an integral geometry operator on a family of curves $\Upsilon\,(x,y)$ (Fig. 1).

Let the function $f\,(x,y)$ be defined for all $(x,y) \in L_H$, the weight function has the form $g\,(x,\xi) = e^{-k(y-\eta)}$, $k > 0$.

Since $x - \xi = (y-\eta)\,tg\theta$ for $\xi < x$ and $\xi - x = (y-\eta)\,tg\theta$ for $\xi > x$ then equation (4) can be written as

$$\frac{1}{\cos\theta}\int\limits_0^y \left[u\,(x - z\,(y-\eta)\,,\eta) + u\,(x + z\,(y-\eta)\,,\eta)\right] e^{-k(y-\eta)} d\eta = f\,(x,y) \tag{5}$$

where $h = y - \eta,\ \ z = tg\theta$.

Let us apply the Fourier transform on the first variable to Eq. (5):

$$\cos\theta \cdot \hat{f}\,(\lambda,\, y) = \frac{1}{\sqrt{2\pi}}\int\limits_{-\infty}^{+\infty} e^{i\lambda x}\int\limits_0^y \left[u\,(x - z\,(y-\eta)\,,\eta) + u\,(x + z\,(y-\eta)\,,\eta)\right] e^{-k(y-\eta)} d\eta dx.$$

We get the equation:

$$\int\limits_0^y \hat{u}\,(\lambda,\eta)\cos\left(z\lambda\,(y-\eta)\right) e^{-k(y-\eta)} d\eta = \frac{\cos\theta}{2}\hat{f}\,(\lambda,y) \tag{6}$$

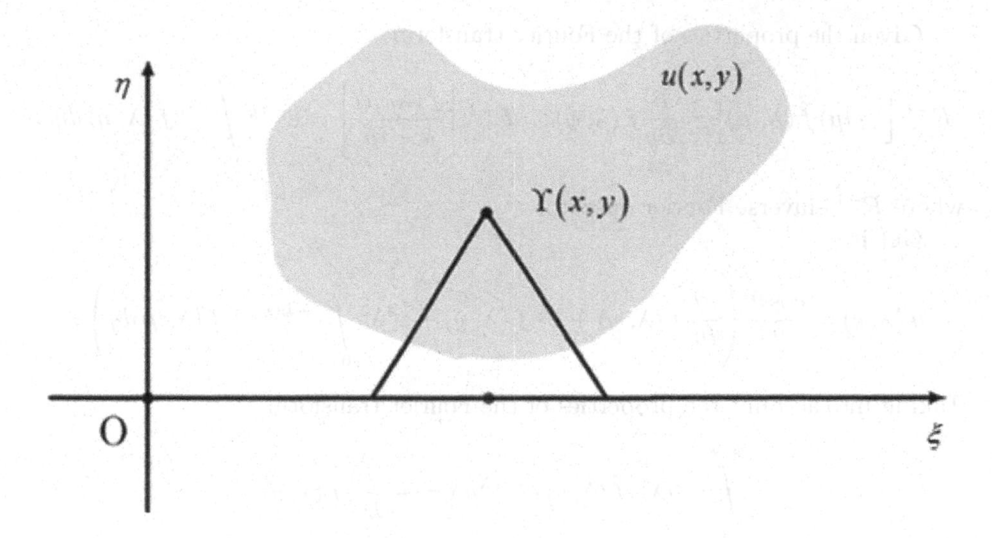

Fig. 1. Geometric illustration of the method.

Let us apply the one-sided Fourier transform to (6) by the variable y:

$$\frac{\cos\theta}{2}\hat{\hat{f}}(\lambda,\mu)=\int_{0}^{+\infty}e^{i\mu y}\int_{0}^{y}\hat{u}(\lambda,\eta)\cos\left(z\lambda\left(y-\eta\right)\right)e^{-k(y-\eta)}d\eta dy, \tag{7}$$

Thus, from Eq. (7) we get:

$$\hat{\hat{u}}(\lambda,\mu)\cdot I(\lambda,\mu)=\frac{\cos\theta}{2}\hat{\hat{f}}(\lambda,\mu) \tag{8}$$

where $\hat{\hat{u}}(\lambda,\mu)=\int_{0}^{\infty}e^{i\mu\eta}\hat{u}(\lambda,\eta)\,d\eta$, $\hat{\hat{f}}(\lambda,\mu)=\int_{0}^{\infty}e^{i\mu y}\hat{f}(\lambda,y)\,dy$.

Given the value of the integral

$$\int_{0}^{+\infty}e^{-px}\cos\left(bx\right)dx=\frac{p}{p^2+b^2}$$

we obtain that the integral of $I(\lambda,\mu)$

$$I(\lambda,\mu)=\int_{0}^{+\infty}e^{-(k-i\mu)\tau}\cos\left(z\lambda\tau\right)d\tau=\frac{k-i\mu}{\left(k-i\mu\right)^2+\left(z\lambda\right)^2}, atk\geq 0$$

From (8) we get

$$\hat{\hat{u}}(\lambda,\mu)=\frac{\cos\theta}{2}\left(k-i\mu+\frac{\left(z\lambda\right)^2}{k-i\mu}\right)\hat{\hat{f}}(\lambda,\mu),$$

Given the properties of the Fourier transform

$$F^{-1}\left[(-i\mu)\,\hat{\hat{f}}(\lambda,\mu)\right]=\frac{\partial}{\partial y}\hat{f}(\lambda,y),\quad F^{-1}\left[\frac{\hat{\hat{f}}(\lambda,\mu)}{k-i\mu}\right]=e^{-ky}\int\limits_0^y e^{k\eta}\hat{f}(\lambda,\eta)\,d\eta,$$

where F^{-1}–inverse Fourier operator.

Get it

$$\hat{u}(\lambda,y)=\frac{\cos\theta}{2}\left(\frac{\partial}{\partial y}\hat{f}(\lambda,y)+k\cdot\hat{f}(\lambda,y)+z^2\lambda^2\int\limits_0^y e^{-k(y-\eta)}\hat{f}(\lambda,\eta)\,d\eta\right),$$

Taking into account the properties of the Fourier transform

$$\int\limits_{-\infty}^{+\infty}(-i\lambda)^2\hat{f}(\lambda,y)\,e^{-ix\lambda}d\lambda=-\frac{\partial^2}{\partial x^2}f(x,y).$$

We arrive at the formula for reversal

$$u(x,y)=\frac{\cos\theta}{2}\left(\frac{\partial}{\partial y}f(x,y)+k\cdot f(x,y)-tg^2\theta\frac{\partial^2}{\partial x^2}\int\limits_0^y e^{-k(y-\eta)}f(x,\eta)\,d\eta\right).$$

We define the Radon transform on the family of broken lines as follows:

$$f(x,y)=\int\limits_{\Upsilon_1(x,y)}u(\xi,\eta)\,ds \tag{9}$$

where $\Upsilon_1(x,y)=\{(\xi,\eta):\,|x-\xi|=y-\eta,\ 0\le y\le H\}$.

Instead of the exact right-hand side $f(x,y)$ let us know an approximate value of f^δ such that $\left\|f^\delta-f\right\|_{L_2}\le\delta$, where δ–is the upper bound of the right-hand side. However, the problem of numerical differentiation of a function measured with errors is ill-posed and it is necessary to use Tikhonov's regularization method for stable differentiation of noisy functions [19,20].

The solution of Eq. (9) is

$$u(x,y)=\frac{1}{2\sqrt{2}}\left(\frac{\partial}{\partial y}f^\delta(x,y)-\frac{\partial^2}{\partial x^2}\int\limits_0^y f^\delta(x,\eta)\,d\eta\right)$$

Denote

$$\psi(x,y)=\frac{\partial}{\partial x}f^\delta(x,y) \tag{10}$$

$$\varphi(x,y)=\frac{\partial^2}{\partial x^2}\mu_1^\delta(x,y) \tag{11}$$

where $\mu_1^\delta(x,y)=\int\limits_0^y f^\delta(x,\eta)\,d\eta$.

Let's introduce a uniform grid in the rectangular area $D = [a, b] \times [c, d]$. Let us rewrite Eq. (10) as

$$A\psi = \int_c^d K(y,\tau)\,\psi(\cdot,\tau)d\tau = f^\delta(\cdot, y),$$

$$\begin{cases} K(y,\tau) = 1, c \le \tau \le y \le d, \\ K(y,\tau) = 0, \tau > y. \end{cases}$$

To ensure the stability of the solution of the last equation, the condition of the minimum of the smoothing functional is introduced

$$\Phi_\alpha[\psi, f^\delta] = \left\|A\psi - f^\delta(\cdot, y)\right\|_{L_2} + \alpha_y\|\psi\|_{L_2},\ \alpha_y > 0 \tag{12}$$

The expansion of (12) leads to the following second kind equation:

$$\alpha_y\psi(t,\cdot) + \int_c^d \bar{K}(t,s)\,\psi(s,\cdot)\,ds = \int_t^d f^\delta(x,\cdot)\,dx \tag{13}$$

where $\bar{K}(t,s) = d - \max\{t,s\}$.

We discretize Eq. (13) on a uniform grid. As a result, we obtain a system of linear algebraic equations of the form

$$\alpha_y\psi_i + \sum_{k=1}^{n_y}\bar{K}(t_i,s_k)\,h_k\psi_i = Q_i,\ Q_i = \int_{t_i}^b \tilde{f}^\delta(x,\cdot)\,dx, \tag{14}$$

$$i = \overline{1, n_y},\ h_1 = h_{n_y} = \frac{h_y}{2},\ h_k = h_y,\ k = \overline{2, n_y - 1}.$$

Let $M-$be a matrix with elements $M_{ik} = \bar{K}(t_i, s_k)\,h_k$. Then the system of Eqs. (14) with respect to the vector with components $(\psi_1, \psi_2, ..., \psi_{n_y})$ can be written as

$$M_{\alpha_y}\psi \equiv M\psi + \alpha_y E\psi = Q, \tag{15}$$

where $Q-$is a vector with components $(q_1/2, q_2, ..., q_{n_y})$, and $E-$is a unit matrix.

As D we can take the following rectangular area $D = [0, 1] \times [0, 1]$.

Now consider the second equation from (11)

$$\varphi(x,y) = \frac{\partial^2}{\partial x^2}\mu_1^\delta(x,y)$$

The equation can be written as

$$\int_0^1 K_1(x,s)\,\varphi(s,\cdot)ds = \mu_1^\delta(x,y),$$

$$K_1(x,s)=\begin{cases}(1-s)\,x,\ 0\le s\le x,\\(1-x)\,s,\ x\le s\le 1.\end{cases}$$

Split the segment as before [0; 1] on the axis Ox and [0; 1] n the axis Oy into n_x-1 and n_x-1 parts, respectively. $x_i=(i-1)\,h_x, y_j=(j-1)\,h_y$.

$$(x_i-1)\int\limits_0^{x_i} s\varphi(s,\cdot)ds + x_i\int\limits_{x_i}^{1}(1-s)\,\varphi(s,\cdot)ds = \mu_1^\delta(x_i,y_j)\,. \qquad (16)$$

After discretization on a grid and approximation of the integral equation by quadrature formulas, Eq. (16) is reduced to the solution of a system of linear algebraic equations. Using the trapezoidal method, Eq. (16) with respect to vector φ with components $(\varphi_1,\varphi_2,...,\varphi_{n_x-1})$ can be written in the form

$$B\varphi=\mu_1^\delta, \qquad (17)$$

Thus, the problem (17) is reduced to the solution of the SLAE. The matrix of this system is symmetric.

$$\left(B^TB+\alpha_xE\right)\varphi=B^T\mu_1^\delta,$$

$$\varphi=\left(B^TB+\alpha_xE\right)^{-1}B^T\mu_1^\delta,$$

where α_x-is the regularization parameter and E-is the unit matrix.

3 Numerical Experiment

Consider the regularized Tikhonov algorithm in a numerical experiment for the model phantom Shepp-Logan. The error function $f(x,y)$ was generated using a normalized random number probe. Figure 2 a shows the original phantom with 256×256 sampling. Figure 2 b, c, d show the results of Tikhonov's regularization algorithm with 5% noise. The results show that the most accurate recovery is obtained when N = 256×256 and $\alpha_x=\alpha_y=10^{-10}$.

For the choice of the regularization parameter α_y, the fitting method described in [8] is used. For the choice of the regularization parameter α_y the dependence of the relative error of the solution $\psi_{\alpha_y}(\cdot,y)$ on the exact solution $\psi(\cdot,y)$ solving SLAE (14) for a number of values of the regularization parameter was constructed.

$$\sigma(\alpha_y)=\frac{\left\|\psi_{\alpha_y}(\cdot,y)-\psi(\cdot,y)\right\|}{\left\|\psi(\cdot,y)\right\|} \qquad (18)$$

Accordingly, the regularization parameter α_x is chosen according to the above scheme using Eq. (18).

Figure 3 shows the results of a section of the Shepp-Logan phantom. Figure 3a shows the results of restoration at N=64×64 with 5% noise. Solid line - exact

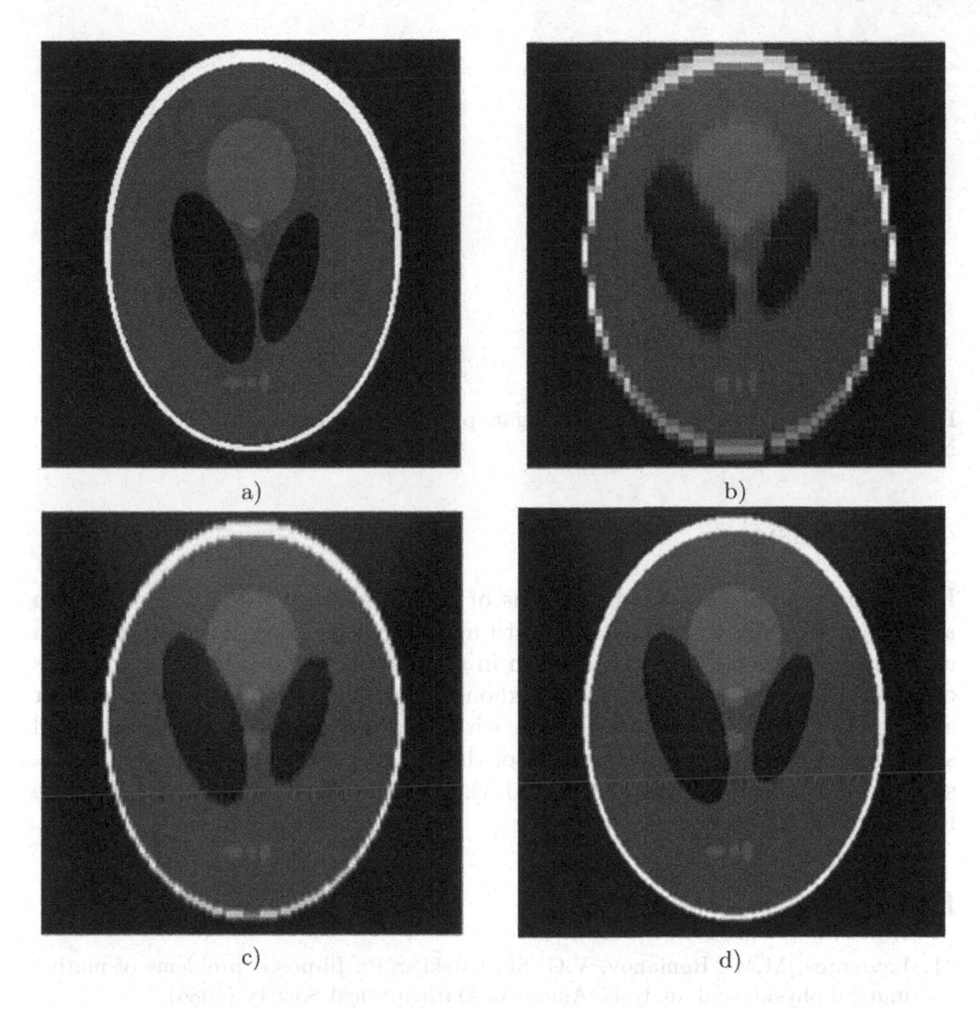

Fig. 2. Sheep-Logan phantom reconstruction. a) Original phantom b) Reconstruction at $N = 64 \times 64$ with noise 5%, c) Reconstruction at $N = 128 \times 128$ with noise 5%, d) Reconstruction at $N = 256 \times 256$ with noise 5%.

solution for, asterisks - restored solution for and $\alpha_x = \alpha_y = 10^{-10}$, discontinuous line -restored solution for $\alpha_x = \alpha_y = 10^{-5}$. Figure 3b shows the results of restoration at N=256 × 256 with 5% noise. Solid line - exact solution, asterisks - restored solution at $\alpha_x = \alpha_y = 10^{-10}$, broken line - restored solution $\alpha_x = \alpha_y = 10^{-5}$.

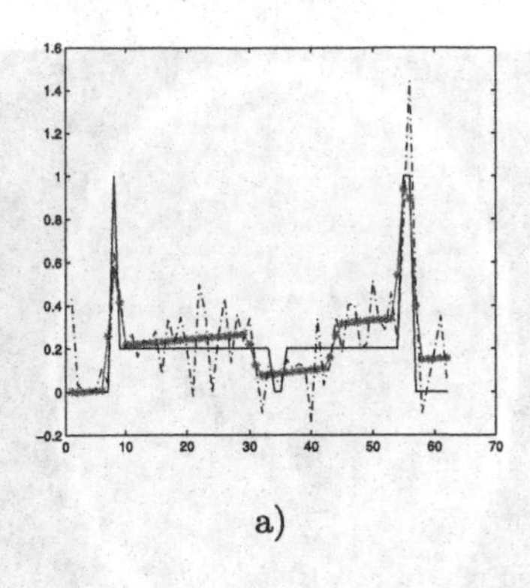

a)

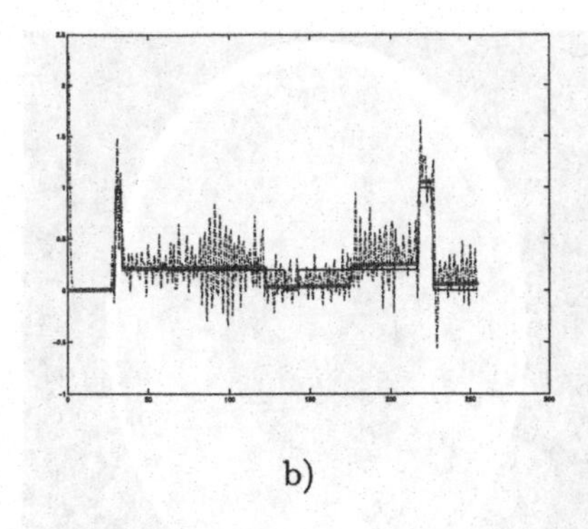

b)

Fig. 3. Cross section of the Shepp-Logan phantom. a) when $N = 64 \times 64$ b) when $N = 256 \times 256$

4 Conclusion

In this paper, we considered solutions of the problem of integral geometry in a strip on a family of broken lines with approximately given integral data. An analytic representation of the solution in the class of smooth finite functions is obtained. An algorithm based on Tikhonov's regularization is constructed for solving an integral geometry problem with an approximately given right-hand side. Numerical and graphical results of the application of this algorithm to the solution of the problem were presented, thereby showing the effectiveness of the results of the developed algorithm.

References

1. Lavrentev, M.M., Romanov, V.G. Shishatski, S.P.: Ill-posed problems of mathematical physics and analysis. American Mathematical Society (1986)
2. Ivanov, V.K., Vasin, V.V., Tanana, V.P.: Theory of linear ill-posed problems and its applications. In: Theory of Linear Ill-Posed Problems and its Applications. De Gruyter (2013)
3. Gelfand, I.M., Gindikin, S.G., Graev, M.I.: Selected topics in integral geometry. American Mathematical Society (2003)
4. Natterer, F.: Inversion of the attenuated Radon transform. Inverse Prob. **17**, 113–119 (2001)
5. Cormack, A.M.: The Radon transform on a family of curves in the plane. Proc. Am. Math. Soc. **83**, 325–330 (1981)
6. Natterer, F.: The mathematics of computerized tomography. Society for Industrial and Applied Mathematics (2001)
7. Radon, J.: Uber die Bestimmung vor Functionen durch ihre Inte-gralwarte langs gewisser Maannigfritigkeiten. Ber. Verh. Sachs. Akad. **69**, 262–277 (1917)
8. Sizikov, V.S.: Sustainable methods for processing measurement results. Spetslit, Saint-Petersburg (1999). [in russian]
9. Herman, G.T.: Fundamentals of Computerized Tomography, 2nd edn. Springer, London (2009). https://doi.org/10.1007/978-1-84628-723-7

10. Tikhonov, A.N., Arsenin, V.Ya., Timonov A.A.: Mathematical problems of computed tomography. Nauka, Moscow (1987). [in russian]
11. Sizikov, V.S.: Inverse Applications and MatLab. Lan, (2011). [in russian]
12. Begmatov, A.K.: The integral geometry problem for a family of cones in then-dimensional space. Sib. Math. J. **37**, 430–435 (1996)
13. Gouia-Zarrad, R., Ambartsoumian, G.: Exact inversion of the conical Radon transform with a fixed opening angle. Inverse Prob. **30**, 045007 (2014)
14. Begmatov, A.H., Djaykov, G.M.: Numerical recovery of function in a strip from given integral data on linear manifolds. In: 2016 11th International Forum on Strategic Technology (IFOST), pp. 478–482 (2016). https://doi.org/10.1109/IFOST.2016.7884159
15. Djaykov, G.M., Arziev, A.D.: Analytical reconstruction of functions from their integral data on a family of straight line segments with a weight function in the form of polynomial. Sci. Educ. Nukus, Uzbekistan **2**, 197–202 (2021)
16. Uteuliev, N.U., Djaykov, G.M., Yadgarov, Sh.A.: Analytical and numerical reconstruction of internal structure of the objects in a family of straight-line segments. In: 2019 International Conference on Information Science and Communications Technologies (ICISCT), pp. 1–4 (2019). https://doi.org/10.1109/ICISCT47635.2019.9011979
17. Begmatov, A.H., Djaykov, G.M.: Linear problem of integral geometry with smooth weight fucntions and perturbation. Vladikavkaz Math. J. **17**(3), 14–22 (2015)
18. Moayedi, F., Azimifar, Z., Fieguth, P., Kazemi A.: A novel coupled transmission-reflection tomography and the V-line Radon transform. In: 2011 18th IEEE International Conference on Image Processing, pp. 413–416 (2011). https://doi.org/10.1109/ICIP.2011.6116537
19. Tikhonov, A.N., Arsenin, V.Ya.: Methods for solving ill-posed problems. Nauka, Moscow (1979). [in russian]
20. Verlan, A.F., Sizikov, V.S., Mosensova, L.V.: Method of computational experiments for solving integral equations in the inverse problem of spectroscopy. In: Electronic modeling, Institute of Modeling Problems in Energy named after. G.E. Pukhova NAS of Ukraine, vol. 33, no. 2, pp. 3–10 (2011). [in russian]

A Framework for Privacy-Preserved Collaborative Learning in Smart Factory Environment

Ericka Pamela Bermudez Pillado, Tori Bukit, Sean Yonathan Tanjung, Hyun-Woo Lim, Ignatius Iwan, Bernardo Nugroho Yahya(✉), and Seok-Lyong Lee(✉)

Hankuk University of Foreign Studies, Yongin, South Korea
{ericka,tori,syt,apu91040,ignatiusiwan,bernardo,sllee}@hufs.ac.kr

Abstract. Integration of artificial intelligence (AI) in a work environment is the key factor of the industry 4.0 revolution. Vertical and horizontal integration among different parties become relevant but privacy could be the issues. Federated learning (FL) has been widely used as a decentralized mechanism to cope with privacy preserving problems. However, the initial setup for FL in the real-world application is difficult and requires a lot of human involvement in daily operation. In addition, project key performance indicator which is important to assess enterprise collaboration accomplishment among partners has been rarely addressed. In this work, we develop a horizontal FL framework and modular dashboard to enable collaborative training among different parties. The module is available for key performance monitoring operational view on both server and clients in three aspects; computer-related indicators, machine learning-related indicators, and manufacturing related indicators. Using the proposed framework, it can provide insight to the user regarding FL results.

Keywords: Modular Dashboard · Federated Learning · Privacy Preserving Mechanism

1 Introduction

In the last few years, factory has been dealing with smart factory transformation demand to utilize AI assistance (digitalization). In addition, digitalization including vertical and horizontal integration are also being pursued to enable information flow between factory floor to enterprise system which can be used to train an AI model. However, enterprises have their own regulation regarding sharing data. Due to the data protection regulations, there should be a privacy preserving mechanism to utilize training among different factories. FL is a solution that enables parties to build AI model together just by sending model updates without sharing any local data. FL is also considered as helping horizontal and vertical integration on which it enables opportunity to establish partnership with different companies and build end-to-end information flow in-house (headquarter and branch factories).

H. Zaynidinov et al. (Eds.): IHCI 2022, LNCS 13741, pp. 428–434, 2023.
https://doi.org/10.1007/978-3-031-27199-1_42

Although it is a promising approach, integrating FL into a system is challenging due to the interpretability for the decision makers. The nature of FL protects viewable data and hinders immediate interpretation. Dashboard can be used as a platform to provide relevant performance indicators for users. Previous works [1–3] already developed FL dashboard on their application. However, the dashboard had been utilized as a hardware resources monitoring and performed in simulation environment. Another popular FL work [4] included the dashboard which requires users to use only their Web Service, leading to the lack of extensibility.

In this work, we propose a framework that utilizes FL as a strategy to collaborate between individual factories and develop visualization dashboards. The dashboard development is to provide relevant information and performance indicators (PI) in three aspects; computer-related PI, machine-learning-related PI, and manufacturing-related PI. The proposed dashboards are considered to be a loosely coupled with FL module so it can be modular and run separately.

This paper is organized as follows. In Sect. 2, we describe the FL application technologies that utilized dashboard as related work. Section 3 explains the proposed framework component for smart factory. Next, Sect. 4 shows the experiment using the proposed framework in real devices. Finally, Sect. 5 summarizes the paper and discusses about the future work.

2 Related Works

FL has been applied into various applications such as smart healthcare, smart keyboard, smart transportation, and smart factory. However, there is no FL implementation for a smart factory that takes into consideration of the dashboard visualization. In the FL for smart factory, the privacy preserving mechanism protects the data for each party and conceals the proper features so that it might puzzle decision makers. Meanwhile, performance indicators are necessary to evaluate the success of collaboration among parties in the FL project.

Several works had addressed the use of dashboard in FL environment. Shariati *et al.* [1] demonstrated the FL implementation in the case of Metro Optical Networks. The authors created a real-time monitoring dashboard using Grafana. The dashboard showed the usage of clients' computational resources. Recently, Lianto *et al.* [2] introduced the web user interface to showcase local differential privacy in a FL setting through simulation. Another work proposed graphical interfaces to illustrate emulated network connections, including the display of training progress with a focus on network parameters [3]. Unfortunately, those previous works only monitored the hardware computational resource and some of the works were dedicated only to the simulated environment.

FedML [4] is a real-life FL Framework that provides a dashboard for FL participants, coined as FedML MLOps. Instead of visualization, the performance of a client in real-time in FedML is shown by logs. In addition, the FedML Ops is limited in terms of the extensibility. The dependency on FedML may limit the extensibility of FL implementation in the real world. Table 1 shows a comparison within FL works that possess a dashboard for visualization.

In summary, there are still some issues to deploy dashboard visualization for the FL implementation in the smart factory. We need to deploy the dashboard that is extensible, easy to setup, and able to monitor the performance of different factories in the FL project. In the next section, we will describe our solution to handle these issues.

Table 1. Comparison with the previous work

Feature	Lianto et al. [2]	Shariati et al. [1]	Conway-Jones [3]	FedML [4]	**Ours**
Real-life setting	No	Yes	No	Yes	**Yes**
Client performance monitoring	No	No	No	Yes*	**Yes**
Hardware resources monitoring	No	Yes	No	Yes	**Yes**
Standalone implementation	Yes	No	Yes	No	**Yes**
Modular	Yes	No	No	No	**Yes**

*Provides experimental interface to show basic stats but the setup requires the interference of an expert. Real-time is only on the logs.

3 Proposed Framework

This section aims to describe the modular horizontal FL framework. Our proposed framework focuses on the type of Horizontal FL, in which each of client's dataset has the same features. Figure 1 shows the overview of the proposed framework. There are three main modules in this framework: Modular dashboard, FL Server, and FL Client. The FL Client module contains two main components, **Client Module** to receive training instruction from the server and **Client's Data** component to pre-process the client data. The following contains the workflow of our framework:

1. The FL server specifies which FL algorithm will be performed during this task.
2. Next, clients join the FL Server host with their previously defined parameters for differential privacy. After that, the FL Server selects the online clients to participate in the training round.
3. FL Server sends the instructions for performing local training and the latest global model to the selected clients.
4. In the next step, the updated model's weights and the statistical data for the server-side dashboard will be sent back to the server from each client. The FL Server aggregates weights using the selected FL algorithm.
5. FL Server sends back the aggregated weights to update to all client's model. The Google Remote Procedure Call (gRPC) communication protocol is used for all communication between a server and clients.

6. For each round, FL client send information such as accuracy, loss, and hardware resources to the backend server for dashboard.
7. At any point of time, FL client can access visualization of their private data in client-side dashboard. Data in client-side dashboard will be shown in real-time.

FL applications on manufacturing have overlooked the user experience perspective. As solutions aimed to improve processes, the user experience is a crucial role in the success of the implementation. Flower Framework [5] has demonstrated to be a promising option for FL in real-life scenarios. However, its configuration and usage require some degree of expertise and external application.

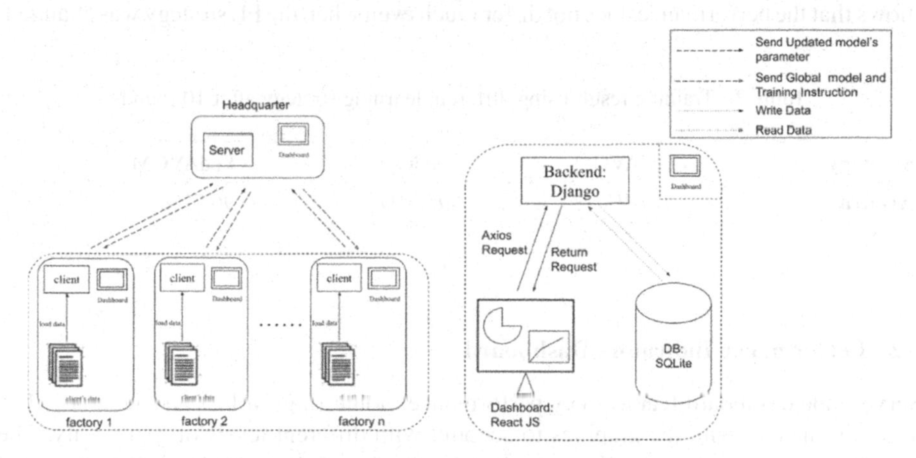

Fig. 1. Proposed framework workflow (left: FL Framework, right: Modular dashboard)

Our framework does not require any third-party software to be implemented, as it can be deployed to any devices. It is a scalable solution, as clients can join the training without human interaction. The dashboard is a modular solution that can be implemented in both client and server. Django is used as backend, and essentially consist of two parts. First, Django ORM creates models, manages database, and queries from a SQLite database. Second, Django REST Framework serializes data from the Django ORM, allows the access, and updates using a RESTful API. React is used for the user interface and utilize GET data from the database using Axios to make the requests.

4 Experiment

In this section, we present the dataset that we used to test the robustness of our framework. We conducted experiment using real world devices as clients that will connect by utilizing our framework. For the dataset, we used Case Western Reserve University (CWRU) bearing dataset for our experiment.

4.1 Framework Evaluation Result

Flower FL framework is used to make the framework development easy [5]. Experiment has been done using 4 devices, 1 device as the server and 3 devices as the clients. This setting was preferred considering the small number of clients that join the training session. FL was carried out for 10 global epochs. From total 17,987 data, we split 5,397 for validation and 12,590 for training samples. The training samples are further divided randomly for each client.

The experiment compares 3 FL strategies available within the Flower FL framework. The evaluation of the model with aggregated weights was done centrally, from the server. It proves that the framework still works well under different types of FL strategies. Table 2 shows that the performance does not differ much even when the FL strategy was changed.

Table 2. Training result using different learning strategy after 10 rounds

Strategy	FedAVG	FedOpt	FedAVGM
Accuracy	91.64%	96.61%	96.72%

4.2 Performance Indicators Dashboard

Server-side dashboard features key performance monitoring for loss and accuracy. The user can interact with the graphics to account with different levels of granularity. The detail information of each client can be observed, and the graphics will be updated accordingly. The dashboard also provides geo-visualization of the status of clients.

We believe that the proposed framework is quite distinct compared to other FL frameworks due to its strengths. The primary strength of this framework is the visualability of the relevant PI in FL, i.e., (a) computer-related PI (e.g., CPU usage, memory usage), (b) machine-learning-related PI (e.g., loss, accuracy), and (c) manufacturing-related PI (e.g., fault detection, fault frequency in the last hours). The listed features for the dashboard can be observed in Fig. 2 and Fig. 3. Despite the strengths presented, the framework also has weaknesses. The cost of read and write operations to the database are yet to be studied.

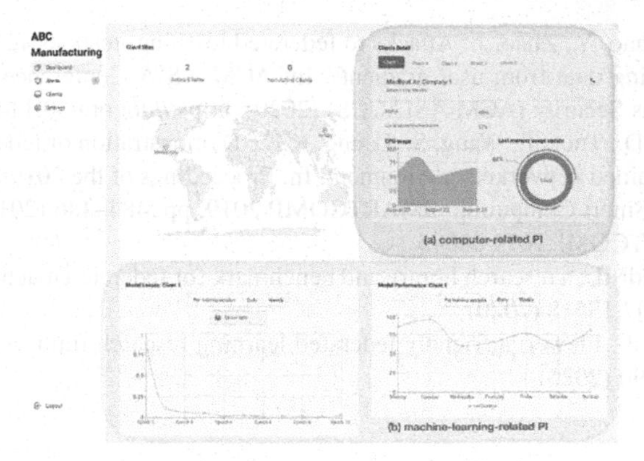

Fig. 2. Server-side Dashboard for monitoring training condition

5 Conclusion and Future Work

We developed a framework that can be utilized by factories to conduct collaborative learning to advance their digital transformation. The framework was developed using robust well-known tools and supported by visualization dashboards for decision making based on result from their collaboration. For future work, our development will specialize on visualization requirement in the manufacturing environment.

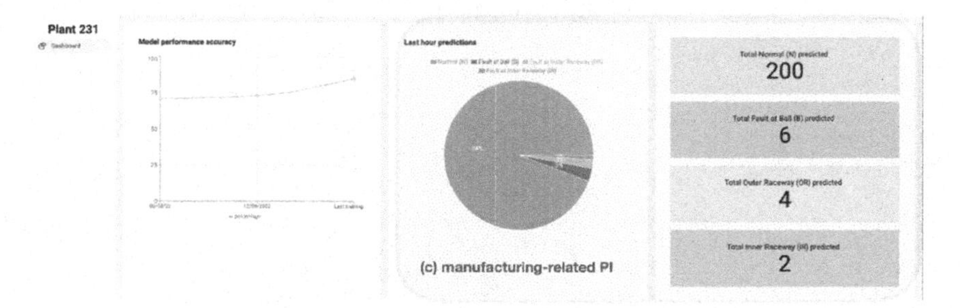

Fig. 3. Client-side dashboard

Acknowledgment. This research was financially supported by the Ministry of Trade, Industry and Energy (MOTIE) and Korea Institute for Advancement of Technology (KIAT) through the International Cooperative R&D program. (Project No. P0022316).

References

1. Shariati, B., Safari, P., Mitrovska, A., Hashemi, N., Fischer, J.K., Freund, R.: Demonstration of Federated Learning over Edge-Computing Enabled Metro Optical Networks (2020). https://doi.org/10.1109/ECOC48923.2020.9333246

2. Lianto, H.A., Zhao, Y., Zhao, J.: Attacks to federated learning: responsive web user interface to recover training data from user gradients. In: ACM ASIA Conference on Computer and Communications Security (ACM ASIACCS) (2020). https://doi.org/10.1145/3320269
3. Conway-Jones, D., Tuor, T., Wang, S., Leung, K.K.: Demonstration of federated learning in a resource-constrained networked environment. In: Proceedings of the 2019 IEEE International Conference on Smart Computing, SMARTCOMP 2019, pp. 484–486 (2019). https://doi.org/10.1109/SMARTCOMP.2019.00095
4. He, C., et al.: FedML: a research library and benchmark for federated machine learning. arXiv Prepr. arXiv:2007.13518 (2020)
5. Beutel, D.J., et al.: Flower: a friendly federated learning research framework. arXiv Preprint arXiv:2007.14390 (2020)

Co-creating Computer Supported Collective Intelligence in Citizen Science Hubs

Aelita Skarzauskiene(✉) and Monika Mačiulienė

Vilnius Gediminas Technical University, Vilnius, Lithuania
aelita.skarzauskiene@vilniustech.lt

Abstract. A Collective Intelligence system is a knowledge network that emerges from human-machine mediated interaction between individuals with personal knowledge. Citizen Science aims to connect society and science by developing a Collective Intelligence ecosystem, which entails collaboration between all QH stakeholders: the public society, researchers and universities, NGOs, governments and funding agencies. The development of crowdsourcing platforms and networks enables volunteers to contribute to different research projects. In addition, artificial intelligence and machine learning technologies extend human intelligence capabilities in the ecosystem. The presented conceptual model was developed based on theoretical insights to understand the complexity of relationships at different levels of human-computer interaction in the Collective Intelligence ecosystem. Citizen Science Hubs are considered an ideal environment for collective intelligence to emerge and bridge the intellectual strengths of humans and machines to take advantage of enormous amounts of data advanced across disciplines.

Keywords: Collective Intelligence · Co-creation · Citizen Science

1 Introduction

1.1 ICT Supported Collective Intelligence Systems

Different types of human groups can be considered a source of collective intelligence. Luo et al. [1] define community as all types of "human groups in which the members share similar characteristics, have common interests or views, work on analogue purposes." Lykourentzou et al. [2] describe an online community, focusing on a critical mass of users in the system. "The individual goals, linked to group actions and supported by technology facilitation, may result in a higher-level intelligence and benefit of the community with real social value" [2]. Like in the case of "swarm intelligence" [4] in natural systems, Collective Intelligence systems incorporate human beings and supporting machine systems. Enabled by information communication technologies (ICT) and under the right circumstances, "the community could generate more beneficial outcomes than a traditional community does" [3]. The innovative capabilities of modern ICT may help communities to complete their information processing tasks because artificial intelligence offers an effective channel for massive data, information and knowledge exchange.

H. Zaynidinov et al. (Eds.): IHCI 2022, LNCS 13741, pp. 435–441, 2023.
https://doi.org/10.1007/978-3-031-27199-1_43

To use collective intelligence potential, several innovative Research Performing and Funding Organisations (RPFOs) in the EU have established interdisciplinary hubs for assisting and stimulating outstanding citizen science. These Citizen Science Hubs are an ideal environment for collective intelligence to emerge and can be considered CI systems. Citizen Science (CS) projects include members of the civic society and other QH stakeholders as active research participants. Citizen scientists usually use crowd-sourced data that would be hard to obtain otherwise due to time, geographic, or resource constraints. Collective efforts of humans outperform automatic procedures in identifying and classifying data collected in various ways. Typical examples include identifying protein or galaxy structures in massively collected relevant images.

This research project presents a conceptual model of the Collective Intelligence ecosystem focusing on human-computer interaction to meet the needs of the QH stakeholders by establishing Citizen Science hubs. The research provides insights into the collaboration specificities and operating models at Micro, Meso and Macro levels supporting Citizen Science communities to deliver intended academic outcomes, leading to inclusive and society-oriented science.

1.2 Citizen Science Hub as Collective Intelligence Ecosystem

A Collective Intelligence system is a knowledge network that emerges from human-machine mediated interaction between individuals with personal knowledge. Citizen Science Hubs aim to connect society with science by developing a Collective Intelligence ecosystem, which entails collaboration between all QH stakeholders: the public society, researchers and universities, NGOs, public authorities and funding organizations [5]. The ecosystems' intellectual capacities emerge in the form of collective intelligence, such as created knowledge and ideas, suggested problem-solving methods, structured and shaped public opinions and discussions, created innovations and prototypes. Networks and crowdsourcing projects offer volunteers possibilities to contribute to different research projects. In addition, artificial intelligence and machine learning technologies extend human intelligence capabilities in the ecosystem. The Structural Model of Community Intelligence [1] explains how the knowledge-related activities of the participants generate community-level intelligence. Firstly, the platform should offer "a memory system that stores collected information and knowledge, similar to the human brain's memory system". Secondly, it is essential for the community to be capable of collective problem-solving and using the stored knowledge to solve problems. Undoubtedly, "the quality of human-computer interaction influences the CI and the system's performance at the current knowledge level" [6]. "The knowledge network consists of a technical media network for information and knowledge transfer, a human network created by community members, and a content network, which hosts knowledge in human-machine systems" [1]. To improve collaboration and communication between citizens and scientists various digital infrastructures, such as mobile applications, sensors, and games, have been developed worldwide. Data collection and processing, information delivery and visualization technologies were adapted to expand the scale and scope of projects and protocol design [7–9]. Citizen science platforms connect the intellectual strengths of humans and machines to take full advantage of data being advanced across disciplines [10].

The conceptual model was developed based on theoretical insights to understand the complexity of relationships at different levels of human-computer interaction in the Collective Intelligence ecosystem. The framework presented in Fig. 1 below provides a holistic view of the Citizen Science Hub as a co-creative collective intelligence ecosystem. In this paper, the Collective Intelligence ecosystem reflects a system in which actors work collectively to co-create beneficial public value. The proposed model has three dimensions – actors, content and processes distributed on three levels – Micro, Meso and Macro between economic and social actors within the networks. The actors can be identified as QH stakeholders participating in the CI ecosystem, including their resources. Hardy et al. [11] suggest that despite the powerful potential of collaboration to create value, not all interactions are successful. Many collaborations do not produce innovative solutions or balanced decisions, and some even fail to start collective action. It is crucial to understand the possible roles of actors involved in ICT-enabled co-creation [12–17]. The identification of roles explains how actors collaborate in service systems [18]. Each participant in the system is a potential resource for other actors within the ecosystem. Interactions happen through resource creation, sharing, obtaining, and integrating. Specific social roles enable co-creation, but there is limited research on such roles and how they function together [19]. Åkesson [20] argues that ecosystem's productivity is based on the heterogeneity of participants and their resources.

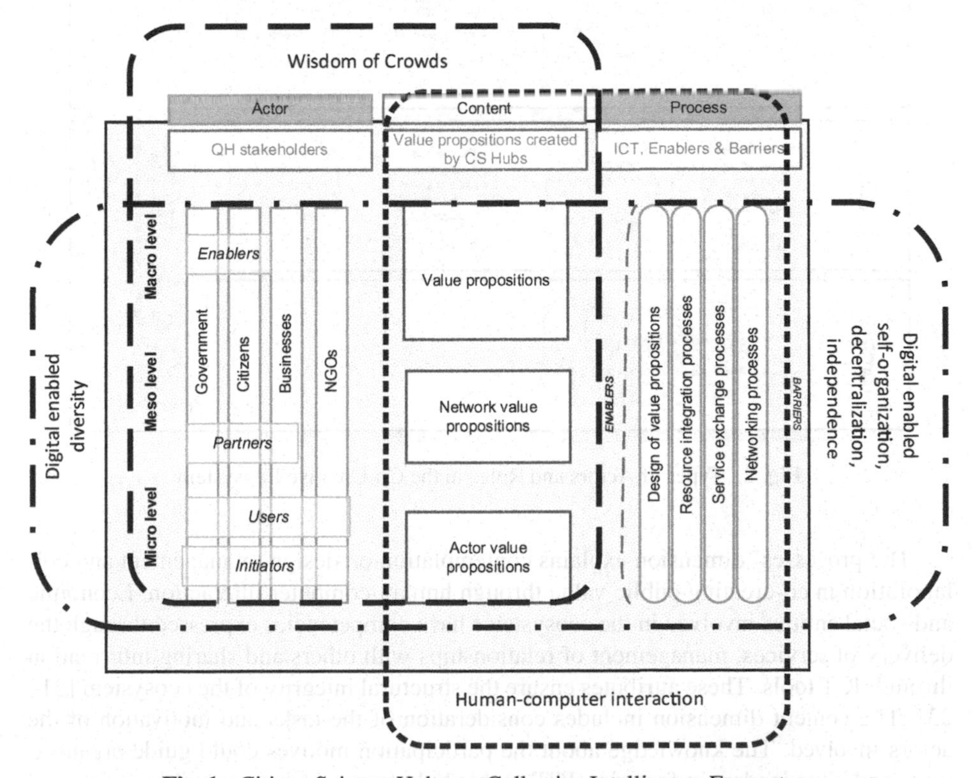

Fig. 1. Citizen Science Hub as a Collective Intelligence Ecosystem

Although different actors are involved in any ecosystem is huge, it is possible to define segments and their relationships. Figure 2 below shows the five types of actors identified in the research literature and the types of roles they can perceive on different levels of the ecosystem. All types of users are considered to be the actors involved in platform activities receiving ICT-supported services. The role of the contributors is similar to the user's role but is more interactive. It is related to such collaborative efforts as suggesting ideas, voting, analysing and reporting issues, and creating content with other contributors beneficial for the active processes of the platform. The partners share operational resources with platform initiators and community developers. This role is performed by developing beneficial relationships in the whole ecosystem, but without losing separate actors‘ autonomy. The role of sponsors is important because they provide financial resources to support platform activities. The sponsoring happens by applying for governmental or business funding. The roles of actors could be switched between actors, which means that any group could initiate, contribute or sponsor the platform.

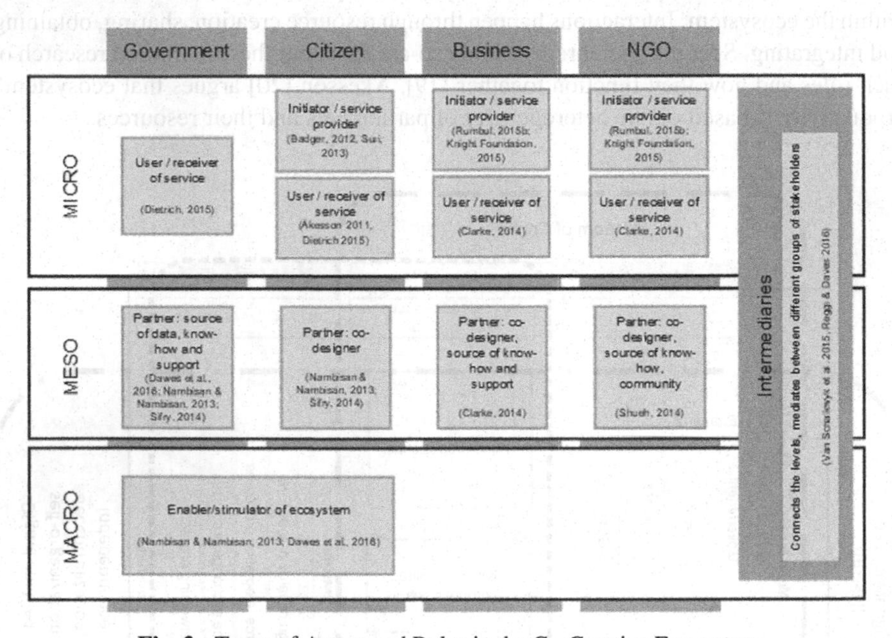

Fig. 2. Types of Actors and Roles in the Co-Creative Ecosystem

The processes' dimension explains contemplation on design, management and collaboration in co-creating public value through human-computer interaction. Economic and social entities involved in the ecosystems have competencies expressed through the delivery of services, management of relationships with others and sharing information through ICT tools. These attributes ensure the structural integrity of the ecosystem [21–23]. The content dimension includes consideration of the tasks and motivation of the actors involved. The knowledge about the participation motives could guide organisations and civic leaders in fostering ICT-enabled platforms. Value propositions are used in integrated systems to connect mutually interested actors. Following the logic of Service Science, the actors able to develop the most compelling value propositions will perform the best. Lusch & Webster [22] stress the importance of constant revision of value propositions responding to changing needs of stakeholders. CS Hubs can be considered testbeds for co-creation of public value because of the involvement of various social groups, the application of social technologies and their social orientation. Due to their small scale, the networks of Citizen Science Hubs are more transparent and open to analysis than the more complex national systems of ICT-enabled governmental services.

Hence, the services offered by the Citizen Science Hubs are only inputs to public value-creating activities in the context of civic society. The Micro level refers to the internal direct service-for-service exchange between users of the platforms. The Meso level explains the indirect service-for-service exchange with the external stakeholders, partners or competitors. The Macro level is related to the complex external relationships between different systems with diverse interests. No stakeholder has all the resources needed to reach their goals, and each actor is a potential source of resources for other actors within the ecosystem. Interactions happen through the digital enhanced creation, sharing, obtainment, and integration of the resources. Collective intelligence in the form of public value emerges in the ecosystem when several actors collectively offer other users access to resources, including people, technologies, and information.

2 Validation of the Collective Intelligence Ecosystem Framework

Collective Intelligence (CI) ecosystem framework was expert-validated and calibrated through a dedicated digital workshop with field experts. In particular, the workshop's purpose was to expert validate the framework by examining its dimensions and collaborations, reflecting on its aspects and providing suggestions for its practical application by establishing Citizen Science Hubs. An information document describing the core structure of the framework and the workshop agenda was sent to the participants in advance of the workshop. The research was initiated to implement the H2020 project INCENTIVE ("Establishing Citizen Science Hubs in European Research Performing and Funding Organisations to drive institutional change and ground Responsible Research and Innovation in society"). The workshops were organised by Vilnius Gediminas Technical University on 20 May 2022 on MS Teams and lasted 2 h. During the workshop, 12 experts were asked to provide their feedback and make suggestions for improving the CI ecosystem framework for adapting it for Citizen Science Hubs. The experts were identified through the author lists European Commission H2020 public deliverables and publications regarding RRI and citizen science-related projects. The workshop with the experts provided information on idyllic co-creation of public value, i.e. the experts discussed the potential and desired roles of governments, citizens and other actors in the ecosystem.

During the expert validation workshop, six groups of actors – governmental organisations, citizens, business companies, NGOs, universities and media, – were confirmed, and three more generalised actor groups – associations, public organisations and international organisations suggested to include. The discussion released insights about increasing society's involvement and openness. Not all citizens and organisations have to be active. However, there is a need for intermediaries, civic leaders, and active citizens who could express the importance of active citizenship and transparency, explain the data and support collaboration between citizens and governments. The intermediary role could be performed chiefly by specialists with skills and knowledge in IT, open data, and governmental processes. Intermediaries explain the complex public sector information and processes to the other groups in the system and make connections easier. Intermediaries serve as the actor connecting the micro, meso and macro levels. In sequence, all experts' comments were considered and incorporated into the final design of the CI ecosystem framework.

3 Conclusions and Insights for Future Research

The long-term vision of CI systems is "to merge the knowledge, experience and expertise residing in the minds of individuals, in order to elevate, through machine facilitation, the optimal information and decisions that will lead to the benefit of the whole community" [24]. Social-technical ecosystems are becoming more complex, with multiply feedback loops connecting humans and machines. The challenging research problem is correlating different factors and finding reachable opportunities for system performance emerging from these causal relationships.

The related fields of Collective Intelligence and Citizen Science require specific attention from academic and practical points of view. Future research could focus on the identification of the structural, organizational and managerial preconditions influencing the development of the ecosystem, prediction of possible evolution scenarios, and definition of risk areas. Particular attention should be paid analyzing the affection of social technologies on the formation of Collective Intelligence, which raises many scientific and practical questions. Research insights related to the problem of social technologies could offer practitioners ideas on how to integrate or create new tools and IT-based solutions oriented toward societal values. Citizen science platforms and other networks face a diversity of various technological tools and solutions. However, technological maturity is not necessarily connected to the growth of civic intelligence. This paper encourages comprehensive research on the phenomenon of Collective Intelligence, formulating holistic conceptions and collecting empirical data to validate conceptual models. A deeper understanding of Collective Intelligence ecosystem is valuable to support Citizen Science communities to reach their tasks and generate high-quality intellectual outcomes.

Funding. This research was funded by the European Union's Horizon 2020 research and innovation program under Grant Agreement No. 101005330 (INCENTIVE).

References

1. Luo, S., Xia, H., Yoshida, T., Wang, Z.: Toward collective intelligence of online communities: a primitive conceptual model. J. Syst. Sci. Syst. Eng. **18**(2), 203–221 (2009)
2. Lykourentzou, I., Vergados, D.J., Kapetanios, E., Loumos, V.: Collective intelligence systems: classification and modelling. J. Emerg. Technol. Web Intell. **3**(3), 217–226 (2011)

3. Stiles, E., Cui, X.: Workings of collective intelligence within open source communities. Adv. Soc. Comput. **6007**, 282–289 (2010)
4. Surowiecki, J.: Wisdom of Crowds. Anchor Books, New York (2005)
5. Haklay, M.M., Dörler, D., Heigl, F., Manzoni, M., Hecker, S., Vohland, K.: What is citizen science? The challenges of definition. In: Vohland, K., et al. (eds.) The Science of Citizen Science, pp. 13–33. Springer, Cham (2021). https://doi.org/10.1007/978-3-030-58278-4_2
6. Mačiulienė, M., Skaržauskienė, A.: Building the capacities of civic tech communities through digital data analytics. J. Innov. Knowl. **5**(4), 244–250 (2020). ISSN 2530-7614. eISSN 2444-569X
7. Newman, G., Wiggins, A., Crall, A., Graham, E., Newman, S., Crowston, K.: The future of citizen science: emerging technologies and shifting paradigms. Front. Ecol. Environ. **10**(6), 298–304 (2012)
8. Bowser, A., Hansen, D.L., Preece, J.: Gamifying citizen science: lessons and future directions. Paper Presented at the Workshop Designing Gamification: Creating Gameful and Playful Experiences, at CHI (2013)
9. Eveleigh, A., Jennett, C., Blandford, A., Brohan, P., Cox, A.L.: Designing for dabblers and deterring dropouts in citizen science. In: Proceedings of the 32nd Annual ACM Conference on Human Factors in Computing Systems, pp. 2985–2994. ACM, New York (2014)
10. Trouille, L., Lintott, C.L., Fortson, L.F.: Citizen science frontiers: efficiency, engagement, and serendipitous discovery with human–machine systems. Proc. Natl. Acad. Sci. **116**(6), 1902–1909 (2019)
11. Hardy, C., Lawrence, T.B., Grant, D.: Discourse and collaboration: the role of conversations and collective identity. Acad. Manag. Rev. **30**(1), 58–77 (2005)
12. Dietrich, A.D.: The Role of Civic Tech Communities in PSI Reuse and Open Data Policies. European Public Sector Information Platform, Brussels (2015)
13. Badger, E.: The next big start-up wave: civic technology (2012). http://www.citylab.com/tech/2012/06/next-big-start-wave-civic-technology/2265/. Accessed 14 May 2021
14. Suri, M.V.: From Crowdsourcing Potholes to Community Policing: Applying Interoperability Theory to Analyze the Expansion of "Open311." Berkman Center Research Publication No. 2013-18, 7641 (2013)
15. Dawes, S., Vidiasova, L., Parkhimovich, O.: Planning and designing open government data programs: an ecosystem approach. Gov. Inf. Q. **33**(1), 15–27 (2016)
16. Nambisan, S., Nambisan, P.: Engaging Citizens in Co-Creation in Public Services - Lessons Learned and Best Practices. Collaboration Across Boundaries Series (2013)
17. Sifry, M.L.: Civic Tech and Engagement: In Search of a Common Language (2014). http://techpresident.com/news/25261/civic-tech-and-engagement-search-common-language. Accessed 14 May 2021
18. Grönroos, C.: Conceptualising value co-creation: a journey to the 1970s and back to the future. J. Mark. Manag. **28**(13/14), 1520–1534 (2012)
19. Akaka, M.A., Chandler, J.D.: Roles as resources: a social roles perspective of change in value networks. Mark. Theory **11**(3), 243–260 (2011)
20. Åkesson, M.: Role constellations in value co-creation: a study of resource integration in an e-government context (Doctoral dissertation). Karlstads University (2011)
21. Evans, P., Wurster, T.S.: Blown to Bits: How the New Economics of Information Transforms Strategy. Harvard Business School Press, Cambridge (2000)
22. Lusch, R.F., Webster, F.E.J.: A stakeholder-unifying, cocreation philosophy for marketing. J. Macromark. **31**, 129–134 (2011)
23. Vargo, S.L., Lusch, R.F.: Service-Dominant Logic: Axioms of Service-Dominant Logic (2016)
24. Kapetanios, E.: Quo Vadis computer science: from turing to personal computer, personal content and collective intelligence. Data Knowl. Eng. **67**, 286–292 (2008)

Gaze Detection Using Encoded Retinomorphic Events

Abeer Banerjee[1,2](✉), Shyam Sunder Prasad[1,2], Naval Kishore Mehta[1,2], Himanshu Kumar[1,2], Sumeet Saurav[1,2], and Sanjay Singh[1,2]

[1] Academy of Scientific and Innovative Research (AcSIR), Ghaziabad, India
abeer.ceeri20a@acsir.res.in

[2] CSIR-Central Electronics Engineering Research Institute (CSIR-CEERI), Pilani, India

Abstract. Event-based gaze detection is a modern problem having several applications and advantages over frame-based techniques. Retinomorphic Event data is logged at a time resolution of microseconds that makes them suitable for the detection of saccadic eye movements. We recorded a new and compact event-based dataset for gaze detection under varying conditions of illumination using a DVS camera. The recorded dataset involved subjects tracking a circle displayed on a screen within a very short duration of time. We propose a novel event encoding technique for encoding event logs resulting from saccadic motion into six channel images. We design a Convolutional Neural Network for the gaze prediction using the encoded events obtained from the retinomorphic sensor. We use multiple evaluation metrics like average distance, average angle, and pixel radius accuracy to validate the reliability of our approach. The recorded dataset will be made available as per request.

Keywords: Gaze Detection · Event Camera · Encoded Events

1 Introduction

1.1 Problem Statement and Motivation

Eye movements are generally classified into three categories, namely fixation, saccades, and smooth pursuits [1,2]. Fixations refer to a steady focus on an object that takes a comparatively longer time compared to smooth pursuits and saccades. Smooth pursuit, as the name suggests, is the smooth continuous movement of the eye, while saccades is a rapid point-to-point movement that changes the eye focus abruptly to track the moving object. Our method focuses on saccades, the most significant type of eye movement observed during fast-action scenarios like games and sports. The analysis of saccadic eye movements can help in the diagnosis of several movement disorders. Capturing saccadic motion can be challenging, considering that rapid changes in gaze are observed during critical scenarios where the ambient light might not be sufficient. Event-based retinomorphic sensors can capture such events owing to the high time resolution and a high dynamic range even under low-light scenarios making them a perfect choice for our experiments.

H. Zaynidinov et al. (Eds.): IHCI 2022, LNCS 13741, pp. 442–453, 2023.
https://doi.org/10.1007/978-3-031-27199-1_44

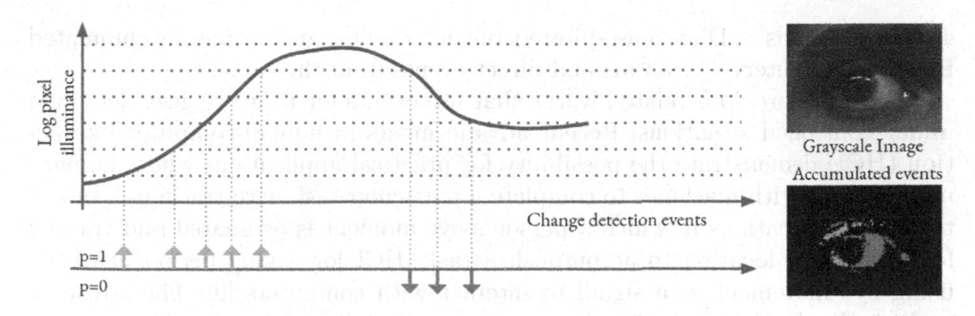

Fig. 1. Working principle of an event-based sensor: Change in illumination intensity generates events with different polarities according to Eq. 1.

Event-based camera systems are becoming a more popular choice for these tasks due to their low latency and high dynamic range [3] since they can address major privacy concerns. Each pixel of an event-based camera operates independently, and the output is a stream of events. These cameras transmit data with events, i.e., changes in illumination. Therefore, it has very low latencies in microseconds (ţs), compared with 50–200 (ms) of standard cameras [4,5]. Conventional cameras fail to capture the environment at such a time resolution making the event-based camera systems superior at handling real-time tasks. Unlike frame-based cameras, motion blur does not affect the data captured by event-based cameras since it only registers data when a change in intensity occurs. Since it does not capture RGB images as frames, there is a significant reduction in data redundancies, and real-time video information processing becomes computationally achievable. With the commercialization of the DVS, a breakthrough vision sensor that mimics some of the capabilities of the human retina, there is a huge potential for deep learning applications that use the DVS output [6–8].

Figure 1 elaborates the working principle of an event-based sensor. Since an event based sensors responds only to changes in intensity, the concept of polarity is used to distinguish between increasing and decreasing intensity. The variation of polarity is expressed by Eq. 1, where the polarity is set to 0 if the rate of change of intensity is positive, and the polarity is set 1 otherwise. The events thus registered are accumulated every 33 ms to form an image named as "Accumulated events" in Fig. 1.

$$p(t) = \begin{cases} 0, & \text{if } \frac{dI}{dt} \geq 0 \\ 1, & \text{otherwise} \end{cases} \tag{1}$$

We use a DVS camera to record a human eye gaze dataset under natural lighting conditions for the detection of rapid changes in eye gaze, especially under low-light situations. The complete dataset was stored in the AEDAT 4.0 format. AEDAT data is composed of positive and negative events detected by the variation of lighting intensities due to the motion of a human. The AEDAT

4.0 file contains a DVS noise-filtered output events, gray frame, accumulated frames at an interval of 33 ms and direct events from the camera.

There are previous related works that use events for near-eye gaze detection under controlled situations. Recent advancements in human-computer interaction (HCI) demonstrate the possibility for practical applications where humans may interact with machines to complete a particular task. Eye-tracking is one of the HCI applications in which a person's eye moment is estimated and tracked for a specified location to accomplish a task. HCI for a long period has been using eye movement as a signal to interact with computers [9]. The advances such as the development of assistance systems for disabled people by removing the dependency of hands to use computers and incorporating the eye-mouse that responds intelligently to eye movement [10]. Further, The Eye-tracking can be used in interactive applications such as virtual learning, computer gaming [11], and augmented reality/virtual reality. The invaluable use cases can be seen in medicine and healthcare such as the diagnosis of concussion, and progressive neurological disorders [12] where the patient impair to move.

Previous work on eye-tracking methods includes electro-oculography, search coils, and other unique approaches [13]. The previously proposed works may be divided into two groups, Model-based and appearance-based gaze estimation. CNN-based approaches [14,15] trained on large-scale datasets-for example, MPIIGaze [15], to estimate eye-gaze direction. The camera-based eye-tracking system progressed from Purkinje reflection-based techniques [16,17] to model-based techniques that track eye movement by extracting frames from a video sequence [18,19]. The research has been expanded to estimate the gaze in 3D coordinates using a single face or eye image. Several additional approaches have been used that are not dependent on the camera, such as limbus trackers based on photodiodes [20], LED-based trackers [21], and display embedded trackers for near-eye displays [22].

[23] proposed a method for creating a hybrid model by fusing events from an event-based camera with frames acquired by a standard camera. The frame-based information is used to initialize pupil fitting functions such as parabola, circle, and parametric ellipse, and event information is used to add weight to frame information so that the frame can estimate the gaze. The model performance is based on frames, which makes it prone to motion blur. To overcome this, [24] proposes a method that works with event cameras and flash lighting conditions. They enhance the specular components of the image by using coded differential lighting causing glints on the cornea. A major drawback is that their approach used strong artificial illumination setup. Our approach aims to address the problems posed by the existing methods by simply using events for deep-learning based prediction under natural lighting. Our contributions in this paper include:

– A novel event encoding scheme for converting sparse volumetric events data to a polarity preserving six-channel image.

- Design and implementation of a deep learning framework for gaze prediction using encoded events that reaches a peak prediction accuracy of 98% and performs in real-time.
- A compact multi-subject dataset captured using retinomorphic sensors under low levels of varying illumination for eye-gaze detection and tracking.

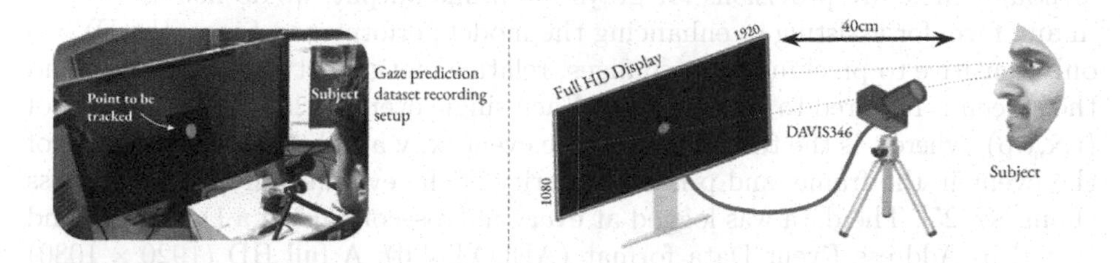

Fig. 2. Recording procedure of event-based eye-gaze detection dataset: The point to be tracked is displayed on a 1920×1080 pixels display. The subject is placed 40 cm away from the screen and is asked to track the point while the DAVIS 346 DVS camera is placed close to the face. A schematic representation is also provided for a better understanding.

2 Methodology

Our method for gaze detection uses a new compact event-based dataset recorded using a DVS camera prepared under varying levels of illumination with the participation of six subjects. The subjects were asked to follow a red circle displayed on a screen that initially starts at the centre and randomly moves to any direction within 500 ms. The corresponding eye-gaze was recorded using a DAVIS346 DVS camera sensor and the recorded train of events was saved in AEDAT 4.0 format and was labeled with the centroids of the displayed red circles. A novel rate encoding technique was used to encode the sparse event data into a coded image such that the complete event information could be processed using a 2D-Convolutional Neural Network. A lightweight 2D-Convolutional Neural Network was designed for gaze detection using the encoded images as an input where the main idea was to predict the centroid of the displayed red circle as a regression task only by using the encoded image. A euclidean distance loss was minimized between the predicted and target centroids.

2.1 Dataset Preparation

Six subjects contributed 560 retinomorphic event sequences of rapid eye movements for our experimentation with different event encoding schemes. We asked the subjects to track a red circle with a diameter of 50 pixels displayed on a

1920 × 1080 pixels resolution display within 500 ms. The event logs thus collected were recorded under low levels of varying illumination levels, and the subjects were not rigorously constrained. The recording procedure and the encoding technique are elaborated on in the subsequent sections.

Recording Procedure: Our approach only uses events for gaze detection and although there are provisions for grayscale frame output, we do not use frames in any form for assisting or enhancing the model performance. Since, the DVS is only sensitive to pixel intensity changes, relative motion between the sensor and the screen is required to capture data. Each single event is classified as a tuple of (t,x,y,p), where t is the time-stamp of the event, x, y are the pixel coordinates of the event in the frame, and p is the polarity of the event(signify the brightness changes) [25]. The data was logged at every microsecond using a DAVIS346 and stored in Address Event Data format (AEDAT 4.0). A full HD (1920 × 1080) monitor was used to display a red circle which is to be tracked by the subject. The screen was placed at a standard reading distance of 40cm and the DVS was placed close to the eye ensuring that it is not an obstruction to the vision of the subject. All recordings were performed under indoor illumination where the subject was asked to track the red circle as fast as they could in order to simulate saccades. Figure 2 elaborates the recording setup.

Algorithm 1. Event rate coded color encoding

$\mathbf{E} \leftarrow \{(x_k, y_k, t_k, p_k)\}$ ▷ **Events**
$\mathbf{T} \leftarrow \{t_1, t_2, ..., t_n\}$ ▷ **Absolute Timestamps**

$\tilde{\mathbf{T}} \leftarrow \frac{t - t_{MIN}}{t_{MAX} - t_{MIN}}$ ▷ **Normalized Timestamps**

$\mathbf{I} \leftarrow \mathbb{R}^{H \times W \times C}$ ▷ **Blank Frame**
for t **in** $\tilde{\mathbf{T}}$ **do**
  if p_t *is* 0 **then**
    I(x,y,0:3) $\leftarrow$ $r[t], g[t], b[t]$
  else
    I(x,y,3:6) $\leftarrow$ $r[t], g[t], b[t]$
  end if
end for
return I

Encoding Procedure: Our method uses a novel rate coded event to image encoding technique. The exhaustive details of the functionality of the encoding algorithm has been provided in Algorithm 1. The idea behind using an event to image encoding scheme is to employ 2D-Convolutional Neural Networks for this regression task. During rapid eye movements, events are spurred randomly, thus generating sparse event logs. A compression technique to encode the time

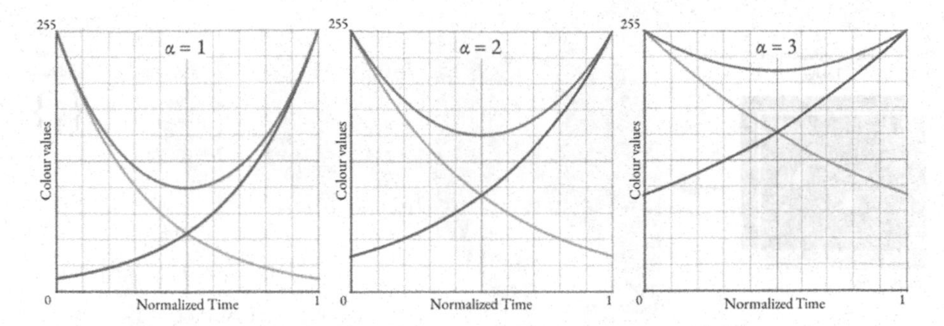

Fig. 3. The encoding functions corresponding to different values of α in Eq. 2. The functions corresponding to different color channels are plotted with their respective colors.The normalized time ranges from $t \in [0, 1]$.

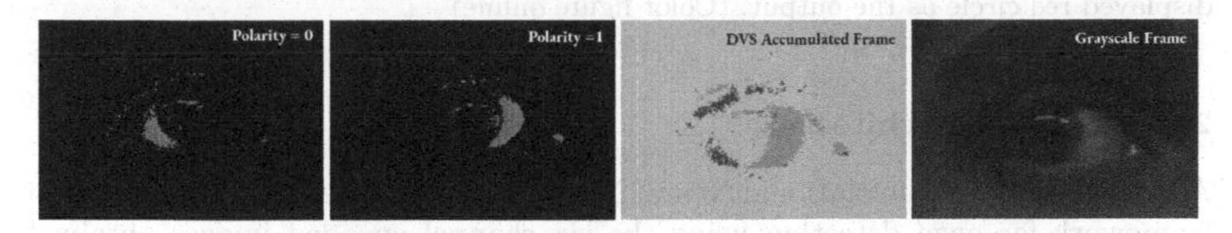

Fig. 4. First (extreme left) image shows the first three channels and the second image shows the rest three channels of the encoded six channel image. The third image is the accumulated output obtained directly from the DVS and the fourth (extreme right) image is the actual grayscale image obtained under low-light situations.

rate of change of information could be helpful to visualize a complete gaze event. Therefore, the timestamps are normalized to encode the rate.

$$r = e^{-\alpha t} \quad b = e^{\alpha(t-1)} \quad g = e^{-\alpha t} + e^{\alpha(t-1)} - \frac{1}{e^{\alpha}} \tag{2}$$

In the image domain, we use concept of color values to distinguish between different event samples where each color value was determined by using monotonic functions within the range of $t \in [0, 1]$ described in Eq. 2. α denotes the factor that determines the rapidity of function decay. Figure 3 plots the encoding functions obtained by varying α. It can be clearly observed that the color values (y-axis of the graph) have a more uniform spread throughout the complete color range for $\alpha = 2$. Therefore, we choose $\alpha = 2$ to perform our experiments. The color values obtained for each normalized time sample $t \in [0, 1]$ is a unique value that encodes the rate of motion during eye movement. A channel separation is maintained for different polarity values resulting in a six channel image, the first three channels for polarity value 0 and the rest for polarity value 1. The computationally formed images using our novel encoding scheme are shown in the top row of Fig. 4.

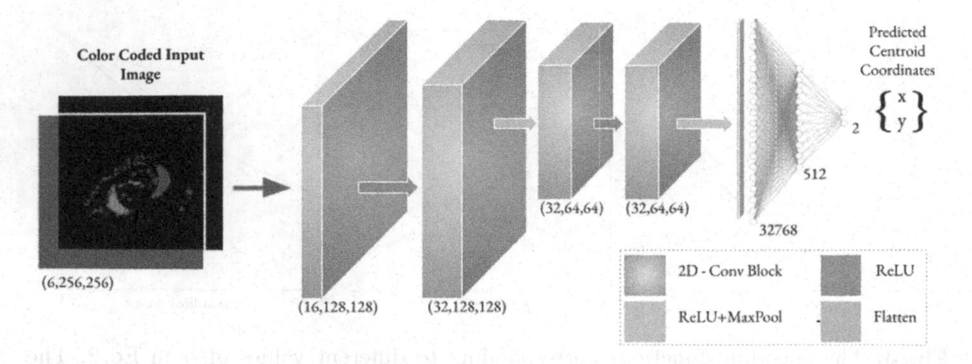

Fig. 5. Our designed 2D-Convolutional architecture for gaze prediction. The model takes the encoded retinomorphic events as its input and predicts the centroid of the displayed red circle as the output. (Color figure online)

2.2 Network Architecture

A lightweight 2D-Convolutional Neural Network was designed using the PyTorch framework for gaze detection using the six channel encoded images obtained from the recorded event logs. The network architecture has been elaborately portrayed in Fig. 5. The network structure consists of four 2D-Convolutional Blocks layers followed by a flatten layer. Two linear layers were used to obtain two value centroid predictions. All the activation functions used in this network were ReLU. Since the accumulated events have been color encoded to form an image, the task of figuring out the centroid becomes relatively simple as CNNs are good at extracting features from the changing gradients in an image. Our network was able to predict the saccadic eye movements with a total parameter count less than 17 million.

2.3 Training

We prepared the training data by generating tuples of the centroid data of the recorded red circle and the corresponding encoded six-channel image. The red circle starts from the center location of the screen and instantaneously moves to a random location within the screen. It is programmatically ensured that the DVS event logging synchronises with the movement of the tracking circle. It depends on the subject's reaction time to track within the event-logging time, and since a large window of 500 ms is sufficient, most of the subjects are easily able to do it. The saccadic movement resulting from tracking the circle can vary for multiple reasons. The number of events triggered within the chosen time interval can vary because the centroid of the tracking point within the screen is randomly assigned, and the eye movements vary from subject to subject. The average count of events within a time interval of 500 ms generally ranges from 8000 to 8500. But, the count may rarely drop as low as 2000 events corresponding to cases where the assigned centroid of the tracking point is close to the center of

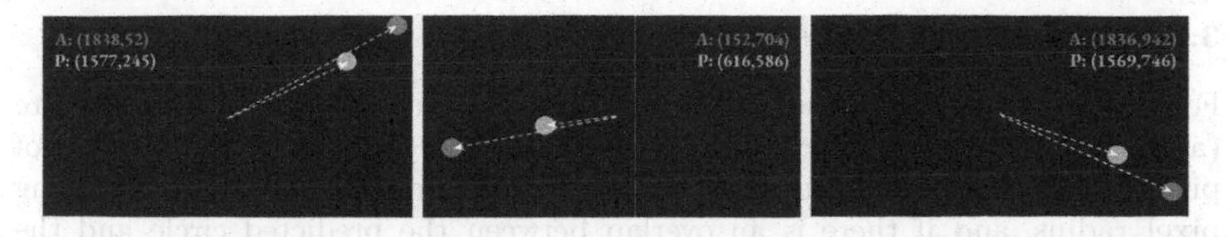

Fig. 6. The predicted circles are shown in green and the target circles are shown in red. It can be seen that the model could follow the direction but could not match the target centroid. (Color figure online)

the screen, resulting in a small range of motion of the pupil. In the rare event of blinking before tracking the point, a huge number of events are triggered within a short range of time resulting in an event count way more than the average. We trained the network on an Alienware R7 desktop with a core i7 8th generation CPU and 8 GB NVIDIA GTX 1080 graphics processor. The model started to converge after 200 epochs of training and validation with a batch size of 10. The execution time of each epoch was less than 5 s, and the L1 loss was minimized using the Adam optimizer.

3 Results

A rigorous and extensive quantitative study was performed to test the reliability of our approach using multiple performance evaluation metrics. In this section, we show the qualitative results where we plot the predicted centroids along with the ground truth centroids to create a more intuitive visualization of the model performance.

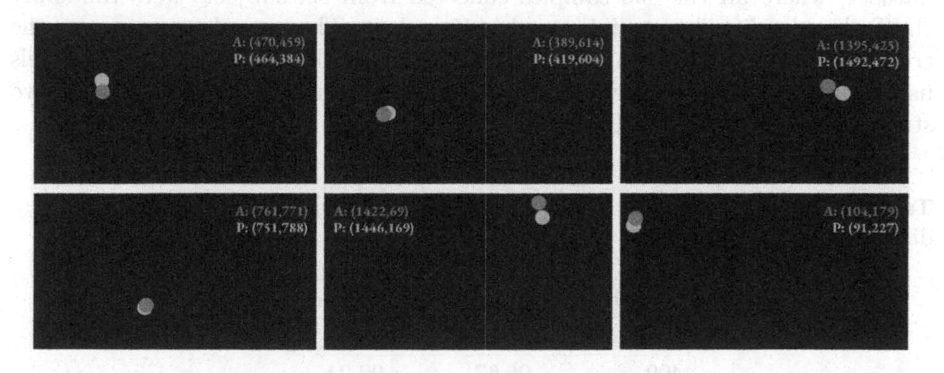

Fig. 7. Obtained results using our best performing model shows the accuracy of the predicted centroids using encoded events.

3.1 Evaluation Metrics

For the quantitative evaluation of our model, we use three metrics: hit rate (accuracy), average angle, and average distance. We define accuracy in terms of pixel radius. A circle is drawn using the ground truth centroid with a varying pixel radius, and if there is an overlap between the predicted circle and the ground truth circle, the prediction is marked as a "hit", otherwise, the prediction is a "miss". A "hit" with a lesser value of pixel radius corresponds to a precise prediction. From the nature of our problem, an acceptable level of precision in pixel radius is determined to be 300 pixels which can easily refer to the zone of interest in the screen.

The nature of the addressed problem includes scenarios that require capturing saccadic eye movements. It was observed that the recorded results might portray a lag due to a variation in the reaction time of the subjects in the rare cases when the target red circle was assigned a location far away from the center of the screen. For subjects with slower reaction times, their saccades might be along the path of the target red circle but they might not be able to track the target within 500 ms.

Figure 6 illustrates some such examples corresponding to subjects with lower reaction times. Therefore, we introduce the metric of average angle to determine the model's ability to predict the direction of the saccade. The angle subtended by the predicted centroid and the target centroid at the center of the screen should be minimal for the direction to be aligned.

3.2 Evaluation Strategy

We employed two strategies to evaluate the generalization performance of our model. The first strategy involves training and testing the model using the whole dataset, where all the 560 samples collected from six subjects were randomly shuffled and splitted to training and testing data. The second strategy uses the training data derived from five randomly selected subjects and tests the models using the remaining subject. The quantitative evaluation scores using the two strategies are reported in Table 1.

Table 1. Testing accuracy obtained using Strategy-1 and Strategy-2 corresponding to different radius (in pixels).

Pixel Radius	S-1 Acc.(%)	S-2 Acc.(%)
500	98.33	96.67
400	96.67	90.24
300	90.00	84.50
200	73.33	73.33
100	45.00	45.00

For Strategy-1, the average angle calculated using 60 test samples is 9.974°, and the average distance calculated between the predicted and target centroids for 60 test samples is 194.032 pixels. For Strategy-2, the average angle obtained for six test-runs corresponding to six different test subjects is 10.836°, and the average distance is 200.046 pixels.

3.3 Qualitative Results

Although the reported average-based quantitative methods are reliable, they often don't perfectly picture the intricacies of model functionality. For example, due to the low reaction time of the subject, the measured distance between the predicted centroid and the target centroid may be high, but the reported angle could be low, indicating that the model could predict the gaze direction successfully. Therefore, to support our claims, we present the qualitative results for visualizing the success rate of our model in Fig. 7. One can observe from Fig. 7 that the predicted centroids overlap the target centroids in most cases, and it is repeatedly confirmed in all the directions spanning across the screen. We performed the test under all possible gaze directions and found the results to follow the same trend.

4 Conclusion and Discussion

We performed gaze detection using encoded events with a 2D Convolutional Neural Network and tested our approach using multiple evaluation metrics. To the best of our knowledge we are the first to record and use encoded events for gaze detection. Our recorded dataset involved multiple subjects to account for the diversity in reaction time while recording saccadic eye motion. The obtained results using the designed lightweight 2D-Convolutional Neural Network are promising since the hit rate of our approach is above 90%. It is to be noted that our approach uses only events for gaze prediction and our designed testing pipeline was able to accurately predict the gaze in real-time. There are a few limitations of our work, a major one being the dependency on the reaction time of the subject. Although a fixed time window for the collection of triggered events resulting from saccadic motion is a straightforward task that works for the general population, the window might not be sufficient for everybody. Therefore, the events are captured while the pupils are still in pursuit before it reaches the target centroid. The accumulated events thus obtained, encode to an image that results in a prediction that might be in the direction of the target centroid, but a few pixels away from it. This drastically reduces the average distance score of the model. To address this, we have used the average angle metric for the analysis of our model performance.

Acknowledgements. All computations were performed using the GPU resources provided by the AI Computing Facility, CSIR-CEERI. The authors sincerely appreciate the willingness of the contributing subjects.

References

1. Rayner, K., Castelhano, M.: Eye movements. Scholarpedia **2**(10), 3649 (2007)
2. Findlay, J., Walker, R.: Human saccadic eye movements. Scholarpedia **7**(7), 5095 (2012)
3. Cheng, W., Luo, H., Yang, W., Yu, L., Chen, S., Li, W.: Det: a high-resolution dvs dataset for lane extraction. In: Proceedings of the IEEE/CVF Conference on Computer Vision and Pattern Recognition Workshops (2019)
4. Gallego, G., et al.: Event-based vision: a survey. IEEE Trans. Pattern Anal. Mach. Intell. **44**(1), 154–180 (2020)
5. Dynamic vision sensor (2022). https://inivation.com/products/customsolutions/videos/ Accessed 13 Apr 2022
6. Baby, S.A., Vinod, B., Chinni, C., Mitra, K.: Dynamic vision sensors for human activity recognition. In: 2017 4th IAPR Asian Conference on Pattern Recognition (ACPR), pp. 316–321. IEEE (2017)
7. Wan, J., et al.: Event-based pedestrian detection using dynamic vision sensors. Electronics **10**(8), 888 (2021)
8. Liao, F., Zhou, F., Chai, Y.: Neuromorphic vision sensors: principle, progress and perspectives. J. Semicond. **42**(1), 013105 (2021)
9. Köles, M.: A review of pupillometry for human-computer interaction studies. Periodica Polytechnica Electr. Eng. Comput. Sci. **61**(4), 320–326 (2017)
10. Lukander, K.: A short review and primer on eye tracking in human computer interaction applications (2016)
11. Corcoran, P.M., Nanu, F., Petrescu, S., Bigioi, P.: Real-time eye gaze tracking for gaming design and consumer electronics systems. IEEE Trans. Cons. Electron. **58**(2), 347–355 (2012)
12. Brunyé, T.T., Drew, T., Weaver, D.L., Elmore, J.G.: A review of eye tracking for understanding and improving diagnostic interpretation. Cogn. Res.: Principles Impl. **4** (2019)
13. Young, L.R., Sheena, D.: Survey of eye movement recording methods. Behav. Res. Methods Instr. **7**, 397–429 (1975)
14. Zhang, X., Sugano, Y., Fritz, M., Bulling, A.: Appearance-based gaze estimation in the wild. In: 2015 IEEE Conference on Computer Vision and Pattern Recognition (CVPR), pp. 4511–4520 (2015)
15. Zhang, X., Sugano, Y., Fritz, M., Bulling, A.: Mpiigaze: real-world dataset and deep appearance-based gaze estimation. IEEE Trans. Pattern Anal. Mach. Intell. **41**(1), 162–175 (2019)
16. Cornsweet, T.N., Crane, H.D.: Accurate two-dimensional eye tracker using first and fourth purkinje images. J. Opt. Soc. Am. **63**(8), 921–928 (1973)
17. Crane, H.D., Steele, C.M.: Generation-v dual-purkinje-image eyetracker. Appl. Opt. **24**(4), 527–537 (1985)
18. Li, Y., Wang, S., Ding, X.: Eye/eyes tracking based on a unified deformable template and particle filtering. Pattern Recogn. Lett. **31**(11), 1377–1387 (2010)
19. Wang, K., Ji, Q.: Real time eye gaze tracking with 3D deformable eye-face model. In: 2017 IEEE International Conference on Computer Vision (ICCV), pp. 1003–1011 (2017)
20. Topal, C., Gerek, Ö.N., Doğan, A.: A head-mounted sensor-based eye tracking device: eye touch system. In: ETRA 2008 (2008)
21. Akşit, K., Kautz, J., Luebke, D.: Gaze-sensing leds for head mounted displays (2020)

22. Vogel, U., et al.: Bidirectional oled microdisplay for interactive see-through hmds: study toward integration of eye-tracking and informational facilities. J. Soc. Inf. Disp. **17**, 03 (2009)
23. Angelopoulos, A.N., Martel, J.N.P., Kohli, A.P., Conradt, J., Wetzstein, G.: Event-based near-eye gaze tracking beyond 10,000 hz. IEEE Trans. Vis. Comput. Graph. **27**(5), 2577–2586 (2021)
24. Stoffregen, T., Daraei, H., Robinson, C., Fix, A.: Event-based kilohertz eye tracking using coded differential lighting. In: 2022 IEEE/CVF Winter Conference on Applications of Computer Vision (WACV), pp. 3937–3945 (2022)
25. Mueggler, E., Forster, C., Baumli, N., Gallego, G., Scaramuzza, D.: Lifetime estimation of events from dynamic vision sensors. In: 2015 IEEE international conference on Robotics and Automation (ICRA), pp. 4874–4881. IEEE (2015)

Extremely Lightweight Skin Segmentation Networks to Improve Remote Photoplethysmography Measurement

Kunyoung Lee[1], Hojoon You[2], Jaemu Oh[2], and Eui Chul Lee[3](✉)

[1] Department of Computer Science, Graduate School, Sangmyung University, Seoul, Republic of Korea

[2] Department of AI and Informatics, Graduate School, Sangmyung University, Seoul, Republic of Korea

toddlf0614@naver.com

[3] Department of Human-Centered Artificial Intelligence, Sangmyung University, Seoul, Republic of Korea

eclee@smu.ac.kr

Abstract. Recently, remote photoplethysmography (rPPG) has been studied and developed not only in a controlled environment but also in a wild environment such as telemedicine and driver monitoring. Although photoplethysmography (PPG) can be measured through a contact sensor, pulse signal can be obtained by remote measuring minute color changes on the skin surface using a camera. This pulse signal is called rPPG and has an advantage of sensing cardiac activity without a contact sensor. The processing pipeline of rPPG can be simply defined as region of interest selection (ROI), pulse signal extraction, signal processing. During this process, in ROI selection, skin segmentation is performed because only the skin pixel region is related to the rPPG signal. We propose extremely lightweight skin segmentation network (ELSNet) for applying deep learning to skin segmentation to measure reliable signals. Our method improved the success rate within 5BPM of heart rate estimation by about 6%, and in the talking environment, an average performance improvement of 9.5% was confirmed. In addition, it was confirmed that *MAPE* was improved by an average of 20%. The ELSNet shows 167 FPS throughput on Intel i9 CPU.

Keywords: Remote photoplethysmography (rPPG) · Skin segmentation · Biomedical monitoring · Image analysis

1 Introduction

Photoplethysmography (PPG) can be obtained by measuring changes in microvascular blood volume, first introduced by Hertzman [1]. It has a lot of information about cardiovascular. PPG is not only easily measurable, but also provides a low-cost solution for health monitoring. Therefore, it is becoming an industry standard [2, 3]. Although PPG can be measured through a contact sensor, Signal can be obtained by measuring minute

H. Zaynidinov et al. (Eds.): IHCI 2022, LNCS 13741, pp. 454–459, 2023.
https://doi.org/10.1007/978-3-031-27199-1_45

color changes on the skin surface using a camera. This signal is called rPPG introduced in 2005 by Wieringa, Mastik & Steen, and in 2007 by Humphreys, Ward & Markham [4, 5]. Representative studies of rPPG are PCA [6], CHROM [7], plane orthogonal to skin (POS) [8], OMIT [9] which are described in related work. The processing pipeline of rPPG can be simply defined as ROI selection, pulse signal extraction and signal processing. During this process, in ROI selection, skin segmentation is performed because only the skin pixel region is related to the rPPG signal [10]. Moreover, skin segmentation can remove elements that can contaminate the rPPG signal, such as glasses or facial hairs and facial features or moving elements, like eyes or lips [6, 7]. In recent years, as deep learning has developed a lot, it has been applied to various tasks. DL based methods do not require hand-crafted features and performance is higher than threshold based or heuristic methods if data is sufficient. Therefore, we propose a ELSNet for fast and reliable measurement of rPPG signal.

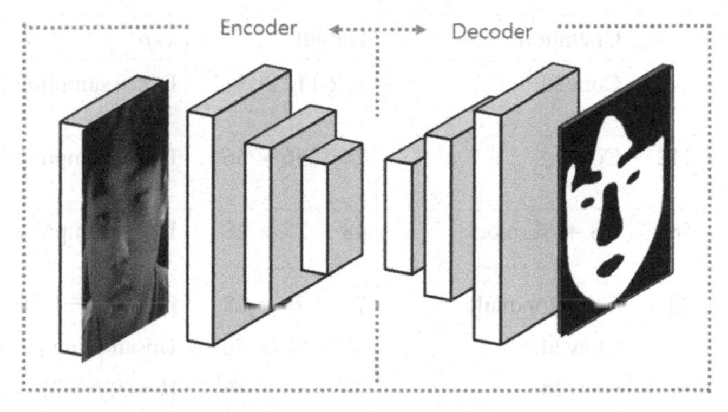

Fig. 1. The architecture of ELSNet. The S2 module consist of bottleneck layer, and the information blocking is a method for fine skin segmentation. DS+SE block means depthwise separable convolution and Squeeze-and-Excitation block

2 Related Work

PCA [7] is an rPPG algorithm using PCA. Only the method of decomposing the signal is different from the Independent Component Analysis (ICA) [12], and the rest is almost the same. The ROI is divided into the entire face area, forehead area, and cheeks. Then, the average and variance of RGB signals in each area were calculated to find out which area was suitable. Based on the above experimental results, it is emphasized that the forehead is the most uniform area. In addition, experiments show that noise increases as the size of the ROI decreases. **CHROM** [8] is an rPPG algorithm that linearly combines color difference signals assuming standardized skin color. A simple face detector and skin-mask were used for ROI selection, and a clean rPPG signal was obtained by removing pixels containing facial hair and facial features that contaminate the rPPG signal. **POS** [9] is an algorithm robust in motion based on CHROM [8]. The method is a method of extracting a pulse signal through a projection surface orthogonal to a skin tone. Support

Vector Machine (SVM)-based classifiers and learning-based object trackers were used for ROI selection. **OMIT** [11] is an rPPG algorithm based on QR decomposition and is robust against compression artifacts. The ROI used two methods: DL-based (uNET) skin segmentation and patch selection based on landmark coordinates. PCA [8] and CHROM [9] sought to improve performance by trying to obtain a clean signal when selecting as many skin areas as possible as ROI, except for areas that could be noisy. Moreover, Bobia's paper [13] emphasizes that ROI selection in rPPG is an important first step in obtaining reliable signals, and [14] states that ROI quality directly affects the quality of rPPG signals. Therefore, we seek to obtain reliable pulse signals through more accurate ROI selection with DL-based skin segmentation.

Table 1. Detailed settings for the ELSNet. *k* Denotes the kernel size of the convolution layer and *p* denotes the kernel size of average pooling layer.

#	Input	Operation	Output	k, p
1	3 × 244 × 244	Conv2d	8 × 112 × 112	Down sampling [k =3, p = 3]
2	8 × 112 × 112	Conv2d	24 × 56 × 56	Down sampling [k =3, p = 3]
3	24 × 56 × 56	DS + SE block	48 × 28 × 28	Down sampling [k =3, p = 3]
4	48 × 28 × 28	6 × S2 module	72 × 28 × 28	[k =3, p =1], [k =5, p =1]
5	72 × 28 × 28	Conv2d	48 × 56 × 56	Upsampling [k =3, p =3]
6	48 × 56 × 56	Conv2d	24 × 56 × 56	[k =3, p =3]
7	24 × 56 × 56	Information blocking operation	2 × 56 × 56	[k =3, p =3]
8	2 × 56 × 56	Conv2d	2 × 112 × 112	Upsampling [k =3, p =3]
9	2 × 112 × 112	Classifier	2 × 224 × 24	Upsampling [k =3, p =3]

3 Method

In this section, we explain our skin segmentation model architecture and the design of experiments that confirmed performance improvement of the rPPG extraction algorithms through our skin segmentation model. The proposed method has been redesigned based on the lightweight and performance enhancement method applied by SINet [15] for real-time portrait segmentation with deep learning approach. In SINet, a network structure that enables feature extraction with less computation through Spatial Squeeze Block (S2 block) and group convolution are used to create a network structure optimized for portrait segmentation. In this paper, we use the S2 block, group convolution and information blocking method to design a network optimized for skin segmentation in the detected face region but change the model architecture and network depth. Also, in 3.2 section, we explain the metrics and experimental methods to measure the rPPG performance improvement. The overall model structure can be seen in Fig. 1. An input image passes through two layers composed of a convolutional layer with 2 stride, a batch normalization layer, and PRelu at the forepart of the encoder. Afterwards, the feature map passes through the DS + SE block composed of depthwise separable convolution and SE block. In the rear part of encoder, the feature map passes through S2 blocks 6 times and the final encoder output is calculated through pointwise convolution. More detailed settings of ELSNet are described in Table 1. Also, we compare the performance of ELSNet and existing skin segmentation methods [16, 17] on the four rPPG methods introduced in Sect. 2 we confirmed through experiments that trough accurate and fast skin segmentation using ELSnet, the rPPG measurement performance was significantly improved in talk and diverse movement environment.

Table 2. rPPG performances measured from PURE dataset (5bpm: *Coverage* with T_γ as 5, 3bpm: *Coverage* with T_γ as 3, Skin seg.: skin segmentation, rPPG alg.: rPPG algorithms).

rPPG alg.	Skin seg.								
	ELSNet (Ours)			YCbCr [16]			HSV [17]		
	MAPE	5 bpm	3 bpm	*MAPE*	5 bpm	3 bpm	*MAPE*	5 bpm	3 bpm
OMIT	**3.8%**	**88%**	**75%**	5.3%	82%	67%	6.1%	82%	69%
CHROME	**3.7%**	**89%**	**74%**	4.8%	83%	69%	6.5%	80%	65%
PCA	**3.7%**	**88%**	**75%**	4.9%	83%	69%	16.4%	56%	46%
POS	**6.7%**	**76%**	**62%**	7.3%	73%	59%	9.1%	75%	**62%**

* The smaller *MAPE* and the larger *Coverage*, the higher the performance. Bold fonts mean best.

Table 3. rPPG performances measured in a talking situation (5bpm: *Coverage* with T_γ as 5, 3bpm: *Coverage* with T_γ as 3, Skin seg.: skin segmentation, rPPG alg.: rPPG algorithms).

rPPG alg.	Skin seg.								
	ELSNet (Ours)			YCbCr [16]			HSV [17]		
	MAPE	5 bpm	3 bpm	*MAPE*	5 bpm	3 bpm	*MAPE*	5 bpm	3 bpm
OMIT	**6.9%**	**67%**	**55%**	10.2%	60%	48%	11.2%	58%	42%
CHROME	**6.7%**	**71%**	**51%**	9%	63%	49%	11.7%	54%	39%
PCA	**6.7%**	**69%**	**54%**	8.7%	65%	49%	20.9%	42%	32%
POS	**14%**	**47%**	**35%**	14.7%	44%	34%	16%	43%	31%

* The smaller *MAPE* and the larger *Coverage*, the higher the performance. Bold fonts mean best.

4 Result

ELSNet trained with the CelebAMask-HQ dataset [18]. We confirmed that the robustness of ELSNet to background and motion leads to improved rPPG performance. The PURE database acquired rPPG data in six different environments, including steady, talking, slow translation, and small rotation [19]. Table 2 shows the results of improving the rPPG measurement performance through ELSNet-based skin segmentation on the PURE dataset. *MAPE* and *Coverage* are used as rPPG performance metrics, and the equations follow (1) and (2), respectively

$$MAPE = \frac{100}{k}\sum_{i=1}^{k}\left|\frac{y_i - \hat{f}(x_i)}{y_i}\right| \tag{1}$$

In *MAPE* Eq. (1), k is an index of 1 s sliding window. y_i is the correct heart rate corresponding to the index. $\hat{f}(x_i)$ is the estimated heart rate corresponding to the index.

$$b_k(T_\gamma) = \begin{cases} 0, \; if \; d(k) > T_\gamma \\ 1, \; if \; d(k) < T_\gamma \end{cases} \tag{2.1}$$

$$Coverage = \frac{\sum_k b_k(T_\gamma)}{K} \tag{2.2}$$

In Eq. (2.1), k is an index of 1 s sliding window. $d(k)$ is a heart rate error function corresponding to the index k. T_γ is the bpm threshold of the binary function $b_k(T_\gamma)$. *Coverage* (2.2) represents the success rate in a time series obtained through a binary function (2.1). As shown in Table 2, performance improvement was confirmed in all four rPPG methods only by applying skin segmentation using ELSNet. Performance improvement of 6% was confirmed on average based on 5BPM coverage. In Table 3, it was confirmed that the performance improvement was distinguished in the talking situation, and it was confirmed that the performance improved in all rPPG methods by an average of 9.5% based on the 5BPM Coverage.

5 Conclusion

In this paper, it is confirmed through experiments that rPPG performance can be improved through skin segmentation corresponding to ROI selection in the rPPG processes. The ELSnet improves the measurement performance of the four rPPG methods by selecting only skin pixels with pulse signals fast and accurately. Based on *MAPE*, the average performance improved by about 20% compared to the existing method. These results show that skin segmentation is an essential component for the widely used of rPPG applications such as telemedicine and driver monitoring.

Acknowledgement. This paper was supported by Field-oriented Technology Development Project for Customs Administration through National Research Foundation of Korea(NRF) funded by the Ministry of Science & ICT and Korea Customs Service(2022M3I1A1095155).

References

1. Herzman, A.B.: Photoelectric plethysmography of the fingers and toes in man. Proc. Soc. Exp. Biol. Med. **37**(3), 529–534 (1937)
2. Schäfer, A., Vagedes, J.: How accurate is pulse rate variability as an estimate of heart rate variability?: a review on studies comparing photoplethysmographic technology with an electrocardiogram. Int. J. Cardiol. **166**(1), 15–29 (2013)
3. Sarkar, P., Etemad, A.: CardioGAN: attentive generative adversarial network with dual discriminator for synthesis of ECG from PPG. In: Proceedings of the AAAI Conference on Artificial Intelligence, vol. 35, pp. 488–496. AAAI Press, California (2021)
4. Wieringa, F.P., Mastik, F., Antonius, F.W.: Contactless multiple wavelength photoplethysmographic imaging: a first step toward 'SpO2 Camera' technology. Ann. Biomed. Eng. **33**(8), 1034–1041 (2005)
5. Humphreys, K., Ward, T., Markham, C.: Noncontact simultaneous dual wavelength photoplethysmography: a further step toward noncontact pulse oximetry. Rev. Sci. Instrum. **78**(4), 044304 (2007)
6. Boccignone, G.: pyVHR: a Python framework for remote photoplethysmography. PeerJ Comput. Sci. **8**, e929 (2022)
7. Lewandowska, M.: Measuring pulse rate with a webcam—a non-contact method for evaluating cardiac activity. In: 2011 Federated Conference on Computer Science and Information Systems (FedCSIS), pp. 405–410. IEEE (2011)
8. De Haan, G., Jeanne, V.: Robust pulse rate from chrominance-based rPPG. IEEE Trans. Biomed. Eng. **60**(10), 2878–2886 (2013)
9. Wang, W., den Brinker, A.C., Stuijk, S., de Haan, G.: Algorithmic principles of remote PPG. IEEE Trans. Biomed. Eng. **64**(7), 1479–1491 (2016)
10. Scherpf, M.: Skin segmentation for imaging photoplethysmography using a specialized deep learning approach. In: 2021 Computing in Cardiology (CinC), vol. 48, pp 1–4. IEEE (2021)
11. Álvarez Casado, C., Bordallo López, M.: Face2PPG: an unsupervised pipeline for blood volume pulse extraction from faces. arXiv e-prints, arXiv:2202.04101 (2022)
12. Poh, M.-Z., McDuff, D.J., Picard, R.W.: Non-contact, automated cardiac pulse measurements using video imaging and blind source separation. Optics Express **18**(10), 10762–10774 (2010)
13. Bobbia, S., Benezeth, Y., Dubois, J.: Remote photoplethysmography based on implicit living skin tissue segmentation. In: 2016 23rd International Conference on Pattern Recognition (ICPR), pp. 361–365. IEEE (2016)
14. Bousefsaf, F., Maaoui, C., Pruski, A.: Continuous wavelet filtering on webcam photoplethysmographic signals to remotely assess the instantaneous heart rate. Biomed. Signal Process. Control **8**(6), 568–574 (2013)
15. Park, H., et al.: SINet: extreme lightweight portrait segmentation networks with spatial squeeze module and information blocking decoder. In: Proceedings of the IEEE/CVF Winter Conference on Applications of Computer Vision (2020)
16. Phung, S.L., Bouzerdoum, A., Chai, D.: A novel skin color model in YCBCR color space and its application to human face detection. In Proceedings. International Conference on Image Processing, vol. 1, p. I. IEEE. (2002)
17. Dahmani, D., Cheref, M., Larabi, S.: Zero-sum game theory model for segmenting skin regions. Image Vision Comput. 103925 (2020)
18. Karras, T., et al.: Progressive growing of GANs for improved quality, stability, and variation. In: International Conference on Representation Learning (ICLR) (2018)
19. Stricker, R., Müller, S., Gross, H.-M.: Non-contact video-based pulse rate measurement on a mobile service robot. In: The 23rd IEEE International Symposium on Robot and Human Interactive Communication. IEEE (2014)

Privacy-Friendly Phishing Attack Detection Using Personalized Federated Learning

Jun Yong Yoon and Bong Jun Choi(✉)

Soongsil University, Seoul, South Korea
{wnsdyd1124,davidchoi}@soongsil.ac.kr

Abstract. Recently, machine learning has been used to detect phishing attacks. However, it is challenging to collect data, such as call recordings and text messages from potential victims, due to privacy issues. Therefore, we introduce a phishing detection technique using federated learning that can preserve user data privacy through collaborative training that keeps local user data private. To improve the detection accuracy of phishing detection, our algorithm groups clients based on their characteristics to recommend personalized data requirements. Our results show that the proposed approach can increase the attack detection accuracy by reducing the impact of data imbalance in the phishing data.

Keywords: Phishing Attack Detection · Federated Learning · Imbalanced Data

1 Introduction

Phishing attacks have become an extensive social problem due to their ability to affect various people using various communication technologies. However, current measures to prevent phishing are only based on the victim's report. Since it detects only the existing attack cases, even if a phishing pattern of a new case appears, it is often not reported as malicious unless the user actively reports it after suffering from the attack. The data collection is based solely on the victim's report because the data includes the user's personal information, such as call history and messenger history. Therefore, we propose a phishing detection algorithm using federated learning that can simultaneously protect and learn personal information so that users can feel safe.

Various algorithms based on machine learning and deep learning models were used to detect voice phishing. However, most existing algorithms are centralized learning algorithms that require a server to collect raw client data to process learning. Centralized learning algorithms generate high communication costs due to data transmission to the server over the Internet. On the other hand, federated learning can reduce communication costs by sending only the learning models to the server without sending the learning data. Therefore, to address

H. Zaynidinov et al. (Eds.): IHCI 2022, LNCS 13741, pp. 460–465, 2023.
https://doi.org/10.1007/978-3-031-27199-1_46

the problem of communication costs and privacy protection, this work proposes a voice phishing detection algorithm using federated learning. To the best of our knowledge, our work is the first attempt at applying federated learning for privacy-protecting phishing detection considering the unique characteristics of the phishing data. We improve the accuracy of the baseline federated learning algorithm by grouping the client features, which reflect the characteristics of the phishing data.

2 Related Works

There are many existing works on detecting phishing attacks in the literature. Some representative works are presented in this section. Gangavarapu et al. [1] proposed a method to extract email content and behavior-based features to detect email. They presented the features suitable for Unsolicited Bulk Emails (UBEs) detection, which selects the most differential feature set. Basnet et al. [2] proposed a method to prevent URL phishing. They presented methods that can be used to develop anti-phishing, such as masking potentially dangerous URLs or alerting users to possible threats in the future. Using federated learning, Thapa et al. [3] detected email phishing and analyzed learning performance in various settings based on Recurrent Neural Network (RNN) and Bidirectional Encoder Representations from Transformers (BERT). This work considers both the balanced and asymmetrical data distribution among clients. The results show that the accuracy of the global model decreases as the data distribution becomes more asymmetric.

3 Research Motivation

3.1 Privacy Preservation Using Federated Learning

In this study, the federated learning approach is used to improve the model's performance in terms of privacy preservation and detection accuracy. Federated learning was first published in 2015 by McMahan et al. from Google [4]. Federated learning can be particularly useful in phishing attack applications because of the following two features: improved data privacy and communication efficiency. First, federated learning allows learning without data leakage in situations where personal privacy must be protected. Second, federated learning can be designed to significantly reduce communication costs by exchanging only updates from local models. On the contrary, sending all data from tens of thousands of local devices to a central server in the centralized learning approach increases network traffic and storage costs.

3.2 Performance Enhancement by Reducing Data Imbalance Problem

The user's personal information is contained in the user's call record. The amount of the user's call record is different for different users, and this is because the amount of calls used per day varies from person to person. Some users take a day to generate 1 Mb of voice text data, while others take a week. Suppose federated learning is naively applied in such a heterogeneous scenario. In that case, the performance of the global model aggregated at the server may reduce due to the data imbalance problem, as discussed in Thapa et al. [3]. Therefore, we propose an algorithm to analyze and group features according to the characteristics of each client to achieve a high-quality model.

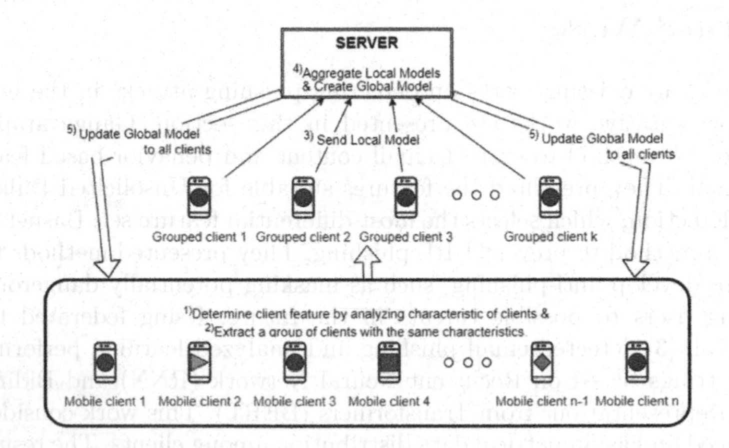

Fig. 1. Proposed federated learning model for phishing attack detection

4 Proposed Method

4.1 Data Setup

The model is learned using Korean Call Content Vishing (KorCCVi) [5]. Since Korean has many phrases in the tokenizing steps compared to English, many studies on Korean natural language processing have been conducted, and various text processing tools have been created. KoNLpy [6], a Python package for Korean natural language processing, is used to learn voice phishing recording data.

4.2 Federated Learning Based Phishing Attack Detection Algorithm

As shown in Fig. 1, the overall process of the proposed algorithm is as follows: 1) the server determines client feature by analyzing the characteristic of client k, 2) the server groups clients according to client feature, 3) the grouped clients send a local model to the server, 4) the server aggregates the received local models and creates a global model, 5) the server sends the updated global model all clients, and 6) steps 1 to 5 are repeated until a satisfactory model is created.

Algorithm 1. *Federated Learning based Phishing Attack Detection Algorithm.* The K clients are indexed by k; B is the local minibatch size, E is the number of local epochs, and η is the learning rate.

1: **1. Client Grouping:**
2: **for** each k **do**
3: determine client feature F_k by analyzing the characteristic of client k
4: send F_k to server
5: server assigns group G to client k based on F_k
6: **end for**
7: **2. Client Update**(k, w)**:**
8: $\mathcal{B} \leftarrow$ (split $\mathcal{P}_k$ into batches of size B based on G recommendation)
9: **for** each local epoch i from 1 to E **do**
10: **for** batch $b \in \mathcal{B}$ **do**
11: $w \leftarrow w - \eta \nabla \ell(w; b)$
12: **end for**
13: **end for**
14: return w to server
15: **3. Server Aggregation:**
16: initialize w_0
17: grouping same client features $F_k = K_{sf}$
18: **for** each round $t = 1, 2, \ldots$ **do**
19: **for** each client $k \in K_{sf}$ **in parallel do**
20: $w_{t+1}^k \leftarrow$Client Update(k, w_t)
21: **end for**
22: $w_{t+1} \leftarrow \sum_{k=1}^{K_{sf}} \frac{n_k}{n} w_{t+1}^k$
23: **end for**

Algorithm 1 shows the details of the proposed algorithm. First, grouping clients part. Each client's dataset has its own characteristics. The server analyzes the client's characteristics and determines the client feature, denoted as K_R. Lines 7 to 14 are the client execution, and lines 15 to 23 are the server execution. The server and client algorithms follow the typical federated learning algorithm, called FedAvg [4]. The server collects and groups the features of each client. For each round t, grouped clients generate and send local gradient w to the server. The server aggregates the local gradients to create an updated global model w_{t+1}. Finally, the updated global is distributed to all clients. This process is repeated until a satisfactory global model is created.

4.3 Experimental Result

As discussed in Duan et al. [7], data imbalance is one of the main factors that lower the accuracy of federated learning. They compared uniform and non-uniform datasets and showed that unbalanced datasets reduce the model's accuracy by approximately 10%. Our algorithm aggregates the training results of the clients that group the same feature clients by analyzing the characteristic of clients to reduce the data imbalance problem caused by heterogeneous clients.

Therefore, by employing the proposed approach, we can improve the overall learning performance by making the dataset of each client more balanced.

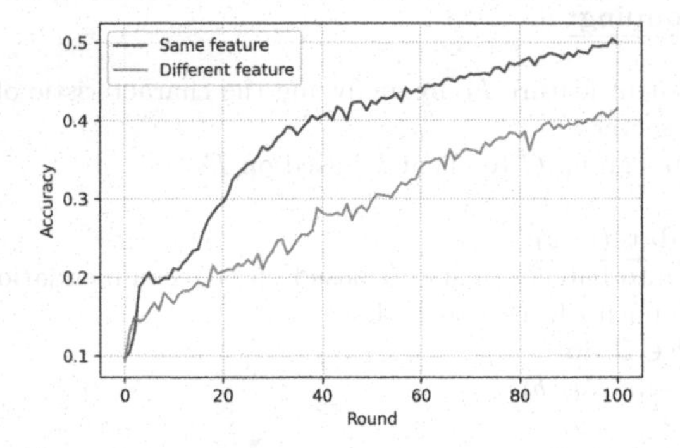

Fig. 2. Comparison of accuracy between balanced and unbalanced data

Figure 2 shows the learning accuracy achieved using the same feature data and different feature data. Here, the same feature data represents a case in the proposed algorithm where the clients are grouped based on their characteristics. The different feature data represents a baseline case where the federated learning algorithm executes with the given data without the grouping. To demonstrate the effectiveness of the personalization based on the grouping of clients, we consider a simple case that assumes similar data sizes for the same feature data. The dataset used in this experiment is CIFAR10. By creating a dataset that divides CIFAR10 evenly by 5000 and randomly divides from 1000 to 9000, we simulated a balanced data environment in which only clients whose data size meets the standard size participate in learning (same feature data) and an unbalanced environment in which they participate in learning regardless of the size of the data (different feature data). FedAvg algorithm of the Flower Framework [8] was used as the baseline of the experiment, and a total of 10 clients were generated and learned using balanced data and randomly split datasets. As a result, the balanced data had an accuracy of 0.5 and the unbalanced data had an accuracy of 0.4. The results of learning using balanced data were better than those of learning using unbalanced data. This result shows that higher accuracy can be obtained by participating clients with the same data characteristics.

5 Conclusion and Further Works

This paper introduces a new phishing attack detection algorithm based on federated learning that can protect user privacy and enhance detection accuracy. The algorithm can improve learning accuracy by mitigating the performance degradation caused by data imbalance between heterogeneous clients, which is one of the inherent problems in federated learning. In the future, we will further design an optimized phishing detection method for each client. We observe that there are various attack patterns as well as personal call patterns, and we plan to create individually optimized models. For example, Hanzely et al. [9] developed a theory to create an individual-optimized FL model by minimizing variants of Local SGD and accelerated coordinate descent/accelerated SVRCD. Anaissi [10] proposed a customized learning algorithm to personalize anomaly detection models. Further, we will consider other available data to increase the detection performance.

Acknowledgement. This research was supported by the MSIT Korea under the NRF Korea (NRF-2022R1A2C4001270), by the MSIT Korea under the India-Korea Joint Programme of Cooperation in Science & Technology (NRF-2020K1A3A1A68093469), and by the ITRC program (IITP-2020-2020-0-01602) supervised by the IITP.

References

1. Gangavarapu, T., Jaidhar, C., Chanduka, B.: Applicability of machine learning in spam and phishing email filtering: review and approaches. Artif. Intell. Rev. **53**(7), 5019–5081 (2020)
2. Basnet, R.B., Doleck, T.: Towards developing a tool to detect phishing URLs: a machine learning approach. In: Proceedings of the IEEE Conference on Computational Intelligence and Communication, pp. 220–223 (2015)
3. Thapa, C., et al.: Evaluation of federated learning in phishing email detection. arXiv:2007.13300 (2020)
4. Brendan McMahan, H., Moore, E., Ramage, D., Hampson, S., y Arcas, B.A.: Communication-efficient learning of deep networks from decentralized data. In: Proceedings of the 20th International Conference on Artificial Intelligence and Statistics (AISTATS) (2017)
5. Milandu, K.M..B., Park, D.J.: A real-time efficient detection technique of voice phishing with AI. In: Proceedings of the Korean Information Science Society Conference, pp. 768–770 (2021)
6. Park, E.L., Cho, S.: KoNLPy: Korean natural language processing in Python. In: Proceedings of the 26th Annual Conference on Human & Cognitive Language Technology, Chuncheon, Korea (2014)
7. Duan, M., Liu, D., Chen, X., Liu, R., Tan, Y.: Self-balancing federated learning with global imbalanced data in mobile systems. IEEE Trans. Parallel Distrib. Syst. **32**(1), 59–71 (2021)
8. Flower Summit 2022, 16 September 2022. https://flower.dev/conf/flower-summit-2022/. Accessed 20 Sept 2022
9. Hanzely, F., Zhao, B., Kolar, M.: Personalized federated learning: a unified framework and universal optimization technique. In: Proceedings of International Conference on Learning Representations (ICLR) (2021)
10. Anaissi, A., Suleiman, B.: A personalized federated learning algorithm: an application in anomaly detection. arXiv preprint arXiv:2111.02627 (2021)

CovidMis20: COVID-19 Misinformation Detection System on Twitter Tweets Using Deep Learning Models

Aos Mulahuwaish[1(✉)], Manish Osti[1], Kevin Gyorick[1], Majdi Maabreh[2], Ajay Gupta[3], and Basheer Qolomany[4]

[1] Department of Computer Science and Information Systems, Saginaw Valley State University, University Center, USA
{amulahuw,mrosti,kpgyoric}@svsu.edu

[2] Department of Information Technology, Faculty of Prince Al-Hussein Bin Abdallah II for Information Technology, The Hashemite University, Zarqa, Jordan
majdi@hu.edu.jo

[3] Department of Computer Science, Western Michigan University, Kalamazoo, USA
ajay.gupta@wmich.edu

[4] Cyber Systems Department, University of Nebraska at Kearney, Kearney, USA
qolomanyb@unk.edu

Abstract. Online news and information sources are convenient and accessible ways to learn about current issues. For instance, more than 300 million people engage with posts on Twitter globally, which provides the possibility to disseminate misleading information. There are numerous cases where violent crimes have been committed due to fake news. This research presents the CovidMis20 dataset (COVID-19 Misinformation 2020 dataset), which consists of 1,375,592 tweets collected from February to July 2020. CovidMis20 can be automatically updated to fetch the latest news and is publicly available at: https://github.com/everythingguy/CovidMis20.

This research was conducted using Bi-LSTM deep learning and an ensemble CNN+Bi-GRU for fake news detection. The results showed that, with testing accuracy of 92.23% and 90.56%, respectively, the ensemble CNN+Bi-GRU model consistently provided higher accuracy than the Bi-LSTM model.

Keywords: COVID-19 · Misinformation · Dataset · Benchmark · Fake News · Twitter · Deep Learning

1 Introduction

Fake news, misinformation and disinformation have been in the media for a long time, with instances cited as far back as 1835 [1]. However, instances of fake news have grown detrimental with the growth of the internet, especially through social media platforms like Twitter, Facebook, etc. Concerns about fake news grew exponentially during the COVID-19 pandemic and the 2020 US Presidential Election. When the pandemic started in 2019,

H. Zaynidinov et al. (Eds.): IHCI 2022, LNCS 13741, pp. 466–479, 2023.
https://doi.org/10.1007/978-3-031-27199-1_47

worry increased, and people went to the internet for answers. Millions of people died from COVID-19, and the global pandemic had lasting economic effects. In addition to COVID-19, there was a US Presidential election in 2020. These two events led to increased fake news led by political biases.

Cases of fake news were especially prevalent on social media platforms, one such social media platform being Twitter. Twitter is a platform where people can share information and images in "Tweets". However, since virtually anybody can post anything, misinformation can also be posted and spread quickly. Therefore, we need to have a way to flag content that could be untrustworthy.

To classify the CovidMis20 dataset, we chose to use Ensemble Convolutional Neural Network (CNN) with the Bi-Directional Gated Recurrent Unit (Bi-GRU) model because Bi-GRU is a less complex architecture than Bidirectional Long Short-Term Memory (Bi-LSTM) [2], which is a good model for text classification purposes in general. CNN is better suited to process the higher dimensional matrices, whereas Bi-GRU is better at temporal sequence data. So, using CNN and Bi-GRU as an ensemble model, we can utilize the power of calculation of higher dimensional matrices and temporal sequence data with less complexity than Bi-LSTM. An alternative is to use Bi-LSTM with Bi-GRU, but we will later see that the ensemble of CNN to capture the high dimensional matrices with Bi-GRU capturing temporal features performed better.

We could also try to use classical machine learning techniques like Support Vector Machine (SVM), k-nearest neighbors (k-NN), Decision Trees, etc. However, we believe deep learning has far more capability in generalizing the various datasets and can be improved with further improvements in the model's architectures.

The main contributions of this work are as follows:

- We built a labeled dataset, CovidMis20, which contains around 1,375,592 tweets; this dataset helps researchers differentiate between fake and real news related to COVID-19 on Twitter; also, CovidMis20 can be automatically updated to fetch the latest news. We collected the Twitter dataset and used Media Bias Fact Check (MBFC) [16], which is a fact-checking page that relies strictly on the International Fact-Checking Network (IFCN) signatories. MBFC portal has one of the most comprehensive databases of fact-checking media sources. Initially, a dataset of URLs identified as fake or real was created using a web crawler tool to identify URLs with COVID-19 information. Based on the Media Bias Fact Check (MBFC) database, these URLs were classified as fake and real.
- We developed and evaluated two deep learning models: ensemble CNN+Bi-GRU and Bi-LSTM for fake and real tweets text prediction with a testing accuracy of 92.23% and 90.56%, respectively.

This paper is organized as follows. Section 2 discusses related works. Section 3 presents the methodology and implementation. Section 4 presents results and discussion. Finally, Sect. 5 provides our conclusions and future work.

2 Related Work

Recently, a significant amount of work has been done in the area of fake news detection. This section reviews recent related works on the different aspects of fake news detection.

Hassan et al. [3] proposed a machine learning approach using supervised machine learning classification models like Support Vector Machine (SVM), Logistic Regression, and Naive Bayes to detect fake online reviews. They have used 1600 examples of the gold standard hotel review dataset, 800 reviews are deceptive, and the other 800 are truthful; as a result, they found that the SVM has given the best accuracy (88.75%) over the remaining models.

Reddy et al. [4] introduced a hybrid detection system for fake data that employs the Multinomial Voting Algorithm based on Naïve Bayes, Random Forest, Decision Tree, Support Vector Machine (SVM), k-Nearest Neighbours (k-NN) models. Their experimental data was collected from the Kaggle datasets (the authors did not mention anything related to the type of the datasets and their sizes or features). They had an accuracy score of 92.58% using above mentioned classical machine learning techniques.

Patwa et al. [5] created a COVID-19 social media and articles dataset for real and fake labels. Fake claims are collected from different fact-checking websites (like Politifact2, NewsChecker 3, Boomlive 4, etc.), and the real labels are collected from Twitter. The dataset vocabulary size is about 37,505. They used classical machine learning techniques like Decision Tree, Logistic Regression, Support Vector Machine (SVM), and Gradient Boost, the SVM model, which performs best with 93.32% accuracy, which is also a very good precision, recall, and accuracy overall. However, SVM gets more complex to work with when increasing the dataset size.

Al Asaad et al. [6] introduced a model to verify the news credibility that used different machine learning techniques for text classification. They used two datasets (fake or real news) with 6335 news articles, 3164 were fake, and 3171 were true, and log data with 32000 titles, 15999 being clickbait and 16001 being non-clickbait. The model's efficiency has been tested on a dataset by using Multinomial Naïve Bayes (MN) and Linear Support Vector Machine classification (LSVC) algorithms; they applied them with Bag-of-Words (BoW 2), Bi-gram (bigram 3), and Term Frequency-Inverse Document Frequency: (TF-IDF). The LSVC classifier performs better with the TF-IDF model.

Yu et al. [7] proposed four hybrid deep learning models (CNN+GRU, CNN+Bi-RNN, CNN+Bi-LSTM, and CNN+Bi-GRU). They applied their models to a real dataset of patients' blood samples for the COVID-19 infection test from Hospital Israelita Albert Einstein in Sao Paulo, Brazil. They included 111 laboratory results from 5644 different patients. The results show that CNN+Bi-GRU performs the best, with an accuracy score of 94%.

Aslam et al. [8] introduced an ensemble Bi-LSTM+GRU dense, deep learning model to detect fake news where Bi-LSTM+GRU was implemented for the textual attribute. In contrast, a dense, deep learning model was used for the remaining attributes. The study used a LIAR dataset with 12.8 K humans labeled from POLITIFACT.COM. The proposed approach achieved an accuracy of 89.8%.

The "COVID-19 Fake News" dataset, which includes 21,379 real and fake news examples for the COVID-19 pandemic and associated vaccines, was used to train and test four deep neural networks for fake news detection. Convolutional Neural Network

(CNN), Long Short Term Memory (LSTM), Bi-directional LSTM, and a combination of CNN and LSTM networks were developed and evaluated to automatically identify fake news content related to the COVID-19 posted on social media platforms. The evaluation's findings demonstrated that the CNN model performed better than the other deep neural networks, with an accuracy rate of 94.2% [9].

In summary, from the literature review, it is evident that social media platforms from the beginning of the pandemic were abuzz with news and information related to COVID-19. It was also clear that information – fake or real – caught the attention of millions of people active on social media, swayed their opinion, and influenced their behavior. To classify a tweet (that is, to analyze its content), many existing works analyze its text content using natural language processing (NLP) techniques and then use machine learning or deep learning algorithms for classifying the tweet's text. However, because the tweet's text is expressed using natural languages, it is hard to detect and extract what the tweet's author means due to the vagueness and the imprecision of the written text, which implies low accuracy.

To aid in those computational efforts, this paper describes the construction of the CovidMis20 (COVID Misinformation 2020), which could be used as a benchmark dataset; CovidMis20 contains confirmed true and fake news from Media Bias Fact Check (MBFC). The size of CovidMis20 is currently about 1,375,592 tweets; CovidMis20 can be automatically updated to fetch and incorporate the latest news. In contrast, most efforts in the literature have used relatively small datasets, and thus scalability remains a question. Obviously, most machine learning models make a good decision when the dataset size is relatively large. The results become more generalizable compared to smaller datasets. Hence, CovidMis20 should contribute to advancing the research on fake news detection. This paper also introduces a technique to build fake news detecting systems by focusing on creating a sequence classification model for fake (non-trustworthy) and real (trustworthy) tweets text (classifying text sequences based on the contextual information present).

3 Methodology and Implementation

3.1 Dataset Collection and Curation

We collected more than 1.5 billion coronavirus-related tweets from more than 40 million users from January 22, 2020, until May 15, 2022, leveraging the Twitter standard search application programming interface (API) and Tweepy Python library. A set of predefined search English keywords were used. These include {"corona", "coronavirus", "Coronavirus", "COVID-19", "stay at home", "lockdown", "Social Distancing", "Epidemic", "pandemic", and "outbreak"}, which are the some of the most widely used scientific and news media terms relating to the novel coronavirus. We extracted and stored the text and metadata of the tweets, such as timestamp, number of likes and retweets, hashtags, language, and user profile information, including user id, username, user location, number of followers, and number of friends.

Typically, it is challenging to label data as "fake" or "real" news since the perception varies from one individual to another. We can, however, focus on the facts that have been established medically or scientifically – and then decipher if these facts were incorrectly shared or twisted by individuals or websites while being posted on a platform. Using a determined list of websites classified as trustworthy or not – obtained from Media Bias Fact Check (MBFC) - a web-crawler tool was deployed to collect an initial URL-based (MBFC dataset) dataset. The web crawler tool used in this work is a Java-based tool to crawl the internet. The tool was executed to search the internet and find and report COVID-19-related sites in an output file.

The keywords used for the search were – Sars-cov-2, COVID-19, Coronavirus, and virus – to only get sites related to the COVID-19 information. Then we looked for COVID articles at each URL (for example, https://achnews.org). If a page contained at least six instances of the key terms: Sars-cov-2, COVID-19, Coronavirus, and virus, it was considered a COVID article. The scraper went through each domain page, searching for these keywords. For the pages considered COVID articles, only the URL was collected. The website content needed to contain any of the keywords mentioned above thrice to ensure a closer hit. This ensured that the words were not present in the infomercial listed or the extra information section. There is an assumption made when we think only these keywords represent sites containing COVID-19 information. Of course, more sites have other keywords present talking about the virus. The URLs from the MBFC dataset were classified as fake and real as per the predetermined list from the MBFC.

For Twitter data analysis, applications were used to clean the data as the tweets that had MBFC were classified, URLs in them were extracted, and fields like tweet date, location, tweet text, and tweet URL were identified and extracted from the data for problem-solving.

Figure 1 shows the data collection process. The first phase of data collection is registering a Twitter application and obtaining a Twitter access key and token. The second phase is to import the Tweepy Python package and write the Python script for accessing Twitter API. The third phase is to connect to Twitter search API and retrieve tweets using some keywords related to COVID-19. The fourth phase reads and processes the data to extract information on tags, agents, and locations for network construction and analysis. The fifth phase filters Tweets containing URLs from a web crawler. The last phase is attaching MBFC labels, including trustworthy and untrustworthy ones.

3.2 Data Preprocessing

We used data mining and machine learning methods to preprocess, classify and analyze the data. Data or text mining, in general, helps identify relevant information in a large corpus, providing qualitative results, and machine learning for text mining or data mining is the process of reviewing the text to help with specific research questions. It is the initial cleaning part of data preprocessing. It helps identify the features and relationships of a given text. Once identified, these features can be used by machine learning methods to find patterns and trends across large data sets, resulting in more quantitative results. We used Natural Language Processing (NLP) which helps machines understand human language in a given context [10].

Initially, Python was used to clean the dataset. Following the standard procedure, it removed all non-English words and converted all text to lower case to ensure the program recognized all the exact words in the same manner. After this, we used Natural

Language Processing Tool Kit (NLTK) library [11] in Python to create a proper corpus using PorterStemmer and Stopword functions.

NLTK in Python has a list of stop words (such as 'the', 'a', 'an') stored in 16 different languages. The Stopword function identifies these words in the corpus. PorterStemmer function is used to stem words; for example, after stemming, a word like "looking" becomes "look". Stemming is done on all words except stop words (removing stop words from the corpus). This process creates a good corpus, saves space in the database, and improves processing time. This corpus is used for classification and prediction purposes.

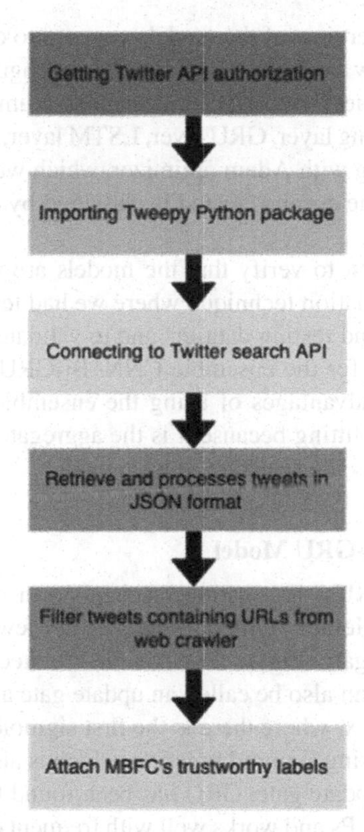

Fig. 1. Process for collecting tweet data

We have used the Embedding Layer from Keras' library in Python [12] (which uses the word embedding technique for text-preprocessing – words with the same meaning will have similar representations). Before the data enters the Embedding Layer, it must be converted to a numerical form categorical, using One-Hot encoding [13]. After this, the timestamp is padded with zeros, either pre or post, to create sentences of equal length. Also, several feature vectors are defined for the Embedding Layer so the model can create vector representations based on the number. This is done so algorithms can generalize the corpus created to make any prediction.

We balanced the data because there were around 10% more true tweets than fake tweets. Thus, we downsampled the true tweets to the same number as fake tweets. We removed URL patterns from the tweets, like https:// and other symbols and characters which resemble the part of a URL to have only normal text without any links to the tweet and then converted all the text to lowercase. The numbers of true and fake news are equally balanced into 289,826. Also, we divided the dataset randomly into 70% training and 30% testing. The total size of the training set is 405,756 (70%), and the testing set is 173,896 (30%).

3.3 Classifications

This section presents an overview of the models we used to classify our dataset.

We used the TensorFlow framework with a Python programming language to build the models. We also used other Python APIs like pandas, NumPy, matplotlib, and NLTK. TensorFlow has an embedding layer, GRU layer, LSTM layer, and dense, fully connected layer readily available along with Adam optimizer, which we used to build the models. After data preprocessing, the corpus created is classified by ensemble CNN+GRU and BiLSTM models.

We used cross-validation to verify that the models are not overfitting. We used a stratified K-Fold cross-validation technique where we had ten folds or splits to balance the classes in the training and testing datasets and to validate the models; We observed a mean accuracy of 91.6% for the ensemble CNN+Bi-GRU and 90.8% for Bi-LSTM models. Also, one of the advantages of using the ensemble CNN+Bi-GRU is that it reduces the chances of overfitting because it is the aggregate of all the model's output. The following sections show the structure of each model.

3.3.1 Ensemble CNN+Bi-GRU Model

Gated Recurrent Units (GRUs) is a gating mechanism in recurrent neural networks, like the Long Short-Term Memory with a forget gate but fewer parameters than LSTM because it lacks an output gate. Figure 2 shows the architecture of GRU, where there are only two gates, which can also be called an update gate and reset gate. From Fig. 2, the first gate from the input x, where there is the first sigmoid function from the left, is the reset gate and the following sigmoid function, which is also connected to the hidden state h to the right, is the update gate. GRU has been found to perform well with tasks like signal modeling and NLPs and works well with frequent datasets. So, a GRU can be generalized as a variation of LSTM because both have a similar design. GRUs use the update gate and the reset gate, the two primary operations in the GRU model. Therefore, these gates control the flow of information, which means that useful information can be kept, and unimportant information can be removed.

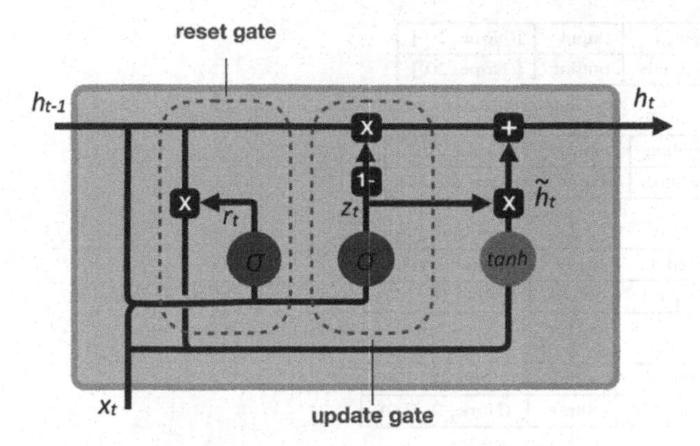

Fig. 2. Gated Recurrent Unit (GRU)

Within deep learning, Convolutional Neural Network (CNN) is used to analyze structured data arrays (such as images) and is mainly used for image and text classification. Figure 3 shows the architecture of a CNN by stacking layers on top of each other in a sequence. These layers are usually convolutional, followed by activation and sometimes pooling layers. Additionally, CNNs are typically made up of 20 to 30 layers, with each layer capable of recognizing something more complex than the last. For example, using 3 to 5 layers, handwritten digits can be recognized, and with 25 layers, a human face can be recognized.

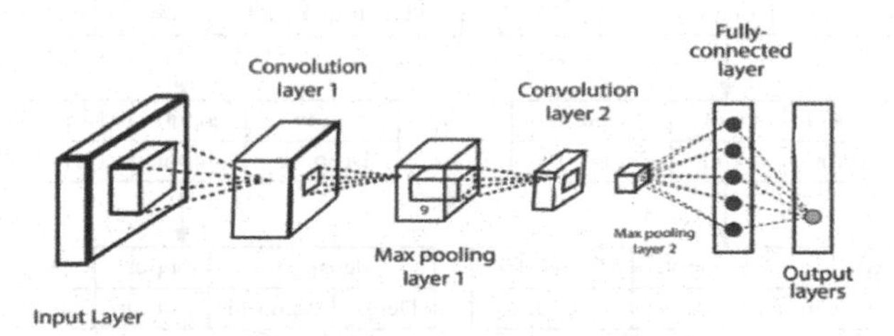

Fig. 3. Convolutional Neural Network (CNN) model

After analyzing how efficient they can be for text classification and NLPs, we decided to use ensemble CNN+Bi-GRU, where we combine 1D CNN with a single Bi-GRU. 1D CNN performs better on text classification [14]. The Bi-GRU model works well on time-series data by looking at the earlier and later information sequences.

Figure 4 shows the layers of the ensemble CNN+Bi-GRU model we used for building CovidMis20 benchmark. There are two inputs for the model training purpose, one for CNN and one for Bi-GRU, followed by the embedding layer, convolutional layer, max pooling, dropout layer, and a dense or a fully connected layer with an activation function (sigmoid function). For GRU, the input layer is followed by the embedding layer, Bi-Directional Gru Layer, dropout layer, and a dense layer with a sigmoid function as an activation function. Then, we merged both the dense layers of CNN and Bi-GRU to form a dense layer (fully connected layer) with a sigmoid function.

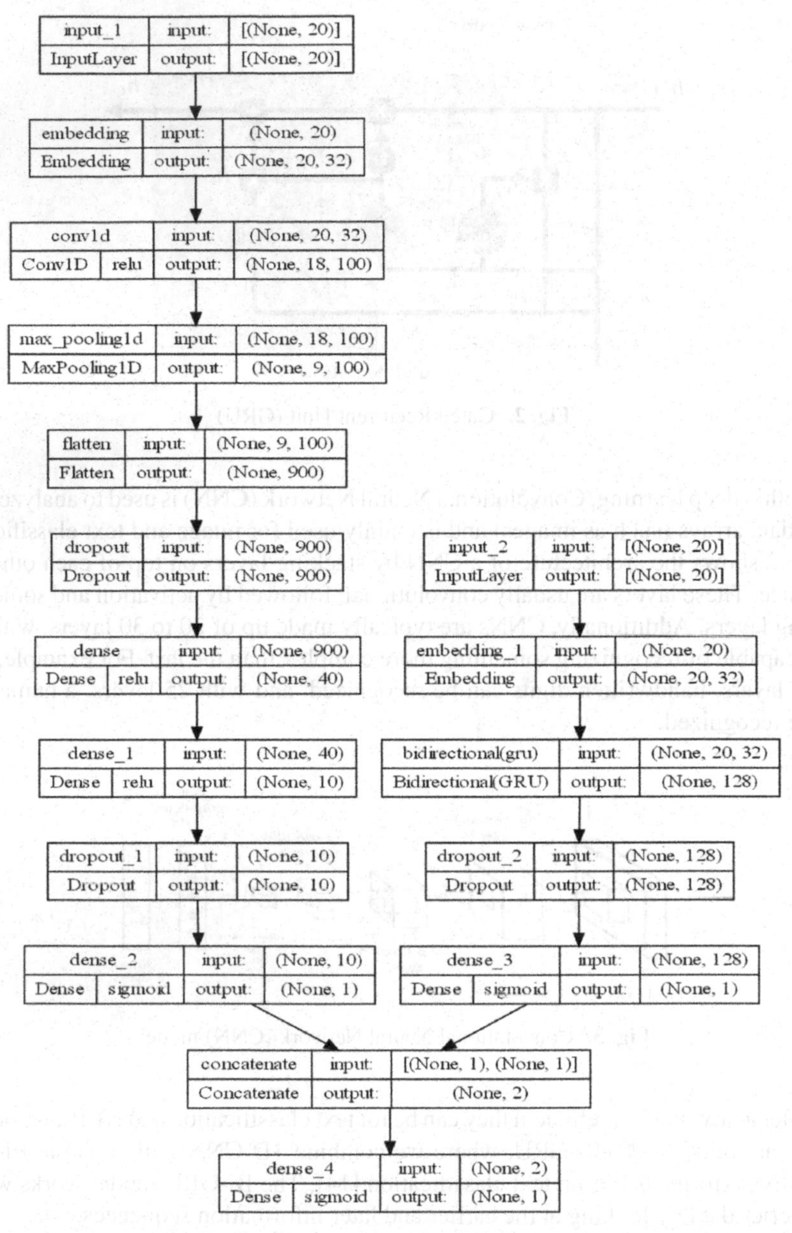

Fig. 4. An ensemble CNN+Bi-GRU model

3.3.2 Bi-LSTM Model

Bi-LSTM is a type of RNN (Recurrent Neural Network [15]) used to avoid vanishing gradient problems. Bi-LSTM models have a cell state representing context information (storing information about past inputs for a specific time). Bidirectional LSTM can look at the data in both directions (left to right and right to left), so it is considered to store contextual information better than LSTM. Bidirectional LSTM is used when prediction depends on previous and future inputs. For example, a sentence like "I say Sam likes eating ___". Eating can be anything, so prediction depends on understanding the future and previous words. The Bi-LSTM model uses the time stamp technique to enter data into the model. The model learns the contextual information for the time stamp entered and tries to predict the next word in the sequence.

Bi-LSTM models and NLP (Natural Language Processing) use distributed representation techniques to describe the same data features across multiple scalable and interdependent layers. It tries to interpret or learn features of the same dataset on different scales (semantic similarities). Another reason to use the Bi-LSTM model is that around 2007, the models started to revolutionize speech recognition outperforming traditional models. Figure 5 shows the structure of an unfolded Bi-LSTM layer that contain the forward and backward LSTM layers.

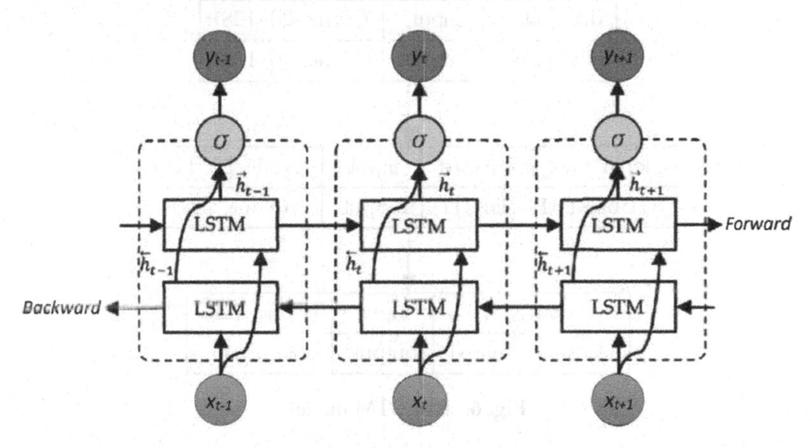

Fig. 5. Unfolded architecture of BiLSTM with three consecutive steps

Figure 6 shows the layers of the Bi-LSTM model we used for this work. There is an input layer, followed by the embedding layer, Bi-Directional LSTM Layer, dropout layer, global max-pooling layer, and a dense layer with a sigmoid function as an activation function.

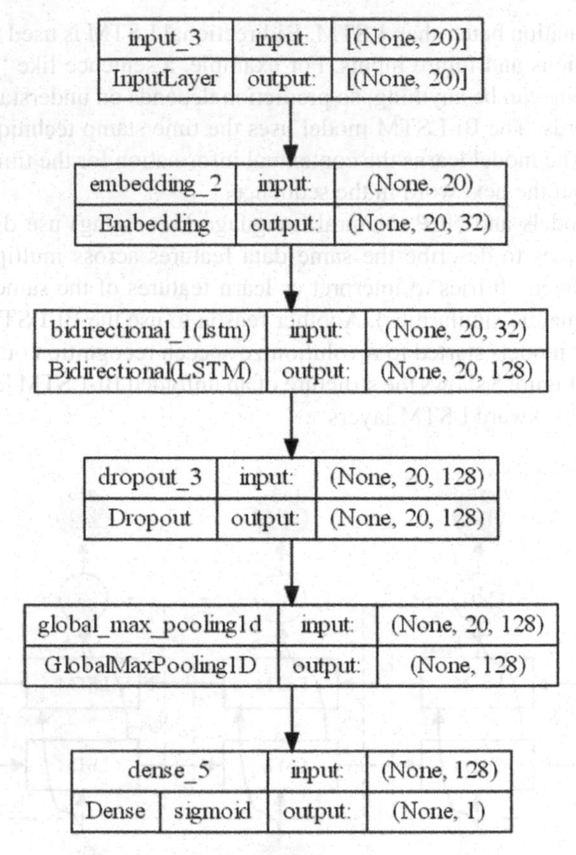

Fig. 6. Bi-LSTM model

4 Results and Discussion

In this section, we provide details on the outcome of running the models on the CovidMis20 dataset.

Figure 7 shows the overall measurement results of the ensemble CNN+Bi-GRU and Bi-LSTM models to detect fake news from the CovidMis20 dataset. We have a 92.23% testing accuracy, 0.9025 precision (specificity), 0.945 recall (sensitivity), and 0.9232 F1-score with the ensemble CNN+Bi-GRU and 91.56% testing accuracy, 0.911 precision (specificity), 0.92 recall (sensitivity), and 0.9154 F1-score with the Bi-LSTM model. Figures 8 show the accuracy and validation accuracy for the ensemble CNN+Bi-GRU and Bi-LSTM, where they trained with a batch size of ten.

Figure 9 shows the confusion matrix for ensemble CNN+Bi-GRU and BiLSTM models for text classification. We analyzed the confusion matrix in the evaluation phase, where 1 is fake and 0 is true. Thus, we can see a high accuracy of True Positive (TP) and True Negative (TN) percentages while detecting fake news from the Twitter dataset. For clarity, the meaning of true positive, true negative, etc. in our context is as follows:

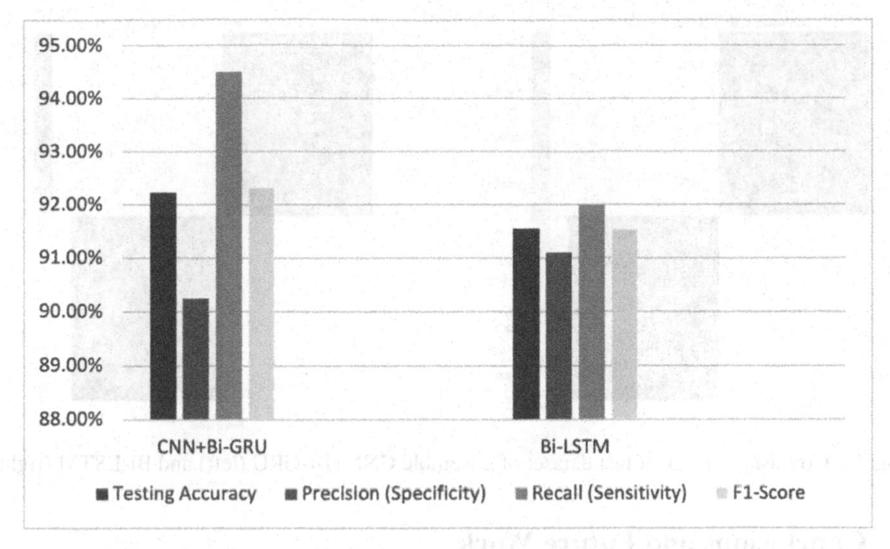

Fig. 7. An ensemble CNN+Bi-GRU and Bi-LSTM performance

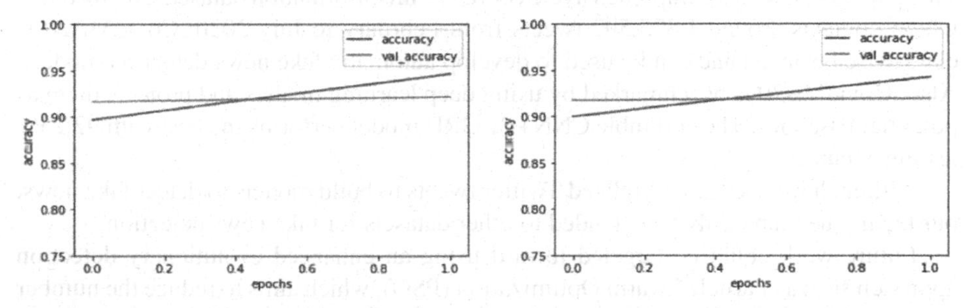

Fig. 8. Evaluation phase performance of ensemble CNN+Bi-GRU (left) and Bi-LSTM (right)

True Positive (TP): when predicted, fake news pieces are annotated as fake news. True Negative (TN): when predicted, true news pieces are annotated as true news. False Negative (FN): when predicted, true news pieces are annotated as fake news. False Positive (FP): when predicted, fake news pieces are annotated as true news.

Based on Fig. 9, the ensemble CNN+Bi-GRU model classified 89.8% of fake tweets correctly and 10.2% incorrectly. At the same time, 5.5% of real tweets text was classified incorrectly by the model as fake, while classifying 94.5% of real tweets text. In contrast, the BiLSTM model correctly identified 91.1% of fake tweets text, misidentifying 8.9% as real. At the same time, 8.0% of real tweets' text was classified correctly and misclassified 92.0% incorrectly.

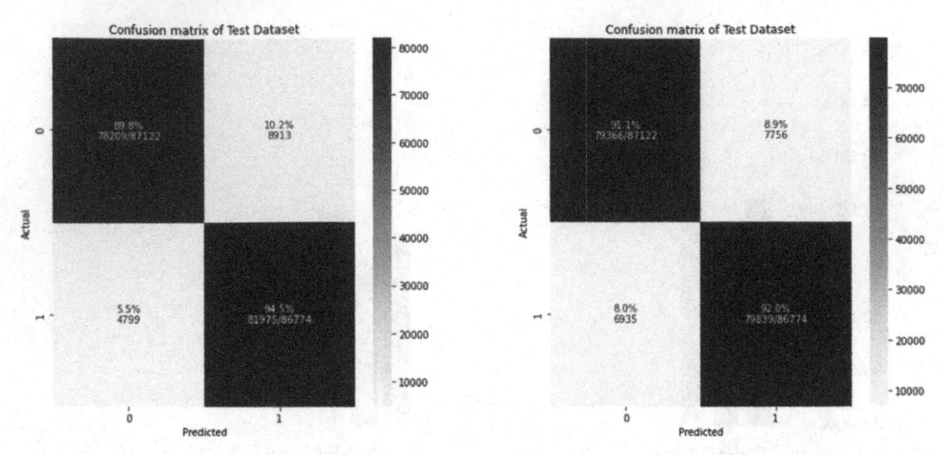

Fig. 9. Confusion matrix of test dataset of ensemble CNN+Bi-GRU (left) and Bi-LSTM (right)

5 Conclusions and Future Work

This paper presented a comprehensive COVID-19 misinformation dataset, CovidMis20, which contains around 1,375,592 tweets from February to July 2020; CovidMis20 is class-wise balanced and can be used to develop automatic fake news detection models. Also, CovidMis20 is benchmarked by using deep learning models and projects them as potential baselines. The ensemble CNN+Bi-GRU model performs the best with 92.23% testing accuracy.

Although we used Covid-related Twitter tweets to build models to detect fake news, our techniques can easily be extended to other datasets for fake news detection.

Future work could be targeted toward using an enhanced evolutionary detection approach such as Particle Swarm Optimization (PSO), which aims to reduce the number of symmetrical features and obtain more accurate models.

Acknowledgments. This work was supported in part by Saginaw Valley State University and the National Science Foundation under Grant OAC-2017289, National Institute of Health under Grant 1R15GM120820-01A1, and WMU FRACAA 2012-22.

References

1. Pennycook, G., Rand, D.G.: The psychology of fake news. Trends Cogn. Sci. **25**(5), 388–402 (2021)
2. Chung, J., Gulcehre, C., Cho, K., Bengio, Y.: Empirical evaluation of gated recurrent neural networks on sequence modeling. arXiv preprint arXiv:1412.3555 (2014)
3. Hassan, R., Islam, M.R.: A supervised machine learning approach to detect fake online reviews. In: 2020 23rd International Conference on Computer and Information Technology (ICCIT), pp. 1–6. IEEE (2020)
4. Reddy, P.B.P., Reddy, M.P.K., Reddy, G.V.M., Mehata, K.M.: Fake data analysis and detection using ensembled hybrid algorithm. In: 2019 3rd International Conference on Computing Methodologies and Communication (ICCMC), pp. 890–897. IEEE (2019)

5. Patwa, P., et al.: Fighting an infodemic: Covid-19 fake news dataset. In: Chakraborty, T., Shu, K., Bernard, H.R., Liu, H., Akhtar, M.S. (eds.) CONSTRAINT 2021. Communications in Computer and Information Science, vol. 1402, pp. 21–29. Springer, Cham (2021). https://doi.org/10.1007/978-3-030-73696-5_3
6. Al Asaad, B., Erascu, M.: A tool for fake news detection. In: 2018 20th International Symposium on Symbolic and Numeric Algorithms for Scientific Computing (SYNASC), pp. 379–386. IEEE (2018)
7. Yu, Z., He, L., Luo, W., Tse, R., Pau, G.: Deep learning for COVID-19 prediction based on blood test. In: IoTBDS, pp. 103–111 (2021)
8. Aslam, N., Khan, I.U., Alotaibi, F.S., Aldaej, L.A., Aldubaikil, A.K.: Fake detect: a deep learning ensemble model for fake news detection. Complexity **2021** (2021)
9. Tashtoush, Y., Alrababah, B., Darwish, O., Maabreh, M., Alsaedi, N.: A deep learning framework for detection of COVID-19 fake news on social media platforms. Data **7**(5), 65 (2022)
10. What is Natural Language Processing and How Does It Work. https://monkeylearn.com/blog/what-is-natural-language-processing/. Accessed 10 Sept 2021
11. Natural Language Toolkit. https://www.nltk.org/. Accessed 2 Sept 2021
12. Keras: the Python deep learning API. https://keras.io/. Accessed 2 Sept 2021
13. Sklearn.preprocessing.OneHotEncoder. https://scikit-learn.org/stable/modules/generated/sklearn.preprocessing.OneHotEncoder.html. Accessed 2 Sept 2021
14. Kowsari, K., Heidarysafa, M., Brown, D.E., Meimandi, K.J., Barnes, L.E.: RMDL: random multimodel deep learning for classification. In: Proceedings of the 2nd International Conference on Information System and Data Mining, pp. 19–28 (2018)
15. Sherstinsky, A.: Fundamentals of recurrent neural network (RNN) and long short-term memory (LSTM) network. Phys. D **404**, 132306 (2020)
16. Media Bias Fact Check (MBFC). https://mediabiasfactcheck.com/. Accessed February 2021

Development of School Library Network Based on Cloud Technologies in Uzbekistan

Marat Rakhmatullaev and Sherbek Normatov(✉)

Tashkent University of Information Technologies named after Muhammad al-Khwarizmi, A.Temur 108, Tashkent, Uzbekistan
shb.normatov@gmail.com

Abstract. Cloud technologies have become an important tool for improving the efficiency of network solutions. This is due not only to the intensive development of information technologies and telecommunications, but also to the active creation of information resources, big databases for collective use, and corporate networks. This article provides the solution based on cloud technologies to the problem of information support for school libraries in the Republic of Uzbekistan. The functional and organizational structure of the MAKTEK system is given, where the main functions performed by the system and organizational decisions for its implementation and operation are given. The problems encountered by developers and the advantages of the system are indicated. The project MAKTEK aims to provide rapid access to electronic educational resources for secondary schools by creating a network of corporate electronic libraries of schools in the Republic of Uzbekistan. The project is implemented with the financial support of the Ministry of public education of the Republic. The implementation of the MAKTEK project has shown some important advantages such as reduced duplication of information resources, improved the quality of bibliographic records in OPAC, fewer errors in these records due to the involvement of catalogers with extensive experience and skills.

Keywords: School libraries · Cloud technology · Digital library · Corporate network · Information library system

1 Introduction

The idea of cloud technologies appeared long ago, even before the era of personal computers. When there were no Internet and Intranet networks, advanced technologies of microprocessor technology, the policy of creating computers was focused on the development and production of big super computers. They had to provide centralized data processing and storage. In addition, the user with his terminal and keyboard was quite far from the head computer. Personal computers have significantly changed people's lives and attitudes to receiving, transmitting, and storing data. Personalization has covered all areas of life. However, the new era has come – the era of active interaction, the era of big data, the need for active data exchange and storage. The rapid development

H. Zaynidinov et al. (Eds.): IHCI 2022, LNCS 13741, pp. 480–492, 2023.
https://doi.org/10.1007/978-3-031-27199-1_48

of telecommunications, the Internet, the ability to quickly transmit large amounts of information over long distances, again returned the old idea of cloud technologies, but to a new level of development.

Cloud technologies are widely used in educational institutions, especially in universities. Each department of a university (departments, libraries, deans, managers, etc.) needs a large set of data about students, teachers, library funds, etc. They can get this data from a shared server of not only one university, but also partner universities, where the necessary information is centrally stored. Sultan, N., (2010) notes that cloud technologies have become an important tool for the collective use of resources in many US educational institutions.

Developers of library systems support the use of cloud technologies, because these technologies are quite useful for protecting and preserving their data and information security. This idea is also shared by librarians who use cloud technologies for easy access to electronic journals, hosting related digital libraries that track statistics and others (Suman & Parminder, 2016). The effectiveness of this technology is that shared resources, software, and other information resources are provided to remote clients over the network. Suciu, Halunga, Apostu, Vulpe and Todoran (2013) they point out that cloud technologies are a service that provides users with secure computing and remote storage capabilities for efficient provision of their services. Some authors as Gireesh, G., Pradeep, G., Gaurav, C., Pooja, V., Gunjan, (2011), point out. Cloud computing is a service where cloud resources are dynamically distributed among multiple users according to predefined conditions. Specialists Zhau, R., Ioan, S. & Lu. I. (2018) believe that cloud computing is an improvement on grid computing, distributed computing, distributed databases, and parallel computing.

There are authors (such as Suciu, Halunga, Apostu, Vulpe and Todoran, 2013) believe that cloud computing provides users who are grouped together with secure computing and efficient data storage capabilities for efficient service delivery. This is also achieved due to the fact that they combine not only information resources (databases), but also software and computing resources. Some researchers (Goldner (2010)) believe that cloud technologies can help libraries get rid of all the technological problems they have with storing and using resources.

In libraries and library networks, the use of cloud technologies began in the early 2000s. Breeding (2012) argues that cloud technologies can significantly improve the automation of library processes, the creation of digital collections, the speed of searching information in databases, and provide other benefits. Professor Nandkishor Gosavi and others (2012) defined advantages of Cloud computing in libraries: 1. Cost saving; 2. Flexibility and innovation; 3. User centric; 4. Openness; 5. Transparency; 6. Interoperability; 7. Representation; 8. Availability anytime anywhere; 9. Connect and Converse10. Create and collaborate. Also authors have given some examples of Cloud libraries:1. OCLC; 2. Library of Congress (LC); 3. Exlibris; 4. Polaris; 5. Scribd; 6. Discovery Service; 7. Google Docs/Google Scholar; 8. Worldcat; 9. Encore.

There are many advantages of using cloud technologies, but as practice shows and as some authors claim, not all associations in a single network solve all problems of access to information and bring the expected effect. Grace (2010) ascertains that the major reasons for adopting cloud computing are scalability, elasticity, virtualization, cost reduction, mobility, and collaboration and risk reduction; however, performance, control, interoperability and security are the primary concern of most organizations. Some researchers, such as Foster, Zhau, Ioan, and Lu (2018) defined Cloud computing as "a large-scale distributed computing paradigm that is driven by economies of scale, in which a pool of abstracted, virtualized, dynamically-scalable, managed computing power, storage, platforms, and services are delivered on demand to external customers over the internet".

Academic libraries are wary of cloud technologies and cloud storage of their resources. These libraries have valuable information material that has been created over the years. Not all authors provide their research papers for open access. Most often, they give their resources only for use in their well-known libraries. Many libraries are afraid that their collections may disappear due to technical problems or actions of the administrators of the main cloud server, or be used in bad faith by other users.

2 Project "MAKTEK"

The use of cloud technologies requires certain conditions that will suit all members of the consortium (corporate network):

1. Reliable technical basis (servers, telecommunications system) that ensures smooth operation of the system;
2. High level of qualification of the staff who will administer cloud databases;
3. Providing a wide range of services for access to information resources (both shared and personal), satisfying both group(corporate network) members and all users of these services;
4. High level of information security and protection of public and private library resources, as well as copyright (Rakhmatullaev M., Normatov Sh. (2018)).

Research and practice show that cloud technologies in libraries are most effective in school corporate library networks. This was shown by the MAKTEK project, which is being implemented in Uzbekistan under a grant from the Ministry of public education of the Republic. MAKTEK is the name of the Information and Library System that operates on the basis of cloud technology. MAKTEK is the name of an information and library system that operates on the basis of cloud technology. This is an abbreviation of three words - MAKTAB (in Uzbek it is "school") and Electron Kutubxona (which means "Digital Library"). The software basis of MAKTAB is the ARMAT system, which was created for the corporate network of academic libraries of Uzbekistan Ministry High Education in 2017.

School libraries in Uzbekistan have a specific feature compared to other types of libraries:

- Most numerous (9698 school libraries), compared to other types of libraries;
- Most often, the librarian is not a specialist a specialist who graduated from a library university or college, but a teacher in literature or other subjects;
- Technical equipment of school libraries is worse than in other types of libraries. Most often, specialized school computer classes are used to work with an digital library;

- The library collection is small (about 20,000 titles, including all textbooks and methodological literature);
- A large number of potential users (more than 6,000,000, including all pupils (more than 4.5 million), as well as teachers and parents);
- The library fund is replenished centrally, with the support of the Ministry of public education. The schools themselves have a little extra money to buy literature;
- There is usually no dedicated maintenance service for servers, databases, telecommunications networks, etc.
- School libraries in different regions of the Republic are not provided with literature evenly and equally. All schools receive the necessary number of textbooks centrally. But in most cases, in large cities and regional centers, they have more information resources, more manuals, electronic textbooks, and additional multimedia materials for teaching and learning.

There is an acute problem of providing schools located in remote areas of the Republic with the necessary literature to improve the effectiveness of secondary education and provide assistance to teachers. It is very important that all school students have equal opportunities to access information resources, gain knowledge and get the necessary education.

The project MAKTEK aims to provide rapid access to electronic educational resources for secondary schools by creating a network of corporate electronic libraries of schools in the Republic of Uzbekistan. The project is implemented with the financial support of the Ministry of public education of the Republic.

The main tasks for implementing the aim are:

- Research, analysis and systematization of available literature for schools in the Republic. Identification of the most popular literature in electronic format for students and teachers to include in the General school electronic library;
- Development of an information and library system for creating electronic libraries and a single electronic catalog (OPAC);
- Creation of software systems for the analysis of demand sources for different categories of users (pupils, teachers, methodologists, parents, etc.). Develop the system of evaluation of sources in databases using the statistical information on the use of sources in various formats (text, multimedia, etc.);
- Information security of the server, databases and library networks.

3 Functional Structure of the Cloud Information Library System

The functional structure of the information system "MAKTEK" is shown in Fig. 1.

System consists of the following subsystems:

"DATA ENTRY". This subsystem is designed for the correct databases creation and includes a number of tasks (Executive modules):

- Creating a friendly interface for convenient work with the system, for organizing operational communication with users, organizing forums, etc.;

- Entering the bibliographic records on MARC21 communicative format and creating an electronic catalog;
- Check records for correctness and the elimination of duplicate records;
- Creating personal accounts for school libraries and users. This allows each library to create a personal electronic library remotely.

«DATA RETRIEVAL». The subsystem is designed to provide data search in databases and includes the following modules:

- Search by author of the work;
- Search by source name;
- Search by subject, key words;
- Advanced search, with restrictions by year, subject, and other criteria;
- Search in the personal account of the school library;
- Search in the OPAC.

"STATISTICS" is a subsystem that performs the functions of generating statistical reports and includes the following modules:

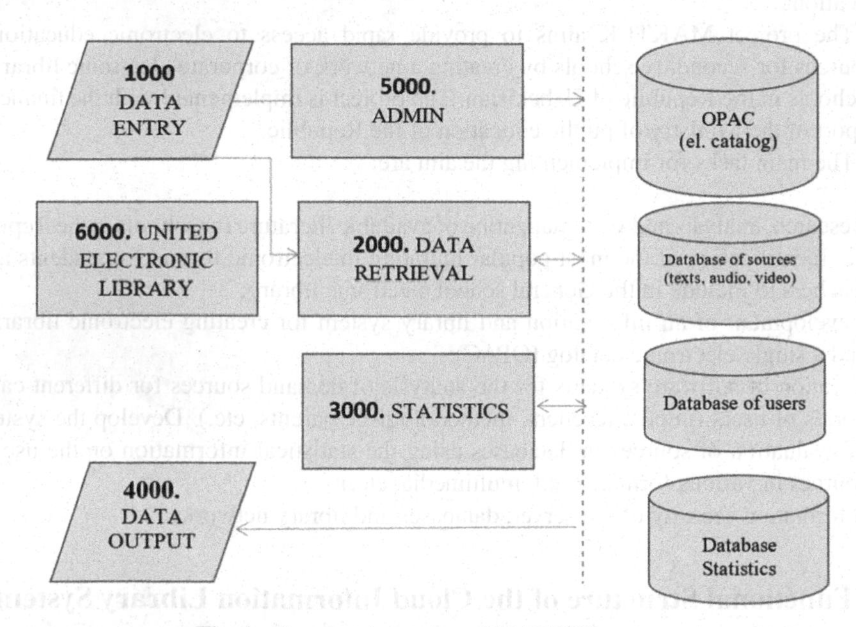

Fig. 1. Functional structure of MAKTEK system

- Formation (calculation) of statistical information on the use of literature (digital library) by schools in different regions of the Republic;
- Generating statistical information for each subject to get reliable information about the demand in the sources;

- Generating statistical information for each area of knowledge;
- Calculating the number of downloads from various sources to determine usage priority and identify demand.

"DATA OUTPUT" is a subsystem that provides output documentation in a form that meets user requirements and includes follow modules:

- Output data in tabular form at the user's request by region and school;
- Output graphical statistics;
- Output relevant information in various formats (text, audio, video).

"ADMIN". This subsystem performs the functions of administration of the head server and ensuring information security of the system and network.

It includes the following modules:

- Providing passwords for access to personal accounts, electronic catalog databases, and full-text resources;
- Organization of user registration;
- Creating personal accounts for school libraries;
- Creating backup copies of information resources;
- Organization of database protection against unauthorized access. Ensuring information security of databases, reserving, providing authorized access.

"UNITED ELECTRONIC LIBRARY" is a subsystem that performs the functions of creation general information resources (electronic catalogs, OPAC, full-text databases) and includes the following modules:

- Identification of duplicate bibliographic data and other information resources. Verification of bibliographic records;
- Creating the general electronic catalog (OPAC) and providing access to it. The function allows you to "look" in electronic catalogs in the personal accounts of school libraries and if it finds new bibliographic records and full-text sources, it enters them into the united electronic library;
- Creating the common database of information resources necessary for school students, as well as methodological literature for teachers. In addition, the database includes multimedia resources that are useful for learning and teaching.

All these subsystems work with General Server.

The main database of corporate information resources in General Server includes the following components:

- Corporate network and database statistics (queries, query categories, services performed, literature used);
- Electronic catalog (OPAC);
- Full-text sources of the corporate network (General sources);
- Links to information sources;

- User data – General information about all users who are members of the consortium.

Personal accounts of school libraries in the General Server space consist of the following parts: Statistics of the i-th library (requests, categories of requests, services performed, activity of literature use, etc.); Electronic catalog of the i-th library; Full-text databases created in the library; Data about users of the i-th library; ID-password of the i-th library.

Databases are updated from various sources. Sources of replenishment of the electronic library are:

- electronic resources created by the Center for Information Technologies(CIT) under the Ministry of public education;
- works (editions) of teachers, doctoral students, researchers and students who work in schools and study at pedagogical and other universities;
- digital books, magazines, articles, and other sources, including copyrights to publications, multimedia sources intended for school education. These resources are often provided by the Republican Children's Library under the Ministry of public education. It is one of the largest libraries in the Republic and has a rich fund of children's literature;
- valuable scientific and educational resources (textbooks, manuals, scientific articles, etc.) from sources of world databases (leading publishers and aggregators, such as Springer Nature, EBSCO Information Services, Oxford University Press, Wiley, ProQuest, e-Library, etc.) by subscription;
- borrowed resources from members of the corporate network, as well as the National Library of Uzbekistan, the corporate network of universities by mutual agreement.

Database storage uses a relational database management system (DBMS) that supports the SQL standard, is industrial, transactional, and fault-tolerant.

When developing the system, the following main requirements for database management systems were defined:

- orientation to DBMS operating in client-server mode;
- ability to process and transmit data in network mode (with data protection);
- ability to duplicate data in case of emergencies;
- ability to exchange data between other software models;
- ability to create relational databases and manipulate highly complex data structures.

Database structures provide:

Combining data of any quantity and volume for sharing (joint use); ability to store information (electronic textbooks and others) in various presentation formats (txt, doc, pdf, rtf, jpg, etc.); minimal data processing time; minimal data redundancy; minimum amount of computer memory for data storage; the most efficient access to the database, etc.

For data recovery in emergency situations, it is possible to back up data about electronic textbooks and other information on servers, register data on storage media in case of failure of the communication channel with remote workstations and the Central server, external memory devices, etc. for data recovery.

4 Hybrid Model

If some libraries are unwilling or unable to share their resources in the cloud, then hybrid mode is provided. The main server contains OPAC, but not all sources themselves can be in the common database and may not be accessible to all users.

The Fig. 2 shows how libraries interact in a cloud environment. There are two categories of libraries that work in the cloud.

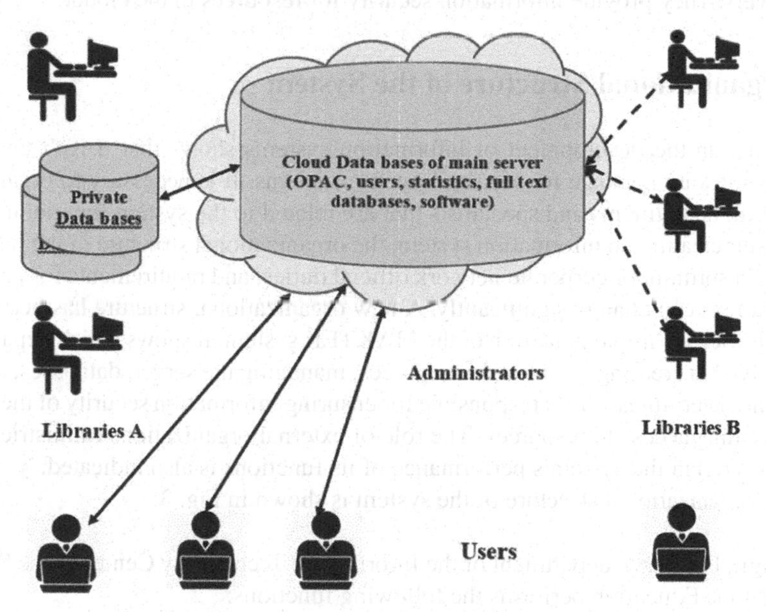

Fig. 2. Library interaction in the Cloud environment

Libraries A – libraries that partially provide their resources to the Cloud. Libraries can only provide links that they have these sources available. In this case, the library can only specify the source URL and access conditions. In other words, some of the resources are located in the Cloud database, and some may be located in the specific library of the corporate network.

These hybrid solutions are used when:

- the library doesn't want to share its resources or part of them;
- the author of the work gave exclusive rights only for this library, but hadn't give access rights for other libraries and their users;
- the information resource is paid.

They give part of their resources only after a request, or registration, or payment for the resource.

Libraries B are those that store all their information resources in the Cloud and, basically, provide them with open access for all members of the corporate network or for their users by passwords.

Administrators are specialists who are responsible for the administration of databases and servers. They provide information security for resources in the cloud.

5 Organizational Structure of the System

Experience in the development of information systems shows that it isn't enough to create a software package for solving system problems. It is necessary to organize the work of all departments and specialists that are related to the system functioning. Very often, after creating an information system, the organizational structure of the library, or library consortium, or corporate network official duties, and requirements for personnel qualifications can change significantly. A new organizational structure has been developed for effective implementation of the MAKTEK system. It shows which departments are involved in creating information resources, managing the server, databases, and network, and determines who is responsible for ensuring information security of the system and providing access to resources. The role of external organizations (ministries, other libraries, etc.) in the system's performance of its functions is also indicated.

The organizational structure of the system is shown in Fig. 3.

1. «Digital Library» department of the Information Technology Center of the Ministry of Public Education performs the following functions:

   - Cataloging using the international communication formats MARC21, DUBLIN CORE. Catalogers are not full-time employees of the ITC. Since cataloging requires high qualifications and skills, catalogers are attracted from libraries where digital libraries are already being created successfully;
   - Selection of literature, acquisition of library funds, methodical assistance in working with the corporate library network, OPAC and etc.;
   - Administration, managing and configuration of the Microsoft Windows network environment; Creation of personal accounts for school libraries; Registration; Ensuring the security of data storage in the databases;
   - Generates statistical information and provides it to MPE and other officials.

2. Republican Children's Library under the MPE performs the following functions:

- Participates in the replenishment of the general digital library and the provision of bibliographic records in the MARC or Dublin Core formats;
- Helps organize training courses for librarians.

3. "Textbook" Department under the MPE provides:

- List of approved by MPE and published literature for schools in the Republic annually;
- Monitoring the completeness of the textbook database in the General server.

4. School libraries

- Participate in the creation of personal electronic libraries and replenishment of the General digital library;
- Provide their own resources and bibliographic records (Library A) to the ITC;
- Conduct training courses for users of their schools (pupils, teachers, parents) on working with the System.
- Participate in the acquisition of school libraries funds. They study the demand for literature in schools and give their suggestions in the MPE.

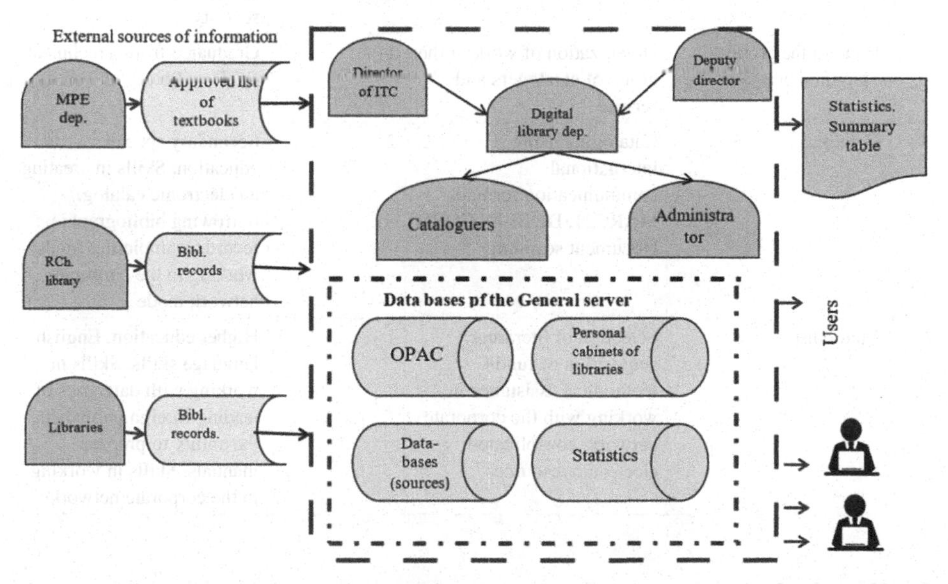

Fig. 3. Organization structure of the System

The requirements for staff to operate the system after the project is implemented are shown in Table 1.

Table 1. Personnel requirements

№	Department	Functions	Staff number	Skill level requirements
1	Deputy director of the CIT	Monitoring the implementation of sector tasks, resolving financial issues, improving the efficiency of development, and servicing	1	Higher education, Knowledge about the functions and tasks of the corporate network and system; management methods in the conditions of working in the corporate network;
2	System administrator (servers and databases)	Administration and configuration of the Microsoft Windows network environment; Creating personal accounts for school libraries; Registration; Generating statistical information Ensuring the security of data storage in the database	1	Graduated from a technical (ICT) university Skills: Microsoft Windows network environment administration and configuration; experience in managing an electronic catalog and full-text databases, as well as obtaining statistical information Knowledge and administration of network security
3	Head of the group «Digital library»	Organization of work of the group of catalogers and service	1	Graduated from a technical (ICT) or library university
4	Cataloguer	Cataloging using international communication formats MARC21, DUBLIN CORE Document scanning	2	Secondary special education. Skills in creating an electronic catalog, borrowing bibliographic records, completing funds, working in the corporate network mode
5	Methodist	Selection of literature, acquisition of funds, methodical assistance in working with the corporate network, consolidated electronic resources	1	Higher education. English language skills. Skills in working with databases of leading foreign publishers, the ability to prepare manuals. Skills in working in the corporate network

6 Conclusion

The development of the project and the study of the functioning of the MAKTEK information system showed the effectiveness of using cloud technologies in creating information and library networks for schools.

In the creation and operation of the system, we are faced with the following problems:

- There are distrust of remote resources and opportunities for hosting their data, storing it, and ensuring information security against unauthorized access and loss of resources;
- There are psychological unpreparedness of users and librarians to work with remote resources. Most of them are used to having resources on their computers and servers, and they can always copy or change them. Working with remote servers and databases takes some time to acquire skills;
- The Ministry of Public Education annually allocates large resources for technical support of schools, but in some regions there are still technical problems, low Internet speed, etc.;
- There is a great need for trainers to train librarians, as well as organizing courses for them and users of the electronic library to train them to work with the information system in the cloud mode.
- There is a great need for highly qualified specialists for database and server administration.

However, despite these difficulties, cloud technologies are increasingly used in library information networks to improve remote service for readers of different categories.

The implementation of the MAKTEK project has shown some important advantages of using cloud technologies in corporate networks for school libraries:

- Reduced duplication of information resources, including bibliographic records and in full-text databases;
- Improved the quality of bibliographic records in the electronic catalog(OPAC), fewer errors in these records due to the involvement of catalogers with extensive experience and skills;
- All users can use valuable educational database resources on the Central server from any location, any region of the republic, 24/7;
- The system generates and shows in a visual form some important statistics: how resources are used; which regions and schools are the most active and which are passive; which literature is in high demand, etc. Such analytical data allows you to effectively plan subscriptions and create library collections.

The perspective of the project is the integration of the MAKTEK system with the UzNEL project of the National library of Uzbekistan. Such cooperation will allow the corporate network of school libraries to access the rich electronic resources of the main library of the Republic automatically, without additional authorization.

References

Sultan, N.: Cloud computing for education: a new dawn? Int. J. Inf. Manage. **30**, 109–116 (2010). http://www.sciencedirect.com/science/article/pii/S0268401209001170

Suman, L., Parminder, S.: Cloud computing in libraries: an overview. Int. J. Digit. Libr. Serv. **6**(1), 121–127 (2016)

Zhau, R., Ioan, S., Lu. I.: Cloud computing and grid computing: 360-degree compared. In: Grid Computing Environments Workshop (2018)

Gireesh, G., Pradeep, G., Gaurav, C., Pooja, V., Gunjan, V.: Library with smart features. Int. J. Sci. Res. Manag. Stud. (IJSRMS) **2**(1), 45–51 (2011)

Suciu, G., Halunga, S., Apostu, A., Vulpe, A., Todoran, G.: Cloud computing as evolution of distributed computing: a case study for SlapOS distributed cloud computing platform. Inf. Econ. **17**(4), 109–122 (2013). https://doi.org/10.12948/issn14531305/17.4.2013.10

Goldner, M.: Winds of change: libraries and cloud computing, multimedia information and technology. Libr. Hi Tech News **37**(3), 24–28 (2010)

Breeding, M.: Cloud computing for libraries Chicago: ALA tech source, pp. 1–8 (2012)

Gosavi, N., Shinde, S.S., Dhakulkar, B.: Use of cloud computing in library and information science field. Int. J. Digit. Libr. Serv. **2**, 52–106 (2012)

Grace. A.L.: Architectures for the cloud: best practices for the adoption of cloud computing (2010). http://files/lu/aers/i45-arif-en.pdf. Accessed 10 Dec 2016

Foster, Y., Zhau, R., Ioan, S., Lu. I.: Cloud computing and grid computing: 360-degree compared. In: Grid Computing Environments Workshop (2018)

Rakhmatullaev M., Normatov Sh.: Analysis of criteria for ensuring information security of scientific and educational resources. In: Proceedings of the International Scientific Conference "Society. Integration. Education", 24th–26th May 2018, vol. V, pp. 430–435. Rezekne, Rezekne Academy of Technologies, Latvia (2018). https://doi.org/10.17770/sie2018vol1.3264

Parallel Resource Defined Fitness Sharing

Blayne Rogers[1(✉)], Ajay Gupta[1], and Pranjal Minocha[2]

[1] Western Michigan University, Kalamazoo, MI 49008, USA
{blayne.a.rogers,ajay.gupta}@wmich.edu
[2] Indian Institute of Technology, Roorkee 247667, Uttarakhand, India
pranjal_m@ph.iitr.ac.in

Abstract. The Resource-defined fitness sharing (RFS) algorithm combines resource and fitness sharing techniques to solve exact cover problems. These problems are defined in terms of resources and cooperating species which attempt to solve the problem via co-evolution. For the testbed, Sudoku is used due to the scalability and complexity of the problem. However, two primary factors limit the performance study of the algorithm. First, the algorithms serial implementation betrays its natural parallel structures. Data analysis is also tedious due the limited number of large Sudoku puzzles and the time required to assess the algorithms performance.

This research presents several solutions aimed at decreasing the compute time of RFS, streamlining testing, and automating the data analysis process. The computation time of RFS is addressed by exploiting the natural parallel structures inherent between RFS and the Sudoku algorithm. Testing and data analysis automation is streamlined by incorporating a user interface (UI). The UI accepts a Sudoku puzzle and provides an easy-to-read view of the current solution state of the RFS algorithm.

Data analysis is interleaved with the UI to show the performance of the algorithm on the current puzzle as well as tracking the performance history of all puzzles introduced to the algorithm. These puzzles are separated by size to address the combinatoric increase in complexity to solve a given puzzle. Once computed on the backend, the UI populates the respective performance graphs.

Keywords: parallel algorithm · user interface · resource sharing · fitness sharing · genetic algorithm · parallel genetic algorithm

1 Introduction

Empirical evidence showed the RFS algorithm is effective in solving exact cover problems [7]. The algorithm is a middle ground between natural resource sharing and fitness sharing techniques. These techniques were introduced in the general

This work was supported in part by the National Science Foundation under Grant OAC-2017289, National Institute of Health under Grant 1R15GM120820-01A1 and WMU FRACAA 2012-22.

H. Zaynidinov et al. (Eds.): IHCI 2022, LNCS 13741, pp. 493–503, 2023.
https://doi.org/10.1007/978-3-031-27199-1_49

genetic algorithm (GA) theories introduced by John Holland [4]. David Goldberg followed by spearheading the utilization and refinement of GA theories, to include these techniques [2,3].

As resource and fitness sharing represent two extremes, research showed the potential usefulness in exploiting the critical boundary which exists between them [6]. This evolved into the RFS algorithm which shows the effectiveness in combining these two techniques [7,8].

Further research showed the scalability of the RFS algorithm by solving large Sudoku puzzles [9]. However, the time required to evaluate these puzzles drastically increases. This comes as no surprise due to the intractability of Sudoku. With this consideration, the serial implementation of RFS was noted to be a major limiting factor.

Parallelization of GAs is by no means new [13,14]. In fact, parallelization of GA algorithms has benefitted from the proliferation of GPGPU computing [15], which offers cost-effective large scale parallelization. However, a review of the literature has shown no attempt to assess the benefits of parallelizing RFS. As such, our research analyzes RFS for its compatibility with GPGPU computing. CUDA is used to exploit natural parallel structures inherent in the algorithm.

Developing parallel RFS (pRFS) to utilize GPGPU computing brings its own challenges [12]. Most notably are the limited memory resources available. Typically, a CUDA graphics card comes equipped with 8 gigabytes of on-board memory. With the restricted memory, simply assessing the parallelization of RFS is not enough. As the scalability of RFS is shown using Sudoku, the memory bound is dictated by the size of the puzzles.

This paper will briefly describe the testing, human-in-the-loop and data analysis automation by incorporating an interactive user interface (UI). The UI will simplify the human interaction with the algorithm by providing a concise view of the problem and solution space. Furthermore, it will automate the both testing and data analysis of the algorithm by providing real time feedback on the current performance of the RFS algorithm. Consequently, such automation will streamline the data analysis and conserve significant time in preparation of future research.

2 Sudoku

Sudoku is a Japanese logic game which is commonly played on a 9×9 grid. This can be abstracted to a $N \times N$ grid. The grid is divided into four overlapping regions: row, column, cell, and square. Each region has one and only one numeral from the set $[1, 2, \ldots, N-1, N]$.

Clues are provided to the player which contain location and numeral information. The clues' location indicates where its numeral is placed w.r.t each of the four regions. All clues provided to the player represent the puzzle to be solved. The players' goal is to fill the remaining cells so there are exactly N copies of each numeral on the grid - distributed between the N copies of each region.

Defining the row and column regions is straight forward since they are the respective parts of the grid. The square regions are a $\sqrt{N} \times \sqrt{N}$ non-overlapping sub grid overlaid on the rows, columns, and cells. Finally, the cell is a unique region as it can only contain one symbol. However, it is representative of all numerals which are available to use, once the clues are considered.

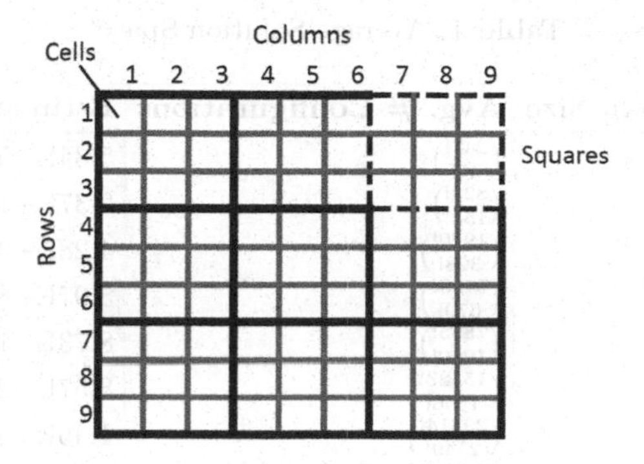

Fig. 1. Restrictions Enforced by Sudoku

As a numeral is fixed in a cell, the remaining numerals of the cell are removed from consideration. The corresponding numeral in each of the conflicting regions are also removed. In the formation of a Sudoku puzzle, each cell starts with $4N - 2\sqrt{N} - 1$ conflicting numerals. This value is greatly reduced for most numerals by applying the clues which are fixed on the grid.

Clues are used in Sudoku to formulate the puzzle. The clue list consists of location and symbol, i.e., cell $[4, 3]$ numeral 4. This removes the numerals $[1 - 3, 5 - N]$ from the $[4, 3]$ cell set and all instances of $[4]$ from the remaining regions in conflict with $[4, 3]$. Applying the clues to a puzzle greatly reduces the number of conflicting numerals. The number of ways in a which the subsequent puzzle can be filled, i.e., the solution space of the puzzle, is described by Eq. 1. This is the number of ways in which the numerals in play can be used to cover the grid. However, there is only one solution to any Sudoku puzzle.

$$\left(\frac{|clues \cap p_{space}|!}{(N^2 - clues)!} \right) \tag{1}$$

Abstracting Sudoku puzzles proves to be efficient for evaluating scalability. To complete this abstraction, let's concisely define an arbitrary space to grow the size and complexity. Since only those puzzles which use natural numbers are considered, a mapping from puzzle size to the natural numbers can be used. In so doing, N can be defined along with the grid.

$$p_{size} := \{n \varepsilon \mathbb{N} | n^2 = N, p_{grid} = N \times N\} \tag{2}$$

From this definition, the problem space of Sudoku can be defined as $p_{space} := N^3$. As described in Eq. 1, the solution space is a combinatoric explosion of possibilities. Refer to Table 1 to see how large this solution can be. These values comprise the average solution space evaluated by RFS at each puzzle size. There is missing data from size 11 to 14 as no puzzles of these sizes were found.

Table 1. Average Solution Space

Sudoku Size	Avg. # Configurations	Estimate
9	$\binom{247!}{57!}$	5.34E+56
16	$\binom{829!}{154!}$	2.37E+171
25	$\binom{1970!}{328!}$	4.28E+383
36	$\binom{5159!}{679!}$	5.97E+870
49	$\binom{7855!}{1060!}$	8.73E+1347
64	$\binom{15483!}{1656!}$	8.57E+2284
81	$\binom{23814!}{2549!}$	1.46E+3517
100	$\binom{39089!}{4000!}$	7.92E+5602
225	$\binom{103384!}{11823!}$	6.27E+15960

3 Resource Defined Fitness Sharing

As briefly discussed, fitness and resource sharing are niching techniques used to emulate evolutionary processes found in nature. Both techniques were derived to exploit the division of resources available [2,4,11]. Naturally, there are finite resources which limits the size of a population. A population's fitness and the resources available to the population are two different ways to view this concept. Likewise, fitness and resource sharing have different uses. Fitness sharing has been used in search and optimization problems citeHorn97. Resource sharing is incorporated into classifiers, immune system models, simple fixed population GAs, etc [3,6].

Resource-defined fitness sharing was derived to balance the advantages of these two techniques. In so doing, the algorithm exploits the boundary between cooperation and competition. The initial application of RFS showed its effectiveness in shape-nesting problems. This was extended further to show its scalability in solving Sudoku puzzles [5,9].

Using Sudoku as a resource-constraint problem, the algorithm begins with the initialization of a population. This is a finite space vector which encapsulates all possible species in a population. The upper bound of this vector is the problem space of the puzzle; generally, it comprises of N^3 species. Each specimen is mapped to a unique numeral within the Sudoku problem space. The species ID is derived from the numeral and its location: [row, column, numeral]. Once initialized, the active population is derived by "terminating" the species which are in direct conflict with the numerals given by the clue.

The pairwise intersection of numerals in the population and clue list is used to remove those species in direct conflict with the clues. This molds the population to a state representative of the specific Sudoku puzzle. In completing the initialization, a simple uniform distribution is used to find how many species exists in the active population. This value is used to weight the initial proportion of the remaining species.

The RFS algorithm uses shared fitness to coevolve these active species into a solution. Looping occurs until a predefined generation limit is reached or a solution to the Sudoku puzzle is found, whichever occurs first. The looping procedure begins by evaluating and storing species and population fitness.

Each specimen's weighted proportion is computed by measuring the level of conflict it has to all species in the active population. To compute this weight, the number of regions in which a specimen conflicts with another is totaled. In the case of Sudoku, there are four regions so this value is divided by four. This preserves each specimen's full proportion while accumulating the weighted proportion of all other species. Performing this computation masks a very sparse activation matrix which reflects the level of conflict all species have within the population.

The fitness for each specimen is computed by the inverse product of its' niche and proportion. The population's average fitness is then computed as the dot product of the species fitness and proportion vectors. These values are used in the computation of the next generation's proportion. This culminates in proportionate selection derived from shared fitness. To apply proportionate selection, a specimen's next generation is computed by the product of its' proportion and shared fitness, weighted by the average fitness of the population.

Once computed, the new proportions are used to advance the population to the next generation. Evaluation is subsequently performed to extract a potential solution from the population array. A simple greedy procedure is used to assess which species has the highest proportion for each numeral on the Sudoku's $N \times N$ grid. Finally, the numerals of the species with the highest proportions are assessed to validate whether a solution has been found.

4 Parallel Resource Defined Fitness Sharing

RFS has shown a competitive ability in finding solutions to constraint problems but is obviously hindered by its serial execution. An earlier version of RFS, pure co-evolutionary shape nesting (PCSN), was tested against three commerical software packages for shape nesting: ArtCAM Insignia, ProNest and OptiNest. The three packages were able to shape nest 10 to 11 circles into a given substrate in 1 or 2 s. PCSN was capable of nesting 12 circles into the same substrate area but required 70 s to converge on a solution [10]. All tests were performed on the same machine — Lenovo R61 laptop with a dual-core 1.8 GHz processor and 2 GB of main memory.

Similarly, the convergence times for serial RFS when applied to Sudoku drastically increases as puzzle size increases. The average time to converge for a 9×9

Sudoku puzzle was 0.18 s. Convergence time increases to 4074 s seconds for a 36×36 Sudoku puzzle. While no puzzles were solved beyond this size, assessment of generation compute time had a similarly profound increase. Using the maximum available puzzle size of 225×225, RFS required 42 s to evolve one generation.

While RFS clearly demonstrates a competitive ability, reducing the computation time is critical to maintaining this competitiveness. Analysis of the computation tasks required of the RFS revealed key parallel structures which could be exploited through GPGPU computing. There are seven tasks used by RFS to coevolve species to a solution for Sudoku. Two auxiliary tasks are used to extract and verify a solution. All of these tasks are near trivially parallelizable.

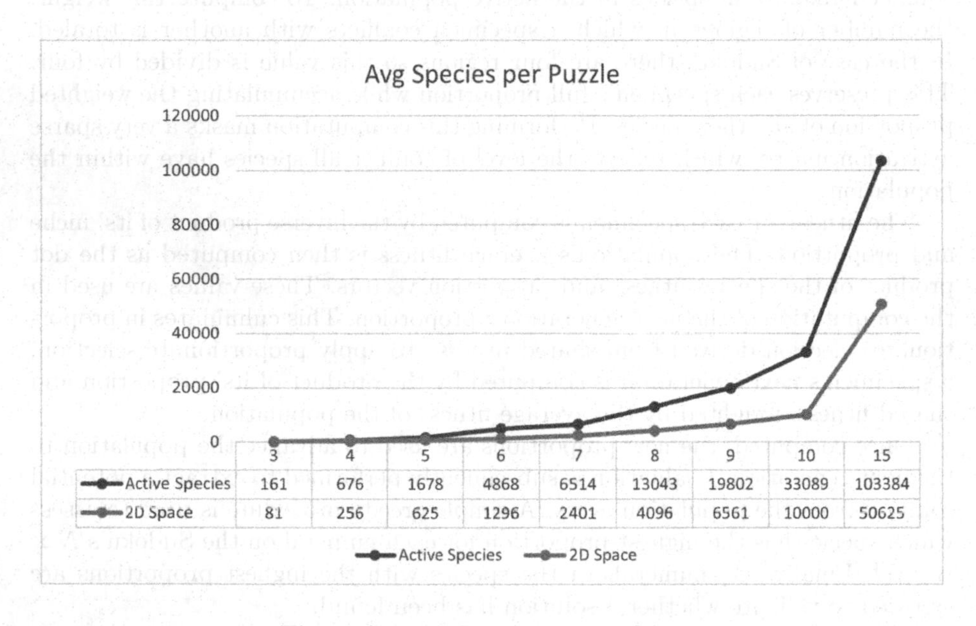

Fig. 2. Average Species per puzzle size

Focusing on these nine tasks forms the basis for studying the parallelization capabilities of RFS. Each of the nine tasks serve to fulfill specific requirements to advance the algorithm forward to the next generation. Nearly all the computations of these tasks can be performed on a species basis. The only exception to this is the dot product computation.

The three procedures for initialization create a population of active species. This begins with a population which matches the size of the puzzle's problem space. The pairwise intersection is used to isolate the non-active population. For any specimen which conflicts with a clue it is removed from the active population. This includes the numerals in the clue list. Each specimen compares itself to each clue, setting its proportion to zero if it conflicts in any region. The average active species assessed for the available puzzle sizes is shown in Fig. 2.

The final step of initialization is counting how many species remain. Once computed, the value is used to assign a uniform distribution to the proportion of all active species. i.e., the sum of the proportions of the active population is 1. This assignment can be performed by each specimen on a dedicated thread. The summation of the active population can be parallelized with an add reduction.

Once the population array is primed, the remaining six tasks are looped to perform coevolution. First, the values used in proportionate selection are computed. This begins with computing the niche of each specimen. Once computed, each specimen uses its niche and proportion to compute its fitness. This procedure can also be computed on a per species basis.

The average fitness is then computed as the dot product of fitness and proportion. While this cannot be done on a per species basis, parallel reduction procedures are common for dot product computations. Each specimen is then able to compute and assign an updated proportion based on its proportion, fitness, and the population's average fitness. This completes the coevolution tasks for the algorithm.

A potential solution to the puzzle is then extracted from the active population. All species within each cell are assessed to find those with the highest proportion. The numeral associated with these "fittest" species are saved as the potential solution. Once all cells have an answer, each region is assessed for multiple copies of numerals. If a numeral is found more than once in a region, the procedure terminates, and the algorithm continues. These validation sections can be performed on a per cell basis.

Whether parallelization is in respect to species, the grid, or some reduction, most of the computation of RFS can be done in parallel. Indeed, RFS falls under those algorithms for which parallelizations are almost trivial. There are still some serial computations - each of the nine tasks must be performed sequentially. However, even the pairwise computations performed by the species have the potential to be parallelized further.

5 User Interface

Current research on RFS is limited by the time requirements of data acquisition and analysis. Our work attempts to resolve these issues by implementing a user interface which addresses two primary areas of human computer interaction: visualization and automation. The UI provides the visualization of the puzzles, metric graphs, and progress of the (p)RFS algorithm. The automation supplements the backend by storing and evaluating the tracked metrics of the algorithm.

Automation will provide two critical components for furthering the evaluation of the algorithm. Any puzzle submitted to the UI must be vetted to ensure it is a real Sudoku puzzle. That is, any puzzle submitted to the UI will be validated to ensure it has one and only one solution. Additionally, large Sudoku puzzles are tediously difficult to come across and translate into a readable format for the algorithm. To resolve this, an abstract Sudoku generator capable of generating $N \times N$ puzzles will be attached.

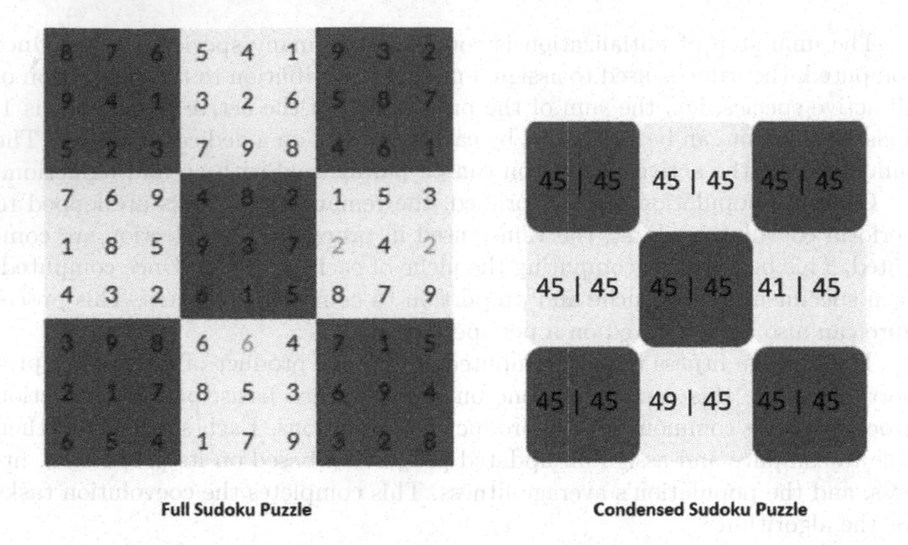

Fig. 3. Full and condensed Sudoku puzzle view

As shown in Fig. 3, the UI presents puzzles with a color coded scheme showing which numerals conflict and in which region they are conflicting. When multiple conflicting regions occur, the conflicted numeral absorbs all colors of the conflicting numerals. Additionally, large Sudoku puzzles become difficult to read. To resolve this, a condensed grid view, Fig. 3, is provided which collapses the values on the square regions of the puzzle. A summation metric will be used to indicate whether the values within a given square are all unique. For example, a size 3 puzzle will have all regions sum to 45. This desired value will be reflected beside the current summation.

Additionally, quality of life features exist and will continued to be added. Generally, a progress graph represents the number of correct numerals found thus far. A database will be used for tracking the historical data of the alogrithm. In addition to populating metric graphs for the current puzzle assessed, graphs for historic view of the RFS algorithm will also be generated.

6 Results

RFS was assessed to resolve the serial bottlenecks. The subsequent pRFS algorithm eliminates the active population search by having each specimen assess its placement in the population. If it conflicts with a clue, then it terminates and the numeral is not placed in active population. This is performed by mapping each of the N^3 specimens to a unique thread on the CUDA architecture. Figure 4 provides visualization of the species converging on a solution. This figure also provides a visualization of the parallel work.

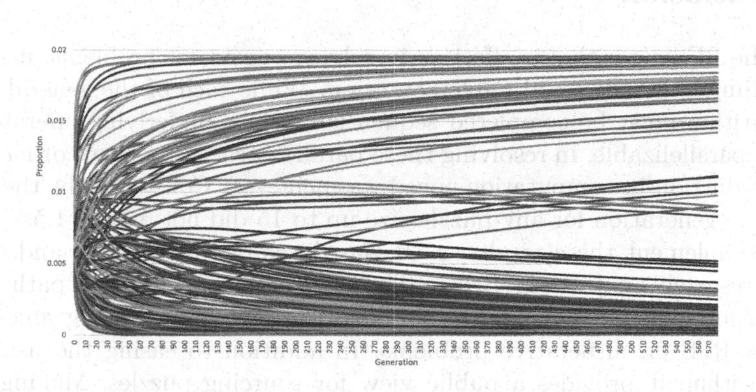

Fig. 4. USA Today 1 Easy Sudoku Puzzle [1] Species Convergence

This gives a point of contention with this form of implementation. Most of the initial N^3 species are inactive in any given puzzle. Consequently, there is a lot of wasted resources being used in retaining the full population array. Still, this one-to-one implementation of pRFS shows significant improvement over the original RFS algorithm as shown in Table 2.

The pRFS algorithm nearly quadrupled the time required to process a size 3 puzzle. This was expected as the associated population size could not saturate the GPGPU enough to be efficiently utilized. This changes with a size 4 puzzle where a 13-fold increase was achieved. While the remaining sizes do not have such a profound increase, pRFS averaged a 2.32-fold increase in performance for every size increase assessed. This culminates in a 41× speed up for a size 15 puzzle and saw an average 13-fold compute time reduction over the cpu for each puzzle size.

Table 2. pRFS generation compute time

Size(N)	$T_{cpu}(s)$	$T_{gpu}(s)$	$cpu\Delta$	$gpu\Delta$	cpu vs gpu
	$\overline{t_{cpu}}(D)$	$\overline{t_{gpu}}(D)$	$\frac{D_{cpu}}{D_{cpu-1}}$	$\frac{D_{gpu}}{D_{gpu-1}}$	$\frac{cpu}{gpu}$
9	0.000303	0.001220	1	1	0.2484
16	0.003968	0.001206	13.0957	0.9885	3.2902
25	0.012145	0.002658	3.0607	2.2039	4.5692
36	0.050122	0.006578	4.1269	2.4747	7.6196
49	0.173802	0.016559	3.4675	2.5173	10.4959
64	0.490754	0.034606	2.8236	2.0898	14.1811
81	1.119670	0.073462	2.2815	2.1228	15.2414
100	2.540480	0.185048	2.2689	2.5189	13.7287
225[a]	61.076566	1.473056	24.0413	7.9603	41.4624

[a] The only puzzle size greater then 10 which was available for analysis

Serial and Parallel RFS timing. Values reflect the average time for pRFS to run at each puzzle size.

7 Conclusion

While the RFS algorithm is effective in solving constraint problems, it is significantly limited by the serial implementation. While each of the general tasks of the algorithm must be completed sequentially, their underlying operations are trivially parallelizable. In resolving these parallel structures, a two order of magnitude reduction in computation speed was achieved. Consequently, the time to compute a generation for any puzzle size up to 15 did not exceed 1.5 s.

To complement these results, a UI was developed to enhance and expedite further research on this algorithm. The UI presents a material path forward for the continued improvement of the algorithm while also being amenable to assessing RFS on altnerative problems. In addition to easing the usability of the algorithm, it provides a public view for sourcing puzzles. Making the UI available online will allow crowd sourcing and tracking of puzzles to compute. As each puzzle is terminated, another can be chosen from the list of puzzles submitted to the UI. This further allows parallel puzzle assessments on multiple GPGPUs.

References

1. Today, U.S.A.: SUDOKU, 200 PUZZLES FOR ALL LEVELS, #22 [1850]. Multi Media International Inc., Tinton Falls, New Jersey (2008)
2. Goldberg, D.E..: In: Genetic Algorithms in Search. Addison-Wesley Professional, Optimization and Machine Learning (1989)
3. Goldberg, D.E., Richardson, J.: Genetic algorithms with sharing for multimodal function optimization. In: Proceedings of the Second International Conference on Genetic Algorithms on Genetic Algorithms and Their Application, pp. 41–49. L. Erlbaum Associates Inc., USA (1987)
4. Holland, J.H.: Adaptation in Natural and Artificial Systems: An Introductory Analysis with Applications to Biology, Control, and Artificial Intelligence. University of Michigan Press, Cambridge (1975). https://books.google.com/books?id=JE5RAAAAMAAJ
5. Horn, J.: Co-evolution of sudoku solutions. Mendel **2014**, 117–122 (2014)
6. Horn, J.: The nature of niching: genetic algorithms and the evolution of optimal, cooperative populations. Technical report, USA (1997)
7. Horn, J.: Resource-based fitness sharing, pp. 381–390, September 2002. https://doi.org/10.1007/3-540-45712-7_37
8. Horn, J.: Coevolving species for shape nesting, vol. 2, pp. 1800–1807, October 2005. https://doi.org/10.1109/CEC.2005.1554906
9. Horn, J.: Solving a large sudoku by co-evolving numerals. In: Proceedings of the Genetic and Evolutionary Computation Conference Companion, pp. 29–30. GECCO 2017, Association for Computing Machinery, New York, NY, USA (2017). https://doi.org/10.1145/3067695.3082050
10. Horn, J.D.: Pure co-evolution for shape nesting. In: IJCCI (2010)
11. PERRY, Z.A.: Experimental study of speciation in ecological niche theory using genetic algorithms. Ph.D. thesis (1984). http://libproxy.library.wmich.edu/login?url=https://www.proquest.com/dissertationstheses/experimental-study-speciation-ecological-niche/docview/303295516/se-2?accountid=15099, copyright

- Database copyright ProQuest LLC; ProQuest does not claim copyright in the individual underlying works; Last updated - 2021-10-04
12. Rogers, B.A.: Parallel resource defined fitness sharing: a study on parallel optimizations for niching algorithms. M.S. thesis (2022). https://scholarworks.wmich.edu/masters_theses/5327/, copyright - Database copyright ProQuest LLC; ProQuest does not claim copyright in the individual underlying works; Last updated - 2022-07-18
13. Sato, Y., Hasegawa, N., Sato, M.: Acceleration of genetic algorithms for sudoku solution on many-core processors. In: Proceedings of the 13th Annual Conference Companion on Genetic and Evolutionary Computation (2011)
14. Tsutsui, S., Collet, P. (eds.): Massively Parallel Evolutionary Computation on GPGPUs. NCS, Springer, Heidelberg (2013). https://doi.org/10.1007/978-3-642-37959-8
15. Wahib, M., Munawar, A., Munetomo, M., Akama, K.: Optimization of parallel genetic algorithms for nVidia GPUs. In: 2011 IEEE Congress of Evolutionary Computation (CEC), pp. 803–811 (2011). https://doi.org/10.1109/CEC.2011.5949701

EEG-Based Key Generation Cryptosystem for Strengthening Security of Blockchain Transactions

Ngoc-Dau Mai, Ha-Trung Nguyen, and Wan-Young Chung(✉)

Department of AI Convergence, Pukyong National University, Busan, South Korea
wychung@pknu.ac.kr

Abstract. This study proposes a key generation cryptosystem based on EEG signals to strengthen security protection for blockchain transactions. For the purpose of gathering brain signals, our system includes an 8-channel custom - designed EEG device. To extract spectral power values over five distinct frequency bands, including the delta band (1–4 Hz), the theta band (4–8 Hz), the alpha band (8–13 Hz), the beta (13–36 Hz), and the gamma band (36–50 Hz), a feature extraction method named the Multitaper spectral power density estimation is performed. Reed-Solomon error correction code is employed to rectify feature set errors to compatible values. Our system utilizes two major algorithms to generate cryptographic keys: enrollment and key creation. The evaluation's results prove the feasibility and efficiency of our EEG-based cryptographic key generation in boosting the security of blockchain transaction with the equal error rate (EER) of 0.368/0.368 (FRR/FAR).

Keywords: Blockchain · Biometrics · EEG · Key Generation Cryptosystem · Transaction Security

1 Introduction

Blockchain has recently drawn a great deal of attention from academic and industry researchers [1]. A distributed ledger that stores data on transactions or events is known as a blockchain. Blockchain is widely applied in a variety of fields, including industry, healthcare, and banking. Bitcoin transactions, that are utilized for online payments to transfer bitcoins, are typical blockchain transactions. In a file known as a Bitcoin wallet, the private and public keys used in the transaction are kept. There are multiple approaches to protect user's wallet. First, users possess a private key that is generated from a password and the Elliptic Curve Cryptography (ECC) [2] encryption algorithm. Another approach is a "cold wallet," which uses a random private key and keeps it on an external computer. Furthermore, using biometrics to generate cryptographic keys is another well-known technique. Traditional biometrics, such as fingerprint image, faces, voices, or signatures, have several advantages over traditional passwords, such as being more challenging to steal, harder to change, and harder to forget. In this study, to guarantee the security of blockchain transactions, we generate cryptographic keys using EEG

H. Zaynidinov et al. (Eds.): IHCI 2022, LNCS 13741, pp. 504–509, 2023.
https://doi.org/10.1007/978-3-031-27199-1_50

as a biometric. The 8-channel EEG device we designed by ourselves is used to gather the EEG signals. EEG signals are known to have extremely small amplitudes and to be accompanied by different varieties of noise during the acquisition process. In this study, we filter out undesirable signals utilizing a bandpass filter. The Multitaper spectral density estimation [3] is then used to extract valuable features from the EEG signals over five different bands. An error-correction process named as Reed-Solomon coding [4] was employed helps cryptographic keys cryptographic to be generated correctly since the EEG signal is unstable and varies over time. Finally, our system employs two algorithms—the enrolment and key generation—to successfully generate a cryptographic key. In this study, we utilize the equal error rate (EER) where the false rejection rate (FRR) is equal to the false acceptance rate (FAR) to assess the system's performance.

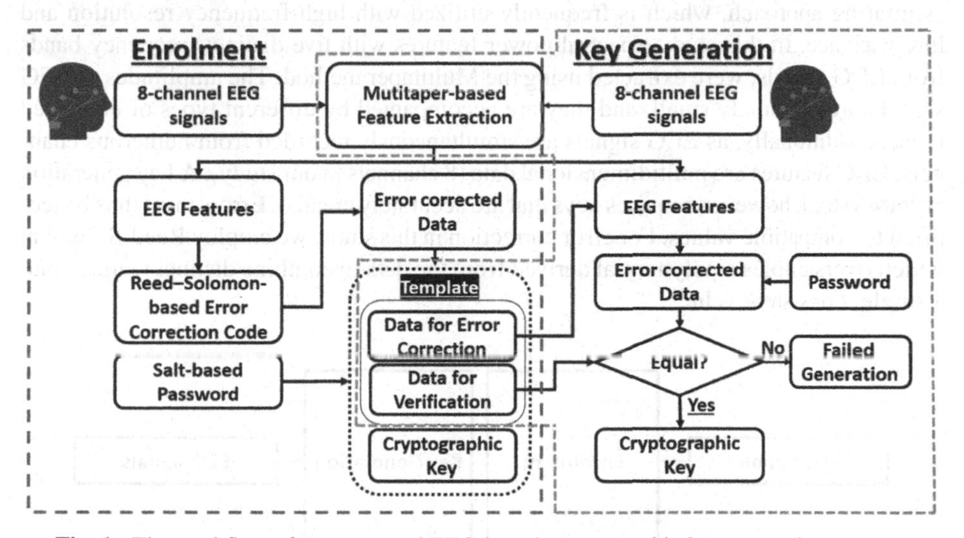

Fig. 1. The workflow of our proposed EEG-based cryptographic key generation system.

2 Methodology

2.1 EEG Acquisition Using a Custom-Designed Device

The EEG data utilized in this study originated from our previous studies [5]. The device comprises of 8 EEG electrodes that are mounted to different locations on the head to gather EEG signals at a 256 Hz sample rate. According to the international 10–20 system, the measuring sites that recorded EEG signals during the experiment were FT7, FT8, T7, T8, TP7, and TP8. In that experiment, EEG signals were gathered from a total of 8 participants as they watched emotional movies (3 categories of emotion: Negative, Neutral, and Positive) to explore emotion. Each participant watched a total of 30 videos, each representing 30 trials. Figure 3 shows our EEG device and the experimental procedure.

2.2 EEG-Based Feature Extraction

The EEG signal is not stationary and continuously varies over time to represent changes in brain activity. However, in a short enough period, the EEG signal is quasi-stationary, allowing the stable extraction of the EEG features, which includes important and distinctive biometric information and is appropriate for the generation of cryptographic keys. In this study, we utilize a bandpass filter to remove undesirable signals such as heartbeat and eye movements while retaining the relevant EEG signals from 1 Hz to 50 Hz. The same sliding window of 5 s with 50% overlap is then applied to split the signals into samples. A spectrum analysis technique called Multitaper was employed to effectively extract useful information from EEG data. The limitations of the conventional methods, such as the classical and Welch's periodograms, are overcome by the Multitaper spectral estimating approach, which is frequently utilized with high-frequency resolution and low variance. In this study, spectral power features with five distinct frequency bands from EEG signals, were extracted using the Multitaper method. The amplitudes of EEG signals are extremely small, and they are accompanied by different types of unwanted noise. Additionally, as EEG signals are simultaneously recorded from numerous channels, EEG features are multidimensional data (8 channels in our study). A key generation cryptosystem, however, requires keys that are accurately created. Errors must thus be rectified to compatible values. For error correction in this study, we employ Reed-Solomon, which oversamples a polynomial derived from the data to combine distinct features into a single, consistent value.

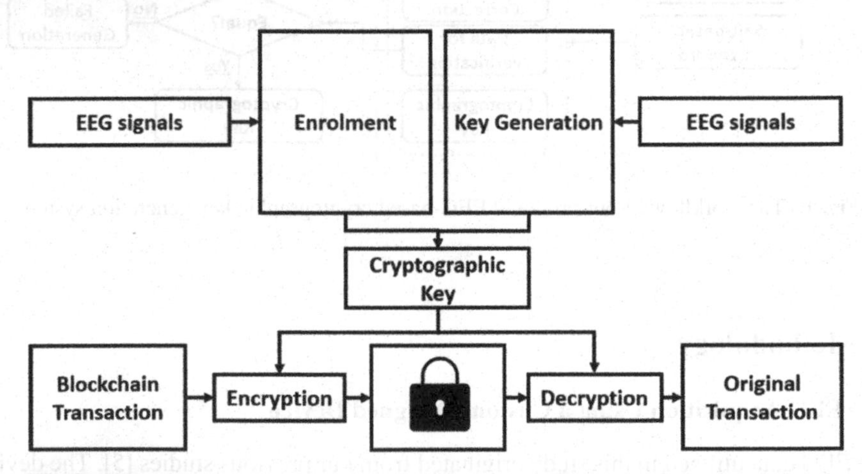

Fig. 2. Cryptographic key generation for Blockchain transactions using EEG signals.

2.3 EEG-Based Enrolment and Key Generation

Figure 1 provides a detailed description of our EEG-based cryptographic key creation's two methods, which include an enrolment algorithm and a key generation algorithm. With the input of EEG samples and an extra password, the enrolment algorithm produces a template and a cryptographic key, which are kept on the system. Without proper authentication, it is impossible to reveal the generated key. The key generation algorithm uses new EEG signals, a password, and the template from the enrolment procedure to determine whether or not to produce a key. The key will be generated if the new EEG samples match the enrolled EEG samples by a given threshold. In this study, thresholds are determined using the multi-threshold estimation algorithm [6]. In summary, a user must possess the following accurately and appropriately in order to successfully unlock the cryptographic key: the template, the EEG samples, and the password.

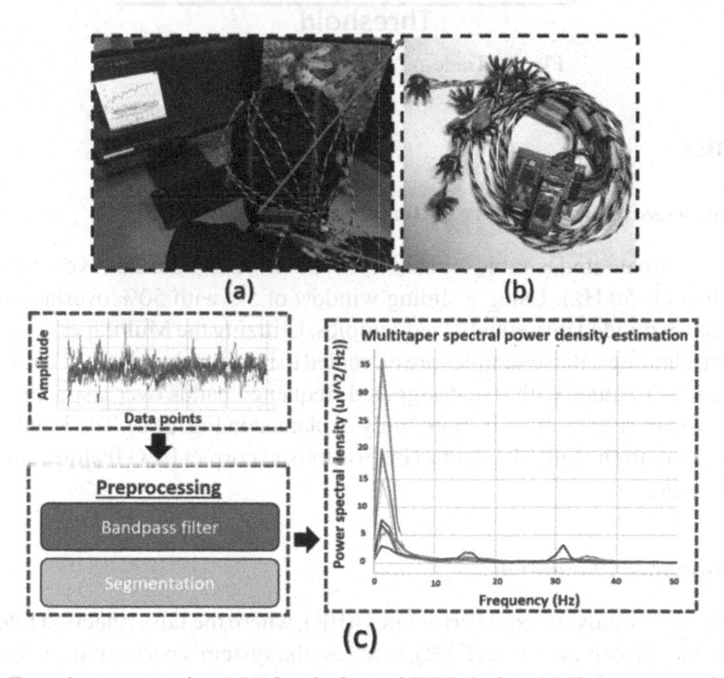

Fig. 3. (a) Experiment procedure, (b) Our 8-channel EEG device, (c) Feature extraction process.

2.4 Performance Evaluation

False rejection and False acceptance errors are employed to assess the performance of our system. False rejection occurs when a user is authentic, but the system is unable to identify the current input EEG data as being sufficiently comparable to the stored template. False acceptance, on the other hand, occurs when the system accepts an unauthorized person because their EEG data is sufficiently close to the template created by our system's enrolment algorithm. If both errors are zero, a system is optimal. However, as shown in Fig. 4, there is always a trade-off between FAR and FRR in a key generation cryptosystem. These two errors have inverse variations. The equal error rate (EER), which is used to assess the system's accuracy, is the position where FAR and FRR are equal.

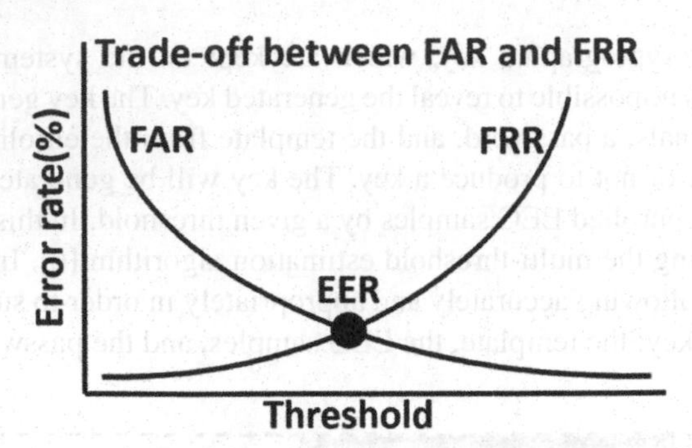

Fig. 4. Trade-off between FAR and FRR.

3 Results

3.1 Pre-processing and Feature Extraction

In order to eliminate undesirable signals from the raw EEG dataset, we first utilize the bandpass filer (1–50 Hz). Using a sliding window of 5 s with 50% overlap, the filtered EEG signals are divided into equal-sized samples. Utilizing the Multitaper spectral power estimation techniques, these samples are extracted to informative values. Features consist of spectral power values with five designated frequency bands over a specific time. The procedure of feature extraction is thoroughly explained in Fig. 3 (c). The Reed-Solomon-based error correction method was used in this study to correct EEG features into a single, repeatable value.

3.2 Performance Evaluation

In this study, we employ the equal error rate (EER), where the false rejection rate (FRR) is equal to the false acceptance rate (FAR), to assess the system's performance. We gathered EEG data from a total of 8 subjects. Each participant in the study underwent 30 trials, or 10 trials for each emotional state. We utilize "leave out 30% of samples cross-validation" to calculate FRR and FAR. This means that we select 30% of each participant's overall samples at random to serve as the testing set. The remaining data is utilized as a training dataset to create unique cryptographic keys and templates for each user. It is determined if the cryptographic key can be generated for each sample from the test set using templates produced from enrolment with the training set. To calculate the FRR and FAR values, this process was done five times and then averaged. Figure 5 displays the outcomes of the system evaluation shown by EER equal to 0.368/0.368, where FRR and FAR cross. The result demonstrates the system's feasibility in generating cryptographic keys from EEG signals to enhance the safety of blockchain transactions. The application of our EEG-based key generation mechanism in blockchain transactions is shown in Fig. 2.

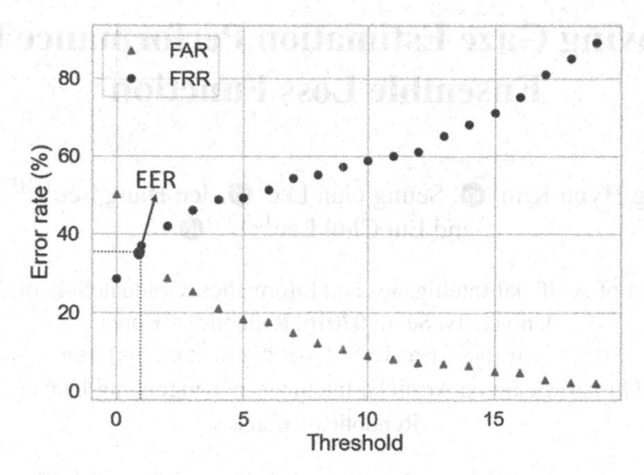

Fig. 5. Performance Evaluation using Equal error rate.

4 Conclusion

This study proposed a key generation cryptosystem utilizing a custom designed EEG device to increase security protection for blockchain transactions. The Multitaper estimation feature extraction method is used to obtain spectral power values over five different frequency bands. The Reed-Solomon error correction code is employed in this study to correct feature set flaws to compatible values. Key generation and enrolment are the two algorithms used to generate cryptographic keys in our study. Evaluation results revealed that our EEG-based system was developed to create cryptographic keys to enhance the security of blockchain transactions with an EER of 0.368/0.368 (FRR/FAR).

References

1. Holotiuk, F., Pisani, F., Moormann, J.: The impact of blockchain technology on business models in the payments industry (2017)
2. Lopez, J., Dahab, R.: An overview of elliptic curve cryptography (2000)
3. Babadi, B., Brown, E.N.: A review of multitaper spectral analysis. IEEE Trans. Biomed. Eng. **61**(5), 1555–1564 (2014)
4. Wicker, S.B., Bhargava, V.K. (Eds.): Reed-Solomon Codes and Their Applications. John Wiley & Sons, Hoboken (1999)
5. Mai, N.D., Lee, B.G., Chung, W.Y.: Affective computing on machine learning-based emotion recognition using a self-made EEG device. Sensors **21**(15), 5135 (2021)
6. Nguyen, D., et al.: Emotional influences on cryptographic key generation systems using EEG signals. Procedia Comput. Sci. **126**, 703–712 (2018)

Improving Gaze Estimation Performance Using Ensemble Loss Function

Seung Hyun Kim[1], Seung Gun Lee[1], Jee Hang Lee[1,2(✉)], and Eui Chul Lee[1,2(✉)]

[1] Department of Artificial Intelligence and Informatics, GraduateSchool, Sangmyung University, Seoul 03016, Republic of Korea
202233053@sangmyung.kr, {jeehang,eclee}@smu.ac.kr

[2] Department of Human-Centered Artificial Intelligence, Sangmyung University, Seoul 03016, Republic of Korea

Abstract. Coupled with deep learning technology, a field of gaze estimation has shown remarkable advances in recent decades. These advances however relied on the large amount of computations and resources due to the complex deep learning architecture or the huge volume of data, respectively. In this paper, we propose a resource efficient approach with ensemble loss function to improve the gaze estimation performance. Since eye gaze estimation is initially a regression problem in a broad sense, we employed the ensemble technique of regression loss functions in the pursuit of estimating gaze coordinates on a higher precision, instead of using additional deep layers or much data. Preliminary experiments on MPIIGaze data showed the improved performance compared to state-of-the-art models, the mean average error of 3.7887 cm which is 10.2%, 50.7%, 42.9% better than that of AFF-Net, Itracker, GazeNet, respectively.

Keywords: Gaze estimation · Deep learning · MPIIGaze · Appearance-based model · Divide and conquer · Ensemble loss function

1 Introduction

Since eye gaze implicitly reflects many facets of human mental states and behavior, gaze estimation has been paid much attention in several disciplines, such as natural user interface (NUI), advertisements, and digital therapeutics and so on. To serve the better user satisfaction, a lot of approaches have been proposed to improve the performance of gaze estimation.

The classical means to predict the eye gaze relied on the numerical calculation based on the facial features, such as the size of the pupil, distances between glints, location of the glints compared to the pupil. However, the classical way revealed a shortcoming especially in the flexibility on the various targets and environments.

Once the numerical model was established to the specific target, it was no more available to other environments.

S.H. Kim and S.G. Lee—Are co-first authors and contributed equally.

H. Zaynidinov et al. (Eds.): IHCI 2022, LNCS 13741, pp. 510–515, 2023.
https://doi.org/10.1007/978-3-031-27199-1_51

A recent decade has witnessed the rise of deep learning-based gaze estimation techniques while there was a remarkable improvement of hardware resources and accumulated huge amounts of datasets (MPIIGaze, GazeCapture, Gaze360). Appearance-based gaze estimation model is one of representative deep learning techniques to achieve a high performance in this research domain. For example, FAZE [1] takes a face image and gaze direction as inputs and predicts the gaze as an outcome, using encoder-and-decoder's latent vector through meta learning. The other example trained with ResNet using a set of images (a face and a pair of eyes) showed better accuracy in gaze than that iTracker [2]. Another model, so called SAGE (faSt Accurate Gaze trackEr) [3] used both eyes' images and landmarks, and achieved the high performance in eye gaze estimation.

The main obstacle to develop a progress on such models is the resource. Usually, the main stream of high performance models employed a complex deep architecture which requires a large amount of computations to fit the many numbers of parameters. In addition, it also necessitates a huge volume of data in subsequent. Thus, it is not realistic to develop high performance models for deploying them into typical applications and services due to the computational cost to train and inference.

Here, we propose a resource efficient computational framework with ensemble loss function to this end. As widely known, gaze estimation is initially a problem of regression which estimates the coordinates of gaze in a broad sense. Inspired by the ensemble approach to design loss function for a simple regression problem [4], we adopted an assemblage of standard regression loss functions at the final layer of the deep neural network. This allows the low cost in computations and resources, so called resource efficiency, for a highly precise estimation of gaze coordinates through a simple tuning of loss functions instead of reconfiguration of entire deep neural networks, training of complex deep neural networks or collecting much data and fitting them with deep neural networks.

We note that research on appearance-based models has also made a lot of progress in two different ways. One is to predict the direction of human gaze in three dimensions (3D) angle, and the other is the prediction of standardized X and Y coordinates in 2D planes, which can be robustly used in laptops, smartphones, and tablet PCs in two dimensions (2D). Since the output coordinates can be seen as a pointer to the mouse or keyboard of the devices as NUI pursues, we conducted our study to predict the user's gaze on the 2d screen.

The remainder of the paper is organized as follows. After the brief introduction in Sect. 1, we described the related works in Sect. 2. It is followed by an experiment illustrating the datasets, baseline models and our proposal in Sect. 3. Afterwards, we showed experimental results in Sect. 4. We conclude the paper with discussions and future works in Sect. 5.

2 Related Work

Adaptive Feature Fusion Network (AFF-Net). [5] is a model which showed the best performance on the problem of predicting X and Y coordinates in the 2D plane [4]. For the input, pictures of the face and both eyes, and coordinate values of the face and both eyes are used. The output, a coordinate of X and Y, was inferred by regression using a single fully connected layer at the prediction stage. We adopted this model as a backbone due to the fact that (i) it showed the highest accuracy in predicting the 2D coordinates of the screen between the appearance-based models, and (ii) the input data can be obtained without additional effort.

L2CS-NET. [6] is a CNN-based model that receives a face as input and predicts the gaze in general situations and predicts the gaze angles of yaw and pitch respectively. The learning of two gaze angles first uses Cross-Entropy-Loss, which is used to classify which range by dividing the entire angle into a certain range and Mean-Squared Error (MSE) between the angle predicted by the model and the Ground-Truth angle to improve the generalization performance of the model and achieve the gaze angle performance of SOTA in the MPIIGaze and Gaze360 datasets. It does not use a unified regression for the output, but there is a limitation in that it is difficult to accurately consider the eye position information and the eye movement because only the face is received as an input.

Squeeze-and-Excitation Network (SENet). [7] is a way of improving the performance of existing deep learning models. Only a slight amount of computational cost increases compared to the remarkable improvement of the performance. Therefore, SE Block is very useful because it can be used to improve performance if there is a convolutional neural network (CNN) based model.

3 Method

3.1 Data Preprocessing

MPIIGaze [8] is an open access database consisting of 213,659 face images from 15 participants. For each image, users were asked to gaze at specific x, y coordinates from the laptop, and the user's face image at this time was acquired through the built in camera of the laptop. We divided the total dataset into 15 folds by ID and used 14 people's data for each fold as train data and one person's data for testing. Using the label value in the database, the face and eye area of the image were cropped, and both eyes were resized to (112, 112) and (224, 224) for faces. In addition, coordinate values of the face and both eyes were added as input information to include distance, location, and angle information between the user and the device.

By inputting the above information together into the deep learning model, various information on the face and eyes can be considered, so that a more accurate gaze can be predicted. Finally, we represent the output of the model as x and y coordinates on the monitor screen rather than the viewing angle.

3.2 Model

We used the structure of AFF-Net as a deep learning-based model to learn features from images and landmarks. Since it was a model showing SOTA performance based on the targeted 2D screen, we thought that the effectiveness of the method we propose, which is the last layer and the loss function, could be shown more clearly. The overall structure of our proposed model is as follows (Fig. 1).

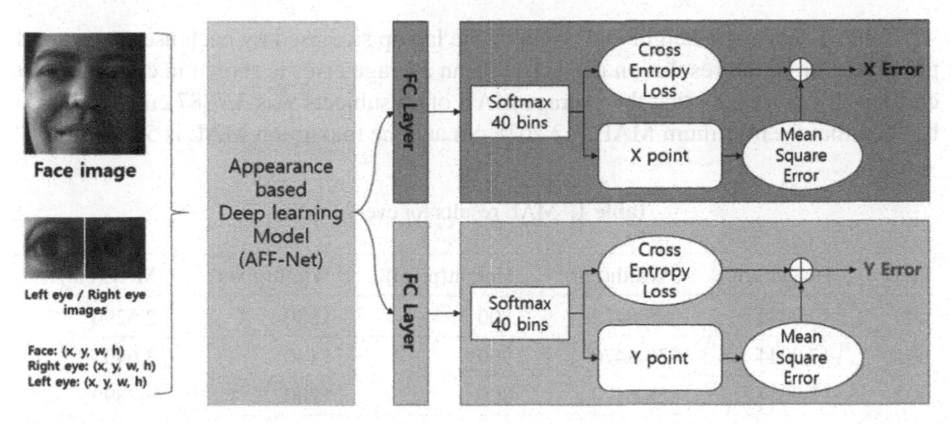

Fig. 1. Overall view of our proposed method. Deep learning model layers were not explained specifically because it was not our contribution point. Any kind of appearance-based model can be used.

3.3 Loss Function

The basic structure of the loss function we used is similar to that of L2CS-Net. We thought that the basic concept borrowed from this method is making prediction from broad range to specific value, which is very well known way of solving in the field of computer science. All existing appearance-based gaze estimation models as well as AFF-Net predict x and y coordinates using the last fully connected layer as a single model. Also, a single result is extracted by the regression method. To improve the performance of the model, we first separated x and y coordinates respectively. This is because we thought that x and y would not interfere with each other on the coordinates. After that, the screen is first divided into 40 blocks on the X-axis and Y-axis, and the block is predicted through softmax and calculated as cross entropy loss. At the same time, it goes through the learning process by combining the mean square error loss with the ground-truth coordinates. The method of splitting the layer and calculating the loss is designed to be attached to the CNN-based eye tracking model like SE-Net. In this study, we attached the proposed function to AFF-Net with SOTA performance to show performance improvement.

4 Experimental Results

First, we check the results of our model with each fold. Every fold uses different set of test data. Since MPIIGaze consists of 15 participants' data in total, we first separated one person for test data and 14 people for training so the total number of 15 folds' results could be made. For the result, we used mean average error (MAE). Equation (1) is the formula for the MAE we used. Here, *xi* 'and *yi* 'are values predicted through the model, and *xi* and *yi*are ground truth values. We measured the error through the Euclidean distance.

$$\mathrm{MAE} = \frac{1}{n}\sum_{i=1}^{n}\left|\sqrt{(x_i' - x_i)^2 + (y_i' - y_i)^2}\right| \quad (1)$$

Table 1 shows the height and width of the laptop size used by each user in cm, and the pixel unit of the resolution used. The mean average error is shown in cm in the last column. It can be seen that the average MAE of 15 subjects was 3.7887 cm. It can also be seen that the minimum MAE is 2.2678 cm and the maximum MAE is 5.7692 cm.

Table 1. MAE results for every fold

Fold	Height(cm)	Width(cm)	Height(pixel)	Width(pixel)	MAE(cm)
p00	17.9044	28.6470	800	1280	2.5290
p01	17.9044	28.6470	900	1440	3.6405
p02	17.9044	28.6470	800	1280	3.2899
p03	17.9044	28.6470	900	1440	2.8647
p04	17.9044	28.6470	800	1280	2.3499
p05	17.9044	28.6470	900	1440	3.1233
p06	20.7314	33.1703	1050	1680	4.0475
p07	17.9044	28.6470	900	1440	4.9137
p08	17.9044	28.6470	900	1440	3.8395
p09	17.9044	28.6470	900	1440	5.3713
p10	17.9044	28.6470	900	1440	5.7692
p11	17.9044	28.6470	800	1280	4.6775
p12	17.9044	28.6470	800	1280	4.6327
p13	17.9044	28.6470	800	1280	3.5137
p14	17.9044	28.6470	900	1440	2.2678
				Average	**3.7887**

Afterwards, we used the results presented in the review paper [9] for performance comparison with existing studies as shown in Table 2. Among them, AFF-Net, which we used following the layer structure of the basic deep learning model, showed an MAE of 4.21 cm, which is superior to other models. After applying our proposed layer, the average MAE of 3.78 cm which is the best between state-of-the-art models. We also computed the ratio on the improvements shown in Eq. (2). Taken together, it is clear that our proposed method improved the performance of gaze estimation by using a simple ensemble approach to the design of regression loss function.

$$Improvement(\%) = \frac{MAE_{old} - MAE_{proposed}}{MAE_{old}} \times 100 \quad (2)$$

Table 2. MAE comparison with existing models

Methods	MAE	Improvements (%)
AFF-Net	4.21 cm	10.2
Itracker	7.67 cm	50.7
GazeNet	6.62 cm	42.9
Proposed Net (fixed num. of blocks)	**3.78** cm	–

5 Conclusion

In this paper, we predicted the user's gaze on the screen through general to specific approach and could confirm the performance of SOTA. This method does not make the model heavier and does not require more data, but it does improve performance. Our future work will be applying our proposed layer and loss function to different kinds of appearance-based models to find out that this method really improves the backbone model. If it does, this can be used to improve the existing deep learningbased gaze estimation models with a little effort like SENet does to other general deep learning models. Also, since MPIIGaze dataset was gathered with a laptop environment only (even though different people and devices were used), we will train and test the model on GazeCapture which gathered 2.5 million images under the mobile phones and tablets with 1,450 people.

References

1. Park, S., et al.: Few-shot adaptive gaze estimation. In: 2019 IEEE/CVF International Conference on Computer Vision (ICCV), Seoul, Korea (South) (2019)
2. Krafka, K., et al.: Eye tracking for everyone. In: 2016 IEEE Conference on Computer Vision and Pattern Recognition (CVPR), pp. 2176–2184 (2016)
3. He, J., et al.: On-device few-shot personalization for real-time gaze estimation. In: 2019 IEEE/CVF International Conference on Computer Vision Workshop (ICCVW), pp. 1149–1158 (2019)
4. Hajiabadi, H., Monsefi, R., Yazdi, H.S.: RELF: robust regression extended with ensemble loss function. Appl. Intell. **49**(4), 1437–1450 (2018). https://doi.org/10.1007/s10489-018-1341-9
5. Bao, Y., Cheng, Y., Liu, Y., Lu, F.: Adaptive feature fusion network for gaze tracking in mobile tablets. In: 2020 25th International Conference on Pattern Recognition (ICPR), pp. 9936–9943 (2021)
6. Abdelrahman, A., Hempel, T., Khalifa, A., Al-Hamadi, A.: L2CS-Net: fine-grained gaze estimation in unconstrained environments (2022)
7. Hu, J., Shen, L., Sun, G.: Squeeze-and-excitation networks. In: 2018 IEEE/CVF Conference on Computer Vision and Pattern Recognition, pp. 7132–7141 (2018)
8. Zhang, X., Sugano, Y., Fritz, M., Bulling, A.: MPIIGaze: real-world dataset and deep appearance-based gaze estimation. IEEE Trans. Pattern Anal. Mach. Intell. **41**(01), 162–175 (2019)
9. Cheng, Y., Wang, H., Bao, Y., Lu, F: Appearance-based gaze estimation with deep learning: a review and benchmark. ArXiv (2021)

Non-overlayed Guidance in Augmented Reality: User Study in Radio-Pharmacy

Yves Simmen(✉), Tabea Eggler, Alexander Legath, Doris Agotai, and Hilko Cords

University of Applied Sciences and Arts Northwestern Switzerland, Bahnhofstrasse 6, 5210 Windisch, Switzerland
yves.simmen@fhnw.ch

Abstract. In many traditional industries, production instructions are usually provided on paper. Past research has shown the effectiveness of Augmented Reality (AR) for virtual user guidance in various cases. Usually, the main focus lies on 3D overlays and spatially anchored tokens to guide the user. Unfortunately, tracking small and moving objects is not always feasible in highly dynamic or complex environments. Additionally, the setup of anchors, 3D models and guidance procedures is often time-consuming and problem-specific. This study addresses such scenarios and provides empirical results of AR user guidance without employing overlays or object tracking. Therefore, we developed two AR concepts that guide the user using either 2D illustrations or 3D models. For evaluation, we designed a user study in the field of radio-pharmaceuticals, assessing quantitative measurements, usability and cognitive load. The conducted user study indicates that AR can also improve the effectiveness of user guidance in scenarios where direct 3D overlays or object tracking approaches are not feasible. The presented 2D and 3D AR concepts performed similarly, while both lead to fewer errors, faster execution time, and lower cognitive load than the paper instructions. Therefore, to reduce the effort required to create 3D instructions, the use of 2D illustrations could often be the more efficient choice.

Keywords: Augmented Reality · Wearable Device · User Guidance · User Interaction · User Studies · Task Guidance · Assistive Systems · Assembly Guidance · Manual Assembly

1 Introduction

Performing tasks according to specific paper-based instructions is a significant challenge in various domains. The creation and maintenance of instructions in highly specified or customized work environments are challenging and expensive. Replacing paper-based instructions with Augmented Reality (AR) user guidance

Y. Simmen and T. Eggler—Both authors contributed equally to this research.

H. Zaynidinov et al. (Eds.): IHCI 2022, LNCS 13741, pp. 516–526, 2023.
https://doi.org/10.1007/978-3-031-27199-1_52

in these scenarios introduces additional complexity due to the difficulty of tracking many small-sized, moving, and partially occluded objects. Further limiting factors are the costs associated with adapting the guidance to different scenarios and environments.

This paper addresses those scenarios and provides empirical results for user guidance where a direct overlay of virtual and physical objects is not feasible. Therefore, we evaluated two methods of non-overlayed guidance in head-mounted augmented reality. As a realistic industrial use case, we have chosen an exemplary production process in radio-pharmaceuticals. In general, the manufacturing process of radio-pharmaceuticals is complex and error-prone while usually being described by paper-based instructions. Laboratory employees often produce different pharmaceuticals on a daily and nightly basis utilizing various materials. Thus, there is an increased need for adequate guidance to avoid costly mistakes. The laboratory instruments used for production in radio-pharmaceuticals, such as syringes or vials, can hardly be tracked. The conducted experiment entails the use of these laboratory instruments and therefore forgoes object tracking, computer vision, or machine learning models. Furthermore, the scenario ensures a certain generality, low overhead, and quick adaptation to other similar use cases in different industries.

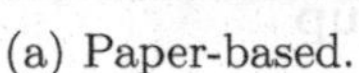

(a) Paper-based. (b) 2D AR guidance. (c) 3D AR guidance.

Fig. 1. Evaluated concepts: state-of-the-art paper-based instructions (a), AR guidance using 2D illustrations (b), AR guidance using 3D models (c). Our AR guidance concepts do not require any object tracking to overlay virtual and physical objects accurately.

Specifically, we evaluated two AR concepts. The first concept (2D concept) contains instructions in the form of 2D illustrations that can be easily created, adapted and applied to many use cases (Fig. 1b). The instructions of the second concept (3D concept) are supported by 3D models. While the creation of 3D models is more costly compared to 2D illustrations, they provide immersive depth information, which can be especially helpful for complex use cases (Fig. 1c).

Based on a user study, we compare the two concepts in contrast to the widely used paper-based instructions. We evaluate indicators such as task completion time, number of errors, usability by using the System Usability Scale (SUS), and cognitive load by using the NASA Task Load Index (NASA-TLX).

2 Related Work

AR guidance systems in work environments have been investigated for several decades [1]. More recent studies on AR enriched user guidance indicate an improvement in effectiveness or error rate compared to paper-based instructions [9,12]. In contrast, varying results can be observed when analysing the efficiency and the cognitive load. The reasons for a decrease in efficiency and increase in cognitive load include limitations of the hardware [11], user interaction design [13] and AR visualization design [8]. Additionally, most of these studies refer to low-complexity assembly tasks [5] with non-professionals [3], which do not scale well to realistic industrial use cases. Furthermore, scenarios are often chosen where the instructions are attached directly to the physical object [3]. Thus, instructions can take full advantage of immersion and spatial localization within AR. Nevertheless, object overlay, tracking, and spatial anchoring remain challenging in complex and dynamic scenarios.

In summary, most recent work in AR User Guidance require the handled objects to be trackable and feature a direct overlay of the virtual 3D models. This paper aims to close the gap in AR User Guidance in which object tracking is not feasible or too cost-intensive.

3 User Study

To validate our two proposed AR guidance concepts, we designed an experiment in which the user is required to perform a selected, complex step in producing radio-pharmaceuticals while being provided with instructions through a head-mounted AR device.

3.1 Experimental Environment and Technical Setup

The experiment was carried out as part of an advanced training seminar at the Institute for Pharmaceutical Sciences at ETH Zurich. During the course, the participants were familiarised with a specific pharmaceutical production process, including preparing a hot cell machine. Being in the sweet spot of complexity and time consumption, we chose the latter task as the use case for the conducted experiment. Figure 2a shows the initial state and materials used in the experiment: a HoloLens 2, a simplified 3D printed model is representing the hot cell machine, and materials (syringes, vials, and tubes). The final state of the experiment is shown in Fig. 2b.

3.2 Conceptual Prototypes for Non-overlaid Guidance in AR

We implemented two augmented reality HMD prototypes based on traditional paper-based instructions. Both prototypes assist users with spatially anchored AR guidance to guide them through the necessary steps to prepare the hot cell machine for the build-up of radioactivity. Both prototypes were intentionally

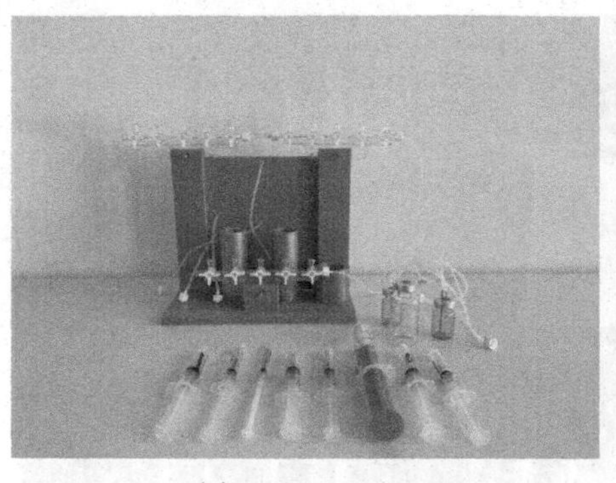

(a) Initial state.

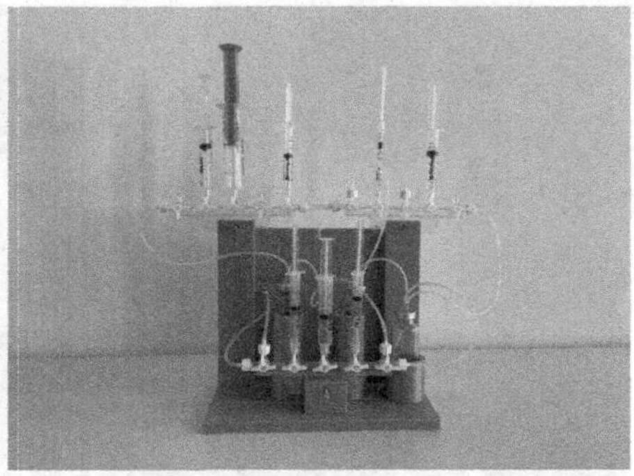

(b) Final state.

Fig. 2. Initial and final states of the experiment.

designed not to require user interaction such as voice recognition or gestures. Thus is guaranteed that the data is collected with regard to instruction visualization and is not disturbed by other factors such as interaction design or hardware limitations. The instructions were split into six sub-steps (Fig. 3):

1. Syringes: Installation of 8 syringes on the cassette,
2. Tubes: Connection of 4 pre-installed tubes on the machine to the cassette,
3. Intermediate vial: Connection of the intermediate vial with 2 tubes on the cassette,
4. Linked vials: Connection of both linked vials on the cassette
5. Levers: Adjustments to levers of all 3 cassettes according to the illustration,
6. Final Check: Check that everything has been implemented correctly.

While the six sub-steps and their used physical materials remain the same in all prototypes, we have varied the structure and placement order within the sub-steps to keep the training effect as low as possible.

Paper-Based Concept. Most laboratories in the pharmaceutical industry use paper-based manufacturing instructions and documentation, and all participants were familiar with this concept. This prototype was used mainly for training purposes so that testers become familiar with the selected use case and the required manipulations of the physical objects. The instructions for this prototype are printed on paper and contain both textual information and illustrations (Fig. 1a).

2D Concept (Using 2D Illustrations). In this prototype, augmented reality is used, and the information is displayed in 2D only, including text and illustrations taken from the paper-based prototype. The content is displayed on a paper-like slate, similar to a traditional desktop interface, which is spatially

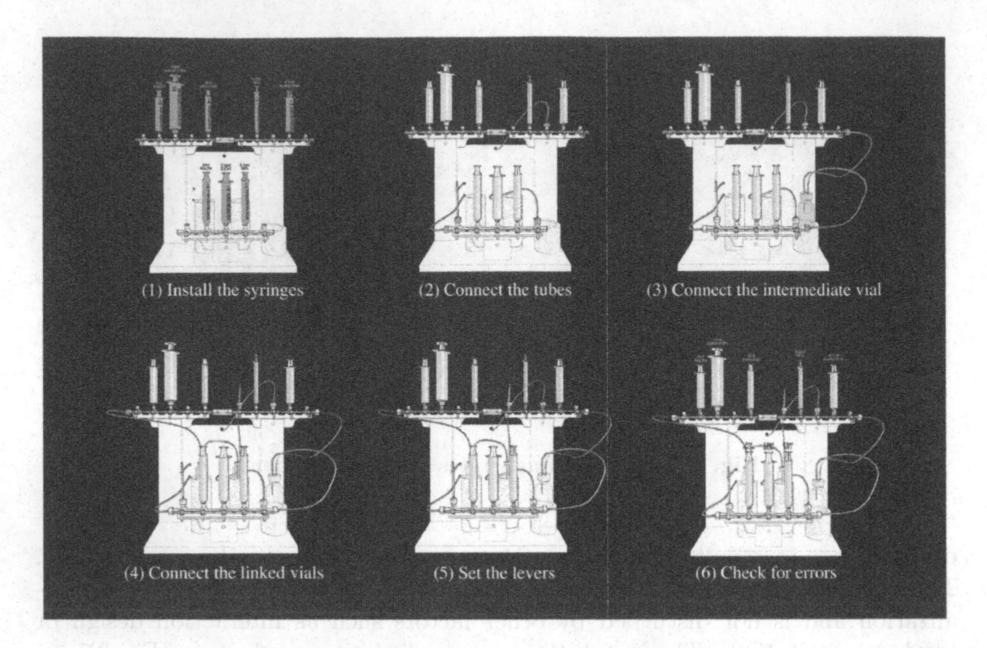

Fig. 3. Illustrations of sub-steps.

anchored above the test setup. It does not require object tracking or the creation of accurate models. The objects in the illustrations were darkly outlined with high contrast between contours and surfaces. Unneeded details have been abstracted (Fig. 1b).

3D Concept (Using 3D Models). This augmented reality prototype consists of 3D models instead of 2D illustrations. The 3D models are spatially anchored above the test setup and hence, do not require object tracking. The 3D models represent the physical object more accurately and with greater detail than 2D illustrations (Fig. 1c).

3.3 Data Collection and Analysis Methods

Before the actual experiment, we conducted the eye calibration of the HMD, asked the participant to fill out a pre-questionnaire, and explained the test setup. Each participant tested all three concepts in a different order. While everyone started with the paper prototype, they continued with either the 2D or the 3D concept. The execution times of all individual steps were measured. To avoid the measurement of interaction issues, the instructor manually switched to the next sub-step as soon as the participant indicated verbally to be done. Instructors observed the users during the task and noted each error that was being made and whether the user corrected it before the next step. Additionally, the final setup was manually checked by the operator after the completion of each procedure and again after all experiments were conducted by photo and video analysis.

We distinguished between self-detected errors, which were noticed and corrected by the tester, and functional errors, which were not detected and could have led to a failed synthesis. We also collected additional data using two different standardized questionnaires: the System Usability Score (SUS) [4] for measuring the perceived usability of the system and the NASA-TLX score [7] for measuring the perceived cognitive load during the task. The participants filled out these two questionnaires after each run. To supplement the quantitative data, we conducted a post-interview to determine the users' preferences and the challenges they faced with the system. Overall, the procedure took an average of 45 min per participant.

3.4 Participants

We recruited 9 participants, most of whom (7) were in the age group of 30 and 49 years. All participants are professionals and have at least an MSc in chemistry or pharmacy, whereas four also have a Ph.D. Four participants reported working in this field for more than six years. Seven participants rated their technical skills related to the versatility with new technologies as beginner or competent and the other two as proficient or expert. None of them had any prior experience with head-mounted AR glasses; only one person was familiar with the concept of AR but had never used an AR application. All the participants reported being skeptical about being accompanied by head-mounted glasses during a pharmaceutical production process.

4 Results

This section summarizes the main features of the user study's results: the indicators of task completion time, number of errors, cognitive load, and usability. The quantitative results are shown in Table 1 and Table 3. Statistical significance p was determined using the Kruskal-Wallis H test [10]. The null hypothesis is defined as follows: different visualizations do not affect task completion time, number of errors, task load, and usability. The following section contains the quantitative analysis of the user study.

4.1 Quantitative Analysis

In order to measure the efficiency and effectiveness of the participants in completing the task, we measured the task completion time, the number of errors they made but corrected during the workflow (self-detected errors), and the number of errors that were not detected and would have lead to a failure in production (functional errors). Additionally, task load and usability were recorded in order to gain insights on whether the system is suited for real-world production environments.

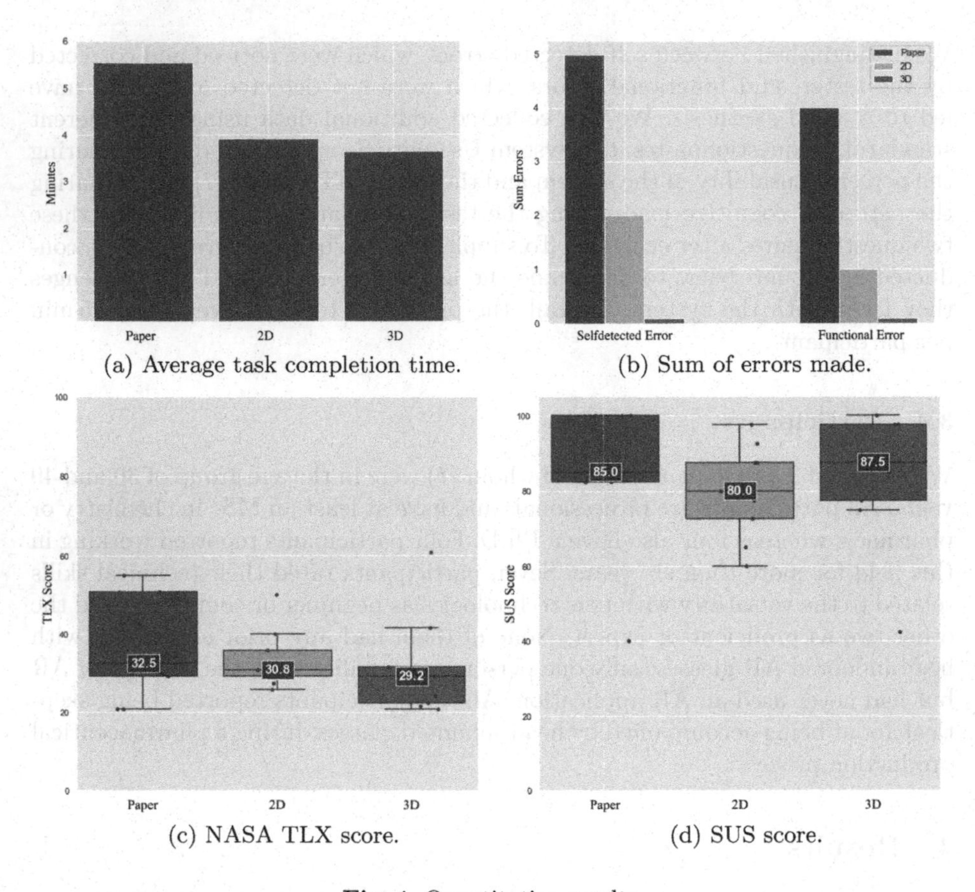

(a) Average task completion time.

(b) Sum of errors made.

(c) NASA TLX score.

(d) SUS score.

Fig. 4. Quantitative results.

Task Completion Time. The execution times for the three concepts are compared in Fig. 4a. The duration of both AR concepts is significantly shorter than the paper concept (1.47 min faster on average, $p < 0.05$). However, since the latter was always tested first, a training effect is to be expected. No significant difference in execution time was found between the 2D and 3D concepts ($p = 0.63$, Table 1).

Number of Errors. A total of 3 errors (2 self-detected, 1 functional error) were made with the 2D concept (Fig. 4b) and none with the 3D concept ($p = 0.07$). In comparison, participants made significantly more errors with the paper instructions (5 self-recognized, 4 functional errors) than with the AR concepts ($p < 0.05$). Table 2 shows the encountered number of errors. The errors made were distributed among almost all testers, and no outliers were detected that distorted the results ($std = 1.15$).

Table 1. Performance results

Concept	Avg. task completion time (mm:ss)	Sum of self-detected errors	Sum of self-detected errors
Paper-based	05:33	5	4
2D	03:54	2	1
3D	03:38	0	0

Table 2. Errors per participant

Participant ID	Sum of self-detected errors	Sum of functional errors
1	0	1
2	3	1
3	1	0
4	0	0
5	0	0
6	1	1
7	2	0
8	0	1
9	0	1

Task Load Reported by Users. Figure 4c shows that the 3D concept has the lowest NASA TLX score with a value of 29.2, indicating "medium" [6] cognitive load. The 2D concept was rated at 30.8 and the paper concept at 32.5, and both fall into the "somewhat high" [6] category ($p = 0.47$). Further evaluations show that the executive order does not play a significant role. Hence, the learning effect does not seem to be significant. On average, the cognitive load tends to be lower with the 3D concept, regardless of whether the 2D concept or 3D concept was tested first.

Table 3. Cognitive load and usability score

Concept	Mean score NASA-TLX	Mean score SUS
Paper-based	32.5	85.0
2D	30.8	80.0
3D	29.2	87.5

Usability Reported by Users. The SUS scores for all three concepts range from "good" [2] (2D concept) to "excellent" [2] (paper and 3D concept) according to Fig. 4d. During the interviews, it was found that the paper received a high score because most participants are familiar with this concept and use it daily in the laboratory. Overall, the 3D concept is preferred in terms of usability. Usability scores show a similar pattern to the NASA TLX scores in terms of order of execution (Table 4).

Table 4. Cognitive load and usability score by order of execution

Order of execution	Mean score NASA-TLX 2D	Mean score NASA-TLX 3D	Mean score SUS 2D	Mean score SUS 3D
2D first	30.8	24.2	87.5	90.0
3D first	32.5	30.4	75.0	75.0

4.2 Qualitative Analysis

In the conducted interviews, we wanted to learn more about the different perceptions of the 2D and 3D concepts. Thus, we did not include the paper version in this section. All participants were optimistic about their first experience with a head-mounted display device. We found that not all people consciously perceived the difference between the 2D and 3D concepts as 4 out of 9 participants did not recall the difference between them. However, all but one person noticed the difference as soon as they were made aware of it. Therefore, this person was unable to answer further questions. In general, the 3D concept was generally preferred. On average, it was the preferred visualization (62.5%), the most secure (75%), most efficient (87.5%), and least cognitively demanding (25%) concept (Fig. 5). These statements are consistent with the quantitative assessments. Although almost half of the participants did not consciously perceive the difference between the two AR concepts, it still seems to have subconsciously impacted, as the majority preferred the 3D concept. Before the participants conducted the experiment, they were skeptical about using AR guidance in the laboratory. However, after the experiment and their first contact with an AR HMD, they stated that they saw the potential for using augmented reality in the pharmaceutical laboratory environment. We found that 66.7% of participants reported that they would wear the HMD device for 1 to 2 h in the lab, 11.1% for up to 5 h, and 22.2% for only a few minutes and specific tasks.

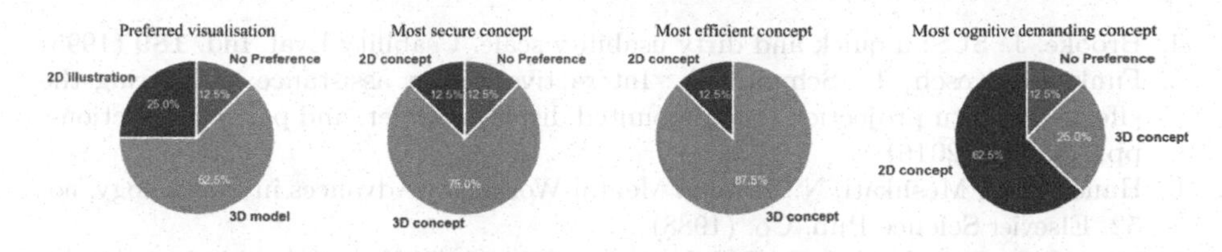

Fig. 5. Interview Questions.

5 Conclusion and Future Work

In this study based work, we presented two concepts of user guidance in AR that do not require object tracking or user interaction. We experimented with professionals and evaluated a concrete scenario in the field of radio-pharmacy that is relevant and involves a certain complexity. Our study indicates that with the 3D approach, the participants performed generally better. There are fewer errors, the execution time is faster, and the best SUS and NASA-TLX scores are obtained. The qualitative analysis supports these findings. However, the proposed two AR concepts performed similarly overall, and the differences are not statistically significant. Since the creation of 3D models is comparatively expensive, using 2D illustrations for guidance in AR might often be the best choice.

Interestingly, our results show that many participants did not consciously perceive a difference between the 2D and 3D concepts during the task. We suspect that the dimension was perceived subconsciously after all, as most of participants preferred the 3D concept. However, due to the relatively small number of participants, we are considering conducting further experiments to quantify these observations with a more extensive set of participants. We will also consider the investigation of other AR concepts such as 2D and 3D overlays in further studies.

Acknowledgements. This work was supported by the project ARIGO funded by the Swiss Innovation Agency Innosuisse (49260.1 IP-LS) and Augmenticon AG. We would like to thank Christian Schmidt, Matthias Friebe and Jamie Gilmartin from Augmenticon AG and Roger Schibli and Annette Krämer from the Institute of Pharmaceutical Sciences at ETH Zurich for the valuable collaboration.

References

1. Azuma, R.T.: A survey of augmented reality. Presence Teleoperators Virtual Environ. **6**(4), 355–385 (1997)
2. Bangor, A., Kortum, P., Miller, J.: Determining what individual SUS scores mean: adding an adjective rating scale. J. Usability Stud. **4**, 114–123 (2009)
3. Blattgerste, J., Pfeiffer, T.: Promptly authored augmented reality instructions can be sufficient to enable cognitively impaired workers. In: Proceedings of DELFI 2020 - Die 2018. Fachtagung Bildungstechnologien der Gesellschaft für Informatik e.V (2020)

4. Brooke, J.: SUS: a quick and dirty usability scale. Usability Eval. Ind. **189** (1995)
5. Funk, M., Kosch, T., Schmidt, A.: Interactive worker assistance: comparing the effects of in-situ projection, head-mounted displays, tablet, and paper instructions, pp. 934–939 (2016)
6. Hancock, P., Meshkati, N.: Human Mental Workload. Advances in Psychology, no. 52. Elsevier Science Pub. Co. (1988)
7. Hart, S.G., Staveland, L.E.: Development of NASA-TLX (task load index): results of empirical and theoretical research. In: Human Mental Workload, Advances in Psychology, vol. 52, pp. 139–183. North-Holland (1988)
8. He, S., et al.: AR assistive system in domestic environment using HMDs: comparing visual and aural instructions. In: Chen, J.Y.C., Fragomeni, G. (eds.) HCII 2019. LNCS, vol. 11574, pp. 71–83. Springer, Cham (2019). https://doi.org/10.1007/978-3-030-21607-8_6
9. Kolla, S., Sanchez, A., Plapper, P.: Comparing effectiveness of paper based and augmented reality instructions for manual assembly and training tasks. In: Proceedings of the Conference on Learning Factories (CLF) 2021 (2021)
10. Kruskal, W.H., Wallis, W.A.: Use of ranks in one-criterion variance analysis. J. Am. Stat. Assoc. **47**(260), 583–621 (1952)
11. Syberfeldt, A., Danielsson, O., Holm, M., Wang, L.: Visual assembling guidance using augmented reality. Procedia Manuf. **1**, 98–109 (2015)
12. Uva, A., Gattullo, M., Manghisi, V., Spagnulo, D., Cascella, G.L., Fiorentino, M.: Evaluating the effectiveness of spatial augmented reality in smart manufacturing: a solution for manual working stations. Int. J. Adv. Manuf. Technol. **94**, 509–521 (2017)
13. Yang, Y., Karreman, J., De Jong, M.: Comparing the effects of paper and mobile augmented reality instructions to guide assembly tasks. In: Proceedings of 2020 IEEE International Professional Communication Conference (ProComm), pp. 96–104 (2020)

Development of a Novel Method for Image Resizing Using Artificial Neural Network

Mukhriddin Arabboev[1(✉)], Shohruh Begmatov[1], Khabibullo Nosirov[1], Shakhzod Tashmetov[1], Saydiakhrol Saydiakbarov[2], Jean Chamberlain Chedjou[3], and Kyandoghere Kyamakya[3]

[1] Tashkent University of Information Technologies Named After Muhammad Al Khwarizmi, Tashkent, Uzbekistan
mukhriddin.9207@gmail.com

[2] UNICON.UZ State Unitary Enterprise Scientific-Engineering and Marketing Research Center, Tashkent, Uzbekistan

[3] Department of Transportation Informatics (TIG), University of Klagenfurt, Klagenfurt, Austria

Abstract. In this paper, we develop an artificial neural network (ANN)-based method for image resizing. The proposed adaptive image interpolation method based on artificial neural networks is compared with neural network and image interpolation-based methods developed by contributed researchers using a USC-SIPI image dataset. The comparison is based on assessment methods such as Mean Squared Error (MSE), Root Mean Square Error (RMSE), Peak Signal-to-Noise Ratio (PSNR) and Structural Similarity Index Measure (SSIM). The comparison clearly shows that the proposed method outperforms its counterparts considered in this work.

Keywords: Artificial Neural Network (ANN) · image resizing · activation function · mean squared error (MSE) · root-mean-square error (RMSE) · peak signal-to-noise ratio (PSNR) · structural similarity index measure (SSIM)

1 Introduction and Literature Overview

In all aspects of processing image signals scientists are concerning to make more effective processing, compression, scaling algorithms in order to improve perspectives of system where the accuracy, mobility and time are taken into account. For instance, bilinear interpolation has been used in an energy-efficient image scaling approach [1]. In addition, hardware-efficient simplified bilinear interpolation is employed [2] in multimedia applications.

Reversible watermarking as medical and military images algorithm [3] based on additive prediction-error has also developed with better PSNR in reversible watermarking systems moreover with same error expansion based reversible data hiding scheme [4] gave positive impacts than state-of-the-art method in terms of embedding capacity too.

One of the main issues in the image processing field is image compression. Most of transform techniques use discrete cosine transform (DCT) therefore, at low bit rates

H. Zaynidinov et al. (Eds.): IHCI 2022, LNCS 13741, pp. 527–539, 2023.
https://doi.org/10.1007/978-3-031-27199-1_53

it makes distortion. Some approaches propose to solve it by the maximum a posteriori way [5] while other introduces another criteria by overlapped-block transform by fusing two prediction values [6]. This kind of block-based DCT methods (BDCT) especially in image compression has been broadly used. Despite of using BDCT to prevent blocking artifacts patch clustering and low-rank minimization [7], maximum a posteriori framework algorithm using Constrained Non-Convex Low-Rank (CONCOLOR) model [11] have been applied. A matrix factorization issues as a convex relaxation of the low rank also exist and it is controlled by typical nuclear norm minimization. Because of denosing problem of it, weighted nuclear norm minimization problem are investigated under various weighing circumstances [10]. On the other hand image decompression although plays crucial role as repeatedly compressed image using DCT gives blocked artifacts after decompression process. And researches tried to reduce it by implementing sparse and redundant representations over a learned dictionary [9]. Key factor here image denoising where in trans-form domain based on an improved sparse representation [8] and it is reached by grouping familiar 2-D image blocks into 3-D ones.

Currently ANN is used in all aspects of image signal processing technologies as it is giving perspective results than other methods. Deep learning implemented researches are giving achievements than other traditional methods. In order to reduce Gibbs effect modified CNN algorithms are demonstrated like Artifact Reduction Convolution Neural Network (AR-CNN) [12–17], Compact Convolution Neural Network (ComCNN) and Reconstruction Convolution Neural Network [18, 19].

The rest of this paper is structured as follows. Section 2 is related to Contributed Neural Network algorithms, Sect. 3 is about the proposed ANN-based model as well as Sect. 4 is about evaluation and experimental results of ANN based method for image resizing.

2 Contributed Neural Network Algorithms

2.1 Convolutional Neural Network (CNN)

Convolutional neural network (CNN, or ConvNet) could be a category of deep neural networks, most ordinarily applied to analyzing visual imagery. They are additionally renowned as shift invariant or area invariant artificial neural networks (SIANN), supported their shared-weights design and translation unchangeable characteristics. They have applications in image and video recognition, recommender systems, image classification, medical image analysis, and language processing.

2.2 Recurrent Neural Network (RNN)

RNN is a class of artificial neural networks in which the connections between nodes form a graph that is oriented in a temporal sequence. This allows it to demonstrate temporary dynamic movement. Coming from neural networks, RNNs are able to process sequences of varying lengths using their internal state (memory) [20]. This applies them to functions such as unrecognized, linked handwriting recognition or speech recognition [21].

2.3 Artificial Neural Network (ANN)

ANNs, usually simply called neural networks (NNs), are computing systems vaguely inspired by the biological neural networks that constitute animal brains. ANN is capable of learning any nonlinear function. Hence, these networks are popularly known as Universal Function Approximates. ANNs have the capacity to learn weights that map any input to the output.

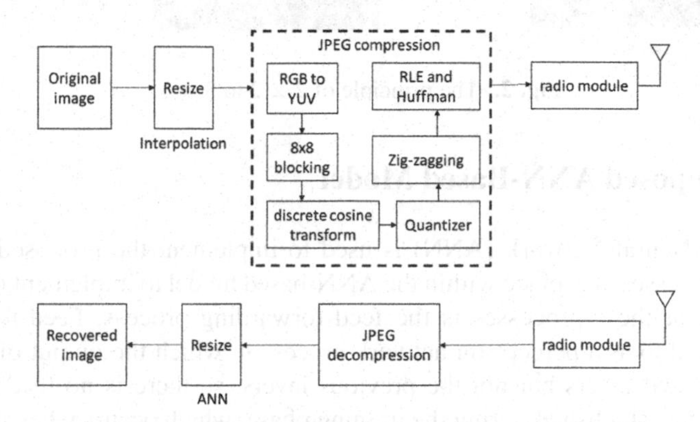

Fig. 1. ANN based method for image resizing

The proposed ANN based method for image resizing consists of the following steps (Fig. 1). First of all, the camera captures the original image. Then, it is resizing using the interpolation method. After that, the JPEG compression process takes place. The compressed image is transmitted to the receiver using a radio module. On the receiving side, the image received using the radio module undergoes JPEG decompression. Then, the next steps occurs choosing convenient neural network model for image resizing. There are several types of neural networks in computing. These include: Convolutional Neural Network (CNN), Recurrent Neural Network (RNN), Artificial Neural Network (ANN), and others. ANN is chosen to organize the image resizing process. Finally, the image is recovered.

ANN based model of image resizing works on the principle of 2×2 to 3×3. A description of the 2×2 to 3×3 principle can be seen in Fig. 2.

In neural networks, there is a hidden layer between the input and output of the model, in which the function applies weights to the data and controls them through the activation function as output. In short, hidden layers perform a nonlinear variation of the inputs included in the network. The latent layers vary depending on the functioning of the neural network, and similar layers may vary depending on their associated weights. These hidden layers are based on the Sigmoid activation function. The first hidden layer consists of 9 neurons based on the Sigmoid activation function. The second hidden layer also consists of 9 neurons based on the Sigmoid activation function.

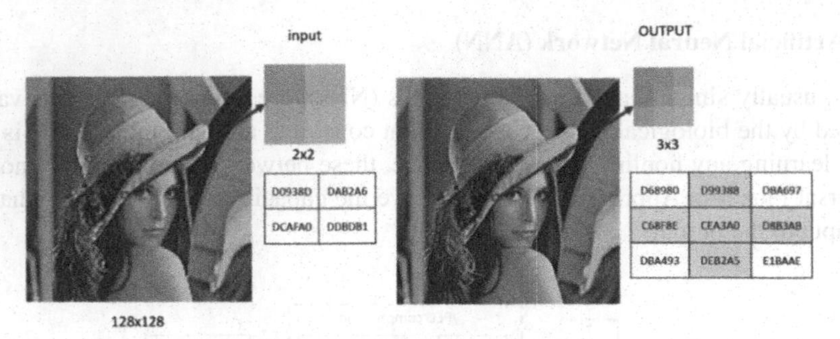

Fig. 2. The principle of 2 × 2 to 3 × 3.

3 The Proposed ANN-Based Model

An Artificial Neural Network (ANN) is used to implement the proposed method. A number of processes take place within the ANN-based model to implement the proposed method. One of these processes is the feed-forwarding process. Feed-forwarding is usually referred to as a perceptron network process in which the output of all neurons passes to the next layers but not the previous layers, so there is no feedback period. Connections are established during the training phase, which occurs when the system is a feedback system.

The feed-forwarding process can be expressed using equations in the following sequence.

First of all, denoting is done in the ANN-based model (see Fig. 3). x_j is equal to $out_{1,j}$. Here j can take numbers from 1 to 12.

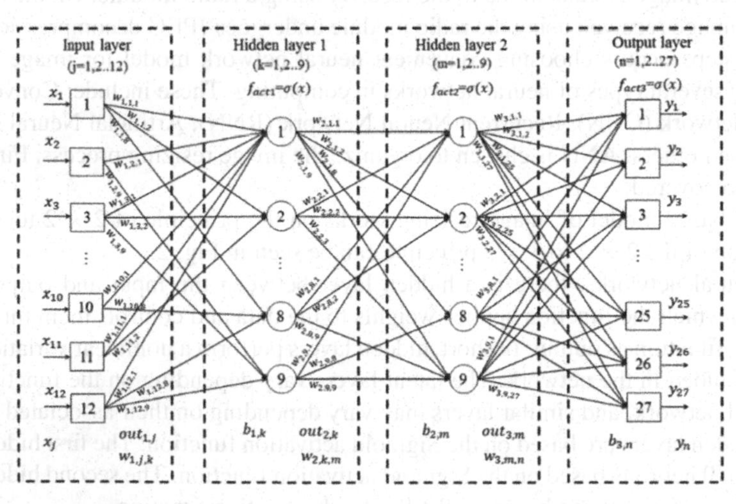

Fig. 3. Feed-forwarding process of the proposed model.

All activation functions, f_{act1}, f_{act2}, f_{act3}, are equal to each other. This equation is denoted by σ (x). σ (x) denotes the sigmoid activation function. For finding the sigmoid activation function is given by the following equation.

$$f_{act1} = f_{act2} = f_{act3} = \sigma(x) = \frac{1}{1+e^{-x}} \tag{1}$$

y_n can be expressed by the action function as follows:

$$y_n = \left[1 + e^{-\left(\left[\sum_{m=1}^{9} \frac{w_{3,m,n}}{1+e^{-\left(\left[\sum_{k=1}^{9} \frac{w_{2,k,m}}{1+e^{-\left(\left[\sum_{j=1}^{12} x_j \cdot w_{1,j,k}\right]+b_{1,k}\right)}}\right]+b_{2,m}\right)}}\right]+b_{3,n}\right)}\right]^{-1} \tag{2}$$

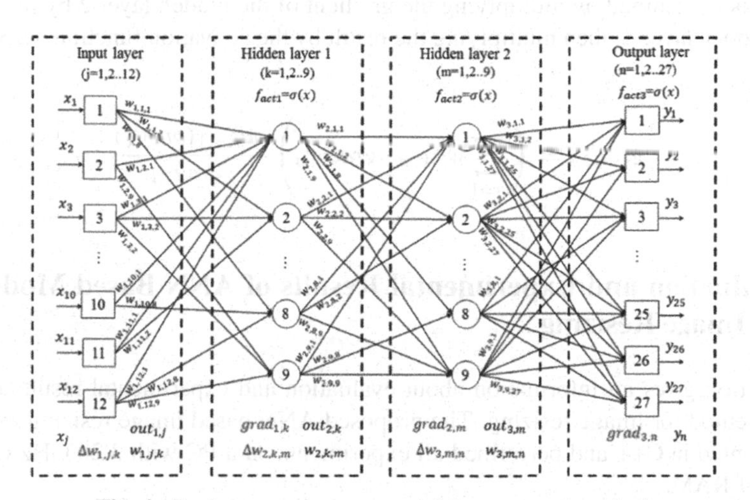

Fig. 4. Backpropagation process of the proposed method.

In Fig. 4, the backpropagation process of the proposed method is shown. Backpropagation is an algorithm that widely used for training feedforward neural networks. Generalization of backpropagation is available for other artificial neural networks (ANNs) and functions in general. These algorithm classes are commonly referred to as "backpropagation". When setting up a neural network, backpropagation calculates the gradient of the loss function relative to the weights of the network for a single input-output example and performs so efficiently that the gradient on each weight is directly as opposed to the correct calculation. This efficiency makes it expedient to train multi-layer networks, using gradient methods to update weights to minimize losses; options such as gradient drop or stochastic gradient drop are commonly used. The backpropagation algorithm is repeated backwards from the last layer to avoid over-calculation of intermediate terms in the chain rule by calculating the gradient of the weight loss function for each weight according to the chain rule; this is an example of dynamic programming [22]. The main purpose of the backpropagation process is to correct output errors.

Back propagation: gradient equation,

The gradient value of the output layer is represented by the following equation:

$$grad_{3,n} = (target_n - y_n) \cdot \frac{df_{act3}(y_n)}{dy_n} \tag{3}$$

where $target_n$ means target value, y_n means the resulting value.

$grad_{2,m}$ is determined by multiplying the gradient of the output layer by the weights of the hidden layer 2, then multiplying the result by the activation function derivative of $out_{3,m}$.

$$grad_{2,m} = \left(\sum_{n=1}^{27} w_{3,m,n} \cdot grad_{3,n}\right) \cdot \frac{df_{act2}(out_{3,m})}{dout_{3,m}} \tag{4}$$

$grad_{1,k}$ is determined by multiplying the gradient of the hidden layer 2 by the weights of the hidden layer 1, then multiplying the result by the activation function derivative of $out_{2,k}$.

$$grad_{1,k} = \left(\sum_{m=1}^{9} w_{2,k,m} \cdot grad_{2,m}\right) \cdot \frac{df_{act1}(out_{2,k})}{dout_{2,k}} \tag{5}$$

4 Evaluation and Experimental Results of ANN Based Model for Image Resizing

This section gives an information about evaluation and experimental results of ANN based method for image resizing. The proposed ANN based image resizing method is implemented in C++ and performed all experiments on a PC with 3.20 GHz CPU and 64 GB of RAM.

The USC-SIPI dataset is a collection of digitized images. It is maintained primarily to support research in image processing, image analysis, and machine vision. This dataset is used to evaluate in order to resize images and to assess the extent to which they affect image quality when resized. In addition, standard images of USC-SIPI dataset were used in this work. In general, the selected ones include: Lena; Baboon (Mandrill); Barbara; Peppers; House; Plane (Airplane); Parrots; Leaves; Cameraman; Butterfly.

4.1 Mean Squared Error (MSE)

MSE is the most common evaluator of image quality measurement metrics. This is a complete reference metric and values close to zero are better. This is the second moment of error. The variance of the estimator and its uncertainty are added by the standard deviation. MSE is the variance of the predictor in the objective estimator. It has units of measurement equal to the square of the magnitude, which is calculated as the variance.

Mean Squared Error (MSE) between two images such as say g (x,y) and ĝ (x,y) is defined as [23],

$$MSE = \frac{1}{MN}\sum_{n=0}^{M}\sum_{m=1}^{N}\left[\hat{g}(n,m) - g(n,m)\right]^2 \tag{6}$$

From Eq. (6), it can be seen that MSE is a representation of absolute error.

The quantitative comparison on MSE is demonstrated in Fig. 5. It is easy to see that the proposed method effectively resizes images. In some cases, our approach and Feng's [18] method show were not very close, but close to similar results. In most images, the proposed method better emphasizes minimal errors. A comparison of the 10 selected images mentioned above with 13 alternative methods on MSE is given in Fig. 5.

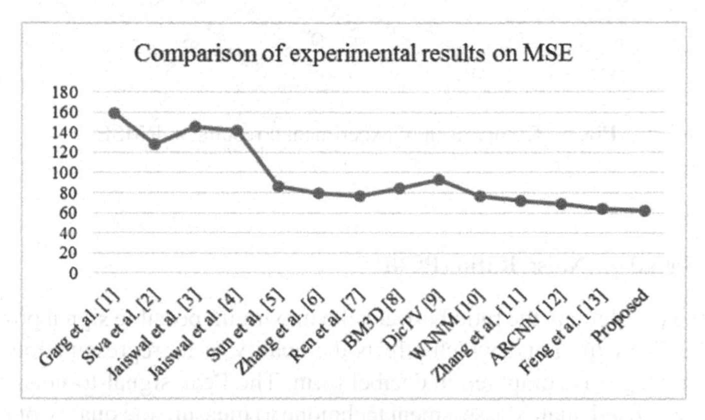

Fig. 5. Comparison of experimental results on MSE.

4.2 Root-Mean Square Error (RMSE)

MSE provides Root-Mean Square Error (RMSE) or Root-Square Deviation (RMSD) and is often referred to as the standard deviation of the variance. The RMSE is another type of error measurement technique that is widely used to measure the difference between the approximate value and the actual value. This estimates the magnitude of the error. This is an excellent measure of the accuracy used to perform the differences in the prediction errors of different predictors to the exact variable [24].

If it is assumed that the estimated parameter given in θ can be a predictor with respect to θ, then the mean square error is actually the square root of the mean square error.

The determination of RMSE is given in the following equation:

$$\mathrm{RMSE}\left(\hat{\theta}\right) = \sqrt{\mathrm{MSE}\left(\hat{\theta}\right)} \tag{7}$$

Comparison on RMSE is demonstrated in Fig. 6. In some cases, our approach and Feng's [18] method show were not very close, but close to similar results. In most images, the proposed method better emphasizes mean square error. A comparison of the 10 selected images mentioned above with 13 alternative methods on RMSE is given in Fig. 6.

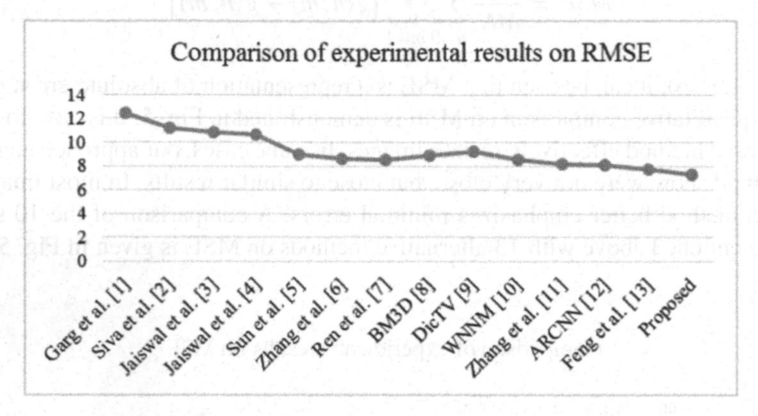

Fig. 6. Comparison of experimental results on RMSE.

4.3 Peak Signal-to-Noise Ratio (PSNR)

PSNR is used to calculate the ratio between the maximum possible signal power and the power of the distorting noise which affects the quality of its representation. This ratio between two images is computed in decibel form. The Peak signal-to-noise ratio is the most commonly used quality assessment technique to measure the quality of reconstruction of lossy image compression codecs. The signal is considered as the original data and the noise is the error yielded by the compression or distortion.

PSNR is expressed as:

$$\mathrm{PSNR} = 10\log_{10}\frac{\mathrm{peakval}^2}{\mathrm{MSE}} \tag{8}$$

where peakval (Peak Value) is the maximal in the image data. If it is an8-bit unsigned integer data type, the peakval is 255 [25]. From Eq. (8), it can be seen that it is a representation of absolute error in dB.

Comparison on PSNR is demonstrated in Fig. 7. In some cases, our approach and Feng's [18] method show were not very close, but close to similar results. In most images, the proposed method better emphasizes mean square error. A comparison of the 10 selected images mentioned above with 13 alternative methods on PSNR is given in Fig. 7.

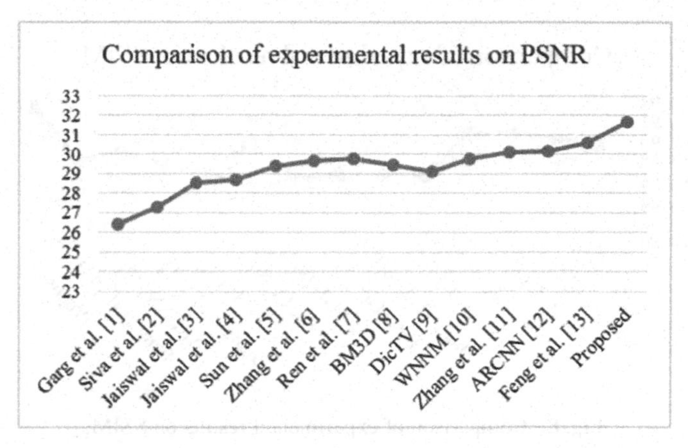

Fig. 7. Comparison of experimental results on PSNR.

4.4 Structural Similarity Index Measure (SSIM)

The structural similarity index method is a model based on this perception. The term structural data refers to interconnected pixels or spatially closed pixels. This interconnected resolution points to a number of important information about objects in the field of images. Lighting masking is a term where the distorted part of the image is less visible at the edges of the image. Contrasting masking, on the other hand, is a term that these distortions are less visible in the image structure. SSIM predicts the perceived quality of images and videos. It measures the similarity between the two images: the original and the restored.

$$\mathrm{SSIM}(x, y) = \frac{(2\mu_x\mu_y + c_1)(2\sigma_{xy} + c_2)}{(\mu_x^2 + \mu_y^2 + c_1)(\sigma_x^2 + \sigma_y^2 + c_2)} \times 100 \tag{9}$$

where μ_x is the average of x and μ_y the average of y; σ_x^2 stands for the variance of x and σ_y^2 the variance of y; σ_{xy} denotes the covariance of x and y.

Comparison on SSIM is demonstrated in Fig. 8. A comparison of the 10 selected images mentioned above with 13 alternative methods on PSNR is given in Fig. 8.

Following our method on MSE, RMSE, PSNR, and SSIM estimation methods show better results than the other 13 alternate methods. The comparative results obtained for these four estimation methods are shown in Fig. 9.

As can be seen from Fig. 9, the results obtained by the PSNR estimation method are expressed in blue, the results obtained by the RMSE estimation method in brown, the results obtained by the MSE estimation method in gray, and the results obtained by the SSIM estimation method in yellow. According to the results, the worst result according to the PSNR method was obtained by Garg et al. [1], and the best result belongs to the proposed method. RMSE best results Feng et al. [18] and belong to the proposed method. Our method is slightly superior to Feng et al. [18]. The results close to the proposed method for all methods belong to Feng et al. [18]. The results for all methods showed our method better results than the closest method. Figure 9 shows images obtained

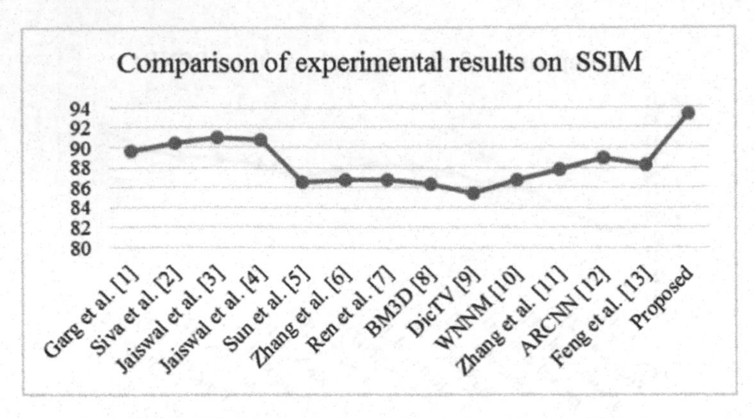

Fig. 8. Comparison of experimental results on SSIM.

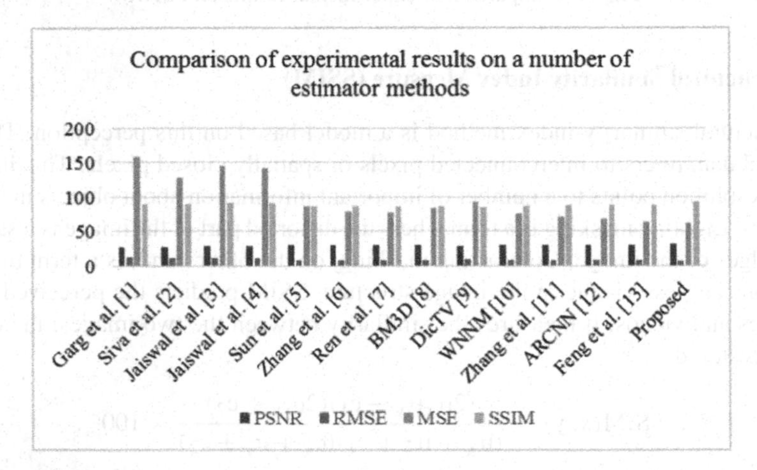

Fig. 9. Experimental results of image resizing on USC-SIPI image dataset.

for comparison in the research work where Fig. 10 represents an example of obtained comparison on Lena image.

Figure 10 shows a comparison of the image named Lena, which is part of the USC-SIPI image dataset given. Here (a) represents the original image, (b) the result obtained on the basis of the nearest method to our method, and (c) the result obtained on the basis of our method. When comparing images, the image (c) obtained based on the proposed method shows a result close to the original image (a). This means that the proposed method better than outperforms its counterparts considered in this work.

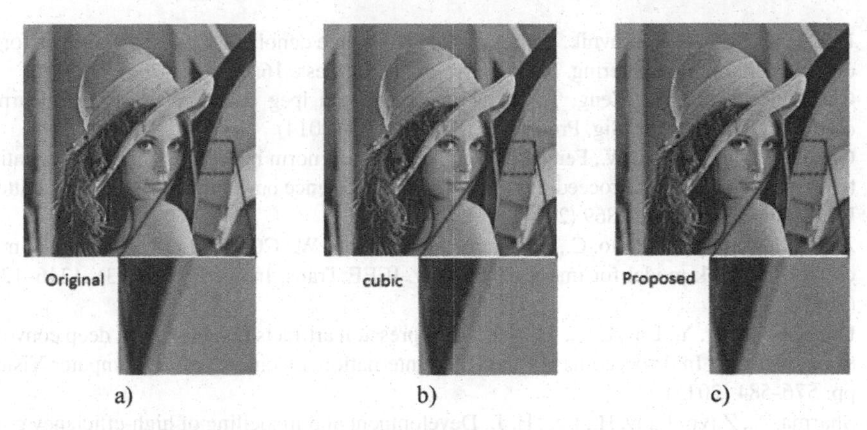

Fig. 10. Example of obtained comparison on Lena image.

5 Conclusion

In this paper, an ANN based image resizing method is developed. Compared to existing neural network and image interpolation-based methods based on evaluation methods such as Mean squared error (MSE), Root-mean-square error (RMSE), Peak signal-to-noise ratio (PSNR), and Structural similarity index measure (SSIM). The comparison clearly showed that the proposed method outperforms each of its counterparts.

In the future work, we will focus on the following: the use of other popular image datasets; getting better results; to apply the proposed method in various fields.

References

1. Garg, B., Goteti, V.N.S.K.C., Sharma, G.K.: A low-cost energy efficient image scaling processor for multimedia applications. In: 20th International Symposium on VLSI Design and Test (VDAT), pp. 1–6 (2016)
2. Siva, M. V., Jayakumar, E. P.: A low cost high performance VLSI architecture for image scaling in multimedia applications. In: 7th International Conference on Signal Processing and Integrated Networks, SPIN, pp. 278–283 (2020)
3. Jaiswal, S.P., Au, O.C., Jakhetiya, V., Guo, Y., Tiwari, A.K., Yue, K.: Efficient adaptive prediction based reversible image watermarking. In: 20th IEEE International Conference on Image Processing (ICIP), pp. 4540–4544. IEEE (2013)
4. Jaiswal, S.P., Au, O., Jakhetiya, V., Guo, A.Y., Tiwari, A.K.: Adaptive predictor structure based interpolation for reversible data hiding. In: Shi, Y.-Q., Kim, H.J., Pérez-González, F., Yang, C.-N. (eds.) IWDW 2014. LNCS, vol. 9023, pp. 276–288. Springer, Cham (2015). https://doi.org/10.1007/978-3-319-19321-2_21
5. Sun, D., Cham, W.-K.: Post processing of low bit-rate block DCT coded images based on a fields of experts prior. IEEE Trans. Image Process. **16**(11), 2743–2751 (2007)
6. Zhang, X., Xiong, R., Fan, X., Ma, S., Gao, W.: Compression artifact reduction by overlapped-block transform coefficient estimation with block similarity. IEEE Trans. Image Process. **22**(12), 4613–4626 (2013)
7. Ren, J., Liu, J., Li, M., Bai, W., Guo, Z.: Image blocking artifacts reduction via patch clustering and low-rank minimization. In: Data Compression Conference (DCC), p. 516. IEEE (2013)

8. Dabov, K., Foi, A., Katkovnik, V., Egiazarian, K.: Image denoising by sparse 3-D transform-domain collaborative filtering. IEEE Trans. Image Process. **16**(8), 2080–2095 (2007)
9. Chang, H., Ng, M.K., Zeng, T.: Reducing artifacts in jpeg decompression via a learned dictionary. IEEE Trans. Sig. Process. **62**(3), 718–728 (2014)
10. Gu, S., Zhang, L., Zuo, W., Feng, X.: Weighted nuclear norm minimization with application to image denoising. In: Proceedings of the IEEE Conference on Computer Vision and Pattern Recognition, pp. 2862–2869 (2014)
11. Zhang, J., Xiong, R., Zhao, C., Zhang, Y., Ma, S., Gao, W.: CONCOLOR: constrained non-convex low-rank model for image deblocking. IEEE Trans. Image Proc. **25**(3), 1246–1259 (2016)
12. Dong, C., Deng, Y., Loy, C.C., Tang, X.: Compression artifacts reduction by a deep convolutional network. In: Proceedings of the IEEE International Conference on Computer Vision, pp. 576–584 (2015)
13. Sharma, A., Zaynidinov, H., Lee, H. J.: Development and modelling of high-efficiency computing structure for digital signal processing. In: International Multimedia, Signal Processing and Communication Technologies, (IMPACT), pp. 189–192 (2009). https://doi.org/10.1109/MSPCT.2009.5164207
14. Arabboev, M., Begmatov, S., Nosirov, K., Shakhobiddinov, A., Chedjou, J.C., Kyamakya, K.: Development of a prototype of a search and rescue robot equipped with multiple cameras: In: International Conference on Information Science and Communications Technologies (ICISCT), pp. 1–5 (2021). https://doi.org/10.1109/ICISCT52966.2021.9670087
15. Zaynidinov, H.N., Mallaev, O.U., Anvarjonov, B.B.: A parallel algorithm for finding the human face in the image. In: IOP Conference Series: Materials Science and Engineering, vol. 862, p. 052004 (2020). https://doi.org/10.1088/1757-899x/862/5/052004
16. Nosirov, K., Begmatov S., Arabboev, M.: Display integrated mobile phone prototype for blind people. In: International Conference on Information Science and Communications Technologies (ICISCT), pp. 1–4 (2019). https://doi.org/10.1109/ICISCT47635.2019.9011919
17. Singh, Dhananjay, Singh, Madhusudan, Hakimjon, Zaynidinov: Evaluation methods of spline. In: Signal Processing Applications Using Multidimensional Polynomial Splines. SAST, pp. 35–46. Springer, Singapore (2019). https://doi.org/10.1007/978-981-13-2239-6_5
18. Jiang, F., Tao, W., Liu, S., Ren, J., Guo, X., Zhao, D.: An end-to-end compression framework based on convolutional neural networks. In: Data Compression Conference (DCC), p. 463 (2017). https://doi.org/10.1109/DCC.2017.54
19. Nosirov, K., Begmatov, S., Arabboev, M., Medetova, K.: Design of a model for disinfection robot system. In: International Conference on Information Science and Communications Technologies (ICISCT), pp. 1–4 (2020). https://doi.org/10.1109/ICISCT50599.2020.9351370
20. Dupond, S.: A thorough review on the current advance of neural network structures. Annu. Rev. Control. **14**, 200–230 (2019)
21. Lee, D., et al.: Long short-term memory recurrent neural network-based acoustic model using connectionist temporal classification on a large-scale training corpus. China Commun. **14**(9), 23–31 (2017)
22. Goodfellow, I., Bengio, Y., Courville, A.: 6.5 Back-propagation and other differentiation algorithms. Deep Learning, pp. 200–220. MIT Press, Cambridge (2016)
23. Søgaard, J., Krasula, L., Shahid, M., Temel, D., Brunnström, K., Razaak, M.: Applicability of existing objective metrics of perceptual quality for adaptive video streaming. In: IS&T International Symposium on Electronic Imaging Science and Technology, pp. 1–7 (2016)

24. Sara, U., Akter, M., Uddin, M.S.: Image quality assessment through FSIM, SSIM, MSE and PSNR—a comparative study. J. Comput. Commun. **7**(3), 8–18 (2019)
25. Deshpande, R., Ragha, G.L.L., Sharma, S.K.: Video quality assessment through PSNR estimation for different compression standards. Indones. J. Electr. Eng. Comput. Sci. **11**(3), 918–924 (2018)

Algorithms for Selecting and Comparing Features of Digital Image Vectors Based on the Analysis of Local Extrema

Mumtozali Tuktasinov(✉)

Namangan Engineering - Construction Institute, I. Karimov Street – 12, Namangan, Uzbekistan
mumtozali@yahoo.com

Abstract. The article proposes new algorithms for extracting recognizable features of one-dimensional vectors obtained from digital images and comparing them. Vectors store one byte, i.e. values in the range $0 \div 255$ taken from a grayscale image. The features of hills located in the intervals of the local minima of the vector were taken as identification features. In particular, the area of the selected hill, its width, the coordinate of the local maximum located on this hill, etc. are taken.

Keywords: digital image · one-dimensional vector · local extremes · vector hills · recognizable features · comparison

1 Introduction

Currently, there are many scientific and practical issues related to the processing and recognition of digital images. In particular, biometric identification of a person [1–3], recognition of license plates on roads [4], automatic reading of texts on images [5], analysis of astronomical images [6], etc. To date, a number of methods and algorithms have been developed in this scientific field, in particular, allowing to recognize digital images [1, 7, 8]. In turn, they improve.

We also conducted research in this direction and achieved certain results. In our study, algorithms for comparing the values of one-dimensional vectors A and B, that is, arrays, were developed. One-dimensional vectors of columns or rows of digital images are obtained as an array. Vectors store one byte, i.e. values in the range $0 \div 255$ taken from the grayscale image. The characteristics of the hills located in the area of the boundaries of the local extreme values of the vector are taken as features. There are different methods and ways for determining extremums [9]. We also used a special version of extrema detection in the algorithm we developed.

H. Zaynidinov et al. (Eds.): IHCI 2022, LNCS 13741, pp. 540–546, 2023.
https://doi.org/10.1007/978-3-031-27199-1_55

2 Formation of Vectors from a Digital Image

Typically, digital images are formed in color, i.e. RGB color scale (red, green, blue). However, to solve scientific and practical problems related to images, in most cases digital images are converted to grayscale [1, 8]. In this case, the color range is reduced from 16.7 million to 256, which simplifies the calculations a bit. We followed the same procedure in our study. As an example, we took the image of a gray face (see Fig. 1).

Fig. 1. Grayscale face image.

Throughout our study, we will work with one-dimensional vectors. Therefore, we extract one-dimensional vectors from columns and rows of digital images. Suppose that the image width is W and the height is H pixels, then we will form W vectors of length H along the columns, i.e. vertically. That is, the 1st vertical vector of length H is generated from the pixels in column 1, the 2nd vertical vector of length H is generated from the pixels in column 2, etc., and finally the W-vertical vector of length H is generated from the pixels in W-column. In the same order, we form H horizontal vectors of length W by rows. Thus, from an image of size W × H, W vertically, H horizontally, total W + H one-dimensional vectors can be formed. For example, a graphical representation of the array values obtained from row 12 (horizontal) of the image in Fig. 1 is shown in Fig. 2.

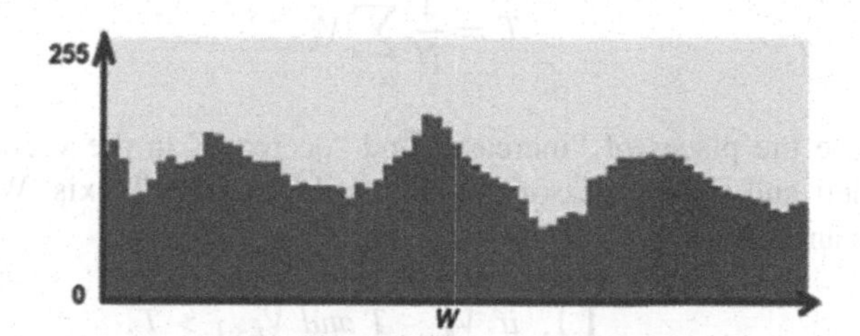

Fig. 2. Graphical representation of array values.

3 An Idea for Extracting the Identification Features of a Vector

As is known, the similarity of the shape of vectors can be determined in different ways. There are several comparison methods, such as calculating the correlation coefficient [1], chi-square [10], etc.

The method we propose considers the possibility of comparing an array (or graph) for similarity in shape. The characteristic features of hills located within the local minimum ($L_V^{\min}$) of the values of the vector V are taken as identification features. In particular, the area of the selected hill is s_h, its width is w_h, and the coordinate of the local maximum located on this hill is $L_h^{\max}$. In Fig. 3, we can visually see these parameters for a single hill extracted as a sample from the vector shown in Fig. 2.

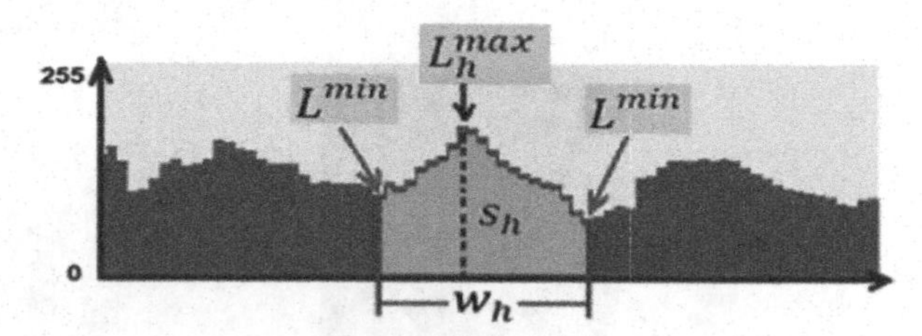

Fig. 3. Features for one selected hill.

For all hills of the vector V, the same features are defined as in this example. In the future, the task of comparing vectors by these features is solved.

4 Algorithm for Extracting Vector Features

To determine the features of the hills of the vector V, we developed the following algorithm.

The *Algorithm for extracting vector features* consists of the following steps:

1) In the N-dimensional vector V, the average value is determined, that is:

$$T = \frac{1}{N}\sum_{i=1}^{N} V_i .$$

2) We determine the places of "increase" and "decrease" in the vector. That is, we define the left and right edges of the hill on the average T axis. We assign these locations to another array C. That is:

$$C_k = \begin{cases} 1, & if\ V_k \le T\ and\ V_{k+1} > T, \\ 2, & if\ V_k > T\ and\ V_{k+1} \le T. \end{cases}$$

where 1 is "increase", 2 is "decrease", $k \in [0, N]$.

3) For each hill, we find local maxima between 1 and 2. We can indicate the location of local maxima with the number 1 in the array $M^{\max}$. That is:

$$M_l^{\max} = 1 ,$$

where $1 \exists \max\{C_l^1, C_l^2\}$, l- hill sequence number.

4) Determine the lowest edges of the hills. This allows for a more complete description of the hill. To find the lower edges of the hill, we find local minima ($L_V^{\min}$) between the previously found local maxima. We can specify the location of local minima with the number 1 in the array $M^{\min}$. That is:

$$M_l^{\min} = 1 \,,$$

where $1 \exists \min\left(C_l^{\max},\ C_{l+1}^{\max}\right)$, l- hill sequence number.

5) Determine the features of hills located between two local minima. That is, the area of the considered hill s_h, its width w_h, the coordinate of the local maximum $L_h^{\max}$ located on this hill. A visual representation of these parameters can be seen in Fig. 3. For the indicated features of the hills, we create a separate array and place the features in it. That is:

$$F_H = \left\{H_1^p,\ H_2^p,\ \ldots,\ H_l^p\right\},$$
$$p \forall \left(s_h,\ w_h,\ L_h^{\max}\right).$$

5 Algorithm for Comparing Identification Features

Suppose we are given two vectors A and B and their feature sets F_H^A and F_H^B. The formula for comparing individual hills from these sets is as follows:

$$D^{A_h,B_h} = \frac{1}{2} \cdot \left(\frac{\min\left(s_h^A,\ s_h^B\right) \cdot 100}{\max\left(s_h^A,\ s_h^B\right)} + \frac{\min\left(w_h^A,\ w_h^B\right) \cdot 100}{\max\left(w_h^A,\ w_h^B\right)}\right)$$
$$-\left(\frac{\left|L_h^{A_{\max}} - L_h^{B_{\max}}\right| \cdot 100}{w} + \left|d_{l,\,l-1}^A - d_{l,\,l-1}^B\right|\right).$$

where $d_{l,\,l-1}^A$ - is the ratio of the distance between two adjacent hills (according to the coordinate of the vertex center) in the vector A. That is:

$$d_{l,\,l-1}^A = \frac{(c_l - c_{l-1}) \cdot 100}{W} \,,$$

where c_l is the coordinate of the hill center at the l-th place, W is the length of the vector.

As a result, the sum of the differences of all features of the hill in the features sets in the vectors A and B is obtained by the following formula:

$$D^{F(A,B)} = \frac{1}{L} \cdot \sum_{h=1}^{L} D^{A_h,B_h}.$$

When compared using the above formula, if two vectors are absolutely similar, the result is 100.

Based on the above formula, if the number of marked hills in vectors A and B is equal to each other, they can be compared by the corresponding sequence number. However, it should be noted that in real problems the number of marked hills - L in vectors A and

B can be different. Therefore, it is recommended to take smaller ones from them, that is, $L = \min(L^A, L^B)$. In this case, if $L^A \neq L^B$ the comparison needs some optimization. That is, for each hill located in a small array, it is necessary to solve the problem of finding the corresponding hill in a large array. For this we have developed the following algorithm.

The "*Algorithm for comparing identification features*" with the case $L^A \neq L^B$ consists of the following steps:

1) If $L^A \neq L^B$, go to the next step.
2) We determine the boundaries of the search for features from a large array for the x-th hills of a small array. That is, we define the Left and Right borders. In first state:

$$x = 1;$$
$$Left = 0;$$
$$Right = L_{\max} - (L_{\min} - Left).$$

where $L_{\max}$ is the length of an array with a large number of hills, $L_{\min}$ is the length of an array with a small number of hills.

3) Comparing the x-th hill's features of the small array between the left and right borders in the large array, we get the maximum result D^{A_h,B_h} of them and label this place as $P_{\max}$. In turn, we update the new search border for the next x + 1st hill, then there is:

$$Left = P_{\max} + 1;$$
$$Right = L_{\max} - (L_{\min} - x) + 1.$$

4) $x = x + 1$; If $x \leq L_{\min}$, go to step 3.
5) We get the final result as $D^{F(A,B)} = \frac{1}{L} \cdot \sum_{h=1}^{L} D^{A_h,B_h}$.
6) We have $L^A \neq L^B$, and therefore we calculate the difference coefficient of the hills and subtract it from the result. That is:

$$D^{F(A,B)} = D^{F(A,B)} - \left(\frac{|L^A - L^B|}{L^B} \cdot 100 \right)$$

6 Experimental Analysis of the Results

Based on the algorithms developed on the basis of our research, a computer program was developed in the C++ programming language, on the basis of which a number of results were obtained and experimentally investigated. The values of the input vectors A and B (in the range $0 \div 255$) were extracted from the real grayscale image (by rows and columns). The following Table 1 shows the numerical values of some of the vectors extracted from images A and B, their corresponding graphs, and their similarity results.

Table 1. Vector comparison results.

N:	Values of the array A and its graph	Values of the array B and its graph	Similarity results of arrays A and B
1.	 111;111;111;111;111;111;114;116;125;125;147;147;152;169;169;147;141;152;133;119;103;97;91;91;114;119;133;130;133;130;125;122;119;116;111;111;116;119;116;114;116;116	 112;112;112;116;116;116;116;116;125;128;128;128;116;112;107;107;107;85;83;83;89;101;101;107;114;116;123;123;123;123;116;107;107;116;116;112;112;107;110;112;114;114	27.32
2.	 116;116;116;125;128;130;136;147;152;152;155;164;164;94;72;61;47;50;44;61;69;86;111;116;136;139;150;139;133;133;130;122;122;122;122;122;119;114;114;119;125;128	 114;119;119;119;121;123;125;128;132;132;123;85;85;92;92;92;83;81;81;81;85;110;112;119;123;125;128;128;128;121;110;103,103,110,110;105;105;105;112;112;112;116	50.58
3.	 188;191;186;67;61;30;30;30;64;47;55;100;144;155;158;164;161;155;155;155;164;141;103;89;64;19;19;64;69;100;111;122;122;122;122;122;125;128;128;122;122;122	 136;136;132;132;101;72;76;78;81;74;58;58;69;112;148;150;148;145;141;141;139;136;101;94;76;65;54;45;54;101;123;130;132;132;128;121;121;121;121;110;110;110	70.01
4.	 175;183;164;139;94;89;83;72;44;39;58;116;119;133;144;150;150;150;155;164;155;128;116;103;75;33;47;58;91;105;116;125;122;119;122;128;128;128;128;128;125;128	 154;161;161;150;105;60;54;51;51;51;47;51;83;94;107;130;134;134;143;148;148;143;94;76;76;58;38;63;78;105;112;121;114;107;107;110;116;121;116;123;123;123	91.07

7 Conclusions

It should be noted that in order to increase the reliability of the results, it is recommended to apply pre-processing methods to input images. For example, scaling, improving image quality, normalizing brightness through filtering, and much more.

The developed algorithms can be used to compare not only images, but also other incoming digital signals.

We are currently conducting further research to improve the methods and algorithms proposed above.

References

1. Kukharev, G.A.: Biometric Systems: Methods and Means of Human Identification, p. 240. Politekhnika, St. Petersburg (2001). (in Russian)
2. Fazilov, S.K., Abdugafarov, I.A., Tukhtasinov, M.T.: Biometric identification of computer system users. In: WCIS –2004, Third World Conference on Intelligent Systems for Industrial Automation, Tashkent, pp. 57–61 (2004)
3. Fazilov, S.K., Tukhtasinov, M.T.: About biometric computer systems. Inf. Energy Prob. Uzb J. (1), 3–8 (2011). (in Uzbek)
4. Obukhov, A.V., Lyasheva, S.A., Shleimovich, M.P.: Methods for automatic recognition of license plates. Bull. Chuvash Univ. (3), 201–208 (2016). (in Russian)
5. Hegghammer, T.: OCR with tesseract, amazon textract, and google document AI: a benchmarking experiment, p. 38 (2021)
6. Zhu, H.J., Han, B.C., Qiu, B.: Survey of Astronomical Image Processing Methods, pp. 420–429. Springer, Switzerland (2015)
7. Gonzalez, R.C., Woods, R.E.: Digital Image Processing, p. 793. Prentice Hall, Upper Saddle River (2002)
8. Pratt, W.K.: Digital Image processing: PIKS Scientific inside, 4th edn, p. 782 (2007)
9. Silverman, R.A.: Essential Calculus with Applications. Dover Publications, New York (2013)
10. Tukhtasinov, M.T., Mirzaev, N., Narzulloev, O.M.: Face recognition on the base of local directional patterns. In: IEEE Conference 2016 Dynamics of Systems, Mechanisms and Machines (Omsk 2016). IEEE (2016). https://doi.org/10.1109/Dynamics.2016.7819101

Calculation of Spectral Coefficients of Signals on the Basis of Haar by the Method of Machine Learning

Yusupov Ibrohimbek[1], Nurmurodov Javohir[1(✉)], Ibragimov Sanjarbek[2], Gofurjonov Muhammadali[1], and Qobilov Sirojiddin[1]

[1] Department of Artificial Intelligence, Tashkent University of Information Technologies Named After Muhammad Al Khwarizmi, Tashkent, Uzbekistan
nurmurodov1994@gmail.com

[2] Department of Information Technology, Andijan Machine-Building Institute, Andijan, Uzbekistan

Abstract. In this article, the spectral analysis of the analytic function based on Haar's piece-polynomial basis was studied. Calculation of spectral coefficients using Haar's piece-square basis is given. A comparative analysis of zero coefficients of spectral values determined using Haar's piece-quadratic bases in different analytical functions is presented in the table. In order to increase the value of the compression coefficients presented in the table, a model of improvement of Haar's piece-quadratic base using the machine learning method was proposed. In determining the hyperparameter, the input values were used based on the problem in the object. The value function was selected and the error was evaluated to check the level of accuracyImprovement was made on the basis of existing algorithms and machine learning method, following the conditions of interpolation of Haar's second-order partial polynomial base using the mentioned sequences. As a result, the number of zero coefficients increases from 81%–85% to 90%–96% when calculating the spectral values of the experimentally obtained functions based on the proposed model, and the number of zero coefficients increases from 75%–80% to 85%–90% when calculating the spectral values of geophysical signals. Percent increase was achieved.

Keywords: Haar base · spectral analysis · machine learning · error · gradient · value function · hyperparameters

1 Introduction

It is known that problems arise in finding a solution to the problems posed in science and technology with clear and easy methods. In many cases, it is necessary to find an approximate solution to the given problem based on certain conditions. The application of convergence and high-speed methods makes a great contribution to increase the practical reliability of science and technology. Today, improved methods are required for solving technical problems. This is due to the introduction of a new concept, artificial

H. Zaynidinov et al. (Eds.): IHCI 2022, LNCS 13741, pp. 547–558, 2023.
https://doi.org/10.1007/978-3-031-27199-1_56

intelligence (IA), into all fields. Machine learning is widely used as a mathematical model for solving problems arising in the field of artificial intelligence. This method is taking a new direction in science and technology. Any mathematical model can be improved based on Kolmogors theorem using the machine learning method. In the course of our research, we will consider the process of calculating spectral coefficients by improving Haars piece-quadratic basis with the help of machine learning method.

2 Construction of Haar Bases

Spectral analysis - this method is considered modern for signal processing and is based on frequency, amplitude, time and spectrum of signals. In spectral analysis, it is necessary to expand the values of the original functions based on other functions and process the values to obtain the signal spectrum values. The values of the spectra correspond to the frequency on the horizontal and the amplitude on the vertical in the graphical view. To date, spectrum analysis of signals is carried out mainly on the basis of Haar bases in the field of technology.

Xaar Base. The Haar base is widely used in various fields of technology and science to solve a large class of theoretical and practical problems. Haar has piece-invariant, piece-linear and piece-quadratic bases. It is distinguished by a number of excellent properties of the basis functions and the presence of spectral analysis calculation algorithms for them. The normalized Haar basis function is considered a multivalued function, and it is more convenient to perform calculations using three values (−1, 0, and 1) for spectral processing. These bases are given analytically by the following expression and are constructed based on the binary number system in the interval [0, 1] or [0, 1)

$$h_k = h_{pj} = \left[\frac{j}{2^{p-1}} \frac{j+1}{2^{p-1}}\right] \tag{1}$$

k-where the Xaar function is the ordinal number. p while $1 \leq p \leq log_2^N$ is equal to the value. N - is the number of incoming values. j = 0,1,……, 2^{p-1} repeated to the power of two. If h_{pj}- we divide the expression into two parts h_{pj}^{+} and h_{pj}^{-} the corresponding left and right parts represent a binary segment [7].

$$\text{(a)}\ h_{pj}^{+} = \left[\frac{j}{2^{p-1}}; \frac{2j-1}{2^{p+1}}\right] \tag{2}$$

$$\text{(b)}\ h_{pj}^{-} = \left[\frac{2j+1}{2^{p+1}}; \frac{j+1}{2^{p}}\right] \tag{3}$$

In this case, the discontinuity of the first type (a) is represented by the continuous reception on the right side (b) in the internal points. Summing these binary segments gives a **piece- invariant** Haar basis [5].

$$har_k(x) = har_{pj}(x) = \begin{cases} +1 & x \in \overset{-}{h_{pj}} \\ -1 & x \in \overset{+}{h_{pj}} \\ 0 & x \in h_{pj} \end{cases} \tag{4}$$

The construction of the piece-invariant basis of Haar in the intervals (a), (b) is formed (4). k-where the Xaar function is the ordinal number. The construction of Haars piece-invariant basis is shown in the figure below.

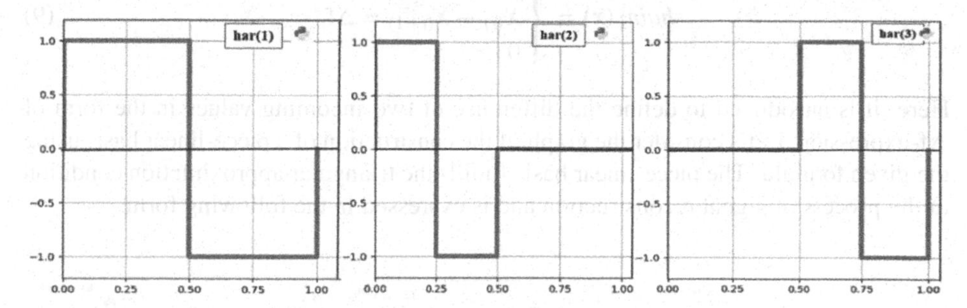

Fig. 1. Haars piece - invariant basis graph.

The graph shown in Fig. 1 fulfills the approximation condition during signal reconstruction. Spectral analysis is widely used today in digital processing of a given signal. In spectral analysis, the calculation of spectrum coefficients and determination of near-zero values is the main factor for solving the given problem. To solve this problem, we consider the fast Haar Endirius method of finding spectral coefficients using the derived expression (4) [6].

$$C_k = \frac{1}{2^P}\sum_{i=0}^{n-1} x(i)\cdot haar(x)dx \tag{5}$$

Here (5) is the formula for calculating spectral coefficients using Haars piece - immutable basis. C_k– spectral coefficient, $x(i)$– incoming values or signal value.

Piece - linear basis - is the result of taking a single integral between two points from a piecewise constant basis in the construction of the basis. For this, it is required to integrate the formula (4) in the interval x(i) and x(i + 1).

$$hain_k(x) = \int_{X(i)}^{X(i+1)} haar_k(x)dx \tag{6}$$

Here, haink(x) is a piecewise linear basis. Haark (x) – if we put formula (4) instead of function, formula (7) is formed.

$$hain_k(x) = \int_{X(i)}^{X(i+1)} (\begin{cases} +1 \ x \in \bar{h}_{pj} \\ -1 \ x \in \overset{+}{h}_{pj} \\ 0 \quad x \in h_{pj} \end{cases})dx \tag{7}$$

Integrating using two point values results in the following form. ki is a free term.

$$hain_k(x) = \begin{cases} X_{(i+1)} - X_{(i)} + k_{(i)} - k_{(i)} = \Delta f \\ -X_{(i+1)} + X_{(i)} + k_{(i)} - k_{(i)} = \Delta f \\ k_{(i)} - k_{(i)} = 0 \end{cases} \tag{8}$$

As a result of shortening the similar expressions in the mentioned formula (8), formula (9) results.

$$hain_k(x) = \begin{cases} X_{(i+1)} - X_{(i)} = \Delta f \\ X_{(i)} - X_{(i+1)} = \Delta f \\ 0 \end{cases} \quad (9)$$

Here, it is introduced to define the difference of two incoming values in the form of Δf-expression. Let's consider the graph of the construction of a piece-linear basis using the given formula. The piece-linear basis fulfills the triangular approximation condition in the process of signal reconstruction and is expressed in the following form.

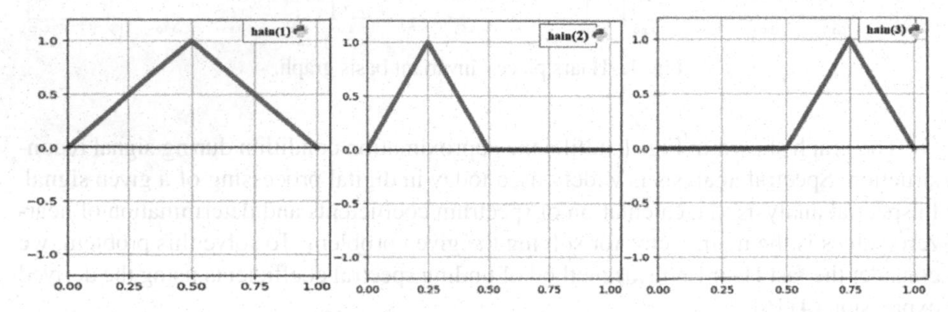

Fig. 2. Haars piece - linear basis plot.

As can be seen in Fig. 2, it approximates better than the piece-invariant basis of Fig. 1. Using the above formula (9), a formula for calculating spectral coefficients based on a piece-linear basis is created [8].

$$C_k = \frac{1}{2^P} \sum_{i=0}^{n-1} \Delta f \cdot haar(x)dx \quad (10)$$

In this created formula, Δf - is equal to the difference of the input values.

Piece-quadratic basis - this basis is formed by taking a double integral from a piecewise-invariant basis or by taking a single integral from a piecewise-linear basis. For this, formula (9) is integrated in the interval x(i) and x(i + 1) and as a result formula (11) is formed.

$$haid_k(x) = \int_0^x hain_k(x)dx \quad (11)$$

Let's consider the graph of the construction of the piece-quadratic basis using the given formula. The piece-quadratic basis fulfills the approximation condition in the parabola method during signal reconstruction and is expressed in the following form (Fig. 3).

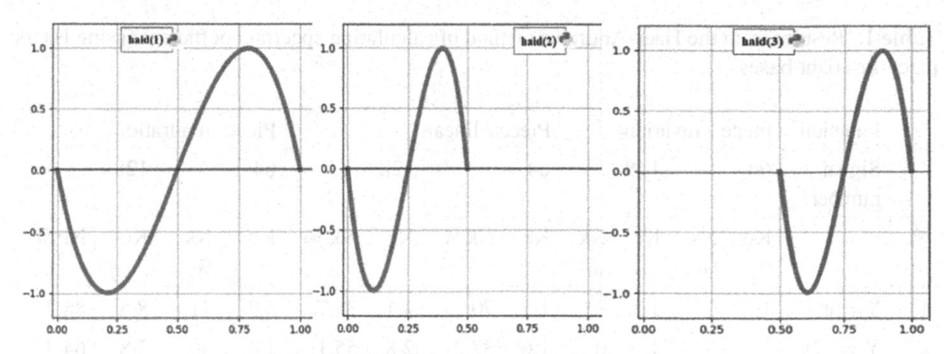

Fig. 3. Haars piece-quadratic basis plot.

From this graph, it approaches more smoothly than the dashed-line basis in Fig. 2. If the spectral coefficients are calculated based on the piece-quadratic basis using the above formula (11), the following expression is obtained.

$$C_k = \frac{1}{2^P} \sum_{i=0}^{n-1} \Delta \mathrm{b_i} \cdot haar(x)dx \tag{12}$$

In this created formula, bi-coefficients can be obtained by calculating based on formulas

3 Point Formula [4]:

$$b_i = (1/6)(-f_{i-1} + 8f_i - f_{i+1}); \tag{13}$$

In this formula, $\mathrm{f_i}$ -is the input value.Lets experiment with several functions using the Haar bases presented as a result of the research. For this, we calculate the spectral coefficients by taking the actual values of the functions and determine the compression coefficients [1].

$$\kappa = N/(N - N1),$$

The number of N-function calculations is N1- the number of zero coefficients obtained as a result of quick changes in Xaar bases.

$\mathrm{K_C}$ - compression coefficient in the table, Nk-number of zero coefficients. Finding the unknown coefficient in the construction of the Haar bases given in the above formula (13) is not widely covered. - since there is no algorithm for finding the coefficient, we use machine learning, one of the widely used methods, to find the value of during the research. This method is a mathematical algorithm of artificial intelligence and is considered the most important factor in creating neural networks [3].

3 Improvement of Haar Bases Based on Machine Learning

Machine Learning– is a new branch of computer programming in which software can read data, learn from that data, and allow computers to learn without being directly programmed. There are also different categories of machine learning. Based on how data is used, we can divide machine learning into the following categories:

Table 1. Results from the Haar-Andrews method of calculating spectral coefficients using Haars piece-invariant bases.

	Function	piece - invariant				Piece - linear				Piece-quadratic			
	Signal number	64		128		64		128		64		128	
№		Kc	Nk %	Kc	Nk %	Kc	Nk %	Kc	Nk %	Kc	Nk %	Kc	Nk %
1	$Y = e^x$	1	0	1	0	1.7	70	4.1	75.7	5.3	81	8.5	88.2
2	Y = e2x coc(4x)	1	0	1	0	1.6	37.2	2.8	55.1	1.7	40	2.8	64.3
3	th(x)	1	0	1	0	1.1	65	2.3	70.5	6.4	85	9.1	89.2

supervised learning;
unsupervised learning;
reinforcement learning;

Supervised learning – the most common method of this type of training is to train a computer program using specified data. Supervised learning itself is divided into 2 categories:

Regression;
Classification.

We use univariate linear regression to solve the fitted mass. One of the main tasks when working with linear regression is to find the best regression line that represents the data. If the regression line is determined based on the following formula (14), we need to find the coefficients wi and b_i.

$$Y_i = w_i \cdot x_i + b_i \tag{14}$$

In this case, the result predicted by these constants must be close to the actual value presented in the data set (Fig. 4).

The 1st line in the graph is the predicted values. Line 2 is the closest value to the actual value based on the predicted values. The difference between the predicted value and the actual value should be as small as possible. This difference is taken as an error and a value function is generated. The main task of the value function is to detect this error. Based on this function, we can determine how correctly the regression coefficients (w_i and b_i) are found, that is, through these coefficients, we can find the error between the predicted and the actual value. For this we use the mean squared error function as the value function. Medium quadratic error is expressed in the form of a graph (Fig. 5).

$$r(loss) = \frac{1}{N}\sum_{i=0}^{N}(\mathrm{f}(x_i) - y_i)^2 \tag{15}$$

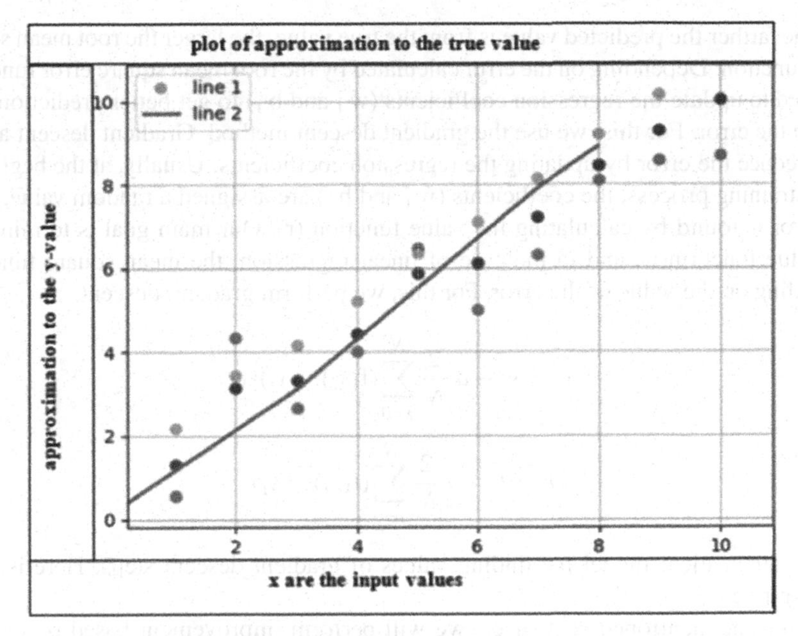

Fig. 4. Finding values closest to actual values using predicted values.

14 is the number of m-datasets in the formula y_1 data set – $f(x_i)$ predicted value

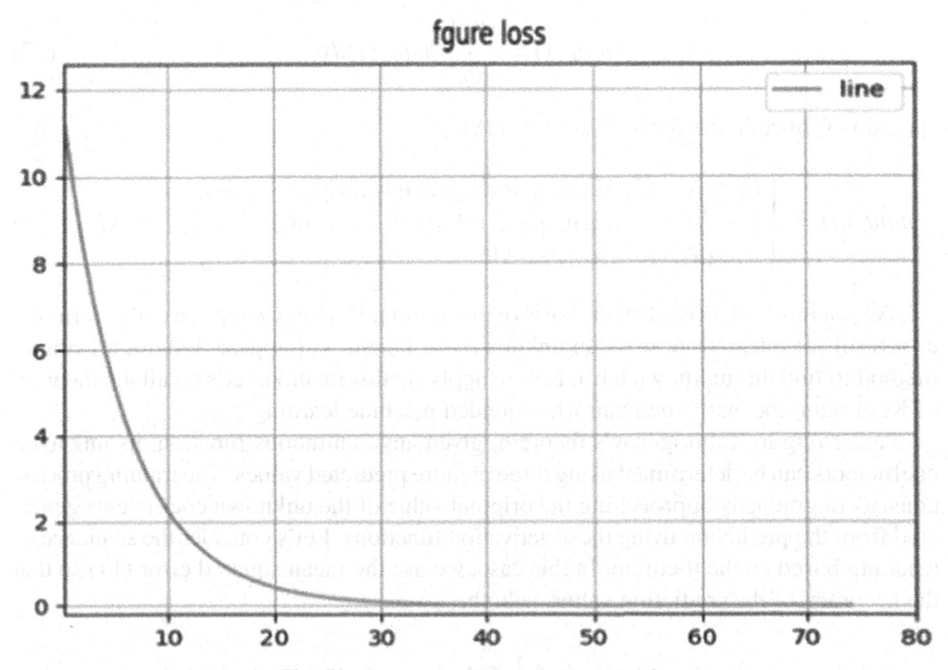

Fig. 5. Mean squared error count graph.

The farther the predicted value is from the true value, the larger the root mean square error function. Depending on the error calculated by the root mean square error function, we need to update the regression coefficients (w_i and b_i) to get better predictions and reduce the error. For this, we use the gradient descent method. Gradient descent allows us to reduce the error by updating the regression coefficients. Usually, at the beginning of the training process, the coefficients (w_i and b_i) are assigned a random value. Then the error is found by calculating the value function (r). Our main goal is to minimize the value function(r), and in the case of linear regression, the mean square function, depending on the value of the error. For this, we perform gradient descent.

$$w = w - a \cdot \frac{2}{N} \sum_{i=0}^{N} (\mathrm{f}(x_i) - y_i) \cdot x_i \tag{16}$$

$$b = b - a \cdot \frac{2}{N} \sum_{i=0}^{N} (\mathrm{f}(x_i) - y_i)$$

are a mathematical model for finding values of gradient descent steps. Here is the a learning rate.

Using the mentioned sequences, we will perform improvement based on existing algorithms and machine learning method, observing the conditions of interpolation of Haar's second-order partial polynomial base. In this case, we determine the b_i coefficient in the formula (10) using the machine controlled learning method [2]. For this, if we integrate between two points,

$$haid_k(x) = \int_{x(i)}^{x(i+1)} hain_k(x)dx \tag{17}$$

spread is formed in the form of formula (18).

$$haid_k(x) = \begin{cases} (X_{(i+1)} - X_{(i)})(X_{(i+1)} + X_{(i)})/2 + w_{(i)} \cdot (X_{(i+1)} - X_{(i)}) = \Delta \mathrm{F} \\ (-X_{(i+1)} + X_{(i)})(X_{(i+1)} + X_{(i)})/2 - w_{(i)} \cdot (X_{(i+1)} - X_{(i)}) = \Delta \mathrm{F} \\ w_{(i)} \cdot (X_{(i+1)} - X_{(i)}) = \Delta \mathrm{F} \end{cases} \tag{18}$$

ΔF_i- a function formed on the basis of integration. If we pay attention to this formula, as a result of integration, w_i - an unknown coefficient will appear. We use a learning method to find this unknown term. Before applying this method, let's recall the theorem of Kolmogor, the mathematician who founded machine learning:

According to Kolmogorov's theorem, given any continuous function, its unknown coefficients can be determined using three or more predicted values. The training process consists in gradually approaching the original value of the unknown coefficients generated from the prediction using these activation functions. Let's consider the sequence of teaching based on the theorem. In this case, we use the mean squared error (14) so that the accuracy of the prediction values is high.

$$r_{loss} = \frac{1}{N} \sum_{i=1}^{N} (\Delta F_i - y_i)^2 \tag{19}$$

Here r is the mean square error *(Mean Square Error)* N- count the number of values. i– step 1, 2, 3,….. N. In the next step, based on the formula (13), we create a mathematical model for finding the values of the gradient steps of the formula (17).

$$w = w - a \cdot \frac{2}{N} \sum_{i=0}^{N} (\Delta F(x_i) - y_i) \cdot x_i \tag{20}$$

Here is the a learning rate. x_i - incoming value. y_i is the original value. (20) is the formula based on teaching done we increase and one of time in itself (19) - using the formula error we count.

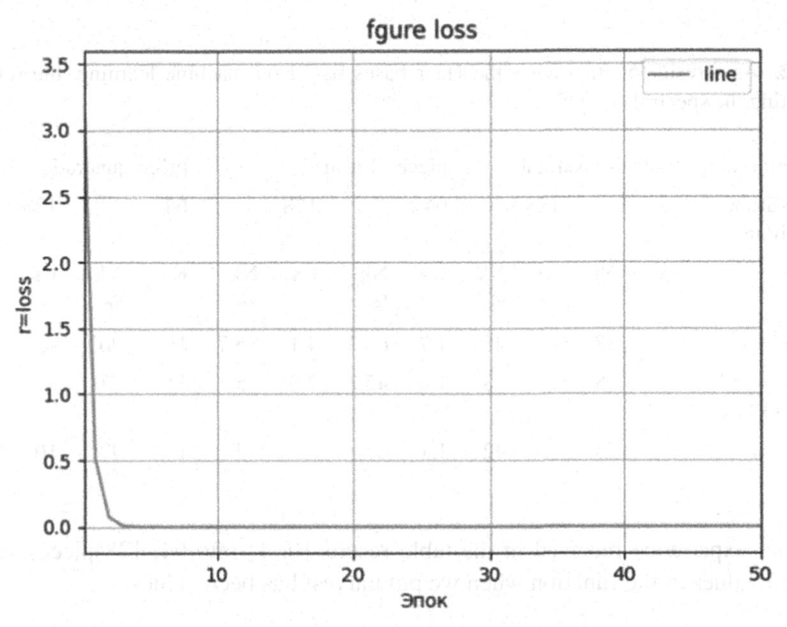

Fig. 6. Mean squared error results graph.

(18) - in order to determine the closest value of the gradient values obtained on the basis of the formula, we carry out training until the error approaches zero. In this case, the

$$haid_k(x) = \begin{cases} (X_{(i+1)} - X_{(i)})(X_{(i+1)} + X_{(i)})/2 + w_{(i)} \cdot (X_{(i+1)} - X_{(i)}) = \Delta F & x \in \bar{h}_{pj} \\ (-X_{(i+1)} + X_{(i)})(X_{(i+1)} + X_{(i)})/2 - w_{(i)} \cdot (X_{(i+1)} - X_{(i)}) = \Delta F & x \in \overset{+}{h}_{pj} \\ w_{(i)} \cdot (X_{(i+1)} - X_{(i)}) = \Delta F & x \in h_{pj} \end{cases} \tag{21}$$

error depends on the number of epochs and appears in the form of Fig. 6. When the optimal value of the gradient is determined as a result of training the above expression, it is multiplied by one input (signal) value and approximation is made between two points.

If the input values in the formula (21) satisfy any condition, calculations are made based on these conditions and the spectral coefficients of the input values are calculated using the following formula (22).

$$C_k = \frac{1}{2^P} \sum_{i=0}^{N-1} \Delta \mathrm{F} \cdot haar(x)dx \tag{22}$$

In the formula above ΔF_i- an expression an expression resulting from machine learning refinement of the piece-quadratic basis. C_k- calculating spectral coefficients to find Based on this formula, the functions in Table 1 on again experience done by increasing we will see.

Table 2. As a result of improving the Haar bases based on machine learning, the results of calculating the spectral coefficients.

	Func-tion	piece - invariant				piece - linear				piece-quadratic			
	Number of signals	64		128		64		128		64		128	
№		Ks	Nk %	Ks	Nk %	Ks	Nk %	Ks	Nk %	Ks	Nk %	Ks	Nk %
1	$Y = e^x$	1	38	1	45	1.7	65.3	4.1	85.7	16	96	43	99.2
2	$Y = e^{2x}$ COS (4x)	1	28	1	33	1.6	47.2	2.8	65.1	11	7 0	9	87.3
3	th (x)	1	35	1	42	1.1	67	2.3	73	13	92	16	94.5

From experience received in the table results [0; 1] into 64, 128 pieces separate received values to the function when we put harvest has been values.

4 Calculation of Spectral Coefficients of Geophysical Signals Using Improved Haar Bases

During the research, we calculate the spectral coefficients of geophysical signals based on the proposed algorithms. When using the actual values of the signal, the accuracy is lower compared to the analytical functions [2] (Table 3).

Table 3. Results of spectral coefficients of geophysical signals using Haar bases.

i	C i	i	C_i	i	C_i	i	C_i
1	0.292493	17	−0.000179	33	−0.000026	49	−0.000026
2	0.000730	18	0.000026	34	0.000013	50	0.000000
3	−0.000128	19	0.000307	35	0.000000	51	−0.000013
4	−0.000883	20	−0.000243	36	0.000000	52	0.000000
5	−0.000179	21	0.000102	37	0.000000	53	0.000000
6	0.000026	22	0.000077	38	0.000026	54	0.000000
7	0.000307	23	0.000038	39	−0.000038	55	−0.000026
8	−0.000243	24	−0.000141	40	0.000000	56	0.000000
9	0.000102	25	0.000307	41	0.000000	57	0.000000
10	0.000077	26	−0.000051	42	0.000000	58	−0.000026
11	0.000038	27	−0.000038	43	0.000026	59	0.000000
12	−0.000141	28	−0.000154	44	0.000026	60	−0.000013
13	0.292493	29	−0.000038	45	0.000000	61	0.000000
14	0.000730	30	−0.000038	46	0.000000	62	0.000000
15	−0.000128	31	−0.000013	47	0.000000	63	0.000000
16	−0.000883	32	−0.000013	48	0.000000	64	−0.000026

Table 4. Results of calculating spectral coefficients of geophysical signals using Haar bases

	Func-tion	piece - invariant				piece - linear				piece-quadratic			
	Number of signals	64		128		64		128		64		128	
№		Ks	Nk %	Ks	Nk %	Ks	Nk %	Ks	Nk %	Ks	Nk %	Ks	Nk %
1	X B	1.7	46.5	2.6	61	2.7	63.4	3.2	68.5	3.7	73.4	4.7	78.5
2	MX B	1.8	47.6	2.6	62.3	2.9	65.5	3.36	70.1	5.8	82.8	6.4	85.1

The above table shows the results of calculating the spectral coefficients of geophysical signals based on the **XB** - Haar Basis. **MXB** – Results from Improving Haar Databases Based on Machine Learning.

5 Conclusion

Signals spectral coefficients in the calculation Haars piece-invariant, piece-linear and piec-quadratic bases analysis done.Haar bases convenience coefficients count for fast algorithms existence. But many practical issues in solving this bases method accuracy enough not Haars piece-polynomial bases teaching the way with accuracy increase possible analysis done. Analysis in the process Haars piece-quadratic basis choose received and 64,128 values for calculations implemented. Based on Haars bases in Table 1 received results. In Table 2, Haar bases from improvement harvest done results. To tables attention giving if we In Table 1, y = e ^{x} function out of 128 values for when used zero coefficients are 88.2%. In Table 2, the function y = e ^{x} is achieving 99.2% when using 128 values. In Table 4, the spectral coefficients of geophysical signals were calculated and increased by 78.5%–85.1%. It can be seen that the accuracy of the values obtained as a result of model improvement is high.

References

1. Xakimjon, Z., Oybek, M.: Definition of synchronization processes during parallel signal processing in multicore processors. In: 2019 International Conference on Information Science and Communications Technologies (ICISCT), pp. 1–4 (2019). https://doi.org/10.1109/ICISCT47635.2019.9012006
2. Xakimjon, Z., Bunyod, A.: Biomedical signals interpolation spline models. In: 2019 International Conference on Information Science and Communications Technologies (ICISCT), pp. 1–3 (2019). https://doi.org/10.1109/ICISCT47635.2019.9011926
3. Zaynidinov, H., Bakhromov, S., Azimov, B., Makhmudjanov, S.: Comparative analysis spline methods in digital processing of signals. Adv. Sci. Technol. Eng. Syst. **5**(6), 1499–1510 (2020). https://doi.org/10.25046/aj0506180
4. Singh, M., Zaynidinov, H., Zaynutdinova, M., Singh, D.: Bi-cubic spline based temperature measurement in the thermal field for navigation and time system. J. Appl. Sci. Eng. **22**(3), 579–586 (2019). https://doi.org/10.6180/jase.201909_22(3).0019
5. Singh, D., Singh, M., Hakimjon, Z.: B-Spline approximation for polynomial splines. In: SpringerBriefs in Applied Sciences and Technology (2019)
6. Singh, D., Singh, M., Hakimjon, Z.: Evaluation methods of spline. In: SpringerBriefs in Applied Sciences and Technology (2019)
7. Zaynidinov, H., Ibragimov, S., Tojiboyev, G., Nurmurodov, J.: Efficiency of parallelization of haar fast transform algorithm in dual-core digital signal processors. In 2021 8th International Conference on Computer and Communication Engineering (ICCCE), pp. 7–12 (2021). https://doi.org/10.1109/ICCCE50029.2021.9467190
8. Zaynidinov, H., Ibragimov, S., Tojiboyev, G.: Comparative analysis of the architecture of dual-core blackfin digital signal processors. In: 2021 International Conference on Information Science and Communications Technologies (ICISCT), pp. 1–4 (2021). https://doi.org/10.1109/ICISCT52966.2021.9670135

Comparison for Polyp Segmentation Models: Focusing on Inference Speed

Seung Gun Lee[1], Seung Min Jeong[1], Chae Lin Seok[1], Jin Man Kim[2], and Eui Chul Lee[3(✉)]

[1] Department of AI and Informatics, Graduate School, Sangmyung University, Seoul, Republic of Korea
{202233053,202132045,202231058}@sangmyung.kr

[2] IoT Convergence & Bigdata Center, Digital Innovation Office, Korea Electronics Association, Seoul, Republic of Korea
jmkim@gokea.org

[3] Department of Human-Centered Artificial Intelligence, Sangmyung University, Seoul, Republic of Korea
eclee@smu.ac.kr

Abstract. Recently, with the development of computer vision technology, models that can detect polyps quickly and accurately have been developed. Algorithms for polyp region segmentation put a lot of effort to accurately find polyps with various patterns in pixel units based on the encoder-decoder structure. However, as this progresses in the direction of increasing the complexity of the model, the time complexity of the model increases exponentially to the extent that it is difficult to implement the model. In this study, the encoders of UNet and DeepLabV3+ models were replaced with lightweight models and compared with high-complexity models. We performed to train and test polyp segmentation on the several polyp segmentation datasets. After that, the processing speed and accuracy of the proposed model were evaluated by comparing the existing models with U-Net and DeepLabv3+ with the changed encoder. As a result of the experiment, proposed model showed that there was no significant difference in accuracy while speeding compared to the existing model.

Keywords: Polyp segmentation · Deep learning · Fine-tuning · Speed · Real-time

1 Introduction

Colorectal cancer refers to a malignant tumor (adenocarcinoma) occurring in the colon or rectum. In the early stages of colorectal cancer, there are no symptoms, and in many cases, it is discovered incidentally. If symptoms of colorectal cancer appear, they have already progressed in many cases, so they should be screened for colorectal cancer regularly. Tests include a digital rectal examination, stool examination, colonography, CT or MRI

S. G. Lee and S. M. Jeong—Co-first authors and contributed equally.

H. Zaynidinov et al. (Eds.): IHCI 2022, LNCS 13741, pp. 559–564, 2023.
https://doi.org/10.1007/978-3-031-27199-1_57

examination, and blood tests. Even after receiving such an examination, the importance of colonoscopy is very high because a biopsy through a colonoscopy must be performed in the end. [1] In the existing colonoscopy, there are polyps that are not found and missed because the doctor directly sees the screen and judges whether polyps are present. To completely detect polyps in the colonoscopy, the insertion time of the endoscope is inevitably prolonged. [2] Recently, with the development of computer vision technology, models that can detect polyps quickly and accurately have been developed.

In the past few years, Convolutional Neural Networks (CNNs) have been variously applied in the field of image processing, and the latest model for medical image segmentation is a variant of the encoder-decoder structure such as the Fully Convolution Network (FCN) that connects these CNNs [3]. Recently, algorithms for polyp region segmentation have made a lot of effort to accurately find polyps with various shapes in pixel units. Accurate prediction in units of pixels increases the complexity of the model, and accordingly, the time complexity of the model increases exponentially to the point where it is intractable to implement the model in real time.

In paper written by Srivastava et al. [4], we confirmed that models such as UNet and DeepLabV3+ showed satisfactory performance at high speed. Accordingly, this paper proposes the need for a balance between accuracy and speed through comparison between the existing method and the model with the encoder changed in UNet and DeepLabV3+.

2 Related Works

2.1 Models

Segmentation Models. Ronneberger et al. [5] is a fully-convolutional-based model of the end-to-end method proposed for the purpose of segmentation and devised a **UNet** model to handle images in the medical field. UNet is a U-shaped model composed of an encoder that reduces the image size and a decoder that increases the image size. UNet has the advantage of using context information well, localizing accurately, and having high speed, but to know the optimal depth of a model that fits a dataset, it requires costly finding or inefficient work of ensemble models. In the case of skip connection, it also has a disadvantage of a limited structure in which only an encoder and a decoder having the same depth are connected [6]. Chen et al. [7] proposes **DeepLab-V3+** as a model to solve the semantic segmentation problem. DeepLab-V3+ is a structured model of Atrous Convolution and Depthwise Separable Convolution with encoder and decoder. DeepLab-V3+ performs calculations through the structure of Atrous Convolution and Xception in the encoder. Atrous convolution performs computation with an empty space inside the convolution filter, so the amount of computation is the same as that of the existing convolution, but it prevents the reduction of detail information by increasing the size of the area that can be seen by one pixel. In addition, in the case of the Xception model, the previous deeplab pooling layer is replaced by a depthwise separable convolution, which acts as an advantage to greatly reduce the cost of the entire model.

Lightweight Deep Neural Networks. Many deep learning models sacrifice time and space complexity to improve performance. This is required not only in the learning process but also in the inference process and causes an increase in latency or a problem in which the model cannot be executed. To solve this problem, several studies have introduced a model focusing on the lightweight model.Howard et al. [8] proposes **MobileNetV3**, a lightweight model while improving performance compared to the previous MobileNetV2 model. In this model, a building block with Squeeze-and-Excite added to the existing bottleneck structure, platform-aware NAS and NetAdapt to find an efficient network structure are applied, and the layer that requires a high amount of computation is redesigned. As a result, it has achieved a significant improvement in performance and latency compared to the previous model. Tan et al. [9] proposes **EfficientNet**, which finds the optimal combination between the depth, width, and resolution of the model through a study on model scaling, and lightens the parameters of the model. In this study, it was confirmed that the increase in accuracy obtained from the increase in depth, width, and resolution of the network rapidly decreased from a certain moment. Based on this, in this paper, we proposed a technique that can optimally balance depth, width, and resolution called compound scaling. As a result, EfficientNet with compound scaling applied to MnasNet shows higher accuracy and significantly improved latency than previous high-complexity models.

3 Method

Network Architecture. In this study, pre-trained neural networks are used as encoders for segmentation models. The pre-trained neural network takes colonoscopy images as input, and it is possible to extract feature maps for the input to the intermediate and final layers of the neural networks. These results replace the encoder's feature map that the existing segmentation model had, and it is used as skip-connection and encoder output. The decoder maintains the structure of the existing segmentation model. In detail, this model uses the structure of UNet or DeepLabV3+ without modification and trains the decoder that receives feature maps and results from the encoder composed of a pre-trained neural network to generate a mask that is as similar as possible to the segmentation masks of the training dataset.

Encoder. In this study, a model whose number of parameters does not exceed 10M was selected. Accordingly, MobileNetV3, EfficientNet, and RegNet models pre-trained with the ImageNet dataset were selected as encoders, and the details are listed in Table 1. The selected models act as encoders in UNet and DeepLabV3+ respectively. In the case of UNet, the feature map of the encoder composed of five layers is replaced with the feature map of the pre-training model of the same size. Meanwhile, the encoder of DeepLabV3+ consists of Deep Convolution Neural Network (DCNN)+ Atrous Spatial Pyramid Pooling (ASPP). In detail, DCNN plays a role of compressing information of input data, and ASPP plays a role of extracting feature maps of various resolutions by inputting DCNN result values.

Decoder. The decoder used in this study uses the UNet and DeepLabV3+ models as they are. The decoder of UNet consists of 5 layers, and each layer compresses the channel through a convolution block, and through transposed convolution between layers and skip

Table 1. Details of the models selected as encoders. For Top1-acc, it is based on the ImageNet dataset.

Models	Detailed Name	Parameters	Top1-acc
MobileNetV3	MobileNetV3-L 0.75	4.0M	73.3%
EfficientNet	EfficientNet-B3 (lite)	8.0M	81.6%
RegNet	RegNetY-1.6GF	10.0M	76.3%

connection with the encoder, it is possible to detect the polyp area using both the context information and local information of the image. On the other hand, DeepLabV3+ has a simplified UNet-style decoder. The DCNN output plays the role of low-level features, and in the case of ASPP, it is concatenated with the low-level feature map after 1×1 convolution and upsampling. After that, a segmentation mask with the same size as the original image is generated through additional convolution and upsampling.

4 Experimental Results

Dataset and Comparison Models. The dataset used in this experiment is the dataset used in the paper of Fan et al. [10], combining the five polyp segmentation datasets: CVC-ColonDB/CVC-612 [11], CVC-Clinic DB [12], EndoScene [13], ETIS [14], and Kvasir [15]. 90% of this dataset is used as data for training, and the remaining 10% is used as a test dataset. In this study, the model is trained with the above dataset, and the test is conducted on CVC-ColonDB and Kvasir.

The models to be compared in this study are DoubleUNet and MSRFNet, which have higher accuracy and lower fps than UNet and DeepLabV3 +. Specially, MSRFNet is a SOTA-level model with a Dice score of about 94% in CVC-ColonDB.

Loss Function. Binary Cross Entropy (BCE) loss and Dice loss are used as loss functions for training the segmentation model in this study. The BCE loss calculates the pixel-wise loss to increase the accuracy of each pixel, and the dice loss calculates the loss of the entire image to measure the similarity between masks. Both the BCE loss and the dice loss have scales from 0 to 1, and as both losses approach 0, it means that the similarity to ground-truth is high. The following equation is the equation of the binary-loss function for ground-truth mask y and prediction mask y^.

$$Loss = Loss_{BCE} + Loss_{Dice} \tag{1}$$

$$Loss_{BCE} = (y - 1)\log(1 - y^{\wedge}) - y \log y^{\wedge} \tag{2}$$

$$Loss_{Dice} = 1 - (2yy^{\wedge} + 1)/(y + y^{\wedge} + 1) \tag{3}$$

Training Settings and Metrics. In this study, learning and testing were conducted with Pytorch as the backend, and RTX 3060 12 GB was used as the graphics card used in the entire process. The validation dataset is further divided, and training is carried out in a ratio of training: validation: test = 8: 1: 1. For model training, the image size was resized to 288 × 384, which is the size of CVC-ClinicDB. In addition, the optimizer used for learning was AdamW, and the learning rate was set to 1e−3, epoch was set to 1000, and patients for early stopping were set to 50. All models used for the encoder were trained with ImageNet, and the data augmentation used for training is as follows: Gaussian Noise (p = 0.3), HorizontalFlip (p = 0.3), VerticalFlip (p = 0.3), CoarseDropout(max_holes = 8, max_height = 10, max_width = 10, fill_value = 0, p = 0.2).

Results. The experimental results are shown in Table 2.

Table 2. Comparison of experimental model and conventional model.

Method	DICE(%)	Parameters(M)	FPS
DoubleUNet	92.72	29.29M	2.39
MSRF-Net	94.20	18.38M	4.09
UNet (MobileNetV3)	92.45	18.64M	18.05
UNet (EfficientNet)	90.86	33.88M	17.57
UNet (RegNet)	89.51	55.27M	9.79
DeepLabV3 + (MobileNetV3)	92.23	6.18M	18.61
DeepLabV3 + (EfficientNet)	92.73	28.25M	16.92
DeepLabV3 + (RegNet)	92.05	45.61M	11.54

5 Discussion

In this paper, we compared the performance and processing speed of the model in which the encoder was changed to segment the polyp quickly and accurately through the neural network and the existing models. All models showed satisfactory performance, but DoubleUNet and MSRFNet models showed processing speed not available in real time. On the other hand, the model that changed the encoder claims that although the detail of the prediction mask is insufficient, it is not lacking in predicting the location of the polyp and shows a reasonable fps for real-time use. Through this study, we suggest that a balance between accuracy and processing speed is necessary for actual field use. In the future, we plan to conduct research with high speed and high accuracy through new model design.

Acknowledgments. This research was supported by the big data analysis and utilization support project for IoT home appliance and smart home (No. 3160–309) of the Ministry of Trade, Industry and Energy of the Republic of Korea.

References

1. National Cancer Information Center, Korea. www.cancer.go, Accessed 24 Apr 2017
2. Ahn, S.B., et al.: The miss rate for colorectal adenoma determined by quality-adjusted, back-to-back colonoscopies. Gut Liver **6**(1), 64 (2012)
3. Long, J., Shelhamer, E., Darrell, T.: Fully convolutional networks for semantic segmentation. In: Proceedings of the IEEE Conference on Computer Vision and Pattern Recognition, pp. 3431–3440 (2015)
4. Srivastava, A., et al.: Msrf-net: A multi-scale residual fusion network for biomedical image segmentation. IEEE J. Biomed. Health Inf. **26**(5), 2252–2263 (2021)
5. Ronneberger, O., Fischer, P., Brox, T.: U-net: convolutional networks for biomedical image segmentation. In: International Conference on Medical Image Computing and Computer-Assisted Intervention, pp. 234–241. Springer, Cham (2015)
6. Zhou, Z., et al.: Unet++: redesigning skip connections to exploit multiscale features in image segmentation. IEEE Trans. Med. Imaging **39**(6), 1856–1867 (2019)
7. Chen, L.-C., Zhu, Y., Papandreou, G., Schroff, F., Adam, H.: Encoder-decoder with atrous separable convolution for semantic image segmentation. In: Ferrari, V., Hebert, M., Sminchisescu, C., Weiss, Y. (eds.) ECCV 2018. LNCS, vol. 11211, pp. 833–851. Springer, Cham (2018). https://doi.org/10.1007/978-3-030-01234-2_49
8. Howard, A., et al.: Searching for mobilenetv3. In: Proceedings of the IEEE/CVF International Conference on Computer Vision, pp. 1314–1324 (2019)
9. Tan, M., Le, Q.: Efficientnet: rethinking model scaling for convolutional neural networks. In: International Conference on Machine Learning, pp. 6105–6114. PMLR (2019)
10. Fan, D.-P., et al.: Pranet: parallel reverse attention network for polyp segmentation. In: International Conference on Medical Image Computing and Computer-Assisted Intervention., pp. 253–273. Springer, Cham (2020)
11. Bernal, J., et al.: WM-DOVA maps for accurate polyp highlighting in colonoscopy: validation vs. saliency maps from physicians. Comput. Med. Imaging Graph. **43**, 99–111 (2015)
12. Tajbakhsh, N., Gurudu, S.R., Liang, J.: Automated polyp detection in colonoscopy videos using shape and context information. IEEE Trans. Med. Imaging **35**(2), 630–644 (2015)
13. Vázquez, D., et al.: A benchmark for endoluminal scene segmentation of colonoscopy images. J. Healthcare Eng. **2017** (2017)
14. Silva, J., et al.: Toward embedded detection of polyps in wce images for early diagnosis of colorectal cancer. Int. J. Comput. Assist. Radiol. Surg. **9**(2), 283–293 (2014)
15. Jha, D., et al.: Kvasir-seg: a segmented polyp dataset. In: International Conference on Multimedia Modeling, pp. 451–462. Springer, Cham (2020)

Custom Object Segmentation by Training R-CNN

Javlon Tursunov[1(✉)], Aziza Narimonova[2], Hamroev Bekzod[2], Djabbarov Bakhtiyor[2], Rustam Rakhmonov[2], and Shakhzod Bobokulov[2]

[1] Tashkent University of Information Technologies Named After Muhammad Al-Khwarizmi, Amir Temur Avenue 108, Tashkent, Uzbekistan
javlontursunov0817@gmail.com

[2] Karshi Branch of TUIT Named After Muhammad Al-Khwarizmi, Karshi-Beshkent 3 Km, Qarshi, Uzbekistan

Abstract. Segmentation of objects in an image is considered crucial as it helps to work with or focus on a specific area of the image. For instance, there is a big image or video which consists of frames and whose resolution is big, making it difficult to work with it since it is computationally heavy to work with all the parts and that is where the image segmentation comes in. So far, various methodologies have been proposed to make this job easier, thus requiring minimum resources to carry out computation. However, there are some shortcomings of the existing works that have been carried out by now. For instance, algorithms and models which are based on the neural networks have some disadvantages such as training them to detect and carry out segmentation. Additionally, they all make use of the common datasets and one of which is COCO dataset which is used by the existing CNN-based algorithms which include only commonly used objects, limiting their capabilities when it comes to detecting and segmenting objects which are not included in those datasets. The proposed method in this work carries out image segmentation by training R-CNN to segment custom objects which are not found in the common datasets, thus making a contribution to image segmentation. In order to train the R-CNN to segment custom objects, overall 1343 images were used in the training process and the obtained result is quite impressive when evaluating the performance of the model.

Keywords: CNN · R-CNN · Training · Accuracy · Semantic segmentation · Instance segmentation · Loss · Annotation · Class

1 Introduction

Segmentation of images into areas according to class objects has been becoming a buzzword in the area of computer vision since it is vital in terms of facilitating the image processing [7]. When it comes to object segmentation in an image, humans outperform machines which, in turn, makes humans push the boundaries of computer vision to compete with humans when carrying out image segmentation, objects detection and objet localization in real time environment. Objects are classified and segmented according to

H. Zaynidinov et al. (Eds.): IHCI 2022, LNCS 13741, pp. 565–574, 2023.
https://doi.org/10.1007/978-3-031-27199-1_58

the classes. Applications of image segmentation is increasing day by day since it helps humans in the tasks which are very difficult to perform.

Since image segmentation is a vital component for visual systems, it requires the images to be partitioned into various objects and segments, thus having a major role to play in a wide range of applications [5] like analysis in the medical images with which, boundaries and size of tumors and cancers can be measured [15], autonomous vehicles for which predicting the suitable path is important and video surveillance which is mainly used in the areas where there is a lot going on. Additionally, it has another applications like detection of road signs [16], colon crypts segmentation [17], and classification of land use and land cover [18].

From the medical perspective, medical image segmentation is considered to be crucial in the field of computer vision as it is extremely helpful in segmenting the exactly needed part in medical images. This, in turn, helps doctors to make the right decision on the case that is being explored without having to spend more time to detect diseased areas. When carrying out image segmentation, there are two types of segmentations: semantic segmentation and instance segmentation. Semantic segmentation sometimes is called pixel-level classification which involves clustering pixels in the image depending on the object of a class [6]. Instance segmentation doesn't make much difference from the semantic segmentation and it just uses different labels for separate instances of objects which belong to the same class. For this reason, instance segmentation can be regarded a task which involves semantic segmentation and immediate solution to detect objects [8]. In other works that have been carried out so far, some different methodologies have been put forward for image segmentation and in the next section called related work, some of those works have been analyzed so as to determine research gap and from this, what is obvious is that making custom model using R-CNN to segment custom objects in the images has not attracted much attention. The methodology section clearly describes what has been done in this work, what kind of specific technique has been used which makes this work distinguishable from other existing works and everything is explained step by step according to the flow chart.

2 Related Work

In order to take image segmentation to the next level, different approaches are being applied and one of them is to use convolutional neural networks. A Convolutional Neural Network (CNN) is a type of artificial neural network which imitates the neural network in the human brain. CNN is commonly a deep learning algorithm that has inputs and can assign importance (biases and learnable parameters) to different aspects or objects which are included in the image and this algorithm can make differentiation among objects [4].

Image segmentation based on monochrome color makes use of either dramatic change in the color intensity of grayscale image or discontinuity of the pixels with the same value intensity. The method which works based on the discontinuity of the pixels divides the images into regions for which, the isolated lines, points and edges are distinguishing local features which can occur when the pixel intensity changes dramatically. Some research papers on image segmentation which are based on the monochrome color covers the most of the available techniques regarding image segmentation [9–11].

According to K.S. Fu et al. [10], image segmentation should be discussed from the point of image processing based on cytology. With the help of that work, all the existing works were categorized into three different classes and they are: a) edge detection, b) region extraction, c) clustering or characteristic feature thresholding. Advantages and disadvantages of each segmentation technique were summarized and some comments were made regarding each approach. Schemes used to choose threshold based on gray level histogram, local features and other techniques based on syntactic techniques were described.

There are also different methods presented which are categorized into two distinctive classes: sequential and parallel techniques [1, 2]. The technique which is based on the parallel edge detection means that whether the set of points are on the edge or not is determined according to the grayscale level and intensity of neighboring pixels [3, 12].

Despite the fact that image segmentation is perceived differently by people, it can be regarded as a prediction which can be made depending on the exiting objects in the image [13, 14].

Accuracy of image segmentation is increasing since the advent and adaptation of DNN (Deep Neural Network) into this field. Prior to the usage of DNN, all the other existing methods were called conventional methods. Before DNN started being used in the field of computer vision, features and classification techniques were regarded important for image segmentation. Feature in the field of computer vision is considered a necessary bit of information relevant to tackle computational tasks. A wide variety of approaches and features are utilized for image segmentation such as color of an each pixel, oriented gradient histogram [19], Scale-invariant feature transform (SIFT) [20], Local Binary Pattern(LBP) [21], SURF [22], Harris Corners [28], Shi-Tomasi [23], Sub-pixel Corner [24], SUSAN [27], Features from Accelerated Segment Test (FAST) [26], FAST- ER [25], AGAST [29] and Multi-scale AGAST Detector [30] and list goes on and on. Methods used in image segmentation are based on two types of segmentations which are unsupervised and supervised. Specifically, the simplest one is based on thresholding which is used for only grayscale images.

Apart from the traditional methods, with the help of the methods based on the CNN and DNN, image segmentation made a huge progress in the field of computer vision.

Artificial Neural Network (ANN) was the result of inspiration from the neural network from the human brain. The basic and main element of ANN is a neuron. Every single neuron differs from each other due to the weighted inputs and those weights are summed up. Artificial neural network has functions called transfer and activation following the input layer. Depending on the structure of the neural network, it can perform a wide variety of tasks and sometimes this neural network is considered complicated with many processing layers, resulting in the DNN.

Neural networks have also the feature of invariance in shift and space based on their architecture that is shared-weight and translation characteristics which are invariant. Because of this remarkable structure, CNN has demonstrated a quite outstanding results in terms of image segmentation, classification and object detection.

3 Methodology

Since the approach used in this work involves using the R-CNN which is based on the CNN (Convolutional Neural Networks), it requires training the model. Unlike any other works which can detect and segment objects which are included in the commonly used datasets like COCO, the model developed in this work can detect and segment the objects which has not been used for artificial neural network training. There are three steps to follow in this work which are shown in the Fig. 1.

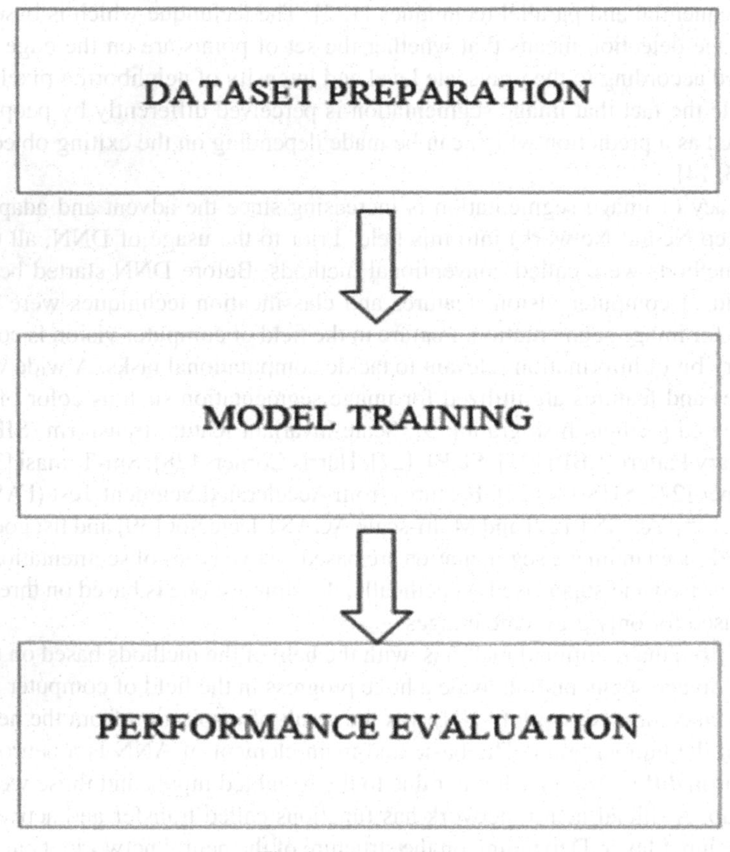

Fig. 1. Steps to follow.

In this work, the custom object is a glass bottle including some other glass object or they could be other objects classified as a glass.

A) In the first step, images with glass bottles are needed and all the images with the glass custom objects were downloaded from the Kaggle dataset which is an open dataset with a wide variety of images for machine learning and model training. Since supervised learning is made use of in this work, it requires the data to be labelled when training the neural network. In machine learning, data annotation is the process of labeling data to show the outcome you want your machine learning model to predict. This process sometimes involves labeling, tagging, transcribing, or processing a dataset with the features you want your machine learning system to learn to recognize. Annotation simply tells the model which features it should focus on while getting trained and of course, when the model put into test in the absence of labels, it can segment objects based on features provided by the annotation and based on those annotation model learns which part should be considered as a reference point in the still image. In order to annotate or label the images, the website called makesense.ai [31] was used and when annotating the images the polygon shape was chosen as it provides more accurate location of objects in the image. Below, the Fig. 2 shows how images were annotated and in the case of this work and there is only one class which is glass. Class is a basically any object which is categorized into one group, distinguishing the one object from the other. For instance, all the bottles are regarded as one type of class and if there were other objects like box or any animal, they would be regarded as a another class in this case. There could have been multiple classes when carrying out annotations but since only a glass needs to be detected and segmented, there is no need for other classes to be detected. In this work, glass is considered one class and since there is no another object classified as another class objcect, we do not need to worry about labeling for other objects.

Fig. 2. Annotations

For annotations, there are a bunch of available tools which can be used when carrying out tasks related to supervised learning. In annotation, extra file is generated which is used along with the raw data to train the model initially.

Once annotation process has finished, all the annotations have been saved in the file with.json extension and for the training, Google Colaboratory was used since it provides free GPU and platform to run python code. During the development of the model, overall 1343 and 200 images were used for the training and validation respectively before the training has started. Figure 3 shows the ground truth depending on the annotations. Ground truth is a metric used to measure how accurately the model is carrying out segmentation. Precisely, ground truth is the perfect segmented part from the image and it can be compared to other segmentations so as to evaluate how model is performing.

B) The training process is time-consuming which continued approximately for 2 h for 5 epochs. What basically epoch means is how many times all the images contained in the dataset are passed through the neural network during the training process. For instance, if dataset contains 200 images, these 200 images should be passed through the neural network for a single epoch during the training process. In this training process, learning rate was 0.001 and in spite of small rate, achieved result is quite remarkable in terms of ground truth which can be inspected visually. In the above Fig. 3, we can see that since there is only one class, only one box is filled with the ground truth and if there had been other classes, they would have been classified as other instances of that second class and illustrated in the other box.

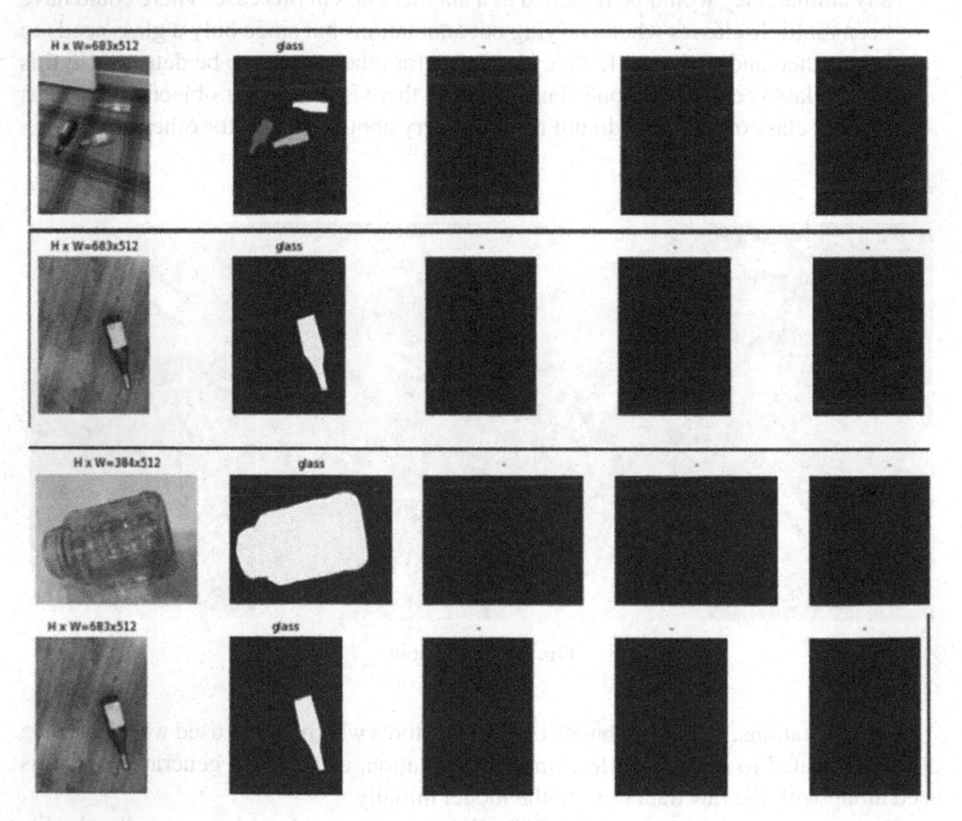

Fig. 3. The ground truth

Ground truth helps to see how well the model is performing. Besides the epoch, there is another factor to keep in mind which is iterations that is the number of times showing how many times weights are updated in the neural network during each epoch.

The training was done with 5 epochs and 500 iterations in each epoch. Below is shown Fig. 4 which illustrates how the accuracy is increasing with the loss decreasing. Loss is calculated by comparing the predicted segmentation with ground truth. Obviously, the more model is trained, the better it performs, thus meaning the decrease of loss and increase of accuracy.

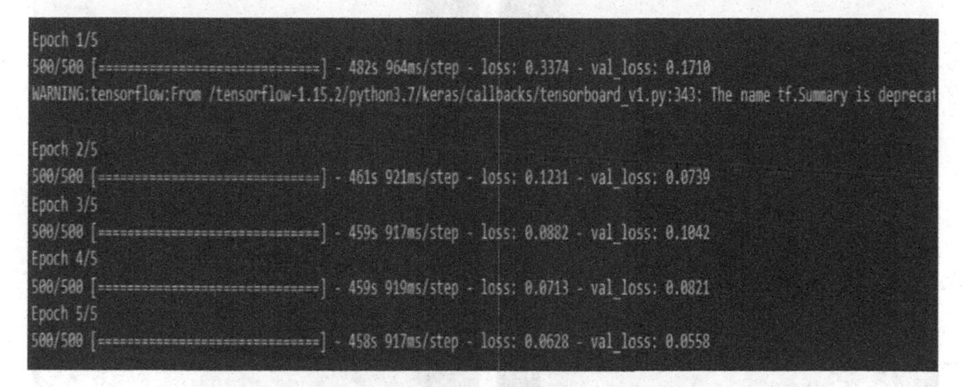

Fig. 4. Training process with loss decreasing.

Since the accuracy is inversely proportional to the loss, meaning that more loss decreases, the more accurately the model classifies the pixels.

C) Once the training is done after all the epochs have been completed, in order to check whether the model is localizing the object accurately in the image with the mask, random images from the validation dataset were shown to model. Some of the images in the Fig. 5 can be seen below while putting the model to the test.

As it can be seen, images from the validation dataset, which were not used in the training process, are being used to check the performance of the model. By making visual comparison among the segmented section by the model and ground truth table made by annotations, it can be said that model is doing pretty well in terms of segmentation of the bottles the images. Besides, the confidence level is high enough to detect other bottles which are new to the model, demonstrating the accuracy of the model with regard to detection, localization and segmentations. On the left side, accuracy level predicted by the model can be seen and it is over 90%, making this model reliable in terms of training it for the detection and segmentation of the any object which a user wants to detect and work on. Given the fact that the dataset is not huge, the over 90% accuracy is quite remarkable. These images were not used in the training process, thus helping to validate the accuracy of the model.

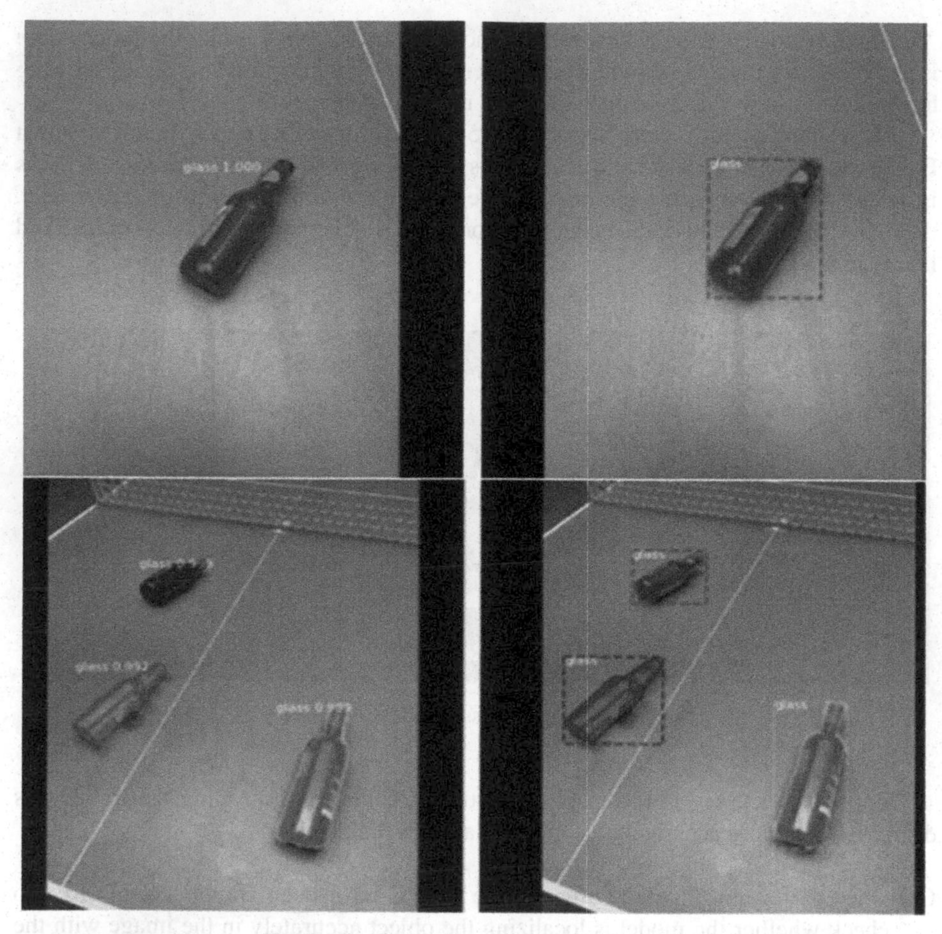

Fig. 5. Checking the performance of model.

4 Conclusion

Segmentation of a particular object in an image is of importance in terms of helping to find and segment that object easily, making the work easier when there are a bunch of unwanted objects surrounding the object of interest. There are some situations where the object of interest is not included in the common datasets which are used for training neural networks, making it challenging for people to detect and segment that object with naked eye with the help of technology. To elaborate on that, algorithms based on the convolutional neural networks make use of datasets used in the training process of the model initially. Pre-trained models strongly rely on the supervised learning, involving the dataset and annotations. In some situations, let's assume we want to detect and segment certain object (car, ball, apple and etc.). But some problems may occur if we want detect and segment object which is not included in the dataset used for pre-trained models and this is where this work that has been done can be helpful. We can train our model which is not pre-trained on any dataset, allowing to make customized model. This research makes a contribution in terms of paving the way to develop a custom model with the CNN-based algorithms to detect and segment objects. To give a practical example of what is meant, there are a bunch of objects on the conveyor and one may want to detect and segment object of interest which has not been detected or segmented by now with the help of the computer vision and that's where this model comes in handy with an extra pair of eyes. Despite the fact that this developed model is still in its infancy, it is performing well to detect and generate mask for the trained object and it has a good potential to be developed to detect and segment object of other classes. As a proof of the achieved result, if we have a look at the Fig. 5, we can see that the trained model in this work is able to detect object (bottle and glass in our case) with more than 90% accuracy and this accuracy rate in terms of segmentation and detection can be increased further if the model is subject to vigorous training. The most important novelty of this research is its contribution to the image segmentation with CNN and supervised learning. The difference that distinguishes this work from other previous works is that other models used in image segmentation were already trained to detect most commonly used objects, with their limitations being obvious. However, what has been done in this work allows to segment almost any object in the still image and video stream as long as the proposed model goes through proper training process.

References

1. Rosenfeld, A., Thurston, M.: Edge and curve detection for visual scene analysis. IEEE Trans. Comput. **C-20**, 562–569 (1971)
2. Rosenfeld, A., Thurston, M., Lee, Y.: Edge and curve detection further experiments. IEEE Trans. Comput. **C-21**, 677–715 (1972)
3. Hueckel, M.: An operator which locates edges in digitized pictures. J. Assoc. Comput. Mach. **18**(1), 113–125 (1971)
4. Gu, J., et al.: Recent advances in convolutional neural networks. Pattern Recogn. **77**, 354–377 (2018)
5. Forsyth, D., Ponce, J.: Computer Vision: A Modern Approach. Prentice Hall, Upper Saddle River (2002)
6. Thoma, M.: A survey of semantic segmentation. arXiv preprint arXiv:1602.06541 VOC2010 preliminary results (2016). http://host.robots.ox.ac.uk/pascal/VOC/voc2010/results/index.html
7. Rosenfeld, A., Kak, A.C.: Digital Picture Processing. Academic Press, Cambridge (1976)
8. Chen, L., Hermans, A., Papandreou, G., Schroff, F., Wang, P., Adam, H.: MaskLab: instance segmentation by refining object detection with semantic and direction features. In: 2018 IEEE/CVF conference on computer vision and pattern recognition, 18–23 June 2018, pp. 4013–4022 (2018). https://doi.org/10.1109/cvpr.2018.00422
9. Pal, S.K., et al.: A review on image segmentation techniques. Pattern Recogn. **29**, 1277–1294 (1993)
10. Fu, K.S., Mui, J.K.: A survey on image segmentation. Pattern Recogn. **13**, 3–16 (1981)
11. Haralick, R.M., Shapiro, L.G.: Image segmentation techniques. Comput. Vision Graph. Image Process. **29**, 100–132 (1985)
12. Hueckel, M.: A local visual operator which recognizes edges and lines. J. Assoc. Comput. Mach. **20**(4), 634–647 (1973)
13. Lin, T.Y., et al.: Microsoft coco: common objects in context. arXiv preprint arXiv:1405.0312 (2014)

14. Li, Y., Qi, H., Dai, J, Ji, X., Wei, Y.: Fully convolutional instance-aware semantic segmentation. In: Computer Vision and Pattern Recognition (CVPR), pp 4438–4446. IEEE (2017)
15. Moon, N., Bullitt, E., van Leemput, K., Gerig, G.: Automatic brain and tumor segmentation. In: Dohi, T., Kikinis, R. (eds.) MICCAI 2002. LNCS, vol. 2488, pp. 372–379. Springer, Heidelberg (2002). https://doi.org/10.1007/3-540-45786-0_46
16. Maldonado-Bascon, S., Lafuente-Arroyo, S., Gil-Jimenez, P., Gomez-Moreno, H., López-Ferreras, F.: Road sign detection and recognition based on support vector machines. IEEE Trans. Intell. Transp. Syst. **8**(2), 264–278 (2007)
17. Cohen, A., Rivlin, E., Shimshoni, I., Sabo, E.: Memory based active contour algorithm using pixel-level classified images for colon crypt segmentation. Comput. Med. Imaging Graph **43**, 150–216 (2015)
18. Huang, C., DavisL, T.J.: An assessment of support vector machines for land cover classification. Int. J. Remote Sens. **23**(4), 725–749 (2002)
19. Bourdev, L., Maji, S., Brox, T., Malik, J.: Detecting people using mutually consistent poselet activations. In: Daniilidis, K., Maragos, P., Paragios, N. (eds.) ECCV 2010. LNCS, vol. 6316, pp. 168–181. Springer, Heidelberg (2010). https://doi.org/10.1007/978-3-642-15567-3_13
20. Lowe, D.G.: Distinctive image features from scale-invariant keypoints. Int. J. Comput. Vis. **60**(2), 91–110 (2004). https://doi.org/10.1023/B:VISI.0000029664.99615.94
21. He, D.C., Wang, L.: Texture unit, texture spectrum, and texture analysis. IEEE Trans. Geosci. Remote Sens. **28**(4), 509–512 (1990)
22. Bay, H., Ess, A., Tuytelaars, T., Van Gool, L.: Speeded-up robust features (surf). Comput. Vis. Image Underst. **110**(3), 346–359 (2008)
23. Shi, J., et al.: Good features to track. In: Proceedings of the 1994 IEEE Computer Society Conference on CVPR'94 Computer Vision and Pattern Recognition, pp. 593–600. IEEE (1994)
24. Medioni, G., Yasumoto, Y.: Corner detection and curve representation using cubic b-splines. Comput. Vis. Graph Image Process **39**(3), 267–278 (1987)
25. Rosten, E., Porter, R., Drummond, T.: Faster and better: a machine learning approach to corner detection. IEEE Trans. Pattern Anal. Mach. Intell. **32**(1), 105–119 (2010)
26. Rosten, E., Drummond, T.: Fusing points and lines for high performance tracking. In: Proceedings of the Tenth IEEE International Conference on Computer Vision. ICCV 2005, vol 2, pp 1508–1515. IEEE (2005)
27. Smith, S.M., Brady, J.M.: Susana new approach to low level image processing. Int. J. Comput. Vis. **23**(1), 45–78 (1997). https://doi.org/10.1023/A:1007963824710
28. Derpanis, K.G.: The Harris Corner Detector. York University, Toronto (2004)
29. Mair, E., Hager, G.D., Burschka, D., Suppa, M., Hirzinger, G.: Adaptive and generic corner detection based on the accelerated segment test. In: Daniilidis, K., Maragos, P., Paragios, N. (eds.) ECCV 2010. LNCS, vol. 6312, pp. 183–196. Springer, Heidelberg (2010). https://doi.org/10.1007/978-3-642-15552-9_14
30. Leutenegger, S., Chli, M., Siegwart, R.Y.: Brisk: binary robust invariant scalable keypoints. In: 2011 IEEE international conference on computer vision (ICCV), pp 2548–2555. IEEE (2011)
31. www.makesense.ai

Optimization of Fractal Structure Pattern Colors in Carpet Design Using Genetic Algorithm

Fakhriddin Nuraliev, Oybek Narzulloyev, and Saida Tastanova(✉)

Tashkent University of Information Technologies Named after Muhammad Al-Khwarizimi, Amir Temur Street 108, Tashkent 100084, Uzbekistan

tasmaat@mail.ru

Abstract. A genetic algorithm is a method of responding to problems that must be used for trial and error that cannot be solved by traditional methods. If there are a lot of parameters, it is difficult to evaluate all the answers. A genetic algorithm can contain a large number of responses and find the most optimal solution by getting feedback from problems. The complex image of the Fractal in the system can be performed several designs of different colors in the design of the carpet. However, many patterns may not have attractiveness and beauty. According to the customer, it is not easy to choose interesting and stylish models from many designs. Having received the necessary feedback from the user, the genetic algorithm can be used to optimize the design and select the perfect patterns. With the help of a genetic algorithm, it is possible to optimize colors by price and select perfect designs taking into account the necessary feedback from the user, which we can use to optimize color in the production of complex image carpets using the proposed algorithm.

Keywords: Genetic algorithm · Fractal · Pascal triangle · Fractal · Mutation · Optimization · Fitness function

1 Introduction

The concept of design is different from the scientific research that we sometimes analyze, design is a design that consists not only of studying the essence of what exists, but of creating innovation. When designing carpet designs in the carpet industry, the choice of color is of great importance.

General customization and automation in the carpet industry, this method of creating a bar mat for the buyer is considered in quality. The genetic algorithm is the direction of optimizing those stimulated by an environment that works in addition, such as interrelated; mutation and the survival of the most suitable one's work for machine learning. This method is widely used in optimization and classification work.

This method was also considered in the textile industry. This article will discuss the application of genetic algorithms in the textile industry. The genetic algorithm allows

H. Zaynidinov et al. (Eds.): IHCI 2022, LNCS 13741, pp. 575–585, 2023.
https://doi.org/10.1007/978-3-031-27199-1_59

you to contain a large number of answers and select perfect designs from them by receiving feedback from assignments. However, many patterns are unpleasant and may not be suitable for customer demand. According to the customer, it is not easy to choose interesting and stylish models from such a variety of designs. To find the optimal price, an interactive genetic algorithm can be used to optimize colors and select ideal patterns based on the user's necessary feedback. We can use it to optimize color in the manufacture of carpets of complex designs [1]. Attempts to eliminate these problems led to the creation of a theory of genetic algorithms. The founder of the theory of genetic algorithms is rightfully considered the American researcher John Holland, who in the late 1960s, it proposed the use of methods and models of the mechanism for the development of the organic world on earth as the principles of combinatorial registration of options for solving optimization problems. In 1975, John Holland published his most famous work "adaptation in natural and artificial systems" [3]. In it, he first introduced the term" genetic algorithm". John Holland's Students Kenneth De Yong and David Goldberg continued his career in the field of genetic algorithm. Goldberg's most famous work is "search optimization and genetic algorithms. in machine learning" [4]. In the carpet industry, several carpet designs of different colors can be created in a complex fractal image system. However, given the time, many patterns may not be attractive and beautiful enough, and choosing patterns among a wide range of customer tastes can be a big problem. To solve such a problem, an optimization and design scheme based on artificial intelligence can be used. This method develops a design system that is consistent with the requirements of customers and is based on market demand, which gradually connects with users, collects feedback and gives optimal results for users. In the carpet weaving system, several designs of different colors can be made. Also, many patterns may not be unpleasant and in demand. According to the buyer, it is not easy to choose interesting and stylish models from such a variety of designs. With the help of an interactive genetic algorithm, it is possible to optimize colors by price and select ideal images, taking into account the necessary feedback from the user. We can use the genetic algorithm to optimize the color of fractals in carpet production with complex image carpet designs [10]. To create an optimal color selection plan, a genetic algorithm is used in the design of carpet production. The interactive genetic algorithm was also used to optimize the color detection of carpet products, and was used in a similar way to previous research databases that included initial reserves to begin the optimization process. This method used two databases; one with basic patterns and the other with colors. This work developed software to optimize the selection of paints in carpet design in the production of complex images of fractal carpets based on a genetic algorithm. As parameters of the genetic algorithm, crossover and mutation probabilities are used. They can be from 0 to 1. Thus, they are considered probability. If the operator is assigned a value of "0", it means that as a result of this operator, no individual will change in this population. Likewise, if "1" is indicated as a parameter, it means that all individuals, in other words, all populations, change. Both of these values are usually not specified, since they do not contribute to the efficiency of the algorithm. So, in this study, we will select a value from 0 to 1 for the crossover and the probability of mutation.

2 Method

The genetic algorithm is similar to an interactive genetic algorithm; however, there is a slight difference where there is primarily a fitness function (target function) and the user sees the level of compatibility for each chromosome. The interactive genetic algorithm can communicate with users and therefore affect the user's emotions and is used in the arts and design fields and is used in production. In the carpet manufacturing industry, the application in color optimization is blurred in accordance with the purpose. For this, too, a program was developed by the Python programming language to express design through complex fractal images. The developed software consists of the following main section. Complex fractal image patterns and colors created by the program are displayed to the user in the main part, which is evaluated by the user. In other parts, the user can view existing carpet design samples from the carpet design sample base and create, edit or delete the desired design samples and place them in the database. The user can do the same for the colors in the software component and finally enter the knitting machine pattern and color boundaries. Using the method of binomial polynomial theory, methods and algorithms for visualizing images in fractal form have been developed, taking into account algebraic structures based on the Pascal triangle and The Theory of prime numbers based on Mod p [7]. Each pattern is shaped on the basis of a Pascal triangle, in which each component is taken into account, the number of rows indicates the number of rows and the number of columns indicates the number of rows in the pattern. For example, the first size in the Pascal triangle indicates the amount of red, the second - green, the third - blue. Genetic algorithm parameters such as population number, number of parents transferred to the next generation, intersection coefficient, mutation rate, number of generations, and time limit are shown. There is no time limit for the algorithm to work, and the chromosome population is considered as a vector, that is, all chromosomes are sent to the fitness function through a matrix, where the number of rows is equal to the number of chromosomes. The number of columns is equal to the number of chromosomes genes. Thus, the fitness function is called only once for each generation, and the speed of execution of the algorithm increases. After determining the required parameters, the task of the genetic algorithm is performed. In the recorded function, the number of genes, the upper and lower limits and previous parameters of each gene are used as inputs, and the final population of chromosomes and their assumptions are used as the final product. Since in the final population some chromosomes may be the same as other chromosomes, samples of a similar design need to be eliminated and a new population with different chromosomes formed, and the best samples are installed on the user. After optimizing the carpet design, the color is optimized in the selected carpet designs. Figure 1 presents the block scheme of the algorithm structure.

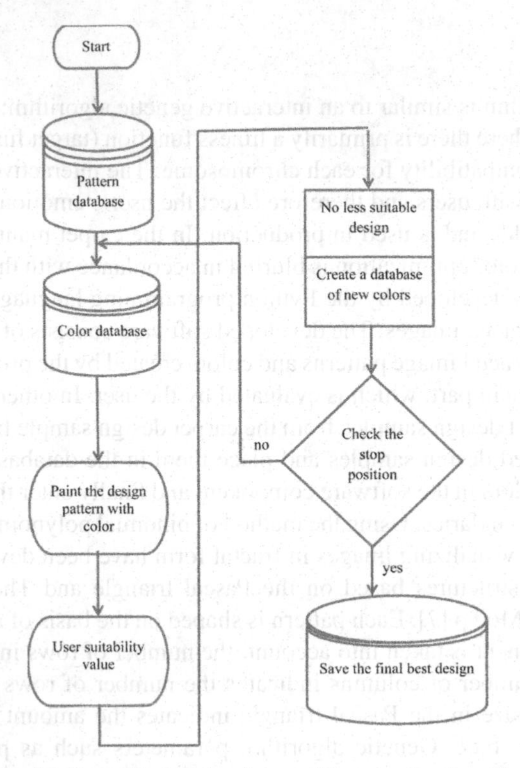

Fig. 1. Operation block scheme of the genetic algorithm

2.1 Genetic Method of Color Optimization of Fractal Structure Images

Like chromosomes, which change in nature due to changes in genes, in the genetic algorithm these elements are constantly changing, turning into full and strong populations. The size of the chromosomes and the number of genes depend on the type of problem. Genes are actually a true determinant of the variables needed to optimize the problem. The suitability of chromosomes determines their effectiveness and the function of solving the problem. Such tasks related to the tastes and feelings of people use a direct definition of fitness instead of its function by a user called an interactive genetic algorithm. The choice of color options for paints for the manufacture of a particular carpet product is a complex issue, since it is necessary, on the one hand, to ensure the established accuracy of painting, and on the other, to be able to produce carpets. The main goal is to maintain the minimum cost and at the same time the highest productivity. To solve these problems, in general, methods of mathematical programming and optimization are used.

In real problems, a connection between objective functions, criteria inevitably arises:

1. criteria can match each other;
2. criteria may contradict each other;
3. criteria can be independent.

The initial expert assessment of the selected criteria [6] allows you to solve the problem of multi-criteria optimization in the simplest, but sometimes most effective ways. The genetic algorithm [5] is based on the theoretical advances of synthetic evolutionary theory and C. Darwin uses the basic principles of evolution theory: heredity, variability, and natural selection. The genetic algorithm works with a set of individuals (population) with rows (chromosomes) that encode one of the solutions to the problem. This genetic algorithm differs from other optimization algorithms in that it only works with one solution and improves it. Each person is evaluated by a measure of his "suitability", depending on how "good" it is to solve the problem that suits him. For this, the fitness function [2] (target function) is used, which highlights the most adapted solutions (which will continue to be used further) and the worst solutions (which will be removed from the population and will not affect the search for the optimal solution). So we strive to increase fitness, and therefore approach the desired solution and approach the desired decision. The operation of the genetic algorithm is an iterative process. Each new iteration over current individuals uses different genetic operators that give birth to new individuals. After that, all individuals are evaluated using the target function, and the most suitable ones are used in the subsequent iteration of the genetic algorithm. This process continues until the desired results are achieved, or the number of iterations exceeds the limit value (the limit of the number of iterations allows you to limit the time of operation of the algorithm from above). Also, one of the symptoms that must stop the iterative process is the approach of the population (the state of the population, all its individuals have been in a certain extreme region for several generations and are almost identical). The convergence of the population usually indicates that a solution closes to the optimal one has been found. Usually, the final solution to the problem is the most adapted person of the last generation. Genetic operators are a means of showing one set to another. They allow the application of the principles of heredity and variability to virtual populations. All genetic operators have probabilistic properties, which brings a certain degree of freedom to the work of the genetic algorithm. The most commonly used genetic operators are the crossing over (cross) operator and mutation operator. The transition Operator models the process of crossing individuals. This genetic operator leads to the creation of new individuals based on existing ones.

2.2 Color Optimization

Unlike design chromosomes with the same number of genes, the number of colored chromosome genes can vary and is determined by the color variety of the selected designs. The number of genes is equal to the number of colors of the selected design. The color chromosome has an auxiliary gene that determines the number of colors, or in other words, the number of genes that are destroyed after the formation of the structure of the color chromosome. The genes of the main color chromosomes will have the desired value depending on the number of colors available in the color database. For example, if there are nineteen colors in the color database, the first gene will have a value from 1 to 19 after the auxiliary gene is excluded. The following genes cannot accept the values of the previous genes, since there may be undesirable changes in the design. In relation to the selected chromosome of the pattern, if the pattern was formed from a single pattern, the number of colors on the color chromosome is equal to the number of colors in the first pattern, if from two patterns, then the number of Colors is equal to the number of patterns. Wider color variety. After the auxiliary gene is removed, the limited color chromosomes are sent to the algorithm. The selected carpet design color chromosome and Color Matrix are sent to the coloring function of the selected design, and this feature paints the selected design in different colors and sends it to the fitness function to show the user when evaluating it. Similar to the automatic evaluation section in the design suitability function, similar colors from different generations of the selected design are included in the color suitability function. This function also imposes restrictions on color chromosome genes for painting images and edits them as needed.

3 Computational Experiment

A fractal pattern of 3 colors of a certain size was selected and the cost of painting the carpet was calculated. The number of each color in the Pascal triangle consists of three colors, for example 63, 78 and 30, here x, y, z are these unknown price natural dyes our function is blurred in the resulting drawing:

1. Constraints that task variables must satisfy:

$$x, y, z \geq 0$$

$n = 18(32, 98, \ldots\ldots)$ line

$P = 3(m)(2, 3, 5, 7\ldots\ldots)$ fuzzy numbers.

Target function of the task.

$$F = \sum_{i=1}^{p_1} x_i y_i \rightarrow \min .$$

Denote F - income from the sale of carpets, then the objective function of the problem is written as follows:

$$F = 63 * x + 78 * y + 30 * z \rightarrow \min \quad (1)$$

Thus, the task is to find $\min F = 63*x+78*y+30*z \rightarrow$ min, under the constraints:

$$11 \leq x \leq 140$$
$$11 \leq y \leq 140$$
$$11 \leq z \leq 140$$

Using the genetic algorithm to solve this optimization problem, it get the following results:

Here we used mutations similar to reproduction, a certain number of individuals are selected from mutants and changed in accordance with predetermined operations.

A mutation was carried out to improve the generation. For each color in the process of solving the problem.

$$F = 63 * x + 78 * y + 30 * z$$

The example below shows the optimization of colors using a genetic algorithm and we got some of the best options (Table 1).

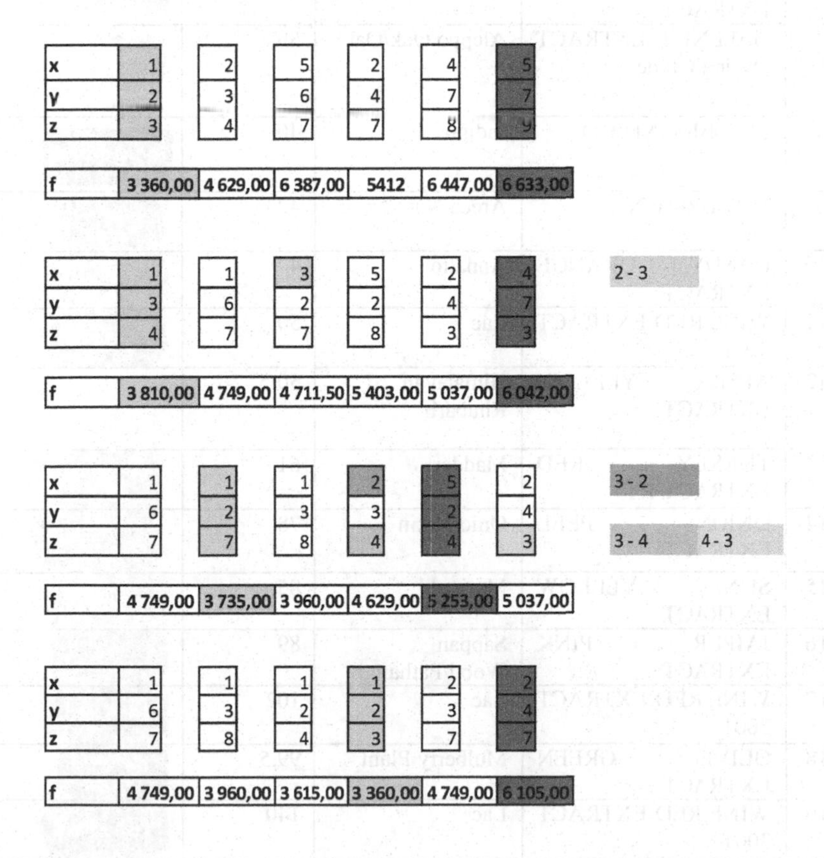

Table 1. We use the values given in

№	Product	English name/commo	Prices/Kg/ USD	Colors
1	MICHIGAN BROWN EXTRACT	Black Catechu/Kattha	11	Michigan Brown
2	CEDAR YELLOW EXTRACT	Myrobalan	24	Cedar Yellow
3	MALLOW GOLD EXTRACT	Pomegranate Peel	26.5	Mallow Gold
4	GARNET BROWN EXTRACT	Bark of Acacia	35	Garnet Brown
5	GALLNUT EXTRACT Tannin Grade	Aleppo Oak/Oak	37	Gallnut
6	CEDAR GREY EXTRACT	Myrobalan	37	
7	GALLNUT EXTRACT Dyeing Grade	Aleppo Oak/Oak	39	Gallnut
8	ESTEBIO INDIGO	Indigo	40	Indigo Blue
9	NUT BROWN	Areca Nut	42	Nut Brown
10	CANDY ORANGE EXTRACT	Annatto	47	Candy Orange
11	WINE RED EXTRACT 4001	Lac	50	Wine Red
12	APSRA YELLOW EXTRACT	Himalayan Rhubarb	50.5	
13	TURKEY RED EXTRACT RT	Madder	61	Turkey Red
14	ONION PEEL EXTRACT	Onion Skin	78	Onion Peel
15	SUN YELLOW EXTRACT	Marigold	85	
16	JAIPUR PINK EXTRACT	Sappan Wood/Pathangi	89	Jaipur Pink
17	WINE RED EXTRACT 2001	Lac	102	Wine Red
18	OLIVE GREEN EXTRACT	Mulberry Plant	99.5	Olive Green
19	WINE RED EXTRACT 3001	Lac	140	Wine Red

Based on the above formula, we Form 6 different chromosomes from three genes. We calculate each chromosome according to the formula and find the maximum and minimum number of chromosomes. We remove the maximum chromosome from the list and choose the chromosome with the minimum value as the best, and when crossing its genes we use twice and the rest one by one. In the new generation, on chromosomes where genes are duplicated, we replace the next gene from the color list. Thus, we carry out four generations of crossing. And for analysis, we calculate the sum of chromosomes in generations. As you can see in Fig. 2, the overall quality has improved, that is, it goes according to the minimum (Fig. 3).

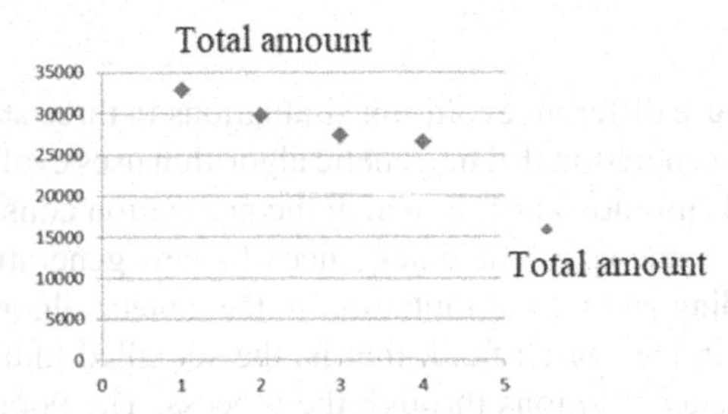

Fig. 2. The average adaptation situation of the generation and the best

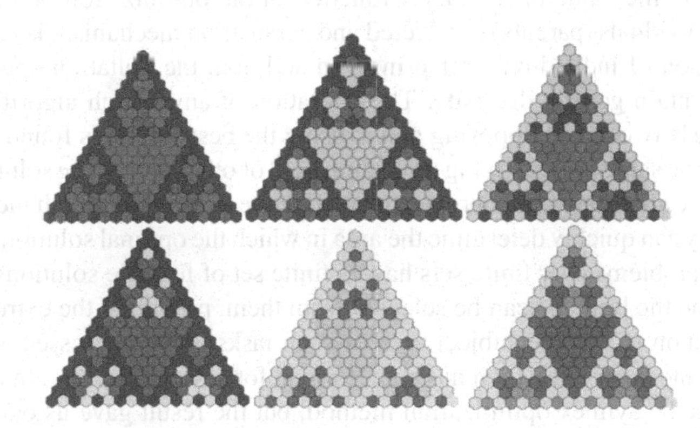

Fig. 3. Mod 3 support complex fractals based on the Pascal triangle

4 Result

Currently, there are many different models and modifications of the genetic algorithm. In test functions, an experimental comparison of all their strengths and weaknesses with each other is impossible to construct within the framework of one article. The size of the chromosomal population is 6, the number of chromosomes transmitted to the next generation is 5, there is activity and 4 generations. The first generation of patterns was created from simple patterns stored in the database to run the genetic algorithm, but they could be modified by the user. The rating of users is from 1 to 9, the more attractive the design, the higher the rating. Subsequently, the first generation was evaluated, the next generations will be created on the basis of user ratings. Selection opera is an interactive operation performed by the user, the number of patterns in the database is the same for each generation, so patterns with a low compatibility value are excluded and will not be passed on to the next generation. The best design in each generation is the one with the highest fitness value or the best fitness value in this generation, and the average fitness value is the average value of all fitness values in this generation. With the production of new generations, the average fitness will be higher, which will show the well-being of the developed design samples and good user ratings.

5 Discussion

Evolution algorithms use different evolution simulations in three stages of genes - chromosomes, human and generational. The genetic algorithm uses evolutionary modeling at the level of genes and chromosomes, in which the population consists of chromosomes with arguments of the same size. The emergence of a new generation is usually caused by chromosomal binding and partial mutation. In the genetic algorithm, the population of possible solutions in the search field, that is, the so-called individuals, the iterative evolution, becomes better solutions through the process. The population in each iteration is called the generation. Each generation assesses the physical fitness of individuals, which is usually the value of the target function in the optimization problem. Then a small set of individuals (parents) is selected and a transition mechanism is used to obtain a new generation of individuals (offspring). In addition, the mutation operator can be applied to maintain genetic diversity. The operation of any search algorithm depends on the balance between two opposing goals: using the best solutions found so far (local search) and at the same time studying the search field of other promising solutions (global search). Genetic algorithms have proven to be effective as a global search method, which means that they can quickly determine the area in which the optimal solution is available. Optimization problems over finite sets have a finite set of feasible solutions that can be enumerated and the best one can be selected from them, providing the extremum of the objective function (CF). The subject area of such tasks is the processes of operations research (IO), the theory of which has been formed for several decades. In our previous work, we used the symlex optimization method, but the result gave us one result. The simplex method has developed an algorithm for optimizing colors in the production of carpets of a given size, it is advisable to use mathematical programming methods to select the optimal set of colors, which can be done with the least cost and maximum benefit. In this paper, a genetic algorithm for optimizing the colors of the carpet industry is considered. One of the optimization methods is the genetic method. This method has also been considered in the textile and carpet industry. The genetic algorithm is a method for solving optimization problems based on the processes of natural selection (mutation, crossing, selection) and is part of a broader direction of artificial intelligence—evolutionary computing. In a genetic algorithm, each possible solution to an optimization problem is called an individual. Individuals form a population. The task is that in the process of evolution, each new generation of individuals (i.e. solutions to the optimization problem) becomes more perfect. In this work, with the help of a genetic algorithm, we got some of the best options.

6 Conclusion

This article analyzes the basic principles of the functioning of search genetic algorithms, presents the most important genetic operators, models and strategies used in the genetic algorithm. Also, a positive experience of practical application of the genetic algorithm for optimizing multi-extreme functions is considered. Carpets are important both artistically and commercially. Research shows that the initial desire to buy a carpet is based on its design pattern and color composition. Therefore, the developed algorithm has optimal capabilities for design and color optimization. The user interface is a powerful application to extract data and tastes from the user, which can adapt to different postures and experiences. This program can create more beautiful and attractive designs and colors from the user's point of view when creating new generations, and will extract different variants of carpets on demand, reducing the cost based on the customer's demand. Compatibility charts also show improvements in color and design.

References

1. Dariush, S., Mehdi, H., Hamed, A., Mohammad, Sh.: Jacquard pattern optimization in weft knitted fabrics via interactive genetic algorithm. Fashion Textiles, 1, 1–9 (2014)
2. Gladkov, L.A., Kureichik, V.V., Kureichik, V.M.: Genetic algorithms. In: FIZMATLIT, 320 p. (2006)
3. Holland, J.H.: Adaptation in Natural and Artificial Systems: An Introductory Analysis, Biology, Control, and Artificial Intelligence, 211 p. Bradford Book, Cambridge (1992)
4. Goldberg, D.: Genetic Algorithms, Optimization and Machine Learning in Search, 432 p. Addison-Wesley Professional, Boston (1989)
5. Darwin, C.: Origin of species through natural selection, 568 p. Nauga, St. Petersburg (2001)
6. Malea, C.I., Nito, E.L.: Optimization of the technological process and equipment of complex profiled parts. In: 2020 IOP Conference Series: Material Science Most, vol. 916, no. 012058, pp. 1–13 (2020)
7. Nuraliev, F.M.N.A., Narzulloev. O.M.: Mathematical and software support of fractal structures from combinatorial numbers. In: 2019 International Conference on Information Sciences and Communication Technologies (ISISST), Tashkent, Uzbekistan, pp. 1–4 (2019)

Algorithm for Digital Processing of Seismic Signals in Distributed Systems

Oybek Mallaev[1], Bunyodbek Azimov[1], Kuchkarov Muslimjon[1](✉), and Ahmadova Kamola[2]

[1] Department of Artificial Intelligence, Tashkent University of Information Technologies named after Muhammad al Khwarizmi, Tashkent, Uzbekistan
muslimjon1010@gmail.com

[2] Department of Information Security, Tashkent University of Information Technologies named after Muhammad al Khwarizmi, Tashkent, Uzbekistan

Abstract. The problem of misallocation of cache lines in processors is studied based on the specific features of memory organization in multi-core and multi-processor systems of distributed systems. Incorrect allocation of cache lines leads to sequential execution of tasks that should be executed in parallel on several cores. It is difficult to determine the problem, and its presence can cause a sharp decrease in the performance of the entire system. In many seismic studies, scientists' efforts are aimed at finding reliable signs of seismic hazard. Abrupt changes in one or another parameter are called abrupt effects, emissions or anomalies, through which the principle of prediction can be implemented - this means predicting the place, strength and time of a future seismic event. The use of distributed systems in the implementation of predictions with the help of machine learning, digital processing of large volumes of digital signals, ensures fast and high-quality determination of results. To increase the speed, it is necessary to develop parallel algorithms. Synchronization of common parallel streams is necessary to solve parallel computing problems and to ignore cache memory in parallel programming. This causes cache lines to be misallocated. Acceleration of parallel algorithms for digital processing of seismic signals and the dependence of the efficiency of using the system's computing resources on the number of parallel flows were analyzed, a parallel algorithm for determining the epicenter using cubic splines and a software tool was created based on this algorithm, and the proposed spline method was used to determine the characteristics of P and S waves. Made it possible to determine the detection time 2.5 times faster than usual.

Keywords: Distributed computing systems · digital signal processing · Spline · multi-processor systems · cache memory · misallocation · multithreaded program

1 Introduction

Over the past decade, the main direction of the development of computing systems, especially distributed computing architecture, has been the design of multi-core multiprocessor systems [1]. This was due to the need to further improve the performance

H. Zaynidinov et al. (Eds.): IHCI 2022, LNCS 13741, pp. 586–593, 2023.
https://doi.org/10.1007/978-3-031-27199-1_60

of computers, as well as to create high-speed algorithms for sorting, searching and processing large amounts of data. The use of multi-processor systems allows software developers to organize parallel computations within distributed computing systems to further accelerate the execution of software tools. In the design of parallel software, in order to effectively distribute the functionality of the program among the available computing resources, it is necessary to take into account the specific characteristics of the hardware on which it is placed [10]. The following factors that may affect the performance of parallel applications should be considered: the number and configuration of processors, the creation of parallel streams for data, changing the cache, improper distribution of processor cache rows.

2 The Main Part

In the last ten years, the main direction of the development of computing systems, especially the architecture of distributed computing systems, has been the design of multi-core multi-processor systems [1]. The reason for this was the need to further increase the performance of computers and, at the same time, to create fast algorithms for sorting, searching and digital processing of large volumes of data. The use of multiprocessor systems allows software developers to organize parallel computations within distributed computing systems to further accelerate the execution of software tools. When designing parallel software tools, it is necessary to take into account the specific characteristics of the hardware on which it will be placed in order to effectively distribute the functionality of the program among the available computing resources. Electromagnetism, anomalous changes in the gravity field, anomalous disturbances in the ionosphere, seismic noise, various acoustic oscillations, and others can be sudden changes. In recent years, dozens of new methods for predicting seismic events have been proposed. In order to understand the results obtained by these methods, the physical processes applied to the seismic event, it is necessary to create physical and mathematical models of the interdependence of the processes necessary for the practical implementation of the principles of forecasting.

Organization of cache memory of local network computers for earthquake source detection. Each processor is equipped with a local cache, which is much faster than the main memory, to reduce the latency of access to the shared memory. Using cache memory avoids the need to access main memory in multiprocessor systems [2]. The cache memory of multi-core processors has several levels [3]:

- first-level cache (L1) works directly with the processor core, has the shortest access time. It is divided into instruction cache (L1i) and data cache (L1d);
- the second-level cache (L2) also belongs to a specific processor core. This cache is larger and slower than the first-level cache;
- third-level cache (L3) is the largest and slowest, but still works much faster than RAM. L3 is shared between all processor cores (Intel).

3 The Problem of Preserving the Integrity of Cached Data When Determining the Epicenter of an Earthquake

Caching shared data requires maintaining its integrity in the cache. If any read operation of a data element performed by any processor returns the last value of this element written by any processor, the memory system preserves the integrity of the data [5]. Thus, there is a need to synchronize cache lines [2]. Changes in one processor cache must be propagated to another processor cache. The value of a shared variable can be repeatedly passed from one cache to another. This phenomenon is called cache modification and can seriously affect the performance of the application [6].

Based on the records (seismogram) recorded at the point, the difference between the arrival times of R and S waves is determined using a hodograph, and on this basis, the distance from the station to the epicenter is determined. This is done as follows: It is known that R-waves arrive at the observation point before S-waves due to their high propagation speed. Let T be the interval of arrival of R- and S-waves measured from the seismogram at the point (determined from the hodograph). The velocities of R and S waves are VP and VS, respectively, and the distance d from the station to the epicenter is determined by the following formula.

For example:

$$\text{For station A: } d = T \cdot \frac{V_p \cdot V_S}{V_p - V_s} \; V_p = 6 \text{ km/cek}; \; V_s = 4 \text{ km/cek}$$

$$T = 25\,cek,\; d_A = \frac{6 \cdot 4}{6 - 4} \cdot 25 = 300 \text{ km}$$

$$\text{For station B: } T = 50\,cek,\; d_B = \frac{6 \cdot 4}{6 - 4} \cdot 50 = 600 \text{ km}$$

A to the epicenter d = 300 km, from station B is d = 600 km. If the observation was made at several stations, that is, d1, d2, d3 are known, then circles are drawn between these distances, and the circles intersect at one point or may not intersect due to inaccuracies in the calculation. . In that case, the center of gravity of the triangle (polygon) formed by the intersection of circles is taken as the location point of the epicenter. During the research, ways to eliminate the problems of efficient allocation of cache memory in the creation of software tools working with parallel streams were identified. Using shared thread synchronization to solve parallel computing problems in multi-threaded software led to misallocation of cache lines. Experiments were conducted on a server equipped with two eight-core Intel Core i5 2.9 GHz processors. Processor cache: L1 - 384 Kb, L2 - 1.5 Mb, L3 - 9 Mb. Operating memory 8 GB. Operating system: Windows 10 Pro. Parallel computing processes were implemented in the Python programming language (Python 3.10.4).

The following indicators are used for analysis. Acceleration is the ratio of the serial execution time of a program to the parallel execution time of the program [8]:

$$S_N = T_1/T_N \tag{1}$$

Efficiency - The program shows how well the system uses computing resources. To calculate the efficiency of a parallel program, the acceleration observed in the experimental processor must be divided by the number of available cores N.

$$\mathrm{E_N} = \mathrm{S_N/N} * 100\% \tag{2}$$

Parallel currents run on different cores, meaning that the number of cores in the processor is the same as the total number of cores.

4 A Method of Parallel Calculation of Spline Function in Distributed Systems

A parallel method of calculating the spline function on a multi-core processor consists of the following sequences. The four pairs of multiplications presented in the spline are converted into an array for parallel computation on individual processor cores.

For example:

$$b_0\ B_0(\mathrm{x}),\ K_j = \sum_{i=0}^{m+1} b_i B_{(i \mod 10)+20}(x),$$

$$b_1\ B_1(\mathrm{x}),\ P_j = \sum_{i=0}^{m+1} b_{i+1} B_{(i \mod 10)+10}(x),$$

$$b_2\ B_2 \text{ is brought to } T_j = \sum_{i=0}^{m+1} b_{i+2} B_{(i \mod 10)}(x).$$

Sums of four arrays are calculated in parallel after one clock cycle of the processor [7].

The problem of predicting the time and place of earthquakes in advance due to its extreme difficulty (the need to obtain information about the processes in the interior of the earth, the need to obtain information about the low speed of differential tectonic movements that lead to earthquakes) is still incomplete. Not resolved internally.

$$P = \max|S(x_i)|, \quad i = 0, 1, \ldots N. \tag{3}$$

S is the maximum value of the digitally processed signal based on the parallel algorithm, which is less than P. S is calculated by the following formula:

$$S = \max|S(x_i)| < P, \quad i = 0, 1, \ldots N. \tag{4}$$

where is a function approximated by a cubic spline.

Figure 1 presents a diagram of a digitally processed seismic signal based on the proposed parallel algorithm, and the maximum and minimum values of this signal are marked with the letters P and S [9]. Using the special parameters of the seismic signal determined by seismologists, parameters P and S are defined to determine the range of changes that occurred in the signal. In order to detect changes in other parts of the signal, apart from the parts defined in the algorithm, the software package has the ability to move the P and S parameters along the x and y coordinate axes. Figure 2 shows the block diagram of the parallel algorithm for determining P and S parameters of seismic signals using cubic splines.

Shared memory allocation for parallel streams in distributed systems. Another approach to parallel programming is to create sub-threads for central parallel threads that run independently in shared memory. The main parallel thread allocates a single block of memory for all structures of these threads before creating sub-parallel threads.

Each sub flow can access elements of its own structure. The structures of all parallel streams are located next to each other in memory, and most likely the elements of these structures fall into the same cache line.

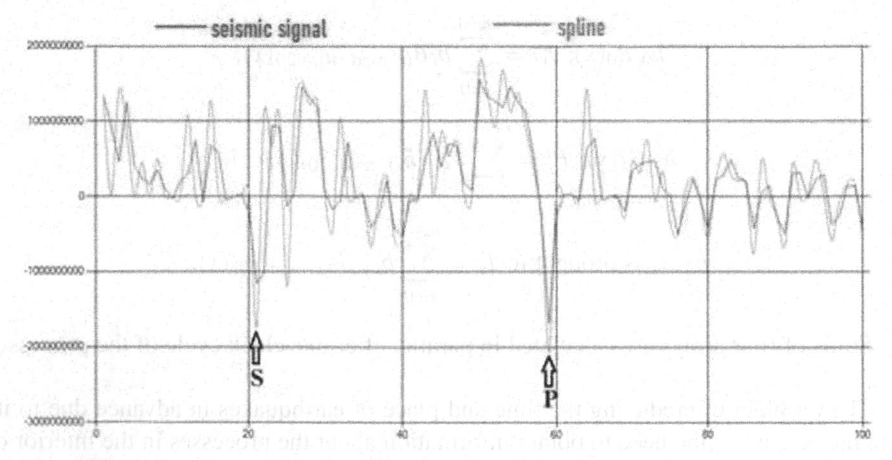

Fig. 1. Results of determination of P and S parameters of the seismic signal

Although parallel threads access unrelated elements of different structures, they cause misallocation of cache lines. In the blue line in Fig. 1, it can be seen that the time characteristics of the program have significantly deteriorated with the increase in the number of parallel streams, and in the red line, the execution time of the program in each parallel stream due to the increase in the number of parallel streams and the correct allocation of the cache line was almost the same.

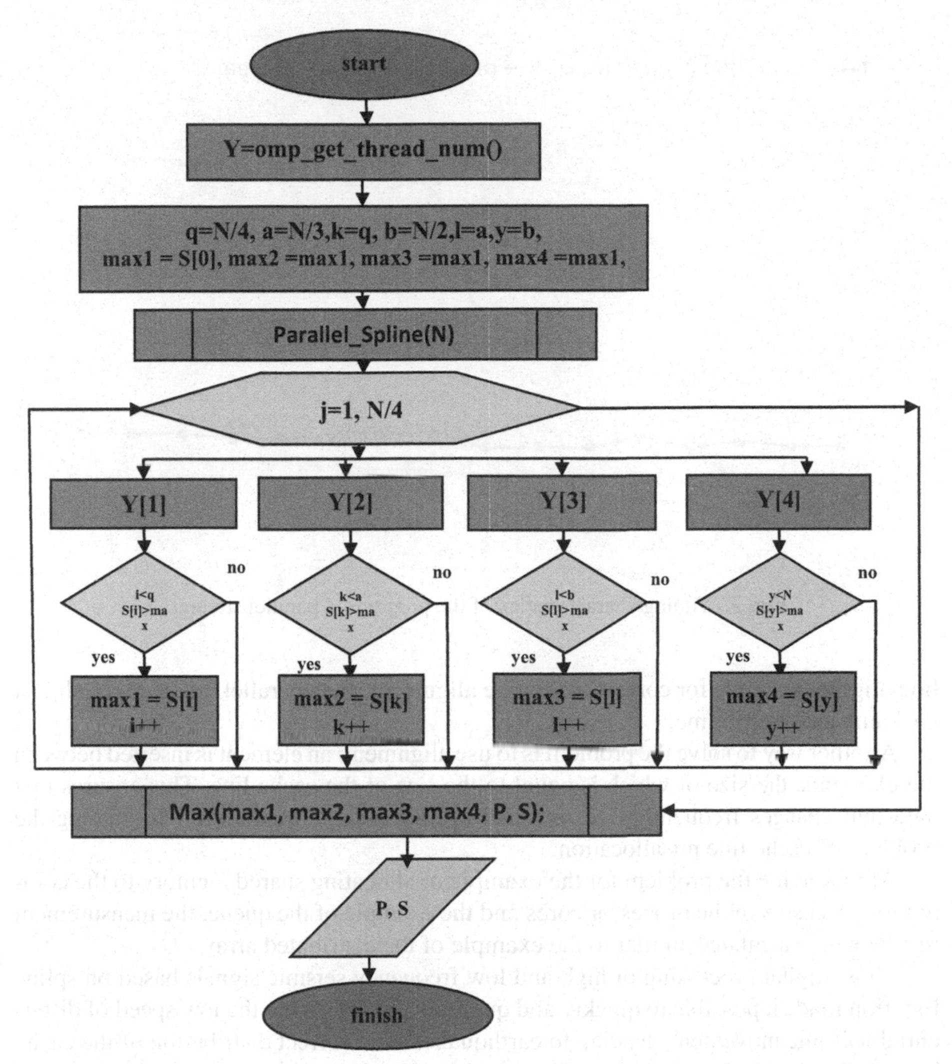

Fig. 2. Block diagram of the algorithm for determining P and S parameters of seismic signals in distributed systems

In Fig. 3, line 1 shows the actual execution time of the program, line 2 shows the execution time of the program after the correct allocation of the cache line. The OY axis of the diagram in Fig. 1 represents the execution time (T) of the program, and the OX axis represents the number of cores (common or main threads) in the processors. This means computing times for distributed arrays in each of the sixteen common streams. For example, the blue line shows the time delay of 0.1 ms to calculate the 1st element of the array in the 1st stream, and the red line shows the time delay of 0.05 ms to calculate the 1st element of the array in the 1st stream. As the number of parallel threads increases, it can be seen that the execution time of the program increases due to incorrect cache

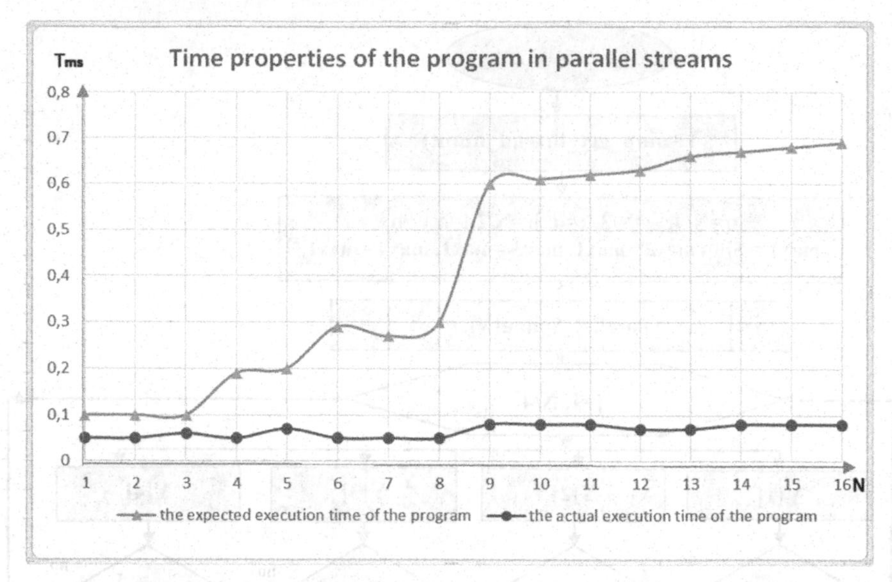

Fig. 3. Timing characteristics of the program in parallel streams.

line alignment, while for correct cache line alignment, each parallel thread takes almost the same amount of time.

Another way to solve the problem is to use alignment: an element is inserted between the elements, the size of which is equal to the size of the cache line. This ensures that data that changes frequently does not end up in the same cache line, eliminating the problem of cache line misallocation.

After solving the problem for the example of allocating shared memory to the computing processes in the processor cores and the example of the queue, the measurement results were calculated similar to the example of the distributed array.

Thus, digital processing of high and low frequency seismic signals based on spline function made it possible to quickly and qualitatively determine the low speed of differential tectonic movements leading to earthquakes. The correct distribution of the cache line of distributed system computers depends on the time characteristics of the parallel algorithms. Running parallel algorithms on dual-core and octa-core CPUs showed that proper cache line allocation can digitize large seismic signals up to 10 times faster than expected. The use of the proposed methods significantly improves the performance of software tools operating on the basis of a parallel algorithm and allows for quick and high-quality identification of seam centers [11].

List of used literatures speed and efficiency of performing parallel algorithms in distributed computing systems.

5 Conclusion

In conclusion, we should note that we have studied the problem of incorrect allocation of cache memory lines in processors due to the peculiarities of the organization of memory in multi-core and multiprocessor systems of distributed systems. The study also identified problems with cache memory in parallel programming, and analyzed the use of common stream synchronization to solve parallel computing problems in software tools leading to incorrect cache memory allocation. Proportional growth was achieved after the time taken to perform large- scale computational processes did not increase proportionally, and after applying the methods proposed in the article to solve this problem. Acceleration of the execution of software algorithms and the dependence of the efficiency of the use of computing resources of the system on the number of parallel flows were analyzed. The study found that misallocation of cache memory had a significant effect on the time characteristics of multi-branch applications, and suggested ways to eliminate it.

References

1. Wu, L., Jing, C., Liu, X., Tian, H.: Diagnostic methods for communication waiting in MPI parallel programs and applications. Guofang Keji Daxue Xuebao/J. Natl. Univ. Def. Technol. **42**(2) (2020). https://doi.org/10.11887/j.cn.202002006
2. Khamdamov, U., Zaynidinov, H.: Parallel Algorithms for bitmap image processing based on daubechies wavelets (2018). https://doi.org/10.1109/ICCSN.2018.8488270
3. Singh, D., Singh, M., Hakimjon, Z.: Geophysical application for splines. In: Singh, D., Singh, M., Hakimjon, Z. (eds.) Signal Processing Applications Using Multidimensional Polynomial Splines. SpringerBriefs in Applied Sciences and Technology, pp. 55–63. Springer, Singapore (2019). https://doi.org/10.1007/978-981-13-2239-6_7
4. Singh, D., Zaynidinov, H., Lee, H.J.. Piecewise quadratic Harmut basis functions and their application to problems in digital signal processing. Int. J. Commun. Syst. **23**(6–7), 751–762 (2010). https://doi.org/10.1002/dac.1093
5. Singh, D., Singh, M., Hakimjon, Z.: Parabolic splines based one-dimensional polynomial. In: Singh, D., Singh, M., Hakimjon, Z. (eds.) Signal Processing Applications Using Multidimensional Polynomial Splines. SpringerBriefs in Applied Sciences and Technology, pp. 1–11. Springer, Singapore (2019). https://doi.org/10.1007/978-981-13-2239-6_1
6. Üncü, Y.A., Mercan, T., Sevym, G., Canpolat, M.: Interpolation applications in diffuse optical tomography system (2018). https://doi.org/10.1109/BIYOMUT.2017.8478855
7. Zayniddinov, H.N., Mallayev, O.U.: Paralleling of calculations and vectorization of processes in digital treatment of seismic signals by cubic spline. In: IOP Conference Series: Materials Science and Engineering, vol. 537, no. 3 (2019). https://doi.org/10.1088/1757-899X/537/3/032002
8. Hennessy, J.L., Patterson, D.A.: Computer Architecture: A Quantitative Approach, 4th edn. (2006)
9. Huang, S.L., Song, W., Wang, Y.Z., Wu, Y.M., Pan, X.M., Sheng, X Q.: Efficient and accurate electromagnetic angular sweeping of rough surfaces by MPI parallel randomized low-rank decomposition. IEEE J. Sel. Top. Appl. Earth Obs. Remote Sens. **13**, 1752–1760 (2020). https://doi.org/10.1109/JSTARS.2020.2981124
10. Rocco, R., Gadioli, D., Palermo, G.: Legio: fault resiliency for embarrassingly parallel MPI applications. J. Supercomput. **78**(2), 2175–2195 (2022). https://doi.org/10.1007/s11227-021-03951-w
11. Zhao, Z., Ma, R., He, L., Chang, X., Zhang, L.: An efficient large-scale mesh deformation method based on MPI/OpenMP hybrid parallel radial basis function interpolation. Chin. J. Aeronaut. **33**(5), 1392–1404 (2020). https://doi.org/10.1016/j.cja.2019.12.025

Author Index

A
Abduganiev, Mukhriddin 18
Abdullaev, Sh. Sh. 100
Abdullaeva, M. I. 39
Agotai, Doris 516
Alimardanov, Shokhzod 251
Amrit, Chintan 373
Arabboev, Mukhriddin 527
Aripov, J. 284
Ariza-Colpas, Paola-Patricia 361
Asif, Mohammad 75
Azimov, Bunyodbek 586
Azimov, Rakhimjon 18

B
Bakhtiyor, Djabbarov 565
Banerjee, Abeer 442
Bashir, Masooda 297
Bedi, Anterpreet Kaur 406
Begmatov, Shohruh 527
Bekzod, Hamroev 565
Bermudez Pillado, Ericka Pamela 428
Bobokulov, Shakhzod 565
Boltayevich, Elov Botir 27
Botirov, F. 230
Bukit, Tori 428
Burman, Puja 140
Butt-Aziz, Shariq 361

C
Chedjou, Jean Chamberlain 527
Cho, Younhee 160
Choi, Bong Jun 213, 460
Chong, Uipil 251
Chung, Wan-Young 504
Cords, Hilko 516

D
Dannemiller, Justin 323
Davronbekov, D. 284
Djaykov, G. M. 417
Djuraev, R. 284
Dutta, Ayan 257

E
Eggler, Tabea 516
Ensing, Sophie 373

F
Fazilov, Shavkat Kh. 177
Ferronato, Priscilla 297

G
Gautam, Ravikant 109
Gupta, Aditya 75
Gupta, Ajay 336, 466, 493
Gyorick, Kevin 466

H
Hegazy, Abdolla 336
Hernandez-Sánchez, Guillermo 361
Hong, Hye-Jin 132

I
Ibrohimbek, Yusupov 547
Ilxomovna, Axmedova Xolisxon 27
Iwan, Ignatius 428

J
Jabbarov, Sh. 284
Jain, Anmol 140
Jain, Sonal 271
Javohir, Nurmurodov 547
Jeon, Su Min 1
Jeong, Seung Min 559
Jung, Eui-chul 160
Jung, Eui-Chul 225, 396
Juraev, D. B. 39
Juturi, Venkateswara Sagar 312

H. Zaynidinov et al. (Eds.): IHCI 2022, LNCS 13741, pp. 595–597, 2023.
https://doi.org/10.1007/978-3-031-27199-1

K
Kakharov, Shukrullo S. 177
Kamel, Ammar 336
Kamola, Ahmadova 586
Kang, Dae-Ki 95
Khalilov, S. P. 230
Khujayarov, Ilyos 51
Kim, Hyewon 160
Kim, Jin Man 559
Kim, Jong-Hoon 201, 323, 386
Kim, Lori Minyoung 396
Kim, Seung Hyun 510
Kim, Seunghyun 154, 348
Kim, So-Hyeon 132
Kim, Sunghwan 239
Ku, Seongyun 239
Kujur, Vinit 406
Kumar, Ajit 213
Kumar, Himanshu 442
Kwon, Jung-Ryun 396
Kyamakya, Kyandoghere 527

L
Lee, Chaewon 348
Lee, Eui Chul 1, 154, 348, 454, 510, 559
Lee, Jee Hang 510
Lee, Jee-Hang 132
Lee, Kunyoung 454
Lee, Seok-Lyong 428
Lee, Seung Gun 510, 559
Legath, Alexander 516
Lewis, Rayan Smith 63
Lim, Hyun-Woo 428

M
Maabreh, Majdi 466
Mačiulienė, Monika 435
Mahapatra, Dwarikanath 271
Mahesh Karimbi, Kavitha 63
Mahmud, Saifuddin 201, 323, 386
Mai, Ngoc-Dau 504
Mallaev, Oybek 586
Mamatov, N. S. 100
Mannapova, M. G. 230
Matar, Sophia 201
Matkurbonov, D. 284
Mehta, Naval Kishore 442
Minocha, Pranjal 493
Mirdjonovna, Hamroyeva Shahlo 27
Mirzaev, Olimjon N. 177
Mishra, Sudhakar 75
Misquith, Shane Christopher 63
Molakatala, Nagamani 109, 312
Morales-Ortega, Roberto-Cesar 361
Muhammadali, Gofurjonov 547
Mulahuwaish, Aos 466
Musaev, Muhammadjon 51
Muslimjon, Kuchkarov 586
Muydinov, Lazizbek 18

N
Narimonova, Aziza 565
Narzulloyev, Oybek 575
Nasrullayev, N. B. 230
Nguyen, Ha-Trung 504
Nirgulkar, Mohit Manohar 166
Niyozmatova, N. A. 100
Normatov, Sherbek 480
Nosirov, Khabibullo 527
Nuraliev, Fakhriddin 575

O
Ochilov, M. M. 39
Ochilov, Mannon 51
Oh, Jaemu 454
Osti, Manish 466

P
Pandey, Sandeep Kumar 166
Park, Karam 225
Parwekar, Pritee 85
Piñeres-Melo, Marlon Alberto 361
Pirimbetov, A. O. 417
Poojary, Sushanth 63
Prabakar, Nagarajan 7
Prasad, Shyam Sunder 442
Priyatham, Nikil 109

Q
Qolomany, Basheer 466

R
Rajender, M. 109
Rakhimov, M. F. 39
Rakhmatullaev, Marat 480
Rakhmonov, Rustam 565
Ridhun, Mohammed 63
Rogers, Blayne 493

S
Saini, Mukesh 271, 406
Samijonov, A. N. 100
Sanjarbek, Ibragimov 547
Saurav, Sumeet 442
Saydiakbarov, Saydiakhrol 527
Seok, Chae Lin 559
Seong, Hyeon Ah 1
Serrano-Torné, Guillermo 361
Shah, Ankita 188
Shaker, Alfred 201
Sharma, Gokarna 257, 386
Sharma, Priyanka 85
Shekhar, Prakash 336
Shekhawat, Hanumant Singh 166
Shenava, Amrth Ashok 386
Shin, Jie Yong 95
Simmen, Yves 516
Singh, Ankit Kumar 213
Singh, Ravi Kumar 121
Singh, Sanjay 442
Singh, Sumit 140
Singh, Varsha 121
Sirojiddin, Qobilov 547
Skarzauskiene, Aelita 435
Soni, Jayesh 7
Sung, Hyeon Ah 154

T
Tambala, Shalem Raju 312
Tang, Liang 297
Tanjung, Sean Yonathan 428
Tashmetov, Shakhzod 527
Tastanova, Saida 575
Teja Kiran, M. 312
Tejaswini, M. 312
Tejo Seshadri, M. 312
Tiwary, Uma Shanker 75, 121, 140, 188
Tuktasinov, Mumtozali 540
Tursunov, Javlon 565

U
Umarov, Azizbek 354
Undru, Vimal Babu 312
Upadhyay, Himanshu 7
Uteuliev, N. U. 417

V
van't Klooster, Jan-Willem 201
Vani, 140
Vasudevan, D. V. K. 109
Vinodbhai, Majithia Tejas 75

W
Whitaker, Mark D. 239
Wimer, Bailey 323

Y
Yahya, Bernardo Nugroho 428
Yoon, Jun Yong 460
You, Hojoon 454
You, Minji 239
Yuldoshev, Yu. Sh. 100
Yusupov, I. 230

Z
Zakirov, Ruslan 354
Zaripov, Oripjon 354

Printed in the United States
by Baker & Taylor Publisher Services

Printed in the United States
by Baker & Taylor Publisher Services